Essential Zebrafish Methods

Reliable Lab Solutions

Essential Zebrafish Methods: Cell and Developmental Biology

Reliable Lab Solutions

Edited by

H. William Detrich, III

Department of Biology, Northeastern University
Boston, Massachusetts 02115

Monte Westerfield

Institute of Neuroscience, University of Oregon
Eugene, Oregon 97403

Leonard I. Zon

Howard Hughes Medical Institute, Children's Hospital
Boston, Massachusetts 02115

ELSEVIER

AMSTERDAM • BOSTON • HEIDELBERG • LONDON
NEW YORK • OXFORD • PARIS • SAN DIEGO
SAN FRANCISCO • SINGAPORE • SYDNEY • TOKYO
Academic Press is an imprint of Elsevier

Academic Press is an imprint of Elsevier
Linacre House, Jordan Hill, Oxford OX2 8DP, UK
30 Corporate Drive, Suite 400, Burlington, MA 01803, USA
525 B Street, Suite 1900, San Diego, CA 92101-4495, USA
32 Jamestown Road, London NW1 7BY, UK

First edition 2009

Material in the work originally appeared in Volume 59 (Academic Press, 1999)
and Volume 76 (Elsevier Inc., 2004) of *Methods in Cell Biology*

Notice
No responsibility is assumed by the publisher for any injury and/or damage to persons
or property as a matter of products liability, negligence or otherwise, or from any use
or operation of any methods, products, instructions or ideas contained in the material
herein. Because of rapid advances in the medical sciences, in particular, independent
verification of diagnoses and drug dosages should be made

ISBN: 978-0-12-374599-6

For information on all Academic Press publications
visit our website at elsevierdirect.com

Printed and bound by CPI Group (UK) Ltd, Croydon, CR0 4YY
Transferred to Digital Print 2011

Working together to grow
libraries in developing countries

www.elsevier.com | www.bookaid.org | www.sabre.org

ELSEVIER   BOOK AID
International   Sabre Foundation

We dedicate this volume to our colleagues in the
zebrafish research community.

CONTENTS

11. Cellular Dissection of Zebrafish Hematopoiesis

David L. Stachura and David Traver

12. Culture of Embryonic Stem and Primordial Germ Cell Lines from Zebrafish

Ten-Tsao Wong, Lianchun Fan, and Paul Collodi

13. Neurogenesis

Prisca Chapouton, Marion Coolen, and Laure Bally-Cuif

CONTRIBUTORS

Numbers in parentheses indicate the pages on which the authors' contributions begin.

R. Craig Albertson (457), Department of Biology, Syracuse University, Syracuse, New York 13244

Courtney Alexander (429), Department of Developmental and Cell Biology, University of California, Irvine, California 92697-2300

James F. Amatruda (183), Departments of Pediatrics and Molecular Biology, and Department of Internal Medicine, UT Southwestern Medical Center, Dallas, Texas 75390-8534

Andrei Avanesov (349), Department of Ophthalmology, Harvard Medical School/ MEEI, Boston, Massachusetts 02114

Laure Bally-Cuif (245), Zebrafish Neurogenetics Department, Institute of Developmental Genetics, Helmholtz Zentrum München, German Research Center for Environmental health, 85764 Neuherberg, Germany

Robert Baker (401), Department of Physiology and Neuroscience, New York University School of Medicine, New York 10016

James C. Beck (401), Department of Physiology and Neuroscience, New York University School of Medicine, New York 10016

Bruce P. Brandhorst (35), Institute of Molecular Biology and Biochemistry, Simon Fraser University, Burnaby, British Columbia, V5A 1S6 Canada

Douglas S. Campbell (159), Laboratory for Developmental Gene Regulation, RIKEN Brain Science Institute, Wako, Saitama 351–0198

Prisca Chapouton (245), Zebrafish Neurogenetics Department, Institute of Developmental Genetics, Helmholtz Zentrum München, German Research Center for Environmental health, 85764 Neuherberg, Germany

Chi-Bin Chien (159), Department of Neurobiology and Anatomy, University of Utah Medical Center, Salt Lake City, Utah 84132

Paul Collodi (233), Department of Animal Sciences, Purdue University, West Lafayette, Indiana 47907

Marion Coolen (245), Zebrafish Neurogenetics Department, Institute of Developmental Genetics, Helmholtz Zentrum München, German Research Center for Environmental health, 85764 Neuherberg, Germany

Mark S. Cooper (103), Department of Biology, University of Washington, Seattle, Washington 98195-1800

Graham E. Corley-Smith (35), Institute of Neuroscience, University of Oregon, Eugene, Oregon 97403-1254

Leonard A. D'Amico (103), Department of Biology, University of Washington, Seattle, Washington 98195-1800

H. William Detrich (1), Department of Biology, Northeastern University, Boston, Massachusetts 02115

Iain A. Drummond (501), Departments of Medicine and Genetics, Harvard Medical School and Nephrology Division, Massachusetts General Hospital, Charlestown, Massachusetts 02129

Lianchun Fan (233), Department of Animal Sciences, Purdue University, West Lafayette, Indiana 47907

Scott E. Fraser (293), Biological Imaging Center, Beckman Institute (139-74), California Institute of Technology, Pasadena, California 91125

Edwin Gilland (401), Department of Anatomy, Howard University School of Medicine, Washington, District of Columbia 20059

Yevgenya Grinblat (73), Departments of Zoology and Anatomy, University of Wisconsin, Madison, Wisconsin 53706

Paul D. Henion (325), Center for Molecular Neurobiology and Department of Neuroscience, Ohio State University, Columbus, Ohio 43210

Clarissa A. Henry (103), Department of Biology, University of Washington, Seattle, Washington 98195-1800

Lara D. Hutson (159), Department of Biology, Williams College, Williamstown, Massachusetts 01267

Yashar Javidan (429), Department of Developmental and Cell Biology, University of California, Irvine, California 92697-2300

Trevor Jowett (51), Department of Biochemistry and Genetics, Medical School, University of Newcastle, Newcastle upon Tyne NE2 4HH, United Kingdom

Reinhard W. Köster (293), Zebrafish Neuroimaging Group, Helmholtz Zentrum München, Institute of Developmental Genetics, 85764 München-Neuherberg, Germany

Donald A. Kane (11), Department of Biological Sciences, Western Michigan University, Kalamazoo, Michigan 49008

John P. Kanki (325), Department of Pediatric Oncology, Dana-Farber Cancer Institute, Boston, Massachusetts 02115

A. Thomas Look (325), Department of Pediatric Oncology, Dana-Farber Cancer Institute, Boston, Massachusetts 02115

Jarema Malicki (349), Department of Ophthalmology, Harvard Medical School/MEEI, Boston, Massachusetts 02114

Barry H. Paw (29), Howard Hughes Medical Institute and Division of Hematology-Oncology, Children's Hospital and Dana-Farber Cancer Institute, Department of Pediatrics, Harvard Medical School, Boston, Massachusetts 02115

John H. Postlethwait (35), Institute of Neuroscience, University of Oregon, Eugene, Oregon 97403-1254

Thomas F. Schilling (429), Department of Developmental and Cell Biology, University of California, Irvine, California 92697-2300

Hazel Sive (73), Whitehead Institute for Biomedical Research and Massachusetts Institute of Technology, Nine Cambridge Center, Cambridge, Massachusetts 02142

Lilianna Solnica-Krezel (133), Department of Biological Sciences, Vanderbilt University, Nashville, Tennessee 37235-1634

David L. Stachura (205), Section of Cell and Developmental Biology, Division of Biological Sciences, University of California, San Diego, La Jolla, California 92093

Didier Y. R. Stainier (479), Department of Biochemistry and Biophysics, Programs in Developmental Biology, Genetics and Human Genetics, University of California, San Francisco, San Francisco, California 94143–0448

Rodney A. Stewart (325), Department of Pediatric Oncology, Dana-Farber Cancer Institute, Boston, Massachusetts 02115

David W. Tank (401), Department of Molecular Biology, and Department of Physics, Princeton University, Princeton, New Jersey 08544

Jacek Topczewski (133), Department of Pediatrics, Children's Memorial Research Center, Northwestern University Feinberg School of Medicine, Chicago, Illinois 60614

David Traver (205), Section of Cell and Developmental Biology, Division of Biological Sciences, University of California, San Diego, La Jolla, California 92093

Le A. Trinh (479), Biology Department, California Institute of Technology, Pasadena, California 91125

Daniel Verduzco (183), Department of Pediatrics and Molecular Biology, UT Southwestern Medical Center, Dallas, Texas 75390-8534

Charline Walker (35), Institute of Neuroscience, University of Oregon, Eugene, Oregon 97403-1254

Monte Westerfield (1), Institute of Neuroscience, University of Oregon, Eugene, Oregon 97403

Ten-Tsao Wong (233), Department of Animal Sciences, Purdue University, West Lafayette, Indiana 47907

Pamela C. Yelick (457), Department of Oral and Maxillofacial Pathology, Tufts University, Boston, Massachusetts 02111

Leonard I. Zon (1, 29), Howard Hughes Medical Institute and Division of Hematology-Oncology, Children's Hospital and Dana-Farber Cancer Institute, Department of Pediatrics, Harvard Medical School, Boston, Massachusetts 02115

PREFACE

Approximately 30 years ago, George Streisinger of the University of Oregon had a prescient insight: analysis of vertebrate development by genetic methods would be facilitated by choosing the zebrafish as a model organism. He recognized that the short generation time of the zebrafish (2–3 months), its high fecundity (100–200 embryos per mating) and oviparous mode of reproduction, and the optical transparency of its early embryo were ideally suited to the rapid screening of the progeny of mutagenized fish for mutations that affect important developmental processes. Before his untimely death in 1984, Streisinger's group cloned the zebrafish, developed methods for mutagenesis, genetic mapping, and clonal analysis of development, and used F1 screens to isolate zygotic recessive mutations with extraordinary embryonic phenotypes. These accomplishments stimulated other laboratories to embrace the zebrafish system, ultimately leading to the publication in December 1996, of a volume of the journal *Development* devoted exclusively to the description of more than 2000 developmental mutations that were obtained from two large-scale mutagenic screens conducted by the laboratories of Christiane Nüsslein-Volhard at the Max-Planck-Institut für Entwicklungsbiologie in Tübingen and of Wolfgang Driever and Mark Fishman at the Massachusetts General Hospital in Boston. This watershed event validated the zebrafish as an important developmental and genetic model and stimulated the explosive growth of the zebrafish research community that continues to this day.

As Editors of four volumes of *Methods in Cell Biology* on the zebrafish, Monte, Len, and I have enjoyed the enthusiastic cooperation of many members of the zebrafish community as we sought to provide practical documentation of technological advances in this system. Those volumes, published in 1999 (vols. 59 and 60) and 2004 (vols. 76 and 77), describe proven methods that form the foundation for much ongoing research. Due to their success our publisher, Elsevier/Academic Press, asked us to identify critically important chapters from the four volumes for republication as part of a new series, *Reliable Lab Solutions*. The present two-volume compilation, *Essential Zebrafish Methods*, contains 39 chapters of proven utility, and many of these have been updated by the original contributors so that they reflect the state-of-the-field as of late 2008. This first volume provides 21 chapters relevant to the cell and developmental biology of the zebrafish, whereas the second is composed of 18 chapters on the genetics and genomics of the system. We believe that *Essential Zebrafish Methods* will provide busy investigators, their postdoctoral fellows, and their students with a cost-effective and time-saving source to the best methods and protocols for the zebrafish system.

We express with pleasure our gratitude to each of our contributors who have worked so diligently on the prior *Methods in Cell Biology* volumes as well as the present *Reliable Lab Solutions* compendium. We also thank the *Methods in Cell Biology* Series Editors, Leslie Wilson and Paul Matsudaira, for their support of our zebrafish volumes. Last, but certainly not least, we thank the staff of Elsevier/Academic Press, especially Jasna Markovac and Tara Hoey, for their help, patience, and encouragement as we developed *Essential Zebrafish Methods*.

H. William Detrich, III
Monte Westerfield
Leonard I. Zon

CHAPTER 1

Overview of the Zebrafish System

H. William Detrich,* Monte Westerfield,† and Leonard I. Zon‡

*Department of Biology
Northeastern University
Boston, Massachusetts 02115

†Institute of Neuroscience
University of Oregon
Eugene, Oregon 97403

‡Howard Hughes Medical Institute
Children's Hospital
Boston, Massachusetts 02115

I. Introduction

A central dogma of developmental biology today is that the fundamental genetic mechanisms that control the development of metazoans have been conserved evolutionarily, albeit frequently modified in their application. For example, invertebrates and vertebrates employ homologous signaling systems that act antagonistically to establish topologically equivalent, but spatially reversed, dorsal/ventral axes (De Robertis and Sasai, 1996). Based on mutant phenotype and protein structure, vertebrate ventralizing signals (e.g., BMP-2, BMP-4) are functionally homologous to *Drosophila* Decapentaplegic, which functions in dorsal determination in the fly, and

ESSENTIAL ZEBRAFISH METHODS:
CELL AND DEVELOPMENTAL BIOLOGY
Copyright 2009, Elsevier Inc. All rights reserved.

DOI: 10.1016/B978-0-12-374599-6.00001-7

the vertebrate dorsalizer Chordin is homologous to the *Drosophila* ventralizing signal, Short gastrulation. Nevertheless, some aspects of development are uniquely vertebrate. The neural crest, for example, is a group of migratory cells that arises in the embryo at the border between neural and nonneural ectoderm. These cells move to many regions of the embryo to form numerous tissues, including part of the cranial skeleton and the peripheral nervous system. The development of complex organ systems, such as the brain, heart, and kidneys, is another hallmark of vertebrates that is not easily studied in invertebrate genetic systems. For developmental analysis of vertebrates, the zebrafish, *Danio rerio*, has arguably emerged as the genetic system *par excellence*.

In December, 1996, the world of biological science witnessed the equivalent of Yogi Berra's "déjà vu all over again." That month's issue of the journal *Development* was devoted entirely to the description, in 37 articles, of approximately 2000 mutations that perturb development of the zebrafish (for highlights, see Currie (1996), Eisen (1996), Grunwald (1996), Holder and McMahon (1996)). This magnificent accomplishment, the result of two independent, large-scale mutagenic screens of the zebrafish genome and phenotypic analysis of embryonic development in the mutants obtained, approximates in a vertebrate the earlier saturation mutagenic screen in *Drosophila* (Nüsslein-Volhard and Wieschaus, 1980). Indeed, two of the investigators leading the zebrafish screens, Christiane Nüsslein-Volhard of the Max-Planck-Institut für Entwicklungsbiologie in Tübingen and Wolfgang Driever of the Massachusetts General Hospital (MGH) in Boston, were veterans of the *Drosophila* program. Working at the European Molecular Biology Laboratory in Heidelberg, "Janni" Nüsslein-Volhard and her colleague Eric Wieschaus (corecipients with Edward Lewis of the 1995 Nobel Prize in Physiology or Medicine) conducted the now legendary *Drosophila* screen, and Driever, as a later member of the Nüsslein-Volhard laboratory, analyzed many of the mutants to determine essential signaling pathways that control development of the fly's body plan. Nüsslein-Volhard in Tübingen, and Driever and his colleague Mark Fishman at the MGH, subsequently applied the conceptual framework of the *Drosophila* screen to the fish. The community of developmental biologists owes these three individuals, and their many colleagues and collaborators, a tremendous debt of gratitude for this repeat performance.

II. History of the Zebrafish System and Its Advantages and Disadvantages

These recent mutagenesis screens provided proof-of-principle that classical, forward genetics can be used to understand vertebrate development. The identification and study of mutations has been extraordinarily successful in providing an understanding of the early development of *Drosophila* and of the nematode worm, *Caenorhabditis elegans*. However, the same level of analysis of early developmental events in vertebrates has been more problematic. In the mouse, historically the species of choice for studies of vertebrate developmental genetics, much of embryogenesis is difficult to follow because it occurs within the mother's uterus. Beginning about 20 years ago at the University of Oregon, George Streisinger recognized the power of genetic analysis for understanding development and the advantages of a small tropical fish with external fertilization as a vertebrate for this approach. Streisinger selected the zebrafish, a freshwater fish commonly available in pet stores, because it has a relatively short generation time (2–3 months), produces large clutches of embryos (100–200 per mating), and provides easy access to all developmental stages. Zebrafish embryos are optically transparent throughout early development, which facilitates a host of embryological experiments and the rapid morphological screening of the live progeny of mutagenized fish for interesting mutations. Before his untimely death in 1984, Streisinger's group cloned the zebrafish (Streisinger *et al.*, 1981) and developed techniques for mutagenesis (e.g., Grunwald and Streisinger, 1992a,b; Walker and Streisinger, 1983), genetic mapping (Streisinger *et al.*, 1986), and clonal analysis of development by genetic mosaics (Streisinger *et al.*, 1989). They also used F1 screens of mutagenized fish to isolate zygotic recessive lethal mutations with wonderfully curious embryonic phenotypes (Felsenfeld *et al.*, 1991; Grunwald *et al.*, 1988).

Streisinger's discoveries, as well as his enthusiasm and generosity, stimulated a number of other laboratories to begin using the zebrafish for developmental and genetic studies. Initially, all of these laboratories were also in Oregon. These groups have extended Streisinger's original studies by isolating and analyzing additional informative mutants (Halpern *et al.*, 1993, 1995; Hatta *et al.*, 1991; Kimmel, 1989; Kimmel *et al.*, 1989) and have developed techniques for production of transgenic zebrafish (Stuart *et al.*, 1988; Westerfield *et al.*, 1993). Moreover, recent work has demonstrated the advantages of zebrafish for cellular studies of vertebrate embryonic development. The embryo is organized very simply (Kimmel *et al.*, 1995) and has fewer cells than other vertebrate species under investigation (Kimmel and Westerfield, 1990). Its transparent cells are accessible for manipulative study. For example, cells can be injected with tracer dyes in intact, developing embryos to track emerging cell lineages (Kimmel and Warga, 1986) or axons growing to their targets (Eisen *et al.*, 1986). Uniquely identified young cells can be ablated singly (Eisen *et al.*, 1989) or transplanted individually to new positions (Eisen, 1991) to address positional influences on development at a level of precision that is unprecedented in any species. The combination of easy mutagenesis and powerful phenotypic screens of the earliest developmental stages eliminates, in principle, the biased detection of mutant phenotypes observed in the mouse, where scoring of mutants is generally restricted to neonatal and adult animals due to intrauterine development of the embryos. The more recent advent of tools for mapping mutations and candidate genes in the zebrafish genome has already begun to facilitate the isolation and functional analysis of genes required for normal development. Even small laboratories can conduct reasonably sized screens for new mutations, and the cost of a fish facility necessary to support such research is significantly lower than for the mouse.

Several disadvantages of the zebrafish system are also apparent. We presently lack in the zebrafish system methods to generate embryonic stem cells for gene "knock-outs" by homologous recombination. In the absence of such methods, we envision a cooperative and synergistic game of "ping pong" between the zebrafish and mammalian research communities. Knock-out analysis of the mouse homologues of genes identified via study of zebrafish mutations should lead to a greater understanding of gene function in vertebrate development. Conversely, conserved syntenies between mammalian and zebrafish genomes, and the numerous mammalian expressed sequence tags (ESTs), will continue to provide candidate genes for zebrafish mutations.

Another potential disadvantage is genetic redundancy in the zebrafish genome, which probably results from duplication of the fish genome subsequent to the phylogenetic divergence of fish and mammals (Chapter 8, Vol. 60). This redundancy may complicate the comparison of homologous developmental pathways in these taxa. Alternatively, extra gene copies may simplify some types of analysis because complex functions in mammals may have been separated and allocated to different gene paralogues in fish.

III. Cell and Developmental Biology, Organogenesis, and Human Disease

The zebrafish embryo (Chapter 2) provides numerous opportunities to examine cellular processes in early development. For example, one can culture cells from embryos (Chapters 3 and 4) *in vitro* for mechanistic studies of cell signaling pathways, analyze gene and protein expression *in situ* (Chapter 6), perturb development using physical and chemical treatments or by ectopic expression of dominant-negative proteins (Herskowitz, 1987; Chapters 7–8), and assay lineage commitment by explant assay (Chapter 9). The roles of cell movements and the cytoskeleton in embryonic axis formation are particularly amenable to analysis (Chapters 10–13). Given that some of these processes occur prior to activation of the zygotic genome, maternal-effect mutants may prove especially informative in revealing the molecular players.

The analysis of vertebrate organogenesis has always been problematic. Thanks to the recent zebrafish genetic screens, mutations that affect virtually all major organ systems are now available for phenotypic and molecular characterization (Chapters 14–20). The hematopoietic mutants, for example, comprise 26 distinct complementation groups that perturb development of the erythroid lineage from the earliest stages of stem cell commitment to terminal differentiation (Ransom *et al.*, 1996; Weinstein *et al.*, 1996; Chapter 17). The cardiovascular mutants include some that affect early development of the heart, vasculature, and blood and others that disrupt the function of an otherwise morphologically normal organ system (Chapters 17–19). Mutations that interfere with development of the central nervous system (Chapter 14), the retina (Chapter 15), and fins (Chapter 16) are also plentiful. We can anticipate a rich harvest of information from the study of these mutants.

Zebrafish mutants will also provide useful models of human diseases. The *one-eyed pinhead* mutation, which disrupts an EGF-signaling pathway in zebrafish (Zhang *et al.*, 1998), phenocopies the human condition holoprosencephaly. *gridlock*, which fails to develop trunk vasculature, resembles the human condition coarctation of the aorta, a common and lethal birth defect (Weinstein *et al.*, 1995). The hematopoietic mutants include representatives of thalassemias, porphyrias, and other human conditions (Chapter 17).

IV. Genetics and Genomics

Although highly successful, the large-scale zebrafish mutant screens (cf. Chapter 2, Vol. 60) failed to achieve saturation. Currie (1996) estimates that the degree of saturation obtained by the combined Tübingen and MGH screens ranges from 50% to 90% of the genes detectable by the methods employed. Furthermore, the morphological parameters of the screens probably precluded identification of many interesting mutants. Thus, it is likely that many investigators will now perform additional screens targeted to particular developmental processes. One exquisite example is the retinotectal projection screen carried out by Friedrich Bonhoeffer and his laboratory (Baier *et al.*, 1996; Karlstrom *et al.*, 1996; Trowe *et al.*, 1996) in conjunction with the Tübingen screen of the Nüsslein-Volhard laboratory. A second is the screen for neural crest mutations (Henion *et al.*, 1996). Maternal effects screens (Chapter 1, Vol. 60) are in progress. The Oregon laboratories have initiated screens based on RNA *in situ* hybridization. Nancy Hopkins (MIT) and her laboratory are now conducting a large-scale insertional mutagenesis screen (Chapter 5, Vol. 60), which promises to ease genetic analysis by providing convenient molecular tags for isolating disrupted genes (Gaiano *et al.*, 1996). Finally, transposable elements (Chapter 6, Vol. 60) and transgenesis with cell-specific promoters that drive the expression of green fluorescent protein (Chapter 7, Vol. 60) (Jessen *et al.*, 1998) will provide important tools for genetic analysis.

Genomic methodologies are advancing rapidly in the zebrafish system. We now have high-density genetic maps (Chapter 8, Vol. 60) that incorporate a variety of markers: randomly amplified polymorphic DNAs (RAPDs) (Chapter 9, Vol. 60), simple-sequence length polymorphisms (e.g., CA repeats (Chapter 10, Vol. 60; Knapik *et al.*, 1998), single-strand conformational polymorphisms (SSLPs)

(Chapter 11, Vol. 60), amplified fragment length polymorphisms (AFLPs) (Chapter 12, Vol. 60), and ESTs (Chapter 13, Vol. 60). Large-insert yeast artificial chromosome (YAC), bacterial artificial chromosome (BAC), and P1 artificial chromosome (PAC) libraries have been produced (Chapter 14, Vol. 60) and are commercially available. With these tools and techniques, we can now map mutations and apply candidate or positional cloning (Chapter 15, Vol. 60) strategies to recover the disrupted genes. At the Third Cold Spring Harbor Conference on *Zebrafish Development and Genetics*, researchers reported the genes for approximately 20 mutants, two identified via positional cloning and the remainder by the candidate approach. Other genomic tools in development include radiation hybrid panels (Chapter 16, Vol. 60), somatic cell hybrids (Chapter 17, Vol. 60), fluorescent *in situ* hybridization (FISH) (Chapter 18, Vol. 60), and a multipurpose zebrafish database (Chapter 19, Vol. 60).

V. Future Prospects

What does the future hold for the zebrafish system? Molecular, cellular, and developmental studies of the extant mutant collections should yield a wealth of new knowledge regarding vertebrate embryogenesis. We envision many more mutant screens directed at particular developmental processes and employing molecular probes (antibodies, antisense RNAs) for phenotypic analysis. Furthermore, the genetics of behavior will certainly be tackled. The genetic epistasis analysis of double mutants will help to establish molecular signaling pathways. The application of suppressor and enhancer screening strategies should reveal gene interactions, and the generation of conditional mutations will contribute to a temporal dissection of gene function. We predict that the zebrafish, with its present and future methodologies and infrastructure, will make important and probably surprising contributions to our understanding of the vertebrate development program.

VI. Conclusions

The consensus of the biologists in attendance at the Third Cold Spring Harbor Conference on *Zebrafish Development and Genetics* is that the zebrafish has now "arrived" as a viable, compelling genetic system for study of vertebrate development. Novel contributions by the zebrafish system have been made and will continue to be made at an accelerating rate. These volumes provide a solid compilation of current methods in zebrafish development and genetics. We trust that they will stimulate further technical development and will attract new scientific converts to the system. Now is certainly the time to fish—not to cut bait.

VII. Epilogue: Volumes 76 and 77 and Technological Advances to Come

In 2004, Monte Westerfield, Leonard I. Zon, and I edited the 2nd Edition of Methods in Cell Biology devoted to The Zebrafish: Vol. 76, Cellular and Developmental Biology, and Vol. 77, Genetics, Genomics, and Informatics. The methodological advances that our contributors described 5 years after publication of the 1st Edition (Vols. 59, 60) were remarkable in scope and sophistication. In 2009, it is clear that previously unimagined methods have been added to the zebrafish technical repertoire while most of the technical shortcomings that we described in our original overview 10 years ago have been overcome. For example, TILLING (Targeting Induced Local Lesions in Genomes) now makes it practical to obtain multiple mutant alleles at almost any known gene locus in the zebrafish genome (Vol. 77; Draper et al., 2004; Weinholds and Plasterk, 2004). Methods for efficient transgenesis of zebrafish have been developed (Dong and Stuart, 2004; Grabher et al., 2004; Hermanson et al., 2004; Kawakami, 2004) and will be further improved.

Recent innovations (post ~2004) include the development of transgenic zebrafish lines that express fluorescent marker proteins either constitutively or inducibly [reviewed by Detrich (2008)]. Reverse-genetic knockout technology using sequence-targeted zinc-finger nucleases is likely to play an increasing role in analysis of gene function in zebrafish development (Doyon et al., 2008; Meng et al., 2008). The zebrafish has also become a premier system for chemical genetics, as shown by screens that have identified small molecules that alter normal embryogenesis (Peterson et al., 2000; Yu et al., 2008), inhibit the cell cycle (Murphey et al., 2006), regulate hematopoietic stem cells (North et al., 2007), and modulate mechanosensory hair cell function (Owens et al., 2008). Furthermore, chemical suppressor screens for molecules that mitigate disease physiology (Peterson et al., 2004) can be performed. We anticipate that the pace of technical advance in the zebrafish system will continue to accelerate and that additional investigators will embrace the zebrafish for molecular genetic analysis of development and disease in vertebrates.

References

Baier, H., Klostermann, S., Trowe, T., Karlstrom, R. O., Nüsslein-Volhard, C., and Bonhoeffer, F. (1996). Genetic dissection of the retinotectal projection. *Development* **123,** 415–425.

Currie, P. D. (1996). Zebrafish genetics: Mutant cornucopia. *Curr. Biol.* **6,** 1548–1552.

De Robertis, E. M., and Sasai, Y. (1996). A common plan for dorsoventral patterning in bilatreia. *Nature* **380,** 37–40.

Detrich, H. W. III (2008). Fluorescent proteins in zebrafish cell and developmental biology. *In* "Methods in Cell Biology, Vol. 85, Fluorescent Proteins 2nd Edition" (K. F. Sullivan, ed.), pp. 219–241. Elsevier Academic Press, San Diego.

Dong, J., and Stuart, G. W. (2004). Transgene manipulation in zebrafish by using recombinases. *In* "Methods in Cell Biology, Vol. 77, The Zebrafish 2nd Edition: Genetics, Genomics, and Informatics" (H. W. Detrich, III, M. Westerfield, and L. I. Zon, eds.), pp. 363–379. Elsevier Academic Press, San Diego, CA.

Doyon, Y., McCammon, J. M., Miller, J. C., Faraji, F., Ngo, C., Katibah, G. E., Amora, R., Hocking, T. D., Zhang, L., Rebar, E. J., Gregory, P. D., Urnov, F. D., *et al.* (2008). Heritable targeted gene disruption in zebrafish using designed zinc-finger nucleases. *Nat. Biotechnol.* **26,** 702–708.

Draper, B. W., McCallum, C. M., Stout, J. L., Slade, A. J., and Moens, C. B. (2004). A high-throughput method for identifying N-ethyl-N-nitrosourea (ENU)-induced point mutations in zebrafish. *In* "Methods in Cell Biology, Vol. 77, The Zebrafish 2nd Edition: Genetics, Genomics, and Informatics" (H. W. Detrich, III, M. Westerfield, and L. I. Zon, eds.), pp. 91–112. Elsevier Academic Press, San Diego, CA.

Eisen, J. S. (1991). Determination of primary motoneuron identity in developing zebrafish embryos. *Science* **252,** 569–572.

Eisen, J. S. (1996). Zebafish make a big splash. *Cell* **87,** 969–977.

Eisen, J. S., Myers, P. Z., and Westerfield, M. (1986). Pathway selection by growth cones of identified motoneurons in live zebrafish embryos. *Nature* **320,** 269–271.

Eisen, J. S., Pike, S. H., and Debu, B. (1989). The growth cones of identified motoneurons in embryonic zebrafish select appropriate pathways in the absence of specific cellular interactions. *Neuron* **2,** 1097–1104.

Felsenfeld, A. L., Curry, M., and Kimmel, C. B. (1991). The *fub-1* mutation blocks initial myofibril formation in zebrafish muscle pioneer cells. *Dev. Biol.* **148,** 23–30.

Gaiano, N., Allende, M., Amsterdam, A., Kawakai, K., and Hopkins, N. (1996). Highly efficient germ-line transmission of proviral insertions in zebrafish. *Proc. Natl. Acad. Sci. USA* **93,** 7777–7782.

Grabher, C., Joly, J.-S., and Wittbrodt, J. (2004). Highly efficient zebrafish transgenesis mediated by the Meganuclease I-SceI. *In* "Methods in Cell Biology, Vol. 77, The Zebrafish 2nd Edition: Genetics, Genomics, and Informatics" (H. W. Detrich, III, M. Westerfield, and L. I. Zon, eds.), pp. 381–401. Elsevier Academic Press, San Diego, CA.

Grunwald, D. J. (1996). A fin-de-siècle achievement: charting new waters in vertebrate biology. *Science* **274,** 1634–1635.

Grunwald, D. J., Kimmel, C. B., Westerfield, M., Walker, C., and Streisinger, G. (1988). A neural degeneration mutation that spares primary neurons in the zebrafish. *Dev. Biol.* **126,** 115–128.

Grunwald, D. J., and Streisinger, G. (1992a). Induction of recessive lethal mutations in zebrafish with ultraviolet light. *Genet. Res.* **59,** 93–101.

Grunwald, D. J., and Streisinger, G. (1992b). Induction of recessive lethal and specific locus mutations in zebrafish with ethylnitrosourea. *Genet. Res.* **59,** 103–116.

Halpern, M. E., Ho, R. K., Walker, C., and Kimmel, C. B. (1993). Induction of muscle pioneers and floor plate is distinguished by the zebrafish *no tail* mutation. *Cell* **75,** 99–111.

Halpern, M. E., Thisse, C., Ho, R. K., Thisse, B., Riggleman, B., Trevarrow, B., Weinberg, E. S., Postlethwait, J. H., and Kimmel, C. B. (1995). Cell-autonomous shift from axial to paraxial meso-dermal development in zebrafish *floating head* mutants. *Development* **121,** 4257–4264.

Hatta, K., Kimmel, C. B., Ho, R. K., and Walker, C. (1991). The *cyclops* mutation blocks specification of the floor plate of the zebrafish central nervous system. *Nature* **350,** 339–341.

Henion, P. D., Raible, D. W., Beattie, C. E., Stoesser, K. L., Weston, J. A., and Eisen, J. S. (1996). Screen for mutations affecting development of zebrafish neural crest. *Dev. Gen.* **18,** 11–17.

Hermanson, S., Davidson, A. E., Sivasubbu, S., Balciunas, D., and Ekker, S. C. (2004). *Sleeping Beauty* transposon for efficient gene delivery. *In* "Methods in Cell Biology, Vol. 77, The Zebrafish 2nd Edition: Genetics, Genomics, and Informatics" (H. W. Detrich, III, M. Westerfield, and L. I. Zon, eds.), pp. 349–362. Elsevier Academic Press, San Diego, CA.

Herskowitz, I. (1987). Functional inactivation of genes by dominant negative mutations. *Nature* **329,** 219–222.

Holder, N., and McMahon, A. (1996). Genes from zebrafish screens. *Nature* **384,** 515–516.

Jessen, J. R., Meng, A., McFarlane, R. J., Paw, B. H., Zon, L. I., Smith, G. R., and Lin, S. (1998). Modification of bacterial artificial chromosomes through Chi-stimulated homologous recombination and its application in zebrafish transgenesis. *Proc. Natl. Acad. Sci. USA* **95,** 5121–5126.

Karlstrom, R. O., Trowe, T., Klostermann, S., Baier, H., Brand, M., Crawford, A. D., Grunewald, B., Haffter, P., Hoffmann, H., Meyer, S. U., Muller, B. K., Richter, S., *et al.* (1996). Zebrafish mutations affecting retinotectal axon pathfinding. *Development* **123**, 427–438.

Kawakami, K. (2004). Transgenesis and gene trap methods in zebrafish by using the *Tol2* transposable element. *In* "Methods in Cell Biology, Vol. 77, The Zebrafish 2nd Edition: Genetics, Genomics, and Informatics" (H. W. Detrich, III, M. Westerfield, and L. I. Zon, eds.), pp. 69–90. Elsevier Academic Press, San Diego, CA.

Kimmel, C. B. (1989). Genetics and early development of the zebrafish. *Trends Genet.* **5**, 283–288.

Kimmel, C. B., Ballard, W. W., Kimmel, S. R., Ullmann, B., and Schilling, T. F. (1995). Stages of embryonic development of the zebrafish. *Dev. Dyn.* **203**, 253–310.

Kimmel, C. B., Kane, D. A., Walker, C., Warga, R. M., and Rothman, M. B. (1989). A mutation that changes cell movement and cell fate in the zebrafish embryo. *Nature* **337**, 358–362.

Kimmel, C. B., and Warga, R. (1986). Tissue specific cell lineages originate in the gastrula of the zebrafish. *Science* **231**, 365–368.

Kimmel, C. B., and Westerfield, M. (1990). Primary neurons of the zebrafish. *In* "Signal and Sense" (G. M. Edelman, W. E. Gall, and M. W. Cowan, eds.), pp. 561–588. Wiley-Liss, New York.

Knapik, E. W., Goodman, A., Ekker, M., Chevrette, M., Delgado, J., Neuhauss, S., Shimoda, N., Driever, W., Fischman, M. C., and Jacob, H. J. (1998). A microsatellite genetic linkage map for zebrafish. *Nat. Genet.* **18**, 338–343.

Meng, X., Noyes, M. B., Zhu, L. J., Lawson, N. D., and Wolfe, S. A. (2008). Targeted gene inactivation in zebrafish using engineered zinc-finger nucleases. *Nat. Biotechnol.* **26**, 695–701.

Murphey, R. D., Stern, H. M., Straub, C. T., and Zon, L. I. (2006). A chemical genetic screen for cell cycle inhibitors in zebrafish embryos. *Chem. Biol. Drug Des.* **68**, 213–219.

North, T. E., Goessling, W., Walkley, C. R., Lengerke, C., Kopani, K. R., Lord, A. M., Weber, G. J., Bowman, T. V., Jang, I. H., Grosser, T., Fitzgerald, G. A., Daley, G. Q., *et al.* (2007). Prostaglandin E2 regulates vertebrate hematopoietic stem cell homeostasis. *Nature* **447**, 1007–1011.

Nüsslein-Volhard, C., and Wieschaus, E. (1980). Mutations affecting segment number and polarity in *Drosophila. Nature* **287**, 795–801.

Owens, K. N., Santos, F., Roberts, B., Linbo, T., Coffin, A. B., Knisely, A. J., Simon, J. A., Rubel, E. W., and Raible, D. W. (2008). Identification of genetic and chemical modulators of zebrafish mechanosensory hair cell death. *PLoS Genet.* **4**(2), e1000020.

Peterson, R. T., Link, B. A., Dowling, J. E., and Schreiber, S. L. (2000). Small molecule developmental screens reveal the logic and timing of vertebrate development. *Proc. Natl. Acad. Sci. USA* **97**, 12965–12969.

Peterson, R. T., Shaw, S. Y., Peterson, T. A., Milan, D. J., Zhong, T. P., Schreiber, S. L., MacRae, C. A., and Fishman, M. C. (2004). Chemical suppression of a genetic mutation in a zebrafish model of aortic coarctation. *Nat. Biotechnol.* **22**, 595–599.

Ransom, D. G., Haffter, P., Odenthal, J., Brownlie, A., Vogelsang, E., Kelsh, R. N., Brand, M., van Eeden, F. J. M., Furutani-Seiki, M., Granato, M., Hammerschmidt, M., Heisenberg, C.-P., *et al.* (1996). Characterization of zebrafish mutants with defects in embryonic hematopoiesis. *Development* **123**, 311–319.

Streisinger, G., Coale, F., Taggart, C., Walker, C., and Grunwald, D. J. (1989). Clonal origins of cells in the pigmented retina of the zebrafish eye. *Dev. Biol.* **131**, 60–69.

Streisinger, G., Singer, F., Walker, C., Knauber, D., and Dower, N. (1986). Segregation analyses and gene-centromere distances in zebrafish. *Genetics* **112**, 311–319.

Streisinger, G., Walker, C., Dower, N., Knauber, D., and Singer, F. (1981). Production of clones of homozygous diploid zebrafish (*Brachydanio rerio*). *Nature* **291**, 293–296.

Stuart, G. W., McMurray, J. V., and Westerfield, M. (1988). Replication, integration and stable germ-line transmission of foreign sequences injected into early zebrafish embryos. *Development* **103**, 403–412.

Trowe, T., Klostermann, S., Baier, H., Granato, M., Crawford, A. D., Grunewald, B., Hoffmann, H., Karlstrom, R. O., Meyer, S. U., Muller, B., Richter, S., Nüsslein-Volhard, C., *et al.* (1996). Mutations disrupting the ordering and topographic mapping of axons in the retinotectal projection of the zebrafish, *Danio rerio. Development* **123**, 439–450.

Walker, C., and Streisinger, G. (1983). Induction of mutations by gamma-rays in pregonial germ cells of zebrafish embryos. *Genetics* **103**, 125–136.

Weinholds, E., and Plasterk, R. H. A. (2004). Target-selected gene inactivation in zebrafish. *In* "Methods in Cell Biology, Vol. 77, The Zebrafish 2nd Edition: Genetics, Genomics, and Informatics" (H. W. Detrich, III, M. Westerfield, and L. I. Zon, eds.), pp. 69–90. Elsevier Academic Press, San Diego, CA.

Weinstein, B. M., Schier, A. F., Abdelilah, S., Malicki, J., Solnica-Krezel, L., Stemple, D. L., Stainier, D. Y. R., Zwartkruis, F., Driever, W., and Fishman, M. C. (1996). Hematopoietic mutations in the zebrafish. *Development* **123**, 303–309.

Weinstein, B. M., Stemple, D. L., Driever, W., and Fishman, M. C. (1995). *Gridlock*, a localized heritable vascular patterning defect in the zebrafish. *Nat. Med.* **1**, 1143–1147.

Westerfield, M., Stuart, G., and Wegner, J. (1993). Expression of foreign genes in zebrafish embryos. *Dev. Ind. Microbiol.* **2**, 658–665.

Yu, P. B., Hong, C. C., Sachidanandan, C., Babitt, J. L., Deng, D. Y., Hoyng, S. A., Lin, H. Y., Bloch, K. D., and Peterson, R. T. (2008). Dorsomorphin inhibits BMP signals required for embryogenesis and iron metabolism. *Nat. Chem. Biol.* **4**, 33–41.

Zhang, J., Talbot, W. S., and Schier, A. F. (1998). Positional cloning identifies zebrafish *one-eyed pinhead* as a permissive EGF-related ligand required during gastrulation. *Cell* **92**, 241–251.

CHAPTER 2

Cell Cycles and Development in the Embryonic Zebrafish

Donald A. Kane

Department of Biological Sciences
Western Michigan University
Kalamazoo, Michigan 49008

By heaven, man, we are turned round and round in this world, like yonder windlass, and Fate is the handspike.

Herman Melville, *Moby Dick* (1851)

I. Introduction

Based on its genetic accessibility, the zebrafish has become the *Drosophila* of vertebrate developmental biology. Zebrafish are easy to grow and care for, and they mature quickly, taking just three months per generation. Each mating produces large numbers of eggs, so that complementation testing and other genetic tests can be accomplished with only a few crosses. Many methods for genetic manipulations now exist (Chakrabarti *et al.*, 1983; Streisinger *et al.*, 1981; Walker and Streisinger, 1983), many of which were developed by George Streisinger in the early 1980s.

Moreover, the zebrafish is also an ideal system for cell biology, as first realized by Roosen-Runge (1937, 1938, 1939), when he used the wonderful transparency of

ESSENTIAL ZEBRAFISH METHODS: CELL AND DEVELOPMENTAL BIOLOGY
Reprinted from *Methods in Cell Biology*, Volume 59 (Academic Press, 1999).
11
DOI: 10.1016/B978-0-12-374599-6.00002-9

the zebrafish to create striking movies of developing embryos, both an artistic as well as a scientific achievement. Zebrafish embryos develop outside the mother allowing the direct observation of all stages of early development, and the optical clarity of the embryos allows individual cells to be followed *in vivo*. Because early development occurs without growth, morphogenesis is due primarily to cell rearrangement, and cell divisions even late in gastrulation are still essentially cleavage divisions. And, since the embryos are available in large numbers on a daily basis year round, experiments are easy to plan and do.

Thus the zebrafish system is especially suited for the study of the cell cycle, as genetic control and cell division can be studied together. Though still incomplete, a complete description of cell division in the zebrafish is emerging; the aim of this chapter is to chronicle these advances in the context of the early normal development of the zebrafish.

The other aim of this chapter is to serve as a ready reference within this volume for early stages of zebrafish development. Note that readers who wish any detail— or stages later than 24 h development—are encouraged to use the excellent staging series of Kimmel *et al.* (1995). Other descriptive and useful works include Hisaola and Battle (1958), Haffter *et al.* (1996), and Karlstrom and Kane (1996).

II. Terminology and the Staging Series

In cold-blooded creatures, the time postfertilization is often a poor description of developmental stage, for the rate of development changes with temperature. Moreover, even at a standard temperature, there are stage variations between clutches of embryos and even among siblings. Thus, in order to describe the developmental time of observations and experiments, it is preferable to stage embryos based on morphological criteria. Observations can be then reported as a stage or, less preferred, as hours of development at some standard temperature.

Zebrafish are staged by comparison to a standard morphological series (Kimmel *et al.*, 1995) incubated at 28.5 °C. The stages are named rather than numbered, making nomenclature easy to remember and making the names more meaningful to biologists working on other systems. There are hazards in the use of a completely morphological-based system. Synonyms confuse and due to variation—and experience—some stages disappear altogether. However, these trivial problems can be smoothed by careful and continuous staging.

III. The Zygote Period

The zygote period encompasses the first cell cycle of the embryo (Fig. 1). At fertilization the egg is a 500- to 600-μm sphere, with the chorion still closely opposed to the cell membrane of the zygote. Unlike the transparent embryos of later development, the zygote is a translucent mixture of yolk and cytoplasm, not

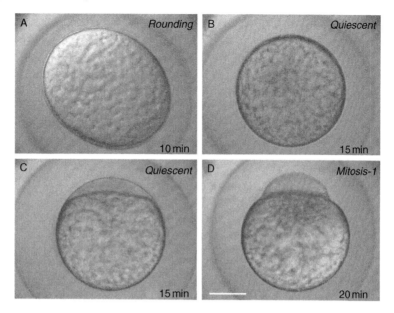

Fig. 1 The zygote period. (A) Mid-rounding stage, 10 min. (B) Animal pole view of mid-quiescent stage, 15 min. (C) Side view of mid-quiescent stage, 15 min. The clear crescent marks the animal pole. (D) Mitosis-1, 20 min. Cytoplasm is streaming toward the animal pole. Scale bar: 200 μm.

transparent at all. At the presumptive animal pole, there is a small divot of clear cytoplasm, the vestige of the germinal vesicle. In this hollow, the maternal nucleus can sometimes be distinguished with careful examination using Nomarski DIC optics.

The zygote period subdivides into four morphological stages. In the first, the flat stage, 6 min long, the zygote resembles a somewhat flattened soccer ball, inside an expanding chorion. In the second stage, the rounding stage, from 6 to 12 min development, the zygote becomes turgid, attaining a diameter of about 600 μm. From 12 to 18 min development, the embryos remain round with little obvious change, in the quiescent stage. By this stage the chorion has reached its final diameter of about 1 mm. Then, at 18 min development, the cytoplasm begins to stream towards the animal pole of the zygote, the first evidence of division-1.

The biology of the first cell cycle was exploited by Streisinger and coworkers in order to produce parthenogenetic diploids from the maternal pronucleus. A common genetic manipulation in zebrafish is to fertilize eggs stripped from an adult female with inactivated sperm, yielding haploid embryos (Streisinger *et al.*, 1981). If high hydrostatic pressure is applied to the eggs during the flat stage, the polar body fails to escape from the egg and is incorporated into the zygote. This produces diploid progeny. Since the female pronucleus and the second polar body share one set of sister chromotids, or in genetic terms, they form a half tetrad, recombination events are uncovered whenever loci are heterozygous.

Thus, the distance from any given locus to the centromere can be estimated by the frequency of heterozygous individuals in a clutch of embryos.

Later in the first cell cycle, other manipulations will produce clonal zebrafish from haploids. At the beginning of the quiescent stage, a short heat shock prevents the first mitotic division, allowing the female pronucleus to undergo a second round of replication before the normal mitosis-2. Thus, the ploidy of the embryo is doubled: Haploid zygotes become clonal diploids and diploid zygotes become tetraploids.

IV. The Cleavage Period

Commencing at the first mitosis, the embryo begins its divisions into blastomeres, divisions which are synchronous and rapid (Fig. 2). The separation of yolk platelets from the cytoplasm, the so-called "bipolar segregation" of Roosen-Runge (1938), clears the embryo, giving the zebrafish embryo its crystalline appearance. Throughout this period the embryo is staged by the number of cells in the embryo, from the 2-cell stage at approximately 30 min of development until the 64-cell stage at 2 h development.

The early cleavages are incomplete, or meroblastic. Cleavage furrows begin at the animal side of the dividing cell and terminate at the condensing body of yolk platelets. Thus, until the 8-cell stage, all of the blastomeres are cytoplasmically continuous with the yolk; there is neither a cell membrane nor any other obvious cellular feature separating the yolk from the forming blastomeres. Whilst the mechanism for the segregation of the yolk from the cytoplasm is unclear, the phenomenon is very useful, for the yolk is the route of choice for early injections into the entire blastoderm.

These early meroblastic cleavages are extremely stereotypic. The cleavage furrow of the second cleavage division aligns at $90°$ to the first to form a 2×2 array of cells at the animal pole. The furrows of the third cleavage align parallel to the plane of first, to form a 2×4 array of blastomeres at the 8-cell stage. And the furrows of the fourth cleavage align parallel to those of the second, to form a 4×4 array of blastomeres at the 16-cell stage. Note in Fig. 2E and F that the blastoderm adopts an ovoid shape, a reflection of the shape of the embryo at the 2-cell stage.

The cell cycle is relatively simple during cleavage, being evenly divided between mitosis and interphase (Fig. 3). This is a brilliant demonstration for students, watching the nuclei disappear, watching the cells round up for mitosis and divide, and watching daughter nuclei reform a few minutes later. Time-lapse microscopy reveals a number of further subdivisions and variations within the cycles. For example, in most cell types, including zebrafish cells after early cleavage, cytokinesis commences shortly after the beginning of anaphase. However, in the huge blastomeres at the 2-cell stage, there is a marked delay between anaphase and cytokinesis. This produces a short "apparent" interphase in the second interphase, with the nuclei sometimes disappearing only 2 or 3 min after the completion of the

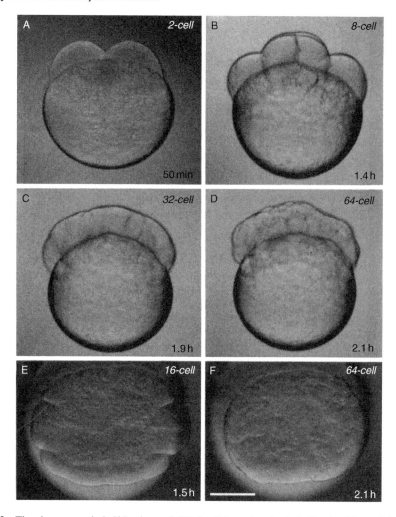

Fig. 2 The cleavage period. Side views of: (A) 2-cell in early mitosis-2, 50 min; (B) 8-cell in early mitosis-4, 1.4 h, showing a view of the long side of the 2 × 4 array; (C) 32-cell in mitosis-6, 1.9 h; and (D) 64-cell stage in interphase-7, 2.1 h. Animal pole views of: (E) 16-cell stage in interphase-5, 1.5 h and (F) 64-cell stage in interphase-7, 2.1 h. The long dimension of the ovoid blastoderm corresponds to the long dimension of the 2-cell stage, perpendicular to the first cleavage plane. Scale bar: 200 μm for A–D; 150 μm for E–F.

first cleavage furrow. With each division this delay diminishes, and it is insignificant by the 64-cell stage.

The separation of cytoplasm and yolk is linked to the cell cycle as well (Fig. 3). As cells round up at the beginning of each mitosis, there is a "loosening" of the cytoplasm, when small particles or organelles begin to move rapidly about the cells in a Brownian style of movement. Coincident with this subcellular movement,

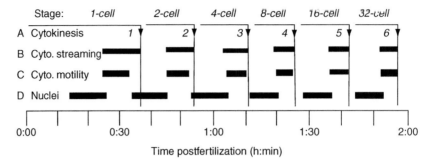

Fig. 3 Time course of cell cycle events during early cleavage from a time-lapse recording. (A) Cytokinesis. Arrows indicate initiation of cleavage furrows. The number next to each arrow indicates the division number, a useful alternative method used for staging parts of cycles, for example, "mitosis-5." (B) Yolk-cytoplasmic streaming, the movement of cytoplasm toward the animal pole. This recording indicates movement at the blastoderm-yolk margin; however, a transition zone of cytoplasmic streaming moves toward the vegetal pole, arriving there 6 to 10 min after initiation. (C) Cytoplasmic motility, the rapid movement of intercellular particles occurring during prophase and metaphase of each mitosis. (D) Record for interphase nuclei indicating when nuclei are visible with Nomarski DIC optics.

cytoplasm streams out of the yolk mass, which, in time-lapse recordings, appears as a pumping of cytoplasm. Then, at anaphase, both movements stop and the cytoplasm remains "gelled" until the next mitosis, some 10 min later.

V. The Blastula Period

This rather disparate period encompasses the late synchronous divisions of cleavage, the midblastula transition to a longer cell cycle, and the early stages of epiboly, covering a time from 2.3 to 5.3 h of development (Fig. 4). During this period the embryo begins the transition from maternal to zygotic control of development.

In early blastula, formal staging is by estimating the number of cells in the blastoderm, sometimes a challenge even for the experienced. Thus, the term "early blastula" is a good synonym for the 128-cell stage through the 2K-cell stage. In the midblastula, the stage is named for the shape of the embryo. The high blastula refers to the time when the blastoderm is perched high on the yolk cell and there is a visual line of yolk syncytial nuclei in the yolk cell; the oblong and sphere stages refer to the rounding of the embryo prior to epiboly.

During the early blastula, the cells of the blastoderm continue to divide in synchrony with a rapid 15 min cycle. In the later divisions there is a loss of global synchrony, with "metasynchronous" waves of mitoses spreading across the embryo. Beginning at cycle 10, at 3 h development, the cycle lengthens and cells lose local synchrony (Kane and Kimmel, 1993), defining the beginning of the midblastula transition. This lengthening and loss of synchrony corresponds to that seen at

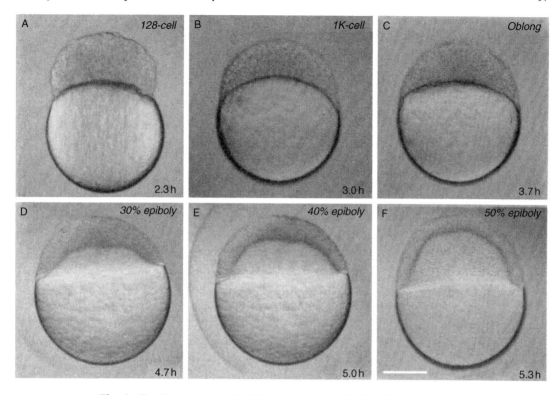

Fig. 4 The blastula period. (A) 128-cell stage, 2.3 h. (B) 1K-cell stage, 3.0 h. (C) oblong, 3.7 h. This stage is also referred to as 2K-cell stage. (D) 30% epiboly, 4.7 h. During early epiboly, the cells deep in the blastoderm radially intercalate with more superficial cells, causing extensive cell mixing. (E) 40% epiboly, 5 h. (F) 50% epiboly, 5.3 h. Note the thinning of the blastoderm. Scale bar: 200 μm.

cycle 12 in *Xenopus* (Newport and Kirschner, 1982) and that seen in *Drosophila* at cycle 10 (Edgar *et al.*, 1986).

As in frogs and flies, the time when lengthening occurs correlates with the attainment of a particular nucleocytoplasmic ratio: Cycle lengthening occurs one cycle early in tetraploid embryos and occurs one cycle late in haploid embryos (Kane and Kimmel, 1993). However, at least in the zebrafish, the volume of the cell—and thus the nucleocytoplasmic ratio—correlates closely with the cell cycle length, suggesting that the nucleocytoplasmic ratio continues to control the cell cycle for the next several cycles. During these cycles there is positive correlation between cycle lengths of daughter cell pairs with that of their mother (Fig. 5). This is expected if control is by the nucleocytoplasmic ratio, for if one assumes no growth during these divisions, the total volume of the daughters will equal that of the mothers.

As in both *Xenopus* (Newport and Kirschner, 1982) and *Drosophila* (Edgar *et al.*, 1986), cells acquire new behaviors as the cell cycle lengthens. Transcription is activated and in fish and *Xenopus*—*Drosophila* is still a syncytial blastoderm—motility is activated. This early motility consists of the cells randomly extending

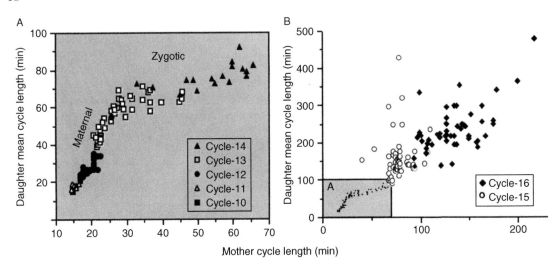

Fig. 5 Relationship of daughter-to-mother cycle length in the blastula. The labeled cycle refers to the daughter's cycle, that is, cycle 12 indicates a cycle-11 mother and its cycle-12 daughters. (A) Cycles 10 through 14, in the blastula period: Mother-to-daughter cycle length in cycles 10 through early 13 is tightly correlated; this correlation is lost in late cycle-13. Compared to subsequent cycles, cycle-14 is more uniform than expected as indicated by the flattening of the curve. Also, compare to Table I. (B) Cycles 15 and 16, in the gastrula and segmentation periods, showing the diverse lengthening of the cell cycle. The inset is data from A. Redrawn with permission from *Nature* (Kane *et al.*, 1992). Copyright 1992 by Macmillan Magazines Ltd.

short bleb-like pseudopods and exhibiting "circus" movements, where the blebs seem to rotate around the cell. These first movements seem quite random, resulting in little if any net cell movement.

During cycles 11 and 12, the blastoderm separates into three domains (Fig. 6). The cells at the yolk-blastoderm margin collapse into the yolk cell (Kimmel and Law, 1985); the nuclei and cytoplasm contributed by these cells is termed the *yolk syncytial layer* (YSL). The outer cells of the blastoderm become epithelial; these cells are termed the *enveloping layer* (EVL). The remaining cells, the *deep cells* of the blastoderm, continue to exhibit blebbing behavior. Each of these cell groups can be distinguished based on their mitotic cycle lengths (Kane *et al.*, 1992): The yolk cell retains a short cycle, the EVL cells acquire a long cycle, and the deep cells acquire an intermediate cycle between the two.

Corresponding to this morphogenetic transition, the midblastula transition ends, defined as the point when the nucleocytoplasmic ratio no longer correlates to the cell cycle, and corresponding roughly to when the cells of the blastoderm reach midcycle 13. Each of the zebrafish mitotic domains acquire unique roles during epiboly: The EVL forms an epithelium covering the blastoderm (Kimmel *et al.*, 1990), the YSL leads the blastoderm vegetal-wards (Kimmel and Law, 1985), and the deep cells of the blastoderm rearrange to form the embryonic anlagen (Warga and Kimmel, 1990).

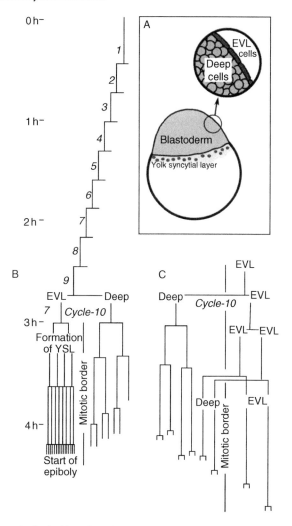

Fig. 6 Mitotic domains in the blastula. (A) Morphological subdivisions at the high blastula stage. The blastoderm consists of a monolayer of flattened EVL cells covering more loosely organized deep cells. The single-celled yolk contains a syncytium of nuclei, the YSL, formed from cells at the margin of the blastoderm. Later, during gastrulation, the deep cells form the germ layers. (B) Early lineage: EVL cells give rise to either a pair of EVL cells or a EVL cell and a deep cell. Thus, EVL cells at the yolk-blastoderm margin are derived from a succession of EVL cells beginning at the 1-cell stage. Late lineage: At division-9, a EVL cell divides to form a EVL and a deep cell. The EVL cell divides to form two EVL daughters, which collapse into the yolk cell after mitosis-10 and become part of the YSL mitotic domain. Note the rapid cycle of the YSL compared to the deep cells. (C) Formation of three separate deep cell lineages and EVL lineage from a single EVL cell. Note that the EVL-deep divisions always result in asymmetric divisions, with the EVL sibling having the longer cycle. Note also that the cells contributed to the deep domain at division-11 have cycles closer in length to their deep cell cousins than their sibling EVL cells. Redrawn with permission from *Nature* (Kane *et al.*, 1992). Copyright 1992 by Macmillan Magazines Ltd.

Morphogenesis begins at 4.5 h with the epibolic spread of the blastoderm vegetal-wards to cover the yolk cell. At this stage, beginning midway through cycle 13, the mother-daughter correlation is lost, suggesting that cycles depart from the nucleocytoplasmic control that defines the cycles of the midblastula transition. In the deep cells, the cell cycle of the daughters becomes shorter than expected in late cycle 13 and cycle 14 (plateau in Fig. 6A). EVL cell cycles lengthen during midcycle 13, to reach a long cycle 14 that extends through gastrulation, and beyond. However, the departure from nucleocytoplasmic control is the most dramatic in the yolk cell, which after cycle lengthening during midblastula, enters mitotic arrest during interphase-14, similar to what has been observed in the teleost *Fundulus* (Trinkaus, 1992, 1993). Recent evidence from UV-treatment of zebrafish embryos at the blastula and epiboly stages suggests that the yolk cell uses microtubular motors to carry the blastoderm (Strähle and Jesuthasan, 1993). Based on the use of microtubules, cell movement and cell mitosis are typically antagonistic behaviors (Trinkaus, 1980); therefore, it is possible that the movement of the YSL forces its mitotic arrest.

The loss of nucleocytoplasmic control may indicate the requirement for zygotic regulation of the cell cycle, as in *Drosophila* cycle 14 (Edgar and O'Farrell, 1989); indeed, in zebrafish embryos that have been treated with alpha-amanitin, cell cycle length abnormalities first appear in cycle 15 (Kane, unpublished observations; see also Kane *et al.*, 1996a).

VI. The Gastrula Period

The gastrula period extends from 5.5 to about 10 h, the time when the germ layers begin to form (Fig. 7). Throughout this time epiboly continues, and percentage epiboly continues as the staging convention.

At 50% epiboly, at 6 h development, the rim of the blastoderm thickens into a bilayered structure termed the germ-ring, consisting of an outer layer, the *epiblast*, and an inner layer, the *hypoblast*. The appearance of the germ-ring marks the beginning of gastrulation, the period when the germ layers of the embryo arise: the epiblast forms the embryonic ectoderm and the hypoblast forms the embryonic mesoderm and endoderm (Warga and Kimmel, 1990). Appearing simultaneously with the germ-ring is the shield, an accentuated thickening of the germ-ring (Fig. 7A). Note that the appearance of the shield is the earliest morphological clue to the zebrafish dorsal side.

As the radial intercalations of the late blastula become less extreme, it is possible to predict the future fate of regions of the embryo (Kimmel *et al.*, 1990). Such a fate map, made at the blastula-to-gastrula transition, overtly resembles the gastrula fate map of *Xenopus* (Keller, 1975, 1976). The cells at the margin of the blastoderm make mesodermal and endodermal structures, and cells further from the margin make ectodermal structures. Cells on the dorsal side of the blastoderm make axial structures and tend to contribute more to anterior structures; cells on the ventral side make lateral and paraxial structures and tend to contribute more to posterior structures.

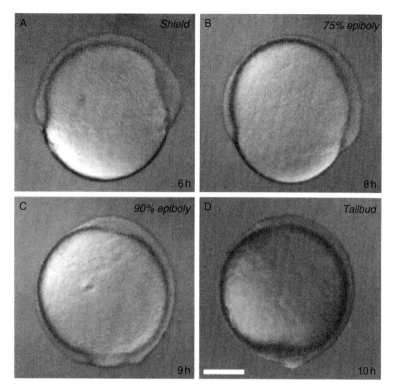

Fig. 7 The gastrulation period. (A) Shield stage, 6 h. The thickening on the dorsal side is the shield; the lesser thickening on ventral side is representative of the germ-ring that completely circles the blastoderm margin. The outer layer of the deep cells at this stage is termed the epiblast, which forms to ectodermal fates; the inner layer is termed the hypoblast, which forms to the mesodermal and endodermal fates. This stage is sometimes referred to as 60% epiboly. (B) 75% epiboly, 8 h. The thickening of the anterior portion of the shield, the anlage of the prechordal plate, has extended halfway from the equator to the animal pole. (C) 90% epiboly, 9 h. Prechordal plate anlage, seen here as a small lens of material under the epiblast, has reached the animal pole. (D) Tailbud stage (or bud stage), 9.5 h. Completion of epiboly; it is thought that gastrulation continues on the ventral side of the tailbud (Kanki and Ho, 1997). The dorsal side is toward right in all panels. Scale bar: 200 μm.

As gastrulation begins, most of the cells of the embryo are in cycle 14 (Fig. 8, see color plate), the first cycle completely released from nucleocytoplasmic control. This cycle is more uniform in length than those before or those afterwards. The standard deviation of cycle 14 is 10% of its total length (Table I), much shorter than the standard deviation of cycles 11, 12, and 13, at about 20% of the cycle length, and of cycles 15 and 16, at about 40% of cycle length. Thus, cycle 14 seems special, and as it connects the blastula to the gastrula, it may have similarities to the 14th cycle in *Drosophila* (Foe, 1989).

The first terminal divisions of differentiating cells occur in the gastrula. Some of these divisions may be at mitosis-14, but most are in mitosis-15. These cells with

Blastula *Gastrulation* **Segmentation**

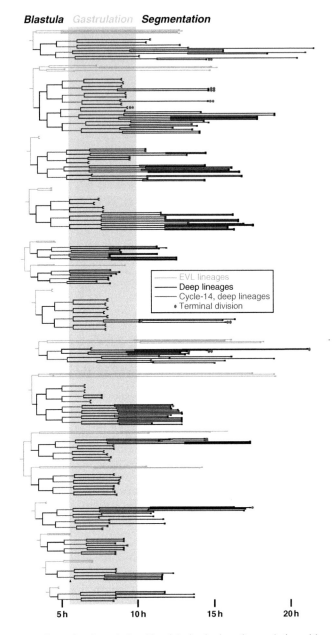

Fig. 8 Lineage recordings, showing relationship of the beginning of gastrulation with cycle-14. Figure is redrawn from Kimmel *et al.* (1994). Enveloping cell lineages are in green; deep cell lineages are in black except for deep cell cycle-14, which is in red. Lineages are from blastomeres injected with florescent tracers at mitosis-10, mitosis-11, or mitosis-12. (See Plate no. 1 in the Color Plate Section.)

Table I

Lengths for Cell Cycles 7 Through 16: Cycle 14 is the Most Uniform Cycle Subsequent to the Midblastula Transition

Cycle	Average length[a] (min)	Standard deviation (min)	Percentage standard deviation (%)
7	14.2	±0.5	3.5
8	14.5	±0.5	3.4
9	15.0	±0.5	3.2
10	17.0	±1.0	6
11	22.5	±5	22
12	33.5	±8	24
13	54	±9	17
14	78	±8	10
15	151	±59	39
16	240	±71	39

[a]Based on Kane and Kimmel (1993), Kimmel *et al.* (1994), and the author's unpublished data.

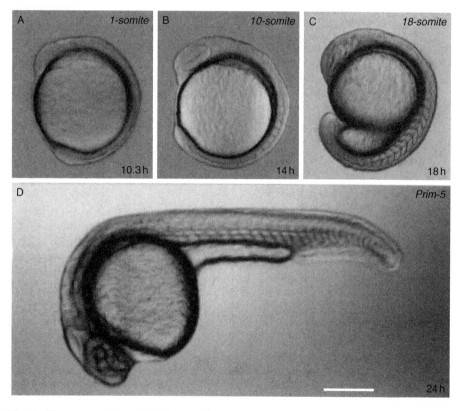

Fig. 9 The segmentation period. (A) 1-somite stage, 10.3 h. (B) 10-somite stage, 14 h. (C) 18-somite stage, 18 h. (D) prim-5 stage, 24 h. The dorsal side is towards right in A-C and towards top in D. Scale bar: 300 μm for A-C and 200 μm for D.

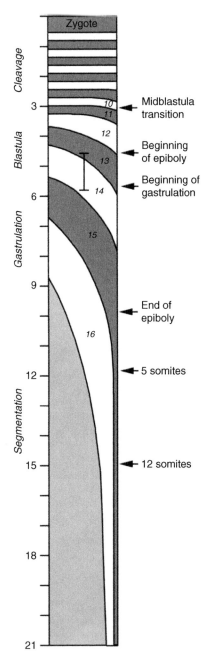

Fig. 10 Distribution of cell cycles during early development, based on Table I. Hours postfertilization and periods are shown on left edge. The cell cycle begins to lose synchrony, indicated by the curvature of the cycle boundary, at the midblastula transition. Due to the early loss of cell synchrony, the embryo is always a mix of two or more cell cycles after cycle 12. For example, at the beginning of epiboly, the

early birthdays include primarily those close to the axial midline of the embryo, such as notochord, floorplate, hatching gland cells, some cells of the somite muscle, and many of the large motor neurons of the brain and spinal cord (Kimmel *et al.*, 1994; Mendelson, 1986).

VII. The Segmentation Period

This long period encompasses the remainder of the first day of development, from 10 to 24 h of development, a period when the embryo begins the subdivision of the body plan (Fig. 9). One of the most conspicuous features of this period is the rhythmic segmentation of the paraxial mesoderm into somites, the means of staging this period.

Using only a dissecting scope, many other features can be seen in the embryo. The eye primordium and the ear primordium appear. The brain neuroectoderm thickens and, by the 18-somite stage, the segmentation of the brain is evident, both in broad subdivisions, such as the forebrain, the telencephalon, the diencephalon, the midbrain, and the hindbrain; and in finer subdivisions, such as the rhombomeres of the hindbrain. The cells of the notochord begin to expand and straighten out the tail of the embryo. With the formation of the horizontal myoseptum, somites take on their herringbone pattern.

At this stage, when much of the embryo is in cycle 16, many cells are leaving the cell cycle as part of a major wave of differentiation (Kimmel *et al.*, 1994). Mutants that arrest early in development, such as *zombie*, *speed bump*, and *ogre* first show their phenotypes in these stages (Kane *et al.*, 1996b). Since many of these mutants produce abnormal nuclei, it is reasonable to suspect they are mutations in genes that necessary for the cell cycle. If so, it is likely that they are producing their mitotic arrests at the end of cycles 15 or 16. Interestingly, and consistent with this hypothesis, cells that have their terminal divisions before cycle 16, such as notochord and hatching gland cells, are unaffected in these early arrest mutants.

At the completion of the first day of development, the embryo has between 80,000 and 100,000 cells (Fig. 10). With notable major exceptions, such as neural crest-derived structures of the jaw and endodermally derived structures of the gut tube, the development of the major systems of the embryo are laid out. The embryos are touch sensitive and their hearts will start beating. Soon, if the fish need be placed under the microscope, they must first be chased and caught.

embryo is a mix of midcycle 13 and early cycle 14; at the beginning of gastrulation, the embryo is a mix of late cycle 13, cycle 14, and early cycle 15. The bar in cycle 14 shows the average length of cycle 14 (from Table I). Many cells have terminal mitoses in cycle 15 and cycle 16, indicated by trailing the cycle boundaries off the bottom of the figure. Not shown are cycles after cycle 16 (diagonal shading); however, many lineages reach cycle 18 by 24 h development.

References

Chakrabarti, S., Streisinger, G., Singer, F., and Walker, C. (1983). Frequency of gamma-ray induced specific locus and recessive lethal mutations in mature germ cells of the zebrafish, *Brachydanio rerio*. *Genetics* **103**, 109–123.

Edgar, B. A., Kiehle, C. P., and Schubiger, G. (1986). Cell cycle control by the nucleo-cytoplasmic ratio in early *Drosophila* development. *Cell* **44**, 365–372.

Edgar, B. A., and O'Farrell, P. H. (1989). Genetic control of cell division patterns in the *Drosophila* embryo. *Cell* **57**, 177–187.

Foe, V. E. (1989). Mitotic domains reveal early commitment of cells in *Drosophila* embryos. *Development* **107**, 1–22.

Haffter, P., Granato, M., Brand, M., Mullins, M. C., Hammerschmidt, M., Kane, D. A., Odenthal, J., van Eeden, F. J. M., Jiang, Y.-J., Heisenberg, C.-P., Kelsh, R. N., Furutani-Seiki, M., *et al.* (1996). The identification of genes with unique and essential functions in the development of the zebrafish, *Danio rerio*. *Development* **123**, 1–36.

Hisaola, K. K., and Battle, H. I. (1958). The normal developmental stages of the zebrafish, *Brachydanio rerio*. (Hamilton-Buchanan). *J. Morphol.* **102**, 311–328.

Kane, D. A., Hammerschmidt, M., Mullins, M. C., Maischein, H. M., Brand, M., van Eeden, F. J., Furutani-Seiki, M., Granato, M., Haffter, P., Heisenberg, C. P., Jiang, Y. J., Kelsh, R. N., *et al.* (1996a). The zebrafish epiboly mutants. *Development* **123**, 47–55.

Kane, D. A., and Kimmel, C. B. (1993). The zebrafish midblastula transition. *Development* **119**, 447–456.

Kane, D. A., Maischein, H. M., Brand, M., van Eeden, F. J., Furutani-Seiki, M., Granato, M., Haffter, P., Hammerschmidt, M., Heisenberg, C. P., Jiang, Y. J., Kelsh, R. N., Mullins, M. C., *et al.* (1996b). The zebrafish early arrest mutants. *Development* **123**, 57–66.

Kane, D. A., Warga, R. M., and Kimmel, C. B. (1992). Mitotic domains in the early embryo of the zebrafish. *Nature* **360**, 735–737.

Kanki, J. P., and Ho, R. K. (1997). The development of the posterior body in zebrafish. *Development* **124**, 881–893.

Karlstrom, R. O., and Kane, D. A. (1996). A flipbook of zebrafish embryogenesis. *Development* **123**, 461.

Keller, R. E. (1975). Vital dye mapping of the gastrula and neurula of *Xenopus laevis*. I. Prospective areas and morphogenetic movements of the superficial layer. *Dev. Biol.* **42**, 222–241.

Keller, R. E. (1976). Vital dye mapping of the gastrula and neurula of *Xenopus laevis*. II. Prospective areas and morphogenetic movements in the deep region. *Dev. Biol.* **51**, 118–137.

Kimmel, C. B., Ballard, W. W., Kimmel, S. R., Ullmann, B., and Schilling, T. F. (1995). Stages of embryonic development of the zebrafish. *Dev. Dyn.* **203**, 253–310.

Kimmel, C. B., and Law, R. D. (1985). Cell lineage of zebrafish blastomeres. II. Formation of the yolk syncytial layer. *Dev. Biol.* **108**, 86–93.

Kimmel, C. B., Warga, R. M., and Kane, D. A. (1994). Cell cycles and clonal strings during the formation of the zebrafish nervous system. *Development* **120**, 265–276.

Kimmel, C. B., Warga, R. M., and Schilling, T. F. (1990). Origin and organization of the zebrafish fate map. *Development* **108**, 581–594.

Mendelson, B. (1986). Development of reticulospinal neurons of the zebrafish. I. Time of origin. *J. Comp. Neurol.* **251**, 160–171.

Newport, J., and Kirschner, M. (1982). A major developmental transition in early *Xenopus* embryos. I. Characterization and timing of cellular changes at the midblastula stage. *Cell* **30**, 675–686.

Roosen-Runge, E. (1937). Observations on the early development of the zebrafish, *Brachydanio rerio*. *Anat. Rec.* **70**, 103.

Roosen-Runge, E. (1938). On the early development—Bipolar differentiation and cleavage—Of the zebra fish, *Brachydanio rerio*. *Biol. Bull.* **75**, 119–133.

Roosen-Runge, E. (1939). Karyokinesis during cleavage of the zebrafish, *Brachydanio rerio*. *Biol. Bull.* **77,** 79–91.

Strähle, U., and Jesuthasan, S. (1993). Ultraviolet irradiation impairs epiboly in zebrafish embryos: Evidence for a microtubule-dependent mechanism of epiboly. *Development* **119,** 909–919.

Streisinger, G., Walker, C., Dower, N., Knauber, D., and Singer, F. (1981). Production of clones of homozygous diploid zebrafish, *Brachydanio rerio*. *Nature* **291,** 293–296.

Trinkaus, J. P. (1980). Formation of protrusions of the cell surface during tissue cell movement. *Prog. Clin. Biol. Res.* **41,** 887–906.

Trinkaus, J. P. (1992). The midblastula transition, the YSL transition and the onset of gastrulation in *Fundulus*. *Development* **116(Suppl.),** 75–80.

Trinkaus, J. P. (1993). The yolk syncytial layer of *Fundulus:* Its origin and history and its significance for early embryogenesis. *J. Exp. Zool.* **265,** 258–284.

Walker, C., and Streisinger, G. (1983). Induction of mutations by gamma-rays in pregonial germ cells of zebrafish embryos. *Genetics* **103,** 125–136.

Warga, R. M., and Kimmel, C. B. (1990). Cell movements during epiboly and gastrulation in zebrafish. *Development* **108,** 569–580.

CHAPTER 3

Primary Fibroblast Cell Culture

Barry H. Paw and Leonard I. Zon

Howard Hughes Medical Institute and Division of Hematology-Oncology
Children's Hospital and Dana-Farber Cancer Institute
Department of Pediatrics
Harvard Medical School
Boston, Massachusetts 02115

I. Introduction

Since the original publication of our chapter detailing the culture of zebrafish-derived cells, a few more zebrafish adherent cells have become available. In addition to the original cell lines, *ZF-4* (Driever and Rangini, 1993), *AB.9*, and *SJD.1* (Paw and Zon, 1999), which are available from the American Type Culture Collection (ATCC, www.atcc.org), an additional embryonic fibroblast line, *ZEM2S* (Collodi *et al.*, 1992), and hepatocyte line, *ZF-L* (Ghosh *et al.*, 1994), have been added to the cell repository. Other zebrafish adherent lines, such as *PAC2* cells, have been reported from Max-Planck-Institute for Developmental Biology, Tubingen, Germany (Vallone *et al.*, 2007). Several of these cells have been successfully adapted for growth in full or partial mixture of

ESSENTIAL ZEBRAFISH METHODS:
CELL AND DEVELOPMENTAL BIOLOGY
Copyright 2009, Elsevier Inc. All rights reserved.

DOI: 10.1016/B978-0-12-374599-6.00003-0

Lebovitz's L-15 medium, which does not require CO_2 for pH buffering (Trede *et al.*, 2007; Vallone *et al.*, 2007). In the near future culture conditions for hematopoietic cells of erythroid and myeloid lineages from zebrafish will become available with the recent derivation of stromal feeder cells from kidney marrow (Stachura *et al.*, 2009) and zebrafish-specific cytokines, such as erythropoietin (Paffert-Lugassay *et al.*, 2007). Much of our basic cell culture conditions based on mammalian cell culture methods have remained the same.

The recent descriptions of mutants of developmental importance that have been recovered from the zebrafish mutagenesis screens attests to the genetic and embryological advantages of this vertebrate model (Driever *et al.*, 1996; Haffter *et al.*, 1996). In contrast to mouse and human model systems, *in vitro* biochemical and molecular methodologies that are widely used to complement the *in vivo* genetic findings are not as developed in zebrafish. The establishment of stable cell lines that can be easily cultured for *in vitro* manipulations is a preliminary step.

Stable cell lines from a variety of fish, such as carp (Shima *et al.*, 1980), bluegill sunfish (Borenfreund *et al.*, 1989), and Chinook salmon (Wolf and Mann, 1980), have been reported. Stable cell lines derived from embryonic zebrafish have also been reported (Collodi *et al.*, 1992; Driever and Rangini, 1993; Sun *et al.*, 1995a). The establishment of these embryonic lines required complex growth media supplements, such as bovine basic fibroblast growth factor (FGF), recombinant leukemia inhibitory factor (LIF), bovine insulin, embryo extract, and trout serum (Sun *et al.*, 1995a). Although stably maintained in the standard Dulbecco's modified Eagle's medium (DMEM) media widely used in mammalian tissue culture, the ZF4 line derived from 1-day-old zebrafish embryos exhibits transformed features such as the loss of contact inhibition, the formation of foci, and hyperdiploid karyotype (Driever and Rangini, 1993). The generation of pluripotential ES-like cell cultures derived from blastula-stage zebrafish embryos have been described by Barnes and his colleagues (Chapter 3; Sun *et al.*, 1995a).

Here we describe a way to culture stable primary fibroblast cell lines from adult zebrafish caudal fins in standard tissue culture medium with bovine serum. These primary fibroblasts retain many features of nontransformed cells such as eudiploidy, contact inhibition, and surface adhesion. *In vitro* applications of these cultured fibroblasts are briefly discussed.

II. Material and Methods

A. Basic Tissue Culture Techniques

We refer readers to an extensive discussion of the general techniques of *in vitro* cell culture in this book (Chapter 3).

B. Zebrafish Strains

AB and SJD zebrafish strains were maintained on a 14–10 h light-dark diurnal cycle according to standard conditions (Westerfield, 1994). The SJD zebrafish strain, a near-isogenic line derived from the Darjeeling strain, was kindly provided by S. L. Johnson (Washington University).

C. Caudal Fin Amputation

Fish were anesthetized in 0.02% tricaine pH 7 (3-aminobenozic acid ethyl ester, Sigma). Caudal fins from the fully anesthetized fish were carefully amputated using sharp dissection scissors (Westerfield, 1994); fins often regenerate within 7 days in healthy donors. The amputated fins were transferred to sterile 50-ml conical tubes containing sterile 0.9 × phosphate buffered saline (PBS) supplemented with 100 units/ml penicillin and 100 μg/ml streptomycin (GIBCO). All aseptic procedures and sterile instruments were used in subsequent steps. To minimize microbial contamination, the fins were extensively washed 10 times in 50-ml volumes of PBS with antibiotics by serial resuspension and centrifugation at 2000–3000 rpm for 5–10 min each. After the last PBS wash, the fins were transferred to "tissue-culture treated" 24-well flat-bottom plates (Corning), ensuring the transfer of one fin piece per well rather than plating fully disaggregated cells. It is advisable to plate several independent wells in the event of microbial contamination in any one well.

D. Cell Culture

All cell culture was performed at 28–29 °C with 5% CO_2 atmospheric pressure. The culture medium used was Dulbecco's modified Eagle's medium (DMEM) low glucose (GIBCO) supplemented with 3 mM glutamine, 100 units/ml penicillin, 100 μg/ml streptomycin, and 15% heat-inactivated fetal bovine serum. During the first 2–3 weeks, fresh medium was added every 7 days to the existing medium, rather than aspirating the existing medium. Care was taken not to aspirate the remnants of the fin pieces. The emergence of tiny fibroblast-like cells was observed immediately adjacent to the fin pieces within the first 3 days. Confluency of the cells was achieved within the first 6–8 weeks. The confluent cells may be transferred into 6-well plates at this stage for an additional 2–4 weeks. A subsequent gradual expansion into 25 and 125-cm^2 flasks was achieved by dissociation in 0.25% trypsin in 0.9 × PBS. The medium was changed every 7 days.

E. Cryopreservation

Cells that achieved confluency were given a fresh medium change 1 day prior to the anticipated date of freezing. The following day, the cells were washed once in 0.9 × PBS and dissociated with 0.25% trypsin. Detached cells were washed in 20 ml of complete DMEM medium to inactivate the trypsin, recentrifuged, and finally

resuspended in 1 ml of complete DMEM medium (as above) with 10% dimethylsulfoxide. After being transferred to cryopreservation vials, the cells were placed in
a prechilled 4 °C StrataCooler insulation container (Stratagene) and then placed
in a −80 °C freezer overnight. There was ∼60–70% viable recovery of frozen
primary fibroblasts.

F. Cell Lines

Primary fibroblast lines from adult caudal fins, AB9 (derived from the AB
zebrafish strain), and SJD1 (derived from the SJD zebrafish strain) have been
deposited in the American Type Culture Collection, Rockville, MD.

III. Results and Discussion

As shown in Fig. 1, one of the stable fibroblast lines, AB9, derived from an AB
strain, exhibits contact inhibition and surface-growth dependency. A karyotype
analysis of these cells showed eudiploid chromosome count (50 chromosomes/
diploid cell) (Kwok *et al.*, 1998). The cells have been maintained in continuous
culture for over 35 passages (over 1 year duration) without change in characteristics. It is uncertain whether these cells retain eudiploid karyotype with the high
passage number.

A continuous source of stable cultured fish cells offers a variety of applications
(Hightower and Renfro, 1988). In addition, high-quality zebrafish genomic DNA
can be easily obtained from these cells. These primary fibroblasts retain the same
genetic polymorphism as the original donor fish when we analyzed a variety of
zebrafish genes; therefore, cultured cells derived from mutant zebrafish could prove
useful in positional cloning and the identification of mutant genes in much the same
way that cultured fibroblasts are in human genetic disease research. This source of
normal fibroblasts has been used to generate high-molecular-weight DNA for one
of the existing zebrafish YAC libraries (Zon *et al.*, 1996) and cell fusion to generate a
radiation hybrid mapping panel (Kwok *et al.*, 1998), both of which are commercially available (from Genome Systems, St. Louis, MO, and Research Genetics,
Huntsville, AL, respectively). Fluorescent *in situ* hybridization (FISH) has been
successfully applied using zebrafish probes on both interphase and metaphase
spreads of cultured fibroblasts (M. Lalande *et al.*, unpublished; Chapter 18,
Vol. 60). These same cells could also serve as feeder layers for coculturing more
demanding cells (Sun *et al.*, 1995b) and as a source for conditioned media in
culturing other cell types (C. Sieff, unpublished). In addition, *in vitro* cytotoxicity
assays using cultured fish cell lines have been well established (Babich *et al.*, 1989;
Borenfreund *et al.*, 1989; Yasuhira *et al.*, 1992). Transfection assays in cultured
zebrafish cells to study transcriptional regulation by enhancer and promoter
elements (Sharps *et al.*, 1992; Driever and Rangini, 1993) have also been described.

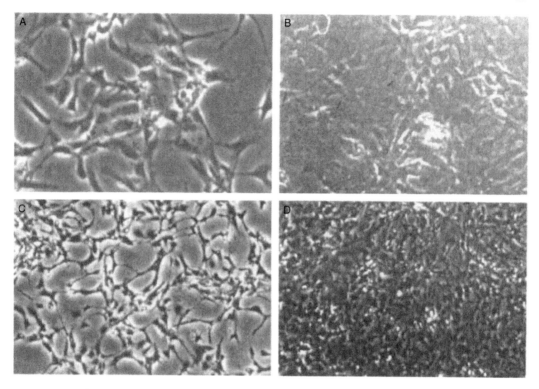

Fig. 1 Phase contrast photography of primary fibroblasts derived from zebrafish caudal fins. Fibroblasts growing at low density: ×40, upper panel (A); ×10, lower panel (B). Fibroblasts growing at high density, showing contact inhibition: ×40, upper panel (C); and ×10, lower panel (D).

Acknowledgments

This work was supported in part by grants from the NIH. B.H.P. is a recipient of a Howard Hughes Medical Institute Postdoctoral Fellowship for Physicians. L.I.Z. is an Associate Investigator of the Howard Hughes Medical Institute.

References

Babich, H., Martin-Alguacil, N., and Borenfreund, E. (1989). Arsenic-selenium interactions determined with cultured fish cells. *Toxicol. Lett.* **45,** 157–164.

Borenfreund, E., Babich, H., and Martin-Alguacil, N. (1989). Effects of methylazoxymethanol acetate on bluegill sunfish cell cultures *in vitro. Ecotoxicol. Environ. Saf.* **17,** 297–307.

Collodi, P., Kamei, Y., Ernst, T., Miranda, C., Buhler, D. R., and Barnes, D. W., (1992). Culture of cells from zebrafish (*Brachydanio rerio*) embryo and adult tissue. *Cell Biol. Toxicol.* **8,** 43–61.

Driever, W., and Rangini, Z. (1993). Characterization of a cell line derived from zebrafish (*Brachydanio rerio*) embryos. *In Vitro Cell. Dev. Biol. Anim.* **29A,** 749–754.

Driever, W., Solnica-Krezel, L., Schier, A., Neuhauss, S., Malicki, J., Stemple, D., Stainier, D., Zwartkruis, F., Abdelilah, S., Rangini, Z., Balak, J., and Boggs, C. (1996). A genetic screen for mutations affecting embryogenesis in zebrafish. *Development* **123**, 37–46.

Ghosh, C., Zhou, Y. L., and Collodi, P. (1994). Derivation and characterization of a zebrafish liver cell line. *Cell Biol. Toxicol.* **10**, 167–176.

Haffter, P., Granato, M., Brand, M., Mullins, M., Hammerschmidt, M., Kane, D., Odenthal, J., van Eeden, F., Jiang, Y., Heisenberg, C., Kelsh, R., Furtani-Seiki, M., *et al.* (1996). The identification of genes with unique and essential functions in the development of the zebrafish. *Danio rerio. Development* **123**, 1–36.

Hightower, L., and Renfro, J. (1988). Recent applications of fish cell culture in biomedical research. *J. Exp. Zool.* **248**, 290–302.

Kwok, C., Kron, R., Davis, M., Burt, D., Critcher, R., Paw, B., Zon, L., Goodfellow, P., and Schmitt, K. (1998). Characterization of whole genome radiation hybrid mapping resources for non-mammalian vertebrates. *Nucleic Acid Res.* **26**, 3562–3566.

Paffett-Lugassy, N., Hsia, N., Fraenkel, P. G., Paw, B., Leshinsky, I., Barut, B., Bahary, N., Caro, J., Handin, R., and Zon, L. I. (2007). Functional conservation of erythropoietin signaling in zebrafish. *Blood* **110**, 2718–2726.

Paw, B. H., and Zon, L. I. (1999). Primary Fibroblast Cell Culture. *Methods Cell Biol.* **59**, 39–43.

Sharps, A., Nishiyama, K., Collodi, P., and Barnes, D. (1992). Comparison of activities of mammalian viral promoters directing gene expression *in vitro* in zebrafish and other fish cell lines. *Mol. Mar. Biol. Biotechol.* **1**, 426–431.

Shima, A., Nikaido, S., Shinohara, S., and Egami, N. (1980). Continued *in vivo* growth of fibroblast-like cells (RBCF-1) derived from the caudal fin of the fish, *Carassius auratus. Exp. Gerontol.* **15**, 305–314.

Stachura, D. L., Reyes, J. R., Bartunek, P., Paw, B. H., Zon, L. I., and Traver, D. (2009). Zebrafish kidney stromal cell lines support multilineage hematopoiesis. *Blood*, in press. (Epublished May 2009).

Sun, L., Bradford, C., Ghosh, C., Collodi, P., and Barnes, D. (1995a). ES-like cell cultures derived from early zebrafish embryos. *Mol. Mar. Biol. Biotechnol.* **4**, 193–199.

Sun, L., Bradford, C., and Barnes, D. (1995b). Feeder cell cultures for zebrafish embryonal cells *in vitro. Mol. Mar. Biol. Biotechnol.* **4**, 43–50.

Trede, N. S., *et al.* (2007). Network of coregulated splicesome components revealed by zebrafish mutant in recycling factor p110. *Proc. Natl. Acad. Sci. USA* **104**, 6608–6613.

Vallone, D., Santoriello, C., Gondi, S. B., and Foulkes, N. S. (2007). Basic protocols in zebrafish cell lines: Maintenance and transfection. *Methods Mol. Biol.* **362**, 429–441.

Westerfield, M. (1994). *In* "The Zebrafish Book," 2nd edn., (M. Westerfield, ed.). Institute of Neuroscience, University of Oregon Press, Eugene, OR.

Wolf, K., and Mann, J. A. (1980). Poikilotherm vertebrate cell lines and viruses: A current listing for fishes. *In Vitro* **16**, 168–179.

Yasuhira, S., Mitani, H., and Shima, A. (1992). Enhancement of photorepair of ultraviolet-induced pyrimidine dimers by preillumination with fluorescent light in the goldfish cell line. The relationship between survival and yield of pyrimidine dimers. *Photochem. Photobiol.* **55**, 97–101.

Zon, L. I., Donovan, A., Paw, B., Thompson, M., Ransom, D., Brownlie, A., Guo, W., Pratt, S., Liao, E., Amemiya, C., and Silverman, G. (1996). *In* "Preparation of a Zebrafish YAC Library. Zebrafish Development and Genetics", p. 211. Cold Spring Harbor Laboratory, New York.

CHAPTER 4

Production of Haploid and Diploid Androgenetic Zebrafish (Including Methodology for Delayed *In Vitro* Fertilization)

Graham E. Corley-Smith,[*] **Bruce P. Brandhorst,**[†]
Charline Walker,[*] **and John H. Postlethwait**[*]

[*]Institute of Neuroscience
University of Oregon
Eugene, Oregon 97403-1254

[†]Institute of Molecular Biology and Biochemistry
Simon Fraser University
Burnaby, British Columbia, V5A 1S6 Canada

ESSENTIAL ZEBRAFISH METHODS: CELL AND DEVELOPMENTAL BIOLOGY
Reprinted from *Methods in Cell Biology*, Volume 59 (Academic Press, 1999).

DOI: 10.1016/B978-0-12-374599-6.00004-2

I. Introduction

The construction of individuals with uniparental inheritance can facilitate genetic analysis. There are two types of uniparental inheritance: gynogenesis, in which the embryo inherits all chromosomes from the mother, and androgenesis in which all chromosomes are inherited from the father. Uniparental inheritance helps overcome a major problem in diploid genetic analysis, including mutation screens and genetic mapping; traits are often masked or confounded by the contribution from the second parent. Unfortunately, in many vertebrate genetic model systems, uniparental progeny cannot be produced. In mice, for example, (McGrath and Solter, 1984) and perhaps in all mammals (Haig and Trivers, 1995), parent-of-origin-specific imprinting (the exclusive expression of paternal alleles of some genes and maternal alleles of others, in specific tissues) prevents the production of uniparental individuals. However, the imprinting of genes essential for development does not limit the production of either androgenotes (animals whose genome comes only from the male parent) (Corley-Smith *et al.*, 1996) or gynogenotes (animals whose genome is solely maternally derived) (Streisinger *et al.*, 1981) in zebrafish, and thus these useful animals can be exploited to solve problems in zebrafish developmental genetics.

There are three types of gynogenotes, all of which are produced by irradiating the sperm to destroy the paternal genome and then using it to fertilize zebrafish eggs *in vitro*. The first category of gynogenotes develops as haploid[1] gynogenotes. They are useful for genetic mapping (Chapter 9, Vol. 60; Postlethwait *et al.*, 1994) and some mutation screens (Kimmel, 1989), but they are of limited use for some applications because haploids die as larvae with some developmental abnormalities (Chapter 4, Vol. 60; Henion *et al.*, 1996; Streisinger *et al.*, 1981, 1986; Walker and Streisinger, 1983). The second category of gynogenotes, homozygous diploids, is produced by inhibiting the first mitotic division of haploid gynogenotes, thereby doubling the haploid genome and making animals homozygous at all loci (Streisinger *et al.*, 1981). The production of the third category, half tetrads, uses the fact that zebrafish eggs are arrested in meiosis II until fertilization; inhibiting the extrusion of the second polar body restores diploidy. These diploids are homozygous for all loci proximal to the first recombination (Streisinger *et al.*, 1986).

The production of androgenotes is similar to that of gynogenotes, except that the maternal, rather than the paternal, genome is destroyed by irradiation; fertilization with normal sperm then initiates the development of an androgenetic haploid (Corley-Smith *et al.*, 1996). Homozygous diploid androgenetic fish can be produced by the inhibition of the first mitotic division of haploid androgenotes. The production of both haploid and diploid androgenetic zebrafish has been substantiated with polymorphic DNA marker evidence (Corley-Smith *et al.*, 1996, 1997).

---

[1] Haploid is used here to designate the set of chromosomes found in one normal gamete. It has been speculated that Pacific salmon may have four sets of chromosomes (Bailey *et al.*, 1969; Klose *et al.*, 1968). Thus, in this chapter, haploid is not necessarily equivalent to one set of chromosomes.

There were two technical obstacles to overcome in the original production of androgenotes. The first was a way to irradiate the eggs to block the transmission of chromosomes without destroying essential cytoplasmic components. The second obstacle was that zebrafish eggs could not be held outside of the zebrafish long enough to irradiate them and still allow them to be fertilized. Exposing freshly oviposited eggs to water activates cytoplasmic streaming and elevation of the chorion, which prevents fertilization. These events are rapid, occurring within approximately 1 min. Even if the eggs are expelled into a dry petri dish or into Hank's solution, they will remain fertilizable for less than 10 min. Except with unusually strong irradiation sources, the irradiation time required to produce androgenotes is longer than 10 min. The discovery that zebrafish eggs could be held in ovarian fluid from coho (*Oncorhynchus kisutch*) or chinook (*O. tshawytscha*) salmon for up to 6 h and still retain viability provided enough time for the eggs to be irradiated for androgenesis.

Using delayed *in vitro* fertilization with salmon ovarian fluid, we performed the Hertwig dose-response experiment and measured survival as a function of the total irradiation dose to optimize the production of haploid androgenotes (Corley-Smith *et al.*, 1996). We irradiated zebrafish eggs with increasing doses of gamma or X-rays, measured the survival to the larval stage, and then plotted survival versus dose. We found that as dosage gradually increased, survival declined to nearly zero; then, as expected from Hertwig's experiments, the survival rate increased again before finally falling once more to zero. These results can be understood if low doses of irradiation lead to more and more aneuploidy and thus to decreased survival. As irradiation increases further, extensive fragmentation of the irradiated chromosome set totally inactivates the maternal genome and, as euploidy (haploidy) is restored, survival again increases and the embryos develop as androgenetic haploids. The final decrease in survival at still higher doses likely is to be due to mitochondrial and cytoplasmic damage from the irradiation. Thus, it is the irradiation dose at the second peak of survival that we used to produce androgenotes. At this dose, the damage to the egg cytoplasm or mitochondria does not prevent the normal development of haploid embryos.

In this chapter, we describe how to produce androgenotes, how they have been used, and their future potential. Section III describes two procedures that are required for androgenesis but which can also be used for other applications: One is the technique of delayed *in vitro* fertilization, and the other is the technique of interfering with the first mitotic division to double the chromosome complement.

Being able to delay the fertilization of zebrafish eggs for periods up to a few hours after they are extruded from the female can facilitate experiments that involve *in vitro* fertilization. It allows manipulations prior to fertilization (e.g., microinjections), it aids studying fertilization itself (Lee *et al.*, 1996), and it also can provide working time for postfertilization manipulations (for example, when injecting 1–4-cell embryos, holding the eggs and fertilizing them in small batches at progressively later times, allows time for injections at a particular stage).

A future use of androgenesis may be in mutagenesis screens and in constructing male genetic maps. Mutant hunts using haploids permits a first generation screen, and this has been exploited quite effectively using gynogenetic haploids (Chapter 3, Vol. 60). While the production of androgenetic haploids is more complicated than gynogenetic haploids, it is advantageous in some situations. For instance, when the germ lines of diploid embryos are mutagenized by high doses of gamma irradiation, most of the embryos develop as males (Walker, 1998; Walker and Streisinger, 1983), which, of course, cannot be used for gynogenetic haploid mutant screens. These males, however, can be used for androgenetic haploid mutant screens and, thus, higher rates of mutagenesis can be achieved. Efficient methods for ENU mutagenesis of adult males are well established (Driever *et al.*, 1996; Haffter *et al.*, 1996) and can be conveniently applied to screens of haploid androgenotes. Cryo-preservation of spermatozoa (Walker and Streisinger, 1995b,c; Appendix 3, Vol. 60) from these mutagenized males permits the recovery of mutations long after the haploid screen is complete without the need to maintain the parents or progeny lines. Another use of androgenesis is in genetic mapping. The production of androgenetic haploids from a diploid male heterozygous for two highly poly-morphic genomes allows the rapid production of a male linkage map based on meiotic recombination frequencies in the male germ line. In many vertebrates, the meiotic recombination frequencies are different in males and females, and when this has been fully documented for zebrafish, this will be useful for the preservation of desired linkages during zebrafish crosses.

II. Equipment and Materials

Androgenetic haploids are produced by the irradiation of eggs to prevent the inheritance of the maternal genome followed by *in vitro* fertilization. Diploids can be produced by blocking the first mitotic division. This section describes the materials needed for these procedures.

A. Collection of Salmonid Ovarian Fluid

Developing a way to keep eggs viable for several hours after ovulation was critical for successful androgenesis. Usually for successful *in vitro* fertilization, zebrafish eggs must be fertilized almost immediately after collection. However, holding eggs in the blood-free ovarian fluid of Pacific salmon (*Oncorhynchus* spp.) extends this time up to 6 h. Ovarian fluid is the extracellular component surrounding the oocytes in the ovary (Corley-Smith *et al.*, 1995).

Diploid embryos produced in this way develop into apparently normal zebra-fish. In one experiment, we held 76 zebrafish eggs for 50 min, from the time the eggs were extruded from the female to the time milt was added to fertilize the eggs. In this group, 72% were fertilized and developed into normal-looking hatched embryos. The remaining 28%, observed after 24 h, showed no embryonic development and appeared to be unfertilized. The technique does not appear to lead to an increased incidence of abnormal development or a decreased fertility of the resulting adults.

Ovarian fluid must be collected carefully from gravid salmon because even slight dilution with water or contamination with blood or ruptured eggs reduces the usefulness of the fluid for holding zebrafish eggs. Fluid from returning salmon can be collected at a hatchery as part of the collection of salmon eggs. Immediately following lethal cranial trauma, the female salmon is suspended by its tail and all the gills are slit to drain the blood from the animal. After 5 min of bleeding, the surface of the fish is dried with paper towels to prevent the exposure of the eggs and ovarian fluid to water as they are collected. As an assistant holds the fish belly downward over a clean dry bowl, it is slit from the anus to the front of the body cavity. The eggs and ovarian fluid are released into the bowl. The eggs are then teased from the skein (ovarian connective tissue) and pieces of skein are removed. A nonabrasive kitchen strainer or colander (stainless steel or plastic) is useful for separating the eggs from ovarian fluid. The eggs should be handled gently at all times to prevent breakage. The ovarian fluid is collected into 50-ml polypropylene centrifuge tubes on ice and the salmon eggs are given to hatchery staff for insemination. Upon return to the laboratory, the fluid is centrifuged at $5500 \times g$ for 5 min at $4\,^\circ$C to sediment cellular debris. The supernate is removed and aliquots are frozen in 1.5-ml screw-cap microfuge tubes and stored at $-20\,^\circ$C. The fluid is robust and can withstand many cycles of freezing and thawing. Each batch of fluid should be tested for its suitability to hold eggs for delayed fertilization.

We have tested fluid from coho salmon (*O. kisutch*), rainbow trout (*O. mykiss*), and chinook salmon (*O. tshawytscha*) for holding eggs. In general, we find chinook salmon ovarian fluid will hold zebrafish eggs the longest with subsequent high fertilization rates. Gibbs *et al.* (1994) has reported success using rainbow trout ovarian fluid for delayed *in vitro* fertilization of zebrafish eggs. Recently, a defined medium has been described for holding eggs for delayed fertilization (Sakai *et al.*, 1997). We have not tested that medium but have had some success with holding eggs in other defined media.

B. Irradiation Source

This section describes an X-ray source we have used to perform androgenesis (Corley-Smith *et al.*, 1996). An X-ray source should produce at least 150 keV. We use a Torrex 150D cabinet-style X-ray inspection system (Faxitron X-Ray Corp., Buffalo Grove, IL., USA). The instrument has a built-in 1.2-mm beryllium window. We use no extra filters, as they extend the time required to deliver the required dose. If an X-ray source with sufficient output is used, a 0.5-mm aluminum or copper filter will selectively remove soft X-rays (low keV), suspected of causing more cytological damage than hard X-rays, which are more selective in targeting DNA. X-ray dosimetry was performed with a MDH1515 dosimeter using a MDH model 10X5-180 ion chamber (paddle chamber). This was calibrated with a known ^{137}Cs source (NBS ^{137}Cs source *47455). Gamma rays have also been used successfully to irradiate eggs (see Section IV.C).

C. Water Baths

Two water baths are needed to maintain water in beakers at constant temperatures: one at $28.5 \pm 0.5\,°C$ and the other at $41.4 \pm 0.05\,°C$. A calibrated thermometer is required for this accuracy (e.g., Fisher Scientific, Cat. No. 15041A, with an uncertainty certified not to exceed $0.03\,°C$). Both water baths contain beakers with fish water (see Section II.D). To promote heat transfer, the water in the beakers is constantly stirred. The temperatures should be measured in the beakers. Since timing of the first mitotic division is temperature-dependent, the accurate temperature control of the cooler water bath is also important for producing diploid androgenotes. To transfer the eggs between the hot water baths and to allow abrupt thermal changes to be applied to the eggs, they are placed in an uncapped 50-ml polypropylene conical tube with the bottom sliced off and a fine nylon mesh (e.g., 153-μm Nitex mesh) melted on.

D. Solutions

The fish and embryos are held in fish water: deionized water filtered through activated charcoal to which 60 mg/l aquarium sea salts (e.g., Instant Ocean) have been added.

The defined medium of Sakai *et al.* (1997) for holding zebrafish eggs contains Hank's saline (0.137 M NaCl, 5.4 mM KCl, 0.25 mM Na_2HPO_4, 0.44 mM KH_2PO_4, 1.3 mM $CaCl_2$, 1.0 mM $MgSO_4$, 4.2 mM $NaHCO_3$) and 0.5% bovine serum albumin (BSA: Sigma fraction V).

To allow easier handling of the small volume of sperm collected from a single zebrafish, the sperm is collected into a sperm-extender solution composed of 80 mM KCl, 45 mM NaCl, 45 mM sodium acetate, 0.4 mM $CaCl_2$, 0.2 mM $MgCl_2$, and 10 mM HEPES; pH 7.7. The solution is filtered through 0.22-μm pores and stored at $4\,°C$.

III. Methods

A. Collection of Zebrafish Eggs for Delayed *In Vitro* Fertilization

In vitro fertilization requires high-quality eggs, which can be obtained in the following way. Shortly after the beginning of the light part of the photocycle, place the gravid female in an 18- or 36-l tank with 1–3 males. Observe the fish and as soon as breeding activity commences remove the female from the tank and squeeze her to obtain the eggs. Good eggs are slightly granular and yellowish in color. Although not always completely spherical, the best eggs look full. The best batches contain few or no broken eggs and no whitish or withered eggs. The chorion of good eggs elevates away from the plasma membrane when ovarian fluid is diluted with water.

The following steps describe the collection and irradiation of zebrafish eggs for androgenesis.

1. Place approximately 100 μl salmon ovarian fluid in 50 mm diameter plastic petri dishes at room temperature. The fluid should form a small dome near the center of the dish.
2. Anaesthetize a female zebrafish in 17 ppm (wv) tricaine (3-aminobenzoic acid ethyl ester) adjusted to pH 7 with sodium bicarbonate (Walker and Streisinger, 1995a; Chapter 3, Vol. 60).
3. Place the female belly up under the dissecting scope, resting the fish in a V-shaped slit of a damp sponge.
4. Carefully dry the belly and genital pore with a facial tissue or Kimwipe.
5. Draw some of the ovarian fluid from the dish up into a silanized glass capillary tube (Kimax-51, Kimble Products Art. No. 34502, ID 0.8–1.1 mm, length 100 mm) and expel it back into the dish. This prewets the tube, reducing friction and lessening the chance of breaking the eggs.
6. Squeeze the fish gently and carefully draw the eggs up into the glass capillary tube.
7. Observe the eggs under the dissecting scope and assess their quality by appearance.
8. Gently expel the eggs into the 100-μl dome of salmon ovarian fluid on the petri dish. Avoid placing the eggs on dry dish, which may reduce subsequent fertilization rates.
9. Return the fish to the water. All the fish should survive, but should not be squeezed again for about a month.
10. Place the lid on the petri dish to reduce evaporation. Although perhaps it is not necessary, we place black plastic sheets over the dishes to shield out light.
11. The eggs are held at room temperature until they are fertilized. The fertilization rate drops the longer the eggs are held. Thus, we try to fertilize the eggs as soon after collection as possible, usually within 30 min.

B. Collection of Sperm for Delayed *In Vitro* Fertilization

1. Put 50 μl sperm extender into 500-μl microcentrifuge tubes on ice.
2. Perform steps in 2–4 in Section III.A on a male fish.
3. Squeeze the fish gently and take up milt into a 2- or 5-μl glass capillary tube (Drummond Microcaps) by capillary action. This tube does not need to be silanized.

4. Gently expel the milt from one male into a microfuge tube containing the sperm extender solution on ice, using about 1–2 μl sperm per 50 μl sperm extender.

5. Gently swirl to mix the sperm and sperm extender.

6. Store on ice until needed.

Sperm are best used fresh. When logistically feasible, we collect sperm while the eggs are being irradiated. Sperm can, however, be collected before the eggs are collected, and stored in extender solution for several hours. When diluted with water, sperm begin to swim actively and are capable of fertilization for about a minute. The quality of the sperm can be assessed microscopically by the dilution of a sample with water; they should swim actively for approximately 1 min following the addition of water and have a normal appearance.

C. Irradiation of Eggs

To prepare eggs for irradiation, about 100 eggs collected in Section III.A are spread in a monolayer in salmon ovarian fluid in a 50-mm diameter plastic petri dish. The eggs should be in the center of the dish in approximately 100 μl of salmon ovarian fluid with a minimum layer covering the eggs. Place the dish 23 cm from the focal point of the X-ray beam of the Torrex 150D (shelf 8). This is the shelf closest to the irradiation source. We used the maximum settings for the instrument: 145 KV and 5 mA, producing a dose of 12.2 R/s. To achieve the desired dose of 10,000 R, irradiate for 820 s. Irradiate at room temperature. After irradiation, fertilize eggs *in vitro* as described in Section III.D.

D. *In Vitro* Fertilization

1. To recycle the ovarian fluid (it can be reused several times), first transfer most of the ovarian fluid from the eggs with a sterile pipette to a clean 0.5-ml microcentrifuge tube on ice. When time permits, centrifuge the fluid at 5500 × g for 5 min at 4 °C and freeze the supernate.

2. Spread 5–15 μl of sperm extender solution containing sperm evenly over all the eggs in the Petri dish.

3. Immediately add 0.5 ml fish water to activate the sperm, and swirl very gently to mix.

4. After 1 min, very gently add 28.5 °C water to fill three-quarters of the petri dish, and leave it undisturbed at 28.5 °C for 1 h. To promote gas exchange, make sure there is a space between the top of the water and the dish cover.

5. After 1 h, add 0.02% stock solution of methylene blue to a final concentration of 0.3 ppm to inhibit fungal growth. To reduce the spread of fungus growing on dead eggs, make sure the eggs do not touch each other.

6. Place the dish in a 28.5 °C incubator, and after 24 h, remove the dead embryos and flush out the methylene blue.

E. Production of Diploid Androgenotes by Heat Shock

Embryos treated according to the above protocol will develop as haploid androgenotes. To produce diploid androgenotes, the first mitotic division should be blocked as described below.

1. Start the timer as soon as 0.5 ml of $28.5 \pm 0.5\,°C$ fish water is added to the milt and eggs. This is time zero.

2. Place the petri dish in a $28.5\,°C$ incubator or on a shallow ledge in a beaker containing $28.5\,°C$ water. After 1.0 min, very gently add $28.5\,°C$ water to fill three-quarters of the petri dish.

3. At 5 min, transfer the eggs to a heat-shocking tube (50-ml tube with net bottom) and suspend the tube in beaker containing $28.5\,°C$ fish water that is in water bath. Tubes should be suspended, not rested on bottom of beakers, and should be left uncapped.

4. At 13.0 min, transfer the heat-shocking tube containing the eggs to a beaker containing $41.4\,°C$ fish water.

5. At 15.0 min, very gently transfer the heat-shocking tube containing the eggs back to the $28.5\,°C$ beaker and leave undisturbed for 1.5 h.

6. After 1.5 h, transfer the eggs very gently into petri dishes three-quarters full of water and place them in a $28.5\,°C$ incubator.

7. At 24 h, inspect the developing embryos under a stereomicroscope.

Three types of control animals help in evaluating the success of the procedure: (1) a normal diploid control group produced by delayed *in vitro* fertilization; (2) irradiated and fertilized, but not heat-shocked, embryos (putative haploid androgenotes); and (3) irradiated, fertilized, and heat-shocked animals (putative diploid androgenotes). Fertilize all the groups of eggs at same time and keep them in a $28.5\,°C$ incubator except during manipulations. The irradiated and nonheat-shocked eggs are included to ensure that the irradiation dose is adequate to prevent the inheritance of the maternal genome in each experiment. If no diploid phenotypes are observed in this group, it is highly likely that the embryos in the irradiated and heat-shocked group that have diploid phenotypes are diploid androgenotes. The confirmation of exclusive paternal inheritance requires investigating the inheritance of parentally polymorphic DNA markers to putative androgenetic progeny. The lack of homozygous maternal-specific markers in the progeny is strong evidence supporting sole paternal inheritance, although it does not rule out the possibility of some maternal genes being inherited by some cells (see Section IV.B).

Haploid androgenetic embryos exhibit the haploid syndrome: a shortened body, kinked neural tube, and small melanocytes in comparison with normal diploid embryos (Fig. 1). The shortened body is noticeable at 24 h and obvious at 48 h, while the difference in melanocyte size is noticeable at 48 h. Haploid larvae rarely feed, do not develop swim bladders, and usually die when 4–5 days old. The development of androgenetic diploid embryos is initially retarded in relation to diploid control embryos (Fig. 1). However, by the end of the first month, the androgenetic diploids achieve approximately the same size as the diploid control fish.

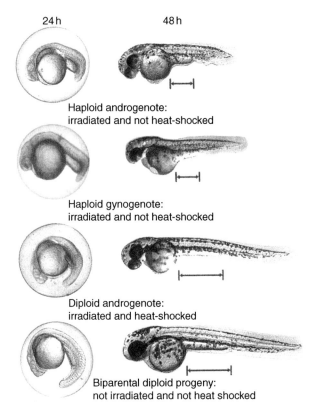

Fig. 1 Putative androgenetic and gynogenetic haploids, putative androgenetic diploid, and putative biparental diploid embryo. Developing embryos of each type were photographed at 24 h (left). The same embryos were photographed again at 48 h (right) after removing the chorion, except for the gynogenote, which is a different embryo. The distance between the posterior yolk-sac margin and the anal pore (as shown by horizontal bars) is greater for the diploid phenotype than for the haploid phenotype. The appearance of two eyes in some embryos and not others is the result of differences in the angle of photography and is not a phenotypic difference. Haploids are distinguishable from diploids at 24 and 48 h, and we did not notice any consistent differences between haploid androgenotes and gynogenotes at either 24 or 48 h (adapted from Corley-Smith *et al.* (1996), Fig. 2, by permission of Genetics Society of America).

IV. Results and Discussion

A. Properties of Androgenetic Embryos

Haploid androgenetic and gynogenetic zebrafish embryos and larvae have been compared using a stereomicroscope (Fig. 1). No consistent differences were observed between these haploids, though we cannot rule out subtle differences. The fraction of androgenetic haploid larvae having optimal haploid appearance ranges from roughly 30–60% in different experiments. This depends on irradiation dose, genotype, and state of the individual fish used as parents. In addition, some haploid androgenotes have minor imperfections such as a bent tail or eye defects. Other embryos have very severe abnormalities.

The utility of androgenetic haploid embryos for genetic screens will depend on the phenotypes sought and the assays used. For instance, screens focused on cellular or tissue differentiation based on *in situ* hybridization or immunostaining assays should be appropriate when applied to haploids having a normal appearance.

Initially we were concerned that recessive lethal mutations would limit the production of androgenetic zebrafish. Most of our investigations have been performed on males from the *AB line, which is prescreened to reduce recessive lethal mutations. However, we have also produced haploid and diploid androgenotes from a modestly inbred line (SFU) of fish acquired from pet stores. These generally produced a higher frequency of haploid androgenotes of normal appearance than did the *AB line. This suggests that recessive lethals are infrequent in this SFU line of fish and that the production of haploid androgenotes can be optimized by the selection of particular males. The efficiency of the production of haploid androgenotes is generally in the range of 5–55% (Corley-Smith *et al.*, 1996). The rate of production of diploid androgenotes is typically less than 2%, and about 10% for diploid gynogenotes. This low survival rate for diploid androgenotes that are produced by heat-shocking haploid androgenotes is likely to be the result of the trauma of the heat shock.

To date, all of the diploid androgenotes produced and raised to maturity have been males. We have crossed some of these diploid androgenotes with wild-type females. Progeny of both sexes have been observed, though the ratios are sometimes highly skewed (our unpublished observations).

B. Are Maternal Genes Transmitted to Androgenotes?

It is necessary to establish the exclusively paternal origin of the genes in putative androgenotes. When investigating putative androgenotes produced by X-rays, we detected no inheritance by putative androgenotes of several maternal-specific DNA markers using PCR and fluorescent detection on an ABI 373A DNA sequencer (Corley-Smith *et al.*, 1996, 1997). The test was sufficiently sensitive to allow us to detect a marker that was present in 2% or more of the cells from which the DNA was extracted (clipped fins or extracts of whole larvae). To learn whether maternal genes are inherited by a small fraction of somatic cells in a mosaic fashion, we used gamma rays (^{137}Cs) and a visible genetic marker. This method is potentially more sensitive than the fluorescent PCR technique. *Golden* (gol^{b1}) is a recessive pigment mutation (Streisinger *et al.*, 1986). Pigment in melanocytes is lacking in homozygous mutant embryos at 48 h, whereas it is clearly visible in wild-type embryos at this stage. Hybrid androgenotes were produced by fertilization of irradiated wild-type eggs with sperm of *golden* males. If the embryos lack pigmentation at 48 h, then melanocytes did not inherit the *gol*[+] allele from the mother. We found that at low doses of irradiation, only morphologically aberrant embryos were produced, as expected from the Hertwig effect; but pigmented cells developed nonetheless (Table I). This shows that low radiation doses do not block transmission of the maternal *gol*[+] allele to all melanocytes. At the dose used to produce androgenotes, 40,000 R for gamma irradiation, no pigmented cells were

Table I
Melanin Rescue in γ-Androgenesis

Treatment (dose of γ-rays)	Visual score[a]	Embryos with no pigmented melanocytes	Embryos containing some pigmented melanocytes (proportion of melanocytes pigmented)
0 R (control), $n = \sim85$	D	0	All (100%)
	AN	0	0
	PH	0	0
	VA	0	0
	MC	0	0
7500 R, $n = 19$	D	0	0
	AN	0	0
	PH	0	0
	VA	7	4 (<10%)
	MC	4	3
10,000 R, $n = 27$	D	0	0
	AN	4	0
	PH	1	0
	VA	15	1 (>25%)
	MC	4	1
13,587 R, $n = 31$	D	0	0
	AN	5	0
	PH	4	0
	VA	13	3 (2 @ 10–25%; 1 @ 1 cell)
	MC	7	0
25,000 R, $n = 53$	D	0	0
	AN	9	1 (>25%)
	PH	6	0
	VA	14	4 (10–25%)
	MC	19	0
40,000 R, $n = 58$	D	0	0
	AN	5	0
	PH	13	0
	VA	27	1 (2 cells)
	MC	11	1 (4 cells)

Embryo scoring adapted from Walker and Streisinger (1995a).

[a]D, diploid embryos; AN, apparent aneuploids, phenotype in between that of diploids and haploids; PH, perfect haploid embryos; VA, very abnormal embryos that have a head, body axis and some sort of tail, but obviously will not survive; MC, mass of cells, embryos that are yolks with masses of cells on them.

observed in perfect or nearly perfect haploid embryos, but a few pigment cells were observed in highly abnormal haploid embryos. While the sample size was small, these observations indicate that fragments of DNA including functional genes can be mosaicly inherited from the irradiated maternal genome. But this is rare at higher doses, and may not occur when the maternal genome has been sufficiently fragmented to allow normal development of androgenotes. The finding that more heavily pigmented animals developed more aberrantly in this dosage range is consistent with the idea that it is the gene dosage and aneuploidy problems arising from partial inactivation of the maternal genome that causes low-dose lethality. Alternative sources of radiation may totally prevent mosaic inheritance of maternal alleles.

C. Alternative Sources of Irradiation

Radiation sources that are useful in blocking the inheritance of chromosome sets include gamma rays (e.g., ^{137}Cs and ^{60}Co), X-rays, and ultraviolet (UV) rays. Gamma rays and X-rays result in the fragmentation of DNA, and UV-rays covalently cross-link DNA. Cross-linking DNA might be preferable if it prevents the mosaic inheritance of fragments of chromosomes. UV has proven useful for irradiating sperm for gynogenesis (Streisinger et al., 1981). However, we did not use it for irradiating eggs for androgenesis because we were concerned that UV irradiation might not penetrate sufficiently to reach the nucleus and might damage the maternal cytoplasmic components required for normal embryonic development, such as axis specification (Black and Gerhart, 1986; Elinson and Rowning, 1988; Jesuthasan and Stahle, 1997; McCrea et al., 1993). We initially chose to use X-rays, as they are a relatively safe source of irradiation and are highly penetrating.

Not all types of radiation result in equivalent DNA damage per unit dose, and thus the location of the second peak of the Hertwig dose curve varies with type of irradiation used; thus this peak must be located empirically for any new type of irradiation. For example, gamma rays which have a higher energy and lower linear energy transfer than X-rays are predicted by the Bragg curve to be less effective in causing radiation damage to living cells. Thus, the second peak of the Hertwig effect is expected to be located at a lower dose for X-rays than for gamma rays. In Hertwig experiments that we have performed with zebrafish eggs (Corley-Smith et al., 1996), rainbow trout (O. mykiss), and chinook salmon eggs (our unpublished results), the location of the second peak for gamma rays (^{60}Co and ^{137}Cs) is at 40,000 R. Survival at the second peak can exceed 80%. We found for X-rays (max keV 145) used on zebrafish eggs that the second peak was located at 10,000 R, which, as expected, is lower than the dose for gamma rays.

A recent report indicates that haploid androgenetic muskellunge fish embryos can be produced by UV irradiation of the egg (Lin and Drabrowski, 1998). These eggs are similar in size and appearance to zebrafish eggs, so the method used may be applicable to zebrafish eggs as well. For muskellunge eggs, the optimal dose of UV radiation was 660–1320 J/m^2, at which 100% putatively haploid larvae were produced at a hatching rate of 22.5 ± 2.8% (Lin and Drabrowski, 1998). In that study, the effectiveness of UV in preventing the transmission of maternal DNA was assessed by flow cytometric measurement of the DNA content of putative haploid androgenotes relative to their diploid parents. It remains to be established if UV irradiation has effectively blocked the inheritance of maternal genetic markers in muskellunge androgenotes. UV irradiation has been used to produce androgenetic haploid zebrafish (Ungar *et al.*, 1998). They used a 254-nm UV fluorescent lamp (model UVG-11, UV Products, San Gabriel, CA). Eggs were irradiated at a distance of 9 cm (dose rate of 600 μW/cm^2) for 4 min (1440 J/m^2). This UV dose of 1440 J/m^2 determined for zebrafish is close to the optimal dose of UV radiation of 660–1320 J/m^2 determined for muskellunge eggs (Lin and Drabrowski, 1998).

The choice of irradiation type depends on what the androgenotes will be used for. UV irradiation has the advantage in that it is more accessible to most zebrafish laboratories and it might prevent mosaic inheritance of chromosome fragments. Furthermore, the resultant embryos at the 2000–4000-cell stage are useful for collecting DNA for PCR detection of gamma ray induced deletions in genes that are heterozygous and are therefore recessive markers in the male parent (Kathryn Helde, 1998, personal communication). In applications where phenotypically normal androgenotes are required, however, gamma irradiation may be advisable.

V. Conclusions and Perspectives

Androgenetic haploid and diploid zebrafish larva can be efficiently produced. The production of both andro- and gynogenetic fish indicates that the irreversible inactivation of genes essential for development by parent-of-origin-genome imprinting does not occur in zebrafish, in contrast to some mammals. The success rate in producing haploid androgenotes with nearly normal morphology should make them suitable for some types of genetic screens. We anticipate that the methods can be refined to a higher level of consistency and success. The use of delayed *in vitro* fertilization facilitates the manipulation of eggs prior to fertilization, as well as the synchronization of batches of eggs manipulated over a period of time. The methods described extend the utility of the zebrafish as a vertebrate genetic system of exceptional power.

Acknowledgments

We thank Chinten (James) Lim for technical assistance. This research was supported by grants from NSERC and NIH.

References

Bailey, G. S., Cocks, G. T., and Wilson, A. C. (1969). Gene duplication in fishes: Malate dehydrogenases of salmon and trout. *Biochem. Biophys. Res. Comm.* **34,** 605–612.

Black, S. D., and Gerhart, J. C. (1986). High-frequency twinning of *Xenopus laevis* embryos from eggs centrifuged before first cleavage. *Dev. Biol.* **116,** 228–240.

Corley-Smith, G. E., Lim, C. J., and Brandhorst, B. P. (1995). Delayed *in vitro* fertilization using coho salmon ovarian fluid. *In* "The Zebrafish Book—A Guide for the Laboratory Use of the Zebrafish (*Brachydanio rerio*)" (M. Westerfield, ed.), pp.7.27–7.26. University of Oregon, Eugene, OR.

Corley-Smith, G. E., Lim, C. J., and Brandhorst, B. P. (1996). Production of androgenetic zebrafish (*Danio rerio*). *Genetics* **142,** 1265–1276.

Corley-Smith, G. E., Lim, C. J., Kalmar, G. B., and Brandhorst, B. P. (1997). Efficient detection of DNA polymorphisms by fluorescent RAPD analysis. *Biotechniques* **22,** 690–696.

Driever, W., Solnica-Krezel, L., Schier, A., Neuhauss, S., Malicki, J., Stemple, D., Stainier, D., Zwartkruis, F., Abdelilah, S., Rangini, Z., Belak, J., and Boggs, C. (1996). A genetic screen for mutations affecting embryogenesis in zebrafish. *Development* **123,** 37–46.

Elinson, R. P., and Rowning, B. (1988). A transient array of parallel microtubules in frog eggs: Potential tracks for a cytoplasmic rotation that specifies the dorso-ventral axis. *Dev. Biol.* **128,** 185–197.

Gibbs, P. D., Peek, L. A., and Thorgaard, G. (1994). An *in vivo* screen for the luciferase transgene in zebrafish. *Mol. Mar. Biol. Biotech.* **3,** 307–316.

Haffter, P., Granato, M., Brand, M., Mullins, M., Hammerschmidt, M., Kane, D., Odenthal, J., van Eeden, F., Jiang, Y., Heisenberg, C., Kelsh, R., Furutani-Seiki, M., et al. (1996). The identification of genes with unique and essential functions in the development of the zebrafish, *Danio rerio*. *Development* **123,** 1–36.

Haig, D., and Trivers, R. (1995). The evolution of parental imprinting: A review of hypotheses. *In* "Genomic Imprinting: Causes and Consequences" (R. Ohlsson, K. Hall, and M. Ritzen, eds.), pp. 17–28. Cambridge University Press, Cambridge.

Henion, P. D., Raible, D. W., Beattie, C. E., Stoesser, K. L., Weston, J. A., and Eisen, J. S. (1996). Screen for mutations affecting development of zebrafish neural crest. *Dev. Genet.* **18,** 11–17.

Jesuthasan, S., and Stahle, U. (1997). Dynamic microtubules and specification of the zebrafish embryonic axis. *Curr. Biol.* **7,** 31–42.

Kimmel, C. B. (1989). Genetics and early development of zebrafish. *Trends Genet.* **5,** 283–288.

Klose, J., Wolf, U., Hitzeroth, H., and Ritter, H. (1968). Duplication of the LDH gene loci by polyploidization in the fish order clupeiformes. *Humangenetik* **5,** 190–196.

Lee, K. W., Baker, R., Galione, N., Gilland, E. H., Hanlon, R. T., and Miller, A. L. (1996). Ionophore-induced calcium waves activate unfertilized (*Danio rerio*) eggs. *Biol. Bull.* **191,** 265–267.

Lin, F., and Drabrowski, K. (1998). Androgenesis and homozygous gynogenesis in muskellunge (*Esox masquinongy*): Evaluation using flow cytometry. *Mol. Reprod. Dev.* **49,** 10–18.

McCrea, P. D., Brieher, W. M., and Gumbiner, B. M. (1993). Induction of a secondary body axis in Xenopus by antibodies to beta-catenin. *J. Cell Biol.* **123,** 477–484.

McGrath, J., and Solter, D. (1984). Completion of mouse embryogenesis requires both the maternal and paternal genomes. *Cell* **37,** 179–183.

Postlethwait, J. H., Johnson, S. L., Midson, C. N., Talbot, W. S., Gates, M., Ballinger, E. W., Africa, D., Andrews, R., Carl, T., Eisen, J. S., *et al.* (1994). A genetic linkage map for the zebrafish. *Science* **264,** 699–703.

Sakai, N., Burgess, S., and Hopkins, N. (1997). Delayed *in vitro* fertilization of zebrafish eggs in Hank's saline containing bovine serum albumin. *Mol. Mar. Biol. Biotechnol.* **6,** 84–87.

Streisinger, G., Singer, F., Walker, C., Knauber, D., and Dower, N. (1986). Segregation analysis and gene-centromere distances in zebrafish. *Genetics* **112,** 311–319.

Streisinger, G., Walker, C., Dower, N., Knauber, D., and Singer, F. (1981). Production of clones of homozygous diploid zebra fish (*Brachydanio rerio*). *Nature* **291,** 293–296.

Ungar, A. R., Helde, K. A., and Moon, R. T. (1998). Production of androgenetic haploids in zebrafish with ultraviolet light. *Mol. Mar. Biol. Biotech.* **7,** 320–326.

Walker, C., and Streisinger, G. (1983). Induction of mutations by gamma-rays in pregonial germ cells of zebrafish embryos. *Genetics* **103,** 125–136.

Walker, C., and Streisinger, G. (1995a). Embryo production by *in vitro* fertilization. *In* "The Zebrafish Book—A Guide for the Laboratory Use of Zebrafish (*Danio rerio*)" (M. Westerfield, ed.), pp. 2.11–2.20. University of Oregon, Eugene, OR.

Walker, C., and Streisinger, G. (1995b). Freezing sperm. *In* "The Zebrafish Book—A Guide for the Laboratory Use of Zebrafish (*Danio rerio*)" (M. Westerfield, ed.), pp. 7.32–7.33. University of Oregon, Eugene, OR.

Walker, C., and Streisinger, G. (1995c). Thawing and using frozen sperm for *in vitro* fertilization. *In* "The Zebrafish Book—A Guide for the Laboratory Use of Zebrafish (*Danio rerio*)" (M. Westerfield, ed.), p. 7.34. University of Oregon, Eugene, OR.

CHAPTER 5

Analysis of Protein and Gene Expression

Trevor Jowett

Department of Biochemistry and Genetics
Medical School
University of Newcastle
Newcastle upon Tyne NE2 4HH
United Kingdom

ESSENTIAL ZEBRAFISH METHODS: CELL AND DEVELOPMENTAL BIOLOGY
Reprinted from *Methods in Cell Biology*, Volume 59 (Academic Press, 1999).
DOI: 10.1016/B978-0-12-374599-6.00005-4

I. *In Situ* Hybridization to RNA and Immunolocalization of Proteins

There are well-established methods for determining the temporal and spatial expression of RNA and proteins, and these can be successfully applied to zebrafish embryos and tissues. The procedures require that the tissues are fixed so that the mRNA and proteins are retained within the cells. The mRNA transcribed by a specific gene is best detected *in situ* by hybridization with an antisense RNA probe labeled so that it can be detected either by chromogenic stains or fluorescence. All that is required is a supply of the appropriate developmental stages of the zebrafish in which the gene of interest is expressed and a cloned fragment of the gene.

The procedure of *in situ* hybridization can be divided into the following component steps.

1. Synthesis of labeled antisense RNA probe(s).
2. Fixation of the tissue and storage of tissue.
3. Pretreatment to permeabilize the tissue and block nonspecific binding of the probe.
4. Hybridization of the probe.
5. Washing to remove the unbound probe.
6. Incubation in blocking solution to prevent nonspecific binding of the antibodies.
7. Incubation in antibody against the probe-hapten.
8. Washing to remove the unbound antibody.
9. Visualization of the bound antibody, either with chromogenic stain or by epifluorescent microscopy.

Modifications to this procedure can allow identification of more than one mRNA and this chapter describes several alternative approaches for identifying multiple transcripts. These methods allow visualization of the transcripts either with chromogenic substrates or different fluorochromes. The latter methods are particularly appropriate when coupled with the use of confocal microscopy.

Localization of proteins can be performed using an antibody raised against the protein. The signal is visualized using a secondary antibody, either labeled with a fluorochrome or conjugated with an enzyme, for which there are chromogenic substrates. Protein localization can also be combined with *in situ* hybridization to mRNA. However, this may not always be possible, since it requires that the protein antigen not be destroyed by the treatments given to the tissue during *in situ* hybridization.

The methods described below are all for performing on whole-mount zebrafish embryos. The same techniques can be applied to tissue sections with minor modifications (Jowett, 1997). However, it is often more practical to perform the *in situ* hybridization or immunolocalization on whole tissues or embryos and then to cut sections. Protocols for making cryostat sections are available in *The Zebrafish Book* (Westerfield, 1995) and on the zebrafish website.

II. Probe Synthesis

A. RNase-Free Conditions

Reasonable care should be taken to prevent contamination with RNases. Most care should be taken in the synthesis of the probe. A single synthesis should be enough for 100–200 hybridizations, and so it is worth storing it in aliquots at −20 °C in the presence of RNase inhibitor. The tissue will normally contain endogenous RNases, which cannot normally be eliminated; rapid fixation is the best way of avoiding potential problems. Solutions are usually made with sterile deionized water and, if possible, autoclaved prior to use. It is recommended that diethyl pyrocarbonate (DEPC)-treated water be used for synthesis and storage of the probe. Glass vials and microcentrifuge tubes used for the fixing and subsequent storage of embryos may be autoclaved, but this is usually not necessary. All solutions used prior to prehybridization are made from sterile stock solutions with deionized water in 50-ml screw-capped polypropylene centrifuge tubes. If these are made up fresh, they should be free from RNase contamination.

B. Antisense RNA Probe Synthesis

There are several plasmids available for cloning cDNA sequences. Many have recognition sequences for bacteriophage RNA polymerases positioned at either end of the multiple cloning site. This allows the plasmids to be linearized at either end of the cDNA sequence and RNA transcripts to be synthesized from either the sense or antisense strand. To make a probe for *in situ* hybridization, the plasmid must be linearized at the 5′ end of the cDNA sequence and the RNA polymerase binding to the recognition site at the 3′ end is used to synthesize antisense RNA. A control often used for *in situ* hybridization with a new probe is a transcript corresponding to the sense strand. However, experience reveals that sometimes what should be a negative control gives a signal through chance cross-hybridization with another transcript. A better control is the use of a probe for a different transcript known to be expressed in the same particular stage of embryo or tissue.

The linearized plasmid is transcribed by the appropriate RNA polymerase in the presence of the four ribonucleotides with a smaller amount of UTP labeled with either digoxigenin or fluorescein. Both of these haptens have the advantage that they are not normally found in animal tissues. This ensures that the immunolocalization of the bound probe does not lead to high backgrounds through the presence of endogenous hapten moieties. In other systems biotin is often used as a means of labeling RNA or DNA, but in zebrafish embryos or tissues this gives rise to very high backgrounds.

The plasmid DNA used for probe synthesis should be as pure as possible. It should be free of RNA and RNases. Plasmid prepared by the various commercial resins (Magic™ or Wizard™ Minipreps, Promega, Qiagen Plasmid Mini or Midi kit) is consistently of high quality. Since the extraction involves the addition of RNase, it is recommended that proteinase K treatment be performed before phenol extraction to remove unwanted proteins.

The amount of DNA used for probe synthesis should be calculated based on the size of the cDNA to be used as template. The reaction detailed below, which is based on reagents available from Boehringer Mannheim, is for the equivalent of $1 \mu g$ of insert cDNA. For a cDNA insert of 1000 bp in Bluescript™ (Stratagene) that is 2.9 kb in size, a total of $3.9 \mu g$ of linearized plasmid should be used per reaction. If the reaction is performed with optimal reagents, the expected yield of RNA should be approximately $10 \mu g$. If the size of the plasmid and insert are not taken into consideration, the yield will vary depending on the size of the insert. It is best to linearize enough DNA to make several probes. For a cDNA insert of 1 kb in Bluescript (2.9 kb), $20 \mu g$ of plasmid DNA is enough for four probe synthesis reactions. Redissolve the final DNA pellet in $20 \mu l$ of TE [10 mM Tris-HCl, 1 mM EDTA ((ethylenedinitrilo) tetra-acetic acid disodium salt)].

Probes for two-color *in situ* hybridization are differentially labeled with digoxigenin and fluorescein. Their synthesis reactions are identical apart from the $10\times$ ribonucleotide mix. One contains 3.5 mM digoxigenin-11-UTP, whereas the other contains 3.5 mM fluorescein-12-UTP. If fluorescein-UTP is used, the tube containing the reaction mix should be kept in the dark by wrapping in aluminium foil.

1. Protocol 1: Linearization of Plasmid and Probe Synthesis

1. Linearize the plasmid equivalent to $1 \mu g$ of insert DNA using a restriction enzyme with a single recognition site at the 5' end of the cDNA insert. Check that the plasmid is fully linearized by running on an agarose minigel before proceeding. Retain some of the reaction to check the yield after phenol/chloroform extraction and precipitation.

2. Add proteinase K to a final concentration of $0.05 \mu g/\mu l$. Incubate for 30 min at 37 °C.

3. Extract the DNA by shaking with an equal volume of phenol equilibrated with TE, pH 8.0. Centrifuge and remove the top aqueous layer to a clean microcentrifuge tube.

4. Extract the DNA by shaking with an equal volume of chloroform. Centrifuge and remove the top aqueous layer to a clean microcentrifuge tube.

5. Add 0.1 volume of 2.5 M sodium acetate and 2.5 volumes of ethanol. Chill at −20 °C or −70 °C.

6. Spin down the precipitated DNA in a microcentrifuge and wash the pellet with cold 80% ethanol. Respin, remove the supernatant, and dry the pellet.

7. Redissolve the DNA pellet in 4 μl TE, pH 8.0. Check the yield by running a small amount on an agarose minigel alongside an equivalent amount of the original restriction digest.

8. Mix together the following:

- 4 μl linearized plasmid (equivalent to 1 μg of insert DNA)
- 2 μl 10× transcription buffer
- 2 μl of 10× nucleotide mix (with digoxigenin-UTP or fluorescein-UTP)
- 20 U of RNase inhibitor
- Water to give a final reaction volume of 20 μl
- 40 U T7, T3, or SP6 RNA polymerase

9. Incubate the mixture for 2 h at 37 °C.

10. Add 40 U of DNase I and incubate at 37 °C for 15 min to remove the plasmid DNA.

11. Stop the reaction by adding 2 μl of 200 mM EDTA pH 8.0.

12. Precipitate the RNA with 2.5 μl of 4 M LiCl and 75 μl prechilled ethanol. The LiCl-ethanol precipitation does not completely remove unincorporated rNTPs, but this is not normally a problem.

13. Spin down the precipitate and redissolve the pellet in 100 μl of RNase-free water containing 40 U of RNase inhibitor.

14. Check the probe by running 2.5 μl on a 0.8% agarose, 1× TBE minigel. Wash the apparatus thoroughly before preparing the gel and run the samples quickly to avoid problems with RNase. In the case of fluorescein probes, the pellet is yellow. Unincorporated fluorescein-UTP runs at the front during electrophoresis and is easily seen on the ultraviolet (UV) transilluminator. Probes are stored in aliquots at −20 °C. There should be a single sharp band migrating behind the front. If it is smeared, then there is a problem with RNase contamination or with the RNA polymerase.

15. The probe can be used without further treatment. It should be split into aliquots and stored at −20 °C.

III. Fixation

Aldehyde fixatives are the most commonly used fixatives for *in situ* hybridizations to both whole-mounts and tissue sections. They cross-link the material retaining the RNA, but still allow penetration of the probe and antibodies. Three commonly used fixatives, in order of increasing cross-linking ability, are

4% paraformaldehyde, 4% formaldehyde, and 1% glutaraldehyde, each in phosphate-buffered saline (PBS). Cross-linking fixatives provide much better retention of nucleic acids than coagulating agents (such as ethanol/acetic acid). Zebrafish embryos of any age are sufficiently fixed in 2 h at room temperature but can be conveniently left overnight at 4 °C. If zebrafish embryos are fixed and dehydrated without dechorionation, a white precipitate forms in the liquid between the chorion and vitelline membrane. This precipitate can be washed away following dechorionation and appears not to interfere with subsequent *in situ* hybridization. Zebrafish embryos can be stored in methanol for 1–12 months without detrimental effects. The methanol dehydration step is always included with zebrafish embryos, as they are usually stored in methanol at least overnight before starting the hybridization. The methanol treatment helps reduce background with zebrafish embryos and provides a convenient way of storing them for prolonged periods at −20 °C.

Fixation prior to immunohistochemistry with antibodies to specific tissue antigens may require modifications to the fixation procedure. Some antigens may be destroyed or otherwise prevented from being recognized by some antibodies. It is also worth noting that assaying for transgenic reporter gene products may also require special consideration. The *lacZ* gene product β-galactosidase may be visualized with X-gal in fixed embryos, but this requires that the embryos are not overfixed and that they are not dehydrated with methanol as this will destroy the enzyme activity.

A. Protocol 2: Fixation of Zebrafish Embryos for *In Situ* Hybridization

1. Fix embryos in their chorions in a 4% solution of paraformaldehyde in PBS for 2 h at room temperature or overnight at 4 °C.
2. Wash twice in PBT (PBS plus 0.1% Tween 20) at 4 °C.
3. Transfer to a glass embryo dish and dechorionate using two pairs of fine watchmaker's forceps.
4. Dehydrate with a series of methanol-PBT solutions (1:3, 1:1, 3:1, methanol:PBT), and then twice with 100% methanol. Incubate 10 min in each solution.
5. The embryos can conveniently be stored at this stage at −20 °C. They are conveniently stored in 7 ml squat glass vials with polyethylene lids.

IV. Hybridization to Whole-Mount Embryos

A. Pretreatments and Hybridization of Zebrafish Embryos

After dehydration, transfer the zebrafish embryos from the glass vials that were used for prolonged storage in methanol to 1.5 ml microcentrifuge tubes. The tubes must have tight-fitting lids, otherwise they may spring open during incubation at 70 °C. Embryos (1–100) may routinely be hybridized per tube. The embryos are easily damaged and so care must be taken when transferring them. A 1-ml blue disposable pipette tip that has had its end cut off with a scalpel is suitable for this purpose. When changing solutions, never withdraw all the liquid from the embryos, otherwise they will be damaged.

Prior to hybridization the tissues must be rehydrated and then treated with proteinase K to further puncture the membranes to allow easy access of the probe and antibodies. For zebrafish embryos up to 24-h old, it is not necessary to treat them with proteinase K. The signals take a little longer to develop, but more embryos remain intact and retain their yolk. The protease digestion causes the tissues to become quite fragile; therefore special care must be taken not to damage the tissue when changing the solutions. After rehydration the embryos must be refixed, otherwise some RNA may be lost. This is irrespective of whether the embryos have been treated with proteinase K.

The embryos must be prehybridized to block potential nonspecific binding sites for the probe. This is done by incubating in hybridization mix without probe at the hybridization temperature. Prehybridization should be for as long as possible. It should be longer than 1 h, and ideally 4–6 h. Most hybridizations are at 65 °C. If background is a problem for a particular probe, it is worth increasing the formamide concentration and hybridization temperature. Many probes will still give a satisfactory signal in 65% formamide at 70 °C. The pH of the hybridization mix may vary from 4.5 to 7.0. However, fluorescein probes are more stable at higher pH.

1. Protocol 3: Pretreatments and Hybridization of Zebrafish Embryos

1. Rehydrate the embryos in a methanol-PBT series (3:1, 1:1, 1:3, methanol: PBT) finishing with three washes of PBT.

2. Treat with proteinase K (10 μg/ml in PBT) to increase the permeability of the membrane. This is performed for 10–20 min at room temperature, depending on the type of embryo and the stage of development. The incubation time in proteinase K may also depend on the particular batch of enzyme.

3. Stop the proteinase K digestion by replacing the solution with 4% paraformaldehyde in PBT for 20 min at room temperature. Earlier protocols stop the proteinase K digestion by replacing the solution with 2 mg/ml glycine in PBT. This is unnecessary for zebrafish embryos if they are immediately placed into fix. This second fixation helps stops the embryos from falling apart and prevents loss of target mRNA.

4. Wash twice for 5 min in PBT.

5. Add 0.2–1.0 ml of hybridization buffer (50% formamide, 5× SSC, 500 μg/ml yeast RNA, 50 μg/ml heparin, and 0.1% Tween-20, brought to pH 7.0 with 1 M citric acid). The volume used depends on the number and size of embryos. 0.2 ml is sufficient for 50–100 embryos that are 1–24-h old.

6. Incubate for 5 min and then add fresh hybridization solution and incubate at 60–70 °C for a minimum of 2 h. Lay the tubes almost horizontally in an upturned empty yellow pipette tip rack. In this way the embryos spread out and do not stick together. The embryos are incubated in a dual hybridization oven (Hybaid), which has a shaking table; gently shake the tubes at the lowest speed. This is not essential for hybridization but does facilitate the washing.

7. Once in the hybridization solution, the embryos can be stored at −20 °C. Short-term storage may be beneficial, but storage for several months in hybridization buffer may be detrimental and lead to reduced signal and higher background. Embryos in hybridization buffer are safe from digestion by RNases.

8. Replace the prehybridization solution with preheated hybridization mix containing probes. The total probe concentration should not exceed 1 μg/ml. As a starting point, use 1/100 of a standard digoxigenin or fluorescein riboprobe reaction in 200 μl hybridization solution (equivalent to a final concentration of 0.5 μg/ml). Probes that give strong signals may be effective at 0.1 μg/ml.

9. Incubate overnight at 60–70 °C. If background proves to be a problem, the temperature and formamide concentration can be increased or the salt concentration in the hybridization buffer reduced.

B. Posthybridization Washes of Zebrafish Embryos

After hybridization it is necessary to wash off unbound probe. This must be done at the temperature of hybridization with prewarmed solutions. The double-stranded RNA formed during the hybridization is very stable, with a melting temperature considerably higher than equivalent RNA/DNA or DNA/DNA hybrids, which is the reason that such stringent hybridization conditions can be used.

1. Protocol 4: Posthybridization Washes for Zebrafish Embryos

All washes and incubations are performed in the same 1.5 ml microcentrifuge tubes used for the hybridizations. Steps 1–7 are performed at the hybridization temperature.

1. Wash the embryos for 10 min in 50% formamide, 5× SSC.

2. Wash for 10 min with 0.5 ml of a 3:1 mixture of 50% formamide, 5× SSC:2× SSC.

3. Wash for 10 min with 0.5 ml of a 1:1 mixture of 50% formamide, 5× SSC:2× SSC.

4. Wash for 10 min with 0.5 ml of a 1:3 mixture of 50% formamide, 5× SSC:2× SSC.

5. Wash for 10 min in 1 ml of 2× SSC.

6. Wash for 30 min in 1 ml of 0.2× SSC, 0.01% Tween-20 at the hybridization temperature. The embryos can be quite sticky at this stage and so should be treated with care. Adding 0.01% Tween-20 to the 0.2× SSC prevents the embryos from sticking to the sides of the tube or to the pipette that is used to withdraw the solutions. This wash is best done on the shaker in the hybridization oven. Alternatively, use a heating block or water bath, but gently invert the tube every 5–10 min.

7. Repeat the stringent wash in step 6.

8. Wash for 10 min in 1 ml of a 3:1 mixture of 0.2× SSC:PBT (1× PBS, 0.1% Tween-20) at room temperature.

9. Wash for 10 min in 1 ml of a 1:1 mixture of 0.2× SSC:PBT at room temperature.

10. Wash for 10 min in 1 ml of a 1:3 mixture of 0.2× SSC: PBT at room temperature.

11. Replace the solution with PBT at room temperature.

V. Immunolocalization of Probes

It is usual to block nonspecific binding sites prior to applying the antibodies. In the past, the Fab fragments used to locate digoxigenin or fluorescein were preabsorbed against zebrafish embryos or a zebrafish acetone powder, but if the antibodies are used at suitably low dilutions then preabsorption is not necessary. All washes are performed by laying the tubes on their side and gently shaking on an orbital shaker. The following protocols are for visualizing signals with Fab fragments conjugated with either horseradish peroxidase or alkaline phosphatase.

There are a large number of alternative substrates available for use in conjunction with alkaline phosphatase- or horseradish peroxidase-conjugated antibodies. If only a single probe is used, in most cases the substrate of choice is that which gives the strongest signal. The most commonly used substrate mix for alkaline phosphatase is nitroblue tetrazolium with 5-bromo-4-chloro-3-indolyl phosphate (NBT/BCIP). This gives a blue/purple precipitate that is insoluble in both aqueous and alcoholic solutions. A stronger signal is obtained with tetranitroblue tetrazolium (TNBT), again in combination with BCIP. With this substrate the precipitate can be almost black.

The sensitivity of horseradish peroxidase-conjugated antibodies is generally less than that of the alkaline phosphatase-conjugated antibodies. Nonetheless, they have been successful on zebrafish embryos. The most common peroxidase substrate used is diaminobenzidine (DAB). This produces a highly insoluble brown precipitate. The color of the precipitate can be changed to grey-black by adding Co or Ni ions in the staining mix. A very much more sensitive alternative to DAB is TrueBlue™. This has the advantage that it is not a carcinogen like DAB and is 10- to 50-fold more sensitive. It produces a bright blue precipitate, which is, however, not stable and may be lost when the staining solution is removed. Because of its great sensitivity, the antibody titer used to identify the probe must be very much (10- to 50-fold) more dilute than that used for DAB staining. A comprehensive discussion of alternative enzyme substrates is provided in Jowett (1997).

A. Protocol 5: Staining Zebrafish Whole–Mounts with NBT/BCIP

This procedure is modified from Jowett and Lettice (1994). After hybridization of a fluorescein-labeled probe, posthybridization washes and equilibration in PBT proceed as follows:

1. Replace the PBT with blocking solution (PBT containing 5% sheep serum, 2 mg/ml BSA, 1% DMSO) and incubate at room temperature for at least 1 h on an orbital shaker.
2. Incubate for 2 h in alkaline phosphatase-conjugated sheep antifluorescein Fab fragments at a dilution of 1:2000 to 1:5000 dilution (0.375–0.15 U/ml).
3. Wash for 2 h with PBT (8×15 min).
4. Equilibrate 3×5 min in freshly made NTMT buffer. Prior to the staining reaction for alkaline phosphatase, it is important to equilibrate the tissue with buffer of pH 9.5. NTMT is buffered weakly with Tris and so with prolonged storage it may develop a lower pH by absorbing carbon dioxide. Therefore make the buffer just prior to use.
5. Stain with NBT/BCIP (4.5 μl of 75 mg/ml NBT in 70% dimethylformamide and 3.5 μl of 50 mg/ml BCIP in dimethylformamide added to 1 ml of NTMT buffer). If the NBT/BCIP signal is reddish-purple, it can be made more blue by increasing Tween-20 to 1% in the staining or wash solutions.
6. Stop the reaction by washing with PBT.
7. Fix the stained embryos in 4% paraformaldehyde in PBS overnight. If the stain is not fixed, prolonged exposure to light can cause a dark background to develop.

Caution: NBT/BCIP solution is potentially harmful and so should be handled with care. All disposable materials should be incinerated, and contaminated glassware should be soaked in a strong sodium hypochlorite solution (6%) before being rinsed thoroughly with water.

B. Protocol 6: Staining Zebrafish Whole–Mounts with Diaminobenzidine

This procedure is modified from Jowett and Lettice (1994). After hybridization of a digoxigenin-labeled probe, posthybridization washes and equilibration in PBT proceed as follows:

1. Replace the PBT with blocking solution (PBT containing 5% sheep serum, 2 mg/ml BSA, 1% DMSO) and incubate at room temperature for at least 1 h on an orbital shaker.

2. Incubate for 2 h in horseradish peroxidase-conjugated sheep antidigoxigenin Fab fragments in blocking solution at a dilution of 1:200 (0.75 U/ml).

3. Wash for 2 h in PBT (8 × 15 min).

4. Incubate for 2 min in 0.5 mg/ml DAB in PBT.

5. Add 1/1000 volume of 3% hydrogen peroxide to each incubation separately. Monitor the staining reaction and stop it by rinsing thoroughly with PBT.

6. Fix the stain in 4% paraformaldehyde in PBS overnight. If the stain is not fixed, prolonged exposure to light can cause a dark background to develop.

Caution: DAB is a potent carcinogen and so should be handled with care. It is supplied in tablet form by Sigma (#D 5905). All disposable materials should be incinerated, and contaminated glassware should be soaked in a strong sodium hypochlorite solution (6%) before being rinsed thoroughly with water.

VI. Two-Color *In Situ* Hybridizations

The availability of differentially labeled probes, antibodies to the different haptens, and alternative chromogenic substrates allows two-color *in situ* hybridization. The antisense probes are labeled with digoxigenin and fluorescein. These are identified with antibodies raised against digoxigenin and fluorescein, respectively, which are conjugated with alkaline phosphatase or horseradish peroxidase.

It makes little difference which hapten is used for each probe, as both fluorescein-12-UTP and digoxigenin-11-UTP label equally efficiently and the antibodies used for the subsequent visualization work with both equally well. However, fluorescein-labeled probes are less stable than digoxigenin-labeled probes. They should be kept in the dark as much as possible and are more stable at higher pH. However, I have used fluorescein-labeled probes up to two years after they were synthesized, having stored them at −20 °C in the dark.

Prior to trying a double-label experiment, it is advisable to check each probe separately, paying attention to the length of time required to obtain the final signal. Ideally, you should adjust the quantities of probe so that they will give equivalent signals in similar times. The most sensitive visualization reaction is achieved with alkaline phosphatase and NBT/BCIP and so this should normally be used for the probe that gives the weakest signal. The Fast Red reaction for alkaline phosphatase and the DAB reaction for horseradish peroxidase are considerably less sensitive.

A. Incubation with Antibodies Conjugated to Horseradish Peroxidase and Alkaline Phosphatase

The aim of this method is to visualize fluorescein- and digoxigenin-labeled probes with antibodies conjugated with horseradish peroxidase and alkaline phosphatase, respectively. There are several alternative ways in which the probes may be visualized. However, through experience, the following rules should be followed to achieve the most consistent results.

The probes should be combined in the hybridization mix.

Since the antibodies are conjugated to different enzymes, they too can be mixed and incubated together with the embryos. However, stronger and more consistent signals are achieved if the antibody incubations and staining are performed sequentially.

The fluorescein-labeled probe should be visualized first because of its lower stability compared to the digoxigenin-labeled RNA.

The horseradish peroxidase is stained with either DAB or TrueBlue™. If DAB is used, it is better to visualize the fluorescein probe first with sheep antifluorescein Fab fragments conjugated with horseradish peroxidase.

If TrueBlue™ is used, it must be the last staining reaction, otherwise the blue precipitate will be lost. The sheep antidigoxigenin Fab fragments conjugated to horseradish peroxidase must be used at a 20- to 50-fold greater dilution than would be used for DAB staining.

The substrate combinations that offer best contrast between the colored precipitates are DAB with NBT/BCIP and Fast Red with TrueBlue™.

The DAB signal generated by horseradish peroxidase is generally weaker than that seen with NBT/BCIP and alkaline phosphatase.

It is important that the antidigoxigenin-horseradish peroxidase antibody be used at a higher concentration (1:200 dilution, 0.75 U/ml) than the alkaline phosphatase-conjugated antibody. The horseradish peroxidase-conjugated antibody is supplied lyophilized. Redissolve the powder in water at 150 U/ml, avoiding excess foaming and air bubbles. This is not the same titer as the AP-conjugated Fab fragments (150 U/200 μl).

1. Protocol 7: Two-Color Whole-Mount Staining with DAB and BCIP/NBT

This procedure is modified from Jowett and Lettice (1994).

1. Replace the PBT with blocking solution (PBT containing 5% sheep serum, 2 mg/ml BSA, 1% DMSO) and incubate at room temperature for at least 1 h on an orbital shaker.

2. Incubate for 2 h in horseradish peroxidase-conjugated sheep antiflucrescein Fab fragments in blocking solution at a dilution of 1:200 (0.75 U/ml).

3. Wash for 2 h in PBT (8 × 15 min).

4. Incubate for 2 min in 0.5 mg/ml DAB in PBT.

5. Add 1/1000 volume of 3% hydrogen peroxide to each incubation. Monitor the staining reaction and stop it by rinsing thoroughly with PBT.

6. Replace the PBT with blocking solution (PBT containing 5% sheep serum, 2 mg/ml BSA, 1% DMSO) and incubate at room temperature for up to 1 h on an orbital shaker (this may not be necessary).

7. Incubate for 2 h in alkaline phosphatase-conjugated sheep antidigoxigenin Fab fragments at a dilution of 1:2000 to 1:5000 dilution (0.375–0.15 U/ml).

8. Wash for 2 h with PBT (8 × 15 min).

9. Equilibrate 3 × 5 min in freshly made NTMT buffer. Prior to the staining reaction for alkaline phosphatase, it is important to equilibrate the tissue with buffer of pH 9.5. NTMT is buffered weakly with Tris base and so with prolonged storage, it may develop a lower pH by absorbing carbon dioxide. Therefore make the buffer just prior to use.

10. Stain with NBT/BCIP (4.5 μl of 75 mg/ml NBT in 70% dimethylformamide and 3.5 μl of 50 mg/ml BCIP in dimethylformamide added to 1 ml of NTMT buffer).

11. Stop the reaction by washing with PBT. The staining reaction can take from 10 min to several hours. If the staining reaction is to take several hours, it is convenient to perform all the antibody washes the day before staining and leave the embryos in PBT overnight. This allows a full working day to monitor the development of the stain.

12. Fix the stained embryos in 4% paraformaldehyde in PBS overnight. If the stain is not fixed, prolonged exposure to light can cause a dark background to develop.

Caution: DAB is a potent carcinogen and so should be handled with care. It is supplied in tablet form by Sigma (#D 5905). All disposable materials should be incinerated, and contaminated glassware should be soaked in a strong sodium hypochlorite solution (6%) before being rinsed thoroughly with water. NBT/BCIP staining solution should be treated with similar caution.

B. Sequential Incubation in Antibodies Conjugated with Alkaline Phosphatase

The aim of this method is to visualize fluorescein-and digoxigenin-labeled probes sequentially using antibodies conjugated with alkaline phosphatase and staining for each antibody with different substrates. Two-color double-label *in situ* hybridizations with chromogens give good results for nonoverlapping signals. The best-contrasting signals are obtained with Fast red substrates in combination with NBT/BCIP or Magenta-phos with BCIP. There are several ways in which the signals may be visualized. However, through experience the following rules should be followed to achieve the most consistent results.

If alkaline phosphatase-staining is to be used for visualizing both probes, the antibodies cannot be added together. They must be added sequentially with an enzyme deactivation step after the development of the first signal.

The fluorescein probe should be visualized first with the weaker staining solution.

The alkaline phosphatase can be inactivated by heat-treating at 65 °C for 30 min when stained with the Vector™ Red precipitate.

Both the Fast Red (Boehringer) and Sigma Fast™ Fast red precipitates are heat-labile and are lost if heat-treated. The alkaline phosphatase activity must be inactivated by incubating in 100 mM glycine-HCl, pH 2.2, for 30 min.

Failure to completely inactivate the first alkaline phosphatase will lead to false signals with the second substrate. If the second substrate is NBT/BCIP, this will overstain the red precipitate.

The Fast Red (Boehringer) staining can be enhanced by adding 0.4 M NaCl to the staining buffer (Chiu *et al.*, 1996).

NBT/BCIP blue stain can be very intense and can easily hide the red stain if the signals are colocalized. However, if the staining reactions are carefully monitored and the weaker red-staining is performed first, then the blue reaction can be stopped before it completely masks the red.

Fast Red, Sigma Fast™, and Vector Red all fluoresce strongly with a rhodamine filter set, so if the tissue is staining only weakly, examination by epifluorescence can enhance the signal.

The following procedure is for two-color *in situ* hybridizations with Fast Red and NBT/BCIP, but these can be replaced with Magenta-phos and BCIP. As its name suggests, the alkaline phosphatase substrate, Magenta-phos (Biosynth AG), gives a magenta-colored precipitate. This contrasts well with the light blue-turquoise color generated by BCIP in the absence of NBT. With both these substrates, the colored precipitates take a long time to develop and staining may take one to two days at 37 °C. If the intensity of each stain is similar, then regions of overlapping signals are a distinct shade of dark blue. Either substrate can be used first in sequential alkaline phosphatase-staining.

1. Protocol 8: Sequential Alkaline Phosphatase Staining with Chromogenic Substrates

The following protocol is modified from that of Jowett and Lettice (1994).

1. Perform the fixation, prehybridization, hybridization, and washes as in Protocols 2–4.
2. Replace the PBT after the posthybridization washes with blocking solution (2 mg/ml BSA, 5% sheep serum, 1% DMSO in PBT) and incubate at room temperature for 60 min.
3. Remove most of the blocking solution and add a 1:5000 dilution (0.15 U/ml) of alkaline phosphatase-conjugated sheep antifluorescein Fab fragments in blocking solution. Incubate for 2 h.
4. Wash for 2 h in blocking solution without sheep serum (8 × 15 min).
5. Equilibrate 3 × 5 min in 100 mM Tris-HCl, pH 8.2.
6. Stain with Vector™ Red, Fast Red (Boehringer), or Sigma *Fast*™ Fast Red.
7. Stop the reaction by washing in PBT.

8. Rinse in PBT and heat to 65 °C for 30 min to inactivate the alkaline phosphatase if stained with Vector™ Red. Alternatively, for Fast Red (Boehringer) and Sigma *Fast*™ Fast Red, incubate in 100 mM glycine-HCl pH 2.2, 0.1% Tween-20 for at least 30 min and then thoroughly wash in PBT.

9. Fix in 4% paraformaldehyde in PBS.

10. Block with blocking solution for 60 min.

11. Incubate for 2 h with a 1:5000 dilution (0.15 U/ml) of alkaline phosphatase-conjugated sheep antidigoxigenin Fab fragments.

12. Wash 8 × 15 min in PBT.

13. Equilibrate 3 × 5 min in NTMT buffer.

14. Stain with NBT/BCIP.

15. Stop the reaction by rinsing in PBT and fixing in 4% paraformaldehyde/PBS.

VII. Double–Fluorescent *In Situ* Hybridization

Chromogenic enzyme substrates are best used for visualizing nonoverlapping signals. When transcripts are colocalized, the heavier and darker precipitate can mask the lighter one. This is particularly so for combinations of NBT/BCIP and Fast Red. These problems can be overcome by using fluorochromes to visualize the signals.

The most sensitive method of fluorescent *in situ* hybridization on whole-mount embryos or tissues sections is to use enzyme substrates that produce insoluble fluorescent precipitates (Jowett and Yan, 1996). The red precipitates produced by the Fast Red products fluoresce strongly when viewed by epifluorescence with a rhodamine filter set. The ELF™ alkaline phosphatase substrate from Molecular Probes initially is nonfluorescent, but when activated by the enzyme it produces a crystalline precipitate that fluoresces yellow-green with a DAPI filter set. By using enzyme substrates there is an amplification of the signal, which greatly increases the sensitivity of the method. The ELF signal is usually greater than that with Fast Red and so it should be used to stain the second antibody.

A. Protocol 9: Sequential Alkaline Phosphatase Staining with Fluorescent Substrates

This procedure is modified from Jowett and Yan (1996).

1. Perform the fixation, prehybridization, hybridization, and washes as in Protocols 2–4.

2. Replace the PBT after the posthybridization washes with blocking solution (2 mg/ml BSA, 5% sheep serum, 1% DMSO in PBT) and incubate at room temperature for 60 min.

3. Remove most of the blocking solution and incubate in a 1:5000 dilution (0.15 U/ml) of alkaline phosphatase-conjugated sheep antifluorescein antibody Fab fragments in blocking solution overnight at 4 °C.

4. Wash the embryos with PBT for 2 h (8×15 min).

5. Equilibrate with 100 mM Tris-HCl pH 8.2 at room temperature by washing 3×5 min. Adding 0.4 M NaCl to this buffer can increase the signal intensity for Fast Red (Boehringer).

6. Stain embryos with Fast Red (Boehringer), Sigma *Fast* Fast Red or Vector Red.

7. Stop the reaction by washing several times with PBT.

8. Inactivate the alkaline phosphatase activity by incubating in 100 mM glycine-HCl pH 2.2, 0.1% Tween-20 for 2×15 min at room temperature.

9. Wash 4×5 min with PBT.

10. Fix in 4% paraformaldehyde in PBS for 20 min.

11. Wash 5×5 min with PBT.

12. Incubate the embryos in blocking solution for 60 min.

13. Incubate in a 1:5000 dilution (0.37–0.15 U/ml) of sheep alkaline phosphatase-conjugated antidigoxigenin antibody in blocking solution overnight at 4 °C.

14. Wash in 0.5% Triton X-100 in PBS for 2 h. DMSO and Tween-20 in the wash solutions cause the final ELF crystals to be large.

15. Wash 3×5 min at room temperature with the ELF prereaction buffer (30 mM Tris-HCl pH 7.5, 150 mM NaCl).

16. For staining incubate in a 1:20 dilution of the ELF substrate reagent at room temperature for 5 h, or a 1:100 dilution of the ELF substrate overnight at room temperature. This is far longer than is recommended by Molecular Probes. With zebrafish embryos, it is not necessary to add 1.0 mM levamisole.

17. Monitor the staining reaction using a UV fluorescent microscope with a DAPI filter set at intervals after starting the reaction.

18. Stop the reaction by washing with 25 mM EDTA, 0.05% Triton X-100 in PBS pH 7.2.

19. Mount the tissue with the special aqueous mounting medium supplied with the kit from Molecular Probes. This mountant preserves the ELF signal better than the usual glycerol-based aqueous mountants for fluorochromes.

VIII. Simultaneous Localization of Transcription and Translation Gene Products

It may be desirable to compare the location of the transcript of a gene with its translation product or that of another gene. This requires an antisense RNA probe to localize the mRNA and an antibody to recognize the protein product. It is best to perform the *in situ* hybridization first, since that reduces the chance of RNases degrading the target transcript. The immunolocalization of the protein requires an antibody, preferably a polyclonal antibody that will recognize the protein antigen after it has gone through the rigors of the hybridization. In this regard, each antibody must be tested separately. Below are protocols for double-labeling using a polyclonal antibody to the zebrafish No tail protein generated by Schulte-Merker *et al.* (1992) and a monoclonal antibody that recognizes the HNK-1 epitope in zebrafish (Trevarrow *et al.*, 1990). The latter is an IgM antibody and so must be identified with a goat antimouse IgM secondary antibody conjugated with horseradish peroxidase. Some protein antigens will be destroyed by the proteinase K permeabilization step after rehydration before refixing and prehybridization. If this is the case, the proteinase K digestion can be left out or replaced with an acetone permeabilization step (incubate for 7 min in acetone at $-20\ ^{\circ}$C). The fixation step may also be critical. Overfixation can mask certain antigens.

A. Protocol 10: Immunolocalization with a Horseradish Peroxidase–Conjugated Secondary Antibody

This is the simplest method of identifying a primary antibody bound to a tissue antigen. It is the least sensitive of the three methods described below.

1. Perform the *in situ* hybridization as described in the previous protocols. Use an antisense RNA probe labeled with digoxigenin and an antibody conjugated with alkaline phosphatase. Stain with either Fast Red or NBT/BCIP.

2. Stop the staining by rinsing in PBT and refix for 20 min in 4% paraformaldehyde in PBS.

3. Incubate in blocking solution (1% DMSO, 2 mg/ml of BSA, and 2% sheep serum in PBT) for 30 min at room temperature.

4. Incubate for 5 h at room temperature in a 1:2000 dilution of monoclonal antibody, which recognizes the HNK-1 epitope in zebrafish. This is an IgM monoclonal antibody (Trevarrow *et al.*, 1990) so the secondary antibody must be one that recognizes IgM rather than the more usual IgG.

5. Wash for 2 h with PBT (8×15 min).

6. Incubate overnight at 4 °C in blocking solution containing a 1:5000 dilution of goat antimouse IgM conjugated to horseradish peroxidase.

7. Wash for 2 h with PBT (8×15 min).

8. Incubate for 2 min in 0.5 mg/ml DAB in PBT. TrueBlue can be used as an alternative substrate to DAB. The TrueBlue horseradish peroxidase substrate is more sensitive than DAB, so decrease the titer of the antibody by 10- to 50-fold. However, the blue precipitate that is formed is less stable, being partially soluble in alcohol and water. If used in a two-color *in situ* hybridization with horseradish peroxidase- and alkaline phosphatase-conjugated antibodies, the phosphatase should be stained before the peroxidase. Otherwise the blue precipitate will be lost or weakened.

9. Add 1/1000 volume of 3% hydrogen peroxide.

10. Monitor the staining and stop it by rinsing thoroughly with PBT.

B. Protocol 11: Immunolocalization by the Peroxidase Antiperoxidase (PAP) Method

The following procedure involves using soluble peroxidase antiperoxidase complex (PAP; Sternberger *et al.*, 1970) to visualize a protein antigen. The PAP complexes are formed from three peroxidase molecules and two antiperoxidase antibodies and are used as a third layer in the visualization step. After incubation with a primary rabbit antibody, a secondary "bridging" antibody (goat antirabbit IgG) is added in excess so that one of its two available identical binding sites binds to the primary antibody and the other binds to the rabbit PAP complex. This method is 100–1000 times more sensitive than indirect methods using fluorochromes or peroxidase-conjugated antibodies.

1. Perform the *in situ* hybridization as described in the previous protocols. Use an antisense RNA probe labeled with digoxigenin and an antibody conjugated with alkaline phosphatase. Stain with either Fast Red or NBT/BCIP.

2. Stop the staining by rinsing in PBT and refix for 20 min in 4% paraformaldehyde in PBS.

3. Incubate in blocking solution (1% DMSO, 2 mg/ml of BSA, and 2% sheep serum in PBT) for 30 min at room temperature.

4. Incubate 5 h at room temperature in a 1:2000 dilution of rabbit antibody directed against the protein product of the zebrafish *no tail* gene (Schulte-Merker *et al.*, 1992).

5. Wash for 2 h with PBT (8 × 15 min).

6. Incubate overnight at 4 °C in blocking solution containing a 1:100 dilution of goat antirabbit IgG (Jackson ImmunoResearch Laboratories). It is important in the PAP technique for the bridging antibody to be applied in excess. This way, one arm of the divalent Fab portion of the immunoglobulin molecule can bind the primary antibody, while the other arm is free to bind the PAP complex.

7. Wash for 2 h with PBT (8 × 15 min).

8. Incubate for 5 h in blocking solution containing a 1:400 dilution of peroxidase-conjugated rabbit antiperoxidase (PAP, Jackson ImmunoResearch Laboratories).

9. Wash for 2 h with PBT (8 × 15 min).

10. Incubate for 2 min in 0.5 mg/ml DAB in PBT. TrueBlue can be used as an alternative substrate to DAB. The TrueBlue horseradish peroxidase substrate is more sensitive than DAB and so the titer of the antibody can be decreased by 10- to 50-fold. However, the blue precipitate that is formed is less stable, being partially soluble in alcohol and water. If used in a two-color *in situ* hybridization with horseradish peroxidase- and alkaline phosphatase-conjugated antibodies, the phosphatase should be stained first and then the peroxidase. Otherwise the blue precipitate will be lost or weakened.

11. Add 1/1000 volume of 3% hydrogen peroxide.

12. Monitor the staining and stop it by rinsing thoroughly with PBT.

C. Protocol 12: Vector Avidin Biotinylated Enzyme Complex (ABC) Antibody Staining of Zebrafish Embryos

This procedure involves a primary antibody followed by a biotinylated secondary antibody and a performed avidin:biotinylated enzyme complex. Avidin has a very high affinity for biotin and the binding of avidin to biotin is effectively irreversible. In addition, avidin has four binding sites for biotin, allowing macromolecular complexes to form with biotinylated enzymes or proteins.

For protein localization following *in situ* hybridization and staining, proceed to step 4 below and transfer the embryos into blocking solution.

1. Perform the *in situ* hybridization as described in the previous protocols. Use an antisense RNA probe labeled with digoxigenin and an antibody conjugated with alkaline phosphatase. Stain with either Fast Red or NBT/BCIP.

2. Stop the staining by rinsing in PBT and refix for 20 min in 4% paraformaldehyde in PBS.

3. Briefly rinse with PBT and then transfer to blocking solution. The composition of the blocking solution depends on how sticky the primary antibody is, but 5% goat serum, 1% DMSO, 0.1% Tween in PBS normally works fine.

4. After blocking for 30 min, transfer the embryos to blocking solution containing the primary antibody (rabbit anti-No tail). Leave overnight in the cold or 4–5 h at room temperature (on a rocker). Perform the incubations in 300 μl in a 24-well dish.

5. Wash 4 × 20 min each with blocking solution: 5% goat serum (supplied with the Vector Elite™ rabbit IgG kit), 1% DMSO, 0.1% Tween-20 in PBS. Biotin is naturally occurring in tissues and serum and, because of the sensitivity of the ABC system endogenous, biotin may lead to background problems. This can be prevented by using a Biotin/Avidin blocking kit (Vector Labs SP-2001). Note also that if serum is used in the blocking solution, this too may be a source of biotin. Either use dialyzed serum (Jackson ImmunoResearch Labs) or serum tested with biotinylated antibody (Vector Labs).

6. Incubate in a 1:2000 dilution of secondary antibody (biotinylated goat anti-rabbit, Vector Elite rabbit IgG kit) for 4–5 h at room temperature or 4 °C overnight.

7. Wash as in step 6.

8. Rinse briefly with PBT (PBS plus 0.1% Tween-20); then proceed with enzymatic detection.

9. During the final 30 min of washing in step 8, prepare the AB Complex. Add 80 μl of Reagent A (Avidin DH) to 10 ml of blocking solution: 5% goat serum (supplied with the Vector Elite™ rabbit IgG kit), 1% DMSO, 0.1% Tween-20 in PBS. Mix and add 80 μl of Reagent B (Biotinylated enzyme); mix and incubate for 30 min at room temperature. Finally, add to the embryos for 45–60 min.

10. Wash 4 × 25 min with 5% goat serum, 1% DMSO, 0.1% Tween in PBS.

11. Wash briefly in PBT and transfer to wherever the stainings are to be done.

12. Incubate for 2 min in 2 ml PBT plus 100 μl of DAB (5 mg/ml in 10 mM Tris-HCl pH 7.0).

13. Add 2 μl 3% H_2O_2. Allow to stain until the signal appears (usually a few minutes). Stop the reaction by washing the specimen in PBS.

14. Dehydrate by putting the specimen in 100% methanol. Change once (2 × 10 min is enough). Transfer to a 2:1 mixture of benzylbenzoate:benzyl alcohol to clear the tissue.

IX. Embedding and Sectioning Whole-Mount Embryos

It can be useful to section the whole-mount embryos after *in situ* hybridization. For this purpose, embryos with strong signals that may even look overstained as whole embryos should be chosen. A little background staining may not be a problem. After fixation, the stain is reasonably stable in organic solvents and so can be embedded in wax without loss of signal. This protocol was designed for *Xenopus* embryos by J. Slack but works well on zebrafish embryos. The original protocol has been modified by replacing tetrahydronaphthalene with Histo-Clear™ (National Diagnostics).

A. Protocol 13: Paraffin-Embedding of Whole Embryos after *In Situ* Hybridization

1. If the embryos are in 70% glycerol, wash them several times with PBT.

2. Transfer to flat-bottomed glass vials (Histo-Clear™ dissolves many plastics). I use 7-ml squat glass vials with polyethylene lids that fit into a specially modified heating block.

3. Wash with methanol for 5 min and with isopropanol for 10 min, and then with Histo-Clear for 15 min. Replace with fresh Histo-Clear.

4. Transfer to 1:1 Histo-Clear: paraffin wax, at 60 °C for 20 min, then to paraffin wax, for 3 × 20 min, also at 60 °C.

5. Transfer to an embryo dish at 60 °C, place at room temperature, orientate under a dissection microscope, and let the wax set.

6. Cut sections.

7. Dewax with Histo-Clear.

8. Mount under DPX-mounting agent.

X. Solutions and Reagents

10× transcription buffer: 400 mM Tris-HCl pH 8.0; 60 mM MgCl$_2$; 100 mM dithiothreitol; 20 mM spermidine; 100 mM NaCl; RNase inhibitor 1 unit/μl. Store in aliquots at −20 °C. (Supplied with RNA polymerase from Boehringer.)

10× digoxigenin-labeling mix: 10 mM ATP, 10 mM CTP, 10 mM GTP, 6.5 mM UTP, 3.5 mM digoxigenin-11-UTP. The pH should be 7.5. Store at −20 °C.

10× fluorescein-labeling mix: 10 mM ATP, 10 mM CTP, 10 mM GTP, 6.5 mM UTP, 3.5 mM fluorescein-12-UTP. The pH should be 7.5. Store at $-20\,°C$.

RNA polymerase SP6 (Boehringer #810 274).

RNA polymerase T3 (Boehringer #1 031 163).

RNA polymerase T7 (Boehringer #881 767).

RNase inhibitor (Boehringer #799 017). Supplied as a solution at 40 U/ml.

10× TBE: dissolve 109 g Tris base, 55 g boric acid, 9.3 g diNaEDTA in 1 l water. pH should be 8.3.1× TBE is 90 mM Tris, 2.5 mM EDTA, 90 mM boric acid.

TE: 10 mM Tris-HCl, 1 mM EDTA, pH 8.0. Make up with RNase-free water and autoclave.

Proteinase K (Boehringer #1 000 144): make up a stock solution of proteinase K at 20 mg/ml in 50% glycerol, 10 mM Tris pH 7.8, and store at $-20\,°C$.

DNase I, RNase-free (Boehringer #776 785): supplied as a solution of 20-50 × 10^3 U/ml.

Paraformaldehyde/PBS fixative: paraformaldehyde is dissolved in PBS at $65\,°C$. If it does not readily dissolve, add a drop or two of 1 M NaOH solution to pH 7.5. It should be cooled to $4\,°C$ and used within two days.

PBS (phosphate-buffered saline): 130 mM NaCl; 7 mM $Na_2HPO_4 \times 2H_2O$; 3 mM $NaH_2PO_4 \cdot 2H_2O$. For 10× PBS mix 75.97 g NaCl, 12.46 g $Na_2HPO_4.2H_2O$, 4.80 g $NaH_2PO_4 \cdot 2H_2O$. Dissolve in less than 1 l distilled water; adjust to pH 7.0 and a final volume of 1 l. Sterilize by autoclaving.

PBT: PBS, 0.1% Tween-20.

Tween-20 (Sigma #P 1379): make a 20% solution in sterile deionized water. Store at room temperature.

20× SSC for hybridization and washing: (20× SSC is 3 M NaCl, 300 mM trisodium citrate). Dissolve 175.3 g NaCl and 88.2 g sodium citrate in 800 ml water. Adjust the pH with 1M citric acid to 4.5 or 6.0 depending on the hybridization solution used. Adjust the volume to 1 l and sterilize by autoclaving.

Hybridization buffer: 50% formamide, 5× SSC, 500 µg/ml yeast RNA, 50 µg/ml heparin and 0.1% Tween-20, brought to pH 6.0 with 1 M citric acid.

Heparin (Sigma #H3393): make a stock solution of 100 mg/ml in deionized water. Store in aliquots at $-20\,°C$.

Yeast RNA (Sigma #R 6750): dissolve in sterile deionized water at 50 mg/ml. Store in aliquots at $-20\,°C$.

Blocking solution for antibodies: 1× PBS, 0.1% Tween-20, 2 mg/ml BSA (BDH #44,155), 5% sheep serum (Gibco BRL #035-6070H), 1% dimethylsulphoxide DMS) (Merck-BDH #28,216).

Alkaline phosphatase inactivation buffer: 0.1 M glycine-HCl pH 2.2, 0.1% Tween-20.

BCIP 5-bromo-4-chloro-3-indolyl-phosphate also known as X-phosphate 4-toluidine salt; (Boehringer #760 994): dissolve at 50 mg/ml in dimethylformamide. Store in aliquots at 20 °C.

ELF™-AP substrate kit (Molecular Probes Inc., Eugene, OR #E-6601): dilute the substrate 1:20 in ELF™ Reaction Buffer supplied with the kit. Use within 30 min. Tissues should be equilibrated with prereaction wash buffer (30 mM Tris, 150 mM NaCl pH 7.5) prior to adding the diluted substrate solution. The reaction is stopped with 25 mM EDTA, 0.05% Triton X-100 in PBS; the final pH should be 7.2 (the addition of 1.0 mM levamisole is optional).

ELF™ stop reaction buffer: 25 mM EDTA, 0.05% Triton X-100 in PBS pH 7.2. Dissolve EDTA in PBS and check pH.

Fast Red tablets (alkaline phosphatase substrate; Boehringer #1 496 549): each tablet contains 0.5 mg naphthol substrate, 2 mg Fast Red chromogen, and 0.4 mg levamisole. Store tablets at −20 °C. Wear gloves and use plastic forceps to handle the tablets. Dissolve one tablet in 2 ml of 100 mM Tris-HCl pH 8.2. Adding 400 mM NaCl can increase the intensity of the staining reaction. Use the solution within 30 min.

NBT (4-nitro blue tetrazolium chloride; Boehringer #1 087 479): dissolve at 75 mg/ml in 70% dimethylformamide. Store in aliquots at −20 °C.

NTMT: 100 mM NaCl, 100 mM Tris-HCl pH 9.5, 50 mM $MgCl_2$, 0.1% Tween-20. Make from concentrated stock solutions on the day of use (the pH will decrease during storage due to absorption of carbon dioxide).

TrueBlue™ peroxidase substrate staining solution (KPL, Kirkegaard & Perry Laboratories, #71-00-64).

References

Chiu, K. P., Sullivan, T., and Bursztajn, S. (1996). Improved in situ hybridization: Color intensity enhancement procedure for the alkaline phosphatase/Fast red system. *Bio Tech.* **20**, 964–968.

Jowett, T. (1997). Alternative enzyme substrates. *In* "Tissue *In Situ* Hybridisation: Methods in Animal Development" (T. Jowett, ed.), pp. 29–32. Spektrum/Wiley, New York.

Jowett, T., and Lettice, L. (1994). Whole-mount *in situ* hybridisation on zebrafish embryos using a mixture of digoxigenin and fluorescein-labeled probes. *Trends Genet.* **10**, 73–74.

Jowett, T., and Yan, Y.-L. (1996). Double fluorescent *in situ* hybridisation to zebrafish embryos. *Trends Genet.* **12**(10), 387–389.

Schulte-Merker, S., Ho, R. K., Herrmann, B. G., and Nüsslein-Volhard, C. (1992). The protein product of the zebrafish homologue of the mouse T gene is expressed in nuclei of the germ ring and the notochord of the early embryo. *Development* **116**, 1021–1032.

Sternberger, L. A., Hardy, P. H., Cuculis, J. J., and Meyer, H. G. (1970). The unlabeled antibody enzyme method of immunohistochemistry: Preparation and properties of soluble antigen-antibody complex (horseradish peroxidase-antiperoxidase) and its use in identification of spirochaetes. *J. Histochem. Cytochem.* **18**, 315.

Trevarrow, W. W., Marks, D. L., and Kimmel, C. B. (1990). Organisation of hindbrain segments in the zebrafish embryos. *Neuron* **4**, 669–679.

Westenfield, M. (1995). "The Zebrafish Book," 3rd edn., pp. 8.1–8.2. University of Oregon Press, Eugene, OR.

CHAPTER 6

Analysis of Zebrafish Development Using Explant Culture Assays

Yevgenya Grinblat* and Hazel Sive[†]

*Departments of Zoology and Anatomy
University of Wisconsin
Madison, Wisconsin 53706

[†]Whitehead Institute for Biomedical Research and Massachusetts Institute of Technology
Nine Cambridge Center
Cambridge, Massachusetts 02142

ESSENTIAL ZEBRAFISH METHODS:
CELL AND DEVELOPMENTAL BIOLOGY

DOI: 10.1016/B978-0-12-374599-6.00006-6

I. Introduction

A. Basic Questions: Lineage Commitment and Inductive Interactions

The most important questions asked by classical embryologists, and ones that must still be considered in any developmental system, are (1) when do cells make the decision to become a particular cell type? and (2) what cell interactions are involved in this decision? The commitment of cells to a particular lineage is referred to as specification or determination, and cell interactions that influence lineage commitment are termed inductive interactions (or induction). Lineage commitment and induction can be assessed directly by explant or transplant assays. These assays led to the discovery of induction and remain the methods of choice to determine when and how cells acquire fate, even amongst genetically tractable organisms (Ang *et al.*, 1994; Duncan, 2003; Spemann and Mangold, 1964; Stern, 1994). Thus, the ability to perform such assays enhances the utility of model organisms for addressing developmental mechanisms.

Explant assays involve removal of groups of cells from the embryo and their culture in a medium believed to be free from inductive factors (Fig. 1A), while

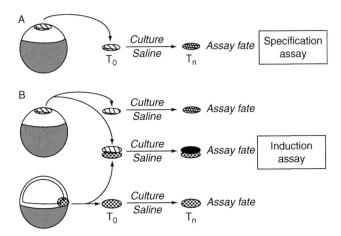

Fig. 1 Analysis of lineage commitment and cell interactions using explant analysis. (A) To assay specification, an explant is isolated from the embryo at time T_0 and cultured for a period of time (T_n) under conditions which support development but do not affect it, for example in saline solution. The fate of the explant after culture, at T_n, is an indication of its lineage commitment at the time of isolation. (B) To assay induction, two different explants are juxtaposed immediately after isolation, at T_0, forming a conjugate, and cultured for a period of time (T_n). Explants of both types are also cultured in isolation alongside the conjugates. The fate of the individual explants within the conjugate at T_n is compared to their fate when cultured in isolation (i.e., to their specification). A change of fate indicates that induction has taken place.

transplant assays involve moving groups of cells or single cells from one part of the donor embryo to a different part of a host embryo. In the host, the transplant is usually placed in a position that is believed to exert little or no influence on the fate of the transplanted cells. In both cases, the fate of explanted or transplanted cells is assayed morphologically, histologically, or molecularly after a period of culture. Transplanted, but not explanted cells must be distinguished from the host embryo, and this is usually achieved by lineage labeling either the host or donor tissue.

Lineage commitment as defined by explant assays is sometimes called "specification," to distinguish it from lineage commitment as assessed by transplantation assays, which is called "determination." In some cases, the timing of specification and determination of the same tissue may be different. Since explanted tissues receive no exogenous inductive signals after isolation, they may display commitment to a particular lineage earlier during development than transplanted cells, which may continue to receive inductive signals in their new surroundings that alter their fate. Thus explant assays of lineage commitment may be a less stringent assay of lineage commitment than transplant assays.

Inductive interactions measured by explant and transplant criteria are likely to be quite similar. In both cases, one measures the fate of test cells after their exposure to a heterologous group of cells believed to be a source of inductive signals. In explants, inductive interactions are tested by juxtaposing the two groups of cells in close apposition in a "conjugation" assay (Fig. 1B). Transplant assays for inductive interactions are similar to those used for lineage commitment, except that the target location for the transplant in the host embryo is chosen near a source of putative inductive signals.

B. Application of Explant Assays to Zebrafish

During the last decade, zebrafish has realized its potential as a powerful, genetically tractable model for vertebrate embryogenesis. Many developmentally important genes have been identified through forward-genetic screens (Driever et al., 1996; Golling et al., 2002; Haffter et al., 1996) and through reverse-genetic methods that include targeted mutagenesis (Doyon et al., 2008; Meng et al., 2008; Moens et al., 2008) and transient gene knock-down assays (Nasevicius and Ekker, 2000; Skromne and Prince, 2008). To fully understand the functions of these genes, it is important to address their roles in inductive cell interactions and lineage commitment decisions. These questions are most readily answered using transplant and explant assays.

Transplant analysis performed over several decades in teleosts have demonstrated the existence of an organizing center analogous to the amphibian Spemann organizer (Oppenheimer, 1936; Shih and Fraser, 1996). In zebrafish, transplant assays have been used successfully in many contexts [reviewed in Carmany-Rampey and Moens (2006)], for example, to analyze the timing of commitment to mesendodermal (Ho and Kimmel, 1993) and neural lineages (Eisen and Pike, 1991; Woo and Fraser, 1997), and to dissect specific mutant phenotypes (Halpern et al., 1995; Schier et al., 1997).

Explant analysis methodology was first developed for use with zebrafish embryonic tissues in the Sive laboratory (Grinblat et al., 1998, 1999; Sagerström et al., 1996). There are a number of advantages offered by explant over transplant analysis. The most important of these is the rigorous control over culture conditions in explants. As discussed in Section I.A, explants are not subject to influences from the host embryo and therefore may reveal earlier, more subtle aspects of lineage commitment. In addition, explants are better suited for testing candidate inducing molecules, for example, growth factors, than are transplants (Roy and Sagerstrom, 2004). Another advantage of the explant method is the relative ease of scoring fate after culture, since transplants always require a means of distinguishing the donor and host tissue. On the other hand, the challenges of explanting zebrafish tissues include small embryo size, which necessitates small explants; the limited amount of yolk contained within the cells of the embryo proper, which suggests that some explants might not survive well in simple nutrient-free culture media; and the fragility of the embryos.

Despite these limitations, we can successfully culture blastula stage animal pole tissue (animal caps) and early gastrula shields for periods of 24 h or more in a simple saline solution with no added nutrients. We showed that blastula stage animal caps were specified primarily as ventral tissue, while shields were specified as a wide array of mesendodermal and ectodermal tissues patterned along the anteroposterior axis (Sagerström et al., 1996). We were also able to demonstrate that the shield was a potent source of neural inducing signals; however, although it was specified to express both anterior and posterior neural markers, it could induce only anterior neural marker expression in juxtaposed animal caps (Sagerström et al., 1996). This indicated that anterior and posterior neural induction were rather different events, as has been suggested for amphibian development (Gamse and Sive, 2000). This initial study demonstrated that zebrafish explant analysis was feasible and capable of yielding new information about zebrafish development (Roush, 1996).

We have analyzed forebrain determination in explant assays (Grinblat et al., 1998), demonstrating that animal cap tissue becomes committed to neurectodermal lineages by early gastrula (shield stage) (Section VI.A). Surprisingly, by this time the presumptive neurectoderm already contains anteroposterior pattern information, with both presumptive telencephalon (anterior forebrain) and diencephalon (posterior forebrain) specified. The source of forebrain inducing and patterning signals appears to be the organizer (Section VI.B), including, but not limited to, the presumptive prechordal plate.

In a separate study, we have used explants to analyze commitment of the outer enveloping layer of the zebrafish embryo (Sagerstrom et al., 2005). We have shown that this layer becomes committed by mid-blastula stages, and that this commitment requires cell interactions. Unexpectedly, unlike in Xenopus, zebrafish ectodermal explants do not auto-neuralize when dissociated and cultured, highlighting the importance of testing developmental mechanisms in multiple model organisms (Sagerstrom et al., 2005).

Thus, our analyses have established explant assays as a robust technique for dissecting lineage commitment and induction at the early stages of zebrafish development. We describe here the methodology of making the explants that we have analyzed, and suggest general guidelines for explant assay design. Finally, we suggest that explant analysis, which yields key information concerning normal fish development, will be particularly powerful when used in conjunction with genetic analysis.

II. Zebrafish Explants: General Considerations

Zebrafish embryonic cells are smaller and less yolky, and therefore more fragile, than those of amphibian embryos. Zebrafish explants are comparable in size to explants made from early gastrula-stage mouse embryos; importantly, unlike mouse explants, zebrafish explants do not require added nutrients for survival. Viability of an explant is influenced by its tissue type, by the embryonic stage at which the explant is made, and by its size.

A. Tissue Choice

The first consideration in choosing the tissue to explant is the question to be addressed. For example, if a researcher wants to assay purified factors on a pluripotent tissue, blastula stage animal caps are a good choice. Isolation of blastula animal caps and other useful embryonic tissues that we have successfully explanted is described in Section IV.B.

B. Reproducibility in Isolating Tissue

It is important that the same group of cells can be identified and isolated from multiple embryos, and this requires embryonic landmarks. In blastula embryos, the blastoderm margin, which marks the vegetal limit of the embryo proper, and the animal pole are useful landmarks. Once gastrulation begins, a dorsal thickening, the shield, can be used as a landmark. As gastrulation proceeds, additional landmarks arise—dorsally, the leading edge of the involuting endomesoderm (prechordal plate), and vegetally the blastoderm margin, which is moving closer to the vegetal pole through epiboly. After the end of gastrulation, the forming somites allow very precise positioning along the anteroposterior axis in the trunk. In the head region, many additional landmarks arise as somitogenesis progresses, including the optic primordium at the 5-somite stage, and the first neuromeres in the developing brain by the 14-somite stage.

As a result of recent dramatic improvements in transgenic technology (Kawakami, 2004, 2007), many transgenic lines that express fluorescent reporter proteins in discrete embryonic domains have been generated. These lines, when available for the tissue of choice, can be used both as landmarks during explant isolation and as a means of monitoring the purity of explanted tissues after dissection.

C. The Size of the Explant

In choosing the size of an explant, it is important to remember that small explants containing fewer cells are better than larger explants, since smaller cell populations are likely to be more homogenous with respect to developmental potential. On the other hand, small explants are less likely than larger explants to survive culture well, and may need to be cultured in groups and/or in methylcellulose (Section IV.A.4) in order to improve survival.

D. The Method of Assay

Culturing zebrafish explants without added nutrients is advantageous since it obviates the possibility that nutrients will interfere with explant development. However, since zebrafish embryonic cells contain little yolk, it also means that explants do not generally survive well for longer than 24 h after removal from the embryo. This relatively short culture period allows time for some morphological differentiation to take place (for example, that seen in isolated shields, described in Section IV.B.3), but molecular assays for changes in gene expression are most useful in general. These include gel based assays (particularly reverse transcriptase-PCR (RT-PCR) assays) and *in situ* hybridization. More discussion of these assays and detailed procedures are given in Section IV.A.5.

III. Materials Required

A. Equipment

1. Warm Room

Dissections and culture of explants should be done between 25 and 31 °C to ensure optimal survival. A small room with a tight door and a standard thermostat which can be set at a constant temperature is adequate for this purpose, unless it becomes critical to control the temperature more precisely.

2. Microscope

A good quality dissecting microscope, with an adjustable range of magnification between 8× and 50×, is required for making explants. It is wise to test several models, looking for ease of positioning and minimal eye and posture stress, before settling on one. Either reflected light, delivered by a fiberlight light source, or transmitted light, delivered by a transmitted light box which fits over the base of the microscope, may be used. Dissecting microscopes fitted with a fluorescent light source are also available (e.g., from Leica). These are used to monitor lineage label incorporation in induction assays (Section V.A) and to visualize fluorescent reporter expression, if it is used to guide microdissection.

3. Microinjectors and Micromanipulators

These are required for embryonic injections to lineage label embryos for induction assays (Section V.A.1). Microinjectors come in a wide range of qualities and prices, and include top of the line models made by Narishige or Medical Systems, and cheaper models made by Drummond. The most important consideration is that an injector can reproducibly deliver 1–10 nl of liquid. Micromanipulators are used to hold and move the needle while injecting and must be able to adjust to an appropriate angle relative to the microscope base. Narishige and Brinkman make good basic models of micromanipulators.

4. Micropipette Puller

This is necessary for making injection needles used for lineage labeling in induction assays (Section V.A.1) and are useful, although optional, for making glass knives for dissecting older embryos (Section III.D.2). Many needle pullers are available, among them those offered by Sutter Instruments, as well as the less expensive models made by Narishige. Micropipettes which are suitable for injection needle and glass knife preparation include 30 μl Drummond micropipette needles from Fisher and 1 mm × 10 cm glass thin-wall capillaries from World Precision Instruments. The latter have an internal filament which facilitates filling of the needle, and are recommended for injections.

B. Solutions and Culture Media

1. Water

We use water purified by the MilliQ Water System (Millipore) with consistent success. The purity of water used to make the solutions listed below is of paramount importance, since we have seen explants die very quickly and dramatically when exposed to lower quality water purified by reverse osmosis.

2. Culture Media

Our lab initially tested several standard culture media used for Xenopus embryonic explants for their ability to support development of zebrafish explants. Modified Barth's Saline (MBS) used at 1× concentration allowed for best survival and was therefore chosen as the standard culture medium for zebrafish explants. We have avoided supplementing this simple saline solution with nutrient additives such as serum, which may contain factors that could affect development of explants during culture. We have, however, found that the addition of chemically defined lipid concentrate (Gibco-BRL) at 1:100 dilution can improve viability of small explants, namely, late blastula animal caps (Sagerstrom and Sive, unpublished observations).

a. 10× MBS Stock Solution

This is the stock buffered saline solution. Sterilize by autoclaving and store at room temperature.

	For 1 l	Final concentration
NaCl	51.3 g	0.88 M
KCl	0.75 g	10 mM
$MgSO_4 \cdot 7H_2O$	2 g	10 mM
HEPES	23.8 g	50 mM
$NaHCO_3$	2 g	25 mM

pH to 7.6 with NaOH. Add H_2O to 1 l.

b. 1× MBS

This is the working solution in which explants are made and cultured. Make fresh from sterile stocks every day.

	For 100 ml	Final concentration
10× MBS	10 ml	1×
1 M $CaCl_2$	0.7 ml	7 mM
H_2O	89.2 ml	
Gentamycin (50 mg/ml)	0.1 ml	50 μg/ml

c. 1 M CaCl₂

Sterilize by filtration. Aliquot into Eppendorf tubes, 1 ml per tube, and store at $-20\,°C$.

	For 100 ml
$CaCl_2$	11.1 g

d. 3% Methyl Cellulose

This is used for culturing small or fragile explants as described in Sections III.C.2 and IV.B. Stored at $-20\,°C$.

Dissolve 3 g of methyl cellulose (Sigma) in 100 ml 1× MBS (without $CaCl_2$ or gentamycin) by stirring at $4\,°C$ for 1–3 days. Centrifuge for 30 min at $12,000 \times g$ to drive out air bubbles and precipitate undissolved solids. Aliquot into 3–5 ml portions and freeze at $-20\,°C$. Just before use thaw the required number of aliquots, transfer to Eppendorf tubes, and centrifuge in a microfuge for 5 min at top speed.

C. Dissection and Culture Dishes

1. Dissection Dishes

Pour a layer of high quality 1% agarose (electrophoresis grade, e.g., Seakem GTG) melted in water into sterile 35 mm Petri dishes. The depth of the agarose layer can vary from 1 to 5 mm, according to preference. Allow agarose to solidify, wrap, and store at 4 °C. Dishes with 5 mm-thick layers of agarose can be stored at 4 °C wrapped in plastic for up to 2 weeks. Dishes with thinner agarose layers should be used sooner, before the agarose dries out and cracks.

2. Culture Dishes

Although explants can be cultured in agarose coated dishes, it is better to transfer them out of the dissection dishes to multiwell tissue culture plates prior to culture, placing one explant into each well. This serves to prevent different expants fusing together during the culture period. Among the commonly available tissue culture grade multiwell plates we have found 4-well plates from Nunc to work best, since they provide a well that is smaller then in most other multiwell dishes, but large enough to allow easy retrieval of explants with a Pipetman. For explants which require extra care, a drop of 3% methyl cellulose should be introduced into the center of each well prior to filling it with $1\times$ MBS. Explants are placed onto the methyl cellulose cushion. Although some methyl cellulose remains associated with the tissue after culture, we have not found it to interfere with subsequent processing, namely, RNA preparation and *in situ* hybridization (unpublished observations).

D. Dissection Tools

The construction of homemade dissection tools is shown in Fig. 2. While the homemade tools work well for gastrula-stage embryos, we recommend investing in a set of Vannas-style spring scissors, available from Fine Science Tools (catalog # 15002–08) for dissections of older embryos. For example, these scissors work very well for retinal primordia explants at mid-somitogenesis stages (N. A. Sanek and Y. Grinblat, unpublished observations). All microdissection tools should be cleaned and sterilized just before and after use by washing with 70% ethanol. It is convenient to keep a squirt bottle with 70% ethanol at the microdissection station for this purpose.

1. Eyebrow Knives

This is the tool of choice for dissecting embryos prior to mid-gastrula (75% epiboly). An eyebrow knife is made by placing a human eyebrow hair into the narrow end of a handle made from a drawn out Pasteur pipette, and fixing it in place with melted beeswax. Eyebrow knives are used by pushing their ends into the

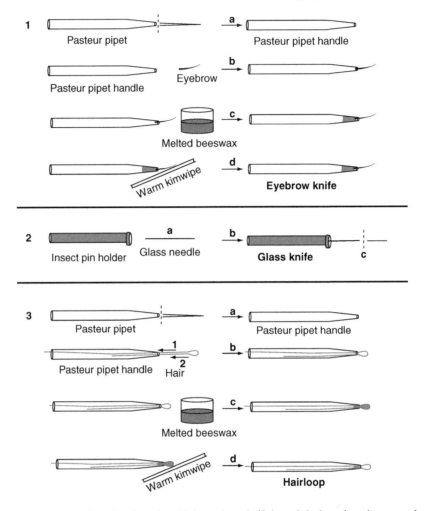

Fig. 2 Making tools for microdissection. (1) An eyebrow knife is made by inserting a human eyebrow hair into a Pasteur pipette handle and fixing it in place with beeswax (see Section III.C.1 for details). (2) A glass knife is prepared by inserting a glass micropipette needle into a commercial insect pin holder and breaking the tip to the desired length (see Section III.C.2 for details). (3) A hairloop is made by inserting a long hair into a Pasteur pipette handle and fixing it with beeswax (see Section III.C.3 for details). Numbers 1 and 2 above the arrows indicate the order in which the two ends of the hair are inserted. In all of the above, letters above arrows correspond to sections within the text which describe these steps.

tissue to be cut and then flicking upward through the tissue to complete the cut. A sharp end allows easy entry, and a smooth surface ensures that the cut does not damage too many cells at the edge of the explant. Therefore, after a period of use, when the sharp end of the eyebrow wears down and its surface becomes visibly rough, the knife should be replaced. A step-by-step guide to the preparation of an eyebrow knife is given below and diagrammed in Fig. 2(1).

a. Preparation

    a. To prepare the handle, a long stemmed Pasteur pipette is heated over a Bunsen burner flame just below the shoulders, and pulled out to about 15 cm in length. The end is broken off carefully by wrapping the drawn-out tip in a Kimwipe, after scoring the pipette with a diamond knife if desired, to produce an opening just large enough to thread an eyebrow or a hair through it.

    b. An eyebrow hair is placed in the opening of the handle, using fine forceps.

    c. The hair is fixed in place with beeswax (obtainable at craft stores). This is done by dipping the knife into a beaker of melted beeswax, which will enter the fine portion of the pipette by capillary action, and will solidify immediately upon removal from the beaker.

    d. Excess wax is removed from the eyebrow hair by touching the hair portion of the knife to a Kimwipe prewarmed on a hotplate. The hotplate should be just warm, not hot, to avoid scorching the eyebrow hair.

    e. Eyebrows from different people can vary in their performance as dissecting tools and testing a range of eyebrow hairs donated by several people before settling on one is highly recommended. The hair of choice should be firm and elastic, and very fine or very coarse eyebrows may not be useful. A newly made eyebrow knife typically has a sharp end and a smooth, shiny surface.

    f. It is a good idea to have at least two good eyebrow knives available at all times, since these are fragile tools prone to accidental damage during routine use. Dull eyebrow knives (or hairloops, Section III.D.3) can be used for moving explants on the dissection plate.

2. Glass Knives

    Eyebrow knives are not suitable for dissections of embryos at mid-gastrula and older because by then the outer enveloping layer of the embryo becomes tougher. Instead, glass knives should be used for these dissections. These can be made from a glass microcapillary set into a commercially available needle holder, as diagrammed in Fig. 2(2), or into a Pasteur pipette as described for eyebrow knives.

a. Preparation

    a. A glass needle is pulled out over a microburner flame (Bunsen burners are not recommended) or on a micropipette needle puller.

    b. The needle is set into a metal needle holder used for holding insect pins (available from Fine Science Tools or Carolina Biological). This is easier then the alternative of setting glass needles into Pasteur pipette handles with beeswax as for eyebrow knives (Section III.D.1).

c. The needle tip is broken off to give a knife of the appropriate length.

d. Since the knife tip is open, glass knives are difficult to keep sterile and should be discarded after one day's use.

3. Hairloops

A hairloop is used to immobilize the embryo while dissecting, and can also be used for cutting. It comprises a loop of hair set into a drawn out Pasteur pipette similar to the procedure for eyebrow knives. A guide to the preparation of a hairloop is given below and diagrammed in Fig. 2(3).

a. Preparation

a. The handle is prepared from a Pasteur pipette as described for eyebrow knives (Section III.D.1).

b. Both ends of a long hair are threaded through the narrow opening of the handle using fine forceps, leaving out a loop. One end of the hair should reach out through the broad end of the Pasteur pipette.

c. The loop size is adjusted by pulling on the long end of the hair. A loop that is approximately 1 mm in diameter is most useful for moving and holding embryos. A smaller loop can be used for moving explants.

d. The hair is fixed in place with beeswax as described for construction of eyebrow knives (Section III.D.1).

4. Forceps

Fine steel forceps are used to remove chorions in preparation for dissections (Section IV.A.1). Dumont #5 or 5A forceps from Fine Science Tools are a good choice. The fine tips of these forceps are very easily damaged; to protect them, avoid contact with hard surfaces by using them only on agarose-coated dishes, and keep the tips covered when not in use. Dulled tips may be sharpened using a fine sharpening stone.

E. Embryos

A healthy explant can only be made from a healthy embryo, which requires that the breeding colony is in top health. The following guidelines are based on The Zebrafish Book (Westerfield, 1995) and tested in our laboratories.

1. Age

The optimal age for egg production is between 4 months and 1 year.

2. Feeding and Density

The egg-producing fish should be fed at least twice a day, alternating between flake and live brine shrimp. Care should be taken not to overfeed, since this reduces egg production. The general rule is to feed just enough food so that it is completely consumed within 5 min. It is very important to avoid over-crowding, both of the juveniles in the nursery and in the adult tanks.

3. Water Quality

Maintaining high standards of water quality is also essential, and this requires good circulation (3–5 volumes exchanged per hour is recommended by many tropical fish breeder manuals), and water purification through mechanical and biological filtration. The optimal pH of the water for egg-laying is 7–8, and the rate of egg-production may drop off dramatically if pH falls below 6.8.

4. Mating

To increase egg production, male and female fish are often maintained in separate tanks and combined only to produce embryos. Crosses are set up in the evening using 1–3 females and 1–4 males per breeding tank, which can be made using a 2 l container (e.g., a plastic mouse cage) with a mesh bottom placed within another 2 l container with a solid bottom. Eggs fall through the mesh bottom and collect on the solid bottom, where they are protected from the adult fish.

After a mating, which usually takes place within 1–2 h of "dawn," males and females are separated again. Mated fish, particularly females, should be allowed to rest for at least one week before being used in a cross. It is often convenient to have a choice of several "dawn" time points, especially when making heterochronic conjugates (see Section V.B below). For this, the fish should be kept and crossed in light-tight photocabinets which have staggered lights on-lights off cycles.

5. Embryo Preparation

Embryos collected from natural crosses should be thoroughly washed to remove any debris, which include fish waste, as soon as possible after they are laid to prevent bacterial and fungal growth. Standard fish water or E3 (Westerfield, 1995) supplemented with methylene blue (final concentration 2 ppm) to prevent bacterial and fungal growth, is used for the washes, and should be kept on hand at the appropriate temperature. In preparation for dissection, washed embryos are transferred to dissection dishes containing culture medium ($1 \times$ MBS) and dechorionated manually (Section IV.A.1). If possible, embryos should not be dechorionated prior to mid-blastula stages (3 hpf), since younger embryos are more easily damaged. Care must be taken not to allow the temperature of the medium to drop below 25 °C while preparing for dissections, which is ensured by doing this in a warm room.

IV. Guide to Explant Isolation and Culture

A. Common Procedures

Dissections of all explants follow certain common procedures or considerations that are described here.

1. Dechorionation

Zebrafish embryos are protected from the environment, and from our dissection tools, by a tough outer covering called the chorion (the vitelline envelope). Chorion must be removed prior to dissection. This should be done manually, since residual quantities of proteases left behind by the enzymatic removal may compromise the health of the explants.

a. Protocol 9.1

a. Use two pairs of forceps to grasp the chorion at two nearby positions on the side of the embryo facing the objective.

b. Pull the forceps apart, creating a tear in the chorion. Do not attempt to enlarge the tear in one continuous movement, since liquid within the chorion will rush toward the newly created opening and in doing so will push the embryo through the opening and damage it. Rather, make several small pulls, repositioning the forceps each time, until the hole is big enough for the embryo to come out without damage.

c. Observe embryos for any sign of damage, that is, torn cell layers or yolk puncture, as these will heal quickly, becoming invisible, but can affect subsequent development of the embryo. Damaged embryos should not be used.

d. When first starting to dechorionate, allow about 1 min per embryo, and be prepared to damage many of them. With practice, 100 embryos can be dechorionated in about 30 min, with very little waste.

e. Once dechorionated, embryos are very fragile. They should not contact air-water interfaces, and should not touch plastic. All dissection and culture dishes should be coated with agarose or methylcellulose. A plastic pipette may be used to transfer dechorionated embryos to a new dish, but its opening should be wide enough to avoid direct contact with the embryo. Alternatively, glass pasteur pipettes, cut to leave a wide opening and then fire polished, can be used for transferring dechorionated embryos.

2. Temperature Control

After dechorionation, all manipulations involving the embryos should be carried out at temperatures ranging from 25 to 31 °C. It is extremely important to prevent sharp temperature changes when transferring embryos from one dish to another. To ensure this, dissection and culture dishes to be used during the day's dissection should be filled with 1× MBS and allowed to warm up to the temperature of the dissection room before use. A bottle of 1× MBS made fresh from stock that day should also be left in the dissection room long enough to come up to temperature.

3. Evaluating Explant Health

Subsequent to explant culture, it is important to evaluate the health of explants before proceeding further. Imperfect explants should be discarded since poor health may give misleading indications concerning their fate. The health of an explant at the end of a culture period can be evaluated in several ways.

a. Under the dissecting microscope, a healthy explant looks translucent and does not contain any white crystalline-like patches. It has lost very few, if any, of its cells (these can usually be seen on the plate surrounding the explant). The surface of a healthy explant looks smooth and shiny, probably due to the presence of the enveloping layer.

b. If more rigorous proof is required that a group of explants is surviving well in culture, sibling explants of the same type can be dissociated, and cell number in an average explant determined before and after culture, to show that cell division has taken place. Dissociation protocols may vary between explants from different tissues. We have found that late sphere animal caps could be dissociated immediately after isolation by pipetting in PhoNaK buffer (50 mM NaPhosphate, 25 mM NaCl, 1 mM KCl) (Godsave and Slack, 1989). After culture for 20 h, addition of collagenase (2 mg/ml) to PhoNaK buffer was required for efficient dissociation. Trypsin, another common enzyme used for cell dissociation, was not as effective. Dissociated cells were fixed by adding formaldehyde (EM grade, EM Sciences) directly to the suspension, to 4% final concentration. After 20 min at room temperature, nuclear stain (DAPI) was added directly to the fixed cell suspension, to 1 μg/ml final concentration. Nuclei in this suspension were counted using a hemocytometer and a fluorescence microscope. Using this protocol we found that the number of cells within a late blastula animal caps increases by 15-fold on average after 20 h in culture (Sagerström *et al.*, 1996). This number is likely to vary greatly between different explants, depending on the type of tissue analyzed and the time in culture, and should be comparable to the number of cell divisions the tissue would undergo if left in the embryo.

4. Suggestions for Improving Survival of Fragile Explants

Very small explants (for example, late blastula animal caps that contain fewer then 50 cells each) can be pooled into groups of 5–10 immediately after the dissection to improve survival. However, since explants are usually oriented randomly within the final conglomerate, this method does not allow axial asymmetries of marker distribution to be evaluated. If pooling explants is not an option, or if it does not improve their viability sufficiently, we recommend culturing them on a thin layer of 3% methyl cellulose (see Section III.C.2). Our lab has found methyl cellulose to make a very dramatic difference in the viability of several explant types as detailed in Section IV.B (M. E. Lane, unpublished observations).

5. Assaying Explant Fate After Culture

The two basic and complementary assays used for explant analysis are *in situ* hybridization and quantitative RNA assays, including Northern, RNAse protection and reverse transcriptase-PCR (RT-PCR) analysis. Gel based assays, particularly RT-PCR (see below) can allow several markers to be assayed on each explant, with quantitative evaluation of relative expression levels. In contrast, *in situ* hybridization allows only one or two markers to be assayed on each individual explant, but unlike gel-based assays, it allows the spatial distribution of gene products to be determined (see Section VI.B for an example of this). In general, specific genes begin to be expressed and reach maximal levels of expression at approximately the same time in explants as in intact sibling embryos, allowing one to time explant harvest to the time of maximal gene expression.

a. RT-PCR Assay

We have previously published a detailed protocol for RT-PCR assays of zebrafish explants which was modified to accommodate the very small quantities of material specific to zebrafish explants (Sagerström *et al.*, 1996). The critical points of this protocol, which has been used extensively and with consistent success in our lab, are detailed below.

1. Explants are harvested by lysis in 200 μl of denaturing solution, aided by vortexing, and processed using a standard acid guanidinium RNA preparation protocol (Chomczynski and Sacchi, 1987). We routinely prepare RNA from pools of 10 late blastula animal caps (approximately equal to 0.3 whole embryos), 5 early gastrula animal caps, or 1–2 early gastrula shields per sample (see Section IV.B for description of these explants). RNA is also prepared from one whole embryo to serve as control. Addition of carrier, for example glycogen, to all precipitation steps during RNA preparation is required to ensure good yields when these small amounts of tissue are being processed.

2. We synthesize cDNA using RNA prepared from 10 late blastula animal caps, 5 early gastrula animal caps, or 1–2 early gastrula shields per reaction and reverse transcriptase (RT) from Gibco, according to manufacturer's instructions. A portion of the RNA should be used in a mock reverse transcription carried out without RT, to control for genomic DNA contamination during RNA preparation. To conserve explant RNA, which is limiting, we routinely use RNA from whole embryos for "no RT" controls.

3. cDNA synthesized as above can be used directly, without additional processing, in standard real-time PCR applications (reviewed in Heid *et al.*, 1996).

4. In order to compare expression levels of a given gene between different samples, expression should be normalized using a ubiquitously expressed gene, e.g., α-tubulin or GAPDH.

b. In situ Hybridization

We have previously published a modified *in situ* hybridization protocol optimized for use with the tiny and fragile zebrafish explants (Sagerström *et al.*, 1996). This protocol contains several critical modifications relative to standard protocols (Hauptmann and Gerster, 1994; Oxtoby and Jowett, 1993). These modifications are outlined below.

1. To minimize transfers, we use commercially available Netwell 15 mm baskets (74 μm mesh, Costar) for all steps, from initial fixation through final staining. In addition to minimizing loss of tissue during transfers, this allows efficient processing of multiple samples.

2. For prehybridization and hybridization steps, explants may be transferred from baskets to eppendorf tubes, in order to minimize the volume of probe used. The transfer should be done using a 20 μl Pipetman and monitored under the microscope. Note that, once placed in prehybridization buffer, explants become transparent and nearly invisible.

3. To avoid losing tissue when replacing prehybridization with hybridization buffer, centrifuge the explant-containing tubes very briefly either in a Picofuge or in a microfuge set at low speed. This brings explants to the bottom of the tube. Remove most of the prehybridization buffer carefully, leaving about 50 μl of it behind, then add hybridization buffer.

4. To transfer explants to baskets at the end of hybridization, briefly spin the tube again to bring explants to the bottom, and remove most of the hybridization buffer, leaving about 50 μl of it behind. Approximately 0.5 ml of wash buffer is then added to the remaining liquid, which contains the explants, mixed gently, and transferred to baskets in wash buffer. Several rinses of the tubes will ensure that all the explants are transferred to baskets. This should be checked by counting the number of explants in the basket under a microscope. Note that used hybridization buffer can be saved and reused at least 3 times with no loss of signal.

5. The mesh size of the Netwell baskets is too large for holding very small explants, such as single late blastula animal caps. For handling such explants, baskets can be hand-made using smaller mesh netting, as described for baskets used to process Drosophila embryos (Wieschaus and Nusslein-Volhard, 1986).

B. Guide to Specific Explants

Timing is a critical factor to consider when planning dissections of these explants, since zebrafish embryos develop quickly and are available at the desired stage for a limited period of time. The length of this period varies from day to day since fish tend to produce embryos over a period of time which can last from 5 min to 1 h. We find that, on average, we have 30–45 min for each of the dissections listed below when using embryos produced by crosses from a single light cycle. Below is a detailed guide to making specific explants, and an estimate of the maximum number of explants which can be generated by one person in one day. This number can be increased by using staggered light cycles.

1. The Late Blastula (Sphere Stage) Animal Cap

An animal cap is the tissue located at and just around the animal pole of an embryo, which at blastula stages is fated to give rise to derivatives of all three germ layers. Small animals cap from late blastula (sphere stage, 4 hpf) embryos contain 20–50 cells, are specified primarily as ventral ectoderm, and lack any neural or mesodermal character. This makes them a good choice as responding tissue in induction assays (see below). Sphere animal caps have been shown to respond to inducing signals by turning on neural and mesodermal molecular markers (Sagerström *et al.*, 1996). A protocol for making these explants is given below and illustrated diagrammatically in Fig. 3A.

a. Protocol 9.2

Up to 100 late blastula animal cap explants can be made per day. You will need this number for an induction assay (see below), but for most experiments fewer explanted caps will be sufficient. When assaying by RT-PCR, we use 20–30 caps (i.e., 2–3 groups of 10), per data point (e.g., for each different time of harvest). Each group of 10 caps is analyzed separately for expression of up to 10 markers. When assaying by *in situ* hybridization, 30–40 caps (3–4 groups of 10) per data point are needed for each marker.

a. Dechorionate the required number of embryos in a prewarmed dissection dish filled with $1\times$ MBS and transfer dechorionated embryos into a clean dish as described in Section IV.A.1.

b. Using a hairloop, turn the embryo so that you are looking directly at the animal pole. Hold it in position with a hairloop and use an eyebrow knife to make a horizontal incision at a distance of approximately 20° from the animal pole. The incision is made with the knife held horizontally, by pushing the tip of it in just under the surface and then flicking it up.

c. Make a second cut, parallel to the first, 20° to the other side of the animal pole. Turn the embryo 90° still facing animal pole up, and make two more cuts, perpendicular to the first two and parallel to each other to make a square.

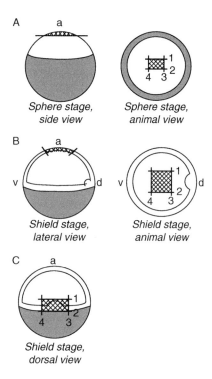

Fig. 3 Preparation of specific explants. (A) Preparation of late blastula animal cap explants (see Section IV.B.1 for details). (B) Preparation of early gastrula animal cap explants (see Section IV. B.2 for details). (C) Preparation of early gastrula shield explants (see Section IV.B.3 for details). In all of the above, positions of incisions are indicated with solid lines and the order in which they are made with numbers next to the lines, that is, 1 indicates the first cut, 2—the second cut etc. Abbreviations: a, animal; d, dorsal; v, ventral.

d. Gently separate the cap from the embryo by passing the eyebrow knife under it and lift it off. This will produce a square explant that is about 3 cell layers thick, and contains 20–50 cells.

e. Blastula animal cap explants can survive for a limited amount of time (up to 4 h) when dissociated and cultured as single cells according to the following protocol (Sagerstrom *et al.*, 2005). To dissociate an explant, place it into an agarose-coated dish containing 1× MBS without Ca^{++} or Mg^{++} and pipette the cells gently to disperse them. For extended cultures the dispersed cells should be pipetted every hour to prevent them from clumping together. To reaggregate the cells, add $CaCl_2$ and $MgCl_2$ to final concentrations of 0.7 and 1.0 mM, respectively, found in standard 1× MBS. Push them together gently with an eyebrow knife to assist reaggregation.

b. Helpful Hints

To improve viability of these very small and fragile explants, and to make them large enough to handle easily, for example, while processing for *in situ* hybridization (see Section IV.A.5), they should be cultured in groups of 5–10. Simply push the explanted caps together right after dissection, and within 1 or 2 min they will form a cohesive group. This group should be allowed to heal in the dissection dish for at least 10–15 min, and then transferred to fresh media for the remaining period of culture.

Transfer explants with a pipetman (P20), using a tip with the end cut to widen the opening somewhat. Use a minimal volume of saline (3–5 μl should be sufficient), and watch the transfer under the microscope, taking care that the explant does not touch the air-liquid interface. To avoid losing explants or contaminating them with other tissue, it is important to keep the explants in a different portion of the dish than the embryos from which they are derived and any embryo fragments. To keep explants from fusing into larger groups during the culture period, they should be transferred to multiwell dishes, such as tissue culture grade plastic 4-well dishes from Costar, filled with 1× MBS and containing a drop of 3% methyl cellulose to improve survival (Section III.D.2).

The major contaminating cell population in this type of explant is derived from the vicinity of the blastoderm margin. These cells are not morphologically distinct but can be identified by the presence of mesendodermal markers such as *no tail* (Schulte-Merker *et al.*, 1992). As these margin-derived cells produce inducing signals, it is strongly recommended that mesodermal contamination be ruled out by routine screening of all explants by RT-PCR for expression of *no tail*.

2. The Early Gastrula (Shield Stage) Animal Cap

Animals cap from early gastrula (shield stage, 6 hpf) embryos are located at the animal pole and are fated to give rise primarily to anterior neural tissues, namely, telencephalon and retina (Woo and Fraser, 1995). They are specified primarily as telencephalon and to a lesser extent as ventral ectoderm, and contain no mesoderm (Grinblat *et al.*, 1998). A protocol for making these explants is given below and diagrammed in Fig. 3B.

a. Protocol 9.3

Up to about 30 early gastrula animal cap explants can be made per session. This produces 6 groups of 5 caps, sufficient to analyze 2–3 different data points (e.g., different harvest times) with up to 10 markers each by RT-PCR, or expression of 1 or 2 markers by *in situ* hybridization.

a. Dechorionate the desired number of embryos in a prewarmed dissection dish and transfer dechorionated embryos to a clean dish as described in Section IV.A.1.

   b. Using a hairloop, turn the embryo so that its animal pole faces the objective. While holding the embryo in that position with a hairloop, use an eyebrow knife to make a horizontal incision at a distance of approximately 30° from the animal pole. The incision is made with the knife held horizontally by pushing the tip between the inner surface of the epiblast and the yolk cell, which by this stage have separated. If done carefully, this incision will not break the yolk cell and will prevent any possibility of contamination of the explant with YSL nuclei.

   c. Make the second cut perpendicular to the first, the third perpendicular to the second, etc., all equidistant from the animal pole, resulting in a square cap which freely lifts off the yolk cell.

b. Helpful Hints

These caps are significantly larger and sturdier then sphere caps and can be cultured singly or in groups of 5. If culturing in groups, allow explants to round up and heal in the dissection dish for at least 10–15 minutes, then transfer them to 4-well plates filled with 1× MBS for the remaining period of culture. If culturing singly, use 3% methyl cellulose to improve survival.

3. The Early Gastrula Shield

The shield is a thickening of the dorsal margin of the blastoderm, which first becomes apparent in early gastrula (shield stage, 6 hpf) embryos. Early gastrula shield is functionally equivalent to the organizer region of amphibians and other vertebrates (Sagerström et al., 1996; Shih and Fraser, 1996). It is fated to give rise to prechordal plate, notochord, and ventral neural tube (Shih and Fraser, 1995). Dorsal mesoderm and neural tissues are specified in the shield, but not anterior-most neural tissue, the prospective telencephalon (Grinblat et al., 1998; Sagerström et al., 1996). A protocol for making these explants is given below and illustrated schematically in Fig. 3C.

a. Protocol 9.4

Up to 30 early gastrula shield explants can be made per session. This is sufficient to analyze six different data points (e.g., different harvest times) with 1 or 2 markers each by in situ hybridization. If assaying by RT-PCR, fewer shield explants will usually be sufficient, since each one can be assayed for expression of 5–10 markers, although pooling RNA prepared from several explants is recommended to control for differences between individual explants.

   a. Dechorionate the desired number of embryos in a prewarmed dissection dish filled with 1× MBS and transfer dechorionated embryos to a clean dish as described in Section IV.A.1.

b. Using a hairloop, turn the embryo so that you are looking at its dorsal side. Holding the embryo in that position with a hairloop, use an eyebrow knife to make a horizontal incision at the border between the yolk cell and the blastoderm margin. This incision will break the yolk cell, which is firmly attached to the blastoderm at this position.

c. Rotate the embryo by 90° while keeping its dorsal side up, and make the second and third cuts perpendicular to the first and parallel to each other, flanking the edges of the shield thickening which should be clearly visible. The fourth cut, made just above the anterior extent of the thickening, will release the shield explant.

b. Helpful Hints

Shield explants are relatively large and can be cultured singly in 1× MBS in 4-well plates. These explants are hardy enough to do well if cultured for short periods (up to 6 h). For longer culture their survival is improved by addition of 3% methyl cellulose. The advantage of culturing shield explants singly is the ability to see anteroposterior pattern within them after just 4–6 h in culture, when they elongate and their posterior ends (equivalent to the notochord) become morphologically distinct from the anterior ones (equivalent to the prechordal plate). Neural and mesodermal markers are distributed along the long axis of the explant in the same order relative to each other as they are distributed in the embryo along its anteroposterior axis (Sagerström *et al.*, 1996).

V. Using Explants to Assay Induction

A. Common Procedures and Considerations

Conjugates are used to assay induction in explants, and are made by placing two explants in close juxtaposition. One of the explants is a candidate source of inducing signals and the other a substrate for the inducing signals to act upon. The common procedures and considerations described for specification assays (Section IV.A) are applicable to induction assays. In addition, the following procedures are specific to induction assays.

1. Lineage Labeling

Because molecular markers used to assay the final fate of the conjugate may be expressed both in the inducing and the responding tissues, complicating the scoring of induction, one of the tissues should be marked with a lineage tracer. Fluresceinated lysine-fixable dextran (FLDX, MW = 10,000 Kd, Molecular Probes), which can be visualized after *in situ* hybridization, is a good choice. 10–20 pg of FLDX injected into the yolk cell of 1–8 cell stage embryos is sufficient to allow detection both directly, by fluorescence, and indirectly, using an anti-FITC antibody. FLDX is nontoxic in fish up to concentrations of at least 100 pg per embryo. Injecting at early cleavages, when all the cells are still connected via cytoplasmic bridges, serves to distribute the label evenly throughout the embryo. As an alternative to labeling by dye injection, embryos from a transgenic line expressing a traceable molecule, for example GFP, in the relevant tissue may be used.

2. Detection of Lineage Label During *In Situ* Hybridization

Both FLDX and GFP lineage labels may be visualized in cultured conjugates directly, by fluorescence, or indirectly, as a second stain for double *in situ* hybridization (Sagerström *et al.*, 1996). In the procedure given, the lineage label is visualized as light blue. Although visualizing lineage label as a color precipitate is more labor intensive, it allows easier and more precise comparison between distributions of gene products and lineage label. A protocol for visualizing FLDX in explants is given below. All steps are carried out in Netwell baskets described in Section IV.A.5.

a. Protocol 9.5

a. At the end of *in situ* hybridization for the gene of interest, which in this procedure works best when visualized as purple color precipitate produced by the action of alkaline phosphatase in the presence of NBT (Research Organics) and BCIP (Sigma), inactivate alkaline phosphatase by immersing in 0.1 M glycine (pH 2.2) for 10–20 min (Hauptmann and Gerster, 1994).

b. After washing and preblocking steps, the explants are incubated with an anti-FITC antibody conjugated to alkaline phosphatase (Boehringer), for 4 h at room temperature or overnight at 4 °C.

c. After the antibody is washed off, the conjugates are stained for alkaline phosphatase using BCIP alone as substrate. This will result in a light blue precipitate, very distinct from the purple precipitate used to detect gene expression. This reaction is very quick, producing strong signals within 10 min, and care should be taken not to overstain.

B. Guide to the Animal Cap/Shield Induction Assay

As described in Section IV.B, late blastula animal caps are specified as nonneural ventral ectoderm and early gastrula shields as dorsal mesoderm and neurectoderm, excluding anterior-most neurectoderm, the prospective telencephalon. When cultured as conjugates, animal caps are respecified to express anterior neural markers, including a telencephalic marker. This indicates that the early gastrula shield is a source of anterior neural inducing signals, and that the late blastula animal cap is competent to receive these signals.

1. Protocol 9.6

1. Collect embryos from crosses with "dawn" points separated by 1–1.5 h. This will allow shield explants to be ready soon after sphere cap explants are made.

2. Inject approximately 50 embryos from the earlier batch (first "dawn"), through the chorion, into the yolk cell at 1–8 cell stage with 10–20 pg of FLDX (see Section V.A.1). Leave to develop in fish water until at least 3 hpf.

3. Dechorionate 100 uninjected embryos from the later batch (the second "dawn"), allowing at least 30 min for this. Carefully transfer dechorionated embryos to a clean dish and cut 100 animal caps when the embryos reach sphere stage, as described in Section IV.B.1. Push into 10 groups of 10, allow explants to heal for at least 10–15 min, and transfer to a clean dissection dish with 1× MBS.

4. Transfer injected embryos to a dissection dish with 1× MBS. Sort them under a microscope equipped with a fluorescent light source, and discard embryos which are not visibly fluorescent.

5. Dechorionate sorted lineage-labeled embryos. Transfer these embryos, which should be approaching shield stage, to the dish with animal cap explants. When shield-donor embryos reach the correct stage, explant two shields as described in Section IV.B.3, then cut one group of 10 caps into two equal halves, and push each half of the cap group together with one shield, with the freshly cut surface of the cap group juxtaposed to the inner (i.e., formerly adjacent to the yolk) surface of the shield. Repeat this procedure, which produces two conjugates, until all shield-donor embryos are past the correct stage. This protocol allows one person to make an average of 10–12 conjugates per session, which lasts about 45 min.

6. Culture and assay the remaining unused cap explants alongside the conjugates. This is a very important control and should be included in every experiment, especially when assaying expression of markers that are already specified at some level in the animal cap. Shields cultured in isolation should also be included as controls, but it may not be necessary to do this in every experiment.

2. Helpful Hints

It is important to culture each conjugate in a separate well to prevent them from fusing together. Methyl cellulose may improve viability of conjugates, especially of their animal cap portions.

C. Assaying Purified Molecules as Inducers

Explants are very useful for assaying the specific effects of purified factors on development. The simplest way to deliver a molecule being tested is to dissolve it, preferably at nearly physiological concentrations, in the medium used to culture the test explants. This approach has been successfully used in zebrafish, for example, to study the effects of activin (Hammerschmidt and Nusslein-Volhard, 1993; Hug *et al.*, 1997), BMPs (Sagerstrom *et al.*, 2005), and Fgf (Roy and Sagerstrom, 2004) on the development of pluripotent tissue, the late blastula animal cap. The caveats of this method are that (1) it may not work for water-insoluble molecules and (2) once the explant has healed and formed an outer layer, the molecule in question may have limited access to the target cells. However, explants will almost always allow better access of the molecule being tested than whole embryos, particularly when explants are made from deep tissues.

VI. Illustrations of Specification and Induction Assays

A. Specification Analysis

1. Aims

We asked when forebrain is first specified in zebrafish. In order to do this we used a novel zebrafish marker of the anterior forebrain (telencephalon), *odd-paired like* (*opl*), a member of the Zic gene family (Grinblat *et al.*, 1998; Kuo *et al.*, 1998). Using specification assays and *opl* we asked whether telencephalon was specified in the ectoderm of late blastula and early gastrula embryos.

2. Procedure

This question was addressed in the experiment outlined schematically in Fig. 4A. Animal caps were explanted at late blastula (sphere) or early gastrula (shield) stages, combined in groups of 10 (sphere caps) or 5 (shield caps), and cultured until control embryos reached tailbud (10 hpf) stage, to allow *opl* expression to reach high levels. Explants were then assayed by *in situ* hybridization for *opl* expression, or by RT-PCR for expression of *opl* and α-*tubulin*, a general marker used as loading control.

3. Outcome

In situ hybridization analysis showed that *opl* is weakly expressed after culture in a small number of cells in late blastula cap explants (Fig. 4B, panel a). In contrast, early gastrula caps activated strong expression of *opl* in the majority of their cells (Fig. 4B, panel b). This indicated that specification of *opl* is strongly upregulated in the animal caps between late blastula and early gastrula stages.

Since *in situ* hybridization is not quantitative, we used an RT-PCR assay to evaluate relative levels of *opl* expression in late blastula and early gastrula caps. Results of a representative experiment are shown in Fig. 4C. Consistent with the findings of analysis by *in situ* hybridization, *opl* is specified weakly in late blastula animal caps (lane 1) and more strongly in early gastrula caps (lane 2). In whole embryos (lane 3) *opl* was expressed at levels similar to those seen in late blastula animal caps. Quantitative analysis showed that *opl* specification increased by an average of 6-fold (Grinblat *et al.*, 1998).

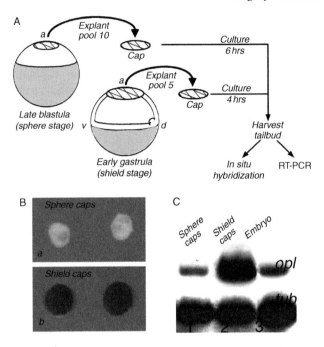

Fig. 4 Assaying the time course of neural specification in late blastula and early gastrula animal caps. (A) Schematic outline of the explant assay. Animal caps were explanted from sphere stage or shield stage embryos. Caps were combined in groups of 5–10 and cultured until control embryos reached tailbud stage (10 hpf), when they were assayed for expression of *opl* (Grinblat *et al.*, submitted) by *in situ* hybridization or by a reverse transcriptase-PCR (RT-PCR) assay. (B) Analysis of *opl* expression in animal cap explants by *in situ* hybridization. (a) Sphere caps showing weak patchy staining. (b) Shield caps showing intense staining throughout the tissue. (C) RT-PCR analysis of *opl* expression in animal cap explants. A ubiquitously expressed marker, α-*tubulin* (G. Conway, unpublished), was used as loading control. Each lane represents a pool of five explants. The data shown here is representative of five experiments. Lane 1: explants from sphere stage embryos, lane 2: explants from shield stage embryos, lane 3: whole embryo controls.

In summary, these specification assays demonstrated that telencephalon was already strongly specified in the ectoderm of early gastrula embryos, at the time when later onset neural markers are first specified (Grinblat *et al.*, 1998).

B. Induction Analysis

1. Aims

We wanted to identify cell interactions which induce forebrain-specific gene expression in zebrafish ectoderm. Specifically, we asked whether the early gastrula shield, a tissue with organizer properties, can induce expression of the telencephalic marker *opl* in nonneural ectoderm.

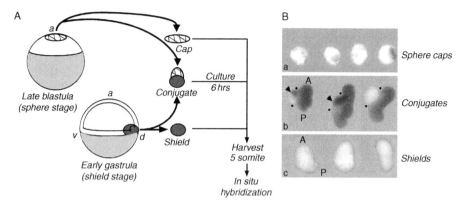

Fig. 5 Assaying induction of anterior neural fates by early gastrula organizer (shield). (A) Schematic outline of the conjugation experiment. Animal cap explants from late blastula (sphere stage) embryos were cultured either in isolation, in groups of 10, or as conjugates with shields explanted from early gastrula (shield stage) embryos. Each conjugate was made with one shield and five animal caps. As a control, shields were cultured alone, singly. For conjugates, embryos from which shields were isolated were lineage labeled with FLDX (pale blue) in order to distinguish inducing from responding tissues. Explants were cultured until shield-donor embryos reached the 5-somite stage, at which time the explants were harvested and stained for *opl* RNA (purple) by *in situ* hybridization. (B) Cultured explants, stained for the presence of *opl* RNA (purple) and lineage label (blue). (a) Four groups of 10 animal caps cultured in isolation. (b) Three conjugates representative of the typical outcomes of the experiment. Dots mark the anterior and posterior edges of the animal cap-derived tissues. A, anterior; P posterior. Arrowheads point to strong *opl* staining. (c) Shield explants cultured in isolation. A total of 38 conjugates were made in seven independent experiments. Induction of *opl* was observed in 58% of them, and was always restricted to the animal cap-derived portion. No detectable *opl* expression has been observed in more then 40 shield explants generated in at least five experiments. Abbreviations: a, anterior; p, posterior. (See Plate no. 2 in the Color Plate Section.)

2. Procedure

We addressed this question, as outlined in Fig. 5A, in conjugates between early gastrula shield and late blastula animal cap explants, which we have previously shown to lack neural specification and to be only weakly specified for *opl* (see previous section). In order to distinguish the inducing and the responding tissue within the conjugates, embryos which shield explants were derived from were lineage labeled dye with FLDX (see Section V.A.1). After culture for 6 h, conjugates were subjected to *in situ* hybridization to visualize *opl* RNA as purple stain and FLDX as light blue stain (Section V.A.2).

3. Outcome

As demonstrated by a representative set of explants shown in Fig. 5B, shields were able to induce strong patterned expression of *opl* in late blastula caps (purple stain, Fig. 5B, panel b). A total of 38 conjugates were made in 7 independent experiments. Induction of *opl* was observed in 22 (58%) of them, and was always restricted to the animal cap-derived portion, while 16 (42%) lacked *opl* expression completely. This indicated the presence of signals within the shield which could induce strong *opl* expression as well as signals that could suppress the low level of *opl* expression seen in cultured caps (Fig. 5B, panel a). Shield explants themselves failed to express *opl*, either when cultured in conjugates or in isolation (Fig. 5B, panel c), suggesting that the ectoderm contained within them is not competent to respond to signals which induce anterior forebrain.

In summary, through use of an induction assay described above, we were able to demonstrate that the early gastrula shield is a source of signals which activate patterned forebrain development in uninduced ectoderm (Grinblat *et al.*, 1998).

VII. Future Directions

Explant assays in zebrafish are feasible and have been used to obtain key information, but have been significantly hampered by several factors. First, until recently there has been a paucity of landmarks that would allow unambiguous identification of tissues of interest in living embryos; second, it is difficult to screen microdissected tissue for contamination with adjacent cells that may be a source of inductive signals during culture; and third, characterization of explant fate after culture has been severely limited by their small size. Recent developments in transgenic technology have produced an explosion of transgenic reporter lines with specific cell populations marked by fluorescent protein expression, and these will largely alleviate the first two limitations. With the advent of ultra high-throughput sequencing methods, transcriptional profiling of even the smallest explants is becoming possible, alleviating the third limitation of the assay. We therefore anticipate that the explant method will be applied much more broadly and effectively in the future, both to understand how cells communicate during normal embryogenesis and to analyze developmental defects caused by genetic lesions and other genetic manipulations, and that this will help us realize the full potential of zebrafish as a model system of choice for vertebrate embryogenesis.

Acknowledgments

Charles Sagerstrom was also pivotal in developing these explant assays. Supported by grants from the NIH to HLS and YG.

References

Ang, S. L., Conlon, R. A., Jin, O., and Rossant, J. (1994). Positive and negative signals from mesoderm regulate the expression of mouse *Otx2* in ectoderm explants. *Development* **120**, 2979–2989.

Carmany-Rampey, A., and Moens, C. B. (2006). Modern mosaic analysis in the zebrafish. *Methods* **39**, 228–238.

Chomczynski, P., and Sacchi, N. (1987). Single-step method of RNA isolation by acid guanidinium thiocyanate-phenol-chloroform extraction. *Anal. Biochem.* **162**, 156–159.

Doyon, Y., McCammon, J. M., Miller, J. C., Faraji, F., Ngo, C., Katibah, G. E., Amora, R., Hocking, T. D., Zhang, L., Rebar, E. J., Gregory, P. D., Urnov, F. D., *et al.* (2008). Heritable targeted gene disruption in zebrafish using designed zinc-finger nucleases. *Nat. Biotechnol.* **26**, 702–708.

Driever, W., Solnica-Krezel, L., Schier, A. F., Neuhauss, S. C., Malicki, J., Stemple, D. L., Stainier, D. Y., Zwartkruis, F., Abdelilah, S., Rangini, Z., Belak, J., and Boggs, C. (1996). A genetic screen for mutations affecting embryogenesis in zebrafish. *Development* **123**, 37–46.

Duncan, S. A. (2003). Mechanisms controlling early development of the liver. *Mech. Dev.* **120**, 19–33.

Eisen, J. S., and Pike, S. H. (1991). The spt-1 mutation alters segmental arrangement and axonal development of identified neurons in the spinal cord of the embryonic zebrafish. *Neuron* **6**, 767–776.

Gamse, J., and Sive, H. (2000). Vertebrate anteroposterior patterning: the Xenopus neurectoderm as a paradigm. *Bioessays* **22**, 976–986.

Godsave, S. F., and Slack, J. M. (1989). Clonal analysis of mesoderm induction in *Xenopus laevis*. *Dev. Biol.* **134**, 486–490.

Golling, G., Amsterdam, A., Sun, Z., Antonelli, M., Maldonado, E., Chen, W., Burgess, S., Haldi, M., Artzt, K., Farrington, S., Lin, S. Y., Nissen, R. M., *et al.* (2002). Insertional mutagenesis in zebrafish rapidly identifies genes essential for early vertebrate development. *Nat. Genet.* **31**, 135–140.

Grinblat, Y., Gamse, J., Patel, M., and Sive, H. (1998). Determination of the zebrafish forebrain: Induction and patterning. *Development* **125**, 4403–4416.

Grinblat, Y., Lane, M. E., Sagerstrom, C., and Sive, H. (1999). Analysis of zebrafish development using explant culture assays. *Methods Cell Biol.* **59**, 127–156.

Haffter, P., Granato, M., Brand, M., Mullins, M. C., Hammerschmidt, M., Kane, D. A., Odenthal, J., van Eeden, F. J., Jiang, Y. J., Heisenberg, C. P., Kelsh, R. N., Furutani-Seiki, M., *et al.* (1996). The identification of genes with unique and essential functions in the development of the zebrafish, *Danio rerio*. *Development* **123**, 1–36.

Halpern, M. E., Thisse, C., Ho, R. K., Thisse, B., Riggleman, B., Trevarrow, B., Weinberg, E. S., Postlethwait, J. H., and Kimmel, C. B. (1995). Cell-autonomous shift from axial to paraxial mesodermal development in zebrafish floating head mutants. *Development* **121**, 4257–4264.

Hammerschmidt, M., and Nusslein-Volhard, C. (1993). The expression of a zebrafish gene homologous to *Drosophila* snail suggests a conserved function in invertebrate and vertebrate gastrulation. *Development* **119**, 1107–1118.

Hauptmann, G., and Gerster, T. (1994). Two-color whole-mount *in situ* hybridization to vertebrate and *Drosophila* embryos. *Trends Genet.* **10**, 266.

Heid, C. A., Stevens, J., Livak, K. J., and Williams, P. M. (1996). Real time quantitative PCR. *Genome Res.* **6**, 986–994.

Ho, R. K., and Kimmel, C. B. (1993). Commitment of cell fate in the early zebrafish embryo. *Science* **261**, 109–111.

Hug, B., Walter, V., and Grunwald, D. J. (1997). Tbx6, a brachyury-related gene expressed by ventral mesodermal precursors in the zebrafish embryo. *Dev. Biol.* **183**, 61–73.

Kawakami, K. (2004). Transgenesis and gene trap methods in zebrafish by using the Tol2 transposable element. *Methods Cell Biol.* **77**, 201–222.

Kawakami, K. (2007). Tol2: A versatile gene transfer vector in vertebrates. *Genome Biol.* **8**(Suppl. 1), S7.

Kuo, J. S., Patel, M., Gamse, J., Merzdorf, C., Liu, X., Apekin, V., and Sive, H. (1998). Opl: A zinc finger protein that regulates neural determination and patterning in xenopus. *Development* **125**, 2867–2882.

Meng, X., Noyes, M. B., Zhu, L. J., Lawson, N. D., and Wolfe, S. A. (2008). Targeted gene inactivation in zebrafish using engineered zinc-finger nucleases. *Nat. Biotechnol.* **26**, 695–701.

Moens, C. B., Donn, T. M., Wolf-Saxon, E. R., and Ma, T. P. (2008). Reverse genetics in zebrafish by TILLING. *Brief. Funct. Genomic. Proteomic.* **7**, 454–459.

Nasevicius, A., and Ekker, S. C. (2000). Effective targeted gene "knockdown" in zebrafish. *Nat. Genet.* **26**, 216–220.

Oppenheimer, J. M. (1936). Transplantation experiments on developing teleosts (*Fundulus* and *Perca*). *J. Exp. Zool.* **72**, 409–437.

Oxtoby, E., and Jowett, T. (1993). Cloning of the zebrafish krox-20 gene (krx-20) and its expression during hindbrain development. *Nucleic Acids Res.* **21**, 1087–1095.

Roush, W. (1996). Zebrafish embryology builds better model vertebrate. *Science* **272**, 1103.

Roy, N. M., and Sagerstrom, C. G. (2004). An early Fgf signal required for gene expression in the zebrafish hindbrain primordium. *Brain Res. Dev. Brain Res.* **148**, 27–42.

Sagerstrom, C. G., Gammill, L. S., Veale, R., and Sive, H. (2005). Specification of the enveloping layer and lack of autoneuralization in zebrafish embryonic explants. *Dev. Dyn.* **232**, 85–97.

Sagerström, C. G., Grinblat, Y., and Sive, H. (1996). Anteroposterior patterning in the zebrafish, *Danio rerio*: An explant assay reveals inductive and suppressive cell interactions. *Development* **122**, 1873–1883.

Schier, A. F., Neuhauss, S. C., Helde, K. A., Talbot, W. S., and Driever, W. (1997). The one-eyed pinhead gene functions in mesoderm and endoderm formation in zebrafish and interacts with no tail. *Development* **124**, 327–342.

Schulte-Merker, S., Ho, R. K., Herrmann, B. G., and Nusslein-Volhard, C. (1992). The protein product of the zebrafish homologue of the mouse T gene is expressed in nuclei of the germ ring and the notochord of the early embryo. *Development* **116**, 1021–1032.

Shih, J., and Fraser, S. E. (1995). Distribution of tissue progenitors within the shield region of the zebrafish gastrula. *Development* **121**, 2755–2765.

Shih, J., and Fraser, S. (1996). Characterizing the zebrafish organizer: Microsurgical analysis at the early-shield stage. *Development* **122**, 1313–1322.

Skromne, I., and Prince, V. E. (2008). Current perspectives in zebrafish reverse genetics: Moving forward. *Dev. Dyn.* **237**, 861–882.

Spemann, H., and Mangold, H. (1964). Part ten: 1924—Induction of embryonic primordia by implantation of organizers from a different species. *In* "Foundations of Experimental Embryology" (B. H. Willier and J. M. Oppenheimer, eds.), pp. 145–184. Prentice-Hall, Englewood Cliffs, NJ.

Stern, C. D. (1994). The avian embryo: A powerful model system for studying neural induction. *FASEB J.* **8**, 687–691.

Westerfield, M. (1995). "The Zebrafish Book." University of Oregon Press, Eugene.

Wieschaus, E., and Nusslein-Volhard, C. (1986). Looking at embryos. *In* "*Drosophila*: A Practical Approach" (D. B. Robers, ed.), pp. 199–227. Oxford University Press, New York, NY.

Woo, K., and Fraser, S. (1995). Order and coherence in the fate map of the zebrafish nervous system. *Development* **121**, 2595–2609.

Woo, K., and Fraser, S. E. (1997). Specification of the zebrafish nervous system by nonaxial signals. *Science* **277**, 254–257.

CHAPTER 7

Confocal Microscopic Analysis of Morphogenetic Movements

Mark S. Cooper, Leonard A. D'Amico, and Clarissa A. Henry

Department of Biology
University of Washington
Seattle, Washington 98195–1800

ESSENTIAL ZEBRAFISH METHODS:
CELL AND DEVELOPMENTAL BIOLOGY
Copyright 2009, Elsevier Inc. All rights reserved.

DOI: 10.1016/B978-0-12-374599-6.00007-8

I. Introduction

During early embryogenesis, cellular ensembles within vertebrate embryos exhibit extraordinary sequences of transient, stereotyped morphogenetic behaviors that are expressed spatiotemporally in a region-specific manner. Recent examinations of such organized cell behavior in various domains of the zebrafish embryo have begun to reveal intricate connections between patterning processes, tissue morphogenesis, and the dynamic cytological activities of the zebrafish's embryonic cells.

Because of its exceedingly rapid rate of development, as well as its optical transparency, the zebrafish embryo provides an ideal system for analyzing the cellular dynamics that underlie early vertebrate morphogenesis. Moreover, the numerous zebrafish strains that have been isolated from recent saturation mutagenesis screens now provide a wide variety of mutant phenotypes from which the patterns of cell division, cell intercalation, cell migration, and coordinate cell shape changes that underlie zebrafish morphogenesis can be experimentally analyzed (Driever *et al.*, 1996; Haffter *et al.*, 1996; Haffter and Nüerlein-Volhard, 1996).

In order to understand the interplay between patterning signals and morphogenesis in specific domains of the embryo, it is necessary to experimentally link these two processes through time. As development proceeds, blastomeres in an embryo undergo profound changes of cell-fate specification in a region-specific and progressive manner. These ongoing processes of cell-fate specification are frequently correlated with distinctive changes in cellular behavior. In this regard, it is very useful to image the patterns of cell behavior that occur during specific morphogenetic events within the interior of a living embryo. New combinations of fluorescence labeling and confocal imaging now make it possible to monitor the collective morphogenetic movements of hundreds of cells simultaneously, allowing the intricate social behavior of embryonic cells within organ-forming primordia to be studied in detail.

In this chapter, we discuss how a variety of fluorescent molecules can be used as vital stains for yolk-containing cytoplasm and interstitial space throughout the entire zebrafish embryo. These fluorescent probe molecules allow all of the cells within a living embryo to be rapidly stained and then visualized *en masse* using a scanning laser confocal microscope. We next describe the use of another fluorescent probe, SYTO-11 (methane, sulfinylbis), which can be used to selectively label and locate a small group of cells (i.e., noninvoluting endocytic marginal (NEM) cells) on the incipient dorsal side (i.e., the organizer region) of late-blastula embryos. We then outline experimental approaches that are useful for making single-level and multilevel confocal time-lapse recordings of vitally stained zebrafish embryos. Finally, we discuss selected applications of visualization technology as applied to confocal imaging. These involve computer-based methods of rendering confocal images of developing zebrafish organ rudiments into uniquely informative three-dimensional (3D) projections or four-dimensional (4D) animations. Such time-compression and visualization approaches can be used to detect and analyze the genetically encoded sequences of cell behaviors that underlie the formation of the zebrafish germ layers and organ rudiments.

II. Confocal Imaging of Embryos

The term "confocal" refers to an imaging system in which the illuminating optic (the condenser) and collecting optic (the objective) are focused on the same volume element of the specimen (Inoue, 1987). In scanning confocal microscopy, the specimen is scanned by a laser beam in a raster-like fashion through an *epifluorescence* objective, which serves as both the illuminating and collecting optical element for the imaging system. Pinhole or slit apertures, which are placed in conjugate image planes to the specimen, prevent fluorescent light that is emitted from above or below the plane-of-focus from impinging on the imaging system's photodetector. By imaging only the in-focus light rays from the specimen plane, the optical configuration of confocal microscopes eliminates much of the blurring that is normally associated with the imaging of thick fluorescent objects. This major improvement in fluorescent imaging permits the morphogenetic movements of numerous cells within a vitally stained zebrafish embryo to be monitored simultaneously with high-spatial resolution.

Confocal imaging of cell movement and cell shape changes within living tissue is intimately linked to the selective placement or accumulation of fluorescent probe molecules (i.e., contrast-enhancing agents) at specific locations within a cell or tissue. Once sufficient numbers of exogenous fluorophores are inserted into specific volume elements or compartments of a living embryo (e.g., interstitial fluid, yolk platelets, or endosomes), these objects of interest stand out from their background and can be easily detected with a confocal microscope. However, to obtain a successful confocal time-lapse recording, the contrast-enhancing fluorescent probe must also be photostable and nontoxic so that multiple images of vitally stained cells or tissues can be acquired over an extended period of time.

Below, we outline an imaging strategy that is based on inserting a large number of photostable fluorescent probe molecules into specific compartments of the living zebrafish embryo. Once these fluorescent probe molecules have been inserted, the embryo can be repeatedly scanned using moderate laser illumination intensity to obtain either single-level or multilevel time-lapse recordings.

III. General Principles of Vital Staining

Vital staining provides a rapid and relatively inexpensive means of enhancing the detectability of cell shape changes and morphogenetic movements within a living embryo. In general, vital staining involves four steps: (1) the solubilization of the vital stain into a labeling medium, (2) the permeation, intercalation, or absorption of the vital stain into (or onto) the embryo, (3) the localization/accumulation of the probe molecule within specific cellular or subcellular compartments, and (4) the washout of the unbound stain.

Many vital stains accumulate in specific cellular compartments of cells and tissues through diffusion-trap mechanisms. As vital stain molecules from the external medium enter a cellular compartment (e.g., the lipid phase of a cell membrane or the lumen of an organelle) by diffusion, physical, and chemical characteristics of the molecules cause them to be retained or "trapped" within the compartment.

Because dimethyl sulfoxide (DMSO) has a low toxicity on living tissues, it is an excellent choice of solvent to solublize and apply vital stains to zebrafish embryos. Vital-staining solutions containing 1–2% DMSO can be applied to zebrafish embryos for up to 1 h without producing toxic or teratogenic effects (Cooper and D'Amico, 1996).

When inserting bath-applied fluorescent probes into embryonic zebrafish tissues, we recommend labeling concentrations on the order of 100 μM. The high-bath concentration produces a large diffusive flux of the vital stain into the embryo, which in turn results in a rapid accumulation of the fluorescent probe into the embryo's constituent tissues.

In the next section, we describe the labeling characteristics of a variety of vital stains/labels that are quickly and easily applied to zebrafish embryos in preparation for confocal imaging. These probes, BODIPY 505/515, BODIPY-ceramide, BODIPY-HPC, and SYTO-11, do not appear to have any teratogenic effects on developing zebrafish embryos. Moreover, extended time-lapse recordings can be obtained using each of the vital stains or vital labels.

A. Vital Stains and Vital Labels for Zebrafish Embryos

1. BODIPY 505/515

"BODIPY" is an abbreviation that refers to a very versatile set of neutral boron-containing fluorophores that are derived from a diazaindacene chemical backbone. Because of their lack of charge, BODIPY fluorophores have a minimal effect on the physiochemical properties of the biomolecules to which they are conjugated (Haugland, 1996).

BODIPY fluorophores are frequently referred to by their excitation and emission maximum. Thus, BODIPY 505/515 indicates that the fluorophore is most strongly excited with visible radiation centered at 505 nm (blue light), and emits a spectrum of longer-wavelength light that peaks at 515 nm. BODIPY 505/515 excitation maximum lies close to the 488 nm line of an Ar/Kr laser. BODIPY 505/515 and its molecular conjugates are ideally suited for scanning laser confocal microscopy due to their relatively low photobleaching rates, as well as their high-quantum yield—approximately 0.9 (Haugland, 1996).

Unconjugated BODIPY 505/515 fluorophore (4,4-difluoro-1,3,5,7-tetramethyl-4-bora-3a,4a-diaza-*s*-indacene; MW 248) is an excellent vital stain for the yolky cytoplasm in zebrafish embryos (Cooper and D'Amico, 1996). It partitions rapidly across cell membranes and accumulates within lipidic yolk platelets. In zebrafish embryos, lipidic yolk platelets are distributed throughout the cytoplasm of the blastoderm and nearly completely fill the volume of the zebrafish's yolk cell. (See Fig. 1) In contrast to its affinity for yolk platelets, BODIPY 505/515 does not stain nucleoplasm, nor does it remain within interstitial space. Vital staining with BODIPY 505/515 thus allows individual cell boundaries and cell nuclei to be imaged clearly with a confocal microscope. Karyokinesis, cytokinesis, and cell rearrangement can be followed in great detail throughout gastrulation, neurulation, and organ rudiment formation (see Fig. 2).

Compared to other fluorescent dyes, BODIPY fluorophores are particularly photostable (Haugland, 1996). This photostability allows multiple confocal micrographs to be acquired without substantial photobleaching. The relatively slow rate of photobleaching of BODIPY fluorophores also dramatically reduces

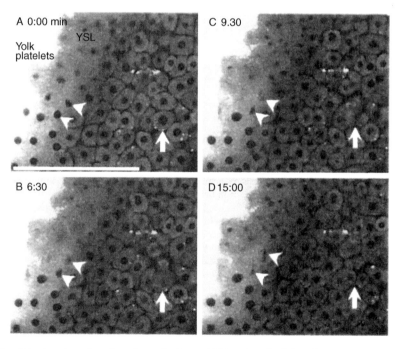

Fig. 1 Cytoplasmic and nuclear dynamics in a mid-blastula zebrafish embryo vitally stained with *BODIPY 505/515*. BODIPY 505/515 preferentially binds to yolk platelets and yolky cytoplasm, leaving nucleoplasm and interstitial space devoid of the fluorophore. Nuclear membranes break down (arrowheads) as a mitotic wave sweeps through the yolk syncytial layer (YSL). A deep cell (arrow) undergoes mitosis in the time-lapse series. Scale bar = 100 μm.

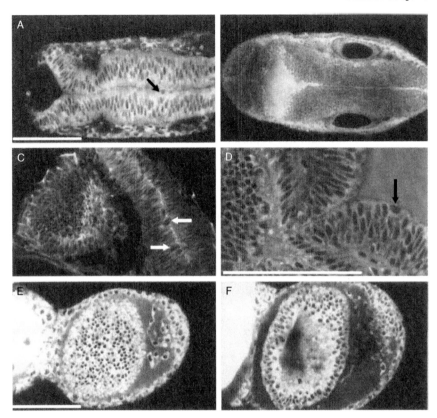

Fig. 2 Spatial structure of cell populations within embryonic organ rudiments (*BODIPY 505/515* staining). Embryos were imaged with a 40×/1.3 NA oil immersion objective. Scale bars = 100 μm. (A) The cell body of a mitotically active cell has migrated to the neural midline in preparation for division (arrow). The neural rudiment is this 18-h embryo is only 1–2 cells wide. (B) In a 48-h embryo, the ventricular walls of the hindbrain are now 8-10 cells thick. In time-lapse, it is possible to detect the location and timing of individual mitotic events within these cortical layers (not shown). Cavitation has occurred along the length of the neural midline, thus forming the central canal (i.e., ependymal canal) of the neural tube. The developing otic vesicles (asterisks) are located adjacent to the neural tube. (C) The outlines of neuroepithelial cells can be revealed in a fixed zebrafish embryo by conventional actin staining. A 24-h embryo was fixed with 4% paraformaldehye, permeabilized with the detergent Triton X-100, and then stained with BODIPY-phallacidin. Arrows point to mitotically active cells located at the neural midline. Mitotically active cells can be detected in live embryos stained with BODIPY 505/515 (see A). (D) A BODIPY-stained embryo showing neuroepithelial cell layers in the anterior region of the midbrain (higher magnification of the same embryo in B). An arrow points to a mitotically active cell located at the ventricular zone. (E, F) Two confocal sections through the left-eye rudiment of a BODIPY-stained 24-h embryo (viewed from the anterior end). Neuroepithelial cells in the optic vesicle, as well as mesenchymal cells in the nascent lens placode, are prominent.

phototoxicity in living embryos (frequently associated with prolonged time-lapse recordings) because fewer free radicals are produced as a consequence of photoillumination.

2. BODIPY-Ceramide

BODIPY-ceramide [full name: BODIPY FL C_5-Cer/C_5-DMB-Cer (N-(4,4-difluoro-5,7-dimethyl-4-bora-3a,4a-diaza-s-indacene-3-pentanoyl)) sphingosine; MW 631] is a fluorescent sphingolipid that has been used for many years as a vital stain for the Golgi apparatus in cultured vertebrate tissue cells (Lipsky and Pagano, 1985a,b; Pagano *et al.*, 1989). When applied to intact zebrafish embryo, BODIPY-ceramide stains the plasma membrane, Golgi apparatus, and cytoplasmic particles within the superficial enveloping layer (EVL) cells of the embryos. However, once the fluorescent lipid percolates through the EVL epithelium, the fluorescent lipid remains localized within the interstitial fluid of the embryo (hereafter referred to as the *interstitium*) and freely diffuses between cells (Fig. 3).

This diffusion of BODIPY-ceramide through the interstitium allows photobleached molecules to be quickly exchanged with unbleached molecules, thus replenishing fluorescence in the scanned field of view. Because there is a large reservoir of mobile fluorescent lipid present in the embryo's segmentation cavity (part of the interstitium), single-level or multilevel time-lapse recordings (Fig. 4) can be acquired over extended periods of time (up to 10 h). Vital staining with BODIPY-ceramide thus allows hundreds of cells to be imaged *en masse* during morphogenetic movements.

3. SYTO-11

During the late-blastula stage, a unique group of cells with increased endocytic activity differentiates at the incipient dorsal margin (organizer region) of the zebrafish blastoderm (Cooper and D'Amico, 1996). The longitudinal position of this cellular domain in late-blastula stage embryos (30–40% epiboly) accurately predicts where the embryonic shield will form at the onset of gastrulation (50% epiboly) (Fig. 5). Unlike other blastomeres around the circumference of the blastoderm, these marginal cells do not involute during germ-ring formation or blastoderm epiboly. Instead, this group of NEM cells remains at the border where the dorsal EVL and dorsal yolk syncytial layer (YSL) come into contact (Fig. 5). During mid- to late-epiboly, deep cells within the NEM cell cluster segregate from neighboring involuting cells and move to the leading edge of the dorsal blastoderm. As the deep NEM cells move into this location, these cells (now visible with Nomarski optics) are referred to as "forerunner cells" (Melby *et al.*, 1996; Cooper and D'Amico, 1996; D'Amico and Cooper, 1997; see Fig. 6).

Using confocal microscopy, both NEM cells and forerunner cells can be easily visualized in late-blastula to late-gastrula stage embryos because they exhibit accelerated endocytic activity and can be selectively labeled by applying membrane-impermeant fluorescent probes, such as SYTO-11 (~400 MW) and SYTO-17, to preepiboly and early-epiboly embryos (Cooper and D'Amico, 1996; D'Amico and Cooper, 1997). Because SYTO-11 (green fluorescence) and SYTO-17 (red fluorescence) labeling can identify the location of the organizer region in a

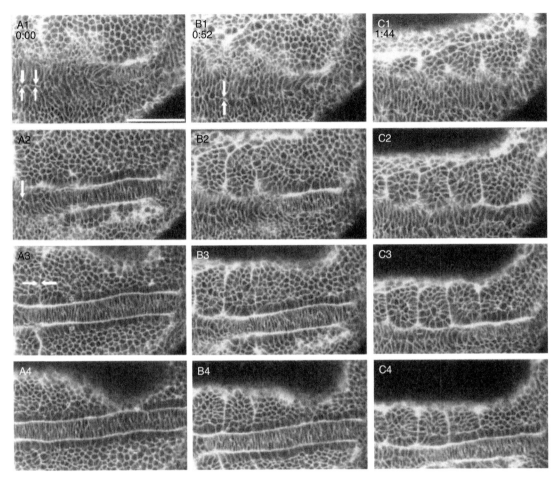

Fig. 3 Morphogenesis of cellular domains in the vicinity of the notochord. A multilevel time-lapse recording of an embryo vitally stained with the fluorescent lipid, *BODIPY-ceramide* (anterior to the left). BODIPY-ceramide remains highly localized to interstitial fluid once it accumulates within a zebrafish embryo, allowing the boundaries of deep cells to be clearly discerned throughout the entire embryo. A large reservoir of this fluorescent lipid in the segmentation cavity also allows extended single-level and multilevel time-lapse recordings to be acquired over extended time intervals. Columns (A–C) are confocal z-series at 0:00, 0:52, and 1:44 h, respectively. The vertical spacing between z-series planes = 7.5 μm. Scale bar = 100 μm.

(A1) Neuroepithelial cells in the presumptive floorplate region become mediolaterally elongated and juxtaposed along the midline of the embryo (arrows). This midline alignment of neuroepithelial cells is the first salient morphological sign of bilateral symmetry within the neural primordium (see also B1).

(A3) On either side of the notochord lie a linear array of mesodermal cells (asterisks) known as *adaxial cells*, a unique set of slow-twitch muscle progenitors (asterisks in A3). Intersomitic furrows form in paraxial mesoderm through selective cell deadhesion. As cells de-adhere from each other, BODIPY-ceramide percolates into the furrow, illuminating the widening gap (arrows in A3). Nascent somites subsequently undergo mediolateral convergence toward the notochord-somitic boundary (A3, C3).

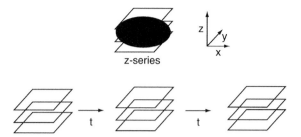

Fig. 4 Acquisition of a multilevel confocal recording. Confocal z-series of the embryo are acquired at sequential time points. These images are digitally stored for postprocessing into 3D projections or 4D animations.

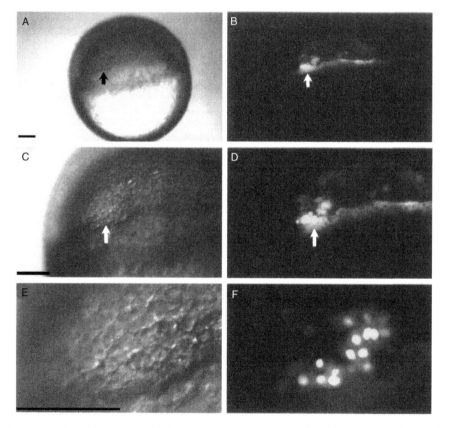

Fig. 5 *SYTO-11* labeling of NEM/forerunner cells. The position of NEM/forerunner cells in early-stage embryos (arrows) correlates with the site of embryonic shield formation. (A, B) A Nomarski confocal image pair of an early-gastrula stage zebrafish embryo showing that the brightly labeled NEM/forerunner cell cluster is located at the leading edge of the dorsal marginal zone (DMZ). (C–F) Higher magnifications of the DMZ and the NEM/forerunner cell cluster. Scale bars = 100 μm.

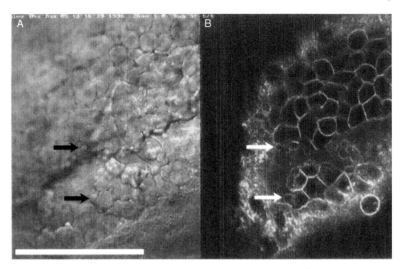

Fig. 6 Protrusive activity of marginal deep cells and forerunner cells in an early-gastrula-stage embryo vitally stained with *BODIPY-sphingomyelin* (Nomarski confocal image pair). Although lobopodia (arrows) can be detected using Nomarski optics (A), these bulbous cellular protrusions are more easily observed using confocal imaging (B). Teleostean deep cells frequently use lobopodia during invasive intercalative cell movements. BODIPY-sphingomyelin associates more closely to the cell membrane than BODIPY-ceramide, and thus delineates cellular protrusions more effectively. BODIPY-ceramide, on the other hand, diffuses more effectively through the interstitium, allowing cellular outlines to be imaged in a given region of the embryo for longer periods of time.

late-blastula stage embryo, these fluorescent probes are potentially useful for experimental studies of dorsal cell-fate specification and embryonic axis formation.

Since SYTO-11 is internalized into intact zebrafish embryos by endocytosis (Cooper and D'Amico, 1996), as opposed to passive permeation, it is more appropriate to refer to SYTO-11 as a vital label than as a vital stain.

4. BODIPY-HPC

BODIPY-HPC [full name: β-BODIPY FL C_{12}-HPC or 2-(4,4-difluoro-5,7-dimethyl-4-bora-3a,4a-diaza-*s*-indacene-3-dodecanoyl)-1-1 hexadecanoyl-*s*-glycerol-3-phosphocholine] is a BODIPY-conjugate of the phospholipid phospholcholine. When it is applied to intact zebrafish embryos, BODIPY-HPC readily intercalates into the membranes of EVL epithelial cells. Most deep cells remain unlabeled by the fluorescent probe, suggesting that it does easily percolate through the outer EVL epithelium and enter the embryo's interstitium.

Vital staining with BODIPY-HPC allows the cellular boundaries of EVL cells be to imaged with great detail. During blastopore closure, the YSL progressively constricts, causing the EVL margin to decrease in circumference. During this process, EVL cells frequently undergo a focal constriction along their leading edge (see arrowhead in Fig. 7) and begin to recede from the EVL to YSL boundary. The leading borders of other EVL cells, however, remain scalloped-shaped (arrows in Fig. 7), as the EVL-YSL margin is constricted by the contracting YSL. Figure 7 illustrates how BODIPY-HPC allows the cytomechanics of EVL epiboly to be studied with subcellular resolution.

5. Vital Staining and Confocal Imaging Procedures

The vital-staining procedures presented in the Section III.C have been previously published elsewhere in more detail. Readers are encouraged to consult Cooper *et al.* (1998) for additional information about dye solubilization, compensations for photobleaching, and computer programs for making single-level and multilevel confocal time-lapse recordings.

B. Materials

1. BODIPY 505/515 (Molecular Probes, Eugene, OR) or other vital stain
2. Anhydrous DMSO
3. 35-mm plastic culture dishes and/or 1.5-ml microfuge tubes
4. Drierite

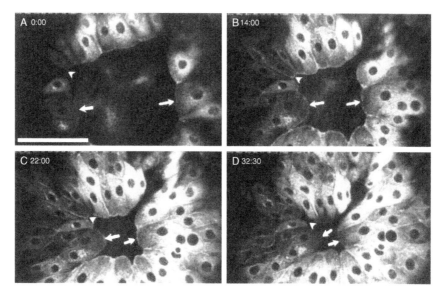

Fig. 7 Constriction of the EVL-YSL boundary (arrows) during blastopore closure in an embryo vitally stained with the fluorescent phospholipid *BODIPY-HPC*. Scale bar = 100 μm. Time points in minutes. Some EVL cells (e.g., arrowhead) undergo a focal contraction at their leading margin and subsequently recede from the EVL-YSL boundary.

5. HEPES-buffered embryo rearing medium (ERM). Add 0.24 g HEPES [*N*-(2-hydroxyethyl)piperazine-*N'*-(2-ethanesulfonic acid)] in 100 ml ERM (see Westerfield, 1995) to obtain a pH buffer capacity of 10 mM in the final solution. Use 1 M NaOH to adjust the solution to pH 7.2. HEPES is added to the ERM in order to stabilize its pH during extended time-lapse recordings.

6. Fire-polished Pasteur pipettes (with bulbs) for transferring embryos

7. 10- or 20-μl micropipettes (VWR Scientific) for molding agarose

8. Hairloop mounted into the tip of a Pasteur pipette with a dab of molten paraffin

9. Agarose

10. Dechorinated zebrafish embryos; see Westerfield (1995), for manual and enzymatic methods. These methods can also be accessed on the at the Fish Net website: http://zfish.uoregon.edu/zf_info/zfbook/zfbk.html.

C. Vital-Staining Procedure

The procedure for vitally staining embryos is quite similar for BODIPY 505/515, BODIPY-ceramide, SYTO-11, or other vital stains. Embryos are placed in a vital-staining solution for 30 min and then washed 3 times with HEPES-buffered ERM in order to remove excess or unbound fluorescent probe molecules. Since the above vital stains are nontoxic, the duration of vital staining can often be increased to achieve a desired staining intensity.

1. Make a stock solution of the unconjugated fluorophore BODIPY 505/515 (4,4-difluoro-1,3,5,7-tetramethyl-4-bora-3a,4a-diaza-*s*-indacene) by dissolving the fluorescent dye in anhydrous DMSO to a stock solution concentration of 5 mM. 20-μl aliquots of the stock solution are placed into microfuge tubes and stored in a light-tight container with Drierite at $-20\,°C$. BODIPY 505/515 aliquots are usually stable for up to 6 months.

2. Prepare labeling chambers for vital staining. Dechorinated zebrafish embryos will stick to plastic surfaces. Therefore, all labeling chambers should be covered with agarose. Molded agarose chambers can also be created to minimize the volume of staining solution needed to vitally stain a group of embryos. For very small volumes (50–100 μl) of labeling solution, a labeling chamber can be constructed from the detached cap of a 1.5-ml microfuge tube. Molten agarose can be dropped into the cap using a Pasteur pipette until the agarose completely covers the bottom of the cap.

3. Embryos can be moved between solutions by using transfer Pasteur pipettes. Care should be taken to avoid having embryos contact air-water interfaces. This is done by tilting the transfer Pasteur pipette to a more horizontal position once embryos have been drawn into the pipette. Transfer pipettes are made by removing the tip, as well as most of the shank, of the glass pipette (a diamond scribe is useful

for this procedure). The remaining 3–5-mm-diameter opening is fire-polished using an alcohol flame.

4. Thaw an aliquot of BODIPY 505/515. The dye is then diluted into HEPES-buffered ERM to a labeling concentration of 100 μM. The final concentration (vol/vol) of DMSO in the labeling solution is 2%.

5. Transfer the embryos to a labeling chamber and stain them with a 100 μM BODIPY 505/515 staining solution for 30 min. As a precaution during the staining procedure, we recommend that the staining chamber be covered with aluminum foil in order to keep vitally stained embryos from being exposed to excess light. However, we have found that exposure to room light for up to 30 min does not harm or overly photobleach vitally stained embryos. During vital staining, the solution in the labeling chamber should be circulated by gentle swirling every 10 min. The yolk cell of zebrafish embryos becomes visibly green within several minutes after the embryo is placed in an aqueous labeling solution containing 100 μM BODIPY 505/515 and 2% DMSO. The yolk cell continues to accumulate dye as the embryo remains in the labeling solution, and becomes intensely green after 30 min of staining. To the naked eye, the blastoderm of the BODIPY-labeled embryo appears colorless. However, when these embryos are examined under the scanning laser confocal microscope, the embryo's blastomeres are well labeled with BODIPY 505/515 (Fig. 1).

6. After vital staining, the embryos are passed through three successive washes of HEPES-buffered ERM. These washes are performed in separate agarose-coated 35-mm culture dishes. After washing, the embryos are ready to be mounted for observation under the confocal microscope. In young embryos (<24 h old), BODIPY 505/515 is retained within the yolk cell and the yolk-containing cytoplasm of cells. In primordia-stage embryos, as in younger embryos, BODIPY 505/515 does not stain the nucleoplasm of cells or fluid-filled organelles, such as vacuoles within the extending notochord. In contrast to younger embryos, BODIPY 505/515 becomes highly concentrated in the blood and cerebral-spinal fluid (data not shown) in later stage embryos (>24 h old). We speculate that this change in staining pattern results from BODIPY 505/515 remaining bound to lipoproteins as they are exported from the yolk cell into the bodily fluids of the embryo.

D. Additional Vital Stains

Embryos can be vitally stained with either BODIPY-ceramide or SYTO-11 using the same procedure as BODIPY 505/515. Recommended labeling concentrations (and labeling times) for BODIPY-ceramide and SYTO-11 are 100 μM (30 min) and 75 μM (15 min), respectively.

It is often advantageous to dual-label embryos with multiple vital stains that illuminate distinct cytological domains. BODIPY 564/591 is excited with the yellow 568 nm line of the Ar-Kr laser and represents an excellent alternative fluorescent vital stain for staining yolk platelets (Cooper and D'Amico, 1996).

BODIPY 564/591 can be used for dual-labeling purposes with other vital stains, such as SYTO-11 (see Fig. 5). BODIPY 564/591 can be obtained from Molecular Probes, Inc. (Eugene, OR) on a custom-synthesis basis. Embryos can be labeled with BODIPY 564/591 using the same procedure as BODIPY 505/515.

A sequential application of the standard vital-staining procedure (Fig. 5) allows embryos to be costained with BODIPY and other fluorophores. Dual-staining with BODIPY 564/591 and SYTO-11, for instance, allows NEM/forerunner cells to be observed along with all other blastomeres. Cell nuclei and endosomes labeled with SYTO-11, or cellular cytoplasm labeled with BODIPY 564/591, can be viewed independently using 488- and 568-nm excitation wavelengths, respectively. When desired, both fluorophores can be viewed simultaneously using dual wavelength excitation (488 and 567 nm) and 585-nm long-pass emission.

BODIPY-sphingomyelin [full name: BODIPY FL C_5-sphingomyelin or N-(4,4-difluoro-5,7-dimethyl-4-bora-3a,4a-diaza-s-indacene-3-pentanoyl) sphingosylphosphocholine] is a fluorescent sphingolipid that exhibits certain vital-staining characteristics that are similar to BODIPY-ceramide. BODIPY-sphingomyelin is able to percolate through the EVL and accumulate in the embryo's interstitium. BODIPY-sphingomyelin, however, associates more closely with the plasma membrane of deep cells than BODIPY-ceramide, allowing certain cellular protrusions, such as lobopodia, to be clearly visualized using a confocal microscope (Fig. 6).

Although we have described the use of a variety of vital stains in this section of the chapter, we would like to comment that there are many other fluorophores and BODIPY-conjugated biomolecules (see Molecular Probes Catalog, Eugene, OR) whose vital-staining properties have yet to be characterized in zebrafish embryos.

IV. Mounting Embryos for Imaging

A. Imaging Chambers

Confocal time-lapse recordings of zebrafish embryos require that the vibration of the specimen be minimized. Live embryos are most easily imaged using an inverted microscope mounted on an airtable, since gravity pulls the embryo toward the coverslip and helps maintain its position with respect to the objective. A useful chamber to image teleost embryos is a 0.7-cm-thick piece of plexiglass with a 24-mm-diameter hole in its center. A 30-mm-diameter circular glass coverslip (#1 thickness) can be secured with high-vacuum silicone grease (Dow Corning, Midland, MI) to serve as the bottom of the bath well. During time-lapse recordings, a plastic culture dish lid can be placed over the well to prevent air currents from producing displacements of culture medium and unsecured embryos.

Here is a new, rapid, and inexpensive method for mounting embryos: take the cap of a 6 ml Falcon polypropylene tube (Becton Dickinson Labware, Lincoln Park, New Jersey) and cut off its end using a pair of large scissors. The resulting plastic ring can now be used as a disposable mounting device for zebrafish embryos. Apply silicon grease to the surface of the plastic ring. Then affix the plastic ring to the surface of a #1 coverslip. Add ERM or culture media. Place embryos inside this volume. Micromanipulations of embryos (e.g., isolating organ primordia) can be performed inside this volume using a dissecting microscope. Isolated tissues can then be oriented and/or immobilized for confocal imaging without having to transfer the isolated tissues to a separate imaging chamber.

Liquid can be added to fill the plastic ring. Following this, a #1 coverslip can be added to the top of the ring. Excess liquid can be dabbed away with the end of a rolled Kimwipe. Surface tension will hold the coverslip in place. Addition of the upper coverslip will reduce convective movements of the embryos or explants while the coverslip sandwich preparation is moved to the confocal microscope for observation.

B. Spatial Orientation of Embryos

To prevent the rolling of embryo during an extended time-lapse recordings, we have found it useful to place fluorescently labeled embryos in an agarose holding well made from ERM plus agarose, Type IX (Sigma, St. Louis, MO) (modified from the procedure in Westerfield (1995)).

1. Apply a thin coat of molten solution of 1.2% agarose in ERM to the coverslip. Let the agarose cool and harden to form a thin layer of agarose gel on the coverslip.

2. To make holding wells in the agarose layer, start with a 10- or 20-μl glass micropipette whose end has been pulled out using an alcohol flame. After breaking off the tip of the pipette, heat the remaining end until it melts into a small bead of glass, approximately 0.8 mm in diameter. Use the rest of the pipette as a handle. After heating this glass bead or ball with the alcohol flame, quickly plunge it into the agarose to melt a hole. The holes should be small enough so that the embryo will not be able to roll, but large enough so that epiboly of the blastoderm is not impeded. Since an embryo will be secured inside this hemispherical-shaped hole, the hole should be as round and smooth as possible. Use room-temperature ERM or water to wash out any melted agarose so that it does not refill the holes. Repeat this procedure to create multiple holes in the agarose layer. You may have to make a new glass bead, since repeated heating will cause the bead to increase in size as more glass melts.

3. After the agarose wells are made, add ERM buffer to cover the agarose hole and then add the embryos. Gently position the embryos into the agarose holes using a nonsticky implement, such as a hairloop mounted on the end of a Pasteur pipette with a dab of molten paraffin. Carefully position the embryo so that the area of interest is facing toward the objective. This operation is most easily accomplished on an inverted microscope with an open holding chamber. However, repositioning the embryos once they are in the wells is difficult and must be done with care. It is useful to secure 5-10 embryos in a given agarose sheet in

preparation for a time-lapse recording. This increases the likelihood that a well-stained embryo can be located in an appropriate orientation for time-lapse imaging.

To prevent muscle twitching during time-lapse recordings, embryos at 20-somite stage or later can be anesthetized with 0.1 mg/ml Tricaine (also known as MS-222; Sigma, St. Louis, MO) dissolved in ERM (Westerfield, 1996).

C. pH Stabilization

One of the most critical experimental variables to control in making time-lapse recordings is the pH of the embryonic medium. Bicarbonate-containing media, such as zebrafish ERM, are subject to major changes in pH with temperature, as carbon dioxide exchange with the atmosphere alters the carbonic acid-bicarbonate equilibrium of the medium (Freshney, 1987). In a small volume, the metabolic activity of several zebrafish embryos can also produce enough CO_2 to substantially alter the pH of the experimental medium. In the absence of CO_2 and temperature control, it is a useful precaution to add a buffering agent at twice the concentration of bicarbonate ion HCO_3^-, to achieve pH stabilization of the experimental solution (Freshney, 1987). Therefore, 10 mM HEPES should be added to all zebrafish salines or media to help stabilize their pH during extended time-lapse recordings.

V. Imaging Procedures

A. Selection of Optics

The choice of objective for imaging zebrafish embryos is determined by several considerations: (1) numerical aperture (NA), (2) working distance, and (3) magnification (M). The brightness (B) of an objective is related to the numerical aperture (NA) and magnification (M) through the following equation (Majlof and Forsgren, 1993):

$$B \cong \frac{(\text{NA})^4}{M^2} \tag{1}$$

Low-magnification objectives (e.g., 10× and 20×) often have comparable brightness values to higher magnification, high-NA objectives. This is fortunate, because low-magnification objectives often provide the required working distance and field of view necessary to observe the collective behavior of cell populations. In addition, low-magnification objectives have a larger depth of field than higher magnification objectives, allowing thicker confocal optical sections to be "in focus." High-NA objectives, however, are usually necessary for imaging applications that require high-spatial resolution.

To examine morphogenetic cell behaviors in zebrafish embryos, we have found a versatile set of microscope objectives to be: (1) a dry 20×/0.75 NA; (2) a dry 40×/0.85 NA; (3) an oil 40×/1.3 NA. A 10×/0.5 NA objective is also useful for locating NEM cells in the organizer region of zebrafish embryos labeled with SYTO-11. It is also particularly useful to configure a confocal microscope for simultaneous (Nomarski) and confocal imaging. This allows macroscopic morphological features (e.g., the embryonic shield) to be easily correlated with the location of specific groups of fluorescently labeled cells. Optical details of such a configuration can be found elsewhere (Simon and Cooper, 1995).

Appropriate filter sets for the confocal are determined from the excitation and emission spectrum of the fluorophores. In Figs. 1–5, embryos stained either with BODIPY 505/515, BODIPY-ceramide or SYTO-11 were imaged using an excitation wavelength of 488 nm. Fluorescent light collected from the specimen was filtered by a 515-nm long-pass filter before it was transmitted to the confocal microscope's photomultiplier tube.

B. Acquiring Confocal Images in Time–Lapse Form

In order to generate a large number of photons for high-quality confocal images with single scans, it is necessary to illuminate the embryo with moderately intense laser light. To obtain optimum contrast for time-lapse recordings, system gain should be increased until saturation begins to occur. At this point, offset is added to bring these pixels below a value of pure white (i.e., 255 on a 0–255 grayscale).

Time-lapse recordings of developing zebrafish embryos are best made from single slow scans of the specimen, as opposed to time-averaged images. Since the scanning laser beam of the confocal microscope passes rapidly over the sample, a stroboscopic illumination of the specimen is produced. This greatly reduces the motional blurring of fluorescent objects that are being displaced within cells by active transport or by diffusion (Cooper et al., 1990). At a slow scan rate, enough photons are collected in a single pass over the specimen to generate adequate signal-noise level in the image. By adjusting neutral density filters, it is useful to use as intense laser light as is possible without producing substantial photobleaching in the specimen over a desired time-lapse interval. For an embryo labeled with BODIPY 505/515 or BODIPY-ceramide, 100–300 frames can be generally recorded with a 300–700 μm field of view over a 1–10-h time period.

Most confocal microscopes now have internal macros for obtaining and transferring single-level and multilevel time-lapse recordings to their host computer's hard disk.

Multilevel time-lapse recordings can be implemented by coupling a stepping motor to the focusing apparatus of the confocal microscope. Bio-Rad command programs for acquiring multilevel time-lapse recordings can be obtained elsewhere

(Cooper *et al.*, 1998; Terasaki and Jaffe, 1993). Software on a variety of commercial confocal microscopes now contain similar programs for acquiring multiple z-series in time-lapse form. Acquired z-series can then be imported into image-processing programs (e.g., NIH-Image) and rendered into 3D projections or 4D animations (see Section VI.C).

OMDRs are now obsolete for time-lapse imaging. The price of digital storage capacity in computer has dramatically decreased in the last 10 years, making the OMDR a data storage instrument of the past.

Computer processors have dramatically increased in speed, allowing much greater playback speeds than 10 years ago. Thus, the advantages of using the OMDR have disappeared.

C. Storage and Analysis of Time-Lapse Recordings

The spatial resolution of a given confocal time-lapse recording is best preserved if the constituent images of the recording are saved as digital image files. These digital image files can then be imported into a suitable program for viewing on a computer. NIH-Image is one such useful program, as it will import Bio-Rad.PIC files using a specific macro (NIH-Image and Bio-Rad macros can be downloaded from the following URL: http://rsb.info.nih.gov/nih-image/more-docs/docs.html). One can load a z-series and scan-through the data set using arrow keys. If it is desired, the scan-through can be animated into a Quicktime movie and saved.

VI. Multilevel Time-Lapse Confocal Analysis

The following procedure was developed by L. D'Amico to determine the individual 3D movements of neighboring cells within a given embryo during a large-scale morphogenetic movement (e.g., germ-ring formation or embryonic shield formation). This procedure employs (1) multilevel, two-wavelength, time-lapse confocal microscopy, (2) image postprocessing-rendering, (3) measurement of cellular coordinates, and (4) a final plotting of cellular trajectories.

A. Embryo Labeling

To determine the morphogenetic movements of multiple cells in a given zebrafish embryo, it is useful to label a large subset of the embryo's cells with a fluorescent lineage tracer. This can be accomplished by injecting a concentrated solution of Texas Red-dextran or BODIPY-dextran into a single blastomere at the 16-128-cell stage. As this labeled cell divides, its numerous cellular progeny will disperse with the onset of morphogenetic movements, resulting in a "scatter-labeling" of cells at a later stage of development. An earlier injection will result in more labeled cells within the embryo. Alternatively, injection into the yolk cell at these stages will result in many more labeled cells, as the dye can diffuse into marginal blastomeres that are still connected to the yolk cell through cytoplasmic bridges. This will result in a greater number of cells labeled in the marginal area of the embryo. The embryo can then be labeled with another fluorescent probe (imaged in a separate wavelength-channel) that labels the boundaries of all the cells of the embryo. In zebrafish, BODIPY-ceramide is an excellent choice for this purpose (see Section III.C for labeling procedures).

One important consideration in time-lapse imaging is the stability of the fluorescent probe. Frequent and prolonged excitation will result in the bleaching of the probe and the eventual phototoxicity and death of the illuminated cells. It is important, therefore, to use the most stable probe available. In addition, it is important to limit the total number of scans taken of the specimen to reduce phototoxicity to the tissue or cells during the total time interval that is of interest. This results in a trade-off: a greater number of scans (or imaged focus levels) at each time point will limit the total number of time points that can be taken. Similarly, shorter time intervals between each imaged time point will limit the total amount of time that one can image a specimen.

B. Multilevel Time–Lapse Data Sets

By using a confocal microscope with a computer controlled focusing motor, it is relatively easy to obtain a z-series or focus-through of a specimen. By repeatedly performing this procedure over an extended time with the same time interval between each z-series, one can obtain a 3D time-lapse data set.

After taking photobleaching/phototoxicity issues into consideration, one has to decide on (1) the desired volume that is to be imaged during a time-lapse recording and (2) the number of focus levels or sections and the spacing between these sections. It is useful to have at least two to three optical sections in a data set that span the thickness of individual cells. This provides more certainty to the position of cells in the z-dimension, as well as giving the observer a general idea of the shape of the cells. If one is particularly interested in cell shape changes, a z-series with much less spacing between focal planes should be made. In the example shown in Fig. 8, imaging of the germ-ring required a total z depth of about 120 μm with 7.75 μm between each focal plane. Since cells at this point in gastrulation are about 15–25 μm in diameter, each cell will span about two to three focal planes. It is especially important to calibrate the focusing motor steps with the actual distance (in μm) traveled between image planes, so that an accurate z-dimension is recorded by the computer.

Another experimental variable to decide on is the time-lapse interval that is required to image the morphogenetic events that are of interest. This depends largely on how fast the cells are moving. This time-lapse interval needs to be (1) short enough so that one can follow the movement of individual cells, but (2) long enough so that excessive photobleaching does not occur in the time scale that you are interested in. The time-lapse interval is likely to be 3–20 min, depending on the stage of develop-ment and speed of cell movement involved in a given morphogenetic movement.

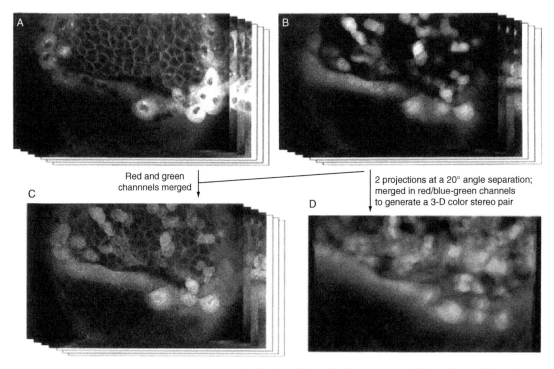

Fig. 8 Two visualization strategies for determining 3D cell movements in deep cell populations: image processing of multilevel time-lapse confocal data sets. A gastrulating zebrafish embryo has been (1) bulk-labeled with Bodipy-ceramide to reveal cell boundaries and (2) scatter-labeled with Texas-Red dextran to fluorescently mark a smaller population of deep cells. Split-screen confocal images (not shown) of the embryo are then acquired using dual-channel imaging. Using a computer-activated stepping-motor on the focus-control of the confocal microscope, a z-series of such dual-channel images is obtained of the embryo. In a multilevel time-lapse recording, a sequential set of z-series would be obtained. Utilizing a macro within NIH-Image, the two color channels of a dual-channel z-series can be separated into two independent image stacks (A-B) to allow for independent adjustment of brightness and contrast. C-D illustrate how the digital information derived from a single dual-channel z-series can be rendered into two types of visualizations that are highly useful for subsequent data analysis. It is possible to combine the separated image stacks into either a two-color merged image stack (C) or a 3D color stereo image-pair (D). (C) Merging of the separated stacks is quite useful for plotting the morphogenetic movements of individual cells. This can be accomplished by displaying multiple time points of merged image stacks on a computer screen (see Fig. 9). The different fluorescent probes can be pseudo-colored so that individual dextran-labeled cells can be easily identified within the entire cell population. The adjusted images are then color merged, resulting in the image stack shown in C. (D) A stack of 3D color stereo image-pairs can be created by generating two projections of the original image stack that deviate from each other by an angle of 10-20°. The two projections are then merged into a single RGB image: the left projection is encoded as red while the right projection is encoded as bluegreen (see text for more detailed information). The above procedure can be repeated for each of the z-series in a multilevel time-lapse series. The rendered images can then be linked together to form a 3D animation. The resulting stereo pair movie allows the relative movements of numerous cells to be visualized within a 3D context.

1. Simultaneous Two–Channel Imaging

Some confocal microscopes (that have two photomultipliers and the appropriate filter sets) have the ability to simultaneously excite in two or more wavelengths and detect the emission of multiple fluorophores. This allows the detection of multiple probes simultaneously. This is very useful for the time-lapse imaging of a specimen that has been labeled with two different fluorescent probes (see embryo labeling discussed previously).

2. Saving the Data Set

Bio-Rad MRC software saves a z-series as a stack of images in a PIC file format. It is convenient to have each time point in the multilevel time-lapse saved as one stack of images (one.PIC file). These files can then be transferred to another computer for image processing. Certain image-processing programs (such as NIH-Image) can readily recognize and import these image stacks for postprocessing.

3. Image Processing

Once a 4D confocal data set (sequential z-series) has been acquired, extensive postprocessing is usually required before cellular trajectories can be traced. NIH-Image (a freeware program developed by Dr Harvey Karten, UCSD) is a very convenient program for working with stacks of confocal images. The program, documentation, and relevant macros are available for downloading from the NIH-Image homepage at http://rsb.info.nih.gov/NIH-Image. The macros that are discussed below are in the confocal-macros.txt file, located in the user-macros directory. Here, we will discuss only the basics on how the images are processed. For more details, the NIH-Image website has very useful documentation dealing specifically with the image processing of fluorescence and confocal microscope generated data.

4. Creating Color–Merged Stacks of the Data Set

When starting with two-channel splitscreen images in stacks, it is convenient to split these stacks in order to work with the two color channels separately. Separating the stacks allows one to optimize the appearance (blackness, contrast, sharpness, etc.) of each color channel independently of one another before merging.

1. Use the *Separate Splitscreen Z Stack* macro in NIH-Image to obtain two separate stacks of the two color channels. Save these separate stacks.
2. Adjust the blackness and contrast as desired for each stack. Once you adjust the blackness and contrast as desired for one image in a stack, you can apply this lookup table (LUT) to a single image (macro: *Apply LUT*) or to the whole stack (macro: *Apply LUT to Stack*), as desired.

3. Now these two stacks can be merged (macro: *Color Merge Two Stacks*) into a single two-color stack (red-green) so that the information in the two channels can be seen simultaneously. All these steps should be repeated for each time point in the data set.

C. Analysis of Cell Movements During Morphogenesis

When analyzing the movement of cells, their cell shape changes and protrusive activity, and their interaction with other cells, we find that using multiple techniques to visualize these processes is helpful. Here we describe some of the techniques that we have found useful.

1. 3D (Stereo Pair) Movie Generation

Generating a 3D movie of the data set is sometimes quite useful in order to get an overall view of the morphogenetic process, as well as watching individual cells move in a 3D context. The basic idea behind this procedure is to project the z-stack at two rotation angles (6–20° apart on the *y*-axis) and then merge these to obtain a color stereo pair. However, if most of the cells are labeled with dextran, or if all the cells (or cell boundaries) are labeled (e.g., with BODIPY-ceramide), the surface cells will obscure any information underneath. As a consequence, a data set with a limited number of labeled cells in the area of interest (e.g., a scatter label) will yield the most useful 3D movies. In addition, only one channel can be viewed in a given color stereo-pair image. To generate red-green or red-blue-green 3D movies of a single labeled data set using NIH-Image, follow these steps:

1. Open one of the separated and adjusted z-stacks (see Section VI.B.4) with NIH-Image.
2. Enter the parameters for Projections: Under the *Stacks* menu item *Project . . .*
   - Enter the *Slice Spacing (Pixels)*: depends on the distance (in μm) between slices and the measured pixel-to-μm ratio (this depends on the objective used and the zoom levels during z-series acquisition).
   - For a separation angle of 12° (adjust to preference) set the *Initial Angle (0–359°)* to −354°; set the *Total Rotation (0–360°)* to 12°; set the *Rotation Angle Increment* to 12°.
   - For *Axis of Rotation*: choose the *Y-axis*; for *Projection Method*: choose *Brightest Point*.
3. Now that there is a stack with the two projections, use the macro: *Add Slice* to add a blank (black) image to the 3/3 slice.
4. To make a merged red-green stereo image for viewing with a pair of red-green stereo glasses, use the *Stacks* menu item *RGB to 8-bit Color*. To make a merged red-blue-green stereo image for viewing with a pair of red-blue stereo

glasses, copy the 2/2 slice and paste it into the 3/3 slice; then use the *RGB to 8-bit Color* function (same as previously).

5. Repeat steps 1–4 for all of the time points. Save each of these with file names that include sequential numbers (*.001, *.002, etc.).

6. Close all windows in NIH-Image and then open all of the stereo images. Use the *Stacks* menu item *Windows to Stack* to make a stack of these images. *Animate* this stack to see the 3D movie (with 3D glasses).

2. 4D-Turnaround and 4D-Viewer

This set of programs makes it relatively easy to follow individual cells, as well as groups of cells, through time and space. The freeware programs are available free at the Integrated Microscopy REsource's 4-D Microscopy website: http:// www.bocklabs.wisc.edu/imr/facility/4D/4d.htm (also see Thomas *et al.*, 1996). 4D-Turnaround converts individual stacks of a multilevel time-lapse data set into one Quicktime movie that contains all the stacks and time points. 4D-Viewer allows the viewing of this Quicktime movie format. What makes 4D-Viewer particularly useful is that it not only allows one to easily pan through the depth of a stack, but one can also advance through the time points at any z-level (or multiple levels). This is easily accomplished by using the arrow keys located on the keyboard. Cells often change their depth in the tissue through time, so the ability to quickly change the level in any stack at any time point provides much needed flexibility for following cell movements.

The 4D-Turnaround and 4D-Viewer programs discussed in this section have been superseded by far more advanced 4D visualization software. In the past 10 years, the processing speeds and RAM of personal computers have greatly increased. As a consequence, many more zebrafish labs are utilizing 4D imaging/ visualization as an experimental approach to analyze GFP-expressing zebrafish. A variety of powerful 4D visualization programs are now commercially available to serve these expanding needs.

Image J is a freeware program that may useful for particular zebrafish imaging and visualization purposes. The URL for this program is: http://rsbweb.nih.gov/ij/.

D. Analysis of Cellular Trajectories

In order to determine the cytomechanics underlying a morphogenetic process, it is often necessary to determine the trajectories of individual cells. A combination of many different programs is useful for extracting this type of information. While 4D-Viewer and 3D stereo pair movies are useful for following cell movements, because of the format of the data, NIH-Image is more practical for finding and recording the x, y, z position of cells, as well as for marking individual cells (with numbers or other labels using the text tool) in each stack (Fig. 9). To have accurate x, y measurements in micrometers instead of pixels, one needs to calibrate the

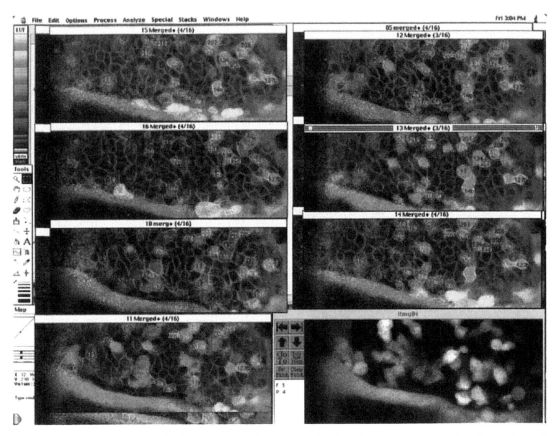

Fig. 9 Multilevel analysis of cell movements involved in zebrafish gastrulation. To analyze cell movements in a multilevel time-lapse data set, one can use the functionality of two freeware programs, NIH-Image and 4D-Viewer, concurrently. With NIH-Image, all of the color-merged stacks representing each timepoint can be opened simultaneously and arranged on the screen to facilitate detection of individual cell movements from one timepoint to the next. NIH-Image can also be used to mark individual cells, as well as to obtain the x,y,z coordinates of cells at each time point. 4D-Viewer (arrows in lower right corner) allows one to rapidly move spatially (up or down) and temporally (back and forth) through a multilevel time-lapse data set.

pixel-to-micrometer ratio by using the *Analyze* menu item *Set Scale*. The *x, y* position (in μm) of cells can then be directly measured by either positioning the cursor in the middle of a cell or, more accurately, outlining the cell boundary and using NIH-Image *Measurement* function to find the *x-y* center. (In order to have the measurement include the *x, y* coordinates, use the *Analyze* menu item *Options* ... and choose *X-Y Center*.) These values can then be exported to a spreadsheet program (e.g., Microsoft Excel) for later analysis. The *z*-position of a cell can be found by panning through the z-stack and manually finding the *z*-position (slice number in stack) of the center of the cell. Since the distance between slices is known, the depth can be easily converted to micrometers. Repeat this process for the same cell at each time point (and then for all the cells that are of interest).

Once the *x, y, z* positions of a cell at each time point have been entered into a spreadsheet, a 3D plot of the trajectory of a cell can be made using a program (such as Igor Pro 3.0) that generates 3D line plots (Fig. 10). Plotting a cell's trajectory is helpful to visualize the cell's movement through 3D space for two reasons. First, it allows one to focus on the movements of a single cell, rather than all the cells at once. Second, it is often much easier to analyze the plotted movement of the cell than to follow the cell in the original data set or 3D stereo-pair movie. This is especially helpful when analyzing the movement of multiple cells, as well presenting the kinematic actions of cellular ensembles to others.

VII. Distribution of Visual Information

The World Wide Web has provided an important new avenue for archiving and disseminating visual information. Confocal time-lapse recordings, in particular, can be compressed into digital movie files in either a QuickTime or a MPEG format. Subsequently, these digital files can be distributed over the Internet. Examples of such compressed time-lapse recordings from our own laboratory, showing zebrafish embryos stained with BODIPY 505/515, BODIPY-ceramide, and BODIPY-dextran are located at the following URL:http://weber.u. washington.edu/~fishscop/.

In 2002, our laboratory assembled a compendium of zebrafish images and time-lapse recordings, and disseminated them to the international zebrafish community as a CD called, "Zebrafish: The Living Laboratory." We described the construction of this CD, as well as the increasing need to disseminate (primary and processed) visual information within the zebrafish research community in Cooper *et al.* (2004).

In the "Zebrafish: The Living Laboratory" CD, we were able to archive a single 4D confocal image dataset of fluorescent YSL nuclei engaging in morphogenetic movements. A recent publication demonstrates that the dissemination of 4D image datasets of developing zebrafish embryos will soon become the norm (Keller *et al.*, 2008).

Open sourcing of primary datasets will allow multiple zebrafish labs to engage in data analysis, without necessarily having to be involved in data acquisition themselves. As this form of visual data exchange increases, a new field of zebrafish bioinformatics will emerge. Gene expression patterns will be linked to the morphogenetic behaviors of cellular ensembles in a more complete and detailed manner. As yet, the subcellular dynamics underlying morphogenetic processes are still largely inferred from a mesoscopic analysis of cell behavior. In the future, the ability to image GFP-fusion proteins at the subcellular level will draw an increasing number of researchers to study dynamic cytological processes within tissues of the living zebrafish embryo.

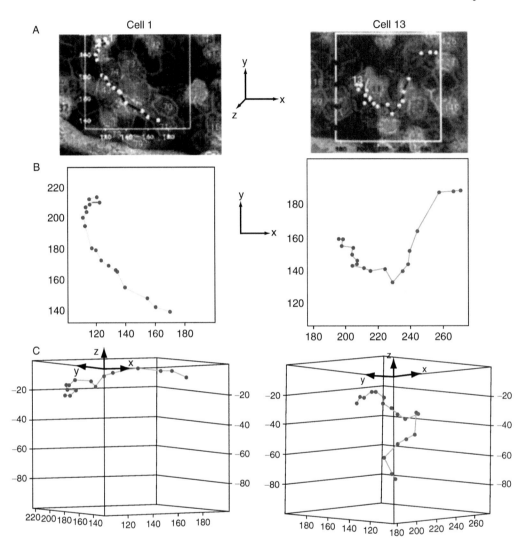

Fig. 10 Morphogenetic cell movements can be plotted as trajectories for detailed analysis. Once the x, y,z coordinates of individual cells have been ascertained and recorded into a digital datafile (see Fig. 9), 3D plots of cellular trajectories can be generated using commercially available software. This allows the three-dimensional movements of individual cells within a multilevel time-lapse recording to be examined and compared. The trajectories of two cells (cell 1 and cell 13) within the marginal zone of a gastrulating embryo are shown here as examples. The axes are labeled in microns. (A) The original image is overlaid with an x,y plot (plot same as in B) of each cell. The z direction is normal to the plane of the image. (B) This x,y plot represents the movement of the cells in the x,y directions. Both cells move toward the right (+x direction). Cell I , however, moves only toward lower y values, while cell 13 moves first toward lower y values then turns and moves in the opposite y-direction (+y direction). (C-D) 3D plots of the same cells.

VIII. Confocal Imaging of Embryos Expressing Green Fluorescent Protein (GFP)

Dramatic advances in the use of GFP-fusion technology within transfected cultured cells and certain vertebrate embryos suggest that GFP will have wide applications as a vital label in living zebrafish embryos. Transgenic mice, for instance, have been made in which GFP is expressed in all tissues except erythrocytes and hair (Okabe et al., 1997). Transgenic mice expressing GFP under an astrocyte-specific promoter have also been generated (Zhuo et al., 1997).

GFP has recently proved useful as a means for rapidly identifying putative transgenic fish. Amsterdam et al. (1995) initially tested the feasibility of using GFP to identify transgenics after injection of a plasmid-encoding GFP from a constitutive promoter. Expression was seen at about 4 h of development, shortly after the mid-blastula transition. Transgenic fish expressing GFP were also generated, although the GFP expression was seen later, at about 20 h of development. Amsterdam et al. (1996) characterized the copy number of GFP in nine transgenic lines. Interestingly, the strongest expressing lines had single-copy integrations of GFP, which demonstrates that single-copy integrations are capable of producing detectable fluorescence.

Transgenic embryos expressing GFP specifically in erythroid cells have also been generated (Long et al., 1997). Fluorescence-activated cell sorting was used to obtain the earliest erythroid progenitor cells. This study demonstrates that GFP can be used to illuminate the origin, lineage, and cell behavior of progenitor cells in developing embryos. It is also likely that GFP technology can be used to study the dynamics and localizations of specific GFP-fusion proteins in vivo.

Owing to the many potential applications of the GFP, confocal imaging of zebrafish embryos vitally labeled with GFP or GFP-fusion proteins will have a wide range of applications. In this regard, it is important to note that certain vital stains, such BODIPY 564/591, have spectral characteristics that do not interfere with GFP imaging. It is therefore feasible to simultaneously image GFP-expressing cells, as well as the cellular organization of organ-forming primordia within an embryo stained with BODIPY 564/591, using two-channel confocal microscopy.

The profound impact of GFP on cell biology and developmental biology led to the 2008 Nobel Prize in Chemistry. Instead of inserting synthetic organic dyes into living cells and tissues (the approach of the 1970s-early1990s), fluorescent fusion proteins can now be selectively expressed in transgenic organisms using genetic techniques. Approximately 300 papers have been published on GFP-expressing zebrafish since the first seminal paper appeared in the mid-1990s (Amsterdam et al., 1995). With expanding uses of GFP, as well as other biological fluorophores, the zebrafish research community should expect to see increasingly novel uses of fluorescent resonance energy transfer (FRET) and fluorescent lifetime imaging (FLIM) in the analysis of embryonic and adult transgenic zebrafish. For example, see Dickinson et al. (2003), Siegel et al. (2003), and Xie et al. (2008).

When this chapter on imaging morphogenetic movements was originally written, it appeared that BODIPY 564/591 would work well as a counterstain for GFP, given its ability to serve as a counterstain for BODIPY 505/515. This turned out not be the case. BODIPY 564/591 showed spectral bleedthrough with GFP fluorescence. To obtain a better fluorescent counterstain, a variety of other BODIPY fluorophores were examined. A dye, known as BODIPY TR methylester, was identified, which had ideal properties as a vital counterstain for GFP in living zebrafish embryos. Embryos can be vitally stained with BODIPY TR methyl ester using vital-staining procedures similar to those outlined in this chapter. For more details on this dye, which has the trademark name CellTrace BODIPY TR methyl ester (Molecular Probes, Eugene, OR), see Cooper *et al.* (2005).

New immobilization methods for imaging organotypic explants have been developed. These methods are especially useful for imaging the morphogenesis and physiology of explanted organ rudiments, after these tissues have been isolated from GFP-transgenic zebrafish embryos (see Langenberg *et al.*, 2003).

IX. Summary

Confocal microscopy is an excellent means of imaging cellular dynamics within living zebrafish embryos because it provides a means of optically sectioning tissues that have been labeled with specific fluorescent probe molecules. In order to study genetically encoded patterns of cell behavior that are involved in the formation of germ layers and various organ primordia, it is possible to vitally stain an entire zebrafish embryo with one or more fluorescent probe molecules and then examine morphogenetic behaviors within specific cell populations of interest using time-lapse confocal microscopy. There are two major advantages to this "bulk-labeling" approach: (1) the applied fluorescent probe (a contrast-enhancing agent) allows all of the cells within an intact zebrafish embryo to be rapidly stained and (2) the morphogenetic movements and shape changes of hundreds of cells can then be examined simultaneously *in vivo* using time-lapse confocal microscopy.

The neutral fluorophore BODIPY 505/515 and its sphingolipid-derivative BODIPY-C_5-ceramide are particularly useful, nonteratogenic vital stains for imaging cellular dynamics in living zebrafish embryos. These photostable fluorescent probes (when applied with 2% DMSO) percolate through the EVL epithelium of the embryo, and localize in yolk-containing cytoplasm and interstitial space, respectively, owing to their different physiochemical characteristics. BODIPY-ceramide, for instance, remains highly localized to interstitial fluid once it accumulates within a zebrafish embryo, allowing the boundaries of deep cells to be clearly discerned throughout the entire embryo. Through the use of either of these fluorescent vital stains, it is possible to rapidly convert a developing zebrafish embryo into a strongly fluorescent specimen that is ideally suited for time-lapse confocal imaging.

For zebrafish embryos whose deep cells have been intentionally "scatter-labeled" with fluorescent lineage tracers (e.g., fluorescent dextrans), sequential confocal z-series (i.e., focus-throughs) of the embryo can be rendered into uniquely informative 3D time-lapse movies using readily available image-processing programs. Similar time-lapse imaging, combined with rapidly advancing computer-assisted visualization techniques, may soon be applied to study the dynamics of GFP-fusion proteins *in vivo*, as well as other types of synthetic probe molecules designed to reveal the cytological processes associated with the patterning and morphological transformations of the zebrafish's embryonic tissues.

Acknowledgments

We dedicate this chapter to Drs Richard A. Cloney and John P. Trinkaus, who independently pioneered the use of time-lapse cinematography for the study of morphogenetic cell behaviors in living embryos. M.S.C. would like to personally acknowledge the following individuals for sharing their theoretical and technical insights on how to image cellular dynamics in living tissues: Raymond E. Keller, Manfred Schliwa, John P. Miller, Scott E. Fraser, Michael V. Danilchik, and Stephen J. Smith. This work was supported by a NSF Presidential Young Investigator Award IBN-9157132, and a University of Washington Royalty Research Fund Grant 65-9926. M. S. C. gratefully acknowledges equipment and software donations from the Bio-Rad Corporation, Meridian Instruments, and the Universal Imaging Corporation through the NSF PYI program. L. A. D. was supported through a NIH Developmental Biology Training Grant 5T32HD07183-18. C. A. H. was supported by a NIH Molecular and Cellular Biology Training Grant PHS NRSA P32 6M07270 from NIGMS.

References

Amsterdam, A., Lin, S., and Hopkins, N. (1995). The *Aequorea victoria* green fluorescent protein can be used as a reporter in live zebrafish embryos. *Dev. Biol.* **171,** 123–129.

Amsterdam, A., Lin, S., Moss, L. G., and Hopkins, N. (1996). Requirements for green fluorescent protein detection in transgenic zebrafish embryos. *Gene* **173,** 99–103.

Cooper, M. S., Cornell-Bell, A. H., Chernjavsky, A., Dani, J. W., and Smith, S. J. (1990). Tubulovesicular processes emerge from *trans*-Golgi cisternae, extend along microtubules, and interlink adjacent *trans*-Golgi elements into a reticulum. *Cell* **61,** 135–145.

Cooper, M. S., and D'Amico, L. A. (1996). A cluster of noninvoluting endocytic cells at the margin of the zebrafish blastoderm marks the site of embryonic shield formation. *Dev. Biol.* **180,** 184–198.

Cooper, M. S., D'Amico, L. A., and Henry, C. A. (1998). Analyzing morphogenetic cell behaviors in living zebrafish embryos. *In* "Protocols in Confocal Microscopy" (S. Paddock, ed.), "Methods in Molecular Biology" series. Humana Press, Totowa, NJ, **122,** 185–204.

Cooper, M. S., Sommers-Herivel, G., Poage, C. T., McCarthy, M. B., Crawford, B. D., and Phillips, C. (2004). The zebrafish DVD exchange project: A bioinformatics initiative. *Methods Cell Biol.* **77,** 439–457.

Cooper, M. S., Szeto, D. P., Sommers-Herivel, G., Topczewski, J., Solnica-Krezel, L., Kang, H.-C., Johnson, I., and Kimelman, D. (2005). Visualizing morphogenesis in transgenic zebrafish embryos using BODIPY TR methyl ester dye as a vital counterstain for GFP. *Dev. Dyn.* **232,** 359–368.

D'Amico, L. A., and Cooper, M. S. (1997). Spatially distinct domains of cell behavior in the zebrafish organizer region. *Biochem. Cell Biol.* **75,** 563–577.

Dickinson, M. E., Simbuerger, E., Zimmermann, B., Waters, C. W., and Fraser, S. E. (2003). Multiphoton excitation spectra in biological samples. *J. Biomed. Opt.* **8,** 329–338.

Driever, W., Solnica-Krezel, L., Schier, A. F., Neuhauss, S. C. F., Malicki, J., Stemple, D. L., Stainier, D. Y. R., Zwartkruis, F., Abdelihah, S., Rangini, Z., Belak, J., and Boggs, C. (1996). A genetic screen for mutations affecting embryogenesis in zebrafish. *Development* **123**, 37–46.

Freshney, R. I. (1987). "Culture of Animal Cells," 2nd edn., Wiley-Liss and Sons, New York.

Haffter, P., Granato, M., Brand, M., Mullins, M. C., Hammerschmidt, M., Kane, D. A., Odenthal, J., van Eeden, F. J. M., Jiang, Y.-J., Heisenberg, C.-P., Kelsh, R. N., Furutani-Seiki, M., *et al.* (1996). The identification of genes with unique and essential functions in the development of the zebrafish, *Danio rerio*. *Development* **123**, 1–36.

Haffter, P., and Nüerlein-Volhard, C. (1996). Large scale genetics in a small vertebrate, the zebrafish. *Int. J. Dev. Biol.* **40**, 221–227.

Haugland, R. (1996). "Handbook of Fluorescent Probes and Research Chemicals." Molecular Probes Inc., Eugene, OR.

Inoue, S. (1987). "Video Microscopy." Plenum Press, New York.

Keller, P. J., Schmidt, A. D., Wittbrodt, J., and Stelzer, E. H. (2008). Reconstruction of zebrafish early embryonic development by scanned light sheet microscopy. *Science* **322**, 1065–1069.

Langenberg, T., Brand, M., and Cooper, M. S. (2003). Imaging brain development and organogenesis in zebrafish using immobilized embryonic explants. *Dev. Dyn.* **228**, 464–474.

Lipsky, N. G., and Pagano, R. E. (1985a). A vital stain for the Golgi apparatus. *Science* **228**, 745–747.

Lipsky, N. G., and Pagano, R. E. (1985b). Intracellular translocation of fluorescent sphingolipids in cultured fibroblasts: Endogenously synthesized sphingomyelin and glucocerebroside analogues pass through the Golgi *en route* to the plasma membrane. *J. Cell Biol.* **100**, 27–34.

Long, Q., Meng, A., Wang, H., Jessen, J. R., Farrell, M. J., and Lin, S. (1997). GATA-1 expression pattern can be recapitulated in living transgenic zebrafish using GFP reporter gene. *Development* **124**, 4105–4111.

Majlof, L., and Forsgren, P.-O. (1993). Confocal microscopy: Important considerations for accurate imaging. *In* "Methods in Cell Biology, Vol. 38, Cell Biological Applications of Confocal Microscopy" (B. Matsumoto, ed.), pp. 79–95. Academic Press, San Diego, CA.

Melby, A. E., Warga, R. M., and Kimmel, C. B. (1996). Specification of cell fates at the dorsal margin of the zebrafish gastrula. *Development* **122**, 2225–2237.

Okabe, M., Ikawa, M., Kominami, K., Nakanishi, T., and Nishimune, Y. (1997). 'Green mice' as a source of ubiquitous green cells. *FEBS Lett.* **407**, 313–319.

Pagano, R. E., Sepanski, M. A., and Martin, O. C. (1989). Molecular trapping of a fluorescent ceramide analogue at the Golgi apparatus of fixed cells: Interactions with endogenous lipids provides a *trans*-Golgi marker for both light and electron microscopy. *J. Cell Biol.* **109**, 2067–2080.

Siegel, J., Elson, D. S., Webb, S. E. D., Lee, K. C. B., Vlandas, A., Gambaruto, G. L., Leveque-Fort, S., Lever, M. J., Tadrous, P. J., Stamp, G. W. H., Wallace, A. L., Sandison, A., *et al.* (2003). Studying biological tissue with fluorescence lifetime imaging: Microscopy, endoscopy, and complex decay profiles. *Appl. Opt.* **42**, 2995–3004.

Simon, J. Z., and Cooper, M. S. (1995). Calcium oscillations and calcium waves coordinate rhythmic contractile activity within the stellate cell layer of medaka fish embryos. *J. Exp. Zool.* **273**, 118–129.

Terasaki, M., and Jaffe, L. A. (1993). Imaging endoplasmic reticulum in living sea urchin eggs. *In* "Methods in Cell Biology, Vol. 38, Cell Biological Applications of Confocal Microscopy" (B. Matsumoto, ed.), pp. 211–220. Academic Press, San Diego, CA.

Thomas, C., DeVries, P., Hardin, J., and White, J. (1996). Four-dimensional imaging: Computer visualization of 3D movements in living specimens. *Science* **273**, 603–607.

Westerfield, M. (1995). "The Zebrafish Book." Oregon University Press, Eugene, OR.

Xie, Y., Ottolia, M., John, S. A., Chen, J. N., and Phillipson, K. D. (2008). Conformational changes of a Ca^{2+}-binding domain of the $Na+/Ca^{2+}$ exchanger monitored by FRET in transgenic zebrafish heart. *Am. J. Physiol. Cell Physiol.* **295**, C388–C393.

Zhuo, L., Sun, B., Zhang, C. L., Fine, A., Chiu, S. Y., and Messing, A. (1997). Live astrocytes visualized by green fluorescent protein in transgenic mice. *Dev. Biol.* **187**, 36–42.

CHAPTER 8

Cytoskeletal Dynamics of the Zebrafish Embryo

Jacek Topczewski[*] and Lilianna Solnica-Krezel[†]

[*]Department of Pediatrics
Children's Memorial Research Center
Northwestern University Feinberg School of Medicine
Chicago, Illinois 60614

[†]Department of Biological Sciences
Vanderbilt University
Nashville, Tennessee 37235-1634

I. Introduction

Since the first publication of this collection of methods impressive progress has been made in the cell biology of zebrafish. Most of these studies employed loss of function methods, such as mutant analysis or antisense morpholino oligonucleotides based interference. Cytoskeletal elements of manipulated zebrafish embryos are usually analyzed by staining fixed samples with specific antibodies or dyes. Several new families of fluorescent dyes, such as Alexa Fluor or cyanine dyes (Cy2, Cy3, and Cy5), afford more photostable and brighter conjugated antibodies, proteins and reagents than described in the original review. However, the basic methods described in the original review are still valid. Here, we have expanded the collection of fixation methods to cover a broad range of conditions that could be tested with a new, difficult antibody. We also added methods that allow testing amounts of actin and tubulin filaments. Finally, we have included a brief overview of the intermediate filaments studies, with examples of antibodies used to detect most common proteins forming this type of cytoskeleton.

Similar to the development of other vertebrates, zebrafish embryogenesis employs multiple morphogenetic processes. Massive intracellular rearrangements are initiated in a fertilized egg, including separation of the cytoplasmic blastodisc from the yolk and directed transport of dorsal determinants. Subsequent synchronous cleavages subdivide the blastoderm in the absence of other morphogenetic changes. Soon after the activation of zygotic transcription at midblastula transition, cellular motility increases and the process of gastrulation is initiated. Gastrulation involves a set of stereotyped cellular rearrangements including epiboly, involution/ingression, convergence, and extension. These concurrent morphogenetic processes create the three-germ layer organization of the embryo, place tissues and organ rudiments in the proper position for further inductive interactions and development, and are the major force underlying the change in the embryonic shape (Kimmel *et al.*, 1995).

While the entire spectrum of morphogenetic cellular behaviors involved in vertebrate embryogenesis is still not known, directed cell migration, cell shape changes, cell intercalation and oriented cell divisions are important components (Trinkaus, 1984). In other systems, both intracellular morphogenetic processes as well as cellular morphogenetic behaviors employ cytoskeletal systems of microfilaments, microtubules, and intermediate filaments (Lauffenburger and Horwitz, 1996; Mitchison and Cramer, 1996). The cytoskeleton of zebrafish has so far received only limited experimental attention. Here, we provide a short overview of the dynamic changes in the organization and function of actin filaments and microtubules during zebrafish development (for more detailed review see Hart and Fluck (1995) and Solnica-Krezel *et al.* (1995)). We also describe selected methods and tools currently available for analysis of cytoskeletal organization and function during the embryogenesis of zebrafish and related fish.

II. Cytoskeleton of the Unfertilized Egg

Unfertilized eggs of teleost fish are enclosed by an acellular chorion that possesses a single micropyle (Brummett and Dumont, 1979; Hart and Donovan, 1983; Kobayashi and Yamamoto, 1981). The egg is compartmentalized into a central yolk mass and a peripheral cortex (Fig. 1). In zebrafish, the cortex is about 15–20 μm thick although it tends to be noticeably thicker at the micropyle (Hart and Yu, 1980). Similar to sea urchin (Chandler, 1991; Spudich *et al.*, 1988) and amphibian (Chow and Elinson, 1993; Merriam and Clark, 1978) eggs, actin is a major component of the teleost egg (Becker and Hart, 1996; Hart *et al.*, 1992; Ivanenkov *et al.*, 1987). Extracts of unfertilized loach and zebrafish eggs resolved by SDS/PAGE electrophoresis display a single polypeptide having a molecular mass of 43 kDa, based on isoelectric focusing (Ivanenkov *et al.*, 1987) and immunoblot analyses (Hart *et al.*, 1992). Zebrafish actin, in Triton X-100-treated egg homogenates, separates principally to the supernatant fraction upon high-speed centrifugation, suggesting that most of the egg actin is soluble and in a nonfilamentous form (see Section VII.B).

The distribution and organization of actin in the unfertilized egg have been investigated with conventional electron microscopy and fluorescent microscopy

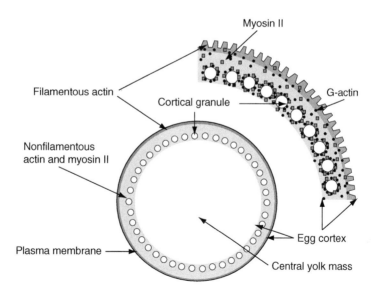

Fig. 1 Schematic representation of spatial distribution of F-actin, G-actin (closed circles), and myosin II (gray squares) in the inactivated zebrafish egg (based on Becker and Hart, 1996). Drawn not to scale.

methods. Ultrathin sectioning of zebrafish (Hart *et al.*, 1992) and chum salmon (Kobayashi and Yamamoto, 1981) cortex reveals a homogeneous-appearing, electron-dense matrix immediately subjacent to the plasma membrane and a deeper, less dense cytoplasm that extends to the yolk mass. The subplasmalemmal layer of the zebrafish egg is about 200 nm thick, except at the site of sperm entry where its thickness increases to about 550 nm. In the electron dense matrix, 5 to 8 nm actin filaments are detected occasionally. Microfilaments appear most prominent in the microplicea and microvilli of the sperm entry site where they tend to run parallel to the long axis of the cytoplasmic extension and course down into the electron-dense layer of the cytoplasm (Hart *et al.*, 1992).

Whole eggs and egg fragments stained with rhodamine-phalloidin (RhPh) exhibit a narrow, continuous rim of fluorescence at the margin of the egg (see Section VII.C.2). The RhPh staining shows only the distribution of filamentous actin (F-actin) in contrast to antibody staining which reveals both filamentous and nonfilamentous actin (Table I) (Becker and Hart, 1996). Cortical patches display an elaborate, three-dimensional meshwork of interconnecting RhPh-staining filaments of variable diameter. Double staining with RhPh and anti-actin antibody has resulted in the identification of a nonfilamentous actin domain juxtaposed to the cellular membrane that holds the cortical granules and other organelles of the cortex (Fig. 1) (Becker and Hart, 1996; Hart *et al.*, 1992).

The zebrafish egg contains a single polypeptide (215–225 kDa) that crossreacts on immunoblots with an antibody against myosin heavy chain (Becker and Hart, 1996; Hart *et al.*, 1992). *In situ* staining with this antibody colocalizes with F-actin at the plasma membrane and within the deeper, nonfilamentous actin domain. Zebrafish extracts also contain a polypeptide, which comigrates with sea urchin spectrin and crossreacts with the anti-sea urchin egg spectrin polyclonal antibody. The spectrin (or spectrin-like) protein colocalizes with F-actin and myosin at the cortex and with F-actin at the sperm entry site. This peripheral distribution of F-actin, myosin, and spectrin suggests the presence of an egg plasma membrane cytoskeleton (Fig. 1). Results of treating zebrafish eggs with cytochalasin B and D, fungal inhibitors of actin polymerization (Table II) (Cooper, 1987), support the view that this cytoskeleton is important in maintaining the structural integrity of the plasma membrane. Specifically, the actin cytoskeleton might stabilize the depression in the egg surface at the sperm entry site (Wolenski and Hart, 1988).

III. Organization and Function of the Cytoskeleton in the Zygote

Fertilization activates several changes in the egg. The chorion swells and lifts away from the newly fertilized egg. Cortical granules are secreted and the fertilization cone forms and serves as a structure for the sperm to enter the egg. Furthermore, the cytoplasm starts to stream toward the animal pole, initiating ooplasm segregation, which will continue during early cleavages (Hart and Fluck, 1995).

Table I

Antibodies and Dyes used for Detection of Cytoskeletal Elements

Cytoskeletal element	Staining/Ab	Concentration/source	Developmental stage/tissue	Reference
Microtubule/ α-tubulin	DM1A (Ab)	1:10,000/ICN Immunobiologicals	1 cell-late gastrula	Solnica-Krezel and Driever (1994)
Microtubule/ β-tubulin	KmX-1 (Ab)	1:500–1:2000/ Boehringer Mannheim	1 cell-late gastrula	Solnica-Krezel and Driever (1994)
Microtubule	Anti-β-tubulin (Ab)	1:500/Amersham	1 cell epiboly	Strahle and Jesuthasan (1993)
Acetylated tubulin	Monoclonal antibody (mAb) 6-11B-1	1:750/Sigma-Aldrich	From early embryonic to adult stages	Piperno and Fuller (1985)
Tubulin/in vivo	5(6)-Carboxyfluorescein labeled porcine tubuline	Molecular Probes	20–25 h/neuronal axons	Takeda et al. (1995)
Total actin	Anti-actin (Ab)	1:50–1:100 Amersham	Egg	Becker and Hart (1996)
Actin	Monoclonal HHF35 to human myocardium	1:500/Biogenex	Hepatic lesion and neoplasms	Bunton (1995)
F-actin	Rhodamine-phalloidin	66 nM/Molecular Probes	Egg	Becker and Hart (1996)
G-actin	Fluorescein-conjugated DNase I	100 μg/ml/Molecular Probes	Egg	Becker and Hart (1996)
Myosin	Human-platelet antibody	1:50/Biomedical Technologies	Egg	Becker and Hart (1996)
Desmin	Polyclonal to chicken gizzard	1:500/Biogenex	Hepatic lesion and neoplasms	Bunton (1995)
Keratin	Mixed monoclonal to human epithelial keratins	1:6000/Boehringer Mannheim	Hepatic lesion and neoplasms	Bunton (1995)

Correlative electron and fluorescent microscopy studies provide insight into the temporal and spatial relationships between the fertilization cone, the fertilizing sperm and the actin cytoskeleton in zebrafish (Hart et al., 1992). The sperm attaches to the microvilli of the sperm entry site within 5-10 s after insemination (Wolenski and Hart, 1987). Membranes of the fused gametes rupture shortly thereafter and the leading edge of the sperm nuclear membrane becomes positioned in direct contact with the subplasmalemmal F-actin meshwork (Hart et al., 1992). The fertilization cone by 1 min postinsemination is an elevated cytoplasmic projection, about 8–10 μm high that contains a partially incorporated spermatozoon at its apex. By 2 min postinsemination, eggs consistently exhibit regression and flattening of the fertilization cone. There is visible thickening and rearrangement of the associated actin meshwork between 30 and 60 s postinsemination.

Table II
Drugs and Physical Treatments Interfering with Cytoskeleton Function

Cytoskeletal element	Drug/action	Concentration/conditions	Reference
Actin filaments	Cytochalasin B/actin filament depolymerization	1–10 μg/ml/1–10 min	Wolenski and Hart (1988)
	Cytochalasin D/actin filament depolymerization	10–50 μg/ml/1–10 min	Wolenski and Hart (1988)
Microtubules	Cold treatment/microtubules depolymerization	18 °C, before 512-cell stage/continuous	Jesuthasan and Strahle (1997)
	UV irradiation/microtubules depolymerization	90 s, 4 mW/cm^2	Jesuthasan and Strahle (1997) Strahle and Jesuthasan (1993)
	Colchicine	100 μM/1 h	Abraham et al. (1993)
	Colcemid (demecolcine)[a]	0.35 μM/1 h	Abraham et al. (1993)
	Nocodazole/microtubule depolymerization	0.6–2 μg/ml/10 min	Strahle and Jesuthasan (1993)
		0.5–20 μg/ml/1 h	Solnica-Krezel and Driever (1994)
		0.17 μM/1 h	Abraham et al. (1993)
	Taxol/microtubule stabilization	10–100 μM/continuous	Solnica-Krezel and Driever (1994)

[a]The inhibitory effects of this poison can be reversed by illuminating the egg with ultraviolet light (360 nm), which photolyzes demecolcine, converting it to lumidemecolcine, a molecule that is not a microtubule poison (Aronson and Inoue, 1970; Webb et al., 1995). Colcemid is trade name for demecolcine.

RhPh staining reveals a gradual, enclosure of the sperm by the actin meshwork during this time. Interestingly, the fertilization cones of zebrafish eggs show normal growth when parthenogenetically activated in tank water, with intense RhPh cortical staining. Treatment of zebrafish eggs with cytochalasins B and D consistently blocks incorporation of the fertilizing sperm, which strongly suggests that the translocation of spermatozoon into the egg requires the assembly of actin filaments (Hart et al., 1992; Wolenski and Hart, 1988).

A. Cortical Granule Exocytosis

Regulated secretion is a dramatic, calcium-dependent response of teleost eggs to either binding with sperm or treatment with an activating agent. During exocytosis, cortical granules (vesicle alveoli) vectorially move to the oolemma, fuse with it, rupture, and discharge their contents at the egg surface (Brummett and Dumont, 1981; Donovan and Hart, 1986). Subplasmalemmal F-actin meshwork has been proposed to act as a barrier to the movement and docking of secretory granules at the plasma membrane of the zebrafish egg. Unstimulated zebrafish eggs treated for 5–10 min with cytochalasin D (50 μg/ml) often spontaneously discharge cortical granules (Wolenski and Hart, 1988). In contrast, injecting loach eggs with phalloidin prevents exocytosis; many granules are frequently displaced deeper into the ooplasm because of a thickened actin network (Ivanenkov *et al.*, 1987).

B. Ooplasmic Segregation

In the zebrafish egg, ooplasm is distributed throughout the ovum and thus intermingles with the yolk. Ooplasmic segregation, following fertilization or egg activation, involves the movement of ooplasm to the animal pole of the zygote and its coalescence into a blastodisc. The movement of ooplasm toward the animal pole can be seen clearly with time-lapse techniques and has been described as bulk flow or streaming (Abraham *et al.*, 1993; Beams *et al.*, 1985). While most ooplasm moves toward the animal pole, some components are left behind near the vegetal pole or appear to be moved actively toward it. These components include the "yolk platelets" that move toward and accumulate at the vegetal pole due to this late-acting counterstream of ooplasm (Roosen-Runge, 1938).

Evidence for the presence and activity of microfilaments during ooplasmic segregation comes from electron microscopy studies and experiments involving the use of cytochalasins (B and D). On the basis of their morphology, microfilaments have been tentatively identified by transmission electron microscopy in the subplasmalemmal electron-dense matrix of zebrafish (Katow, 1983), as well as in loach zygotes (Ivanenkov *et al.*, 1987). Scanning electron microscopy reveals a filamentous appearance that may be due to the presence of F-actin (Beams *et al.*, 1985). Actin, by RhPh staining and molecules crossreacting with antiplatelet heavy-chain myosin is found both in the peripheral ooplasm and in the blastodisc of zebrafish zygotes (Hart and Becker, 1994). Treatment of zebrafish zygotes with cytochalasins causes the meshwork of subplasmalemmal microfilaments to detach from the plasma membrane and prevents formation of the blastodisc and subsequent cleavage (Hart and Becker, 1994; Katow, 1983). In medaka, cytochalasin D inhibits cytoplasmic streaming and formation of the blastodisc, but has no effect on oil-droplet movement or saltatory movement (Webb and Fluck, 1995).

The role of microtubules in ooplasm segregation is not clear. Anti microtubule drugs—colchicine (100 μM), demecolcine (Colcemid; 0.35 μM), or nocodazole (0.17 μM)—slow the rate of growth of the blastodisc and inhibit oil-droplet movement toward the vegetal pole, saltatory movement toward both poles, and pronuclear movement in medaka zygotes (Abraham *et al.*, 1993; Webb and Fluck, 1995; Table II). By contrast, colchicine had no apparent effect on ooplasmic segregation in zebrafish (Katow, 1983). It is not clear if this difference is due to the 40-fold lower concentration of the drug used in experiments involving zebrafish zygotes or to a lower permeability of the zebrafish egg to the molecule. However, the possibility that ooplasmic segregation in the zebrafish does not require microtubules cannot be excluded.

IV. Cleavage and Blastula Period

After the zygotic period ends with the first cleavage about 40 min after fertilization, blastomeres divide synchronously at about 15 min intervals. These divisions are meroblastic; they only incompletely undercut the blastodisc, and initially (2–8-cell stage) all blastomeres and later (16–512-cell stage) the marginal blastomeres, remain connected to the yolk cell by cytoplasmic bridges. During the end of the 512-cell stage and particularly as they enter the next, 10th mitosis, the marginal blastomeres collapse into the yolk cell and form the yolk syncytial layer (YSL) (Kimmel *et al.*, 1995). The second cell lineage forms within the blastodermal cap. This lineage comprises the superficial blastomeres that will give rise to an epithelial monolayer, called the enveloping layer (EVL). Separation of YSL and EVL from the remaining deep cell layer roughly coincides with midblastula transition (Kimmel *et al.*, 1990).

A. The Role of a Vegetal Array of Parallel Microtubules in the Directed Transport of Dorsal Determinants at the Zygote and Cleavage Stages

Dorsal maternal determinants in teleost embryos are thought to be present in the vegetal mass of the yolk cell soon after fertilization (Mizuno *et al.*, 1997). Recent work implicates microtubules in the transport of dorsal determinants from the vegetal pole of the zygote toward the blastodisc. This microtubule-dependent transport might play a key role in the establishment of embryonic polarity in the zebrafish (Jesuthasan and Strahle, 1997). Approximately 20 min after fertilization (at 28.5 °C) a parallel array of microtubules was observed at the vegetal pole, at a shallow location, less than 2 μm from the surface (Fig. 2A; See Section VII.C.3). This array has not been detected in other regions of the yolk cell, and was absent up to 10 min after fertilization. Microtubules adjacent to the forming blastodisc and at the equator were present but disorganized. At 30 min postfertilization, a number of aligned fibrils could be seen offset from the vegetal pole, whereas fibrils near the blastodisc were not aligned. During and after the first cleavage, aligned fibrils were not detected at the vegetal pole. At the 4-cell stage, short fibrils with no obvious orientation were detected at the equator (Jesuthasan and Strahle, 1997). However, at the 8-cell stage, microtubules appear to emerge from blastomeres and extend along the animal-vegetal axis (Solnica-Krezel and Driever, 1994). In embryos examined from the 32-cell stage up to the 256-cell stage, long arrays of microtubules extending beyond the equator, were observed (Fig. 2B) (Jesuthasan and Strahle, 1997; Solnica-Krezel and Driever, 1994).

The yolk microtubule array has been proposed to transport substances from the vegetal region of the yolk cell into the blastoderm. Treatments that disrupt this array, cold (18 °C), UV irradiation (90 s, 4 mW/cm^2), or microtubule poison (nocodazole 0.1 μg/ml), were shown to prevent axis formation (Table II) (Jesuthasan and Strahle, 1997; Strahle and Jesuthasan, 1993). To assess whether substances localized in the vegetal hemisphere could be transported to the

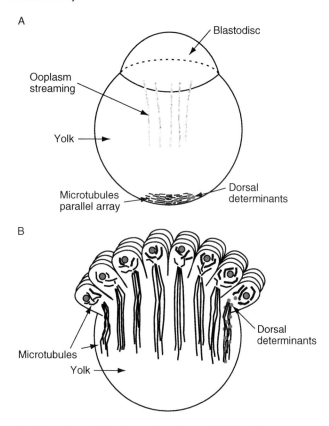

Fig. 2 Schematic illustration of the changes in the organization of the cytoskeletal microtubules during zygote period and cleavage stage. (A) 30 min-old zygote, the animal pole at the top. Yolk-free cytoplasm segregated to the animal pole. The array of parallel microtubules is formed near the vegetal pole and dorsal determinants start to be asymmetricaly transported toward the blastodisc. (B) 32-cell stage; long arrays of microtubules originating from the marginal blastomeres and extending beyond the equator can be observed. Dorsal determinants reached the blastomeres. Drawn not to scale.

blastoderm, 0.2 μm polystyrene beads were injected into the vegetal pole region at 1–4-cell stage of zebrafish development (see Section VII.D). Such beads have been shown to mimic the movement of organelles in a variety of cell types (Adams and Bray, 1983; Beckerle, 1984). Injected beads were able to enter the blastoderm and moved only toward the animal pole, in contrast to the uniform spread of coinjected dextran. Several experimental observations support the notion that the movement of beads is microtubule dependent. First, treatment of bead-injected embryos with nocodazole or cold, arrested the movement of the particles. Second, incubation of the embryos in 1 μg/ml nocodazole 5 min after bead injection, completely prevented translocation of beads in the cortex. The beads could be detected only in the vegetal hemisphere when examined after 1.5 or 4.5 h. When embryos were treated with nocodazole 30 min after injection, cortical beads only reached the equator, while in embryos treated 60 min after injection beads advanced into blastomeres. Similar results were obtained using 10 μg/ml nocodazole.

In contrast, treatment with 0.1 μg/ml nocodazole or incubation at 18 °C, conditions under which an axis fails to form, did not prevent beads movement toward the animal pole, although beads moved slower. In embryos treated with the lower dose of nocodazole immediately after injection at the 2–4-cell stage, cortical beads reached only the equator when those in untreated siblings already translocated into blastomeres (1 h after injection). After 1 h, beads reached the blastoderm, but they stopped in enlarged yolk syncytium. As a result, embryos treated prior to the 32-cell stage, approximately 1 h and 15 min after vegetal microtubule array formation, fail to accumulate β-catenin in the nuclei, to express *goosecoid* and *foxa3* genes, and to gastrulate properly. Treatment at later stages did not have these effects (Jesuthasan and Strahle, 1997). The role of the parallel array of microtubules at the vegetal pole in the asymmetric distribution of dorsal determinants has been also documented in medaka (Trimble and Fluck, 1995) and *Xenopus* (Rowning *et al.*, 1997) zygotes. Saltatory motion of subcellular particles in the vegetal pole of medaka zygotes is oriented along the microtubules in the vegetal array. The motion is absent from medaka zygotes treated with microtubule poisons such as demecolcine (Webb *et al.*, 1995). Notably, the vector of saltatory motion along microtubules points directly from the ventral surface to the future dorsal surface of the embryo (Trimble and Fluck, 1995).

V. Yolk Cell Microtubules during Epiboly

During epiboly, the deep cells, the EVL, and the YSL lineages expand vegetally to cover the yolk sphere completely 10 h postfertilization. The mechanisms involved in epibolic movements of these three cell types are, however, distinct. At the beginning of epiboly, the YSL, a thickened cytoplasmic layer populated by the yolk nuclei, is localized in the animal part of the yolk cell and is partially covered by blastoderm. The remainder of the yolk cortex is a thin anuclear layer of cytoplasm called the yolk cytoplasmic layer (YCL). The YSL exhibits a network of mitotic and interphase microtubules, while an array of microtubules aligned along the animal-vegetal axis exists in the YCL. Microtubules of the YCL appear to emanate from the centers associated with the syncytial nuclei. Since these centers also radiate mitotic spindle microtubules during the mitotic divisions of the YSN, they probably correspond to true microtubule organizing centers. It is thought that the YCL microtubules exhibit uniform polarity: the minus ends are in the YSL and the plus ends point toward the vegetal pole. During the period of mitotic divisions of the YSN, the external YSL expands and the YSN spread toward the vegetal pole while maintaining a very regular distribution. The astral microtubules of neighboring spindles in the YSL overlap and interdigitate. It has been hypothesized that microtubule-dependent forces generated by plus-end directed microtubule motors might act between antiparallel astral microtubules of adjacent spindles to push nuclei apart (Solnica-Krezel and Driever, 1994).

Changes in the organization of the yolk cell microtubules correlate with both the process of crowding of the YSN at the beginning of epiboly, and with their subsequent vegetal movements. During epiboly the yolk cell is equipped with two distinct microtubule arrays (Fig. 3). One array, an extensive network of

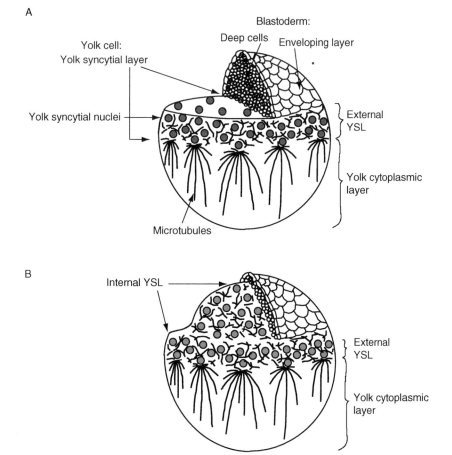

Fig. 3 Organization of yolk cell microtubules during epiboly. (A) The sphere stage just before the onset of epiboly. The blastoderm, composed of the internal deep cells and the superficial enveloping layer (EVL), is positioned atop of the syncytial yolk cell. Most of the yolk syncytial nuclei (YSN) are in the external yolk syncytial layer (YSL) positioned vegetal to the blastoderm. The microtubules of the external YSL form a network. The microtubules of the anuclear yolk cytoplasmic layer (YCL) radiate from the organizing centers associated with the vegetal-most YSN and are aligned along the animal-vegetal axis. (B) 30% epiboly. The external YSL has contracted in a doming of yolk cell and exhibits densely packed YSN and a dense network of microtubules. The external YSL is partially covered by the expanding vegetally blastoderm (based on Solnica-Krezel and Driever, 1994). Drawn not to scale.

intercrossing microtubules, is part of the YSL and expands as epiboly of the YSL proceeds. By contrast, another array of microtubules oriented along the animal-vegetal axis within the YCL becomes shorter as this layer diminishes (Solnica-Krezel and Driever, 1994). Treatment of sphere-stage embryos with 5–20 μg/ml nocodazole dramatically affected the organization of microtubules, cell divisions, epiboly, and gastrulation (Solnica-Krezel and Driever, 1994). In embryos fixed 30 min after the addition of 10 μg/ml nocodazole, microtubule arrays of the blastoderm and the yolk cell were completely disorganized (Table III). Blastoderm cells remained large throughout the experiment, indicating that cell divisions were inhibited. Both the germ ring and the embryonic shield failed to form, indicating that both involution and convergence toward the dorsal side were either greatly repressed or completely blocked. Distinct aspects of epiboly were variably impaired in the absence of microtubules. The external YSL remained as a wide belt below the blastoderm ring and no crowding of the YSN was observed. Embryos treated at the sphere stage (4 h) for 1 h with 0.5 μg/ml nocodazole exhibited numerous long microtubules in the yolk cell. In these embryos, in contrast to the embryos treated with higher concentrations of nocodazole, the YSN became concentrated close to the blastoderm rim. These observations are consistent with a function of the yolk cell microtubules in the compaction of the YSN and contraction of the YSL at the beginning of epiboly. In the later stages of epiboly, the

Table III
Survey of Fixation Methods Used for *In Situ* Detection of Cytoskeletal Components

	Fixation stage	Fixing solution	Time	Reference
Actin	Egg	3.7–5% formaldehyde diluted in actin stabilizing buffer (ASB)	15 min RT or 1–3 min on ice	Becker and Hart (1996) Hart *et al.* (1992)
		3.7% formaldehyde in ASB	4–8 h cold	Hart and Fluck (1995)
Myosin	Egg	0.45 M sodium acetate buffer pH 6.2	4–6 h/cold fresh buffer	Hart *et al.* (1992)
Tubulin	1 cell-late gastrulation	Formaldehyde-glutaraldehyde-taxol	2–4 h/25–28.5 °C	Solnica-Krezel and Driever (1994)
	1 cell-epiboly	MAB pH 6.5 (Schroeder and Gard, 1992)	5 h/RT	Strahle and Jesuthasan (1993)
	Late gastrulation	4% PFA in PBS	2–4 h/RT	McMenamin *et al.* (2003)
	2–4 days postfertilization	Prefer fixative (Anatech, Battle Creek, MI)	4 h/RT	Pathak *et al.* (2007)
	2–4 days postfertilization	Dent's fixative (80% methanol:20% dimethyl sulfoxide)	Overnight/4 °C	Pathak *et al.* (2007)
Pankreatin	2–4 days postfertilization	Dent's fixative	Overnight/−20 °C	Webb *et al.* (2008)

movements of the YSN toward the vegetal pole were shown to be blocked in the nocodazole-treated embryos, while the vegetal expansion of the EVL and deep blastoderm cells was only partially inhibited.

Several observations argue that the inhibition of nuclear movements in the above experiments occurred due to the loss of microtubules. First, some treatments with nocodazole that led to the inhibition of nuclear movements were initiated at the sphere stage or later during epiboly, after the cessation of mitotic divisions of the syncytial nuclei. Thus, observed effects are rather unlikely to be an indirect consequence of interference with proliferation of the YSN. Second, when sphere-stage embryos were treated with a low concentration of nocodazole such that disruption of microtubules was delayed, then the YSN became densely packed near the blastoderm rim. Finally, epiboly of the blastoderm proceeded further than epiboly of the YSN. However, the possibility that some aspects of epiboly inhibition in nocodazole-treated embryos were secondary to the disruption of microtubules cannot be excluded. Indeed, other cytoskeletal elements—actin and intermediate filaments—were not monitored in these experiments.

Taxol treatment experiments also support the involvement of microtubules in epiboly (Table II). Thirty minutes after incubation of sphere-stage embryos (4 h) in 100 μM taxol, the YSN were covered by the blastoderm, and only a belt of the dense network of the YSL microtubules was visible vegetal to the blastoderm rim. Thus, the contraction of the YSL was not inhibited. In taxol-treated embryos both the YSL and the YCL microtubule arrays had a denser appearance than in control embryos, with the microtubule arrays completely covering the vegetal pole and exhibiting a higher resistance to nocodazole. However, the movements of the YSN, the EVL, and deep cells toward the vegetal pole in the later stages of epiboly were delayed in taxol-treated embryos. In contrast to nocodazole-treated embryos, in embryos treated with taxol, epibolic movements of the YSN, the EVL, and deep cells were affected to a similar extent. Taxol treatment affected only epiboly, but not other aspects of gastrulation or morphogenesis. Notably, at 10.5 h of development, both the control and taxol-treated embryos exhibited two somites and a forming notochord. However, most of the taxol-treated embryos had not completed epiboly at this stage. Instead, they formed abnormally shaped gastrulae, with a portion of the yolk cell protruding from contracted blastopore lips (Solnica-Krezel and Driever, 1994).

A similar separation of epiboly and gastrulation was observed in UV-treated embryos (254 nm, 1.8–3.6 J/cm^2) (Strahle and Jesuthasan, 1993). Epiboly started later and proceeded slower in UV-treated embryos. The blastoderm spread with a speed of 15.2 $\pm$ 0.6% epiboly per hour over the yolk in controls, whereas the speed of epiboly was reduced to 7.5 $\pm$ 0.4% epiboly per hour in UV-treated embryos. β-Tubulin antibody staining (Table I) revealed that most of UV-treated embryos contained abnormal microtubules. They were either undetectable or shorter than normal and were either not aligned along the A-V axis in the YCL or took the form of a "comet-tail." Nevertheless, other aspects of embryo development were not affected (Strahle and Jesuthasan, 1993).

VI. Tubulin Dynamics in Neuronal Axons of Living Zebrafish Embryos

The translucent character of the zebrafish embryo, its rapid development, and accessibility for injections create a unique opportunity for analysis of cytoskeletal dynamics within cells of a living vertebrate *in situ*. Takeda and coworkers demonstrated the feasibility of such experiments in zebrafish addressing the mechanisms of slow axonal transport using fluorescence recovery after photobleaching (FRAP) (Takeda *et al.*, 1995; Table I).

In these experiments, purified porcine tubulin labeled with 5(6)-carboxyfluorescein succinimide was microinjected into single nonmarginal blastomeres of 16–32-cell stage zebrafish embryos. About 5% of the surviving embryos exhibited fluorescently labeled neurons. These embryos were subsequently dechorionated and mounted in 2% methyl cellulose, and a 488 nm argon laser was applied for 1 s to a small portion of one of the labeled neurites. The recovery of fluorescence after photobleaching was then analyzed by collecting serial images using a digital camera to monitor changes in the relative fluorescence intensity of the bleached area. Such photobleaching experiments were performed on a total of 11 Rohon-Beard cells and 25 motorneurons in the spinal cord. The results were similar for both cell types studied: the bleached marks did not translocate in any direction while the fluorescence recovered gradually. The average recovery half-time for the two cell types was determined to be 44.2 ± 11.2 min. Based on these results it has been proposed that in zebrafish neurons, tubulin is transported down the axon in the form of a heterodimer or a small oligomer. Furthermore, tubulin molecules once incorporated into the preexisting microtubules turn over with a relatively short half-time (Takeda *et al.*, 1995).

The growth rate of Rohon-Beard neuron axons containing fluorescently labeled microtubules was shown to be comparable to that reported previously for this cell type. This indicated that fluorescently labeled tubulin does not significantly alter cellular morphogenetic behaviors. Therefore, such technology should prove invaluable to study the changes in the organization of microtubules and other cytoskeletal components *in situ* at various stages of zebrafish development.

VII. Intermediate Filaments in Zebrafish

Intermediate filaments are the third major fibrous polymer of the cell that function as a cytoskeletal scaffold in the nucleus and cytoplasm. In contrast to microtubules and actin-based filaments, the protein composition of intermediate filaments is much more complex, and their basic attributes, such as the lack of structural polarity, are unique (Coulombe and Wong, 2004). The first insights into complex functions of intermediate filaments during zebrafish development are only beginning to emerge. Several keratin genes were identified in zebrafish and other fish species (Krushna Padhi *et al.*, 2006; Imboden *et al.*, 1997). Sequence conservation allowed using antibodies against chick and mammalian intermediate filaments for analysis of the neoplastic lesions in bass (*Morone saxatilis*), and 6-month-old medaka (*Oryzias latipes*) (Bunton, 1993), as well as for the localization of cytokeratins in rainbow trout (*Salmo gairdneri*) tissues (Markl and Franke, 1988). More recently, several antikeratin antibodies have been used in zebrafish. An extensive comparison of keratins and their immunoreactivity between carp *Cyprinus carpio*, goldfish *Carassius auratus* and zebrafish was conducted by Conrad *et al.* (1998) and Garcia *et al.* (2005). A commercial pankeratin antibody Ks pan1–8 (Progen Biotechnik) was used by Webb *et al.* (2008). Schaffeld *et al.* (2003) detected specific keratins 8 and 18 with monoclonal antibodies C10 and C04, respectively. Both biochemical properties and the expression pattern of zebrafish vimentin were characterized (Cerda *et al.*, 1998) and the protein was detected in the zebrafish retina with a specific antibody (Immunotech) (Lee *et al.*, 2005). A gene encoding Desmin, a muscle-specific protein and a constitutive subunit of the intermediate filaments, was cloned by Loh *et al.* (2000) and the protein visualized in zebrafish muscle (Costa *et al.*, 2002) by use of an antidesmin antibody (Sigma). Glial fibrillary acidic protein (GFAP), a member of the same class of intermediate filaments protein as vimentin and desmin, is expressed in astrocytes and glia, and can be visualized with anti GFAP antibody G-A-5 (Sigma) (Zupanc *et al.*, 2005). The *gfap* gene was isolated and the basic biochemical properties of its product were characterized (Nielsen and Jorgensen, 2003).

VIII. Methods

A. Collecting Oocytes, Eggs, and Embryos

1. Protocol 12.1

Collecting Oocytes

Ovary tissue is usually obtained by dissection of the pair of organs from a gravid female immobilized on ice (Becker and Hart, 1996; Kessel *et al.*, 1984). Isolated tissue is kept in FBSS buffer (see Table IV).

2. Protocol 12.2

Collecting Eggs and In Vitro Fertilization (Driever et al., 1996)

1. To induce oocyte maturation set up crosses, in the late afternoon, in breeding traps (two females, six to eleven months old, and one male).
2. The following morning, immediately after the light is turned on, separate the males and females. Perform *in vitro* fertilization within the next 2 h.
3. Anesthetize females with Tricaine (Westerfield, 1996).

Table IV
Egg Buffers

Buffer name	Buffer composition	Reference
Fish balanced saline solution (FBSS)[a]	137 mM NaCl, 5.36 mM KCl, 0.98 mM $MgCl_2 \cdot 6H_2O$, 0.81 nM $NaHPO_4 \cdot 7H_2O$, 1.27 mM $CaCl_2 \cdot 2H_2O$, 0.44 mM KH_2PO_4, 1.34 mM $Na_2HPO_4 \cdot 7H_2O$, and 5.5 mM dextrose	Hart *et al.* (1992)
Fish ringer solution	128 mM NaCl, 4.2 mM $MgSO_4$, 3.5 mM $NaHPO_4$, 3.6 mM KCl, 1.9 mM $NaHCO_3$, and 2.7 mM $CaCl_2$	Wolenski and Hart (1987)
Ginsburg fish ringer	111 mM NaCl, 3.6 mM KCl, 2.7 mM $CaCl_2$, and 1.9 mM $NaHCO_3$	Hart and Fluck (1995)

[a]FBSS is particularly useful because eggs in this medium can be maintained for 15–25 min without autoactivation (Becker and Hart, 1996). For longer incubation eggs can be kept in coho salmon ovarian fluid (Westerfield, 1996).

4. Rinse fish well after Tricaine treatment, remove excess of water with a paper towel.

5. Transfer females to a 6-cm Petri dish. Females should not bring any water into the dish that could activate the eggs.

6. Squeeze females by applying gentle pressure with dampened finger tips to both sides of the abdomen.

7. Carefully remove the eggs from the body of the female with a very fine (2.5 mm) paintbrush. The paintbrush should be damp, with no excess water. Use only good egg clutches (these tend to be yellowish and stick to side of female) with more than 150 good eggs.

8. Transfer female to a recovery tank.

9. Eggs can be fertilized immediately or can be kept several minutes in one of the buffers given in Table IV.

10. For sperm collection, rinse male fish well after tricaine treatment, dry as above.

11. Place the male, ventral side up, in a slit cut in a polyfoam sponge. Blot excess of water from anal fin area once fish is positioned on the sponge.

12. Obtain sperm directly from the genital pore, by applying gentle pressure with Millipore forceps and collect the sperm into a glass capillary. Volumes may range from 0.25 to 2.0 μl for each fish.

13. Place the sperm in a 70 μl drop of I-buffer (116 mM NaCl, 23 mM KCl, 6 mM $CaCl_2$, 2 mM $MgSO_4$, 29 mM $NaHCO_3$, and 0.5% fructose pH 7.2; filtered through 0.22 μm filter) in a Petri dish next to the eggs to be fertilized. Combine the egg clutch with sperm sample and incubate for 15–30 s.

14. Activate fertilization upon addition of 750 μl of 0.5% fructose in egg water.

15. After 2 min add more 0.5% fructose to fill the dish. The success of fertilization can be evaluated by examining eggs for cleavage furrow formation between 40 and 50 min postinsemination.

(Alternative protocol for *in vitro* fertilization can be found in Hart and Fluck (1995).)

Eggs, unfertilized or fertilized, can be either homogenized and prepared for protein gel electrophoresis, or fixed and processed for cytoskeletal protein localization (see below). Other buffers used for egg storage are given in Table IV.

3. Protocol 12.3

Collecting Embryos

Normal procedure for collecting embryos from natural spawnings can be used according to Westerfield (1996).

B. Preparation of Egg Extracts (Becker and Hart, 1996)

Transfer eggs ($n = 200$) or ovary tissue in FBSS (Table IV) to a 1 ml micro-homogenizer (Wheaton Instruments, Millville, NJ) chilled on ice. All subsequent manipulation should be carried out at 4 °C unless otherwise indicated.

1. Protocol 12.4

Total Actin Preparation (Becker and Hart, 1996)

Homogenize eggs or ovaries in 500 μl of lysis buffer A (100 mM HEPES, pH 7.4, 50 mM NaCl, 20 mM KCl, 5 mM MgCl$_2$, 5 mM EGTA, 10 mM benzamidine, 50 μg/ml aprotinin, 1 mM dithiothreitol, 10 μg/ml leupeptin, and 1 mM phenyl-methylsulfonyl fluoride) and then combine the homogenate directly with hot SDS-PAGE sample buffer.

2. Protocol 12.5

Isolation of Actin Soluble and Insoluble Fraction (Becker and Hart, 1996)

Homogenize eggs in lysis buffer A (see above Section VII.B.3) containing 0.5% Triton X-100 and centrifuge at 12,000 × g for 20 min Combine pellet (Triton insoluble) and supernatant (Triton soluble) fractions with SDS-PAGE sample buffer, boil for 1–2 min and use immediately or store at −70 °C.

3. Protocol 12.6

Myosin Preparation (Becker and Hart, 1996)

Homogenize eggs in buffer B (50 mM HEPES, pH 7.4, 150 mM NaCl, 1 M KCl, 1 mM $MgCl_2$, 5 mM EGTA, 2.5 mM *p*-tosyl-L-arginine methyl ester, 2.5 mM benzamidine, 50 μg/ml aprotinin, 1 mM dithiothreitol, 25 μg/ml leupeptin, 2 mM ATP, and 1% Triton X-100) followed by centrifugation at 86,000 $\times$ g for 30 min using a TLA-45 fixed angle rotor in a Beckman TL-100 tabletop ultracentrifuge (Beckman Instruments Co., Palo Alto, CA). Resuspended the pellet in buffer B. Recentrifuge supernatant and pellet fractions at 86,000 $\times$ g for 30 min. Process the final pellet and supernatants as described above (see Section VII.B.2).

4. Protocol 12.7

SDS-polyacrylamide Gel Electrophoresis and Immunoblotting of Egg Extracts (Becker and Hart, 1996)

1. Resolve egg or ovary sample (approximately 60 μg protein per well) in a standard 7.5% or 10% SDS-polyacrylamide minigel (e.g., Hoefer Scientific Instruments, San Francisco, CA), with molecular weight standard proteins.

2. Stain gel with Coomassie brilliant blue R-250 or blot to nitrocellulose paper (NC, BioRad) using the Genie Transfer apparatus (Idea Scientific Co., Minneapolis, MN) containing 25 mM Tris, pH 8.3, 192 mM glycine, 20% methanol, and 0.05% SDS (Towbin *et al.*, 1979).

3. Block the NC sheet overnight in TTBS (20 mM Tris, pH 7.5, 150 mM NaCl, and 0.1% Tween-20) containing 5% nonfat dry milk and 3% bovine serum albumin (BSA).

4. Wash the NC sheet several times in TTBS.

5. Incubate the NC sheet for either 4 h with antichicken monoclonal actin antibody (Amersham Life Science, Arlington Heights, Il) or for 12–18 h with human antiplatelet myosin-II antibody (Biomedical Technologies Inc., Stoughton, MA).

6. Wash blot in TTBS and TBS (20 mM Tris, pH 7.5, 150 mM NaCl), and incubate for 2 h in 1:2000 dilution of alkaline phosphatase conjugated with either goat antirabbit (myosin) or goat antimouse (actin) secondary antibodies (Hyclone Laboratories, Logan, UT).

5. Protocol 12.8

Filamentous Actin Assay (Becker and Hart, 1999)

To compare the F-actin content of unactivated and activated eggs, Becker and Hart (1999) used a fluorometric assay based on the affinity of rhodamine-phalloidin (RhPh) to bind with actin filaments (Greenberg *et al.*, 1991; Howard and Oresajo, 1985).

1. Fix 60 unactivated eggs and 60 eggs activated for 2 or 5 min for 6 h in 5.0% formaldehyde in actin stabilizing buffer (ASB).

2. Incubate eggs for 3 h in 150 mM glycine-ASB buffer and wash intensively with ASB.

3. Label with 165 nM RhPh (Molecular Probes, Eugene, OR). (Incubate control eggs in unlabeled phalloidin (20 mg/ml) for 50 min prior to RhPh staining.) Wash several times in ASB.

4. Extract samples in the dark with methanol (1 ml) for 36 h at 10 °C.

5. Measure fluorescence of the extract in a spectrofluorometer using an excitation wavelength of 565 nm and an emission wavelength of 580 nm.

6. Protocol 12.9

Microtubule Abundance Assay (Hsu et al., 2006)

A very interesting finding of the role of the cholesterol derivative pregnenolone in microtubule stability during the epiboly process was supported by an assay testing the amount of polymerized tubulin in manipulated embryos. This assay was based on a commercially available microtubules/tubulin *in vivo* assay kit (Cytoskeleton, Inc. Denver). Hsu *et al.* (2006) also describe a method for synthesis of fluorescein-conjugated pregnenolone and colocalization of this reagent with microtubules. Based on the same principle of ultracentrifugation of polymerized tubulin, microtubule associated proteins can be isolated as follows.

1. Homogenize 20 embryos per experimental condition in 200 μl lysis buffer with 1 mM GTP, 10 mM ATP, and 20 μl proteinase inhibitor cocktail.

2. Transfer 180 μl-aliquot into prewarmed at 30 °C tube and centrifuge at 100,000 × g, 30 °C for 30 min.

3. Dissolve pellet in 180 μl of cold 200 μM CaCl$_2$.

4. Use 15 μl of the pellet sample and 2.5 μl of the total lysate for western blot analysis.

5. The pellet can be directly stained with anti α-tubulin antibody and imaged, after spotting onto glass slide.

C. Whole Mount Cytoskeleton Staining

1. Protocol 12.10

Actin and Myosin Staining in Zebrafish Eggs (Hart and Fluck, 1995)

1. Fix eggs in cold 3.7% formaldehyde in ASB (1 mM PMSF, 10 mM EGTA, 10 mM PIPES pH 7.3, 5 mM MgCl$_2$, 100 mM KCl) for 4–8 h.

2. Rinse in ASB with several solution changes and dechorionate manually.

3. Transfer to cold quenching buffer (150 mM glycine in ASB) for 30 min.

4. Rinse briefly in ASB.

5. Transfer to cold 0.5% Triton X-100 in ASB for 30 min.

6. Rinse several times in ASB and incubate in cold blocking buffer (1–3% BSA or 1% BSA and 2% Normal Goat Serum).

7. Incubate in cold primary antibody for about 4 h (actin) or 12–18 h (myosin). For actin, dilute antichicken monoclonal antibody (Amersham Life Science Inc.) 1:50 or 1:100 in ASB. For myosin, dilute antiplatelet myosin-II antibody (Biomedical Technologies Inc.) 1:50 in ASB.

8. Rinse thoroughly with several changes of cold ASB.

9. Incubate in FITC-conjugated secondary antibody diluted 1:25 or 1:50 in ASB, at 4 °C for 2 h.

10. Control eggs should be incubated in secondary antibody alone or with primary antibody preabsorbed with antigen followed with secondary antibody.

11. Mount whole eggs on acid-cleaned slides using 2% *n*-propyl gallate in 50% glycerol. A thin layer of Vaseline can be applied to the edges of the coverglass before mounting.

2. Protocol 12.11

F-Actin Staining with Rhodamine-Phalloidin (Becker and Hart, 1996)

1. Block overnight fixed and dechorionated eggs (Section VII.C.1 steps 1 and 2) in cold 1% fetal calf serum or 5% nonfat dry milk plus 3% BSA in ASB, and then wash extensively in buffer.

2. Incubate in the dark with 66 nM rhodamine-phalloidin in ASB for 45–60 min.

3. Wash the eggs in ASB and mount on slides as Section VII.C.1 step 11.

(Similar protocol can be used for monomeric G-actin staining with 100 μg/ml fluorescein-conjugated DNAse I in single- and double-label with RhPh.)

3. Protocol 12.12

Microtubule Staining in Zebrafish Embryos (Gard, 1991) with later modification (Solnica-Krezel and Driever, 1994)

1. Fix embryos in microtubule assembly buffer (MAB; 80 mM KPIPES, pH 6.5, 5 mM EGTA, 1 mM $MgCl_2$) containing 3.7% formaldehyde, 0.25% glutaraldehyde, 0.5 μM taxol, and 0.2% Triton X-100 for 2–4 h.

2. Incubate for postfixation in absolute methanol at -20 °C overnight. (PIPES pH 6.5, magnesium, and taxol improve stabilization of microtubules. Microtubules can be preserved in embryos in the absence of taxol).

3. Rehydrate in PBS and incubate for 6–16 h at RT in PBS containing 100 mM $NaBH_4$. Rinse embryos extensively in Tris-buffered saline (TBS: 155 mM NaCl, 10 mM Tris-Cl pH 7.4, 0.1% NP-40).

4. Block embryos for 0.5 h at RT in TBS containing 2% BSA and 5% normal goat serum.

5. Rehydrate embryos and incubate with primary antibody (1:500–1:2000) (KMX-1, Boerhinger Mannheim) in TBS containing 2% BSA and 5% normal goat serum for 16–24 h at 4 °C.

6. Wash with several changes of TBS for 24 h with gentle agitation at 4 °C (or alternatively wash with TBS 1 × 5 min and 3 × 30 min at RT).

7. Incubate with secondary antibodies in TBS/BSA/normal goat serum for 16–24 h. (As a secondary, antimouse IgG antibodies conjugated to Texas Red from Jackson Immunoresearch Laboratories can be used. Alternatively, use secondary antibody conjugated with biotin, followed by avidin/horse radish peroxidase-biotin complexes. Follow the manufacturer's manual (Elite Vectastain ABC staining kit from Vector Laboratories.). Reduce secondary antibody concentration to 1/2500 and that of the AB reagents to 10 μl/ml of buffer and finally develop with diaminobenzidine and H_2O_2.

8. Wash with TBS for 24–36 h as above.

9. Dehydrate with several changes of absolute methanol (3 × 30 min). Embryos can be cleared in benzyl benzoate-benzyl alcohol (2:1 v/v). (Alternatively embryos washed in TBS can be cleared in 100% glycerol.)

10. Mount in clearing solution using standard microscope slides with no.1 coverslips (Westerfield, 1996) and seal with nail polish (or Ladd-O-Lac from LADD Inc.).

A comparison of different fixation techniques by McMenamin *et al.* (2003) indicated satisfactory staining with the use of standard 4% PFA fixative (4% paraformaldehyde in PBS). At the same time, these authors explored alternative methods of fixation that may be useful at later stages of development. No difference in the staining was observed in embryos that were not treated with methanol.

A modified version of McMenamin *et al.*'s (2003) method was used by Redd *et al.* (2006) to stain microtubules in macrophages of 22 h postfertilization embryos.

1. Fix dechorionated embryos in 3.7% formaldehyde, 0.25% glutaraldehyde, 1 mM $MgCl_2$, 5 mM EGTA, 0.2% Triton X-100 for 30 min at room temperature.

2. Wash embryos several times with PBS, 0.5% Triton X-100 and quench with 1 mg/ml $Na_2B_4O_7$ for 20 min at room temperature.

3. Rinse embryos intensively in PBS, 0.5% Triton X-100 and block in 10% goat serum and 1% DMSO in PBS for1 h.

4. Incubate embryos in anti-α-tubulin antibody (Sigma) diluted 1:200 in blocking solution for 12 h at 4 °C.

5. Wash in the block solution 4 times for 20 min at RT and incubate with secondary antibody diluted in the block solution for 12 h at 4 °C.

6. Wash in PBS, 0.5% Triton X-100 4 times for 30 min at RT and transfer to 70% glycerol in PBS for observation.

4. Protocol 12.13

Bead/Nile Red Injection (Jesuthasan and Strahle, 1997)

1. Dilute a suspension of 0.2 μm fluorescent polystyrene beads (1μ, #17151, Polysciences, Warrington, PA) in 23 μl water, together with 1 μl of a Nile Red (Molecular Probes, Eugene, OR) stock solution (1 μg/ml in acetone).

2. Backfill the suspension into the capillaries (1.0 mm OD with filament) and microinject into the yolk cell of dechorionated embryos using a standard gas pressure injector.

3. To record the position of beads/Nile Red mount embryos in methyl cellulose, image with confocal microscope using 10× objective.

(As a result of the fragility and distorted morphology of early embryos, less than 15 min old after dechorionation, beads could not be targeted to the vegetal pole of young embryos.)

Acknowledgments

We thank Diane Sepich, Alex Schier, Kim Fekany, and Florence Marlow for comments on the manuscript. Work in the LSK laboratory is supported by NIH RO1 grants GM55101, GM77770, and the Human Frontiers in Science Program and in JT laboratory by NIH R01 grant DE016678.

References

Abraham, V. C., Gupta, S., and Fluck, R. A. (1993). Ooplasmic segregation in the medaka *Oryzias latipes* egg. *Biol. Bull.* **184,** 115–124.

Adams, R. J., and Bray, D. (1983). Rapid transport of foreign particles microinjected into crab axons. *Nature* **303,** 718–720.

Aronson, J., and Inoue, S. (1970). Reversal by light of the action of N-methyl N-desacetyl colchicine on mitosis. *J. Cell Biol.* **45,** 470–477.

Beams, H. W., Kessel, R. G., Shih, C. Y., and Tung, H. N. (1985). Scanning electron microscope studies on blastodisc formation in the zebrafish, *Brachydanio rerio. J. Morph.* **184,** 41–50.

Becker, K. A., and Hart, N. H. (1996). The cortical actin cytoskeleton of unactivated zebrafish eggs: Spatial organization and distribution of filamentous actin, nonfilamentous actin, and myosin-II. *Mol. Reprod. Dev.* **43,** 536–547.

Becker, K. A., and Hart, N. H. (1999). Reorganization of filamentous actin and myosin-II in zebrafish eggs correlates temporally and spatially with cortical granule exocytosis. *J. Cell Sci.* **112**(Pt 1), 97–110.

Beckerle, M. C. (1984). Microinjected fluorescent polystyrene beads exhibit saltatory motion in tissue culture cells. *J. Cell Biol.* **98,** 2126–2132.

Brummett, A. R., and Dumont, J. N. (1979). Initial stages of sperm penetration into the egg of *Fundulus heteroclitus. J. Exp. Zool.* **210,** 417–434.

Brummett, A. R., and Dumont, J. N. (1981). Cortical vesicle breakdown in fertilized eggs of *Fundulus heteroclitus. J. Exp. Zool.* **216,** 63–79.

Bunton, T. E. (1993). The immunocytochemistry of cytokeratin in fish tissues. *Vet. Pathol.* **30,** 418–425.

Bunton, T. E. (1995). Expression of actin and desmin in experimentally induced hepatic lesions and neoplasms from medaka (*Oryzias latipes*). *Carcinogenesis* **16,** 1059–1063.

Cerda, J., Conrad, M., Markl, J., Brand, M., and Herrmann, H. (1998). Zebrafish vimentin: Molecular characterization, assembly properties and developmental expression. *Eur. J. Cell Biol.* **77**, 175–187.

Chandler, D. E. (1991). Multiple intracellular signals coordinate structural dynamics in the sea urchin egg cortex at fertilization. *J. Electron Microsc. Tech.* **17**, 266–293.

Chow, R. L., and Elinson, R. P. (1993). Local alteration of cortical actin in *Xenopus* eggs by the fertilizing sperm. *Mol. Reprod. Dev.* **35**, 69–75.

Cooper, J. A. (1987). Effects of cytochalasin and phalloidin on actin. *J. Cell Biol.* **105**, 1473–1478.

Conrad, M., Lemb, K., Schubert, T.,and Markl, J. (1998). Biochemical identification and tissue-specific expression patterns of keratins in the zebrafish Danio rerio. *Cell Tissue Res.* **293**, 195–205.

Costa, M. L., Escaleira, R. C., Rodrigues, V. B., Manasfi, M.,and Mermelstein, C. S. (2002). Some distinctive features of zebrafish myogenesis based on unexpected distributions of the muscle cytoskeletal proteins actin, myosin, desmin, alpha-actinin, troponin and titin. *Mech. Dev.* **116**, 95–104.

Coulombe, P. A., and Wong, P. (2004). Cytoplasmic intermediate filaments revealed as dynamic and multipurpose scaffolds. *Nat. Cell Biol.* **6**, 699–706.

Donovan, M. J., and Hart, N. H. (1986). Cortical granule exocytosis is coupled with membrane retrieval in the egg of *Brachydanio*. *J. Exp. Zool.* **237**, 391–405.

Driever, W., Solnica-Krezel, L., Schier, A. F., Neuhauss, S. C., Malicki, J., Stemple, D. L., Stainier, D. Y., Zwartkruis, F., Abdelilah, S., Rangini, Z., Belak, J., and Boggs, C. (1996). A genetic screen for mutations affecting embryogenesis in zebrafish. *Development* **123**, 37–46.

Garcia, D. M., Bauer, H., Dietz, T., Schubert, T., Markl, J.,and Schaffeld, M. (2005). Identification of keratins and analysis of their expression in carp and goldfish: Comparison with the zebrafish and trout keratin catalog. *Cell Tissue Res.* **322**, 245–256.

Gard, D. L. (1991). Organization, nucleation, and acetylation of microtubules in *Xenopus laevis* oocytes: A study by confocal immunofluorescence microscopy. *Dev. Biol.* **143**, 346–362.

Greenberg, S., el Khoury, J., di Virgilio, F., Kaplan, E. M.,and Silverstein, S. C. (1991). Ca(2+)-independent F-actin assembly and disassembly during Fc receptor-mediated phagocytosis in mouse macrophages. *J. Cell Biol.* **113**, 757–767.

Hart, N. H., and Becker, K. A. (1994). Ooplasmic segregation in the zebrafish egg. *Mol. Biol. Cell.* **5**, 100a.

Hart, N. H., Becker, K. A., and Wolenski, J. S. (1992). The sperm entry site during fertilization of the zebrafish egg: Localization of actin. *Mol. Reprod. Dev.* **32**, 217–228.

Hart, N. H., and Donovan, M. (1983). Fine structure of the chorion and site of sperm entry in the egg of *Brachydanio rerio*. *J. Exp. Zool* **227**, 277–296.

Hart, N. H., and Fluck, R. A. (1995). Cytoskeleton in teleost eggs and early embryos: Contributions to cytoarchitecture and motile events. *Curr Top Dev Biol* **31**, 343–381.

Hart, N. H., and Yu, S. F. (1980). Cortical granule exocytosis and cell surface reorganization in eggs of *Brachydanio*. *J Exp Zool* **213**, 137–159.

Howard, T. H., and Oresajo, C. O. (1985). A method for quantifying F-actin in chemotactic peptide activated neutrophils: Study of the effect of tBOC peptide. *Cell Motil.* **5**, 545–557.

Hsu, H. J., Liang, M. R., Chen, C. T.,and Chung, B. C. (2006). Pregnenolone stabilizes microtubules and promotes zebrafish embryonic cell movement. *Nature* **439**, 480–483.

Imboden, M., Goblet, C., Korn, H., and Vriz, S. (1997). Cytokeratin 8 is a suitable epidermal marker during zebrafish development. *C R Acad. Sci. III* **320**, 689–700.

Ivanenkov, V. V., Minin, A. A., Meshcheryakov, V. N., and Martynova, L. E. (1987). The effect of local cortical microfilament disorganization on ooplasmic segregation in the loach (*Misgurnus fossilis*) egg. *Cell Differ.* **22**, 19–28.

Jesuthasan, S., and Strahle, U. (1997). Dynamic microtubules and specification of the zebrafish embryonic axis. *Curr. Biol.* **7**, 31–42.

Katow, H. (1983). Obstruction of blastodisk formation by cytochalasin B in the zebrafish, *Brachydanio rerio*. *Dev. Growth Differ.* **25**, 477–484.

Kessel, R. G., Beams, H. W., and Tung, H. N. (1984). Relationships between annulate lamellae and filament bundles in oocytes of the zebrafish, *Brachydanio rerio*. *Cell Tissue Res.* **236**, 725–727.

Kimmel, C. B., Ballard, W. W., Kimmel, S. R., Ullmann, B., and Schilling, T. F. (1995). Stages of embryonic development of the zebrafish. *Dev. Dyn.* **203,** 253–310.

Kimmel, C. B. Warga, R. M., and Schilling, T. F (1990). Origin and organization of the zebrafish fate map. *Development* **108,** 581–594.

Kobayashi, W., and Yamamoto, T. S. (1981). Fine structure of the micropylar apparatus of sperm entry in the chum salmon egg. *J. Exp. Zool.* **243,** 311–322.

Krushna Padhi, B., Akimenko, M. A., and Ekker, M. (2006). Independent expansion of the keratin gene family in teleostean fish and mammals: An insight from phylogenetic analysis and radiation hybrid mapping of keratin genes in zebrafish. *Gene* **368,** 37–45.

Lauffenburger, D. A., and Horwitz, A. F. (1996). Cell migration: A physically integrated molecular process. *Cell* **84,** 359–369.

Lee, E., Chang, B. S., Mun, G. H., Chung, Y. H., Kim, J.,and Shin, D. H. (2005). An ultramicroscopic study on the distribution of Muller cell processes in the outer retinal layers of the zebrafish. *Ann. Anat.* **187,** 43–50.

Loh, S. H., Chan, W. T., Gong, Z., Lim, T. M.,and Chua, K. L. (2000). Characterization of a zebrafish (Danio rerio) desmin cDNA: An early molecular marker of myogenesis. *Differentiation* **65,** 247–254.

Markl, J., and Franke, W. W. (1988). Localization of cytokeratins in tissues of the rainbow trout: Fundamental differences in expression pattern between fish and higher vertebrates. *Differentiation* **39,** 97–122.

McMenamin, S., Reinsch, S., and Conway, G. (2003). Direct comparison of common fixation methods for preservation of microtubules in zebrafish embryos. *BioTechniques* **34,** 468–472.

Merriam, R. W., and Clark, T. G. (1978). Actin in *Xenopus* oocytes. II. Intracellular distribution and polymerizability. *J. Cell Biol.* **77,** 439–447.

Mitchison, T. J., and Cramer, L. P. (1996). Actin-based cell motility and cell locomotion. *Cell* **84,** 371–379.

Mizuno, T., Yamaha, E., and Yamazaki, F. (1997). Localized determinant in the early cleavage of the goldfish, *Carassius auratus.* *Dev. Genes Evol.* **206,** 389–396.

Nielsen, A. L., and Jorgensen, A. L. (2003). Structural and functional characterization of the zebrafish gene for glial fibrillary acidic protein, GFAP. *Gene* **310,** 123–132.

Pathak, N., Obara, T., Mangos, S., Liu, Y.,and Drummond, I. A. (2007). The zebrafish *fleer* gene encodes an essential regulator of cilia tubulin polyglutamylation. *Mol. Biol. Cell* **18,** 4353–4364.

Piperno, G., and Fuller, M. T. (1985) Monoclonal antibodies specific for an acetylated form of alpha-tubulin recognize the antigen in cilia and flagella from a variety of organisms. *J. Cell Biol.* **101,** 2085–2094.

Redd, M. J., Kelly, G., Dunn, G., Way, M.,and Martin, P. (2006). Imaging macrophage chemotaxis *in vivo*: Studies of microtubule function in zebrafish wound inflammation. *Cell Motil. Cytoskeleton* **63,** 415–422.

Roosen-Runge, E. (1938). On the early development—Bipolar differentiation and cleavage—of the zebra fish, *Brachydanio rerio.* *Biol. Bull.* **75,** 119–133.

Rowning, B. A., Wells, J., Wu, M., Gerhart, J. C., Moon, R. T., and Larabell, C. A. (1997). Microtubule-mediated transport of organelles and localization of beta- catenin to the future dorsal side of *Xenopus* eggs. *Proc. Natl. Acad. Sci. USA* **94,** 1224–1229.

Schaffeld, M., Knappe, M., Hunzinger, C.,and Markl, J. (2003). cDNA sequences of the authentic keratins 8 and 18 in zebrafish. *Differentiation* **71,** 73–82.

Schroeder, M. M., and Gard, D. L. (1992). Organization and regulation of cortical microtubules during the first cell cycle of *Xenopus* eggs. *Development* **114,** 699–709.

Solnica-Krezel, L., and Driever, W. (1994). Microtubule arrays of the zebrafish yolk cell: Organization and function during epiboly. *Development* **120,** 2443–2455.

Solnica-Krezel, L., Stemple, D. L., and Driever, W. (1995). Transparent things: Cell fates and cell movements during early embryogenesis of zebrafish. *Bioessays* **17,** 931–939.

Spudich, A., Wrenn, J. T., and Wessells, N. K. (1988). Unfertilized sea urchin eggs contain a discrete cortical shell of actin that is subdivided into two organizational states. *Cell Motil. Cytoskeleton* **9,** 85–96.

Strahle, U., and Jesuthasan, S. (1993). Ultraviolet irradiation impairs epiboly in zebrafish embryos: Evidence for a microtubule-dependent mechanism of epiboly. *Development* **119,** 909–919.

Takeda, S., Funakoshi, T., and Hirokawa, N. (1995). Tubulin dynamics in neuronal axons of living zebrafish embryos. *Neuron* **14,** 1257–1264.

Towbin, H., Staehelin, T., and Gordon, J. (1979). Electrophoretic transfer of proteins from polyacrylamide gels to nitrocellulose sheets: Procedure and some applications. *Proc. Natl. Acad. Sci. USA* **76,** 4350–4354.

Trimble, L. M., and Fluck, R. A. (1995). Indicators of the dorsoventral axis in medaka (*Oryzias latipes*) zygotes. *Fish Biol. J. Medaka* **7,** 37–41.

Trinkaus, J. P. (1984). "Cells into Organs," 2nd edn., Prentice-Hall, Englewood Cliffs.

Webb, A. E., Driever, W.,and Kimelman, D. (2008). *psoriasis* regulates epidermal development in zebrafish. *Dev. Dyn.* **237,** 1153–1164.

Webb, T. A., and Fluck, R. A. (1995). The spatiotemporal pattern of microtubules in parthenogenetically activated medaka fish eggs *Oryzias latipes*. *J. Pa. Acad. Sci.* **68,** 197.

Webb, T. A., Kowalski, W. J., and Fluck, R. A. (1995). Microtubule-based movements during ooplasmic segregation in the medaka fish egg (*Oryzias latipes*). *Biol. Bull.* **188,** 146–156.

Westerfield, M. (1996) "The Zebrafish Book," 3rd edn., University of Oregon Press, Eugene.

Wolenski, J. S., and Hart, N. H. (1987). Scanning electron microscope studies of sperm incorporation into the zebrafish (*Brachydanio*) egg. *J. Exp. Zool.* **243,** 259–273.

Wolenski, J. S., and Hart, N. H. (1988). Effects of cytochalasins B and D on the fertilization of zebrafish (*Brachydanio*) eggs. *J. Exp. Zool.* **246,** 202–215.

Zupanc, G. K., Hinsch, K.,and Gage, F. H. (2005). Proliferation, migration, neuronal differentiation, and long-term survival of new cells in the adult zebrafish brain. *J. Comp. Neurol.* **488,** 290–319.

CHAPTER 9

Analyzing Axon Guidance in the Zebrafish Retinotectal System

Lara D. Hutson,* Douglas S. Campbell,† and Chi–Bin Chien‡

*Department of Biology
Williams College
Williamstown, Massachusetts 01267

†Laboratory for Developmental Gene Regulation
RIKEN Brain Science Institute
Wako, Saitama 351–0198

‡Department of Neurobiology and Anatomy
University of Utah Medical Center
Salt Lake City, Utah 84132

I. Introduction

The developing visual system has been the subject of intensive study in many model organisms except *Caenorhabditis elegans*, which unfortunately lacks eyes. There are several reasons for its popularity. It is experimentally accessible, its normal anatomy is understood extremely well, and its function is understood equally well. Thus one can observe how the visual system develops, use perturbations to test the mechanisms of its development, and link alterations in development to changes in mature anatomy and, ultimately, to visual behavior. Many laboratories, including ours, have been led by these advantages to study the projections of retinal ganglion cells (RGCs) to the optic tectum. The retinotectal projection in the zebrafish is an especially good system for imaging because of the larva's transparency. In addition, it is a useful system for genetic analysis because of the large number of known mutants that affect its development (reviewed in Culverwell and Karlstrom (2002) and Hutson and Chien (2002b)). Here we describe the strategies that have been used to observe and to perturb retinotectal development in the zebrafish.

Retinal axons originate from the RGCs, which are the primary cell type in the innermost cellular layer of the retina. The axons first navigate radially to the optic nerve head near the center of the retina and exit the eye. This occurs at 32 hpf in zebrafish (Burrill and Easter, 1994). The retinal axons then cross the midline and enter the contralateral optic tract, coursing dorsalward in the diencephalon until they reach the pretectal nuclei and the optic tectum at 48 hpf. The tectum, which is located at the dorsal roof of the midbrain, is the primary visual center in nonmammalian vertebrates and is likely primarily responsible for high spatial-resolution vision. Axons find their topographic targets on the tectum, then arborize and begin to form synapses. The axonal arbors are refined over time via activity-dependent mechanisms (Schmidt, 2004). The first retinal axons reach the tectum before visual function begins, but by about 3 days postfertilization (dpf), larvae begin to show visually evoked behavior, and by 5 dpf, behaviors such as the optokinetic response are robustly displayed (Easter and Nicola, 1996; Neuhauss, 2003).

In the next section, we begin with a brief overview of the genetic control of retinal axon guidance. We then describe methods for labeling and observing retinal axons. We follow with methods for perturbing retinotectal development. Finally, we end with a brief discussion of methods likely to be important in the future. The methods described here for the retinotectal system are also useful, either directly or with some modifications, for studying the development of other axon tracts.

II. Retinotectal Mutants

Among the many zebrafish mutants affecting specific developmental processes that were isolated in large-scale genetic screens (Neumann, 2002), at least 27 are known to affect development of the retinotectal projection (Table I; Haffter *et al.*, 1996; reviewed in Culverwell and Karlstrom (2002) and Hutson and Chien (2002b)). Most of these were discovered in a landmark screen carried out in Friedrich Bonhoeffer's laboratory, which used as an assay the direct labeling of retinal axons with lipophilic dyes (Baier *et al.*, 1996; Karlstrom *et al.*, 1996; Trowe *et al.*, 1996). Screens for visual behavioral defects are a promising avenue for

Table I
Retinotectal Pathfinding Mutants

Mutant	Region in which pathfinding affected	Gene	Brain defect?	References
acerebellar (ace)	Chiasm, anterior projections, optic tract, topography	fgf8	Yes	Picker et al. (1999) and Shanmugalingam et al. (2000)
astray (ast)	Chiasm, anterior projections, optic tract	robo2	No	Fricke et al. (2001) and Karlstrom et al. (1996)
bashful (bal)	Retinal exit, anterior projections	?	Yes	Karlstrom et al. (1996)
belladonna (bel)	Midline crossing	?	Yes	Karlstrom et al. (1996)
blowout (blw)	Midline crossing	?	Yes	Karlstrom et al. (1996), Nakano et al.(2004)
blumenkohl (blu)	Expanded terminations	?	No	Trowe et al.(1996)
boxer (box)	Tract sorting, crossing in posterior commissure	?	No	Karlstrom et al. (1996), Trowe et al.(1996)
chameleon (con)	Retinal exit, midline crossing	dispatched 1	Yes	Karlstrom et al. (1996), Nakano et al. (2004)
cyclops (cyc)	Midline crossing	Nodal related-2 (ndr2)	Yes	Karlstrom et al. (1996), Rebagliati et al. (1998), Sampath et al. (1998)
dackel (dak)	Tract sorting	?	No	Karlstrom et al. (1996), Trowe et al. (1996)
detour (dtr)	Midline crossing	gli1	Yes	Karlstrom et al. (1996, 2003)
esrom (esr)	Midline crossing, termination	?	No	Karlstrom et al. (1996) and Trowe et al. (1996)
gnarled (gna)	Tectal entry, tectalmisrouting	?	No	Trowe et al. (1996) and Wagle et al. (2004)
grumpy (gup)	Anterior projections, midline crossing	laminin β1	Yes	Karlstrom et al. (1996) and Parsons et al. (2002)
iguana (igu)	Midline crossing	?	Yes	Karlstrom et al. (1996)
macho (mao)	Expanded terminations	?	No	Gnuegge et al. (2001) and Trowe et al. (1996)
nevermind (nev)	Tract sorting, D-V topography	?	No	Trowe et al. (1996)
no isthmus (noi)	Chiasm, anterior projections, tectal bypass	pax2a	Yes	Brand et al. (1996), Macdonald et al. (1997) and Trowe et al. (1996)

(continues)

Table I (*continued*)

Mutant	Region in which pathfinding affected	Gene	Brain defect?	References
parachute (*pac*)	Ipsilateral projections, entering chiasm area	*N-cadherin*	Yes	Lele *et al.* (2002) and Masai *et al.* (2003)
pinscher (*pic*)	Tract sorting, crossing in posterior commissure	?	No	Karlstrom *et al.* (1996) and Trowe *et al.* (1996)
sleepy (*sly*)	Anterior projections, midline crossing	*laminin γ1*	Yes	Karlstrom *et al.* (1996) and Parsons *et al.* (2002)
smooth muscle omitted (*smu*)	Midline crossing	*smoothened* (*smo*)	Yes	Chen *et al.* (2001) and Varga *et al.* (2001)
sonic-you (*syu*)	Retinal exit, midline crossing	*sonic hedgehog* (*shh*)	Yes	Brand *et al.* (1996) and Schauerte *et al.* (1998)
space cadet (*spc*)	Retinal exit, midline crossing	?	No	Karlstrom *et al.* (1996) and Lorent *et al.* (2001)
umleitung (*uml*)	Midline crossing	?	Yes	Karlstrom *et al.* (1996)
who cares (*woe*)	Tract sorting, D-V topography	?	No	Trowe *et al.* (1996)
you-too (*yot*)	Midline crossing	*gli2*	Yes	Karlstrom *et al.* (1996, 1999)

discovering additional mutants (reviewed in Baier (2000) and Neuhauss (2003)), although behavioral defects do not necessarily reflect axon guidance defects, nor vice versa (Neuhauss *et al.*, 1999).

In most of the retinotectal mutants, the axon guidance defect seems to be a secondary consequence of disrupted brain patterning that results in abnormally patterned axon guidance cues (Barresi and Karlstrom, personal communication, 2004; Culverwell and Karlstrom, 2002; Hutson and Chien, 2002b). In others, brain patterning appears to be normal, suggesting that the mutation affects retinal axon guidance more specifically. Of course, one of the most interesting questions is, What are the mutated genes? Eleven retinotectal mutants have been cloned and published as of 2004 (Table I), and most of the remaining have either been cloned or are in the process of being cloned (personal communication from several authors). To date, most of the cloned genes clearly affect brain patterning, either through the Hedgehog pathway [*syu*(*shh*), *smu*(*smo*), *dtr*(*gli1*), *yot*(*gli2*)] or other pathways [*cyc* (*ndr2*), *noi* (*pax2a*), *ace* (*fgf8*)]. The only clear "axon guidance gene" to emerge so far is *astray* (*ast*), which encodes the axon guidance receptor Robo2, a homolog of the *Drosophila* Roundabout receptor first found in *Drosophila* (Fricke *et al.*, 2001; Kidd *et al.*, 1998). RGC axons in *ast* mutants are presumably "blind" to members of the Slit family of axon guidance signals, three of which are expressed in or near the optic pathway (Hutson and Chien, 2002a; Hutson *et al.*, 2003). The lack of Slit-Robo signaling in the retinal axons causes defects in the formation of the optic chiasm and the optic tract. At the current pace, cloning of (nearly) all of the retinotectal mutants will be complete by 2006. This should give us an idea of a large part of the genetic toolkit used to build the visual system. This information will be important for understanding human visual development, since many of these genes have conserved functions in mammals. However, the identification of the genes is only half the battle. We will still need to understand how these genes control retinal axons. It is here that the experimental accessibility of the retinotectal system—the ease of labeling, observing, and perturbing retinal axons—will be especially important.

III. Labeling the Retinotectal System

A. Overview

The discovery of many mutants with specific retinotectal phenotypes (Karlstrom *et al.*, 1996; Trowe *et al.*, 1996) has provided the impetus to develop an array of techniques to visualize retinal axons, using antibody staining, lipophilic dyes, or DNA constructs encoding green fluorescent protein (GFP) or other fluorescent proteins. Each technique is best suited to a particular part of the pathway (inside the retina, near the chiasm, in the optic tract, or on the tectum) and a particular type of sample (few or many axons, live or fixed embryos). A rogue's gallery of labeled embryos is shown in Fig. 1.

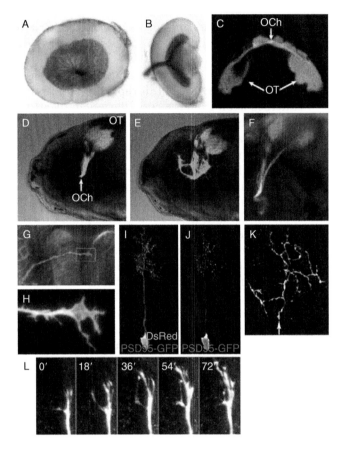

Fig. 1 A variety of methods for labeling the zebrafish retinotectal system. (A, B) 48 hpf zebrafish eye labeled with the zn-8 antibody, which recognizes the cell-surface molecule Alcam/Neurolin/DM-GRASP, expressed in all RGCs. Courtesy of Arminda Suli. (A) Lateral view, rostral to the right. (B) Dorsal view, rostral up. (C) Dorsal view of 5 dpf retinotectal projection labeled by intraocular injection (Method 1) with DiO (*green*, left eye) and DiI (*red*, right eye). The eyes have been dissected away. Rostral is up. (D, E) Lateral views of optic tracts labeled with DiO in 5 dpf wild type (D) and *astray* (E) larvae. The eye contralateral to the injected eye was removed in order to facilitate imaging. (F) Lateral view of optic tract and tectum of 6 dpf wild-type larva, after topographic labeling of the eye. DiI (*red*) was injected into dorsonasal retina, and DiO (*green*) was injected into ventrotemporal retina. Courtesy of Jeong-Soo Lee. (G, H) Leading growth cone labeled by intraocular injection of DiI (whole eye fill, Method 1). The *boxed area* in G is shown at higher magnification in H. (G) 20 × dry objective. (H) 60 × water-immersion objective. (I, J) A single wild-type tectal neuron at 100 hpf, labeled by coinjection of pBSKα-tubulin:GAL4 and UAS:PSD95-GFP; UAS:DsRedExpress-1 (Niell *et al.*, 2004; Method 3). The DsRed labels the cytoplasm of the soma and apical dendrite, while PSD95-GFP is preferentially localized to postsynaptic sites. 60×/1.2 water-immersion objective. (I) Merged image. (J) PSD95-GFP image alone. (K) A single retinal axon arbor on the tectum, labeled by injecting plasmid DNA containing the Brn3c promoter driving expression of membrane-targeted GFP using GAL4/UAS amplification (Method 3). (L) Time-lapse imaging of a cluster of zebrafish retinal growth cones, labeled with DiI, as they grow through the developing brain over a 72-min in period (Method 2). 60× water-immersion objective. (A, B) Obtained on an Olympus BX50WI compound microscope. (C–L) Captured on Olympus FV200 or FV300 confocal microscopes. OCh, optic chiasm; OT, optic tectum. (See Plate no. 3 in the Color Plate Section.)

B. Antibody Labeling

Two antibodies have been widely used to label retinal axons, using standard whole-mount antibody staining techniques. Anti-acetylated tubulin (Sigma St. Louis, MO) recognizes a form of tubulin preferentially found in stable micro-tubules, and thus labels all axons. This staining has been used to look at the earliest retinal axons crossing the optic chiasm (Karlstrom *et al.*, 1996), but at later stages it becomes hard to distinguish retinal axons amongst the rest of the axon scaffold. Two monoclonal antibodies from the panel generated by Trevarrow *et al.* (1990), zn-5 and zn-8 (Zebrafish International Resource Center, Developmental Studies Hybridoma Bank) both recognize Alcam (previously known as either neurolin or DM-GRASP) (Fashena and Westerfield, 1999), Zn-5 and zn-8 are likely duplicate isolates derived from the same hybridoma. Alcam is a cell-surface protein expressed by newly born RGCs and their axons (Kanki *et al.*, 1994; Laessing and Stuermer, 1996), which are added in a ring around the periphery of the retina. Since central RGCs have started to turn off Alcam expression by 48 hpf (Laessing and Stuermer, 1996), zn-5 or zn-8 staining is especially useful for showing RGC axons navigating within the retina to the optic nerve head (Fig. 1A and B).

C. Lipophilic Dye Labeling

Lipophilic dyes such as DiI, DiO, or DiD are wonderful tools for labeling axons (Honig and Hume, 1989). Structurally, these dyes consist of a fluorophore attached to two long aliphatic tails, which make them lipophilic so that they dissolve easily into cell membranes. Lipophilic dyes applied to live neurons become incorporated into intracellular membrane pools and then are transported both anterogradely and retrogradely by fast axonal transport. In addition, these dyes have the unique ability to label axons in fixed tissue. Since aldehyde fixation cross-links proteins, but leaves membranes untouched, lipophilic dye applied to a neuron's cell body can dissolve into the plasma membrane, then diffuse through the plane of the membrane, lighting up the entire axon. Because the zebrafish larva is so small, its axons are relatively short, and the required diffusion times for lipophilic dye labeling are only a few hours.

We use mainly DiO and DiI, but sometimes DiD, in several different ways—labeling either the entire retina or specific retinal regions in fixed larvae, or labeling live axons in living embryos. The original retinotectal screen analyzed fixed 5 dpf larvae, using a vibrating-needle injection apparatus (which works much like a tiny tattoo needle), to label the dorsonasal and ventrotemporal quadrants of one eye with DiO and DiI (described in detail by Baier *et al.* (1996)). We have replicated this apparatus and routinely use it for topographic injections (Fig. 1F).

Here we describe a simpler method that can be easily duplicated in any laboratory that uses a pressure injector to label eyes in fixed embryos or larvae (Method 1). Using DiI or DiO dissolved in chloroform, it is quite simple to label the entire retinotectal projection (whole eye fills) in order to analyze axon pathfinding (Fig. 1C–E) and growth cone morphology (Fig. 1G and H). This method can be modified slightly to give reliable topographic labeling of regions of the eye by injecting less dye solution and dissolving the dye in ethanol or dimethylformamide (which gives more localized labeling because the dye precipitates immediately at the point of injection, rather than spreading over the entire RGC layer). Individual ganglion cells can also be retrogradely labeled with DiI, allowing the visualization of their dendritic arbors (Mangrum *et al.*, 2002). We also describe a method for labeling RGCs in live embryos in order to observe axon growth using time-lapse microscopy (Method 2; Hutson and Chien, 2002a).

D. Using GFP or Other Fluorescent Proteins for Labeling

As useful as lipophilic dyes are (1) in our hands, it is difficult to label single axons using pressure injections (although others have had better success, e.g., Schmidt *et al.* (2004)). Single axon labeling is particularly important for visualizing arbors on the tectum, as neighboring arbors frequently overlap. (2) For large-scale experiments such as screens, lipophilic dye labeling is very time-consuming. These two problems can be overcome by transient expression and stable transgenic expression, respectively, of constructs encoding fluorescent proteins such as GFP. DNA-based labeling methods can also be used to express fusion proteins, for instance to visualize synaptogenesis (Niell *et al.*, 2004). Furthermore, DNA-based methods are not restricted to reporter proteins, but can also be used to express proteins that may perturb axonal development. In Method 3, discussed later in this chapter, we describe how we use these methods to label individual tectal neurons or retinal arbors (Fig. 1I-K).

Transient expression of DNA constructs is comparatively easy in the zebrafish. One can simply inject plasmid DNA (either supercoiled or linearized) for a construct with an enhancer and promoter (usually just abbreviated as "promoter") placed upstream of the desired coding sequence. DNA constructs are expressed mosaically, perhaps because the injected DNA is packaged into "micronuclei," which are inherited at random by a subset of cells in the embryo (Forbes *et al.*, 1983; Westerfield *et al.*, 1992). Several groups have carried out time-lapse imaging of neuronal development using transient GFP expression (Dynes and Ngai, 1998; Köster and Fraser, 2001a) or stable transgenic lines (Bak and Fraser, 2003). For a given experiment, it is necessary to optimize both the promoter and the coding sequence. By using a cell type-specific promoter, it is possible to get expression specifically in the tissue of interest. Promoters from at least four RGC-specific genes have been characterized in zebrafish: *nAChRβ3* (*PAR*) (Tokuoka *et al.*, 2002), *brn3c* (Roeser and Baier, personal communication, 2003), *isl3* (Pittman and Chien, unpublished data, 2003), and *ath5* (Masai, personal communication, 2004). Each of these drives expression starting at a slightly different time and each also drives in a different set of cells outside the retina.

Single axons must express at relatively high levels in order to be visible. Two methods are available for increasing the brightness of GFP expression. First, by changing the coding sequence, GFP localization can be biased to the axon using a membrane-targeting sequence such as the N-terminal palmitoylation sequence from GAP-43 (Moriyoshi *et al.*, 1996). Second, by changing the promoter, either by using a stronger promoter or by using the GAL4/UAS amplification system, in which a tissue-specific promoter drives expression of the strong transcriptional activator GAL4-VP16, which then drives expression from a 14×UAS control element (Köster and Fraser, 2001b). We have successfully used both of these methods to improve labeling of retinal axons. Note however that it is important to control for potential disruptions of cell behavior or viability that may be caused by high levels of either cytoplasmic or membrane-targeted GFP.

Stable transgenic lines can be generated by raising injected founders to sexual maturity, crossing them, and examining their offspring to find those founders with germline integration of the injected DNA construct (Meng *et al.*, 1999). Several stable transgenic lines now exist that express GFP under the control of RGC-specific promoters (Table II; see also review by Udvadia and Linney (2003)). Some of these lines express GFP in the optic pathway or the optic tectum, making it difficult to distinguish retinal axons from surrounding tissue. However, several lines allow clear visualization of the retinal projection, making it possible to label large numbers of embryos simply by crossing transgenic carriers. Indeed, we are beginning to use transgenics in new screens for retinotectal mutants.

Table II
Transgenic Lines that Label Retinal Ganglion Cells

Transgenic line	Retinal expression	Other expression	References
ath5:GFP	RGCs, cytoplasmic GFP	Forebrain, tectum	Masai *et al.* (2003)
brn3c:mGFP	Subset of RGCs, membrane-targeted GFP	Inner ear, lateral line neuromasts	Roeser and Baier (personal communication, 2003)
isl3:GFP	All RGCs, cytoplasmic GFP	Trigeminal ganglion, Rohon-Beard neurons, a few cells in dorsal midbrain	Pittman and Chien (unpublished data, 2003)
nAChRβ3:GFP	RGCs, cytoplasmic GFP	Trigeminal ganglion, Rohon-Beard neurons some tectal cells	Tokuoka *et al.* (2002) and Yoshida and Mishina (2003)
shh:GFP	RGCs, cytoplasmic GFP	Notochord, floor plate, pharyngeal arch endoderm, ventral forebrain	Neumann and Nuesslein-Volhard (2000) and Roeser and Baier (2003)
[RARE + GATA2 basal promoter]: nGFP	Dorsal retina (weak), ventral retina (strong), nuclear GFP	Caudal hindbrain, spinal cord, notochord	Perz-Edwards *et al.* (2001)
deltaD:GAL4	Drives expression throughout retina from UAS reporter constructs	Not described	Scheer *et al.* (2001)

mGFP, membrane-targeted GFP; nGFP, nuclear-localized GFP; RGC, retinal ganglion cell.

1. Method 1: Whole Eye Fills in Fixed Embryos

a. Summary

This method uses lipophilic dyes (DiI, DiO, or DiD) to label the entire retino-fugal projection in fixed embryos. The dyes are made up at approximately 1% w/v in chloroform (originally suggested by Suresh Jesuthasan). The chloroform evaporates quickly, but the concentration is not critical in chloroform. Since these dyes are very soluble in chloroform, pipette clogging is not usually a problem. When the dye is injected just right into the gap between retina and lens, it spreads out quickly to cover the RGC layer (since chloroform is immiscible with water), not stopping until the dye has dissolved into cell membranes. This labels all of the retinal axons, often so strongly that the optic nerve can even be seen in bright field.

b. Solutions Needed

4% PFA	4% paraformaldehyde in 0.1 M phosphate buffer, pH 7.4
PBS	Phosphate-buffered saline

- 1.0% DiI, DiO, or DiD in chloroform (Molecular Probes, Eugene, OR)
- 1.5% agarose in 1/3 × PBS*
- 1.5% low-melt agarose in 1/3 × PBS*

c. Protocol

1. If the embryos or larvae are pigmented, expose them to bright light from a fiberoptic illuminator for a few minutes before fixation to contract the melanophores so that they do not obscure the tectum. Alternately, raise the embryos in 0.1 mM phenylthiourea (PTU) to prevent melanin formation. PTU also makes it easy to see into the eye being injected and is therefore a good idea when learning the procedure.

2. For fixation, remove as much of the embryo medium as possible and add ice-cold fix (4% PFA), then leave for a few hours at room temperature or overnight at 4 °C. For imaging growth cones and arbors, however, do not use cold fix as it depolymerizes microtubules. Rather, fix for about 30 min at room temperature, then store at 4 °C. For lipophilic dye labeling, the length of fixation is not critical—we often store embryos in fix at 4 °C for months and then have successful injections.

3. For a quick mounting system, make up at least 50 ml of 1.5% agarose in 1/3 × PBS. Do this in an Erlenmeyer flask, so that it will stay hot long enough to work with. Carefully microwave the agarose to melt it. Pipette 1–2 ml onto a microscope slide (Fig. 2A).

---

* (At one point, it was thought that 1/3 × PBS gave better labeling than 1 × PBS, but we have not seen a great difference and, at this point, its use is historical.)

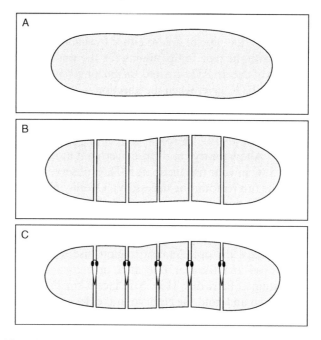

Fig. 2 Embedding of zebrafish embryos or larvae for intraocular injections (whole eye fills), Method 1. (A) Drop of agarose on microscope slide. (B) Agarose with slices made with a scalpel or razor blade. (C) Embryos/larvae inserted into slices. The edges of the cuts hold the tail fin for a perfect dorsal view of the embryo.

4. After the agarose cools, use a scalpel or razor blade to slice it into slabs, and separate the slabs slightly (Fig. 2B). Each slab will hold an embryo.

5. Working under a dissecting microscope at 20× magnification, use a pipette or forceps to place the embryos on the agarose, then orient the embryo with forceps or a pin (Fig. 2C). The vertical sides of the cut hold the embryo's tail fin to give a perfect dorsal orientation. Push the agarose slabs together to hold the embryos, then pipette on a bit more agarose to seal the embryos in place. Using 1.5% low-melt agarose for sealing is recommended for easier penetration.

With a large number of embryos to analyze, an alternate method is to fill the lid of a 60-mm Petri dish with 1.5% agarose in 1/3 × PBS. Once the agarose cools, make several long cuts and arrange embryos head-to-tail in each slab. Each slab can hold up to 25 embryos. Seal in the embryos with 1.5% low-melt agarose as described earlier. Overlay the entire dish with 1/3 × PBS to prevent desiccation.

6. Fill a pulled glass micropipette with 1-2 μl of lipophilic dye solution (we use Eppendorf Micro loader tips to back fill). We use a Picospritzer pressure injector for injections, set at 40 psi pressure and 30 ms pulse time (adjust pulse time to change drop size). Working under a dissecting microscope at 40× magnification, eject test droplets into the medium and carefully break back the tip of the pipette with forceps until you get the desired drop size.

The eye will be easiest to reach if there is not too much agarose covering it. Holding the pipette holder either freehand or with a micromanipulator, pierce the eye with the pipette tip, aiming for the small gap between retina and lens. Inject a drop of dye. In PTU-treated, *albino*, or *golden* embryos, you will see the dye instantly fill the RGC layer when the injection is correct. In pigmented embryos, you may see the eye swell briefly when dye is injected. If the injection goes awry, dye may go through the eye, back out of the eye, etc. Stray drops of dye should be teased away with a pin before they cause background labeling.

7. Allow the dye to diffuse overnight at room temperature or for a few hours at 28.5 °C in your fish incubator. Then observe on an upright fluorescence dissecting scope (no remounting necessary). Overlaying the slides or dishes with a little 1/3 × PBS can improve the optics.

8. Use a compound microscope, ideally a confocal microscope, to attain higher resolution imaging (Fig. 1C–H). It may be necessary to remount your samples. To image the optic chiasm region, use forceps to tease the embryos out of the agarose and transfer one at a time to a 24 × 50 mm coverslip or coverslip-bottomed Petri dish (Fig. 3). Place a small droplet of low-melt agarose over the embryo and hold the embryo in the desired orientation until the agarose hardens. To image the optic tract, remove the contralateral eye by using an electrolytically sharpened tungsten needle to cut away the surrounding skin, sever the optic nerve, and very carefully release the eye. We prefer an inverted microscope, but for an upright microscope, it helps to sandwich the mounted samples with another coverslip. A 20× objective lens is sufficient to view the entire projection, while a 40× or 60× objective lens is necessary to image individual growth cones or arbors.

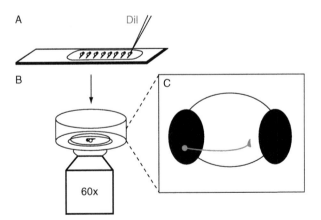

Fig. 3 Time-lapse imaging of retinal axons, Method 2. (A) Injection of DiI into eyes of embryos while immobilized in methylcellulose on a microscope slide. (B) Embryo with DiI-labeled RGCs embedded in agarose in observation chamber (a drilled Petri dish with a coverslip glued on the bottom). (C) Schematic view of ventral diencephalic region of the optic pathway with labeled axon and growth cone. Dorsorostral is up.

2. Method 2: Live Axon Labeling for Time-Lapse

a. Summary

Time-lapse imaging of RGC axons and their arbors has been performed by many investigators (Gnuegge *et al.*, 2001; Hutson and Chien, 2002a; Kaethner and Stuermer, 1992; Schmidt *et al.*, 2000). We have obtained better labeling with DiI than with DiO. However, DiI has been reported to cause greater phototoxicity than DiO (Kaethner and Stuermer, 1992), so it may be advisable to test both. DiD also works well, but its long-red emission makes it difficult to screen for labeling by eye.

b. Solutions and Materials Needed

E2 embryo medium	15.0 mM NaCl, 0.5 mM KCl, 1.0 mM CaCl$_2$, 1.0 mM MgSO$_4$, 0.15 mM KH$_2$PO$_4$, 0.70 mM NaHCO$_3$
E2/GN	10 μg/ml gentamycin in E2
Tricaine stock	0.4% tricaine (MESAB, MS222), 10 mM HEPES, pH 7.4
Methylcellulose	3% methylcellulose, 0.016% tricaine in E2/GN
DiI, DiO, or DiD	0.25% dissolved as well as possible in 95% EtOH
1.5% agarose	1.5% low-melt agarose, 0.016% tricaine in E2/GN

- Picospritzer or other microinjection apparatus
- Microinjection needles
- Microscope slide
- Observation chamber (coverslip-bottomed Petri dish, see Fig. 3.)

c. Protocol

1. Label the embryos 3–6 h before imaging. This allows time for axons to become well-labeled, without being long enough for dye to enter the blood stream and give high background fluorescence. The first cohort of axons enters the ventral diencephalon at approximately 32 hpf, enters the contralateral optic tract at approximately 38 hpf, and reaches the optic tectum at approximately 48 hpf. Anesthetize the embryos by adding tricaine stock to a final concentration of 0.016%. Carefully transfer the embryos, a few at a time, into a large pool of 3% methylcellulose solution on a microscope slide. Transfer as little embryo medium as possible and remove extra medium using the corner of a tissue.

2. Use forceps or a large-caliber dissecting needle to *carefully* orient the embryos with one eye facing up. The methylcellulose is extremely viscous and can shear the embryos.

3. Microcentrifuge the dye solution briefly to pellet precipitated dye since 0.25% is near saturation for the lipophilic dyes. Fill a micropipette with dye solution, taking care not to transfer any crystals.

Pressure-inject a very small volume of DiI solution into one eye of each embryo (Fig. 3A). Do not inject between the eye and the lens, as you did for whole eye fills, but instead aim for the body of the neural retina. (The ethanol solution does not penetrate as well as chloroform.) Figure 3 shows the best angle of approach. When injections are complete, place the slide containing the embryos into a Petri dish that contains E2, and transfer the embryos to 28.5 °C for at least 3 h. The E2 will slowly dissolve the methylcellulose, making it easier to liberate the embryos.

4. At the desired stage, release embryos, using a transfer pipette to "blow" away any remaining methylcellulose.

5. Add tricaine to a final concentration of 0.016%. Screen under a fluorescence dissecting microscope for nicely labeled axons.

6. Mount 6–8 of your best-labeled embryos. Liquefy the low-melt agarose solution in a glass test tube, then hold it at 40 °C in a heat block. This temperature is close to the maximum that embryos can withstand, but high enough to keep the agarose liquid. Using a fire-polished Pasteur pipette, transfer one embryo at a time into the agarose, then onto a slide (if using an upright microscope) or into the bottom of an observation chamber (if using an inverted microscope; see Fig. 3B). Before the agarose cools, orient the embryo using forceps. If using an upright microscope, you need not coverslip the embryo, and it can be imaged directly using a dipping lens. If using an inverted microscope, partially fill the dish with E2 containing tricaine to keep it wet.

7. Use confocal microscopy because it works well for time-lapse imaging (see Fig. 1L). Phototoxicity is not generally a problem in our hands, although we do try to minimize laser intensity and dwell time. We have been able to take as many as 22 slices per Z-series every 3 min, for as long as 8 h (Hutson and Chien, 2002a).

3. Method 3: Single Axon Labeling Using Xfp Constructs

a. Summary

This is a method to label small numbers of RGCs by injecting embryos at the 1–4-cell stage with plasmids encoding a fluorescent protein (e.g., GFP or DsRed) under the control of an RGC-specific promoter. This gives mosaic expression from the DNA, so that by chance one or a few RGCs will express the construct. The first published report using this method to label single zebrafish RGCs used the nicotinic acetylcholine receptor $\beta3$ subunit promoter (PAR) to drive the expression of soluble enhanced green fluorescent protein (EGFP) to label arbors, and a vesicle associated membrane protein (VAMP)-EGFP fusion construct to label presynaptic sites (Tokuoka *et al.*, 2002). However, in our hands the GFP expression from these constructs is very dim, making it difficult to visualize individual arbors and sites of synapse formation in living embryos (Campbell and Chien, unpublished data, 2004). We find that using a GAL4/UAS amplification cassette (Köster and Fraser, 2001b; Brn3c constructs, gift of Tong Xiao) overcomes this problem and gives sufficiently bright expression in retinal axons (see Fig. 1K). Using a similar set of constructs described by Niell *et al.* (2004) gives very nice labeling of tectal neurons (Fig. 1I and J).

b. Solutions and Materials Needed

DNA constructs	pG1 Brn3c:GAL4;UAS:GAP43-GFP (gift of T. Roeser, T. Xiao, M. Smear, and H. Baier). Drives expression of membrane-targeted GFP in RGCs
	pBSKα-tubulin:GAL4 (Köster and Fraser, 2001b). Drives expression in differentiated neurons including tectal cells
	UAS:PSD95-GFP;UAS:DsRedExpress-1 (Niell *et al.*, 2004). Dual construct that labels tectal dendrites with DsRedExpress, and postsynaptic sites with PSD95-GFP, when coinjected with pBSKα-tubulin:GAL4
Phenol red stock	1% phenol red in deionized water
E2 embryo medium	15.0 mM NaCl, 0.5 mM KCl, 1.0 mM CaCl$_2$, 1.0 mM MgSO$_4$, 0.15 mM KH$_2$PO$_4$, 0.70 mM NaHCO$_3$
E2/GN	10 μg/ml gentamycin in E2
E3 medium	5 mM NaCl, 0.17 mM KCl, 0.33 mM CaCl$_2$, 0.33 mM MgSO$_4$
E3/PTU	0.1 mM phenylthiourea (phenylthiocarbamide) in E3
1.5% agarose	1.5% low-melt agarose in E2/GN for injection plates
1% agarose	1% low-melt agarose in E2/GN for mounting embryos

- Picospritzer or other microinjection apparatus
- Microinjection needles
- Coverslip-bottomed dishes

c. Protocol

1. Prepare DNA solution to be injected in deionized water, with a final concentration of 0.1% phenol red to help visualize the injected bolus. For labeling single RGC axons and arbors, the DNA concentration should be approximately 35 ng/μl of pG1 Brn3c:GAL4;UAS:GAP43-GFP. For labeling single tectal cells, the DNA concentration should be approximately 25 ng/μl pBSKα-tubulin:GAL4 and 25 ng/μl UAS:PSD95-GFP;UAS:DsRed Express-1. The injected bolus should be approximately 1 nl, or a sphere 120 μm in diameter. This can be measured with an eyepiece micrometer or estimated very roughly as one-sixth the diameter of the yolk cell.

2. Arrange the embryos in grooves in agarose injection trays overlaid with E2/GN and inject DNA into the cell(s) at the 1–4-cell stage, using a Picospritzer or similar microinjector.

3. Once injected, transfer the embryos to E3 medium and raise at 28.5 °C.

4. Between 12 and 22 hpf, replace the medium with E3/PTU to prevent melanin formation.

5. At approximately 3 dpf, screen the embryos under a fluorescence dissecting microscope for GFP or DsRed expression. Mount labeled embryos dorsal side down in 1% agarose in a coverslip-bottomed Petri dish for imaging on an inverted confocal microscope. Labeled arbors can be visualized using a $60\times$ water-immersion objective.

6. Amplify the signal (should the GFP appear too dim) by fixing the embryos and staining them with an anti-GFP polyclonal antibody (Molecular Probes, Inc., Eugene, OR) followed by an Alexa 488-labeled antirabbit secondary antibody.

7. This method can be extended to examine the effects of perturbing gene function on RGC arbor formation *in vivo*, by coinjection of the pG1 Brn3c:GAL4;UAS:GAP43-GFP construct together with morpholino antisense oligonucleotides.

IV. Perturbing the Retinotectal System

A. Overview

Simple observation of normal embryos can tell us much about development. However, to rigorously test hypotheses about developmental mechanisms, one must perform perturbations. The strategies for perturbing retinotectal development are, broadly, pharmacological manipulation, generation of mosaics by transplantation, loss-of-function perturbation, and gain-of-function perturbation. For pharmacological experiments, drugs have been applied by injection into the eyes (Schmidt *et al.*, 2004; Stuermer *et al.*, 1990), or by bath application (Masai *et al.*, 2003; Schmidt *et al.*, 2000). The penetration of bath-applied agents is presumably aided by the small size of zebrafish larvae.

Transplantation of cells from labeled donors to unlabeled hosts at the blastula or gastrula stage is the classic zebrafish method for creating genetic mosaics to test cell autonomy (Ho and Kane, 1990), and can be used to target the retina quite effectively (Moens and Fritz, 1999). Indeed, using a stable transgenic line expressing GFP under an RGC-specific promoter, it is possible to test cell autonomy of axon guidance mutations (Suli and Chien, unpublished data, 2002). However, such mosaics often have clones of cells in the brain that may make the results difficult to interpret. A method that is more difficult, but perhaps more elegant, is to transplant entire eye primordia (Fig. 4; Method 4). This yields mosaics in which the entire eye comes from the donor, and the rest of the embryo is completely host derived. This method has been used to show that *astray* acts eye autonomously in retinal axon guidance, while *ace* (*fgf8*) has both eye autonomous and nonautonomous actions (Fricke *et al.*, 2001; Picker *et al.*, 1999).

Two main methods are used for loss-of-function experiments. The first is, obviously, mutant analysis. The second, injection of stable, nontoxic antisense morpholino oligonucleotides (MOs) into 1-cell stage embryos, is well established

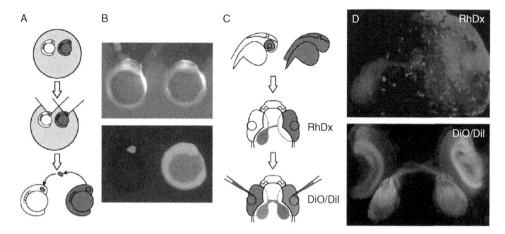

Fig. 4 Eye transplant procedure, Method 4. (A) At 5-somite stage, host embryo (unlabeled) and donor embryo (labeled with rhodamine dextran) are embedded in a small drop of low-melt agarose. Scalpel cuts are used to remove wedges of agarose, giving access to eye primordium, then the host eye is replaced with the donor eye. (B) Bright field (*top*) and rhodamine fluorescence (*bottom*) micrographs of embedded embryos after removing agarose wedges. (C) Host and donor embryos are raised until 4-5 dpf. Rhodamine fluorescence is used to check that the entire eye has been transplanted without any accompanying brain tissue. Lipophilic dye labeling is then used to assay the retinal projections. (D) Confocal images of transplants. *Top*, rhodamine dextran labeling shows the transplanted eye and retinal axons that have grown out into the host brain. Notice a few cells that have migrated out of the transplanted eye. *Bottom*, lipophilic dye labeling shows the host (*red*) and donor (*green*) projections in a wild-type to wild-type transplant. Note that a few donor axons project ipsilaterally (*yellow* labeling in right tectum). (See Plate no. 4 in the Color Plate Section.)

for the short-term disruption of gene function (Nasevicius and Ekker, 2000). MOs directed near a start codon can inhibit protein translation. MOs directed against an exon/intron or intron/exon boundary can inhibit splicing of pre-mRNAs (Draper *et al.*, 2001), and have the advantage that their efficacy can be tested by RT-PCR. A general problem with MOs is that they lose efficacy at later stages, as they become diluted by cell divisions and titrated out by new mRNA synthesis. This is a particular concern for the retinotectal system, which develops relatively late. We have used both translation-blocking and splice-blocking morpholinos to successfully disrupt development of the retinotectal pathway at 48 hpf and in some cases up to 4 dpf (Hutson and Chien, unpublished results, 2002). Another problem is that, for genes with dual roles in early development, as well as in retinotectal development, a potential retinotectal phenotype may be obscured by earlier defects. The discovery that morpholinos appear to be effective when injected into the yolk at later stages suggests a possible method to circumvent this problem (Stenkamp *et al.*, 2000). Blastula- or gastrula-stage cell transplants are another way to avoid early lethality.

It is becoming quite simple to perform gain-of-function experiments (i.e., to incorrectly express genes in particular cells of interest) by using transient or stable transgenic expression driven by cell-specific, tissue-specific, or inducible promoters. The nicotinic acetylcholine receptor-β3 subunit promoter has been used to drive incorrect expression in RGCs and to analyze the arborization of retinal axons (Tokuoka *et al.*, 2002; Yoshida and Mishina, 2003). The *hsp70* promoter, derived from the heat shock protein 70 gene, allows genes to misexpress either globally using heat shock or locally using focal laser induction (Halloran *et al.*, 2000). Mary Halloran's group (Liu *et al.*, 2004) has used heat shock controlled incorrect expression of Sema3D in order to study its effects on retinal axon guidance. The temporal control provided by the *hsp70* promoter makes it an especially powerful tool, sure to be used widely in the future.

1. Method 4: Eye Transplants

a. Summary

This procedure is tricky, but is the most convincing way to determine whether a gene acts eye autonomously or brain autonomously during retinotectal development (Fricke *et al.*, 2001). Wild-type to wild-type transplantations can result in certain retinal axon guidance defects (especially ipsilateral projections), so it is important to do transplants in both directions (i.e., wild type into mutant and mutant into wild type) and to evaluate the results quantitatively.

b. Solutions and Materials

Fish Ringer's	116 mM NaCl, 3 mM KCl, 4 mM CaCl$_2$, 1 mM MgCl$_2$, 5 mM HEPES, 20 μg/ml gentamycin sulfate. This has been slightly modified from the Ringer's in the Zebrafish Book, (Westerfield, 2000) by addition of magnesium and the raising of calcium to promote healing
E2 embryo medium	15 mM NaCl, 0.5 mM KCl, 1 mM Ca Cl$_2$, 1 mM Mg SO$_4$, 0.15 mM KH$_2$PO$_4$, 0.05 mM Na$_2$HPO$_4$, 0.70 mM NaHCO$_3$
E2/PTU	0.1 mM phenylthiourea (phenylthiocarbamide) in E2
Rhodamine dextran	Lysine-fixable rhodamine dextran, 3000 MW, 50 mg/ml in water
Low-melt agarose	1.2% low-melt agarose in Fish Ringer's

- Mineral oil
- Picospritzer or other microinjector
- 18 °C incubator
- Agarose injection trays for single-cell injections
- Tungsten dissecting needles, electrolytically sharpened
- Flame-polished large-bore Pasteur pipette

- Scalpel, #11 blade
- Nylon crystal-picking loop (0.1–0.2 mm, Hampton Research, Aliso Viejo, CA)

c. Protocol

1. Label donor embryos at the 1–2-cell stage with rhodamine dextran. We use Tübingen-style agarose injecting dishes with long grooves (Gilmour *et al.*, 2002). Hold the pipette holder freehand and, using a Picospritzer, inject the embryos. Use medium forceps to orient the embryos and hold them when withdrawing the needle. It is gentlest to inject dextran into the yolk just below the cell(s). The dextran will be carried upwards by cytoplasmic streaming. Inject just enough to turn the embryo slightly pink. Occasional batches of dextran are said to be much more toxic than usual, presumably because of impurities.

2. Carry out the transplants at 5–10 somites (about 12–13 hpf), when the eye anlage has just formed a distinct pocket. Before this, the tissue is not firm enough to manipulate; after this, healing is excellent but retinal axons never enter the brain. The reason seems to be that the optic stalk is still quite wide at this stage. At later stages the stalk thins down rapidly, making it impossible to match up the ends of the donor and host stalks. Unless you have fish on a delayed light cycle, the most convenient schedule is to inject dextran the first morning, raise the embryos at 28.5 °C until about 70-80% epiboly that afternoon (about 5 p.m.), then shift them to 18 °C to slow their development. They will then reach 5 somites around 9 or 10 a.m. the next morning. At this stage, transfer about 20 donors and 20 hosts to separate 50-mm glass Petri dishes filled with E2. It is important to match stages well so that the donor and host eyes are the same size. Dechorionate embryos manually with fine forceps. Hereafter, transfer them using a cut-off, flame-polished Pasteur pipette. It is easy to damage the yolks of embryos at this stage, so always use clean tools.

3. Microwave 0.12 g of low-melt agarose in 10 ml of Ringer's in a glass test tube. Hold this at 38–40 °C (no hotter) in a heat block. To mount embryos, pick up one donor and one host in the same pipette, drop them briefly in the agarose to wash off the E2, then place them back in the pipette and put them in a drop of agarose on an inverted 35-mm Petri dish lid. Orient them quickly with forceps: lateral view, right eye up, noses in the same direction (see Fig. 4A and B).

4. After orienting several pairs of donors and hosts (typically two or four pairs in each lid), use a clean #11 scalpel to cut wedges in the agarose drops, pointing at the heads of the embryos. Next, flood the dishes with Ringer's. Use a sharpened tungsten needle to flip the wedges of agarose out, starting just underneath the embryo so that no film of agarose is left over the eye (see Fig. 4A and B).

5. Make an entry through the tough embryonic skin. This is a key step. Use a sharpened tungsten needle to nick the skin just dorsal to the eyes of both donor and host. Apply a small drop of mineral oil here. To apply the oil, we use a broken-off micropipette attached to a piece of silastic tubing whose end has been plugged,

mounted using a 1-ml tuberculin syringe as a handle. Squeezing the silastic with an index finger gives good control over oil ejection. After the oil droplet wets the skin, it can be removed immediately. After about 2 min, a small bubble of dead, white skin should rise up. This can be flicked away with a needle, leaving a window through which to reach the eye.

6. Use a very sharp tungsten needle with a hooked end to loosen the eye anlage, cutting through the optic stalk and also separating it from overlying skin. Loosen the donor eye first, then the host.

7. Use a fine nylon loop to remove the host eye. (These loops are designed to handle protein crystals for X-ray crystallography. Search for one that is just the right size for the eyes.) It is important to leave the overlying skin to hold the transplanted eye in place. The donor eye should then be moved into place as soon as possible using the loop, and pushed into the pocket formed by the overlying skin.

8. Carefully free the donor and host from the agarose after healing occurs (within 20–30 min). They can then be raised in E2/PTU to inhibit pigment formation.

9. Label eyes by whole eye fill at 4-5 dpf. It is important to use the rhodamine dextran lineage label to confirm that the entire eye is donor derived, and also to check that no donor cells have been inadvertently transplanted into the brain (see Fig. 4C and D).

V. Future Directions

Two techniques stand out as ones that will undoubtedly be adapted to the study of the zebrafish retinotectal system: *in vivo* electroporation and multiphoton imaging. In *Xenopus* embryos, *in vivo* electroporation is a powerful method for delivering both DNA constructs (Haas *et al.*, 2001) and dextran-coupled indicators (Edwards and Cline, 1999). There seems to be no reason that it should not be just as useful in zebrafish (Teh *et al.*, 2003). Electroporation is especially attractive because it would allow both spatial and temporal control when expressing reporters, gain-of-function constructs, or dominant negative constructs. Indeed, with effective electroporation, it should be possible to use short hairpin constructs driven by Pol III promoters to generate small interfering RNAs (siRNAs) in RGCs, thus specifically inhibiting gene function (see McManus and Sharp (2002) for review).

Multiphoton excitation has two main advantages (reviewed in Denk and Svoboda (1997) and Helmchen and Denk (2002)): (1) reduced photobleaching and phototoxicity, since multiphoton absorption is restricted to the plane of focus and (2) improved brightness and resolution in thick or scattering tissues, since infrared light scatters less than visible light and nondescanned detectors can make use of scattered emitted light. Thus, it is ideally suited to imaging cell behavior in live thick samples such as the intact zebrafish larva and in particular in the retinotectal system.

Combining these emerging techniques with established methods in the zebrafish retinotectal system promises to make significant contributions to our understanding of the *in vivo* control of axon guidance.

Acknowledgments

We dedicate this chapter to Friedrich Bonhoeffer for his farsighted support of the retinotectal screen. We thank Arminda Suli for Fig. 1A and B, Jeong-Soo Lee for Fig. 1F, and Andrew Pittman and Ichiro Masai for allowing us to mention unpublished transgenic lines. We thank many colleagues for generously providing reagents: Tobias Roeser, Tong Xiao, and Herwig Baier for unpublished Brn3c constructs and transgenic fish; Reinhard Köster for GAL4 and UAS constructs; and Cris Niell and Stephen Smith for PSD95-GFP constructs. Thanks to Nick Marsh-Armstrong and Pam Kainz for suggestions on the eye transplant procedure. This work was supported by NIH F32 EY07017 to LDH, an EMBO Long-Term Fellowship to DSC, and NSF IBN-021385 and NIH R01 EY12873 to CBC.

References

Baier, H. (2000). Zebrafish on the move: Towards a behavior-genetic analysis of vertebrate vision. *Curr. Opin. Neurobiol.* **10**, 451–455.

Baier, H., Klostermann, S., Trowe, T., Karlstrom, R. O., Nusslein-Volhard, C., and Bonhoeffer, F. (1996). Genetic dissection of the retinotectal projection. *Development* **123**, 415–425.

Bak, M., and Fraser, S. E. (2003). Axon fasciculation and differences in midline kinetics between pioneer and follower axons within commissural fascicles. *Development* **130**, 4999–5008.

Brand, M., Heisenberg, C. P., Jiang, Y. J., Beuchle, D., Lun, K., Furutani-Seiki, M., Granato, M., Haffter, P., Hammerschmidt, M., Kane, D. A., Kelsh, R. N., Mullins, M. C., *et al.* (1996). Mutations in zebrafish genes affecting the formation of the boundary between midbrain and hindbrain. *Development* **123**, 179–190.

Burrill, J. D., and Easter, S. S., Jr. (1994). Development of the retinofugal projections in the embryonic and larval zebrafish (Brachydanio rerio). *J. Comp. Neurol.* **346**, 583–600.

Chen, W., Burgess, S., and Hopkins, N. (2001). Analysis of the zebrafish smoothened mutant reveals conserved and divergent functions of hedgehog activity. *Development* **128**, 2385–2396.

Culverwell, J., and Karlstrom, R. O. (2002). Making the connection: Retinal axon guidance in the zebrafish. *Semin. Cell Dev. Biol.* **13**, 497–506.

Denk, W., and Svoboda, K. (1997). Photon upmanship: Why multiphoton imaging is more than a gimmick. *Neuron* **18**, 351–357.

Draper, B. W., Morcos, P. A., and Kimmel, C. B. (2001). Inhibition of zebrafish fgf8 pre-mRNA splicing with morpholino oligos: A quantifiable method for gene knockdown. *Genesis* **30**, 154–156.

Dynes, J. L., and Ngai, J. (1998). Pathfinding of olfactory neuron axons to stereotyped glomerular targets revealed by dynamic imaging in living zebrafish embryos. *Neuron.* **20**, 1081–1091.

Easter, S. S., Jr., and Nicola, G. N. (1996). The development of vision in the zebrafish (Danio rerio). *Dev. Biol.* **180**, 646–663.

Edwards, J. A., and Cline, H. T. (1999). Light-induced calcium influx into retinal axons is regulated by presynaptic nicotinic acetylcholine receptor activity *in vivo*. *J. Neurophysiol.* **81**, 895–907.

Fashena, D., and Westerfield, M. (1999). Secondary motoneuron axons localize DM-GRASP on their fasciculated segments. *J. Comp. Neurol.* **406**, 415–424.

Forbes, D. J., Kirschner, M. W., and Newport, J. W. (1983). Spontaneous formation of nucleus-like structures around bacteriophage DNA microinjected into Xenopus eggs. *Cell* **34**, 13–23.

Fricke, C., Lee, J. S., Geiger-Rudolph, S., Bonhoeffer, F., and Chien, C.-B. (2001). Astray, a zebrafish roundabout homolog required for retinal axon guidance. *Science* **292**, 507–510.

Gilmour, D. T., Jessen, J. R., and Lin, S. (2002). Manipulating gene expression in the zebrafish. *In* "Zebrafish: A Practical Approach" (C. Nüsslein-Volhard and R. Dahm, eds.), pp. 121–144. Oxford University Press, New York.

Gnuegge, L., Schmid, S., and Neuhauss, S. C. (2001). Analysis of the activity-deprived zebrafish mutant macho reveals an essential requirement of neuronal activity for the development of a finegrained visuotopic map. *J. Neurosci.* **21,** 3542–3548.

Haas, K., Sin, W. C., Javaherian, A., Li, Z., and Cline, H. T. (2001). Single-cell electroporation for gene transfer *in vivo. Neuron* **29,** 583–591.

Haffter, P., Granato, M., Brand, M., Mullins, M. C., Hammerschmidt, M., Kane, D. A., Odenthal, J., van Eeden, F. J., Jiang, Y. J., Heisenberg, C. P., Kelsh, R. N., Furutani-Seiki, M., *et al.* (1996). The identification of genes with unique and essential functions in the development of the zebrafish, Danio rerio. *Development* **123,** 1–36.

Halloran, M. C., Sato-Maeda, M., Warren, J. T., Su, F., Lele, Z., Krone, P. H., Kuwada, J. Y., and Shoji, W. (2000). Laser-induced gene expression in specific cells of transgenic zebrafish. *Development* **127,** 1953–1960.

Helmchen, F., and Denk, W. (2002). New developments in multiphoton microscopy. *Curr. Opin. Neurobiol.* **12,** 593–601.

Ho, R. K., and Kane, D. A. (1990). Cell-autonomous action of zebrafish spt-1 mutation in specific mesodermal precursors. *Nature* **348,** 728–730.

Honig, M. G., and Hume, R. I. (1989). DiI and diO: Versatile fluorescent dyes for neuronal labeling and pathway tracing. *Trends Neurosci.* **12,** 333–335, 340-341.

Hutson, L. D., and Chien, C.-B. (2002a). Pathfinding and error correction by retinal axons: The role of astray/robo2. *Neuron* **33,** 205–217.

Hutson, L. D., and Chien, C.-B. (2002b). Wiring the zebrafish: Axon guidance and synaptogenesis. *Curr. Opin. Neurobiol.* **12,** 87–92.

Hutson, L. D., Jurynec, M. J., Yeo, S. Y., Okamoto, H., and Chien, C.-B. (2003). Two divergent slit1 genes in zebrafish. *Dev. Dyn.* **228,** 358–369.

Kaethner, R. J., and Stuermer, C. A. (1992). Dynamics of terminal arbor formation and target approach of retinotectal axons in living zebrafish embryos: A time-lapse study of single axons. *J. Neurosci.* **12,** 3257–3271.

Kanki, J. P., Chang, S., and Kuwada, J. Y. (1994). The molcular cloning and characterization of potential chick DM-GRASP homologs in zebrafish and mouse. *J. Neurobiol.* **25,** 831–845.

Karlstrom, R. O., Talbot, W. S., and Schier, A. F. (1999). Comparative synteny cloning of zebrafish you-too: Mutations in the Hedgehog target gli2 affect ventral forebrain patterning. *Genes Dev.* **13,** 388–393.

Karlstrom, R. O., Trowe, T., Klostermann, S., Baier, H., Brand, M., Crawford, A. D., Grunewald, B., Haffter, P., Hoffmann, H., Meyer, S. U., Müller, B. K., Richter, S., *et al.* (1996). Zebrafish mutations affecting retinotectal axon pathfinding. *Development* **123,** 427–438.

Karlstrom, R. O., Tyurina, O. V., Kawakami, A., Nishioka, N., Talbot, W. S., Sasaki, H., and Schier, A. F. (2003). Genetic analysis of zebrafish gli1 and gli2 reveals divergent requirements for gli genes in vertebrate development. *Development* **130,** 1549–1564.

Kidd, T., Brose, K., Mitchell, K. J., Fetter, R. D., Tessier-Lavigne, M., Goodman, C. S., and Tear, G. (1998). Roundabout controls axon crossing of the CNS midline and defines a novel subfamily of evolutionarily conserved guidance receptors. *Cell* **92,** 205–215.

Köster, R. W., and Fraser, S. E. (2001a). Direct imaging of *in vivo* neuronal migration in the developing cerebellum. *Curr. Biol.* **11,** 1858–1863.

Köster, R. W., and Fraser, S. E. (2001b). Tracing transgene expression in living zebrafish embryos. *Dev. Biol.* **233,** 329–346.

Laessing, U., and Stuermer, C. A. (1996). Spatiotemporal pattern of retinal ganglion cell differetiation revealed by the expression of neurolin in embryonic zebrafish. *J. Neurobiol.* **29,** 65–74.

Lele, Z., Folchert, A., Concha, M., Rauch, G. J., Geisler, R., Rosa, F., Wilson, S. W., Hammerschmidt, M., and Bally-Cuif, L. (2002). Parachute/n-cadherin is required for morphogenesis and maintained integrity of the zebrafish neural tube. *Development* **129,** 3281–3294.

Liu, Y., Berndt, J., Su, F., Tawarayama, H., Shoji, W., Kuwada, J. Y., and Halloran, M. C. (2004). Semaphorin3D guides retinal axons along the dorsoventral axis of the tectum. *J. Neurosci.* **24,** 310–318.

Lorent, K., Liu, K. S., Fetcho, J. R., and Granato, M. (2001). The zebrafish space cadet gene controls axonal pathfinding of neurons that modulate fast turning movements. *Development* **128,** 2131–2142.

Macdonald, R., Scholes, J., Strahle, U., Brennan, C., Holder, N., Brand, M., and Wilson, S. W. (1997). The Pax protein Noi is required for commissural axon pathway formation in the rostral forebrain. *Development* **124,** 2397–2408.

Mangrum, W. I., Dowling, J. E., and Cohen, E. D. (2002). A morphological classification of ganglion cells in the zebrafish retina. *Vis. Neurosci.* **19,** 767–779.

Masai, I., Lele, Z., Yamaguchi, M., Komori, A., Nakata, A., Nishiwaki, Y., Wada, H., Tanaka, H., Nojima, Y., Hammerschmidt, M., Wilson, S. W., and Okamoto, H. (2003). N-cadherin mediates retinal lamination, maintenance of forebrain compartments and patterning of retinal neurites. *Development* **130,** 2479–2494.

McManus, M. T., and Sharp, P. A. (2002). Gene silencing in mammals by small interfering RNAs. *Nat. Rev. Genet.* **3,** 737–747.

Meng, A., Jessen, J. R., and Lin, S. (1999). Transgenesis. *Methods Cell Biol.* **60,** 133–148.

Moens, C. B., and Fritz, A. (1999). Techniques in neural development. *Methods Cell Biol.* **59,** 253–272.

Moriyoshi, K., Richards, L. J., Akazawa, C., O'Leary, D. D., and Nakanishi, S. (1996). Labeling neural cells using adenoviral gene transfer of membrane-targeted GFP. *Neuron* **16,** 255–260.

Nakano, Y., Kim, H. R., Kawakami, A., Roy, S., Schier, A. F., and Ingham, P. W. (2004). Inactivation of dispatched 1 by the chameleon mutation disrupts Hedgehog signaling in the zebrafish embryo. *Dev. Biol.* **269,** 381–392.

Nasevicius, A., and Ekker, S. C. (2000). Effective targeted gene 'knockdown' in zebrafish. *Nat. Genet.* **26,** 216–220.

Neuhauss, S. C. (2003). Behavioral genetic approaches to visual system development and function in zebrafish. *J. Neurobiol.* **54,** 148–160.

Neuhauss, S. C., Biehlmaier, O., Seeliger, M. W., Das, T., Kohler, K., Harris, W. A., and Baier, H. (1999). Genetic disorders of vision revealed by a behavioral screen of 400 essential loci in zebrafish. *J. Neurosci.* **19,** 8603–8615.

Neumann, C. J. (2002). Vertebrate development: A view from the zebrafish. *Semin. Cell Dev. Biol.* **13,** 469.

Neumann, C. J., and Nuesslein-Volhard, C. (2000). Patterning of the zebrafish retina by a wave of sonic hedgehog activity. *Science* **289,** 2137–2139.

Niell, C. M., Meyer, M. P., and Smith, S. J. (2004). *In Vivo* imaging of synapse formation on a growing dendritic arbor. *Nat. Neurosci.* **7,** 254–260.

Parsons, M. J., Pollard, S. M., Saude, L., Feldman, B., Coutinho, P., Hirst, E. M., and Stemple, D. L. (2002). Zebrafish mutants identify an essential role for laminins in notochord formation. *Development* **129,** 3137–3146.

Perz-Edwards, A., Hardison, N. L., and Linney, E. (2001). Retinoic acid-mediated gene expression in transgenic reporter zebrafish. *Dev. Biol.* **229,** 89–101.

Picker, A., Brennan, C., Reifers, F., Clarke, J. D., Holder, N., and Brand, M. (1999). Requirement for the zebrafish mid-hindbrain boundary in midbrain polarisation, mapping and confinement of the retinotectal projection. *Development* **126,** 2967–2978.

Rebagliati, M. R., Toyama, R., Haffter, P., and Dawid, I. B. (1998). Cyclops encodes a nodal-related factor involved in midline signaling. *Proc. Natl. Acad. Sci. USA* **95,** 9932–9937.

Roeser, T., and Baier, H. (2003). Visuomotor behaviors in larval zebrafish after GFP-guided laser ablation of the optic tectum. *J. Neurosci.* **23,** 3726–3734.

Sampath, K., Rubinstein, A. L., Cheng, A. M., Liang, J. O., Fekany, K., Solnica-Krezel, L., Korzh, V., Halpern, M. E., and Wright, C. V. (1998). Induction of the zebrafish ventral brain and floorplate requires cyclops/nodal signalling. *Nature* **395,** 185–189.

Schauerte, H. E., van Eeden, F. J., Fricke, C., Odenthal, J., Strahle, U., and Haffter, P. (1998). Sonic hedgehog is not required for the induction of medial floor plate cells in the zebrafish. *Development* **125**, 2983–2993.

Scheer, N., Groth, A., Hans, S., and Campos-Ortega, J. A. (2001). An instructive function for Notch in promoting gliogenesis in the zebrafish retina. *Development* **128**, 1099–1107.

Schmidt, J. T. (2004). Activity-driven sharpening of the retinotectal projection: The search for retrograde synaptic signaling pathways. *J. Neurobiol.* **59**, 114–133.

Schmidt, J. T., Buzzard, M., Borress, R., and Dhillon, S. (2000). MK801 increases retinotectal arbor size in developing zebrafish without affecting kinetics of branch elimination and addition. *J. Neurobiol.* **42**, 303–314.

Schmidt, J. T., Fleming, M. R., and Leu, B. (2004). Presynaptic protein kinase C controls maturation and branch dynamics of developing retinotectal arbors: Possible role in activity-driven sharpening. *J. Neurobiol.* **58**, 328–340.

Shanmugalingam, S., Houart, C., Picker, A., Reifers, F., Macdonald, R., Barth, A., Griffin, K., Brand, M., and Wilson, S. W. (2000). Ace/Fgf8 is required for forebrain commissure formation and patterning of the telencephalon. *Development* **127**, 2549–2561.

Stenkamp, D. L., Frey, R. A., Prabhudesai, S. N., and Raymond, P. A. (2000). Function for Hedgehog genes in zebrafish retinal development. *Dev. Biol.* **220**, 238–252.

Stuermer, C. A., Rohrer, B., and Munz, H. (1990). Development of the retinotectal projection in zebrafish embryos under TTX-induced neural-impulse blockade. *J. Neurosci.* **10**, 3615–3626.

Teh, C., Chong, S. W., and Korzh, V. (2003). DNA delivery into anterior neural tube of zebrafish embryos by electroporation. *Biotechniques* **35**, 950–954.

Tokuoka, H., Yoshida, T., Matsuda, N., and Mishina, M. (2002). Regulation by glycogen synthase kinase-3beta of the arborization field and maturation of retinotectal projection in zebrafish. *J. Neurosci.* **22**, 10324–10332.

Trevarrow, B., Marks, D. L., and Kimmel, C. B. (1990). Organization of hindbrain segments in the zebrafish embryo. *Neuron* **4**, 669–679.

Trowe, T., Klostermann, S., Baier, H., Granato, M., Crawford, A. D., Grunewald, B., Hoffmann, H., Karlstrom, R. O., Meyer, S. U., Muller, B., Richter, S., Nüsslein-Volhard, C., *et al.* (1996). Mutations disrupting the ordering and topographic mapping of axons in the retinotectal projection of the zebrafish, Danio rerio. *Development* **123**, 439–450.

Udvadia, A. J., and Linney, E. (2003). Windows into development: Historic, current, and future perspectives on transgenic zebrafish. *Dev. Biol.* **256**, 1–17.

Varga, Z. M., Amores, A., Lewis, K. E., Yan, Y. L., Postlethwait, J. H., Eisen, J. S., and Westerfield, M. (2001). Zebrafish smoothened functions in ventral neural tube specification and axon tract formation. *Development* **128**, 3497–3509.

Wagle, M., Grunewald, B., Subburaju, S., Barzaghi, C., Le Guyader, S., Chan, J., and Jesuthasan, S. (2004). EphrinB2a in the zebrafish retinotectal system. *J. Neurobiol.* **59**, 57–65.

Westerfield, M., Wegner, J., Jegalian, B. G., DeRobertis, E. M., and Puschel, A. W. (1992). Specific activation of mammalian Hox promoters in mosaic transgenic zebrafish. *Genes Dev.* **6**, 591–598.

Westerfield, M. (2000). The zebrafish book. A guide for the laboratory use of zebrafish (*Danio rerio*). 4th ed., Univ. of Oregon Press, Eugene, OR.

Yoshida, T., and Mishina, M. (2003). Neuron-specific gene manipulations to transparent zebrafish embryos. *Methods Cell. Sci.* **25**, 15–23.

CHAPTER 10

Analysis of Cell Proliferation, Senescence, and Cell Death in Zebrafish Embryos

Daniel Verduzco* and James F. Amatruda*,†

*Departments of Pediatrics and Molecular Biology
UT Southwestern Medical Center
Dallas, Texas 75390-8534

†Department of Internal Medicine
UT Southwestern Medical Center
Dallas, Texas 75390-8534

I. Introduction: The Cell Cycle in Zebrafish

In multicellular organisms, the cell cycle is a fundamental feature of cellular physiology that is critical for normal development, organogenesis and tissue homeostasis. Reflecting this central role, the molecular pathways that regulate cell division in eukaryotes are evolutionarily conserved. Aberrations in the control of the cell cycle are common in degenerative diseases and cancer. Therefore, analysis of the cell cycle in nonmammalian organisms can illuminate the processes underlying human development and disease. Forward-genetic screens in yeast and

ESSENTIAL ZEBRAFISH METHODS:
CELL AND DEVELOPMENTAL BIOLOGY
183
DOI: 10.1016/B978-0-12-374599-6.00010-8

Drosophila have been invaluable for gene discovery and have made important contributions to understanding pathways regulating cell proliferation. Importantly, it has been found that the human orthologs of some genes identified in these organisms are misexpressed in human tumors (Hariharan and Haber, 2003). Zebrafish have proven to be an excellent model of early vertebrate development (Driever *et al.*, 1996; Haffter *et al.*, 1996) and also of a wide variety of human diseases such as cancer, anemia, cardiovascular defects, neuromuscular conditions, kidney disease, and host-pathogen interaction, to name a few examples (Ackermann and Paw, 2003; Bassett and Currie, 2003; Drummond, 2005; Goessling *et al.*, 2007; Hsu *et al.*, 2007; Lambrechts and Carmeliet, 2004; Miller and Neely, 2004).

The particular advantages that make zebrafish ideal for developmental embryology—including external fertilization of oocytes, transparent embryos, and rapid embryonic development—also provide the opportunity to study early cell divisions, tissue-specific cellular proliferation and, more broadly, the role of cell-cycle genes in development and disease. A number of methods and markers have been successfully applied to investigate the cell cycle in zebrafish embryos, including video microscopy (Kane, 1999; Kane *et al.*, 1992), histone-GFP fusions (Pauls *et al.*, 2001), BrdU labeling (Baye and Link, 2007; Link *et al.*, 2000), proliferating cell nuclear antigen (PCNA) RNA and protein expression (Koudijs *et al.*, 2005; Wullimann and Knipp, 2000), phosphohistone H3 (pH3) immunohistochemistry (Shepard *et al.*, 2005), and minichromosome maintenance protein expression (Ryu and Driever, 2006).

Studies of the developing zebrafish embryo have revealed similarities to the early cell divisions of other vertebrates, such as *Xenopus*. In the zebrafish, the first seven cell divisions are synchronous and cycle rapidly between DNA replication (S phase) and mitosis (M phase) without the intervening gap phases, G1 and G2 (Kimmel *et al.*, 1995). The mid-blastula transition (MBT) ensues during the 10th cell division, which is approximately 3 hpf. MBT is accompanied by loss of division synchrony, increased cell-cycle duration, activation of zygotic transcription, and the onset of cellular motility (Kane and Kimmel, 1993). Embryonic cells first exhibit a G1 gap phase between the M and S phases during MBT. Recently, Dalle Nogare *et al.* demonstrated that during cycles 11–13, embryonic cells acquire a G2 phase in a transcription-independent fashion, through inhibition of Cdk1 and its activating phosphatase, Cdc25a (Dalle Nogare *et al.*, 2008).

Further understanding of cell-cycle regulation in zebrafish embryos was obtained by studying their responses to various cell-cycle inhibitors, including aphidocolin, hydroxyurea, etoposide, camptothecin, and nocodazole (Ikegami *et al.*, 1997a,b, 1999). Exposure to these agents after MBT induces cell-cycle arrest, sometimes accompanied by initiation of an apoptotic program. However, prior to MBT, the embryonic cells continue to divide, often with deleterious effects, after exposure to cell-cycle inhibitors. These studies indicate that zebrafish embryos do possess cell-cycle checkpoints, but they are not functional until after MBT.

Later developmental stages of zebrafish embryogenesis provide the opportunity to study the cell cycle in distinct tissue types. Studies of cell-cycle regulation in older embryos (10–36 hpf) have focused on the developing eyes and central nervous system. Lineage analysis of CNS progenitor cells revealed a correlation between morphogenesis and cell-cycle number, implying that the nervous system development may be at least partially regulated by the cell cycle (Kimmel *et al.*, 1994). Whereas most developing vertebrate embryos exhibit a constant lengthening of the cell-cycle duration throughout development, meticulous analysis of cell number in the developing zebrafish retina revealed a surprising mechanism of modulated cell-cycle control. Li *et al.* (2000) reported that the retinal cell-cycle duration temporarily slows between 16 and 24 hpf, followed by an abrupt change to more rapid cell divisions.

Several studies have elucidated a role for the zebrafish cell-cycle machinery in tissue differentiation during development and in the regenerative response to injury. Bessa *et al.* found that Meis1, a marker of the eye primordium, promotes G1-S progression and a block of differentiation in the zebrafish eye through regulation of Cyclin D1 and c-Myc expression (Bessa *et al.*, 2008). Fischer and coworkers showed that loss of caf1b in zebrafish (by mutation or MO injection) leads to an S-phase arrest and eventual apoptosis that can be rescued by p53 deficiency. However, loss of caf1b also leads to a block in differentiation in tissues that express caf1b, implicating caf1b in the switch from proliferation to differentiation (Fischer *et al.*, 2007). The effect of loss of early mitotic inhibitor 1 (emi1) on somite formation was evaluated by Zhang *et al.* These authors found that cell-cycle progression was required for proper somite morphogenesis, but not for formation of the segmentation clock (Zhang *et al.*, 2008). The role of the cell cycle in regeneration has also been assessed. Certain traumas result in loss of hair cell precursors, which results in deafness in vertebrates. Hernández and coworkers used BrdU labeling and transgenic GFP reporter lines to study hair cell regeneration, identifying proliferation-dependent and -independent mechanisms of hair cell renewal (Hernandez *et al.*, 2007).

A. Forward-Genetic Screens

Several groups have carried out forward-genetic screens to identify mutations that alter cell proliferation in embryos. Shepard *et al.* used phosphohistone H3 (pH3) as a marker of cell proliferation in a two-generation haploid genetic screen. They identified seven mutant lines with different alterations in pH3 immunoreactivity. At least two of these lines demonstrate aneuploidy and increased cancer susceptibility as heterozygotes (Shepard *et al.*, 2005, 2007). Using a similar screening strategy, Pfaff *et al.* identified a further set of genes required for cell-proliferation mutants, among which was Scl-interrupting locus (SIL), which was identified as a novel, vertebrate-specific regulator of mitotic spindle assembly (Pfaff *et al.*, 2007). Koudijs and coworkers used proliferating cell nuclear antigen (PCNA) expression in the CNS as a readout to identify new mutations in repressors of the hedgehog (Hh) signaling pathway (Koudijs *et al.*, 2005). Finally, another screen for genes that control eye growth uncovered two zebrafish lines mutant for the anaphase-promoting complex/cyclosome (APC/C) (Wehman *et al.*, 2006). Loss of APC/C results in a loss of mitotic progression and apoptosis; in this study, colabeling with BrdU and pH3 revealed cells undergoing mitotic catastrophe.

In this chapter, we provide protocols to characterize the various phases of cell division in zebrafish embryos and protocols to detect DNA damage, senescence, and cell death. Assays discussed in this chapter include: DNA content analysis by flow cytometry, whole-mount embryonic antibody staining, mitotic spindle analysis, BrdU incorporation, cell death analysis, and *in situ* hybridization with cell-cycle regulatory genes. Each assay targets different phases of the cell cycle and in total create a detailed picture of zebrafish embryo cell proliferation. Although our studies have focused on embryonic assays for cell-cycle characterization, it is likely that these protocols can be modified to study adult tissues. These protocols can be applied to a variety of experiments, such as characterization of the cell-cycle phenotypes of mutants or the analysis of RNA overexpression and morpholino knockdown of cell cycle regulatory genes. Furthermore, the genetic tractability of the zebrafish system (Patton and Zon, 2001) makes it an excellent organism in which to pursue forward-genetic screens for mutations or chemical screens for novel compounds that alter cell division using one or more of these cell-cycle assays.

II. Zebrafish Embryo Cell–Cycle Protocols[1]

A. Analysis of Cell Proliferation and Mitosis

1. DNA Content Analysis

A profile of the cell cycle in disaggregated zebrafish embryos or adult tissue can be obtained through DNA content analysis. In this technique, cells are stained with a dye that fluoresces upon DNA binding, such as Hoechst 33342 or propidium iodide. The intensity of fluorescence is proportional to the amount of DNA in each cell (Krishan, 1975). Analysis by fluorescence activated cell sorting (FACS) generates a histogram showing the proportion of cells that have an unreplicated complement of DNA (G1 phase), those that have a fully replicated complement of DNA (G2 or M phase) and those that have an intermediate amount of DNA (S phase).

a. Protocol

All steps are performed on ice except for the dechorionation (step 1) and RNAse incubation (step 9).

1. Dechorionate embryos and wash with **E3**. Analysis of single embryos is possible, though in practice we typically pool approximately 40 embryos/tube.

---

[1] Items in boldface indicate reagents and supplies listed in Section V.

2. Disaggregate embryos (using small **pellet pestle**) in 500 μL of DMEM (or other tissue culture medium) + 10% fetal calf serum in a matching **homogenizing tube**.

3. Bring volume to 1 mL with DMEM/serum and remove aggregates by passing cell suspension sequentially through **105** and **40 μm mesh**.

4. Count a sample using a hemocytometer.

5. Place volume containing at least 2×10^6 cells in a 15 mL conical tube, and bring volume to 5 mL with $1\times$ **PBS**.

6. Spin at 1200 rpm for 10 min at 4 °C.

7. Carefully aspirate off liquid and gently resuspend cell pellet in 2 mL **Propidium Iodide solution**.

8. Add 2 μg of DNAse-free RNAse (Roche). This step is necessary to remove double-stranded RNA, which binds propidium iodide.

9. Incubate in the dark at room temperature for 30 min.

10. Place samples on ice and analyze on FACS machine.

Note: Samples can also be fixed in ethanol, allowing multiple samples or time points to be collected for subsequent analysis.

1. Harvest cells and prepare single-cell suspension in DMEM/serum as above, steps 1–4.

2. Wash cells in PBS and resuspend at 1–2×10^6 cells/mL.

3. To 1 mL cells in a 15 mL polypropylene, V-bottom tube add 3 mL ice-cold absolute EtOH. To avoid clumping, add the ethanol dropwise while vortexing the sample.

4. Fix cells for at least 1 h at 4 °C. Cells may be stored for several weeks at −20 °C before undergoing PI staining.

5. Wash cells twice in $1\times$ PBS. We typically increase the speed of centrifugation to 2500 rpm because the cells do not pellet as readily after EtOH fixation.

6. Resuspend the pellet in 1 mL **Propidium Iodide solution**. Add 2 μg of DNAse-free RNAse (Roche) and incubate 3 h at 4 °C.

7. Place samples on ice and analyze on FACS machine.

2. Whole-Mount Immunohistochemistry with Mitotic Marker Phosphohistone H3

Histone H3 phosphorylation is considered to be a crucial event for the onset of mitosis and this antibody has been widely used in *Drosophila* and mammalian cell lines as a mitotic marker (Hendzel *et al.*, 1997). Two members of the Aurora/AIK kinase family, Aurora A and Aurora B, phosphorylate histone H3 at the serine 10 residue (Chadee *et al.*, 1999; Crosio *et al.*, 2002). Increased serine 10 phosphorylation of histone H3 has been seen in transformed fibroblasts (Chadee *et al.*, 1999), suggesting that this antibody could make an excellent marker for cell proliferation in the zebrafish as well as detecting cell-cycle mutations that may result in transformed phenotypes. In zebrafish, the phosphohistone H3 antibody (pH3) stains mitotic cells throughout the embryo (Fig. 1A). pH3 staining in developing organs like the nervous system increases as they undergo proliferation during distinct developmental stages.

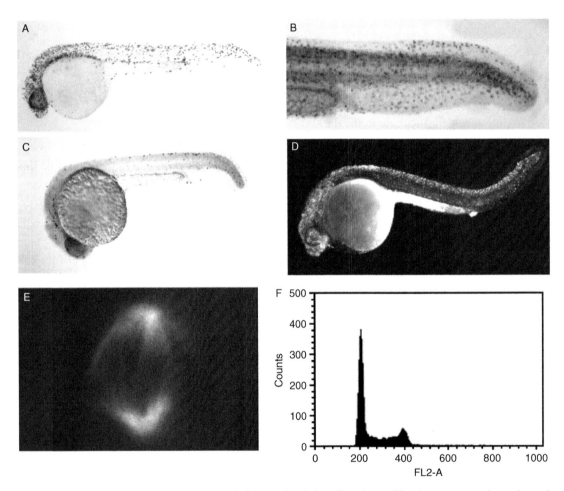

Fig. 1 Useful techniques for the study of the cell cycle, proliferation or apoptosis as shown in zebrafish embryos. (A) Antibody staining against phosphorylated histone H3 in wild-type 24 hpf embryos. (B) BrdU incorporation to mark cells in S phase in the tail of a 28 hpf wild-type embryo. (C, D) Apoptotic cells can be visualized by TUNEL (C; wild-type 24 hpf embryos) or acridine orange (D, 24 hpf *crash&burn* mutant embryo). (E) Anti-alpha tubulin can be used to examine mitotic spindle formation. (F) DNA content analysis shows the population of embryonic cells present in all phases of the cell cycle.

a. Protocol

1. Fix embryos overnight at 4 °C in 4% paraformaldehyde (**PFA**).
2. Permeabilize embryos for 7 min in −20 °C acetone.
3. Wash embryos in H_2O followed by 2× 5 min washes in **PBST**.
4. Incubate for 30 min at room temperature in **block**.
5. Incubate overnight at 4 °C in rabbit anti-phosphohistone H3 at a concentration of 1.33 mg/mL in **block**. Two different sources of antibody have been used: Santa Cruz Biotechnology and an anti-phosphopeptide polyclonal antibody to the sequence (ARKS[PO_4]TGGKAPRKQLC) made and affinity purified by Genemed Synthesis.
6. Wash 4× 15 min in **PBST**.
7. Incubate 2 h at room temperature in horseradish peroxidase-conjugated secondary goat anti-rabbit IgG (Jackson Immunoresearch) at a concentration of 3 μg/mL in **block**.
8. Wash 4× 15 min in **PBST**.
9. Develop in the dark for 3–5 min at room temperature in diaminobenzidine (DAB) solution (0.67 mg/mL DAB in 15 mL of **PBST** to which 12 μL of 30% H_2O_2 has been added).
10. Wash in **PBST** and store embryos at 4 °C in **PFA**.

3. Mitotic Spindle/Centrosome Detection

Study of the mitotic spindle and centrosomes is an important step in understanding mutants with cell-cycle defects, particularly those whose phenotypes appear to be related to problems in mitosis. Genomic instability is one of the main alterations seen in human cancers and such unequal segregation of chromosomes can be caused by problems in mitotic spindle formation or centrosome number (Kramer *et al.*, 2002). In this protocol, anti-α-tubulin labels the mitotic spindle, anti-γ-tubulin the centrosome, and DAPI the DNA.

a. Protocol

1. Fix embryos in **PFA** for 4 h at room temperature.
2. Dehydrate in methanol at −20 °C for at least 30 min.
3. Rehydrate embryos in graded methanol:**PBST** series (3:1, 1:1, 1:3) for 5 min each.
4. Wash 1× 5 min in **PBST**.
5. Place in −20 °C acetone for 7 min.
6. Wash 3× 5 min in **PBST**.
7. Incubate 1 h at room temperature in **block**.

8. Incubate in monoclonal mouse α-tubulin antibody (Sigma) at a concentration of 1:500 and in polyclonal rabbit γ-tubulin antibody (Sigma) at a concentration of 1:1000 (both diluted in block) at 4 °C overnight.

9. Wash 4× 15 min in **PBST**.

10. Incubate in rhodamine-conjugated goat anti-mouse secondary (Molecular Probes) at 1:600 dilution and fluorescein-conjugated goat anti-rabbit secondary (Jackson Immunoresearch) at 1:600 dilution for 2 h room temperature.

11. Wash 2× 15 min in **PBST**.

12. Include a 1:500 dilution of 100 μM **DAPI** during the third wash to stain DNA.

13. Wash 2× 15 min in **PBST**.

14. For observation by epifluorescence microscopy, embryos are mounted on glass slides with VectaShield mounting media (Vector Labs) and coverslipped. To permit the specimen to lie flat, is helpful to remove the yolk using forceps or a tungsten needle. Alternatively, for embryos >18 hpf, the tail can be cut off from the embryo and mounted on the slide.

4. BrdU Incorporation

5-Bromo-2-deoxyuridine (BrdU) is a nucleoside analog that is specifically incorporated into DNA during S phase (Meyn *et al.*, 1973) and can subsequently be detected with an anti-BrdU-specific antibody. This technique has been used to label replicating cells in zebrafish embryos (Larison and Bremiller, 1990) and adults (Rowlerson *et al.*, 1997). The following protocol is designed to label a fraction of proliferating cells in zebrafish embryos, to allow comparison of the replication fraction of different embryos (Fig. 1B). If the embryos are chased for varying amounts of time after the BrdU pulse, then fixed and stained for both BrdU and pH3 (Section II.B), the transit of cells from S phase into G2/M can be assessed. This is useful in analyzing mutants with mitotic phenotypes.

a. Protocol

1. Dechorionate embryos and chill 15 min on ice in **E3**.

2. Prepare cold 10 mM BrdU/15% dimethylsulfoxide in E3 and chill on ice. Place embryos in BrdU solution and incubate 20 min on ice to allow uptake of BrdU.

3. Change into warm **E3** and incubate exactly 5 min, 28.5 °C. *Note*: longer incubation times will result in more cells being labeled.

4. Fix 2 h, room temperature in **PFA**. Longer fixation may decrease the staining.

5. Transfer to methanol at −20 °C. overnight. All subsequent steps are performed at room temperature unless otherwise noted.

6. Rehydrate in graded methanol:**PBST** series (3:1, 1:1, 1:3) for 5 min each.

7. Wash 2× in **PBST**, 5 min.

8. Digest embryos in 10 μg/mL proteinase K, 10 min.

9. Wash **PBST**. Refix in **PFA** for not more than 20 min.

10. Wash quickly 3× in H_2O, then 2× in 2 N HCl.

11. Incubate 1 h in 2 N HCl. This step denatures the labeled DNA to expose the BrdU epitope.

12. Remove the 2 N HCl solution from the embryos and neutralize in 0.1 M borate buffer, pH 8.5, 20 min, room temparature.

13. Rinse several times in **PBST**. Block for 30 min in **BrdU blocking solution**.

14. Incubate in monoclonal **anti-BrdU** antibody at a dilution of 1:100 in **BrdU block** for 2 h at room temperature or overnight at 4 °C. (If carrying out simultaneous BrdU/pH3 staining, add the primary anti-pH3 antibody as described in Section II.B, except that **BrdU block** is used.)

15. Wash 5× 10 min in **PBST**.

16. Incubate 2 h room temperature with horseradish peroxidase or fluorophore-conjugated anti-mouse secondary antibody. (For simultaneous BrdU/pH3 stain, add a fluorescent anti-rabbit antibody as well.)

17. Wash 5× 10 min in **PBST**. If using fluorescent secondary, mount embryos as described in Section II.C, step 14.

18. If using HRP-conjugated secondary antibody, develop in the dark for 3–5 min at room temperature in diaminobenzidine (DAB) solution (0.67 mg/mL DAB in 15 mL of **PBST** to which 12 μL of 30% H_2O_2 has been added). When staining is complete, wash 3× 5 min in **PBST**, then fix in **PFA**.

B. Analysis of DNA Damage, Senescence, and Apoptosis

1. COMET Assay

The COMET Assay, also known as the single-cell microgel electrophoresis (SCGE) assay, is a highly sensitive technique that is used to detect DNA damage at the single-cell level (Singh *et al.*, 1988). Cells are embedded into a thin agarose gel, through which a current is run allowing for migration of DNA. Smaller fragments of DNA, resulting from DNA damage, will travel more quickly and appear as a tail to the nucleus "comet head." The comets can be visualized using a nuclear stain, such as SYBR green, and visualized under a fluorescent microscope (Figure 2). The following protocol is designed to isolate cells from zebrafish embryos and detect any kind of DNA damage. Variations of this technique allows for specific detection of double-stranded breaks.

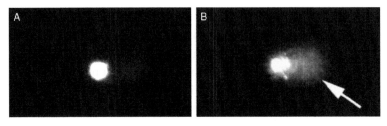

Fig. 2 The COMET assay reveals double-strand DNA breaks. Images of zebrafish embryo cells from a single-cell microgel electrophoresis experiment. (A) Unirradiated control cell. (B) A cell after exposure to 2000R (20 Gy) gamma irradiation. The arrow indicates the comet tail, composed of fragmented DNA (Verduzco and Amatruda, unpublished).

a. Protocol

1. Dechorionate embryos and wash with **E3**. Typically, about 25–50 embryos are used.
2. Disaggregate embryos (using small **pellet pestle**) in 500 μL of DMEM (or other tissue culture medium) + 10% fetal calf serum or lamb serum in a matching **homogenizing tube**.
3. Bring volume to 1 mL with DMEM/serum.
4. Count the samples using a hemocytometer.
5. Spin down cells at 3000 rpm and resuspend in **PBS** to a concentration of 1×10^5 cells/mL.
6. Combine 10 μL of cells with 90 μL of molten LMAgarose (Trevigen) prewarmed to 37 °C. Pipette 75 μL of the cell agarose mixture onto a CometSlide (Trevigen) prewarmed to 37 °C.
7. Incubate the slide flat at 4 °C for 30 min in the dark to allow the gel to solidify.
8. Immerse the slide in Lysis solution (Trevigen) containing 9% DMSO. After this point, it is very important to retain your slide in low-light conditions.
9. Dry off the slide, and immerse it in **alkaline solution** for 30 min.
10. Prepare a large horizontal electrophoresis apparatus by filling the chamber with fresh **alkaline electrophoresis buffer** and adjusting the volume of the alkaline electrophoresis buffer such that the current is 300 mA when the voltage is set to 25 V. Additionally, the chamber should be prepared and used in a 4 °C room.
11. Place the slide in the electrophoresis apparatus. Run for 30 min at 4 °C in the dark.
12. Dry off the slide. Rinse by dipping in ddH$_2$O.
13. Incubate the slide in 70% EtOH for 5 min at RT in the dark.
14. Air dry the slide for 1 hour.

15. Pipette 50 μL of SYBR green staining solution (Trevigen) onto the microgel on the slide.

16. View the slide using epifluorescence microscopy under a fluorescein filter.

17. Comets can be analyzed using CometScore by Tritek Corp or another similar software program.

2. Detection of Senescence-Associated Beta Galactosidase

The study of cellular senescence was initiated by Hayflick and Moorhead (1961). Cellular senescence pertains to the cessation of cell replication and certain morphological and transcriptional changes that occur when cells permanently cease dividing. Although it is unclear whether the events that occur during *in vitro* cellular senescence also occur during organismal aging (Hayflick, 2007; Masoro, 2006) studies have revealed strong connections between cellular senescence, cancer, and age-related diseases (Campisi, 2005). Cellular senescence most likely arose evolutionarily as a mechanism to defend against tumorigenesis (Shay and Roninson, 2004). When a cell is afflicted by stress that may result in transformation (such as oxidative stress, DNA damage, or overepxpression of oncogenes) tumor-suppressor genes such as p53 may force the cell to undergo senescence-induced arrest. Arrested cells are functional but are not a risk for tumor initiation. Senescence also occurs as the ends of chromosomes, the telomeres, shorten. During reach replication cycle, if no active telomerase is present (Bodnar *et al.*, 1998), the telomeres shorten, leading eventually to critically short telomeres which may interfere with gene expression and genomic stability (Shay and Wright, 2006). Normal cells senesce before telomeres shorten to the point of causing genomic instability, therefore instilling a counting mechanism which confirms Hayflick's observation in 1961 (Shay and Wright, 2006).

Senescent cells lose sensitivity to mitogens or growth factors, repress cell-cycle genes such as cdk2, and become insensitive to apoptotic signals. Morphological changes occur resulting in an enlarged shape and flattened body (Ben-Porath and Weinberg, 2005), as well as expression of unique markers, many of unknown function, such as β-galactosidase activity at pH 6.0 (Dimri *et al.*, 1995). Kishi and coworkers have used senescence associated as β-galactosidase staining in several studies to characterize senescence in normal and mutant zebrafish embryos and during aging of zebrafish adults (Kishi, 2004; Kishi *et al.*, 2003, 2008; Tsai *et al.*, 2007).

a. Protocol

We have used the Senescence-Associated Beta-Galacotsidase Detection Kit from Sigma (CS 0030). The following protocol adapts the manufacturer's instructions specifically for use with zebrafish embryos, and is kindly provided by Jenny Richardson and Dr. Elizabeth Patton, Edinburgh Cancer Research Centre:

1. Dechorionate embryos and add 1.5 mL of 1× fixation buffer (prepared from **10× Sigma Senescence Fixation Buffer**). Incubate overnight at 4 °C.

2. Wash embryos 4 times in 1× PBS, 1 h each wash.

3. Make up the **Senescence Staining Mixture** as per the manufacturer's protocol. Add 1 mL to embryos and incubate for 24 h at 37 °C.

4. Wash embryos 3 times in 1× PBS, 10 min each wash.

5. Embryos can be stored at 4 °C in 1× PBS and 0.1% NaN_3 or in 70% glycerol at 4 °C.

An alternative protocol was described by Dr. Shuji Kishi and coworkers in a recent paper describing a senescence-based genetic screen (Kishi *et al.*, 2008). The following protocol is adapted from Kishi *et al.* (2008):

1. Fix embryos or adult zebrafish in 4% paraformaldehyde in 1× PBS at 4 °C (for 3 days in adults and overnight in embryos).

2. Wash 3× for 1 h in PBS-pH 7.4 and for a further 1 h in PBS-pH 6.0 at 4 °C.

3. Stain the samples overnight at 37 °C in 5 mM potassium ferrocyanide, 5 mM potassium ferricyanide, 2 mM $MgCl_2$, and 1 mg/mL X-gal in PBS adjusted to pH 6.0.

3. Apoptosis Detection by TUNEL Staining

Apoptosis is a form of programmed cell death that eliminates damaged or unneeded cells. It is controlled by multiple signaling pathways that mediate responses to growth, survival, or death signals. Cell-cycle checkpoint controls are linked to apoptotic cascades and these connections can be compromised in diseases, including cancer. The defining characteristics of apoptosis are membrane blebbing, cell shrinkage, nuclear condensation, segmentation, and division into apoptotic bodies that are phagocytosed (Wyllie, 1987). The DNA strand breaks that occur during apoptosis can be detected by enzymatically labeling the free ends with modified nucleotides which can then be detected with antibodies (Gavrieli *et al.*, 1992).

a. Protocol

1. Embryos are fixed overnight at 4 °C in **PFA**.

2. Wash in **PBS** and transfer to methanol for 30 min at −20 °C.

3. Rehydrate embryos in a graded methanol:**PBST** series (3:1, 1:1, 1:3) for 5 min each.

4. Wash 1× 5 min in **PBST**.

5. Digest embryos in proteinase K (10 μg/mL) at room temperature (1 min for embryos younger than 16 hpf, 2 min for embryos older than 16 hpf).

6. Wash twice in **PBST**.

7. Postfix in **PFA** for 20 min room temperature.
8. Wash 5× 5 min in **PBST**.
9. Postfix for 10 min at −20 °C with prechilled ethanol:acetic acid 2:1.
10. Wash 3× 5 min in **PBST** at room temperature.
11. Incubate for 1 h at room temperature in 75 μL equilibration buffer (TdT-Apop Tag Peroxidase *In Situ* Apoptosis Detection Kit from Serologics Corporation).
12. Add small volume of working strength TdT (reaction buffer and TdT at a ratio of 2:1 plus 0.3% Triton) (Serologics Corporation).
13. Incubate overnight at 37 °C.
14. Stop reaction by washing in working strength stop/wash buffer (1 mL concentrated buffer from Serologics Kit with 34 mL water) for 3–4 h at 37 °C.
15. Wash 3× 5 min in **PBST**.
16. Block with 2 mg/mL BSA, 5% sheep serum in **PBST** for 1 h at room temperature.
17. Incubate in anti-digoxigenin peroxidase antibody included in kit (full strength).
18. Wash 4× 30 min **PBST** at room temperature.
19. Develop in the dark for 5 min at room temperature in diaminobenzidine (DAB) solution (0.67 mg/mL in 15 mL of **PBST**) and 12 μL 30% H_2O_2.
20. Wash in **PBST** and store embryos at 4 °C in **PFA**.

4. Apoptosis Detection by Acridine Orange

Another method of apoptotic cell detection that can be performed on living embryos is acridine orange staining. The basis of this method is that the ATP-dependent lysosomal proton pump is preserved in apoptotic but not necrotic cells; therefore apoptotic cells will take up the acridine orange dye whereas living or necrotic cells will not (Darzynkiewicz *et al.*, 1992). This method is useful for identifying mutants based on an apoptotic phenotype in order to further characterize them in living assays.

a. Protocol

1. Live dechorionated embryos are incubated in a 2 μg/mL solution of acridine orange (Sigma) in 1× **PBS** for 30 min at room temperature.
2. Embryos are washed 5× quickly in **E3**, then 5 × 5 minutes in **E3** and then visualized on a stereo dissecting microscope equipped for FITC epifluorescence.

C. *In Situ* Hybridization

RNA expression analysis by *in situ* hybridization of antisense probes in whole-mount zebrafish embryos is a commonly used technique to localize expression of developmental regulatory genes. While the technique is not exceptionally quantitative, it can reveal stark differences in gene expression. More quantitative analysis of gene expression, such as Northern blotting, RT-PCR, or real-time PCR do not permit the examination of alterations in tissue-specific expression or an expression pattern.

Cell division is a highly controlled process that involves regulation at both the transcriptional and posttranslational stages. Cyclins are a class of proteins that play critical roles in guiding cells through the G1, S, G2, and M phases of the cell cycle by regulating the activity of the cyclin-dependent kinases. The name cyclin alludes to the fact that their expression levels oscillate between peaks and nadirs that are coordinated with particular phases of the cell cycle (reviewed in Murray, 2004). The tightly regulated expression of these important cell-cycle genes incorporates transcriptional, translational, and posttranslational controls. Many genes involved in cell-cycle regulation are specifically expressed during the cell-cycle phase in which they act.

Zebrafish orthologs of cell cycle regulatory genes such as PCNA and cyclins have been found to possess similar expression patterns throughout the proliferative tissues of developing zebrafish embryos (C. Thisse and B. Thisse, unpublished and www.zfin.org). *In situ* hybridization for cell cycle regulatory genes can be performed using previously published *in situ* hybridization protocols (Jowett, 1999; Thisse *et al.*, 1993, 1994).

III. Screening for Chemical Suppressors of Zebrafish Cell-Cycle Mutants

Another way to probe the cell cycle is via chemical agents. Chemical screens could identify novel compounds that are useful tools for studying the cell cycle. Furthermore, mutations in cell-cycle genes are commonly found in human cancer. Given the need to improve upon current cancer therapy, one approach is to identify small molecule suppressors that bypass the consequences of specific cell-cycle gene mutations. Akin to the use of genetic modifier screens to identify secondary mutations that enhance or suppress a primary defect (St Johnston, 2002), chemical suppressor screens would directly identify small molecules that rescue a genetic phenotype. If the phenotype is disease-related, such compounds might represent lead therapeutic agents.

Zebrafish have recently been utilized in chemical screens to identify compounds that perturb specific aspects of development (Anderson *et al.*, 2007; Bayliss *et al.*, 2006; den Hertog, 2005; Khersonsky *et al.*, 2003; Peterson *et al.*, 2000, 2004). The zebrafish system offers several advantages for chemical screens, providing information on tissue specificity and toxicity, and accounting for compound activation via drug metabolism. Furthermore, cells are not transformed and are in their normal physiological milieu of cell-cell and cell-extracellular matrix interactions. Murphey and coworkers carried out a high-throughput chemical screen to detect small molecules capable of perturbing the cell cycle during zebrafish development, identifying several compounds that were not previously detected in cell-based screens of the same library (Murphey *et al.*, 2006). As another application of this technique, Stern *et al.* screened a 16,000-compound library to identify small molecules capable of suppressing the cell-proliferation defect in the *crash and burn* cell-cycle mutant (Stern *et al.*, 2005). This technology could easily be applied to other cell-cycle mutants and could be modified to use cell-cycle assays other than pH3 staining. In addition, such chemical suppressor screens could be applied to any zebrafish model of human disease (Dooley and Zon, 2000). For these reasons, we provide a detailed protocol below. The original version of this protocol was designed by Ryan Murphey and Howard M. Stern in the Zon laboratory.

The following protocol can be repeated weekly giving a throughput of over 1000 compounds per week for a recessive lethal mutation. In the case of homozygous viable mutants, the throughput could be improved by using fewer embryos (3–5) per well in 96-well plates.

A. Protocol

1. For a chemical screen, large numbers of embryos at approximately the same developmental stage need to be generated. Set up 100 heterozygote pairwise matings with fish separated by a divider. The next morning, remove the divider, allow the fish to mate, and collect the embryos.

2. Dilute chemicals into **screening medium**. The screen is conducted in 48-well plates with a volume of 300 mL per well. Individual chemicals could be added to each well, but to improve throughput, we devised a matrix pooling strategy: The chemical library (courtesy of the Institute of Chemistry and Cell Biology, Harvard Medical School) was arrayed in 384-well plates with the last four columns empty, thus containing 320 compounds per plate. Given this plate geometry, 8×10 matrix pools were created. A hit detected in both a horizontal and a vertical pool identified the individual compound (Fig. 3A).

   a. Transfer 80 mL of screening medium to each well of four 384-well plates using a TECAN liquid handling robot.

   b. Pin transfer 1 mL of each compound (arrayed at 5 mg/mL in **DMSO**) into each well of screening medium by performing 10 transfers with a 100 nL 384-pin array for each of the four 384-well plates (total of $320 \times 4 = 1280$ compounds).

   c. Pooling was performed with a TECAN liquid handling robot by pipeting the diluted chemicals from the 384- to 48-well plates. For vertical pools, 30 mL was transferred from each of eight wells plus an additional 60 mL of screening medium to bring the total volume to 300 mL. For horizontal pools, 30 mL was transferred from each of 10 wells.

3. Aliquot embryos to the 48-well plates at 50% epiboly.

A

1	2	3	4	5	6	7	8	9	10	A
11	12	13	14	15	16	17	18	19	20	B
21	22	23	24	25	26	27	28	29	30	C
31	32	33	34	35	36	37	38	39	40	D
41	42	43	44	45	46	47	48	49	50	E
51	52	53	54	55	56	57	58	59	60	F
61	62	63	64	65	66	67	68	69	70	G
71	72	73	74	75	76	77	78	79	80	H
I	J	K	L	M	N	O	P	Q	R	

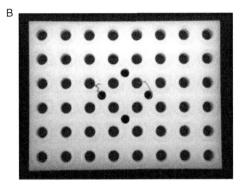

Fig. 3 Chemical screen methods. (A) Example of matrix pooling. In an 8 by 10 matrix pool, 80 compounds (numbered 1–80) are screened in 18 pools (A–R). Each compound is thus tested in two distinct pools. Individual active compounds are identified by deconvoluting the matrix. For example, if a hit is identified in pools B and P, the active compound would be number 18. (B) Photograph of a 48-well staining grid.

    a. Prior to aliquoting embryos to wells, examine them under a dissecting microscope and discard all dead, delayed or deformed embryos.

    b. Pool embryos in a single 100 mm tissue culture dish or a 50 mL conical tube.

    c. Decant the embryo medium and remove as much liquid from the embryo suspension as possible with a transfer pipet. Pressing the transfer pipet tip to the bottom of the tube or dish allows most liquid to be removed without aspirating the embryos.

    d. Add approximately 20 embryos to each well by scooping them with a small chemical weighing spatula. With 20 embryos per well and a Mendelian recessive inheritance, there is a 0.3% chance of a well having no mutants. Since a hit requires detection in both a horizontal and a vertical pool, each with 20 embryos, the false-positive rate for identification of complete suppressors is 0.001%.

4. Place 48-well plates into an incubator at 28.5 °C.

5. One to two hours later, clean out any dead embryos from each well using a long glass Pasteur pipet bent at a 90 °.

6. Incubate at 28.5 °C overnight.

7. Dechorionate embryos by adding 150 mL of a 5 mg/mL pronase solution to each well. After 10 min, gently shake plates until embryos come out of the chorions.

8. Using a transfer pipet fitted with a 10 mL tip, remove as much of the pronase/chemical mixture as possible from each well.

9. Rinse the embryos once in fresh embryo medium and remove as in step 8.

10. Add 500 mL of **PFA** to each well.

11. Parafilm the edges of the plates to prevent evaporation and fix at 4 °C at least overnight but not longer than a week.

12. Using a transfer pipet, move embryos to 48-well staining grids made of acetone resistant plastic with a wire mesh bottom (Fig. 3B).

13. Perform the pH3 staining protocol by placing staining grids into 11 × 8.5 cm reservoirs containing 20–30 mL of the appropriate solution. To change solutions, the grid can be lifted out of one reservoir and placed into another reservoir with the next solution. For overnight antibody incubations, the reservoir should be sealed with parafilm to prevent evaporation.

14. After staining is complete, move embryos with a transfer pipet back into 48-well plates that have been precoated with 100 mL of 1% agarose in 1× **PBS**. The agarose forms a meniscus that keeps embryos in the center of the well where they are easier to score.

15. Score for absence of mutants or for partial suppression without effect on wild types. In addition to suppressors and enhancers, one can identify compounds that affect both wild types and mutants, thus having a more general effect.

16. Deconvolute matrix pool to identify individual chemicals.

IV. Conclusions

Given the power of zebrafish in forward-vertebrate genetics and organism-based small molecule screens, the system will nicely complement traditional model organisms for studying the cell division cycle. Many of the assays that are commonly used to probe the cell cycle in systems such as yeast, *Drosophila*, and mammalian cells can be used in the zebrafish. The protocols outlined in this chapter can be utilized to characterize known mutants for alterations in cell proliferation or, alternatively, can be used to screen for more cell-cycle mutants. Given that zebrafish embryos are amenable to gene knockdown via antisense morpholino-modified oligonucleotides and overexpression by mRNA injection, these protocols can also be used to study cell-cycle genes in the zebrafish without generating a mutant.

V. Reagents and Supplies

Alkaline solution	0.6 g NaOH, 250 μL 200 mM EDTA pH 10, to 50 mL ddH$_2$O
Alkaline electrophoresis Buffer	12 g NaOH, 2 mL 500 mM EDTA pH 8, to 1 L ddH$_2$O
Anti-BrdU	Roche cat # 1170 376
Block	2% blocking reagent (Roche 1096 176), 10% fetal calf serum, 1% dimethylsulfoxide in PBST

BrdU Block	0.2% blocking reagent (Roche 1096 176), 10% fetal calf serum, 1% dimethylsulfoxide in PBST. The lower concentration of blocking reagent improves detection
DAPI	4′,6-Diamidino-2-phenylindole
DMSO	Dimethylsulfoxide
E3	5 mM NaCl, 0.17 mM KCl, 0.33 mM $CaCl_2$, 0.33 mM $MgSO_4$
Mesh	Small Parts, inc. 105 μm mesh is cat # U-CMN-105D. 40 μm mesh is cat # U-CMN-40D
PBS	Phosphate-buffered saline, pH 7.5
PBST	1× PBS with 0.1% (v/v) Tween-20
Pellet pestle and tubes	Fisher, cat # K749520-0090
PFA	4% paraformaldehyde buffered with 1× PBS
Propidium iodide solution	0.1% sodium citrate, 0.05 mg/mL propidium iodide, 0.0002% Triton X-100 (added fresh)

Sigma senescence fixation buffer, 10× (catalog number F1797)

Contains 20% formaldehyde, 2% glutaraldehyde, 70.4 mM Na_2HPO_4, 14.7 mM KH_2PO_4, 1.37 M NaCl, and 26.8 mM KCl

Sigma senescence staining mixture
(Prepare just prior to use)

Mix the following for preparation of 10 mL of the staining mixture

1 mL of staining solution 10× buffer (catalog number S5818)
125 μL of reagent B (catalog number R5272)
125 μL of reagent C (catalog number R5147)
0.25 mL of X-gal solution (catalog number X3753)
8.50 mL of ultrapure water

| E3 Screening medium | Supplemented with 1% DMSO, 20 μM metronidazole, 50 units/mL penicillin, 50 mg/mL streptomycin, and 1 mM Tris pH 7.4 |

Acknowledgments

We thank Len Zon and members of the Zon laboratory for useful discussions. Original versions of many of these protocols were worked out by Jennifer L. Shepard, Ryan Murphey, Howard M. Stern, Kathryn L. Pfaff, and J. F. A. D. V. was supported by NIH Training Grant 5 T32 GM008203 and J. F. A. was supported by grants from the Lance Armstrong Foundation, the Amon G. Carter Foundation, the Welch Foundation and NIH/NCI grant 1R01CA135731.

References

Ackermann, G. E., and Paw, B. H. (2003). Zebrafish: A genetic model for vertebrate organogenesis and human disorders. *Front. Biosci.* **8**, d1227–d1253.

Anderson, C., Bartlett, S. J., Gansner, J. M., Wilson, D., He, L., Gitlin, J. D., Kelsh, R. N., and Dowden, J. (2007). Chemical genetics suggests a critical role for lysyl oxidase in zebrafish notochord morphogenesis. *Mol. Biosyst.* **3**(1), 51–59.

Bassett, D. I., and Currie, P. D. (2003). The zebrafish as a model for muscular dystrophy and congenital myopathy. *Hum. Mol. Genet.* **12**(Spec No 2), R265–R270.

Baye, L. M., and Link, B. A. (2007). The disarrayed mutation results in cell cycle and neurogenesis defects during retinal development in zebrafish. *BMC Dev. Biol.* **7**, 28.

Bayliss, P. E., Bellavance, K. L., Whitehead, G. G., Abrams, J. M., Aegerter, S., Robbins, H. S., Cowan, D. B., Keating, M. T., O'Reilly, T., Wood, J. M., Roberts, T. M., and Chan, J. (2006). Chemical modulation of receptor signaling inhibits regenerative angiogenesis in adult zebrafish. *Nat. Chem. Biol.* **2**(5), 265–273.

Ben-Porath, I., and Weinberg, R. A. (2005). The signals and pathways activating cellular senescence. *Int. J. Biochem. Cell Biol.* **37**(5), 961–976.

Bessa, J., Tavares, M. J., Santos, J., Kikuta, H., Laplante, M., Becker, T. S., Gomez-Skarmeta, J. L., and Casares, F. (2008). meis1 regulates cyclin D1 and c-myc expression, and controls the proliferation of the multipotent cells in the early developing zebrafish eye. *Development* **135**(5), 799–803.

Bodnar, A. G., Ouellette, M., Frolkis, M., Holt, S. E., Chiu, C. P., Morin, G. B., Harley, C. B., Shay, J. W., Lichtsteiner, S., and Wright, W. E. (1998). Extension of life-span by introduction of telomerase into normal human cells. *Science* **279**(5349), 349–352.

Campisi, J. (2005). Senescent cells, tumor suppression, and organismal aging: Good citizens, bad neighbors. *Cell* **120**(4), 513–522.

Chadee, D. N., Hendzel, M. J., Tylipski, C. P., Allis, C. D., Bazett-Jones, D. P., Wright, J. A., and Davie, J. R. (1999). Increased Ser-10 phosphorylation of histone H3 in mitogen-stimulated and oncogene-transformed mouse fibroblasts. *J. Biol. Chem.* **274**(35), 24914–24920.

Crosio, C., Fimia, G. M., Loury, R., Kimura, M., Okano, Y., Zhou, H., Sen, S., Allis, C. D., and Sassone-Corsi, P. (2002). Mitotic phosphorylation of histone H3: Spatio-temporal regulation by mammalian Aurora kinases. *Mol. Cell. Biol.* **22**(3), 874–885.

Dalle Nogare, D. E., Pauerstein, P. T., and Lane, M. E. (2008). G2 acquisition by transcription-independent mechanism at the zebrafish midblastula transition. *Dev. Biol.* **326**(1), 131–142.

Darzynkiewicz, Z., Bruno, S., Del Bino, G., Gorczyca, W., Hotz, M. A., Lassota, P., and Traganos, F. (1992). Features of apoptotic cells measured by flow cytometry. *Cytometry* **13**(8), 795–808.

den Hertog, J. (2005). Chemical genetics: Drug screens in Zebrafish. *Biosci. Rep.* **25**(5–6), 289–297.

Dimri, G. P., Lee, X., Basile, G., Acosta, M., Scott, G., Roskelley, C., Medrano, E. E., Linskens, M., Rubelj, I., Pereira-Smith, O., Peacocke, M., and Campisi, J. (1995). A biomarker that identifies senescent human cells in culture and in aging skin *in vivo*. *Proc. Natl. Acad. Sci. USA* **92**(20), 9363–9367.

Dooley, K., and Zon, L. I. (2000). Zebrafish: A model system for the study of human disease. *Curr. Opin. Genet. Dev.* **10**(3), 252–256.

Driever, W., Solnica-Krezel, L., Schier, A. F., Neuhauss, S. C., Malicki, J., Stemple, D. L., Stainier, D. Y., Zwartkruis, F., Abdelilah, S., Rangini, Z., Belak, J., and Boggs, C. (1996). A genetic screen for mutations affecting embryogenesis in zebrafish. *Development* **123**, 37–46.

Drummond, I. A. (2005). Kidney development and disease in the zebrafish. *J. Am. Soc. Nephrol.* **16**(2), 299–304.

Fischer, S., Prykhozhij, S., Rau, M. J., and Neumann, C. J. (2007). Mutation of zebrafish caf-1b results in S phase arrest, defective differentiation, and p53-mediated apoptosis during organogenesis. *Cell. Cycle* **6**(23), 2962–2969.

Gavrieli, Y., Sherman, Y., and Ben-Sasson, S. A. (1992). Identification of programmed cell death *in situ* via specific labeling of nuclear DNA fragmentation. *J. Cell. Biol.* **119**(3), 493–501.

Goessling, W., North, T. E., and Zon, L. I. (2007). New waves of discovery: Modeling cancer in zebrafish. *J. Clin. Oncol.* **25**(17), 2473–2479.

Haffter, P., Granato, M., Brand, M., Mullins, M. C., Hammerschmidt, M., Kane, D. A., Odenthal, J., van Eeden, F. J., Jiang, Y. J., Heisenberg, C. P., Kelsh, R. N., Furutani-Seiki, M., *et al.* (1996). The identification of genes with unique and essential functions in the development of the zebrafish, Danio rerio. *Development* **123**, 1–36.

Hariharan, I. K., and Haber, D. A. (2003). Yeast, flies, worms, and fish in the study of human disease. *New. Engl. J. Med.* **348**(24), 2457–2463.

Hayflick, L. (2007). Biological aging is no longer an unsolved problem. *Ann. NY Acad. Sci.* **1100**, 1–13.

Hayflick, L., and Moorhead, P. S. (1961). The serial cultivation of human diploid cell strains. *Exp. Cell Res.* **25**, 585–621.

Hendzel, M. J., Wei, Y., Mancini, M. A., Van Hooser, A., Ranalli, T., Brinkley, B. R., Bazett-Jones, D. P., and Allis, C. D. (1997). Mitosis-specific phosphorylation of histone H3 initiates primarily within pericentromeric heterochromatin during G2 and spreads in an ordered fashion coincident with mitotic chromosome condensation. *Chromosoma* **106**(6), 348–360.

Hernandez, P. P., Olivari, F. A., Sarrazin, A. F., Sandoval, P. C., and Allende, M. L. (2007). Regeneration in zebrafish lateral line neuromasts: Expression of the neural progenitor cell marker sox2 and proliferation-dependent and -independent mechanisms of hair cell renewal. *Dev. Neurobiol.* **67**(5), 637–654.

Hsu, C. H., Wen, Z. H., Lin, C. S., and Chakraborty, C. (2007). The zebrafish model: Use in studying cellular mechanisms for a spectrum of clinical disease entities. *Curr. Neurovasc. Res.* **4**(2), 111–120.

Ikegami, R., Hunter, P., and Yager, T. D. (1999). Developmental activation of the capability to undergo checkpoint-induced apoptosis in the early zebrafish embryo. *Dev. Biol.* **209**(2), 409–433.

Ikegami, R., Rivera-Bennetts, A. K., Brooker, D.L., and Yager, T. D. (1997a). Effect of inhibitors of DNA replication on early zebrafish embryos: Evidence for coordinate activation of multiple intrinsic cell-cycle checkpoints at the mid-blastula transition. *Zygote* **5**(2), 153–175.

Ikegami, R., Zhang, J., Rivera-Bennetts, A.K., and Yager, T. D. (1997b). Activation of the metaphase checkpoint and an apoptosis programme in the early zebrafish embryo, by treatment with the spindle-destabilising agent nocodazole. *Zygote* **5**(4), 329–350.

Jowett, T. (1999). Analysis of protein and gene expression. *Methods Cell Biol.* **59**, 63–85.

Kane, D. A. (1999). Cell cycles and development in the embryonic zebrafish. *Methods Cell Biol.* **59**, 11–26.

Kane, D. A., and Kimmel, C. B. (1993). The zebrafish midblastula transition. *Development* **119**(2), 447–456.

Kane, D. A., Warga, R. M., and Kimmel, C. B. (1992). Mitotic domains in the early embryo of the zebrafish. *Nature* **360**(6406), 735–737.

Khersonsky, S. M., Jung, D. W., Kang, T. W., Walsh, D. P., Moon, H. S., Jo, H., Jacobson, E. M., Shetty, V., Neubert, T. A., and Chang, Y. T. (2003). Facilitated forward chemical genetics using a tagged triazine library and zebrafish embryo screening. *J. Am. Chem. Soc.* **125**(39), 11804–11805.

Kimmel, C. B., Ballard, W. W., Kimmel, S. R., Ullmann, B., and Schilling, T. F. (1995). Stages of embryonic development of the zebrafish. *Dev. Dyn.* **203**(3), 253–310.

Kimmel, C. B., Warga, R. M., and Kane, D. A. (1994). Cell cycles and clonal strings during formation of the zebrafish central nervous system. *Development* **120**(2), 265–276.

Kishi, S. (2004). Functional aging and gradual senescence in zebrafish. *Ann. NY Acad. Sci.* **1019**, 521–526.

Kishi, S., Bayliss, P. E., Uchiyama, J., Koshimizu, E., Qi, J., Nanjappa, P., Imamura, S., Islam, A., Neuberg, D., Amsterdam, A., and Roberts, T. M. (2008). The identification of zebrafish mutants showing alterations in senescence-associated biomarkers. *PLoS Genet.* **4**(8), e1000152.

Kishi, S., Uchiyama, J., Baughman, A. M., Goto, T., Lin, M. C., and Tsai, S. B. (2003). The zebrafish as a vertebrate model of functional aging and very gradual senescence. *Exp. Gerontol.* **38**(7), 777–786.

Koudijs, M. J., den Broeder, M. J., Keijser, A., Wienholds, E., Houwing, S., van Rooijen, E. M., Geisler, R., and van Eeden, F. J. (2005). The zebrafish mutants dre, uki, and lep encode negative regulators of the hedgehog signaling pathway. *PLoS Genet.* **1**(2), e19.

Kramer, A., Neben, K., and Ho, A. D. (2002). Centrosome replication, genomic instability and cancer. *Leukemia* **16**(5), 767–775.

Krishan, A. (1975). Rapid flow cytofluorometric analysis of mammalian cell cycle by propidium iodide staining. *J. Cell. Biol.* **66**(1), 188–193.

Lambrechts, D., and Carmeliet, P. (2004). Genetics in zebrafish, mice, and humans to dissect congenital heart disease: Insights in the role of VEGF. *Curr. Top. Dev. Biol.* **62**, 189–224.

Larison, K. D., and Bremiller, R. (1990). Early onset of phenotype and cell patterning in the embryonic zebrafish retina. *Development* **109**(3), 567–576.

Li, Z., Hu, M., Ochocinska, M. J., Joseph, N. M., and Easter, S. S., Jr. (2000). Modulation of cell proliferation in the embryonic retina of zebrafish (Danio rerio). *Dev. Dyn.* **219**(3), 391–401.

Link, B. A., Fadool, J. M., Malicki, J., and Dowling, J. E. (2000). The zebrafish young mutation acts non-cell-autonomously to uncouple differentiation from specification for all retinal cells. *Development* **127**(10), 2177–2188.

Masoro, E. J. (2006). Dietary restriction-induced life extension: A broadly based biological phenomenon. *Biogerontology* **7**(3), 153–155.

Meyn, R. E., Hewitt, R. R., and Humphrey, R. M. (1973). Evaluation of S phase synchronization by analysis of DNA replication in 5-bromodeoxyuridine. *Exp. Cell Res.* **82**(1), 137–142.

Miller, J. D., and Neely, M. N. (2004). Zebrafish as a model host for streptococcal pathogenesis. *Acta Trop.* **91**(1), 53–68.

Murphey, R. D., Stern, H. M., Straub, C. T., and Zon, L. I. (2006). A chemical genetic screen for cell cycle inhibitors in zebrafish embryos. *Chem. Biol. Drug Des.* **68**(4), 213–219.

Murray, A. W. (2004). Recycling the cell cycle: Cyclins revisited. *Cell* **116**(2), 221–234.

Patton, E. E., and Zon, L. I. (2001). The art and design of genetic screens: Zebrafish. *Nat. Rev. Genet.* **2**(12), 956–966.

Pauls, S., Geldmacher-Voss, B., and Campos-Ortega, J. A. (2001). A zebrafish histone variant H2A.F/Z and a transgenic H2A.F/Z:GFP fusion protein for *in vivo* studies of embryonic development. *Dev. Genes Evol.* **211**(12), 603–610.

Peterson, R. T., Link, B. A., Dowling, J. E., and Schreiber, S. L. (2000). Small molecule developmental screens reveal the logic and timing of vertebrate development. *Proc. Natl. Acad. Sci. USA* **97**(24), 12965–12969.

Peterson, R. T., Shaw, S. Y., Peterson, T. A., Milan, D. J., Zhong, T. P., Schreiber, S. L., MacRae, C. A., and Fishman, M. C. (2004). Chemical suppression of a genetic mutation in a zebrafish model of aortic coarctation. *Nat. Biotechnol.* **22**(5), 595–599.

Pfaff, K. L., Straub, C. T., Chiang, K., Bear, D. M., Zhou, Y., and Zon, L. I. (2007). The zebra fish cassiopeia mutant reveals that SIL is required for mitotic spindle organization. *Mol. Cell. Biol.* **27**(16), 5887–5897.

Rowlerson, A., Radaelli, G., Mascarello, F., and Veggetti, A. (1997). Regeneration of skeletal muscle in two teleost fish: Sparus aurata and Brachydanio rerio. *Cell Tissue Res.* **289**(2), 311–322.

Ryu, S., and Driever, W. (2006). Minichromosome maintenance proteins as markers for proliferation zones during embryogenesis. *Cell Cycle* **5**(11), 1140–1142.

Shay, J. W., and Roninson, I. B. (2004). Hallmarks of senescence in carcinogenesis and cancer therapy. *Oncogene* **23**(16), 2919–2933.

Shay, J. W., and Wright, W. E. (2006). Telomerase therapeutics for cancer: Challenges and new directions. *Nat. Rev. Drug. Discov.* **5**(7), 577–584.

Shepard, J. L., Amatruda, J. F., Finkelstein, D., Ziai, J., Finley, K. R., Stern, H. M., Chiang, K., Hersey, C., Barut, B., Freeman, J. L., Lee, C., Glickman, J. N., *et al.* (2007). A mutation in separase causes genome instability and increased susceptibility to epithelial cancer. *Genes Dev.* **21**(1), 55–59.

Shepard, J. L., Amatruda, J. F., Stern, H. M., Subramanian, A., Finkelstein, D., Ziai, J., Finley, K. R., Pfaff, K. L., Hersey, C., Zhou, Y., Barut, B., Freedman, M., *et al.* (2005). A zebrafish bmyb mutation causes genome instability and increased cancer susceptibility. *Proc. Natl. Acad. Sci. USA* **102**(37), 13194–13199.

Singh, N. P., McCoy, M. T., Tice, R. R., and Schneider, E. L. (1988). A simple technique for quantitation of low levels of DNA damage in individual cells. *Exp. Cell Res.* **175**(1), 184–191.

St. Johnston, D. (2002). The art and design of genetic screens: Drosophila melanogaster. *Nat. Rev. Genet.* **3**(3), 176–188.

Stern, H. M., Murphey, R. D., Shepard, J. L., Amatruda, J. F., Straub, C. T., Pfaff, K. L., Weber, G., Tallarico, J. A., King, R. W., and Zon, L. I. (2005). Small molecules that delay S phase suppress a zebrafish bmyb mutant. *Nat. Chem. Biol.* **1**(7), 366–370.

Thisse, C., Thisse, B., Halpern, M. E., and Postlethwait, J. H. (1994). Goosecoid expression in neurectoderm and mesendoderm is disrupted in zebrafish cyclops gastrulas. *Dev. Biol.* **164**(2), 420–429.

Thisse, C., Thisse, B., Schilling, T. F., and Postlethwait, J. H. (1993). Structure of the zebrafish snail1 gene and its expression in wild-type, spadetail and no tail mutant embryos. *Development* **119**(4), 1203–1215.

Tsai, S. B., Tucci, V., Uchiyama, J., Fabian, N. J., Lin, M. C., Bayliss, P. E., Neuberg, D. S., Zhdanova, I. V., and Kishi, S. (2007). Differential effects of genotoxic stress on both concurrent body growth and gradual senescence in the adult zebrafish. *Aging Cell* **6**(2), 209–224.

Wehman, A. M., Staub, W., and Baier, H. (2006). The anaphase-promoting complex is required in both dividing and quiescent cells during zebrafish development. *Dev. Biol.* **303**(1), 144–156.

Wullimann, M. F., and Knipp, S. (2000). Proliferation pattern changes in the zebrafish brain from embryonic through early postembryonic stages. *Anat. Embryol. (Berl.)* **202**(5), 385–400.

Wyllie, A. H. (1987). Apoptosis: Cell death in tissue regulation. *J. Pathol.* **153**(4), 313–316.

Zhang, L., Kendrick, C., Julich, D., and Holley, S. A. (2008). Cell cycle progression is required for zebrafish somite morphogenesis but not segmentation clock function. *Development* **135**(12), 2065–2070.

CHAPTER 11

Cellular Dissection of Zebrafish Hematopoiesis

David L. Stachura and David Traver

Section of Cell and Developmental Biology
Division of Biological Sciences
University of California
San Diego, La Jolla, California 92093

I. Introduction

Over the past decade, the development of forward genetic approaches in the zebrafish model system has provided unprecedented power in understanding the molecular basis of vertebrate blood development. Establishment of cellular and hematological approaches to better understand the biology of resulting blood mutants, however, has lagged behind these efforts. In this chapter, recent advances in zebrafish hematology will be reviewed, with an emphasis on prospective

ESSENTIAL ZEBRAFISH METHODS:
CELL AND DEVELOPMENTAL BIOLOGY
DOI: 10.1016/B978-0-12-374599-6.00011-X

isolation strategies for both embryonic and adult hematopoietic stem cells (HSCs) and the development of assays with which to rigorously test their function.

II. Zebrafish Hematopoiesis

Developmental hematopoiesis, in both mammals and teleosts, occurs in sequential waves (Fig. 1). The first is termed primitive, and is characterized by rapid commitment of embryonic mesoderm to monopotent hematopoietic precursors (Keller *et al.*, 1999; Palis *et al.*, 1999). Next are definitive hematopoietic precursors, which are multipotent hematopoietic stem and progenitor cells that give rise to the full repertoire of mature blood cells throughout adulthood (Cumano and Godin, 2007).

A. Primitive Hematopoiesis

Primitive hematopoiesis has been extensively studied in the mouse, where primitive macrophages and erythroid cells are generated in the extraembryonic yolk sac (YS). In the zebrafish, primitive macrophages develop in an anatomically distinct area known as the rostral blood island (RBI) (Fig. 1A). Transcripts for *scl, lmo2, gata2,* and *fli1a* are found in the RBI between the 3- and 5-somite stage (Brown *et al.*, 2000; Liao *et al.*, 1998; Thompson *et al.*, 1998). This is quickly followed by expression of *pu.1* in a subset of these precursors (Bennett *et al.*, 2001; Lieschke *et al.*, 2002). Between 11 and 15 somites, *pu.1*⁺ macrophages are detectable, and they migrate toward the head midline (Herbomel *et al.*, 1999; Lieschke *et al.*, 2002; Ward *et al.*, 2003), and across the yolk ball (Fig. 1C). Some of these precursors enter circulation, while others migrate into the head (Herbomel *et al.*, 1999). By 28–32 hpf macrophages are found in circulation and dispersed throughout the embryo.

Primitive erythroid cell generation begins in the YS blood islands at day 7.5 postcoitum (E7.5) (Fig. 1B). These blood islands consist of nucleated erythroid cells that express embryonic *globin* genes surrounded by endothelial cells. The zebrafish has an equivalent anatomical site, known as the intermediate cell mass (ICM), where two stripes of mesodermal cells expressing *scl, lmo2,* and *gata1* converge to the midline of the zebrafish embryo and are surrounded by endothelial cells that become the cardinal vein (Al-Adhami and Kunz, 1977; Detrich *et al.*, 1995) (Fig. 1A and B). Although the ICM is intraembryonic in zebrafish, it has a similar cellular architecture to the mammalian YS blood islands (Al-Adhami and Kunz, 1977; Willett *et al.*, 1999).

Importantly, the development of transgenic zebrafish expressing fluorescent markers under the control of early hematopoietic promoters such as *lmo2* (and others, see Table I) now allow testing of early fate potentials by prospective isolation strategies and functional assays outlined below.

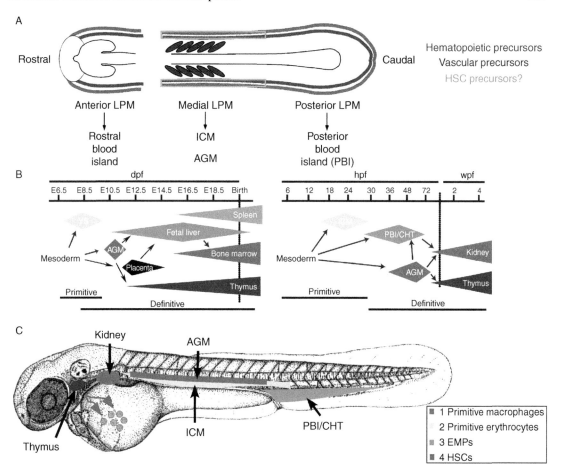

Fig. 1 Model of hematopoietic ontogeny in the developing zebrafish embryo. (A) Different regions of lateral plate mesoderm (LPM) give rise to anatomically distinct regions of blood cell precursors. Anatomical regions of embryo responsible for generation of hematopoietic precursors (red), vasculature (blue), and pre-HSCs (green) are highlighted. Cartoon is a 5-somite stage embryo, dorsal view. (B) Timing of mouse and zebrafish hematopoietic development. In mouse, primitive hematopoiesis initiates in the yolk sac (YS; yellow), producing primitive erythroid cells and macrophages. Later, definitive EMPs emerge in the YS. HSCs are specified in the aorta, gonad, and mesonephros (AGM, teal) region. These HSCs eventually seed the fetal liver (orange), the main site of embryonic hematopoiesis. Adult hematopoiesis occurs in the thymus (blue), spleen (green), and bone marrow (red). Zebrafish hematopoiesis is similar: temporal analogy to mouse hematopoiesis shown in (B), spatial locations shown in (C). Numbers in (C) correspond to timing of distinct precursor waves. (C) Embryonic hematopoiesis occurs through four independent waves of precursor production. First, primitive macrophages arise in cephalic mesoderm, migrate onto the yolk ball, and spread throughout the embryo (purple, 1). Then, primitive erythrocytes develop in the intermediate cell mass (ICM; yellow, 2). The first definitive progenitors are EMPs, which develop in the posterior blood island (PBI; orange, 3). Later, HSCs arise in the AGM region (teal, 4), migrate to the CHT (later name for the PBI, orange), and eventually seed the thymus and kidney (blue; red). Similar hematopoietic events in mouse and fish are color-matched between right and left panels of (B). Hematopoietic sites (B) and locations (C) are also color-matched. (Hpf, hours postfertilization; dpf, days postfertilization; wpf, weeks postfertilization; E, embryonic day.) (See Plate no. 5 in the Color Plate Section.)

Table I

List of Relevant Transgenic Zebrafish Lines Currently Available for Hematopoietic Studies, Indicating the Promoter Driver, the Specific Fluorophore, and Cell Population Identified

Transgene	Tissue	Transgene	Tissue
lmo2:gfp (Zhu et al., 2005)	Prehematopoietic	*flk1:gfp* (Cross et al., 2003)	Prehematopoietic, vasculature
lmo2:mCherry	Prehematopoietic	*flk1:rfp*	Prehematopoietic, vasculature
lmo2:DsRed (Lin et al., 2005)	Prehematopoietic	*fli1:gfp* (Lawson and Weinstein, 2002)	Prehematopoietic, vasculature
cd41:gfp (Lin et al., 2005)	EMPs, HSCs, thrombocytes	*fli1:rfp*	Prehematopoietic, vasculature
cd41:mCherry	EMPs, HSCs, thrombocytes	*gata3:AmCyan* (Bertrand et al., 2008)	Kidney
cd41:cfp	EMPs, HSCs, thrombocytes	*rag2:gfp* (Langenau et al., 2003)	T cells
cd45:DsRed (Bertrand et al., 2008)	Pan-leukocyte	*lck:gfp* (Langenau et al., 2004)	T cells
cd45:AmCyan	Pan-leukocyte	*mhcII:mOrange*	Antigen-presenting cells
gata1:gfp (Long et al., 1997)	Red blood cells	*mhcII:AmCyan*	Antigen-presenting cells
gata1:DsRed (Traver et al., 2003b)	Red blood cells	*lysc:gfp* (Hall et al., 2007)	Neutrophils
cmyb:gfp (Bertrand et al., 2008)	HSCs, neural	*lysc:DsRed* (Hall et al., 2007)	Neutrophils
mpx:gfp (Renshaw et al., 2006)	Neutrophils	*Runx1P1:gfp* (Lam et al., 2009)	EMPs
		runx1P2:gfp (Lam et al., 2009)	HSCs

Other transgenics are constantly being generated, but these are a few of the essential tools currently being used in zebrafish hematopoiesis laboratories.

B. Definitive Hematopoiesis

Similar to primitive hematopoiesis, definitive hematopoiesis initiates through two distinct precursor subsets. In the mouse, multilineage hematopoiesis is first evident in the YS (Bertrand et al., 2005b; Palis et al., 1999; Yoder et al., 1997a,b) and placenta (Gekas et al., 2005; Ottersbach and Dzierzak, 2005) by E9.5. Multilineage precursors in either tissue can be isolated and distinguished by the expression of CD41, an integrin molecule that labels early hematopoietic progenitors. CD41$^+$ cells differentiate into both myeloid and erythroid lineages, but conspicuously lack lymphoid potential (Bertrand et al., 2005a; Yokota et al., 2006). These studies suggested that the definitive hematopoietic program in the developing mouse begins with committed erythromyeloid progenitors (EMPs). Recent studies in the zebrafish have confirmed these findings. EMPs can be isolated from the zebrafish posterior blood island (PBI) at 30–36 hpf (Fig. 1C) by their coexpression of fluorescent transgenes driven by the *lmo2* and *gata1* promoters (Fig. 2A) (Bertrand et al., 2008). *In vitro* differentiation experiments (Fig. 2B and C), *in vivo*

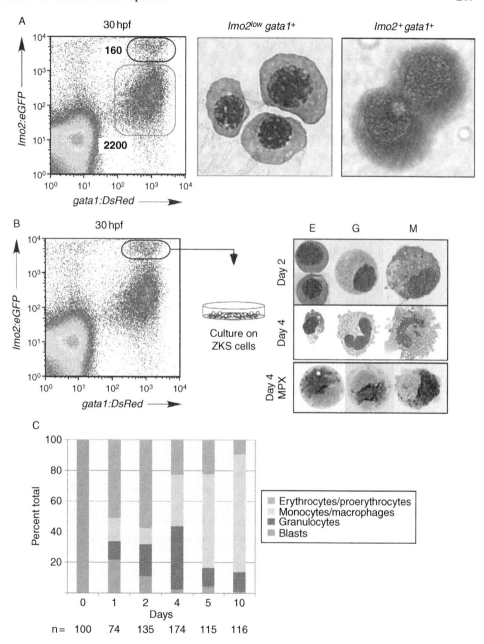

Fig. 2 Functional *in vitro* differentiation studies demonstrate that *gata1⁺ lmo2⁺* cells are committed erythromyeloid (EMP) progenitors. (A) Purified erythromyeloid progenitors at 30 hpf (*lmo2⁺ gata1⁺*, black gate) have the immature morphology of early hematopoietic progenitors. As a comparison, purified primitive erythroblasts are shown (*lmo2^low gata1⁺*, red gate). Magnification, 1000×. (B) Short term *in vitro* culture of *lmo2⁺ gata1⁺* cells atop zebrafish kidney stromal (ZKS) (Stachura *et al.*, 2009)

lineage tracing experiments, and transplantation assays have shown that these cells give rise only to erythroid and myeloid lineages (Bertrand *et al.*, 2008), and not lymphoid cells.

HSCs have the ability to not only self renew, but to give rise to all definitive blood cell lineages, including cells of the lymphoid lineage. It was initially believed that HSCs first originated in an area of the mid-gestation mouse bounded by the aorta, gonads, and mesonephros (AGM) at E10-10.5 (Fig. 1B) (Cumano and Godin, 2007; Dzierzak, 2005). Many studies have subsequently demonstrated that transplantable HSCs are present in the YS on E9 (Lux *et al.*, 2008; Weissman *et al.*, 1978; Yoder *et al.*, 1997a,b), the para-aortic splanchnopleura (P-Sp; precursor of the AGM) on E9 (Cumano and Godin, 2001; Yoder *et al.*, 1997a,b), and later in the placenta by E11 (Gekas *et al.*, 2005; Ottersbach and Dzierzak, 2005). These results suggest that HSCs arise in distinctly different locations in the developing mouse embryo. HSCs are only present in each of these locations transiently; by E11 the fetal liver (FL) is populated by circulating HSCs (Houssaint, 1981; Johnson and Moore, 1975) and becomes the predominant site of blood production during mid-gestation, producing the first full complement of definitive, adult-type effector cells. Shortly afterwards, hematopoiesis is evident in the fetal spleen, and occurs in bone marrow throughout adulthood (Keller *et al.*, 1999).

The zebrafish possesses an anatomical site that closely resembles the mammalian AGM (Fig. 1B and C). Between the dorsal aorta (DA) and cardinal vein between 28 and 48 hpf, *c-myb*$^+$ and *runx-1*$^+$ blood cells appear in intimate contact with the DA (Burns *et al.*, 2002; Kalev-Zylinska *et al.*, 2002; Thompson *et al.*, 1998). Lineage tracing of CD41$^+$ HSCs derived from this ventral aortic region show their ability to colonize the thymus (Bertrand *et al.*, 2008; Kissa *et al.*, 2008) and pronephros (Bertrand *et al.*, 2008; Murayama *et al.*, 2006), which are the sites of adult hematopoiesis (Jin *et al.*, 2007; Murayama *et al.*, 2006). After 48 hpf, blood production appears to shift to the caudal hematopoietic tissue (CHT) (Fig. 1C), and later the pronephros, which serves as the definitive hematopoietic organ for the remainder of life.

The development of transgenic zebrafish expressing fluorescent markers under the control of definitive hematopoietic promoters such as *cd41*, *cmyb*, *runx1P1*, and *P2* (and others, see Table I) now allow testing of fate potentials by prospective isolation strategies and functional assays outlined later in this chapter.

C. Adult Hematopoiesis

Previous genetic screens in zebrafish were extremely successful in identifying mutants in primitive erythropoiesis. The screening criteria used in these screens scored visual defects in circulating blood cells at early time points in

---

cells demonstrate erythroid (E), granulocytic (G), and monocytic/macrophage (M) differentiation potentials. Cultured cells were stained with May-Grünwald Giemsa and for myeloperoxidase (MPX) activity. *lmo2*^{low}*gata1*$^+$ cells only differentiated into erythroid cells (not shown). Magnification, 1000×. (C) Lineage differentials of cell types produced from cultured EMPs. *n*, number of cells counted from each time point. (See Plate no. 6 in the Color Plate Section.)

embryogenesis. Mutants defective in definitive hematopoiesis but displaying normal primitive blood cell development were therefore likely missed. Current screens aimed at identifying mutants with defects in the generation of definitive HSCs in the AGM should reveal new genetic pathways required for multilineage hematopoiesis. Understanding the biology of mutants isolated using these approaches, however, first requires the characterization of normal, definitive hematopoiesis and the development of assays to more precisely study the biology of zebrafish blood cells. To this end, we have established several tools to characterize the definitive blood-forming system of adult zebrafish.

Blood production in adult zebrafish, like other teleosts, occurs in the kidney, which supports both renal functions and multilineage hematopoiesis (Zapata, 1979). Similar to mammals, T lymphocytes develop in the thymus (Trede and Zon, 1998; Willett *et al.*, 1999) (Fig. 3A), which exists in two bilateral sites in zebrafish (Hansen and Zapata, 1998; Willett *et al.*, 1997). The teleostean kidney is a sheath of tissue that runs along the spine (Fig. 3B and E); the anterior portion, or head kidney (HK), shows a higher ratio of blood cells to renal tubules than does the posterior portion (Zapata, 1979), termed the trunk kidney (TK) (Fig. 3B and C). All mature blood cell types are found in the kidney and morphologically resemble their mammalian counterparts (Figs. 3G and 4), with the exceptions that erythrocytes remain nucleated and thrombocytes perform the clotting functions of platelets (Jagadeeswaran *et al.*, 1999). Histologically, the zebrafish spleen (Fig. 3D) has a simpler structure than its mammalian counterpart in which germinal centers have not been observed (Zapata and Amemiya, 2000). The absence of immature precursors in the spleen, or any other adult tissue, suggests that the kidney is the predominant hematopoietic site in adult zebrafish. The cellular composition of whole kidney marrow (WKM), spleen, and blood are shown in Fig. 3F–H. Morphological examples of all kidney cell types are shown in Fig. 4.

Analysis of WKM by fluorescence activated cell sorting (FACS) showed that several distinct populations could be resolved by light scatter characteristics (Fig. 5A). Forward scatter (FSC) is directly proportional to cell size, and side scatter (SSC) proportional to cellular granularity (Shapiro, 2002). Using combined scatter profiles, the major blood lineages can be isolated to purity from WKM following two rounds of cell sorting (Traver *et al.*, 2003b). Mature erythroid cells were found exclusively within two FSC^{low} fractions (Populations R1 and R2, Fig. 5A and D), lymphoid cells within a $FSC^{int} SSC^{low}$ subset (Population R3, Fig. 5A and E), immature precursors within a $FSC^{high} SSC^{int}$ subset (Population R4, Fig. 5A and F), and myelomonocytic cells within only a $FSC^{high} SSC^{high}$ population (Population R5, Fig. 5A and G). Interestingly, two distinct populations of mature erythroid cells exist (Fig. 5A, R1, R2 gates). Attempts at sorting either of these subsets reproducibly resulted in approximately equal recovery of both (Fig. 5D). This likely resulted due to the elliptical nature of zebrafish red blood cells, since sorting of all other populations yielded cells that fell within the original sorting gates upon reanalysis. Examination of splenic (Fig. 5B) and

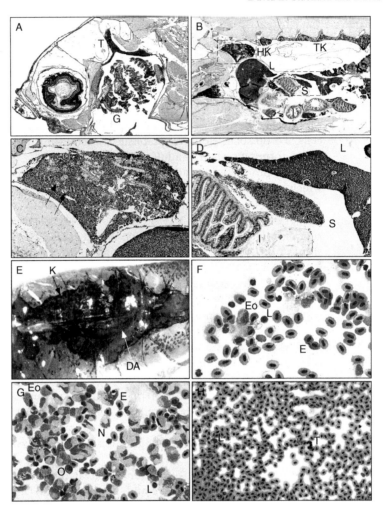

Fig. 3 Histological analyses of adult hematopoietic sites. (A) Sagittal section showing location of the thymus (T), which is dorsal to the gills (G). (B) Midline sagittal section showing location of the kidney, which is divided into the (HK), and trunk kidney (TK), and spleen (S). (C) The head kidney shows a higher ratio of blood cells to renal tubules (black arrows), as shown in a close up view of the HK. (D) Close up view of the spleen, which is positioned between the liver (L) and the intestine (I). (E) Light microscopic view of the kidney (K), over which passes the dorsal aorta (DA, white arrow). (F) Cytospin preparation of splenic cells, showing erythrocytes (E), lymphocytes (L), and an eosinophil (Eo). (G) Cytospin preparation of kidney cells showing cell types as noted above plus neutrophils (N) and erythroid precursors (O, orthochromic erythroblast). (H) Peripheral blood smear showing occasional lymphocytes and thrombocytes (T) clusters amongst mature erythrocytes. (A–D) Hematoxylin and Eosin stains, (F–H) May-Grünwald/Giemsa stains.

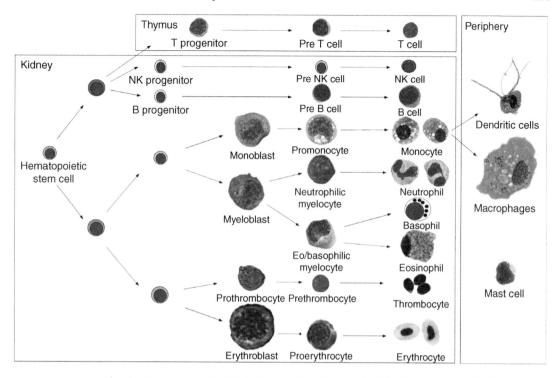

Fig. 4 Proposed model of definitive hematopoiesis in zebrafish. Shown are actual cells types from adult kidney marrow. All cells were photographed with a 100× oil objective from cytospin preparations. Proposed lineage relationships are based on those demonstrated using clonogenic murine progenitor cells.

peripheral blood (Fig. 5C) suspensions showed each to have distinct profiles from WKM, each being predominantly erythroid. It should be noted that, due to differences in the fluidics and beam size, that erythroid cells are not discretely detectable on BD FACScan, FACS Caliber, or FACS Aria I flow cytometers. Sorting of each scatter population from spleen and blood showed each to contain only erythrocytes, lymphocytes, or myelomonocytes in a manner identical to those in the kidney. Immature precursors were not observed in either tissue. Percentages of cells within each scatter population closely matched those obtained by morphological cell counts, demonstrating that this flow cytometric assay is accurate in measuring the relative percentages of each of the major blood lineages.

Many transgenic zebrafish lines have been created using proximal promoter elements from genes that demonstrate lineage-affiliated expression patterns in the mouse. These include *gata1:gfp* (Long *et al.*, 1997), *gata2:egfp* (Jessen *et al.*, 1998; Traver *et al.*, 2003b), *rag2:egfp* (Langenau *et al.*, 2003), *lck:egfp* (Langenau *et al.*, 2004), *pu.1:egfp* (Hsu *et al.*, 2004; Ward *et al.*, 2003), and *cd41:egfp* (Lin *et al.*, 2005; Traver *et al.*, 2003b) stable transgenic lines. In the adult kidney, we

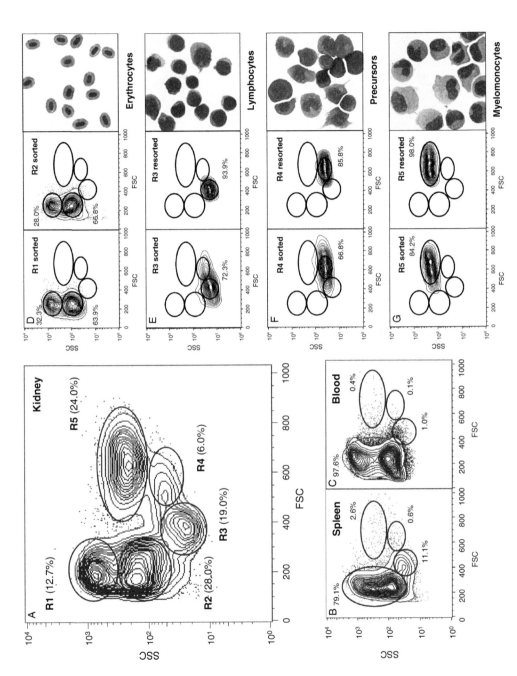

Fig. 5 Each major blood lineage can be isolated by size and granularity using flow cytometry. (A) Scatter profile for WKM. Mature erythrocytes are found within R1 and R2 gates, lymphocytes within the R3 gate, immature precursors within the R4 gate, and myeloid cells within the R5 gate. Mean percentages of each population within WKM are shown. Scatter profiling can also be used for splenocytes (B) and peripheral blood (C). Purification of each WKM fraction by FACS (D–G). (D) Sorting of populations R1 or R2 yields both upon reanalysis. This appears to be due to the elliptical shape of erythrocytes (right panel). (E) Isolation of lymphoid cells. (F) Isolation of precursor fraction. (G) Isolation of myeloid cells. FACS profiles following one round of sorting are shown in left panels, after two rounds in middle panels, and morphology of double sorted cells shown in right panels (E–G).

have demonstrated that each of these animals expresses GFP in the expected kidney scatter fractions (Traver *et al.*, 2003b). For example, all mature erythrocytes express GFP in *gata1:gfp* transgenic animals, as do erythroid progenitors within the precursor population. High expression levels of *gata2* are seen only within eosinophils that are contained within the myeloid population, *rag2* and lck only within cells in the lymphoid fraction, and *pu.1* in both myeloid cells and rare lymphoid cells. The development of *cd41:egfp* transgenic animals has demonstrated that rare thrombocytic cells are found within the kidney, with thrombocyte precursors appearing in the precursor scatter fraction and mature thrombocytes in the lymphoid fraction. Without fluorescent reporter genes, rare populations such as thrombocytes cannot be resolved by light scatter characteristics alone. By combining the simple technique of scatter separation with fluorescent transgenesis, specific hematopoietic cell subpopulations can now be isolated to a relatively high degree of purity for further analyses.

FACS profiling can also serve as a diagnostic tool in the examination of zebrafish blood mutants. The majority of blood mutants identified to date are those displaying defects in embryonic erythrocyte production (Traver *et al.*, 2003a). Most of these mutants are recessive and many are embryonic lethal when homozygous. Most have not been examined for subtle defects as heterozygotes. Several heterozygous mutants, such as *retsina, riesling*, and *merlot* showed haploinsufficiency as evidenced by aberrant kidney erythropoiesis (Traver *et al.*, 2003b). All mutants displayed anemia with concomitant increases in erythroid precursors. These findings suggest that many of the gene functions required to make embryonic erythrocytes are similarly required in their adult counterparts at full gene dosage for normal function.

III. Hematopoietic Cell Transplantation

In mammals, transplantation has been used extensively to functionally test putative hematopoietic stem and progenitor cell populations, precursor/progeny relationships, and cell autonomy of mutant gene function. To address similar issues in zebrafish, we have developed several different varieties of hematopoietic cell transplantation (HCT; Fig. 6).

A. Embryonic Donor Cells

While scatter profiling has proven very useful in analyzing and isolating specific blood lineages from the adult kidney, it cannot be used to enrich blood cells from the developing embryo. In order to study the biology of the earliest blood-forming cells in the embryo, we have made use of transgenic zebrafish expressing fluorescent proteins. As discussed above, hematopoietic precursors appear to be specified from mesodermal derivatives that express *lmo2, flk1*, and *gata2*. The proximal promoter elements from each of these genes have been shown to be sufficient to

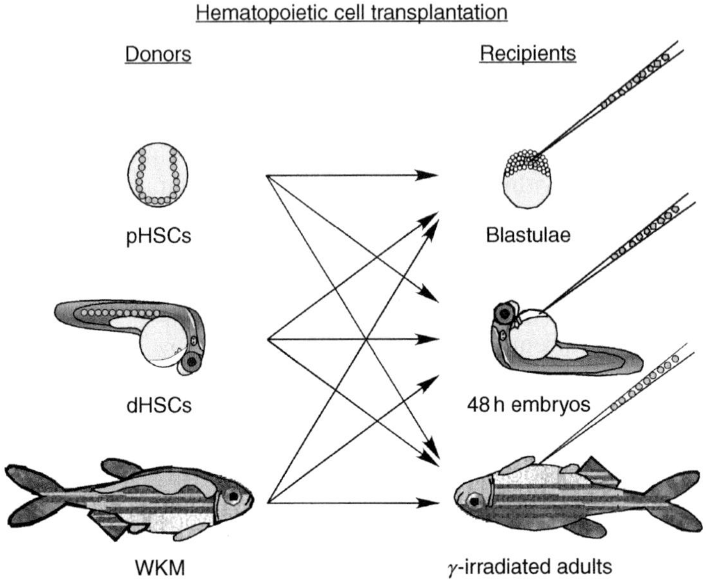

Fig. 6 Methods of hematopoietic cell transplantation in the zebrafish. See text for details.

recapitulate their endogenous expression patterns. Using germline transgenic animals expressing GFP under the control of each of these promoters, blood cell precursors can be isolated by flow cytometry from embryonic and larval animals for transplantation into wild-type recipients. For example, GFP$^+$ cells in *lmo2:egfp* embryos can be visualized by FACS by 8–10 somites (Traver, 2004). These cells can be sorted to purity and tested for functional potential in a variety of transplantation (Fig. 6) or *in vitro* culture assays (Fig. 2).

We have used two types of heterochronic transplantation strategies to address two fundamental questions in developmental hematopoiesis. The first is whether cells that express *lmo2* at 8–12 somites have hemangioblastic potential, that is, can generate both blood and vascular cells. We reasoned that purified cells should be placed into a relatively naive environment to provide the most permissive conditions to read out their full fate potentials. Therefore, we attempted transplantation into 1000-cell stage blastulae recipients. Transplanted cells appear to survive this procedure well and GFP$^+$ cells could be found over several days later in developing embryos and larvae. By isolating GFP$^+$ cells from *lmo2:egfp* animals also carrying a *gata1:dsred* transgene, both donor-derived endothelial and erythroid cells can be independently visualized in green and red, respectively. Using this approach, we have shown that *lmo2*$^+$ cells from 8- to 12-somite stage embryos can generate robust regions of donor endothelium and intermediate levels of circulating erythrocytes (D. Traver, C. E. Burns, H. Zhu, and L. I. Zon, unpublished results). We are currently generating additional transgenic lines that express dsRED under ubiquitous promoters to test the full fate potentials of *lmo2*⁺ cells upon transplantation. Additionally, while these studies demonstrate that *lmo2*⁺ cells can generate at least blood and endothelial cells at the population level, single-cell fate-mapping studies need to be performed to assess whether clonogenic hemangioblasts can be identified *in vivo*.

The second question addressed through transplantation is whether the earliest identifiable primitive blood precursors can generate the definitive hematopoietic cells that arise later in embryogenesis. It has been previously reported that the embryonic lethal *vlad tepes* mutant dies from erythropoietic failure due to a defect in the *gata-1* gene (Lyons *et al.*, 2002). This lethality can be rescued by transplantation of WKM from wild-type adults into mutant recipients at 48 hpf (Traver *et al.*, 2003b). We therefore tested whether cells isolated from 8- to 12-somite stage *lmo2:egfp* embryos could give rise to definitive cell types and rescue embryonic lethality in *vlad tepes* recipients. Following transplantation of GFP⁺ cells at 48 hpf, approximately half of the cells in circulation were GFP⁺ and the other half was dsRED⁺ 1 day posttransplantation. Analyses of the same animals, 3 days later, showed that the vast majority of cells in circulation were dsRED⁺, apparently due to the differentiation of *lmo2*⁺ precursors to the erythroid fates. Compared to untransplanted control animals which all died by 12 dpf, some mutant recipients survived for 1–2 months following transplantation. We observed no proliferation of donor cells at any time point following transplantation, however, and survivors analyzed over 1 month posttransplantation showed no remaining cells in circulation (D. Traver, C. E. Burns, H. Zhu, and L. I. Zon, unpublished results). Therefore, these data indicate that mutant survivors were only transiently rescued by short-lived, donor-derived erythrocytes. Thus, within the context of this transplantation setting, it does not appear that primitive hematopoietic precursors can seed definitive hematopoietic organs to give rise to enduring repopulation of the host blood-forming system.

1. Protocol for Isolating Hematopoietic Cells from Embryos

This simple physical dissociation procedure is effective in producing single cell suspensions from early embryos (8–12-somite stage) as well as from embryos as late as 48 hpf.

1. Stage and collect embryos. We estimate that approximately 200 cells can be isolated per 10–12-somite stage *lmo2:egfp* embryo. It is recommended that as many embryos can be collected as possible since subsequent transplantation efficiency depends largely upon cell concentration. At least 500–1000 embryos are recommended.

2. Transfer embryos to 1.5-ml Eppendorf centrifuge tubes. Add embryos until they sediment to the 0.5 ml mark. Remove embryo medium since it is not optimal for cellular viability.

3. Wash 2× with 0.9× Dulbecco's PBS (Gibco; 500 ml 1× Dulbecco's PBS + 55 ml ddH₂O).

4. Remove $0.9\times$ PBS and add 750 μl ice-cold staining media (SM; $0.9\times$ Dulbecco's PBS + 5% FCS). Keep cells on ice from this point onward.

5. Homogenize with blue plastic pestle and pipette a few times with a p1000 tip.

6. Strain resulting cellular slurry through a 40-μm nylon cell strainer (Falcon 2340) atop a 50-ml conical tube. Rinse with additional SM to flush cells through the filter.

7. Gently mash remaining debris atop strainer with a plunger removed from a 28-gauge insulin syringe.

8. Rinse with more SM until conical is filled to 25 ml mark (helps remove yolk).

9. Centrifuge for 5 min at 200g and 4 °C. Remove supernatant until 1–2 ml remain.

10. Add 2–3 ml SM, resuspend by pipetting.

11. Strain again through 40 μm nylon mesh into a 5-ml Falcon 2054 tube. It is important to filter the cell suspension at least twice before running the sample by FACS. Embryonic cells are sticky and will clog the nozzle if clumps are not properly removed beforehand.

12. Centrifuge again for 5 min at 200g and 4 °C. Repeat steps 10–12 if necessary.

13. Remove supernatant, resuspend with 1–2 ml SM depending upon number of embryos used.

14. Propidium Iodide (PI) may be added at this point to 1 μg/ml to exclude dead cells and debris on the flow cytometer. When using, however, bring samples having PI only and GFP only to set compensations properly. Otherwise, the signal from PI may bleed into the GFP channel resulting in false positives.

Embryonic cells are now ready for analysis or sorting by flow cytometry. It is often difficult to visualize GFP$^+$ cells when the expression is low or target population is rare, so one should always prepare age-matched GFP negative embryos in parallel with transgenic embryos. It is then apparent where the sorting gates should be drawn to sort *bona fide* GFP$^+$ cells. If highly purified cells are desired, one must perform two successive rounds of sorting. In general, sorting GFP$^+$ cells once yields populations of approximately 50–70% purity. Two rounds of cell sorting generally yields >90% purity as observed with 10-somite stage *lmo2:egfp* cells (Traver, 2004). Cells should be kept ice-cold during the sorting procedure.

2. Transplanting Purified Cells into Embryonic Recipients

After sorting, centrifuge cells for 5 min at 200g and 4 °C. Carefully remove all supernatant. Resuspend cell pellet in 5–10 μl of ice-cold SM containing 3 U Heparin and 1 U DnaseI to prevent coagulation and lessen aggregation. Preventing the cells from aggregating or adhering to the glass capillary needle used for transplantation is critical. Mix the cells by gently pipetting with a 10-μl pipette tip. Keep on ice. For transplantation, we use the same needle-pulling parameters used to make needles for nucleic acid injections, the only difference being the use of filament-free capillaries to maintain cell viability. We also use the standard air-powered injection stations used for nucleic acid injections.

3. Transplanting Cells into Blastula Recipients

1. Stage embryonic recipients to reach the 500–1000-cell stage at the time of transplantation.
2. Prepare plates for transplantation by pouring a thin layer of 2% agarose made in E3 embryo medium into a 6-cm Petri dish. Drop transplantation mold (similar to the embryo injection mold described in Chapter 5 of The Zebrafish Book (Westerfield, 2000) but having individual depressions rather than troughs) atop molten agarose and let solidify.
3. Dechorionate blastulae in 1–2% agarose-coated Petri dishes by light pronase treatment or manually with watchmaker's forceps.
4. Transfer individual blastulae into individual wells of transplantation plate that has been immersed in $1 \times$ HBSS (Gibco). Position the animal pole upward.
5. Using glass, filament-free, fine-pulled capillary needles (1.0 mm OD) back-load 3–6 μl of cell suspension after breaking needle on a bevel to an opening of $\sim$20 μm. Load into needle holder and force cells to injection end by positive pressure using a pressurized air injection station.
6. Gently insert needle into the center of the embryo and expel cells using either very gentle pressure bursts or slight positive pressure. Transplanting cells near the marginal zone of the blastula leads to higher blood cell yields since embryonic fate maps show blood cells to derive from this region in later gastrula stage embryos.
7. Carefully transfer embryos to agarose-coated Petri dishes using glass transfer pipettes.
8. Place into E3 embryo medium and incubate at 28.5 °C. Many embryos will not survive the transplantation procedure, so clean periodically to prevent microbial outgrowth.
9. Monitor by fluorescence microscopy for donor cell types.

4. Transplanting Cells into 48 hpf Embryos

1. All procedures are performed as above except that dechorionated 48 hpf embryos are staged and used as transplant recipients.
2. Fill transplantation plate with $1 \times$ HBSS containing $1 \times$ Penicillin/Streptomycin and $1 \times$ buffered tricaine, pH 7.0. Do not use E3 as it is suboptimal for cellular viability. Anesthetize recipients in tricaine then array individual embryos into individual wells of transplantation plate. Position head at bottom of well, yolk side up.

3. Load cells as above. Insert injection needle into the sinus venosus/duct of Cuvier and gently expel cells by positive pressure or gentle pressure bursts. Take care not to rupture the YS membrane. A very limited volume can be injected into each recipient. It is thus important to use very concentrated cell suspensions in order to reconstitute the host blood-forming system. If using WKM as donor cells, concentrations of 5×10^5 cells/μl can be achieved if care is taken to filter and anticoagulate the sample.

4. Allow animals to recover at 28.5 °C in E3. Keep clean and visualize daily by microscopy for the presence of donor-derived cells.

B. Adult Donor Cells

Whereas the first HSCs appear to arise within the embryonic AGM, multilineage hematopoiesis is not fully apparent until the kidney becomes the site of blood cell production. The kidney appears to be the only site of adult hematopoiesis, and we have previously demonstrated that it contains HSCs capable of long-term repopulation of embryonic (Traver et al., 2003b) and adult (Langenau et al., 2004) recipients. For HSC enrichment strategies, both high-dose transplants and limiting dilution assays are required to gauge the purity of input cell populations. In embryonic recipients, we estimate that the maximum number of cells that can be transplanted is approximately 5×10^3, and precise quantitation of transplanted cell numbers is difficult. To circumvent both issues, we have developed HCT into adult recipients.

For transplantation into adult recipients, myeloablation is necessary for successful engraftment of donor cells. We have found γ-irradiation to be the most consistent way to deplete zebrafish hematopoietic cells. The minimum lethal dose (MLD) of 40 Gy specifically ablates cells of the blood-forming system and can be rescued by transplantation of one kidney equivalent (10^6 WKM cells). Thirty-day survival of transplanted recipients is approximately 75% (Traver et al., 2004). An irradiation dose of 20 Gy is sublethal, and the vast majority of animals survive this treatment despite having nearly total depletion of all leukocyte subsets 1 week following irradiation (Traver et al., 2004). We have shown that this dose is necessary and sufficient for transfer of a lethal T-cell leukemia (Traver et al., 2004), and for long-term (>6 months) engraftment of thymus repopulating cells (Langenau et al., 2004). We do not yet know the average relative chimerism of donor to host cells when transplantation is performed following 20 Gy. That this dose is sufficient for robust engraftment, for long-term repopulation, and yields extremely high survival suggests that 20 Gy may be the optimal dose for myeloablative conditioning prior to transplantation. Improvement in short-term engraftment and long-term survival of transplant recipients will also likely require matching of MHC loci between donor and host genotypes.

1. Protocols for Isolating Hematopoietic Cells from Adult Zebrafish

Anesthetize adult animals in 0.02% tricaine in fish water.

For blood collection, dry animal briefly on tissue then place on a flat surface with head to the left, dorsal side up. Coat a 10-μl pipette tip with heparin (3 μm/μl) then insert tip just behind the pectoral fin and puncture the skin. Direct the tip into the heart cavity, puncture the heart and aspirate up to 10 μl blood by gentle suction. Immediately perform blood smears or place into 0.9× PBS containing 5% FCS and 1 μm/μl heparin. Mix immediately to prevent clotting. Blood from several animals may be pooled in this manner for later use by flow cytometry. Red cells may be removed using red blood cell hypotonic lysis solution (Sigma; 8.3 g/l ammonium chloride in.01 M Tris-HCl, pH 7.5) on ice for 5 min. Add 10 volumes of ice-cold SM then centrifuge at 200g for 5 min at 4 °C. Resuspended blood leukocytes can then be analyzed by flow cytometry or cytocentrifuge preparations.

For collection of other hematopoietic tissues, place fish on ice for several minutes following tricaine. Make a ventral, midline incision using fine scissors under a dissection microscope.

For spleen collection, locate spleen just dorsal to the major intestinal loops and tease out with watchmaker's forceps. Place into ice-cold SM. Dissect any nonsplenic tissue away and place on a 40-μm nylon cell strainer (Falcon 2340) atop a 50-ml conical tube. Gently mash the spleen using a plunger removed from a 28-gauge insulin syringe and rinse with SM to flush cells through the filter. Up to 10 spleens can be processed through each filter. Centrifuge at 200g for 5 min at 4 °C. Filter again through 40-μm nylon mesh if using for FACS.

For kidney collection, remove all internal organs using forceps and a dissection microscope. Take care in dissection as ruptured intestines or gonads will contaminate the kidney preparation. Using watchmaker's forceps, tease the entire kidney away from the body wall starting at the HK and working toward the rear. Place into ice-cold SM. Aspirate vigorously with a 1-ml pipetteman to separate hematopoietic cells (WKM) from renal cells. Filter through 40-μm nylon mesh, wash, centrifuge, and repeat. Perform last filtration step into Falcon 2054 tube if using for FACS. It is important to filter the WKM cell suspension at least twice before running the sample. PI may be added at this point to 1 μg/ml to exclude dead cells and debris on the flow cytometer. When using, however, compare to samples having PI only and GFP only (if using) to set compensations properly. Otherwise, the signal from PI may bleed into the GFP channel resulting in false positives.

2. Transplanting Whole Kidney Marrow

After filtering and washing WKM suspension three times, centrifuge cells for 5 min at 200g at 4 °C. Carefully remove all supernatant. Resuspend cell pellet in 5-10 μl of ice-cold SM containing 3 U Heparin and 1 U DnaseI to prevent coagulation and lessen aggregation. Preventing the cells from aggregating or adhering to the glass capillary needle used for transplantation is critical. Mix the cells by gently pipetting with a 10-μl pipette tip. Keep on ice. For blastulae and embryo transplantation, perform following previous protocols. Between 5×10^2 and 5×10^3 cells can be injected into each 48 hpf embryo if the final cell concentration is approximately $5 \times 10^5/\mu l$.

3. Transplanting Cells into Irradiated Adult Recipients

For irradiation of adult zebrafish, we have used a ^{137}Cesium source irradiator typically used for the irradiation of cultured cells (Gammacell 1000). We lightly anesthetize five animals at a time then irradiate in sealed Petri dishes filled with fish water (without tricaine). We performed careful calibration of the irradiator using calibration microchips to obtain the dose rate at the height within the irradiation chamber nearest to the ^{137}Cesium point source. We found the dose rate to be uniform amongst calibration chips placed within euthanized animals under water, under water alone, or in air alone, verifying that the tissue dosage via TBI was accurate.

Transplantation into circulation is most efficiently performed by injecting cells directly into the heart. We perform intracardiac transplantation using pulled filament-free capillary needles as above, but break needles at a larger bore size of approximately 50 μm. The needle assembly can be handheld and used with a standard gas-powered microinjection station. We have also had limited success transplanting cells intraperitoneally using a 10-μl Hamilton syringe. Engraftment efficiency for WKM is only marginal using this method, but transplantation of T-cell leukemia or solid tumor suspensions is highly efficient following irradiation at 20 Gy (Traver *et al.*, 2004).

4. Irradiation

1. Briefly anesthetize adult zebrafish in 0.02% tricaine in fish water.
2. Place five at a time into 60×15 mm Petri dishes (Falcon) containing fish water. Wrap dish with parafilm and irradiate for length of time necessary to achieve desired dose.
3. Return irradiated animals to clean tanks containing fish water. We have successfully transplanted irradiated animals from 12 to 72 h following irradiation. Using a 20 Gy dose, the nadir of host hematopoietic cell numbers occurs at approximately 72 h postirradiation.

5. Transplantation

1. Prepare cells to be transplanted as above, taking care to remove particulates/contaminants by multiple filtration and washes. When using WKM as donor cells, we typically make final cell suspensions at 2×10^5 cells/μl. Keep cells on ice.
2. Anesthetize an irradiated animal in 0.02% tricaine in fish water.

3. Transfer ventral side up into a well cut into a sponge wetted with fish water. Under a dissection microscope, remove scales covering the pericardial region with fine forceps.

4. Fill injection needle with ~20 μl of cell suspension. Force cells to end of needle with positive pressure and adjust pressure balance to be neutral. Hold needle assembly in right hand then place gentle pressure on the abdomen of the recipient with left index finger. This will position the heart adjacent to the skin and allow visualization of the heartbeat. Insert needle through the skin and into the heart. If the needle is positioned within the heart, and the pressure balance is neutral, blood from the heart will enter the needle and the meniscus will rise and fall with the heartbeat. Inject approximately 5-10 μl by gentle pressure bursts.

5. Return recipient to fresh fish water. Repeat for each additional recipient. Do not feed until the next day to lessen chance of infection.

IV. Enrichment of HSCs

The development of many different transplantation techniques now permits the testing of cell autonomy of mutant gene function, oncogenic transformation, and stem cell enrichment strategies in the zebrafish. For HSC enrichment strategies, fractionation techniques can be used to divide WKM into distinct subsets for functional testing via transplantation. The most successful means of HSC enrichment in the mouse has resulted from the subfractionation of whole bone marrow cells with monoclonal antibodies (mAbs) and flow cytometry (Spangrude *et al.*, 1988). We have attempted to generate mAbs against zebrafish leukocytes by repeated mouse immunizations using both live WKM and purified membrane fractions followed by standard fusion techniques. Many resulting hybridoma supernatants showed affinity to zebrafish WKM cells in FACS analyses (Fig. 7A). All antibodies showed one of two patterns, however. The first showed binding to all WKM cells at similar levels. The second showed binding to all kidney leukocyte subsets but not to kidney erythrocytes, similar to the pattern shown in the left panel of Fig. 7A. We found no mAbs that specifically bound only to myeloid cells, lymphoid cells, etc. when analyzing positive cells by their scatter profiles. We reasoned that these nonspecific binding affinities may be due to different oligosaccharide groups present on zebrafish blood cells. If the glycosylation of zebrafish membrane proteins is different from the mouse, then the murine immune system would likely mount an immune response against these epitopes. To test this idea, we removed both O- and N-linked sugars from WKM using a deglycosylation kit (Prozyme), then incubated these cells with previously positive mAbs. All mAbs tested in this way showed a time-dependent decrease in binding, with nearly all binding disappearing following 2 h of deglycosylation (Fig. 7A). It thus appears that standard immunization approaches using zebrafish WKM cells elicit a strong immune response against oligosaccharide epitopes. This response is

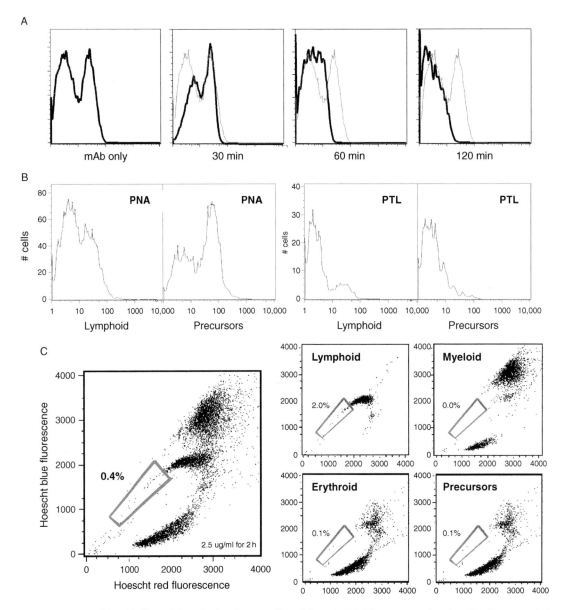

Fig. 7 Potential methods of stem cell enrichment. (A) Mouse monoclonal antibodies generated against zebrafish WKM cells react against oligosaccharide epitopes. Deglycosylation enzymes result in time-dependent loss of antibody binding (bold histograms) compared to no enzyme control (left panel and grey histograms). (B) Differential binding of lectins to WKM scatter fractions. Peanut agglutinin splits both the lymphoid and precursor fraction into positive and negative populations (left panels). Potato lectin shows a minor positive fraction only within the lymphoid fraction (right panels). (C) Hoechst 33342 dye reveals a side population (SP) within WKM. 0.4% of WKM cells appear within the verapamil-sensitive SP gate (left panel). Only the lymphoid fraction, where kidney HSCs reside, contains appreciable numbers of SP cells (right panels). (See Plate no. 7 in the Color Plate Section.)

likely extremely robust, since we did not recover any mAbs that reacted with specific blood cell lineages. Similar approaches by other investigators using blood cells from frogs or other teleost species have yielded similar results (L. du Pasquier and M. Flajnik, personal communication). In an attempt to circumvent the glycoprotein issue, we are now preparing to perform a new series of immunizations using kidney cell membrane preparations that have been deglycosylated.

Previous studies have shown that specific lectins can be used to enrich hematopoietic stem and progenitor cell subsets in the mouse (Huang and Auerbach, 1993; Lu et al., 1996; Visser et al., 1984). In preliminary studies, we have shown that FITC-labeled lectins such as peanut agglutinin (PNA) and potato lectin (PTL) differentially bind to zebrafish kidney subsets. As shown in Fig. 7B, PNA binds to a subset of cells both within the lymphoid and precursor kidney scatter fractions. Staining with PTL also shows that a minor fraction of lymphoid cells binds PTL, whereas the precursor (and other) scatter fractions are largely negative (Fig. 7B). We are currently testing both positive and negative fractions in transplantation assays to determine whether these differential binding affinities can be used to enrich HSCs.

We have previously demonstrated that long-term HSCs reside in the adult kidney (Traver et al., 2003b). We therefore isolated each of the kidney scatter fractions from gata1:egfp transgenic animals and transplanted cells from each into 48 hpf recipients to determine which subset contains HSC activity. The only population that could generate GFP$^+$ cells for over 3 weeks in wild-type recipients was the lymphoid fraction. This finding is in accordance with mouse and human studies that have shown purified HSCs to have the size and morphological characteristics of inactive lymphocytes (Morrison et al., 1995).

Another method that has been extremely useful in isolating stem cells from whole bone marrow is differential dye efflux. Dyes such as rhodamine 123 (Mulder and Visser, 1987; Visser and de Vries, 1988) or Hoechst 33342 (Goodell et al., 1996) allow the visualization and purification of a "side population" (SP) that is highly enriched for HSCs. This technique appears to take advantage of the relatively high activity of multidrug resistance transporter proteins in HSCs that actively pump each dye out of the cell in a verapamil-sensitive manner (Goodell et al., 1996). Other cell types lack this activity and become positively stained, allowing isolation of the negative SP fraction by FACS. Our preliminary studies of SP cells in the zebrafish kidney have demonstrated a typical SP profile when stained with 2.5 μg/ml of Hoechst 33342 for 2 h at 28 °C (Fig. 7C). This population disappears when verapamil is added to the incubation. Interestingly, the vast majority of SP cells appear within the lymphoid scatter fraction (Fig. 7C). Further examination of whether this population is enriched for HSC activity in transplantation assays is warranted.

Finally, there are many other ways that hematopoietic stem and progenitor cells can be enriched from WKM including sublethal irradiation, cytoreductive drug treatment, and use of transgenic lines expressing fluorescent reporter genes (see Table I). We have shown following 20 Gy doses of γ-irradiation that nearly all hematopoietic lineages are depleted within 1 week (Traver *et al.*, 2004). Examination of kidney cytocentrifuge preparations at this time shows that the vast majority of cells are immature precursors. That this dose does not lead to death of the animals demonstrates that HSCs are spared and are likely highly enriched 5-8 days following exposure. We have also shown that cytoreductive drugs such as Cytoxan and 5-Fluorouracil have similar effects on kidney cell depletion, although the effects were more variable than those achieved with sublethal irradiation (A. Winzeler, D. Traver, and L. I. Zon, unpublished results). Since HSCs are contained within the kidney lymphoid fraction, they can be further enriched by HSC-specific or lymphocyte-specific transgenic markers. Possible examples of transgenic promoters are *lmo-2, gata-2,* or *c-myb* to positively mark HSCs and *rag-2, lck,* or B-cell receptor genes to exclude lymphocytes from this subset (see Table I).

V. *In Vitro* Culture of Hematopoietic Progenitors

In vitro culture of hematopoietic cell subtypes in mammals has proven to be an essential tool for elucidating the regulation of hematopoietic cell maintenance, proliferation, and differentiation. Such assays have proven to be invaluable for functionally testing the developmental potential of progenitors, the autonomy of mutant gene functions, and the stages in development where these mutant gene functions are critical. Up until recently, no such *in vitro* assays existed for zebrafish. To address this issue, we generated zebrafish stromal cell lines that support hematopoietic cell growth, expansion, and differentiation (Stachura *et al.*, 2009).

A. Zebrafish Kidney Stroma (ZKS)

To create a suitable *in vitro* environment for the culture of zebrafish hematopoietic cells, we isolated the stromal fraction of the zebrafish kidney, which is the main site of hematopoiesis in the adult fish (Zapata, 1979). Similar strategies have been used for years in the mammalian and murine systems, allowing the culture of all definable mature blood cell lineages. The benefit of utilizing hematopoietic stromal layers is twofold. First, performing culture assays in the zebrafish has been hampered by a paucity of defined and purified hematopoietic cytokines. Most zebrafish cytokines have poor sequence homology to their mammalian counterparts, and as a consequence, have not been well described, characterized, or rigorously tested. The other reason to utilize a stromal cell layer is that some hematopoietic cell types require physical cell-cell interaction for differentiation.

ZKS cells have recently been utilized to demonstrate the bipotentiality of embryonic EMPs *in vitro* (Bertrand *et al.*, 2007). Culturing $lmo2^{GFP+}$ $gata1^{dsRed+}$ cells from 30 hpf transgenic animals showed differentiation down both the myeloid and erythroid pathways (Fig. 2).

B. Protocol for Generation, Culture, and Maintenance of ZKS Cells

To create ZKS cells, kidney was isolated from AB* wild-type fish as described above. The kidney tissue was sterilized by washing for 5 min in 0.000525% sodium hypochlorite (Fisher Scientific, Pittsburgh, PA), then rinsed in sterile 0.9× Dulbecco's PBS. Tissue was then mechanically dissociated by trituration then filtered through a 40-μm filter (BD Biosciences, San Jose, CA). Flow through cells (WKM) were discarded, and the remaining kidney tissue was cultured in vacuum plasma treated 12.5 cm^2 vented flasks (Corning Incorporated Life Sciences, Lowell, MA) at 32 °C and 5% CO_2.

ZKS cells are maintained in tissue culture media consisting of 10% ES cell qualified FBS (American Type Culture Collection, Manassas, VA), 55% L-15, 32.5% DMEM, and 12.5% Ham's F-12 (Mediatech, Herndon, VA). Media is supplemented with 150 mg/L sodium bicarbonate, 2% penicillin/streptomycin (10 U/ml stock), 1.5% HEPES, 1% L-glutmanine and 0.1 mg/ml gentamycin (all from Mediatech, Herndon, VA).

Cells are grown until 60-80% confluent before splitting. Medium is then removed, and enough trypsin (0.25%; Invitrogen, Grand Island, NY) is added to cover the stromal cells. Allow cells to incubate for 5 min at 32 °C. Add medium to cells to stop trypsinization, pipetting up and down to achieve a single cell suspension. Split the cells 1:10 to passage.

As with all tissue culture, strict attention to sterility and cleanliness should be adhered to at all times.

C. Protocols for *In Vitro* Proliferation and Differentiation Assays

Purify prospective progenitors by FACS as described above. Plate cells on confluent ZKS at a density of 1×10^4 cells/well in a 12-well tissue culture plate, using 2 ml complete medium per well. Lower density of progenitors is not recommended. If using fewer cells, reduce the size of the tissue culture well and volume of medium.

A. Morphological assessment of hematopoietic cells after *in vitro* culture

   a. Gently aspirate hematopoietic cells from the ZKS cultures, taking care not to disturb the stromal underlayer.

   b. Cytocentrifuge up to 200 μl of the hematopoietic cells at 250g for 5 min on to glass slides using a Shandon Cytospin 4 (Thermo Fischer Scientific, Waltham, MA). It is possible to concentrate the cells, if needed, before cytocentrifugation at 300g for 10 min. Cytocentrifugation of over 200 μl of cell suspensions is not recommended.

   c. Perform May-Grünwald Giemsa staining by allowing slides to air dry briefly. Then, submerge slide in May-Grünwald staining solution (Sigma Aldrich, St. Louis, MO) for 10 min. Transfer slide to 1:5 dilution of Giemsa stain (Sigma Aldrich, St. Louis, MO) in dH_2O for an additional 20 min.

Rinse slide in dH_2O, and allow to air dry. Coverslip slide with cytoseal XYL mounting medium (Richard-Allan Scientific, Kalamazoo, MI) and Corning no.1 18-mm square cover glass (Corning Incorporated, Lowell, MA). Allow slides to completely dry before visualization on upright microscope, especially if using an oil immersion lens.

B. Proliferation assessment of hematopoietic cells after *in vitro* culture

    a. Gently aspirate hematopoietic cells as above.

    b. Count cells with use of a bright line hemacytometer (Hausser Scientific, Horsham, PA) using Trypan Blue dye (Invitrogen, Philadelphia, PA) exclusion to assess viability.

C. RT-PCR analysis of hematopoietic cells after *in vitro* culture

    a. Gently aspirate hematopoietic cells as above.

    b. Isolate RNA from hematopoietic cells using either Trizol (Invitrogen, Philadelphia, PA) or RNAeasy kit (Qiagen, Valencia, CA).

    c. Generate cDNA with oligo dT primers and the Superscript RT-PCR kit (Invitrogen, Philadelphia, PA).

    d. Perform PCR with desired zebrafish DNA primers.

D. Cell labeling and cell division determination of hematopoietic cells after *in vitro* culture

    a. Prior to plating cells on ZKS monolayer, wash cells twice with 0.1% bovine serum albumin (BSA) to remove FBS from the media.

    b. Resuspend cells in 0.1% BSA and 2 μl/ml of 5 mM carboxyfluorescein succinimidyl ester (CFSE) (Invitrogen, Philadelphia, PA) at room temperature for 10 min, in the dark.

    c. Wash cells with complete media supplanted with an additional 10% FBS twice.

    d. Save 1/10 of the culture and perform FACS (Day 0 time point). Culture remaining cells in complete media as described above.

    e. For analysis, remove hematopoietic cells from culture at desired time points as described above and FACS. CFSE is read in the FL-1/GFP channel, and will decrease in fluorescence intensity as cells divide. Compare divisions to Day 0 time point with FloJo software (TreeStar, Ashland, OR).

VI. Conclusions

Over the past decade, the zebrafish has rapidly become a powerful model system in which to elucidate the molecular mechanisms of vertebrate blood development through forward genetic screens. In this review, we have described the cellular characterization of the zebrafish blood-forming system and provided detailed protocols for the isolation and transplantation of hematopoietic cells. Through the development of lineal subfractionation techniques, transplantation technology, and *in vitro* hematopoietic assays, a hematological framework now exists for the continued study of the genetics of hematopoiesis. By adapting these experimental approaches that have proven to be powerful in the mouse, the zebrafish is uniquely positioned to address fundamental questions regarding the biology of hematopoietic stem and progenitor cells.

References

Al-Adhami, M. A., and Kunz, Y. W. (1977). Ontogenesis of haematopoietic sites in Brachydanio rerio. *Dev. Growth Differ.* **19**, 171–179.

Bennett, C. M., Kanki, J. P., Rhodes, J., Liu, T. X., Paw, B. H., Kieran, M. W., Langenau, D. M., Delahaye-Brown, A., Zon, L. I., Fleming, M. D., and Look, A. T. (2001). Myelopoiesis in the zebrafish, Danio rerio. *Blood* **98**, 643–651.

Bertrand, J. Y., Giroux, S., Golub, R., Klaine, M., Jalil, A., Boucontet, L., Godin, I., and Cumano, A. (2005a). Characterization of purified intraembryonic hematopoietic stem cells as a tool to define their site of origin. *Proc. Natl. Acad. Sci. USA* **102**, 134–139.

Bertrand, J. Y., Jalil, A., Klaine, M., Jung, S., Cumano, A., and Godin, I. (2005b). Three pathways to mature macrophages in the early mouse yolk sac. *Blood* **106**, 3004–3011.

Bertrand, J. Y., Kim, A. D., Violette, E. P., Stachura, D. L., Cisson, J. L., and Traver, D. (2007). Definitive hematopoiesis initiates through a committed erythromyeloid progenitor in the zebrafish embryo. *Development* **134**, 4147–4156.

Bertrand, J. Y., Kim, A. D., Teng, S., and Traver, D. (2008). CD41 + cmyb + precursors colonize the zebrafish pronephros by a novel migration route to initiate adult hematopoiesis. *Development* **135**, 1853–1862.

Brown, L. A., Rodaway, A. R., Schilling, T. F., Jowett, T., Ingham, P. W., Patient, R. K., and Sharrocks, A. D. (2000). Insights into early vasculogenesis revealed by expression of the ETS-domain transcription factor Fli-1 in wild-type and mutant zebrafish embryos. *Mech. Dev.* **90**, 237–252.

Burns, C. E., DeBlasio, T., Zhou, Y., Zhang, J., Zon, L., and Nimer, S. D. (2002). Isolation and characterization of runxa and runxb, zebrafish members of the runt family of transcriptional regulators. *Exp. Hematol.* **30**, 1381–1389.

Cross, L. M., Cook, M. A., Lin, S., Chen, J. N., and Rubinstein, A. L. (2003). Rapid analysis of angiogenesis drugs in a live fluorescent zebrafish assay. *Arterioscler. Thromb. Vasc. Biol.* **23**, 911–912.

Cumano, A., and Godin, I. (2001). Pluripotent hematopoietic stem cell development during embryogenesis. *Curr. Opin. Immunol.* **13**, 166–171.

Cumano, A., and Godin, I. (2007). Ontogeny of the hematopoietic system. *Annu. Rev. Immunol.* **25**, 745–785.

Detrich, H. W., 3rd, Kieran, M. W., Chan, F. Y., Barone, L. M., Yee, K., Rundstadler, J. A., Pratt, S., Ransom, D., and Zon, L. I. (1995). Intraembryonic hematopoietic cell migration during vertebrate development. *Proc. Natl. Acad. Sci. USA* **92**, 10713–10717.

Dzierzak, E. (2005). The emergence of definitive hematopoietic stem cells in the mammal. *Curr. Opin. Hematol.* **12**, 197–202.

Gekas, C., Dieterlen-Lievre, F., Orkin, S. H., and Mikkola, H. K. (2005). The placenta is a niche for hematopoietic stem cells. *Dev. Cell* **8**, 365–375.

Goodell, M. A., Brose, K., Paradis, G., Conner, A. S., and Mulligan, R. C. (1996). Isolation and functional properties of murine hematopoietic stem cells that are replicating *in vivo*. *J. Exp. Med.* **183**, 1797–1806.

Hall, C., Flores, M. V., Storm, T., Crosier, K., and Crosier, P. (2007). The zebrafish lysozyme C promoter drives myeloid-specific expression in transgenic fish. *BMC Dev. Biol.* **7**, 42.

Hansen, J. D., and Zapata, A. G. (1998). Lymphocyte development in fish and amphibians. *Immunol. Rev.* **166,** 199–220.

Herbomel, P., Thisse, B., and Thisse, C. (1999). Ontogeny and behaviour of early macrophages in the zebrafish embryo. *Development* **126,** 3735–3745.

Houssaint, E. (1981). Differentiation of the mouse hepatic primordium. II. Extrinsic origin of the haemopoietic cell line. *Cell Differ.* **10,** 243–252.

Hsu, K., Traver, D., Kutok, J. L., Hagen, A., Liu, T. X., Paw, B. H., Rhodes, J., Berman, J., Zon, L. I., Kanki, J. P., and Look, A. T. (2004). The pu.1 promoter drives myeloid gene expression in zebrafish. *Blood* **104,** 1291–1297.

Huang, H., and Auerbach, R. (1993). Identification and characterization of hematopoietic stem cells from the yolk sac of the early mouse embryo. *Proc. Natl. Acad. Sci. USA* **90,** 10110–10114.

Jagadeeswaran, P., Sheehan, J. P., Craig, F. E., and Troyer, D. (1999). Identification and characterization of zebrafish thrombocytes. *Br. J. Haematol.* **107,** 731–738.

Jessen, J. R., Meng, A., McFarlane, R. J., Paw, B. H., Zon, L. I., Smith, G. R., and Lin, S. (1998). Modification of bacterial artificial chromosomes through chi-stimulated homologous recombination and its application in zebrafish transgenesis. *Proc. Natl. Acad. Sci. USA* **95,** 5121–5126.

Jin, H., Xu, J., and Wen, Z. (2007). Migratory path of definitive hematopoietic stem/progenitor cells during zebrafish development. *Blood* **109,** 5208–5214.

Johnson, G. R., and Moore, M. A. (1975). Role of stem cell migration in initiation of mouse foetal liver haemopoiesis. *Nature* **258,** 726–728.

Kalev-Zylinska, M. L., Horsfield, J. A., Flores, M. V., Postlethwait, J. H., Vitas, M. R., Baas, A. M., Crosier, P. S., and Crosier, K. E. (2002). Runx1 is required for zebrafish blood and vessel development and expression of a human RUNX1-CBF2T1 transgene advances a model for studies of leukemogenesis. *Development* **129,** 2015–2030.

Keller, G., Lacaud, G., and Robertson, S. (1999). Development of the hematopoietic system in the mouse. *Exp. Hematol.* **27,** 777–787.

Kissa, K., Murayama, E., Zapata, A., Cortes, A., Perret, E., Machu, C., and Herbomel, P. (2008). Live imaging of emerging hematopoietic stem cells and early thymus colonization. *Blood* **111,** 1147–1156.

Lam, E. Y., Chau, J. Y., Kalev-Zylinska, M. L., Fountaine, T. M., Mead, R. S., Hall, C. J., Crosier, P. S., Crosier, K. E., and Flores, M. V. (2009). Zebrafish runx1 promoter-EGFP transgenics mark discrete sites of definitive blood progenitors. *Blood* **113,** 1241–1249.

Langenau, D. M., Traver, D., Ferrando, A. A., Kutok, J., Aster, J. C., Kanki, J. P., Lin, H. S., Prochownik, E., Trede, N. S., Zon, L. I., and Look, A. T. (2003). Myc-induced T-cell leukemia in transgenic zebrafish. *Science* **299,** 887–890.

Langenau, D. M., Ferrando, A. A., Traver, D., Kutok, J. L., Hezel, J. P., Kanki, J. P., Zon, L. I., Look, A. T., and Trede, N. S. (2004). *In vivo* tracking of T cell development, ablation and engraftment in transgenic zebrafish. *Proc. Natl. Acad. Sci. USA* **101,** 7369–7374.

Lawson, N. D., and Weinstein, B. M. (2002). *In vivo* imaging of embryonic vascular development using transgenic zebrafish. *Dev Biol.* **248,** 307–318.

Liao, E. C., Paw, B. H., Oates, A. C., Pratt, S. J., Postlethwait, J. H., and Zon, L. I. (1998). SCL/Tal-1 transcription factor acts downstream of cloche to specify hematopoietic and vascular progenitors in zebrafish. *Genes Dev.* **12,** 621–626.

Lieschke, G. J., Oates, A. C., Paw, B. H., Thompson, M. A., Hall, N. E., Ward, A. C., Ho, R. K., Zon, L. I., and Layton, J. E. (2002). Zebrafish SPI-1 (PU.1) marks a site of myeloid development independent of primitive erythropoiesis: Implications for axial patterning. *Dev. Biol.* **246,** 274–295.

Lin, H. F., Traver, D., Zhu, H., Dooley, K., Paw, B. H., Zon, L. I., and Handin, R. I. (2005). Analysis of thrombocyte development in CD41-GFP transgenic zebrafish. *Blood* **106,** 3803–3810.

Long, Q., Meng, A., Wang, H., Jessen, J. R., Farrell, M. J., and Lin, S. (1997). GATA-1 expression pattern can be recapitulated in living transgenic zebrafish using GFP reporter gene. *Development* **124,** 4105–4111.

Lu, L. S., Wang, S. J., and Auerbach, R. (1996). *In vitro* and *in vivo* differentiation into B cells, T cells, and myeloid cells of primitive yolk sac hematopoietic precursor cells expanded >100-fold by coculture with a clonal yolk sac endothelial cell line. *Proc. Natl. Acad. Sci. USA* **93,** 14782–14787.

Lux, C. T., Yoshimoto, M., McGrath, K., Conway, S. J., Palis, J., and Yoder, M. C. (2008). All primitive and definitive hematopoietic progenitor cells emerging before E10 in the mouse embryo are products of the yolk sac. *Blood* **111,** 3435–3438.

Lyons, S. E., Lawson, N. D., Lei, L., Bennett, P. E., Weinstein, B. M., and Liu, P. P. (2002). A nonsense mutation in zebrafish gata1 causes the bloodless phenotype in vlad tepes. *Proc. Natl. Acad. Sci. USA* **99,** 5454–5459.

Morrison, S. J., Uchida, N., and Weissman, I. L. (1995). The biology of hematopoietic stem cells. *Annu. Rev. Cell Dev. Biol.* **11,** 35–71.

Mulder, A. H., and Visser, J. W. M. (1987). Separation and functional analysis of bone marrow cells separated by rhodamine-123 fluorescence. *Exp. Hematol.* **15,** 99–104.

Murayama, E., Kissa, K., Zapata, A., Mordelet, E., Briolat, V., Lin, H. F., Handin, R. I., and Herbomel, P. (2006). Tracing hematopoietic precursor migration to successive hematopoietic organs during zebrafish development. *Immunity* **25,** 963–975.

Ottersbach, K., and Dzierzak, E. (2005). The murine placenta contains hematopoietic stem cells within the vascular labyrinth region. *Dev. Cell* **8,** 377–387.

Palis, J., Robertson, S., Kennedy, M., Wall, C., and Keller, G. (1999). Development of erythroid and myeloid progenitors in the yolk sac and embryo proper of the mouse. *Development* **126,** 5073–5084.

Renshaw, S. A., Loynes, C. A., Trushell, D. M., Elworthy, S., Ingham, P. W., and Whyte, M. K. (2006). A transgenic zebrafish model of neutrophilic inflammation. *Blood* **108,** 3976–3978.

Shapiro, H. M. (2002). "Practical Flow Cytometry." Wiley-Liss, New York.

Spangrude, G. J., Heimfeld, S., and Weissman, I. L. (1988). Purification and characterization of mouse hematopoietic stem cells. *Science* **241,** 58–62.

Stachura, D. L., Reyes, J. R., Bartunek, P., Paw, B. H., Zon, L. I., and Traver, D. (2009). Zebrafish kidney stromal cell lines support multilineage hematopoiesis. *Blood.* Doi:10.1182/blood-2009-02-203638.

Thompson, M. A., Ransom, D. G., Pratt, S. J., MacLennan, H., Kieran, M. W., Detrich, H. W., III, Vail, B., Huber, T. L., Paw, B., Brownlie, A. J., Oates, A. C., Fritz, A., *et al.* (1998). The cloche and spadetail genes differentially affect hematopoiesis and vasculogenesis. *Dev. Biol.* **197,** 248–269.

Traver, D. (2004). Cellular dissection of zebrafish hematopoiesis. *Methods Cell Biol.* **76,** 127–149.

Traver, D., Herbomel, P., Patton, E. E., Murphy, R. D., Yoder, J. A., Litman, G. W., Catic, A., Amemiya, C. T., Zon, L. I., and Trede, N. S. (2003a). "The zebrafish as a model organism to study development of the immune system." Academic Press, London.

Traver, D., Paw, B. H., Poss, K. D., Penberthy, W. T., Lin, S., and Zon, L. I. (2003b). Transplantation and *in vivo* imaging of multilineage engraftment in zebrafish bloodless mutants. *Nat. Immunol.* **4,** 1238–1246.

Traver, D., Winzeler, E. A., Stern, H. M., Mayhall, E. A., Langenau, D. M., Kutok, J. L., Look, A. T., and Zon, L. I. (2004). Biological effects of lethal irradiation and rescue by hematopoietic cell transplantation in zebrafish. *Blood* **104,** 1298–1305.

Trede, N. S., and Zon, L. I. (1998). Development of T-cells during fish embryogenesis. *Dev. Comp. Immunol.* **22,** 253–263.

Visser, J. W., and de Vries, P. (1988). Isolation of spleen-colony forming cells (CFU-s) using wheat germ agglutinin and rhodamine 123 labeling. *Blood Cells* **14,** 369–384.

Visser, J. W., Bauman, J. G., Mulder, A. H., Eliason, J. F., and de Leeuw, A. M. (2004). Isolation of murine pluripotent hemopoietic stem cells. *J. Exp. Med.* **159,** 1576–1590.

Ward, A. C., McPhee, D. O., Condron, M. M., Varma, S., Cody, S. H., Onnebo, S. M., Paw, B. H., Zon, L. I., and Lieschke, G. J. (2003). The zebrafish spi1 promoter drives myeloid-specific expression in stable transgenic fish. *Blood* **102,** 3238–3240.

Weissman, I., Papaioannou, V., and Gardner, R. (1978). Fetal hematopoietic origins of the adult hematolymphoid system. *In* "Differentiation of Normal and Neoplastic Cells" (B. Clarkson, P. A. Marks, and J. E. Till, eds.), pp. 33–47. Cold Spring Harbor Laboratory Press, New York.

Westerfield, M. (2000). "The Zebrafish Book. A guide for the Laboratory Use of Zebrafish (Danio rerio)," 4th edn., University of Oregon Press, Eugene.

Willett, C. E., Zapata, A. G., Hopkins, N., and Steiner, L. A. (1997). Expression of zebrafish rag genes during early development identifies the thymus. *Dev. Biol.* **182,** 331–341.

Willett, C. E., Cortes, A., Zuasti, A., and Zapata, A. G. (1999). Early hematopoiesis and developing lymphoid organs in the zebrafish. *Dev. Dyn.* **214,** 323–336.

Yoder, M., Hiatt, K., and Mukherjee, P. (1997a). *In vivo* repopulating hematopoietic stem cells are present in the murine yolk sac at day 9.0 postcoitus. *Proc. Natl. Acad. Sci. USA* **94,** 6776.

Yoder, M. C., Hiatt, K., Dutt, P., Mukherjee, P., Bodine, D. M., and Orlic, D. (1997b). Characterization of definitive lymphohematopoietic stem cells in the day 9 murine yolk sac. *Immunity* **7,** 335–344.

Yokota, T., Huang, J., Tavian, M., Nagai, Y., Hirose, J., Zuniga-Pflucker, J. C., Peault, B., and Kincade, P. W. (2006). Tracing the first waves of lymphopoiesis in mice. *Development* **133,** 2041–2051.

Zapata, A. (1979). Ultrastructural study of the teleost fish kidney. *Dev. Comp. Immunol.* **3,** 55–65.

Zapata, A., and Amemiya, C. T. (2000). Phylogeny of lower vertebrates and their immunological structures. *Curr. Top. Microbiol. Immunol.* **248,** 67–107.

Zhu, H., Traver, D., Davidson, A. J., Dibiase, A., Thisse, C., Thisse, B., Nimer, S., and Zon, L. I. (2005). Regulation of the lmo2 promoter during hematopoietic and vascular development in zebrafish. *Dev. Biol.* **281,** 256–269.

CHAPTER 12

Culture of Embryonic Stem and Primordial Germ Cell Lines from Zebrafish

Ten–Tsao Wong, Lianchun Fan, and Paul Collodi

Department of Animal Sciences
Purdue University
West Lafayette, Indiana 47907

Although zebrafish ES cells exhibit characteristics of pluripotency in culture, when they are transplanted to a host embryo, the cells contribute at a low frequency to the germ cell lineage of the recipient embryo. As a result, germ line chimeras have been successfully generated by transplanting ES cells taken from an entire culture, but the small percentage of germ line competent ES cells in the culture has precluded the production of germ line chimeras using individual ES cell colonies that were grown from a single cell. The production of germ line chimeras from individual ES cell colonies will be required in order to generate lines of fish derived from selected colonies that carry a specific genetic alteration. The low frequency germ line transmission by the ES cells is most likely due to the small number of germ line-specified cells present in the early-stage zebrafish embryos that are used to initiate ES cell cultures (Yoon *et al.*, 1997). To circumvent this problem and improve the efficiency of germ line chimera production, we have recently focused on establishing cultures of primordial germ cells (PGCs) initiated from late-stage zebrafish embryos (Fan *et al.*, 2008). Since the PGCs are committed to the germ cell lineage in the embryo, a culture consisting entirely of PGCs would be expected to generate germ line chimeras at a high frequency when transplanted into a host. The general methods used to initiate PGC cultures are similar to those for ES cells that are described in this chapter. One key difference is that feeder cells that express zebrafish kit ligand A are used to initiate and maintain the PGC cultures. Also, in order to obtain homogeneous cultures of PGCs, the cultures are initiated from transgenic embryos that express genes conferring drug resistance and encoding the red fluorescent protein (RFP) specifically in the germ cell lineage (Fan *et al.*, 2008). Homogenous cultures of PGCs have been initiated and maintained for several months using a combination of fluorescence-activated cell sorting (FACS) and/or drug selection. Detailed methods for the derivation of zebrafish PGC cultures are described on our website (http://www.ag.purdue.edu/ansc/Pages/pcollodi.aspx).

Key differences between the methods used to derive zebrafish ES cell versus PGC cultures are

- PGC cultures are initiated from zebrafish embryos at the 12–15-somite stage instead of the blastula or early-gastrula stages that are used for ES cells.
- The PGCs are grown on RTS34st feeder cells that express zebrafish kit-ligand A.
- PGCs are initiated from transgenic embryos that express genes encoding RFP and drug resistance. PGC cultures have also been initiated from wild-type embryos that were injected at the 1-cell stage with mRNA encoding RFP and drug resistance along with the nanos 3'-UTR to direct expression in the PGCs of the developing embryo.
- Drug selection is used to obtain homogeneous PGC cultures.

I. Introduction

Despite its many advantages for studies of embryo development and human disease, one deficiency of the zebrafish model has been the lack of methods for targeted mutagenesis using embryonic stem (ES) cells. In mice, ES cell-mediated gene targeting has provided a powerful approach to the study of gene function (Lui *et al.*, 1993; Zhang, 1995). A similar strategy applied to zebrafish would complement other genetic methods currently available such as large-scale random mutagenesis (Currie, 1996; Golling *et al.*, 2002; Van Eeden *et al.*, 1999), antisense-based gene knockdown (Nasevicius and Ekker, 2002), and target selected mutagenesis (Wienholds *et al.*, 2002) approaches to increase the utility of this model system. To address this problem, our laboratory has been working to establish zebrafish ES cell lines that are suitable for use in a gene targeting approach (Fan *et al.*, 2004; Ghosh and Collodi, 1994; Ma *et al.*, 2001). Successful ES cell-mediated gene targeting requires the use of pluripotent ES cell lines that possess the capacity to contribute to the germ-cell lineage of a host embryo (Evans and Kaufman, 1981;

Gossler *et al.*, 1996). The germ-line competent ES cells are genetically altered in culture by targeted incorporation of foreign DNA by homologous recombination followed by *in vitro* selection of cell colonies that have undergone the targeting event (Fig. 1). The selected colonies are expanded in culture and introduced into host embryos to generate germ-line chimeras carrying the targeted mutation (Capecchi, 1989; Doetschman *et al.*, 1987). Once the chimeras are sexually mature they are used to establish the knockout line. This chapter describes methods for the derivation of zebrafish ES cell cultures along with a protocol for the efficient introduction of plasmid DNA into the cells by electroporation and *in vitro* selection of homologous recombinants.

II. Methods

A. General Characteristics of Zebrafish ES Cell Cultures

Although pluripotent, germ-line competent ES cell cultures have been derived from both blastula- and gastrula-stage zebrafish embryos (Fan *et al.*, 2004), the blastula, which is comprised of nondifferentiated cells, is the optimal stage for use in initiating the cultures. To maintain pluripotency and germ-line competency, the ES cell cultures are initiated and maintained on a feeder layer of growth-arrested rainbow trout spleen cells derived from the established RTS34st cell line (Ma *et al.*, 2001; Gannassin and Bols, 1999). In the presence of the feeder layer, the ES cell cultures remain germ-line competent for at least six passages (6 weeks) in culture, a sufficient period of time for electroporation and selection of homologous recombinants (Fan *et al.*, 2004). When introduced into a recipient embryo the cultured ES cells contribute to multiple tissues including the germ cell lineage of the host (Fan and Collodi, 2002). Zebrafish germ-line chimeras have been generated from ES cells maintained for multiple passages in culture (Fan *et al.* 2004).

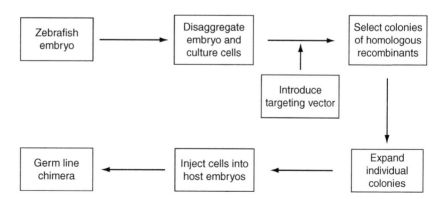

Fig. 1 Strategy for ES cell-mediated gene targeting in zebrafish.

When a primary cell culture is initiated from zebrafish blastulas and maintained on an RTS34st feeder layer, within 24 h after plating, the embryo cells form dense homogeneous aggregates present throughout the culture (Fig. 2A). As the zebrafish embryo cells proliferate, the aggregates become larger while continuing to exhibit a homogeneous appearance without showing morphological indication of differentiation. The primary culture must be passaged at days 4–6 to prevent the cell aggregates from becoming too large and differentiating. During the first passage, the aggregates are partially dissociated and added to a fresh feeder layer of growth-arrested RTS34st cells. The aggregates become easier to dissociate with each passage and eventually grow to form a monolayer by the fourth passage (Fig. 2B).

The same strategy can be used to derive pluripotent ES cell cultures initiated from embryos at the germ-ring stage of development (approximately 6 hpf) (Ma *et al.*, 2001). Since cell differentiation has begun to occur at this stage of development, the majority of the cell aggregates in the primary culture will be comprised of differentiated cell types including pigmented melanocytes, neural cells, and fibroblasts. Even though cell differentiation is pervasive in the gastrula-derived primary culture, pluripotent ES cells can be obtained by manually selecting the small number of cell aggregates that possess an ES-like morphology characterized by a compact and homogeneous appearance. The ES-like cell aggregates are dissociated and passaged to initiate the long-term culture.

Plasmid DNA is efficiently introduced into the ES cells by electroporation and drug selection is used to isolate colonies of stable transformants. Gene targeting in the ES cells is accomplished by introducing a vector containing sequences that are homologous to the gene that is being targeted for disruption (Fig. 3A) (Fan *et al.*, 2006). Zebrafish ES cells that have incorporated the plasmid DNA in a targeted

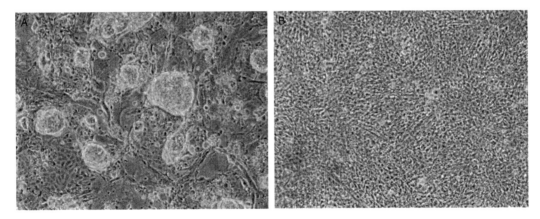

Fig. 2 Zebrafish embryo cell culture initiated from blastula-stage embryos on a feeder layer of RTS34st cells. (A) The primary culture is comprised of ES-like cell aggregates growing on the monolayer of feeder cells. (B) By passage 4 the embryo cells grow to form a confluent monolayer in close association with the feeder cells.

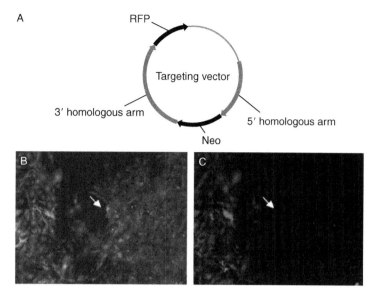

Fig. 3 (A) The general design of a targeting vector that contains *neo* flanked by sequences that are homologous to the gene being targeted along with *RFP* located outside of the homologous region. The targeting vector was introduced into an ES cell line that constitutively expresses the enhanced green fluorescent protein (EGFP). (B) After G418 selection, two GFP⁺ colonies are shown by fluorescence microscopy. (C) The same two colonies shown in (B) are visualized using a rhodamine filter to detect RFP expression. The colony on the right (arrow) is RFP⁻ indicating that it is a homologous recombinant.

fashion by homologous recombination are selected using a strategy that is based on the positive/negative selection method commonly employed with mouse ES cells (Capecchi, 1989). To use this selection strategy, a targeting vector is constructed that contains *neo* flanked on each side by 3 kb or longer arms that are homologous to the gene being targeted (Fig. 3A) (Fan *et al.*, 2006). The vector also contains the red fluorescent protein (RFP) gene with its own promoter located outside of the homologous region. After electroporation, the ES cells that have incorporated the vector are selected in G418 and drug resistant colonies are examined by fluorescence microscopy. The colonies that have incorporated the plasmid by random insertion will possess the entire vector sequence including RFP whereas the colonies of homologous recombinants will lack RFP (Fig. 3B). Approximately 5 weeks after electroporation, individual RFP negative colonies are manually picked from the plate using a drawn out Pasteur pipet or a micropipetor and transferred to individual wells of a 24-well plate (one colony/well) and eventually expanded into 25 cm² flasks. Disruption of the targeted gene is confirmed by PCR and southern blot analysis. We have found that the frequency of homologous recombination in the ES cells is approximately 1 out of 400 G418 resistant colonies. Using this selection strategy approximately 50% of the manually selected RFP-negative colonies were confirmed to be homologous recombinants.

B. Derivation of ES Cell Cultures from Blastula-Stage Embryos

1. Collect the embryos in a fine mesh net and rinse several times with water to remove debris. Transfer the embryos into a Petri dish containing egg water and incubate at 28 °C until they reach the blastula-stage of development (approximately 4 hpf). Divide the embryos into groups of approximately 50 individuals, and transfer each group into a 2.5-ml Eppendorf microfuge tube that had the conical shaped bottom cut off and replaced with fine mesh netting attached with a rubber band. Submerge the bottom of the tube containing the embryos sitting on the net into 70% ethanol for 10 s and then immediately submerge the embryos into a beaker of sterile egg water. Transfer the embryos from the tube into a 60-mm Petri dish containing egg water and remove any dead individuals. Remove the egg water and rinse the embryos 3 times with LDF medium (Collodi *et al.*, 1992). Remove the LDF and add approximately 2 ml of bleach solution to the dish. Incubate the embryos for 2 min in the bleach solution and then remove the bleach and immediately rinse with LDF. Repeat the bleach treatment and rinse two additional times. It is important not to expose the embryos to the bleach solution for periods longer than 2 min without rinsing. Following the final bleach treatment, rinse the embryos three additional times with LDF medium.

2. Remove the chorions by incubating each group of embryos in 3 ml of pronase solution for approximately 15 min or until the chorions begin to break apart. Gently swirl the suspended embryos in the dish to release them from the digested chorion. Use a pipettor to remove the pronase solution containing the floating chorions and gently rinse the dechorionated embryos with LDF medium. Incubate each group of embryos in 3 ml of trypsin/EDTA for 1–2 min and gently pipet to dissociate the cells. Collect the cells by centrifugation ($500 \times g$; 5 min) and resuspend the cell pellet obtained from each group of approximately 50 embryos in 1.8 ml LDF medium. Transfer the cell suspension (1.8 ml) to a single well of a 6-well tissue culture plate (Falcon) containing a confluent monolayer of growth-arrested RTS34st cells. Let the plate sit undisturbed for 30 min to allow the embryo cells to attach to the RTS34st monolayer. After the cells have attached, add the following factors to each well: 150 μl of fetal bovine serum (FBS), 15 μl of zebrafish embryo extract, 30 μl of trout plasma, 30 μl of insulin stock solution, 15 μl of epidermal growth factor (EGF) stock solution, 15 μl of basic fibroblast growth factor (bFGF) stock solution, and 945 μl of RTS34st conditioned medium. If the cells are not attached after 30 min the plate can be incubated for 1–2 h before adding the factors. Also, EGF can be added immediately after plating the cells to enhance cell attachment. Incubate the culture for 5 days (22 °C). During this time the cell aggregates should increase in size while maintaining a homogeneous,

nondifferentiated appearance. Although zebrafish cell cultures are normally propagated at 26 °C (Collodi *et al.*, 1992), the embryo cell cultures are maintained at 22 °C to accommodate the feeder layer of trout spleen cells.

3. To passage the primary culture, harvest the cells from each well by adding 2 ml of trypsin/EDTA solution per well and incubating 30 s before transferring the cell suspension to a 15-ml polypropylene centrifuge tube (Corning). Pipet the cell suspension up and down several times to partially dissociate the cell aggregates and add 0.2 ml of FBS to stop the action of the trypsin. The cell aggregates cannot be completely dissociated during the first passage. Collect the cells by centrifugation (500 × g; 5 min) and resuspend the pellet in 3.6 ml of LDF medium. Add 1.8 ml of the cell suspension to each of two wells of a 6-well plate (Falcon) containing a confluent monolayer of growth-arrested RTS34st cells and add the factors listed in step 2 to each well. Incubate the 6-well plate for 5 days (22 °C) and harvest the cells as described in step 2. Combine the cells harvested from two wells, collect the cells by centrifugation and resuspend the cell pellet in 3.6 ml of LDF medium. The suspension will still contain a large number of cell aggregates.

4. Add the cell suspension to a 25-cm^2 tissue culture flask (Falcon) containing a confluent monolayer of growth-arrested RTS34st cells. Let the flask sit undisturbed for 1–3 h to allow the cells to attach to the feeder layer.

5. Add the following factors to the flask: 300 μl of FBS, 30 μl of zebrafish embryo extract, 60 μl of trout serum, 60 μl of bovine insulin stock solution, 30 μl of EGF stock solution, 30 μl of bFGF stock solution, and 1.890 μl of RTS34st conditioned medium.

6. Incubate the flask for 7 days (22 °C) and then harvest the cells in trypsin/EDTA as described above and seed them into two flasks that each contains a confluent monolayer of growth-arrested RTS34st cells. With each passage the cell aggregates become easier to dissociate and fewer aggregates are present in the culture. Continue to passage the culture approximately every 7 days.

7. The cultures can be cryopreserved beginning at passage 4 when the cells begin to grow as a monolayer. A portion of the culture is frozen at each passage. Harvest the cultures in trypsin/EDTA as described in step 3 and resuspend the cell pellet obtained from one 25 cm^2 flask in 1.2 ml freezing medium. Transfer the cell suspension to a cryovial (Nalgene), place the vial in Styrofoam insulation and incubate the vial at 4 °C for 10 min followed by −80 °C for at least 1 h and then submerge and store the vial in liquid nitrogen.

C. Derivation of ES Cell Cultures from Zebrafish Gastrula-Stage Embryos

1. Initiate primary cell cultures from embryos at the germ-ring stage of development (approximately 6 hpf) (Westerfield, 1995) using the methods described in Section B. After the primary culture has been growing for approximately 5 days,

use a drawn out Pasteur pipet or a micropipetor (Rainin) to remove aggregates of densely packed cells that appear homogeneous without morphological indications of differentiation. Combine 30–50 of the isolated cell aggregates in a sterile 2 ml centrifuge tube containing LDF medium.

2. Collect the cell aggregates by centrifugation (500 × g, 5 min), resuspend the pellet in 1.0 ml of trypsin/EDTA solution and incubate 2 min while occasionally pipeting the cell suspension through a 5-ml pipet to partially dissociate the aggregates. Add 0.1 ml of FBS to stop the action of the trypsin and collect the cells by centrifugation (500 × g, 5 min).

3. Resuspend the cell pellet in 1.8 ml of LDF medium and add to a single well of a 6-well plate containing a monolayer of growth-arrested RTS34st feeder cells.

4. Let the plate sit undisturbed for 5 h to allow the embryo cells to attach and add the factors listed Section B, step 2. The culture consists of small cell aggregates and some single embryo cells attached to the RTS34st cells.

5. Incubate the plate (22 °C) for 7 days. As the cells proliferate the aggregates should become larger without exhibiting morphological indications of differentiation. The culture is passaged every 7 days as described in Section B. The cell aggregates will become easier to dissociate and eventually grow as a monolayer after approximately four passages.

D. Electroporation of Plasmid DNA into the ES Cell Cultures

1. Once the ES cell culture begins to grow as a monolayer (passage 4) (Fig. 2B), the cells can be efficiently transformed with plasmid DNA by electroporation and colonies of stable transformants selected. To prepare the ES cells for electroporation, harvest the cells by trypsinization, wash two times with phosphate buffered saline (PBS) and suspend 6×10^6 cells in 0.75 ml of PBS in a 0.4-cm electroporation cuvette.

2. Add 50 μg of sterile, linearized, plasmid DNA dissolved in 50 μl TE buffer. In addition to the gene of interest, the plasmid should contain a selectable marker gene such as *neo* under the control of a constitutively expressed promoter.

3. Electroporate the cells (950 μF, 300 V) and measure cell mortality by trypan blue staining 0.5 h after electroporation. Cell mortality should be approximately 50%. Plate the cells into two 100-mm diameter culture dishes containing a confluent layer of growth-arrested RTS34st(*neo*) cells and add the medium and supplements described in Section B. The next day, add 5 μl/ml of the G418 stock solution and change the medium every 2 days adding fresh G418. Colonies will begin to appear 2–3 weeks after G418 selection is initiated.

4. Gene targeting by homologous recombination can be accomplished in the cells using a targeting vector containing *neo* flanked by 5′ and 3′ arms that are homologous to the targeted gene along with *RFP* located outside of the

homologous region (Fig. 3A) (Fan *et al.*, 2006). The targeting vector is electroporated into the cells and the cells are selected in G418 as described in Step 3. Following G418 selection the colonies are examined by fluorescence microscopy and the homologous recombinants are identified by the absence of RFP expression (Fig. 3B). The RFP negative colonies are manually removed from the plate using a Pipetman micropipetor (Rainin) approximately 5 weeks after the start of G418 selection. The individual selected colonies are transferred to single wells of a 24-well plate containing growth arrested RTS34st(*neo*) feeder cells. The individual colonies are cultured for 2–3 weeks before passaging into single wells of a 12-well plate. During passage, a portion of the cells from each colony are harvested for PCR analysis to confirm homologous recombination.

III. Materials

A. Reagents

1. *Cell culture media*: Leibowitz's L-15 (catalog no. 41300-039), Ham's F12 (catalog no. 21700-075) and Dulbecco's modified Eagle's media (catalog no. 12100-046) are available from Invitrogen Corporation, Carlsbad, CA. One liter of each medium is prepared separately by dissolving the powder in ddH$_2$O and adding HEPES buffer (final concentration 15 mM, pH 7.2), penicillin G (120 μg/ml), ampicillin (25 μg/ml), and streptomycin sulfate (200 μg/ml). LDF medium is prepared by combining Leibowitz's L-15, Dulbecco's modified Eagles and Ham's F12 media (50:35:15) and supplementing with sodium bicarbonate (0.180 g/l) and sodium selenite (10^{-8} M). The medium is filter sterilized before use.

2. PBS (catalog no. 21600-010) obtained from Invitrogen.

3. TE buffer: 10 mM Tris-HCl, 1 mM EDTA, pH 8.0.

4. FBS (catalog no. BT-9501-500) is available from Harlan Laboratories, Indianapolis, IN.

5. Calf serum (catalog no. 26170-043) is available from Invitrogen.

6. Trout plasma (SeaGrow) is available from East Coast Biologics, Inc., North Berwick, ME. The plasma is sterile filtered and heat treated (56 °C, 25 min) and centrifuged (10,000 $\times$ *g*, 10 min) before use.

7. Trypsin/EDTA solution (2 mg/ml trypsin, 1 mM EDTA) is prepared in PBS. The solution is filter sterilized before use. Trypsin (catalog no. T-7409) and EDTA (catalog no. E-6511) are available from Sigma, St. Louis, MO.

8. Human EGF (catalog no. 13247-051) is available from Invitrogen, Carlsbad, CA. Stock EGF solution is prepared at 10 μg/ml in ddH$_2$O.

9. Human bFGF (catalog no. 13256-029) is available from Invitrogen. Stock bFGF solution is prepared at 10 μg/ml in 10 mM Tris-HCl, pH 7.6.

10. Bovine insulin catalog no. I-5500 is available from Sigma. Stock insulin is prepared at 1 mg/ml in 20 mM HCl.

11. Bleach (Chlorox) solution is prepared fresh at 0.5% in ddH$_2$O from a newly opened bottle.

12. Zebrafish embryo extract is prepared by homogenizing approximately 500 embryos in 0.5 ml of LDF medium and centrifuging (20,000 × g, 10 min) to remove the debris. The supernatant is collected, filter sterilized and the protein measured. The extract is diluted to 10 mg protein/ml and stored frozen (-20 °C) in 0.2 ml aliquots.

13. Geneticin (G418 sulfate, catalog no. 11811-031) is available from Invitrogen. G418 stock solution is prepared at 100 mg/ml in ddH$_2$O and filter sterilized before use.

14. Pronase (catalog no. P6911) is available from Sigma and is prepared at 0.5 mg/ml in Hanks solution.

15. *Egg water*: 60 μg/ml aquarium salt.

16. *Freezing medium*: 80% FD medium (1:1 mixture of Ham's F12 and DMEM), 10% FBS, 10% DMSO.

B. Feeder Cell Lines

1. Growth-Arrested Feeder Cells

RTS34st cells (Gannassin and Bols, 1999) are cultured (18 °C) in Leibowitz's L-15 medium (Sigma, catalog no. L5520) supplemented with 30% calf serum. To prepare growth-arrested cells, a confluent culture of RTS34st cells contained in a flask (25 cm^2) or dish (100 mm diameter) is irradiated (3000 RADS) and then harvested by trypsinization and frozen in liquid nitrogen within 24 h after irradiation. To recover the frozen growth arrested cells, the vial is thawed briefly in a water bath (37 °C) and the cells are collected by centrifugation and resuspending in L-15 medium. The cells from one frozen vial are distributed into two 25 cm^2 flasks or 4 wells of a 6-well plate. After the cells have attached to the culture surface, the medium is supplemented with calf serum (30%). After 24 h the growth-arrested cells should be spread on the culture surface and are used immediately as feeder layers. Before using the growth-arrested cells as a feeder layer, the L-15 medium is removed and the cells are rinsed one time.

2. Drug Resistant Feeder Cells

A feeder cell line that is resistant to G418 was prepared by transfecting RTS34st with the pBKRSV plasmid which contains the aminoglycoside phophotransferase gene (*neo*) under the control of RSV promoter. Colonies of cells that stably express *neo* were selected in G418 (500 μg/ml). The *neo* resistant cell line, RTS34st(*neo*), is cultured in L15 medium supplemented with 30% calf serum plus 200 μg/ml G418. Growth arrested RTS34st(*neo*) cells are prepared using the same methods described for RTS34st.

3. RTS34ST Cell-Conditioned Medium

Conditioned medium is prepared by adding fresh L-15 plus 30% FBS to a confluent culture of RTS34st cells and incubating for 3 days (18 °C). The medium is removed, filter sterilized and stored frozen (−20 °C).

Acknowledgments

This work was supported by grants from the U.S. Department of Agriculture NRI 01-3242, Illinois-Indiana SeaGrant R/A-03-01 and the National Institutes of Health R01-GM069384.IV.

References

Capecchi, M. (1989). Altering the genome by homologous recombination. *Science* **244**, 1288–1292.

Collodi, P., Kamei, Y., Ernst, T., Miranda, C., Buhler, D., and Barnes, D. (1992). Culture of cells from zebrafish embryo and adult tissues. *Cell Biol. Toxicol.* **8**, 43–61.

Currie, P. D. (1996). Zebrafish genetics: Mutant cornucopia. *Curr. Biol.* **6**, 1548–1552.

Doetschman, T., Gregg, R. G., Maeda, N., Hooper, M. L., Melton, D. W., Thompson, S., and Smithies, O. (1987). Targeted correction of a mutant HPRT gene in mouse embryonic stem cells. *Nature* **330**, 576–578.

Evans, M. J., and Kaufman, M. H. (1981). Establishment in culture of pluripotential cells from mouse embryos. *Nature* **292**, 154–156.

Fan, L., Moon, J., Wong, T.-T., Crodian, J., and Collodi, P. (2008). Zebrafish primordial germ cell cultures derived from vasa::RFP transgenic embryos. *Stem Cells Dev.* **17**, 585–597.

Fan, L., Moon, J., Crodian, J., and Collodi, P. (2006). Homologous recombination in zebrafish ES cells. *Transgenic Res.* **15**, 21–30.

Fan, L., Alestrom, A., Alestrom, P., and Collodi, P. (2004). Development of zebrafish cell cultures with competency for contributing to the germ line. *Crit. Rev. Eukaryot. Gene Expr.* **14**, 43–51.

Fan, L., and Collodi, P. (2002). Progress towards cell-mediated gene transfer in zebrafish. *Brief. Funct. Genomic. Proteomic.* **1**, 131–138.

Ganassin, R., and Bols, N. C. (1999). A stromal cell line from rainbow trout spleen, RTS34st, that supports the growth of rainbow trout macrophages and produces conditioned medium with mitogenic effects on leukocytes. *In Vitro Cell Dev. Biol. Anim.* **35**, 80–86.

Ghosh, C., and Collodi, P. (1994). Culture of cells from zebrafish blastula-stage embryos. *Cytotechnology* **14**, 21–26.

Golling, G., Amsterdam, A., Sun, Z., Antonelli, M., Maldonado, E., Chen, W., Burgess, S., Haldi, M., Artzt, K., Farrington, S., Lin, S-Y., Nissen, R., *et al.* (2002). Insertional mutagenesis in zebrafish rapidly identifies genes essential for early vertebrate development. *Nat. Genet.* **31**, 135–140.

Gossler, A., Doetschman, T., Korn, R., Serfling, E., and Kemler, R. (1986). Transgenesis by means of blastocyst-derived embryonic stem cell lines. *Proc. Natl. Acad. Sci. USA* **83**, 9065–9069.

Lui, J., Baker, J., Perkins, A. S., Robertson, E. J., and Efstratiadis, A. (1993). Mice carrying null mutations of the genes encoding insulin-like growth factor 1 (Igf-1) and type 1 IGF receptor (Igf1r). *Cell* **75**, 59–72.

Ma, C., Fan, L., Ganassin, R., Bols, N., and Collodi, P. (2001). Production of zebrafish germ-line chimeras from embryo cell cultures. *Proc. Natl. Acad. Sci. USA* **98**, 2461–2466.

Nasevicius, A., and Ekker, S. C. (2000). Effective targeted gene "knockdown" in zebrafish. *Nat. Genet.* **26**, 216–220.

Van Eeden, F. J. M., Granato, M., Odenthal, J., and Haffter, P. (1999). Developmental mutant screens in the zebrafish. *In* "Methods in Cell Biology" (H. W. Detrich, M. Westerfield, and L. I. Zon, eds.), Vol. 60, pp. 21–41. Academic Press, San Diego.

Westerfield, M. (1995). "The Zebrafish Book," 3rd edn., , University of Oregon Press, Eugene, OR

Wienholds, E., Schulte-Merker, S., Walderich, B., and Plasterk, R. (2002). Target-selected inactivation of the zebrafish rag1 gene. *Science* **297**, 99–101.

Yoon, C., Kawakami, K., and Hopkins, N. (1997). Zebrafish *vasa* homologue RNA is localized to the cleavage planes of 2- and 4-cell-stage embryos and is expressed in the primordial germ cells. *Development* **124**, 3157–3166.

Zhang, W., Behringer, R. R., and Olson, E. N. (1995). Inactivation of the myogenic bHLH gene MRF4 results in up-regulation of myogenin and rib anomalies. *Genes Dev.* **9**, 1388–1399.

CHAPTER 13

Neurogenesis

Prisca Chapouton, Marion Coolen, and Laure Bally-Cuif

Zebrafish Neurogenetics Department
Institute of Developmental Genetics
Helmholtz Zentrum München
German Research Center for Environmental health
85764 Neuherberg, Germany

I. Introduction

We will adopt here a broad definition of neurogenesis, shared by most authors, as the multistep process that brings from neural induction to the differentiation of functional neurons (Appel and Chitnis, 2002). Along this process, cells transit from a precursor state (further named "precursor cell" or "progenitor cell" or "neuroblast") to a committed and freshly postmitotic state ("early differentiating neuron" or "early postmitotic neuron") to the differentiated state. We will restrict our review to the building of the zebrafish central nervous system, but include current knowledge on the formation of glia (sensu stricto "gliogenesis"), which can influence neuronal selection, birth and differentiation as well as take an active part in late neurogenesis (Alvarez-Buylla and Garcia-Verdugo, 2002; Gotz *et al.*, 2002; Taupin and Gage, 2002).

Neurogenesis in zebrafish, as in other lower vertebrates, is classically considered to occur in two successive ("primary" and "secondary") waves (Kimmel, 1993) (see our discussion below). Primary neurogenesis is initiated at late gastrulation and continues during embryogenesis to produce early-born big neurons with long axons such as the brain epiphyseal and postoptic clusters, Mauthner cells, Rohon-Beard (RB) sensory neurons, and the three types of primary spinal motoneurons (CaP, MiP, RoP) (Kimmel and Westerfield, 1990). Axonogenesis starts between 14 and 24 hpf and primary neurons build the first functional embryonic and early larval neuronal scaffold. Secondary neurogenesis occurs massively from postembryonic stages (2 dpf) onward at all rostro-caudal levels, to take over the primary system by a refined and increasingly complex network (Mueller and Wullimann, 2003). Zebrafish neurogenesis also extends into adulthood.

The last few years of the neurogenesis field have seen an increased interest for characterizing late neurogenesis steps and determining whether these follow intrinsically different mechanisms compared to early neuronal production. In particular, the formation of oligodendrocytes and late spinal cord neurons has been a topic of focus (see Section IV.B), and several studies characterized neurogenesis domains in the adult brain and spinal cord (see Section IV.C). Stimulated by the maintenance of numerous active neurogenic niches during adulthood, the mechanisms of regeneration recently began to be addressed, with first publications in the adult retina and spinal cord (Section IV.C). Recent years were also marked by the generation of hundreds of transgenic lines labeling selective neuronal populations or neural territories with fluorescent reporters (see tables). In addition to efforts directed toward specific gene candidates, several gene- or enhancer-trap screens have been conducted, producing extremely useful libraries (see Table V).

II. The Primary Neuronal Scaffold

The initiation of neuronal differentiation in the zebrafish embryo has originally been revealed using acetylcholinesterase (AChE) activity (Hanneman and Westerfield, 1989; Ross *et al.*, 1992; Wilson *et al.*, 1990) or antibodies against acetylated tubulin (Chitnis and Kuwada, 1990) or HNK1 (Metcalfe *et al.*, 1990). Axon tracts have also been visualized following DiI applications, facilitated by the large size of zebrafish primary neurons (Myers *et al.*, 1986; Wilson and Easter,

1991a,b). These techniques, later complemented by *in situ* hybridization approaches with neurogenesis markers, and by the production of large sets of antibodies (Trewarrow *et al.*, 1990), highlighted defined differentiation centers and stereotyped axonal pathways within the early neural tube (Kimmel, 1993).

The embryonic brain neuronal network (Ross *et al.*, 1992; Wilson and Easter, 1991a,b) is composed of several nuclei (dorsorostral, ventrorostral, epiphyseal, ventrocaudal, nucleus of the posterior commissure) that extend longitudinal or transverse (commissural) axon tracts connecting the clusters with each other, with the spinal cord or with the peripheral system.

Reticulospinal neurons are long projection interneurons whose cell bodies are located in a rhombomere-specific pattern within the hindbrain or within the basal mesencephalon (nMLF), and whose axons extend into the spinal cord (Mendelson, 1986a,b). These neurons can be labeled by backfilling with fluorescent dyes from the spinal cord at trunk levels. An extensively studied reticulospinal neuron is the Mauthner cell, located in rhombomere 4, which links sensory information originating from trigeminal ganglia to contralateral motoneurons, mediating the escape response (Kimmel *et al.*, 1981, 1982; O'Malley *et al.*, 1996). The specific projections of other reticulospinal neurons (Fame *et al.*, 2006) and their diverse roles in locomotor control is beginning to be appreciated (Gahtan *et al.*, 2005; Nakayama and Oda, 2004) (Fig. 1).

Spinal cord neurons (Bernhardt *et al.*, 1990; Eisen *et al.*, 1990; Kuwada *et al.*, 1990; Westerfield *et al.*, 1986): comprise moto-, intermediate, and sensory neurons, arranged in a ventral to dorsal sequence. Three types of primary motoneurons (CaP, MiP, RoP), regularly arranged opposite each somite, differ in their position and projection pattern to the somite-derived fast muscles (Eisen *et al.*, 1990; Westerfield *et al.*, 1986). Eight types of interneurons have been identified (Bernhardt *et al.*, 1990; Hale *et al.*, 2001) that, at least in part, are differentially recruited during motor behavior control (Gahtan *et al.*, 2005; McLean *et al.*, 2007; Ritter *et al.*, 2001). Among those, at least CoPA and VeLD have both primary and secondary components. RB mechanosensory neurons are the only type of primary sensory neurons. Their central axons join the lateral longitudinal fascicle and extend into the hindbrain and spinal cord; their peripheral axons exit the spinal cord and innervate the skin. Of all six primary spinal neurons, RBs are the only transient cell type. They gradually die by apoptosis around 3 dpf, coinciding with the emergence of the dorsal root ganglia (Reyes *et al.*, 2004; Svoboda *et al.*, 2001; Williams *et al.*, 2000).

Together, these results highlight a number of interesting features. First, at all locations, the basal (ventral) plate generally matures earlier than the alar (dorsal) plate. Second, the location and connection pattern of all zebrafish early clusters/ neuronal groups is very reminiscent of the early neuronal organization observed in other vertebrates, such as the mouse or chick (Easter *et al.*, 1994). This suggests that establishment of the early neuronal differentiation profile in vertebrates is controlled in time and space in a strict and evolutionarily conserved manner. Finally, the position of these clusters is generally related to the neuromeric organization of the brain. Thus patterning cues that regulate neuromere's development

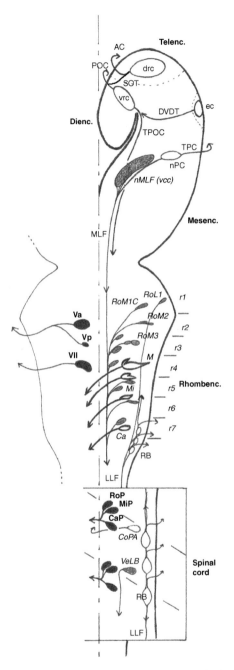

Fig. 1 Earliest clusters of zebrafish central primary neurons, schematized on an "open" preparation of the neural tube at 24–48 hpf (anterior up, spinal cord—boxed—represented for clarity at a higher magnification). Brain clusters in the tel-, di-, and mesencephalon (drc, ep, nMLF, nPC, vrc) build a scaffold of orthogonally oriented tracts (AC, DVDT, POC, SOT, TPC). More posterior neuronal

might also play a role in the spatial regulation of neurogenesis onset (see for instance in the rhombencephalon: Amoyel *et al.* (2005) and Cheng *et al.* (2004)).

III. Early Development of the Zebrafish Neural Plate

A. Morphogenesis

Direct lineage tracings using fluorescent markers and time-lapse video-microscopy have localized presumptive neural plate cells to the central four blastomeres of the 16-cell embryo (Helde *et al.*, 1994; Wilson *et al.*, 1995) and to the dorsal aspect of the blastoderm at the shield stage (Fig. 2A) (Woo and Fraser, 1995). Starting at gastrulation, neural plate cells are then rearranged along the antero-posterior and mediolateral axes (Fig. 2B) by convergence-extension movements (Solnica-Krezel and Cooper, 2002) and/or active migration (Varga *et al.*, 1999). Zebrafish neurulation has been studied in detailed histological and lineage analyses (Kimmel and Warga, 1986; Papan and Campos-Ortega, 1994; Schmitz *et al.*, 1993). It is initiated at early somitogenesis, when characteristic thickenings form along the lateral and medial aspects of the neural plate (Fig. 2C). The lateral thickenings progressively converge toward the midline while the neural plate folds inward, giving rise around the 6–10-somite stage to a solid neural keel (Fig. 2D), that later detaches from the adjacent epidermis and forms a neural rod (Fig. 2E). The neural lumen starts becoming visible around the 17-somite stage (Fig. 2F) and has been hypothesized to result from secondary cavitation (Kimmel, 1993; Papan and Campos-Ortega, 1994; Schmitz *et al.*, 1993). Several important molecular mechanisms have been unraveled that control neurulation. Polarized cell movements of convergence and intercalation toward the midline within the

---

groups can be subdivided into moto- (black cell bodies, boldface labeling), intermediate (italics), and sensory (regular labeling) neurons. Primary motoneurons of the rhombencephalon first differentiate in rhombomeres 2–4 and produce cranial nerves V and VII that innervate the branchial arches. Primary motoneurons of the spinal cord are of three types (RoP, MiP, CaP) that differ in their location relative to somitic boundaries (diagonal lines) and axonal arborization. Primary interneurons comprise the reticulospinal system of the mes- (nMLF) and rhombencephalon (Ca, M, Mi, Ro) (for a full nomenclature see Mendelson (1986a,b), and CoPA and VeLP spinal neurons. They project ipsilaterally (dark gray cell bodies) or contralaterally (light grey cell bodies). Finally, the primary sensory system is composed of the RB cells, which axons form the LLF and also innervate the skin. *Abbreviations: clusters/neurons*: Ca, caudal reticulospinal neurons; CaP, caudal primary motoneurons; CoPA, commissural primary ascending interneuron; drc, dorsorostral cluster; ec, epiphyseal cluster; M, Mauthner cell; Mi, middle reticulospinal neurons; MiP, middle primary motoneuron; nMLF (or vcc), nucleus of the MLF; nPC, nucleus of the PC; RB, RB neurons; RoM, rostral medial reticulospinal neurons; RoP, rostral primary motoneurons; RoL, rostral lateral reticulospinal neurons; VeLD, ventral longitudinal descending interneurons; vcc (or nMLF), ventro-caudal cluster; vrc; ventro-rostral cluster; *tracts*: AC, anterior commissure; DVDT, dorsoventral diencephalic tract; LLF, lateral longitudinal fascicle; MLF, medial longitudinal fascicle; POC, postoptic commissure; TPC, tract of the posterior commissure; TPOC, tract of the postoptic commissure; V, fifth cranial nerve; VII, seventh cranial nerve; *territories*: dienc., diencephalon; mesenc., mesencephalon (midbrain), telencephalon.

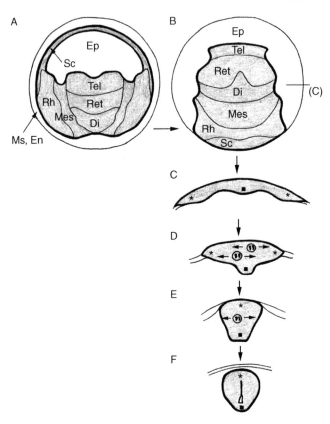

Fig. 2 Fate map and morphogenesis of the zebrafish embryonic neural plate. (A, B) Fate map of the neural plate (gray shading) at the shield (A) and tail-bud (B) stages (dorsal views), schematized from (Woo and Fraser, 1995). The different neural territories are organized into coherent (and partially overlapping) domains that align along AP during gastrulation. (C–F) Morphogenesis of the neural tube (gray shading) at tail-bud (C), 5 somites (D), 14 somites (E) and 20 somites (cross sections at the level indicated in B, schematized from Schmitz *et al.* (1993)). Lateral neural plate bulges converge toward the midline, which itself folds inward, leading to the formation of a compact neural keel/rod that later cavitates. However, as in other vertebrates, prospective dorsal neural tube cells originate from lateral neural plate domains (black stars) while ventral cells originate at the midline (black square). At the keel/rod stage, dividing cells often contribute progeny to both sides of the future neural tube (arrows in D, E). *Abbreviations*: Di, diencephalon; En, endoderm; Ep, epidermis; Mes, mesencephalon; Ms, mesoderm; Ret, retina; Rh, rhombencephalon; Sc, spinal cord; Tel, telencephalon.

neural plate control neural keel formation, and neural tube closure. Implicated molecules comprise Ncadherin, Nodal, Hh, and the planar cell polarity pathway (Aquilina-Beck *et al.*, 2007; Ciruna *et al.*, 2006; Hong and Brewster, 2006; Takamiya and Campos-Ortega, 2006). An important specific feature of zebrafish neurulation is the fact that dividing cells frequently contribute progeny to both sides of the neural tube (Papan and Campos-Ortega, 1997). This bilateral distribution is permitted by the long-lasting keel/rod structure and the perpendicular orientation of mitoses at that stage (Fig. 2D and E) (Geldmacher-Voss *et al.*, 2003). Polarity molecules such as Numb1 and Par3 are asymmetrically distributed at this stage. In daughter cells distributing bilaterally, the localization of Par3 at the apical pole is transmitted to both daughters in a new, mirror-symmetric mode of division (Tawk *et al.*, 2007). A causal implication of Numb1 and Par3 in the orientation of cell divisions remains unclear (Reugels *et al.*, 2006; von Trotha *et al.*, 2006), but Par3 is involved in midline crossing following mirror-symmetric division, and in the maintenance of an apical surface directing the formation of a neural tube lumen (Tawk *et al.*, 2007). Sister cells distributing bilaterally are not related in fate and mostly give rise to different neuronal or glial types (Papan and Campos-Ortega, 1997). Outside these specific features, zebrafish neurulation is comparable to that of other vertebrates, in particular with the distribution of medial versus lateral cells of the neural plate to the ventral versus midline, respectively, of the mature neural tube (Fig. 2C–F).

B. Neural Induction

Markers of neural induction, or "preneural" genes, best established in *Xenopus* and chicken, include the Sox family members *Sox2* and *Sox3*, and *ERNI*. Likewise, expression of several zebrafish *sox* genes delimits the early neural plate at gastrulation (de Martino *et al.*, 2000; Rimini *et al.*, 1999; Vriz *et al.*, 1996), and zebrafish *sox3* was recently shown to determine the neural fate (Dee *et al.*, 2008). Traditionally, early markers of neural AP and DV patterning are used to assess neural induction in zebrafish. These include *otx2*, expressed within the anterior neural plate (presumptive fore- and midbrain), and *hoxb1b* (previously *hoxb1, hoxa-1*) (rhombomere 4 and beyond), from early gastrulation (Koshida *et al.*, 1998; Li *et al.*, 1994; Prince *et al.*, 1998). Induction of the anterior neural plate border (placodal fields) is revealed by *dlx3b* (previously *dlx3*)/*dlx5a* (previously *dlx4*) and *foxi1* at early somitogenesis stages (Akimenko *et al.*, 1994; Lee *et al.*, 2003; Nissen *et al.*, 2003; Quint *et al.*, 2000; Solomon *et al.*, 2003). Finally, the posterior neural plate border (future neural crests and RB neurons) (Cornell and Eisen, 2002) expresses *foxd3* (previously *fkd6*), *sox9b, sox10* (*cll*), or *snail1b* (previously *snail2*) at late gastrulation (Dutton *et al.*, 2001; Li *et al.*, 2002; Odenthal and Nusslein-Volhard, 1998; Thisse *et al.*, 1995) and is delimited by the action of Midkine-b (Liedtke and Winkler, 2008). As in other vertebrates, the initial specification of the zebrafish neural plate is anterior in character, later progressively posteriorized by signals originating from the nonaxial mesoderm (Erter *et al.*, 2001; Lekven *et al.*, 2001; Woo and Fraser, 1997, 1998).

Until recently, largely based on studies in amphibians, neural induction was viewed as a default pathway, where ectodermal cells would become neural unless driven toward an epithelial fate by BMP signaling at gastrulation (Hemmati-Brivanlou and Melton, 1997). In this model, neural development on the dorsal embryonic side was permitted by the diffusion of BMP inhibitors (such as Noggin and Chordin) from the Spemann organizer (or its equivalents the mouse node, chicken Hensen's node and zebrafish shield). Along this line, the size of the neural plate is impaired in zebrafish mutants affected in Bmp2/4 signaling (Hammerschmidt and Mullins, 2002) or when organizer activity is weakened (e. g., in *bozozok* (*dharma*) (Fekany *et al.*, 1999; Koos and Ho, 1999; Leung *et al.*, 2003; Yamanaka *et al.*, 1998), *squint;cyclops* (*ndr1;ndr2*) or maternal zygotic *one-eyed pinhead* (*oep*), mutants (Gritsman *et al.*, 1999) or in *ichabod* mutants (showing lowered levels of beta-catenin2) (Bellipanni *et al.*, 2006). However, even in these cases, a neural plate forms, suggesting that blocking Bmp is neither required nor sufficient for neural induction.

Initial findings in chicken and mouse first pointed to two major other signaling pathways as the prime neural inducers, acting before gastrulation: Fgf and Wnt (Bainter *et al.*, 2001; Sheng *et al.*, 2003; Stern, 2002; Streit *et al.*, 2000; Wilson and Edlund, 2001; Wilson *et al.*, 2000). Early Fgf and Wnt signaling in zebrafish have been primarily implicated in neural patterning (Furthauer *et al.*, 1997; Kudoh *et al.*, 2002; Reifers *et al.*, 1998; Shiomi *et al.*, 2003; Erter *et al.*, 2001). Recently however, a role for Fgf in neural induction independently of BMP repression was demonstrated (Rentzsch *et al.*, 2004), although an implication of early Wnt signaling is still lacking. BMP signaling remains also important in the definition of the neural plate border and dorsal neural cell types at all AP levels of the zebrafish neural plate (Barth *et al.*, 1999; Houart *et al.*, 2002; Wilson *et al.*, 2002).

C. Delimitation of Proneural Fields by Prepattern Factors

Our understanding of neurogenesis in vertebrates, including zebrafish, stems largely from key original findings in *Drosophila* (Campos-Ortega, 1993). The definition of neurogenesis-competent domains ("proneural clusters" or "proneural fields") within the fly is achieved by the combinatorial expression of neurogenesis activators ("proneural" factors) and neurogenesis inhibitors. The former comprise Achaete-Scute proteins, the latter Hairy-like factors. All belong to the bHLH class of transcription factors, and are expressed in direct response to the early embryonic patterning machinery, to establish a neurogenesis "prepattern" within the neuroectoderm.

As discussed above, neurogenesis in the early vertebrate neural plate occurs at stereotyped loci, and avoids others, suggesting the existence of a prepattern of proneural/incompetent fields similar to that observed *Drosophila*. Immediately downstream of neural induction, several transcription factors that promote the neural fate have been recently identified. The most studied of these belong to the Sox, Gli, POU, and Iroquois families, and are expressed across broad domains of the neural plate (Bainter *et al.*, 2001; Bally-Cuif and Hammerschmidt, 2003). In zebrafish, *iro1, iro7,* and *pou5f1* (previously *pou2*) are expressed across the presumptive mid- and hindbrain areas (Hauptmann and Gerster, 1995; Itoh *et al.*, 2002; Lecaudey *et al.*, 2005, 2001; Parvin *et al.*, 2008). Iro1, 7 and Pou5f1 are required for *neurog1* (previously *ngn1*) expression in their respective expression domains, and at least Iro1 and 7 are sufficient, when misexpressed, to induce ectopic *neurog1* expression within nonneural ectoderm (Belting *et al.*, 2001; Itoh *et al.*, 2002; Lecaudey *et al.*, 2004). More anteriorly, expression of *flh* (*znot*) defines the epiphyseal proneural field and permits expression of the proneural factors Neurog1 and Ash1a, driving neurogenesis in this area (Cau and Wilson, 2003).

Recent evidence in zebrafish and *Xenopus* highlight that proneural fields are also defined as the domains that do not express active neurogenesis inhibitors. These domains include the anterior neural plate (prospective telencephalon, diencephalon, and eyes), the longitudinal spinal cord stripes that separate the columns of sensory, moto- and interneurons, and the midbrain-hindbrain boundary. They are generally characterized by the expression of transcription factors such as Zic1-3, Iro3, Anf, and BF1, as well as members of Hairy/E(Spl) family (so-called Hes, Hey, Her, or Hairy in different species), which have been best studied in *Xenopus* (Andreazzoli *et al.*, 2003; Bellefroid *et al.*, 1998; Brewster *et al.*, 1998; Hardcastle and Papalopulu, 2000). In zebrafish, *zic1* (previously *opl*) and *zic3* expression highlight, respectively anterior and posterior neural plate domains (Grinblat and Sive, 2001; Grinblat *et al.*, 1998), *zic1* and *zic4* are expressed later in the dorsal neural tube (Elsen *et al.*, 2008), *hesx1* (previously *anf*) and *foxg1* (previously *bf1*) label the anterior neural plate (Houart *et al.*, 2002), *her5* and *her11* the midbrain-hindbrain boundary (Müller *et al.*, 1996), and *her3* and *her9* inter-proneural stripes of the presumptive spinal cord (Bae *et al.*, 2005; Hans *et al.*, 2004). Gain- and loss-of-function experiments have established Her3/5/9/11 as crucial inhibitors of neurogenesis in their expression domains, acting upstream of the definition of proneural fields (Bae *et al.*, 2005; Geling *et al.*, 2003, 2004; Hans *et al.*, 2004; Ninkovic *et al.*, 2005). A new player negatively defining proneural zones is d-Asb11, a ubiquitin ligase involved in the control of Notch signaling (Diks *et al.*, 2008). *d-asb11* is expressed along the margin of the neural plate and maintains progenitor cells in this location, as its loss of function leads to premature neurogenesis and progenitor depletion (Diks *et al.*, 2006).

In all examples studied, expression of neurogenesis prepattern factors is directly controlled by the early embryonic patterning machinery, thus links early patterning information to the definition of the first neurogenesis sites, in a manner analogous to the definition of proneural fields in *Drosophila*. The molecular pathways activated or inhibited by these factors in vertebrates remain however poorly understood. These factors might directly inhibit expression of proneural genes or control the cell-cycle machinery, or both.

As a result of prepatterning, early neurogenesis occurs at discrete sites, which can be revealed at the 3-somite stage by the characteristic expression profile of the proneural genes *neurog1* (see Section IV.A and Blader *et al.*, 2004) (Fig. 3A), *elavl3* (Park *et al.*, 2000b), or *zash1* (Allende and Weinberg, 1994). Anteriorly, prominent *neurog1* expression highlights precursors of the olfactory neurons, the vcc, and trigeminal ganglia. Posteriorly, longitudinal columns of presumptive moto-, inter-, and sensory neurons are visible. These proneural clusters are organized around zones where inhibitory prepatterning takes place, such as the anterior neural plate,

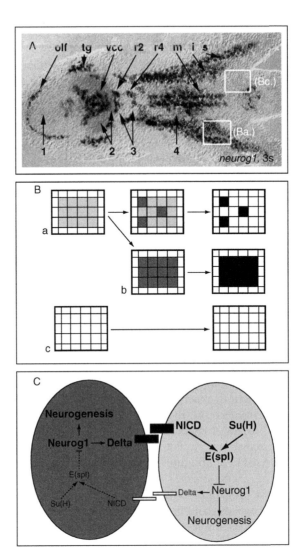

Fig. 3 Early proneural clusters and the principles of lateral inhibition. (A) Expression of *neurog1* revealed at the 3-somite stage by whole-mount *in situ* hybridization, dorsal view of a flat-mounted embryo, anterior left. *neurog1* expression highlights a discrete distribution of early proneural clusters (olf, olfactory neuron precursors; tg, trigeminal ganglion; vcc, ventro-caudal cluster; r2, motoneuron precursors in rhombomere 2; r4, motoneuron precursors in rhombomere 4; m, spinal motoneuron precursors; i, spinal interneuron precursors; s, spinal sensory neuron precursors). These clusters are separated by prepatterned zones of neuronal inhibition (1, ANP; 2, MHB; 3, longitudinal stripes in r2 and r4; 4, longitudinal stripes in the spinal cord). (B, C) Lateral inhibition selects a few competent precursors for further neurogenesis. B(a) Precursors selection within a proneural cluster (boxed in A.) occurs in three steps: (i) weak and ubiquitous expression of an early proneural factors, for example, Neurog1 (light gray) (left panel), (ii) reinforcement of stochastic differences in the expression of *neurog1* between adjacent cells by Notch-Delta interactions (C), leading to the juxtaposition of strongly and weakly *neurog1*-positive cells (dark and light gray, respectively) (middle panel), and (iii) further commitment of strongly *neurog1*-expressing cells (black) and extinction of *neurog1* expression in their neighbors (white) (right panel). B(b) In the absence of lateral inhibition, all early precursors commit to differentiation. B(c) In prepatterned domains of lateral inhibitions (as boxed in A), early expression of proneural genes is blocked and lateral inhibition does not take place.

the midbrain-hindbrain boundary, and domains separating the posterior longitudinal stripes in the presumptive hindbrain and spinal cord (Fig. 3A, numbers
1 to 4).

D. Neural Tube Organizers as Neurogenesis Signals?

The expression of early neurogenesis prepatterning genes, and the positioning of
neurogenic zones at later stages, is mainly influenced by local secreted factors
produced within or adjacent to the neural plate. Members of the Wnt, Fgf, BMP,
and Hh families are involved in this process. In addition to patterning control,
these factors have been directly implicated in the control of cell proliferation/
differentiation in many systems. Thus, in addition to positioning the expression
of prepatterning genes, these factors may also play a direct role on neurogenesis by
influencing cellular status.

Along the AP axis, several local signaling centers have been identified that influence neural plate patterning and correlate with specific neurogenesis status. In zebrafish, at gastrulation stages, a signal originating from the nonaxial blastoderm margin
and possibly encoded by *wnt8* posteriorizes the neural plate, and may partly account
for the anterior to posterior gradient of CNS maturation. At late gastrulation, the
first cell row of the neural plate acts as a signaling source, the "anterior neural
border," to permit telencephalic maintenance (Houart *et al.*, 1998). Anterior neural
border activity is encoded by Sfrp (previously Tlc), a secreted Wnt inhibitor of the
Frizzled family, proposed to antagonize Wnt8b signaling from the posterior diencephalon (Houart *et al.*, 2002). A role of the anterior neural border on neurogenesis
has not been studied but its position is suggestive: it lies at the junction between
prospective placodal fields, which undergo early neurogenesis, and the anterior
neural plate, where neurogenesis is actively prevented by prepatterning inhibitors.
At later stages, forebrain neurogenesis is influenced by Fgf3 and Fgf8 (Walshe and
Mason, 2003). The midbrain-hindbrain boundary is the source of secreted factors of
the Wnt (Wnt1, Wnt10b, Wnt8, Wnt3) and Fgf (Fgf8, Fgf15, Fgf17, Fgf18) families
(Kelly and Moon, 1995; Kelly *et al.*, 1995; Lekven *et al.*, 2003; Reifers *et al.*, 1998,
2000), which at least in part encode activity of the isthmic organizer (Martinez, 2001;
Rhinn and Brand, 2001; Wurst and Bally-Cuif, 2001) and are required for growth and
patterning of the mid- and anterior hindbrain. It positions expression of the prepatterning neurogenesis inhibitor Her5 (Geling *et al.*, 2003). A possible direct role of
midbrain-hindbrain boundary secreted factors in neurogenesis control has not been
studied, however both Wnt1 and Fgf8 have been shown to influence cell proliferation
within the midbrain-hindbrain area in the mouse (Panhuysen *et al.*, 2004; Xu *et al.*,
2000). Finally, Wnt1 is expressed at rhombomere boundaries in response to the
Notch pathway and orchestrates the repeated pattern of rhombencephalon neurogenesis by activating proneural genes in adjacent, nonboundary regions (Amoyel
et al., 2005; Cheng *et al.*, 2004).

Signals influencing neurogenesis along the DV axis include BMP, Wnt, and Hh.
BMPs are expressed by the dorsal ectoderm and dorsal neural tube, and can

distinctly affect neurogenesis in different contexts. In the mouse embryo, BMP signaling sequentially promotes proliferation then cell-cycle exit of dorsal neural tube cells via alternative BMP receptor usage (Panchision *et al.*, 2001). The zebrafish mutant *narrowminded*, carrying a loss of function of the transcription factor Prdm1 (Hernandez-Lagunas *et al.*, 2005), lacks dorsal neuronal types including sensory neurons and neural crests (Artinger *et al.*, 1999). Expression of *prdm1* responds to BMP signaling (Hernandez-Lagunas *et al.*, 2005). Expression of *Wnt1* and *Wnt3a* at the dorsal midline has also been proposed to establish a dorsoventral gradient promoting proliferation (and blocking differentiation) across the neural tube (Megason and McMahon, 2002). Interestingly, sensitivity to this gradient might be influenced by ventrally derived Shh (Ishibashi and McMahon, 2002). Hh signaling itself was associated with several events of late neural proliferation (Ruiz i Altaba *et al.*, 2002; Ruiz *et al.*, 2002), as well as with the regulation of expression of Zic and Gli factors to define longitudinal stripes of neurogenesis in the *Xenopus* spinal cord (Brewster *et al.*, 1998). The role of BMP, Wnt, and Hh signals in controlling the neural proliferation/differentiation process in zebrafish has not been yet intensely studied. Hh signaling controls the mediolateral pattern of *neurog1* expression (Blader *et al.*, 1997), but a direct regulation of the *neurog1* enhancer (as opposed to general neural plate repatterning) by Gli factors has not been reported (Blader *et al.*, 2003). Recently, Gli1 and Gil2 factors were implicated in the choice of cellular status in the ventral neural plate (Ke *et al.*, 2008; Ninkovic *et al.*, 2008). However, these activities appear in part Hh-independent.

IV. Lateral Inhibition and the Neurogenesis Cascade

A. Primary Neurogenesis

Competent neurogenic domains (or "proneural fields") express low levels of proneural factors, which need to be reinforced to drive neurogenesis, and only some precursors will further commit to neuro- or gliogenesis. In *Drosophila*, the selection of these precursors relies on the process of lateral inhibition, where cells expressing high levels of the Notch ligand Delta will commit to differentiation and at the same time, via Delta-Notch binding, inhibit their neighboring cells from doing so (Simpson, 1997). After binding of Delta, the intracellular domain of Notch (NICD) translocates to the nucleus and activates transcription of downstream effectors, among which bHLH transcriptional repressors of the enhancer-of-split (E(Spl)) family. E(Spl) proteins prevent expression or activity of proneural factors. In parallel, cells expressing high levels of Delta will maintain expression of proneural factors and Delta transcription. Thus, initial differences in Delta expression are amplified, leading to the reinforcement of neurogenic predispositions in a selection of precursors (Bray and Furriols, 2001; Mumm and Kopan, 2000) (see Fig. 3C).

Current evidence suggests that similar molecular and cellular mechanisms are at play in vertebrates (Appel and Chitnis, 2002; Lewis, 1998). Studies in zebrafish have revealed expression of *notch* (*notch1a, notch1b, notch3* (previously *notch5*), *notch2* (previously *notch6*)) (Westin and Lardelli, 1997), delta (*dlA, B, C, D*) (Haddon *et al.*, 1998; Julich *et al.*, 2005), *jagged* (*jgd1a* and *jgd2*) (Zecchin *et al.*, 2005), and *E(spl)* (*her4*) (Takke *et al.*, 1999) genes, as well as of proneural genes (*neurog1, neurog3* (previously *ngn2, ngn3*), *ash1a, ash1b, atoh1* (previously *ath1*), *neurod4* (previously *ath3*), *coe2, elavl3*) (Allende and Weinberg, 1994; Bally-Cuif *et al.*, 1998; Blader *et al.*, 1997; Koster and Fraser, 2001; Liao *et al.*, 1999a; Park *et al.*, 2000b, 2003; Wang *et al.*, 2001) in proneural domains of the neural plate. The relative expression of these factors obeys the mechanics of lateral inhibition: for instance, NICD induces *her4* and prevents *neurog1* expression (Takke *et al.*, 1999), while increase of Delta function maintains high levels of *neurog1* throughout early proneural fields (Appel and Eisen, 1998; Appel *et al.*, 2001; Dornseifer *et al.*, 1997; Haddon *et al.*, 1998). In addition, a number of zebrafish mutants with altered *neurog1* expression proved invaluable to our understanding of vertebrate neurogenesis and Notch/Delta function in this process. Deficiencies in Neurog1 function were produced by insertional mutagenesis into the *neurog1* locus (*neurod3^{hi1059}*) (Golling *et al.*, 2002) and injection of *neurog1* morpholino into zebrafish eggs (Cornell and Eisen, 2002; Park *et al.*, 2003). Resulting embryos display a severe reduction of cranial ganglia and of the number of nMLF and spinal sensory neurons, while spinal moto- and interneurons, and epiphysial neurons are less affected (Cau and Wilson, 2003; Cornell and Eisen, 2002; Geling *et al.*, 2004; Golling *et al.*, 2002; Park *et al.*, 2003). These results point to Neurog1 as a crucial proneural factor of zebrafish primary neurogenesis, and highlight differential spatial requirements for Neurog1 function in this process. *after eight* (*aei*) (*delD*), *deadly seven* (*des*) (*notch1a*), *delA^{dx2}* (*dlA*), and beamter (*delC*) mutations all directly affect Notch/Delta signaling (Appel *et al.*, 1999; Holley *et al.*, 2002; Jiang *et al.*, 1996; Julich *et al.*, 2005; van Eeden *et al.*, 1996), and *aei, des, delA*, and *delC* mutants all display some excess of primary neurons, as predicted from a lateral inhibition defect (Gray *et al.*, 2001; Jiang *et al.*, 1996; Julich *et al.*, 2005). Redundancy between Notch and Delta proteins within the zebrafish neural plate might explain the relatively mild effects of these mutations. Recent work emphasized the role of Notch/Delta regulators, such as factors involved in promoting ubiquitylation and endocytosis of Delta to permit Notch signaling. Mind-bomb (Mib) and Mib2 encode ring E3 ubiquitin ligases that interact with the intracellular domain of Delta to promote Notch signaling and neurogenesis inhibition in neighboring cells (Itoh *et al.*, 2003; Zhang *et al.*, 2007a,b), and *mib* mutants suffer from a severe neurogenic phenotype. Likewise, D-asb11 ubiquitylates Delta to permit Notch signaling, although this function is limited to the lateral neural plate (Diks *et al.*, 2006, 2008). Finally, Delta interactors of the MAGI family have been cloned. These are scaffolding proteins that bind the intracellular domain of zebrafish Delta C and D (Wright *et al.*, 2004). Their role in neurogenesis control remains to be demonstrated. Regulators of Notch/Delta signaling downstream of the Notch and Delta molecules include Histone deacetylase-1 (Hdac-1). Hdac-1 antagonizes Notch signaling in particular by directly or indirectly inhibiting expression of E(Spl) genes such as *her6* and *her4* in the neural tube and retina, respectively (Cunliffe, 2004; Yamaguchi *et al.*, 2005). Notch/Delta signaling and lateral inhibition also control the definition of the neural crest progenitor compartment along the sides of the neural plate. Manipulations of Notch/Delta influence development of crest versus RB identities: deficiencies in Notch signaling expand the RB population at the expense of neural crests (Cornell and Eisen, 2000; Jiang *et al.*, 1996) (reviewed in Cornell and Eisen, 2005). A possibly downstream or collaborating factor promoting expression of proneural genes in dorsal interneuron precursors is Olig3 (Filippi *et al.*, 2005). Recently, a zebrafish non-HLH-containing factor expressed in primary neuron progenitors, Onecut1 (previously Onecut), was also isolated (Hong *et al.*, 2002). Its function in the neuronal specification process is unknown.

Following lateral inhibition, the progenitors displaying increased levels of Delta and of proneural expression will further commit to differentiation. In zebrafish, as in *Drosophila* and other vertebrates, this is accompanied by cell-cycle exit and the transcription of a new set of proneural bHLH genes, such as Neurod2 (previously Ndr2) (Korzh *et al.*, 1998; Liao *et al.*, 1999b). In addition, a late Delta factor, DlB (Haddon *et al.*, 1998), and RNA-binding proteins of the Hu family (Mueller and Wullimann, 2002a; Park *et al.*, 2000a), are expressed in committed progenitors. The transition from proliferating to postmitotic cells involves a number of cell-cycle regulators, which have been best studied in *Xenopus* (Ohnuma *et al.*, 2001, 2002). A recent report, however, demonstrated the role of Cdkn1, a p57 homolog, in controlling cell-cycle exit and neuronal differentiation in the zebrafish neural plate, in a manner epistatic to Notch/Delta signaling (Park *et al.*, 2005). Most studies to date focus on retinal differentiation, perhaps because cell-cycle exit in this system is technically easier to score. Several zebrafish mutants have been isolated where the transition from proliferation to differentiation in this system is impaired. In *young* mutants, affecting the *smarca4* gene, cell-cycle withdrawal across the retina is initiated normally but is slower than in wild type, and retinal cells do not fully differentiate (Gregg *et al.*, 2003; Link *et al.*, 2000). *perplexed* and *confused* mutations respectively cause generalized or restricted apoptosis of retinal precursors during the transition phase from proliferating to postmitotic cells (Link *et al.*, 2001). *disarrayed* mutants display small eyes (and reduced forebrain) due to reduced cell-cycle exit (and thereby reduced neurogenesis) (Baye and Link, 2007). The retina is also interesting as it displays a clear link between interkinetic nuclear migration (INM) and the occurrence of neurogenic divisions: the latter tend to occur when progenitors exhibit less pronounced INM (Baye and Link, 2007). This phenomenon has been linked to increased Notch signaling at the basal side of the retinal neuroepithelium (Del Bene *et al.*, 2008). Molecular identification of the mutations above promises to provide great insight into the mechanisms linking INM, cell-cycle arrest and neuronal differentiation in vertebrates.

B. Molecular Control of Secondary Neurogenesis and Gliogenesis

A classical distinction is made between early-developing primary neurons and late-developing secondary neurons (Kimmel and Westerfield, 1990). Nevertheless it should be noted that neurogenesis in the embryo is a continuous process, not obviously divided in two distinct phases (Lyons *et al.*, 2003). Early findings suggest that at least some of these processes might differ from primary neurogenesis control, possibly at late stages of neuronal maturation (Grunwald *et al.*, 1988). Molecular data however strongly support a view where, at least for early steps of the neurogenesis cascade, the same early processes are used reiteratively throughout embryonic and larval development to generate both primary and secondary neurons. Indeed, the early neurogenesis program (*notch1a, ash1a, ash1b, neurog1, neurod2, dlA*) is expressed in the postembryonic zebrafish brain, in a similar relative distribution than at embryonic stages, and with a comparable correlation to proliferation zones (Mueller and Wullimann, 2002a,b, 2003, 2005; Wullimann and Knipp, 2000; Wullimann and Mueller, 2002). For instance, *notch1a* expression generally coincides with proliferating cells, *neurod2* with immediately postmitotic neurons, and *neurog1* transiently covers both populations. Analyses in the zebrafish spinal cord indicate that lateral inhibition through Notch signaling is continuously required during development to maintain a pool of neural progenitors, by limiting the number that exit cell cycle and differentiate into neural or glial cells (Itoh *et al.*, 2003; Kim *et al.*, 2008; Park and Appel, 2003; Shin *et al.*, 2007; Yeo and Chitnis, 2007). Throughout development distinct neuronal subtypes then emerge from this pool of progenitors in a distinct spatiotemporal order (Park *et al.*, 2004).

During vertebrate neural development, neuroepithelial cells acquire features of radial glial cells, such as expression of the glial fibrillary acidic protein (GFAP). These radial glial cells were considered initially as a structural supporting element of the CNS. However it has recently been shown that they generate new neurons and glial cells during embryogenesis, and also may become neural stem cells maintained throughout adulthood (reviewed in Gotz and Huttner, 2005). In zebrafish, analysis of a GFAP reporter line has demonstrated that GFAP + radial glia arise during zebrafish development at the time when the first neurons are produced (Kim *et al.*, 2008). As previously shown in rodents, some of these radial glial cells are neural progenitors in the zebrafish spinal cord, maintained throughout development via Notch signaling (Kim *et al.*, 2008).

Precursors of oligodendrocytes (OPCs), the myelinating glial cells of the CNS, originate from olig2 + progenitors located in the ventral spinal cord, that also give rise to secondary motoneurons and interneurons (Park *et al.*, 2004). Blocking Notch signaling just before their formation results in a decrease of the number of OPCs and a parallel increase of the number of secondary motoneurons (Kim *et al.*, 2008).

C. Adult Neurogenesis

In contrast to the telencephalon-restricted neurogenic activity of adult mammals, neurogenesis in adult zebrafish and other teleosts is by far more widespread throughout the brain, happening along ventricular areas along the whole rostrocaudal axis, with exception of the nonventricular neurogenesis in the cerebellum and valvula cerebelaris. A number of distinct regions harboring neurogenesis have been listed (Adolf *et al.*, 2006; Chapouton *et al.*, 2006; reviewed in Chapouton *et al.*, 2007; Grandel *et al.*, 2006) and out of these, mainly the telencephalon has been studied, in order to be compared to its mammalian counterpart.

Proliferative activity, as defined by the expression of PCNA, MCM5, or by the permanent incorporation of the thymidine analogon BrdU, takes place within regions containing the somata of radial glial cells, which themselves cover the whole ventricular surface. The majority of dividing cells are radial glial cells, and only a small proportion of all radial glial cells are dividing.

BrdU tracing experiments show that cell division is followed within the time frame of 1–3 days by the formation of Hu C/D positive postmitotic neuroblasts (Adolf *et al.*, 2006; Pellegrini *et al.*, 2007), which then give rise in the telencephalon and in the olfactory bulb to gabaergic and few tyrosine hydroxylase positive interneurons. The proliferation pattern and generation of olfactory bulb interneurons is reminiscent to the mammalian rostral migratory stream: a stripe of proliferating cells expressing the migrating neuroblast marker PSA-NCAM follows a posterior to anterior path along the ventricle of the telencephalon toward the olfactory bulb, and new born neurons do not originate within the olfactory bulb itself.

Cell fate tracing by BrdU reveals also BrdU-positive radial glial cells several days after division (Adolf *et al.*, 2006; Grandel *et al.*, 2006), which could either represent *de novo* generated progenitors or postmitotic glial cells. The generation of oligodendrocytes has been described only in the postnatal spinal cord and in the midbrain (Chapouton *et al.*, 2006a; Park *et al.*, 2007), but has not been examined in the telencephalon.

As the study of adult neurogenesis in zebrafish is at its beginning, the molecular mechanisms controlling the steps of stem cell maintenance, entry into proliferation, cell-cycle exit, maturation, and differentiation of new born neurons, have barely been studied so far. Components of the FGF signaling cascade are expressed in radial glial cells, without being associated with the proliferative activity of these (Topp *et al.*, 2008). As well the testosterone converting enzyme aromatase is expressed in the whole radial glial population (Pellegrini *et al.*, 2007), suggesting a regulation of this population by sex steroid hormones.

Besides the telencephalon, the adult retina with its *de novo* genesis of photoreceptors, and with its strong regenerative capacity, is a very informative model contributing to the elucidation of adult neurogenesis mechanisms. The ciliary marginal zone at the rim of the retina is a source of cone photoreceptors and other retinal neurons, whereas rod photoreceptors are generated in a scattered manner throughout the retina by divisions of Müller glial cells (Raymond *et al.*, 2006). Experimental lesions of the retinal tissue and regeneration, which most likely represents an enhanced picture of the naturally occurring neurogenesis, strongly suggest that Müller glia are a multipotent source of all retinal neuronal subtypes (Bernardos and Raymond, 2006; Fausett and Goldman, 2006; Thummel *et al.*, 2008). Expression studies of the ciliary marginal zone or microarray analysis of the regenerating retina, have shown the implication of several developmental genes, such as pax6, ngn1, stat3, olig2, vsx2, and rx1 (Fausett and Goldman, 2006; Kassen *et al.*, 2007; Raymond *et al.*, 2006). Thummel *et al.* (2008) have shown in a functional study the requirement of Ascla in the adult retinal neurogenic process.

As a last examined system in the process of adult neurogenesis, the lesioned spinal cord also revealed the ability for generation of new motoneurons, most probably generated from olig2 expressing progenitors (Reimer *et al.*, 2008). These new motoneurons are surrounded by the presynaptic marker SV2, indicating an integration into the neuronal networks. However, the general integration of adult born neurons into the preexisting circuitries of the zebrafish brain, and their contribution to specific brain functions have not been approached yet.

V. Establishment of Neuronal Identity

Important parameters in the choice of neuronal identity are combinations of positional cues and differentiation timing. Positional cues are extrinsic (e.g., distance to a signaling factor) and/or intrinsic (e.g., expression of a given set of transcriptional regulators by the precursor cell), and are significant in relation to a specific differentiation status, generally the moment of cell-cycle exit (birth date) or of the last few rounds of cell division that precede cell-cycle exit. Mutants and functional tests in zebrafish made important contribution to this field in vertebrates, by leading to the identification of a number of pathways or factors attributing specific CNS neuronal identities.

Major signaling pathways are thus involved in the specification of neuronal identities. At gastrulation, as discussed above, BMP levels determine the extent of nonneural versus neural ectoderm. However, extensive studies of zebrafish BMP signaling mutants revealed an additional role of BMPs in the specification of intermediate and dorsal neural cell types. Indeed, embryos where BMP signaling is strongly reduced, such as *swirl* (*bmp2b*) or strong alleles of *somitabun* (*madh5*, previously *smad5*), develop an excess of interneurons at the expense of neural crests/RBs. In contrast, weak *somitabun* alleles or *snailhouse* (*bmp7*) mutants, where BMP signaling is less reduced, display an excess of dorsal most neuronal types (Barth *et al.*, 1999; Nguyen *et al.*, 2000; Schmid *et al.*, 2000). Thus, graded BMP levels might be required *in vivo* for the specification of neuronal identities along the neural plate border. In a series of single-cell RNA injection experiments, canonical Wnt signaling from the dorsal midline (possibly encoded by Wnt1 or Wnt3a) was also directly implicated in neural versus pigment cell fate choice of neural crests (Dorsky *et al.*, 1998). In the vertebrate spinal cord, the Hh morphogen is a key determinant of neural cell fates along the dorsoventral axis (reviewed in Dessaud *et al.*, 2008).

Extrinsic position signals result in the induction of specific transcription factors in progenitors. Studies in chick and mice thus show that Hh signaling from the floor plate and BMP/Wnt signaling from the roof plate regulate the expression of particular combinations of transcription factors along the dorsoventral axis of the spinal cord (Dessaud *et al.*, 2008). The combination of factors expressed in the progenitor domain defines their potency to generate a restricted number of cell types. The dorsoventral patterning of the spinal cord progenitor domains appears to be highly conserved in zebrafish (Gribble *et al.*, 2007; Lewis *et al.*, 2005; Park *et al.*, 2004). The set of transcription factors in the differentiating cells then further resolves the final phenotype of the neural or glial cell (see Table II for a list of neuronal identity markers). Interestingly, recent evidence implicate lateral inhibition via Notch signaling for the determination of the final identity of cell type emerging from the same progenitors (Cau *et al.*, 2008; Kimura *et al.*, 2008; Shin *et al.*, 2007).

Concomitant to or following withdrawal from mitosis and specification to particular cell types by multipotent progenitors, key steps in the generation of functional neurons include their migration to the appropriate position, their molecular and morphological differentiation, with the establishment of characteristic projections, and their survival. These aspects will not be considered within the frame of this review.

VI. Lab Methods to Study Adult Neurogenesis

A. BrdU Injections

- Dissolve 10 mg of BrdU into 4 ml saline (110 mM NaCl pH 7.2) with a bit of methylene blue to color the solution. Vortex for about 5 min to dissolve completely the BrdU.
- Weigh the fish and inject intraperitoneally 5 μl solution per 0.1 g body weight.
- Drop the required injection volume onto a parafilm piece, aspirate into a small synringe (insulin syringe). Take the fish into a net, lay the fish on its side into a wet petridish and inject the solution intraperitoneally.
- Store BrdU solution at 4 °C for not more than 1 week.

B. Staining Methods

1. Immunohistochemistry on Vibratome Sections

- Anesthetize the fish on an ice/water mix for 3–5 min and decapitate with a scalpel.
- Either dissect the brain immediately in PBS and fix for 4–6 h at 4 °C in 4% PFA (paraformaldehyde) or fix the whole head at 4 °C in 4% PFA overnight on a shaker and dissect the brain the next day. After washing out the PFA in PBS, proceed to a methanol series and freeze the brain in 100% methanol at −20 °C for at least 1 h. The brain can be conserved for several months at −20 °C. Some

antigens do not tolerate a methanol treatment, if they are lipid soluble for example. Some antibodies do not require any methanol treatment.

• Go back the methanol series to 100% PBS, and embed the brain in 3% agarose in PBS.

• Cool down the agarose and cut a block around the brain.

• Section the block at the vibratome with a section thickness ranging from 70 to 100 μm.

• Some antibodies will then require a special pretreatment, such as a second methanol series, an HCl, or a citrate buffer treatment (see antigene retrieval).

• Block the section in PBS + 0.5% Tx + 10% normal goat serum. Incubate primary antibody 2 h at room temperature or O/N at 4 °C, shaking.

• Wash 2–3× for 5 min in PBS.

• Incubate the secondary antibody (at a 1:1000 dilution, Alexafluor coupled antibodies from, i.e., Molecular Probes-Invitrogen) in PBS 0.5%Tx, 10% NGS for 30–45 min at RT in the dark.

• Wash 2–3× in PBS.

a. Pretreatments, antigen retrieval

• *HCl pretreatment*: BrdU revelation requires a pretreatment for opening the DNA, which might destroy other antigens. Therefore, it should be performed after the completion of the immunohistochemistry for the other markers. Incubate the slices with fresh 2 M HCl (1:4.4 of the 32% HCl solution) 30 min at RT. Wash once quickly, and twice 5 min with PBS. The BrdU antibody should be diluted in PBS + 0.5% Tx, without goat serum.

• *Citrate retrieval*: Some antigens require a retrieval step in citrate buffer. Slices are incubated at 85 °C for 30 min in 10 mM sodium citrate in PBS, pH 6, and washed 3 times in PBS.

2. *In Situ* Hybridization Starting on Whole–Mount Adult Brains

• Place a fixed brain (after MeOH gradient back to PBS) into a 48-well plate.

• Treat with proteinase K (10 μg/ml) for 30 min at room temperature. Proceed then for the *in situ* hybridization after the standard zebrafish embryo protocol, until all posthybridization washes have been performed. Embed the brain in 3% agarose and cut sections of 70–100 μm at the vibratome. Block the sections in blocking buffer and incubate them into the antidig antibodies, and continue following the standard embryo *in situ* protocol.

a. In Situ Hybridization on Gelatine Albumine Sections

• For weak *in situ* probes which do not reveal a strong signal after whole-mount *in situ* hybridization, gelatine albumine sections are better suited. However, fluorescence immunohistochemistry after the *in situ* hibrydization is not practicable on these sections, due to the high level of autofluorescence of the embedding mixture.

- Solve 4.5 g gelatine up to 250 ml **PBS** upon heating at 50 °C.
- Solve 270 g albumin + 180 g sucrose up to 750 ml **PBS** overnight at room temperature, filter the solution.
- Mix both solutions after the gelatine has cooled down. The mix can be aliquoted at −20 °C.

b. Immediately Before Embedding the Brain

- Give 200–500 μl glutaraldehyde to 5 ml of albumin/gelatine mixture and embed the brain very quickly, as the gelatine albumin polymerises within 30 s to 1 min.
- Cut out a block around the brain and sections at a vibratome, with a thickness ranging from 70 to 100 μm. Process the sections into a methanol series and freeze in 100% methanol at −20 °C in for a minimum of 1 h. Reverse the methanol series to 100% PBS and process for *in situ* hybridization, after a standard zebrafish embryo protocol, without proteinase K pretreatment.

C. DNA Electroporation of the Adult Brain

Dissect brains in cold ACSF (see Section VI.D). Position the freshly dissected brains ventral to the bottom onto a Petri dish containing a 2% agarose layer, immobilize the brains with thin metal needles (insect needles). Backfill an injection capillary (same capillaries as for embryo injections at one-cell stage) with DNA solution (1–4 mg/ml DNA, containing 1 mg/ml Fast Green FCF). Fix the capillary into the injector holder onto a micromanipulator. Break the tip of the capillary with a forcep and pressure-inject the DNA solution into the brain ventricles (for telencephalon electroporation). The capillary should be introduced through the meninges, at the level of the epiphysis or between the olfactory bulbs, in order to target the telencephalic ventricle, or at the posterior midbrain between the two tectal hemispheres in order to target the midbrain ventricles. The DNA solution colored with fast green spreads into the brain ventricles. Immediately after the injection, place the electrodes on both sides of the brain and give 5–7 electroshocks (voltage 99.9 V, pulse width 50 ms, space 1000 ms). After every pulse-series, clean the electrodes to remove air bubbles.

Process the brains further for slice cultures (see below).

Material:

- *Injector*: Pneumatic Pico Pump (PV 820, World Precision Instruments; Sarasota, USA) connected to a compressor (3–4/547994, Jun-Air).
- Ovodyne electroporator and current amplifier (TSS20 and EP21, Intracel; Herts, UK)
- *Electrodes*: Tweezertrodes 520, Harvard Apparatus, Inc.; San Diego, USA.

D. Slice Culture

Culture medium (modified ACSF) to be prepared freshly for 1 l

100 mM	NaCl	5.84 g
2.46 mM	KCl	0.183 g
1 mM	$MgCl_2 \cdot 6H_2O$	0.203 g
0.44 mM	$NaH_2PO_4 \cdot H_2O$	0.060 g
1.13 mM	$CaCl_2$	0.166 g
5 mM	$NaHCO_3$	0.420 g
10 mM	Glucose	1.802 g
pH 7.2		
Sterile filter the solution		

Brain Embedding, Cutting, and Culture: Dissect brains in cold ACSF and embedded in 2% low-geling agarose cooled down to 28 °C. Cut sections of 270 μm at the vibratome. Collect the sections gently on a spatula or in a pipette tip with a large opening. Lay the slices on a Millicell-CM (Milipore) culture plate insert placed in 6-well plate filled with 1.5 ml culture medium (containing penicillin/streptomycin/well). The slices should not float, as they are only covered by a thin layer of medium. Culture at 28 °C, in a normal egg incubator, change the medium every second day. Slices can be cultured for about 4–5 days.

Fixation: Remove the medium and fill the wells with about 3 ml of 4% PFA, drop some more PFA gently onto the slices. Fix for 1 h at 4 °C. Wash the slices in PBS and proceed to a methanol series, freeze in 100% methanol at −20 °C for at least 1 h, before proceeding for immunohistochemistry.

VII. Useful Tools for the Study of Zebrafish Neurogenesis

See Tables I–V

VIII. Conclusion

Our understanding of zebrafish early neurogenesis has made enormous progress in the last years, with the unraveling of several patterning mechanisms, a large number of proneural factors, and some of their interactions. Interestingly however, in most cases, expression of different proneural factors are complementary and/or only partially overlapping rather than identical. This suggests that these factors might exert shared as well as specific functions, achieved either by each factor alone or as a combinatorial "code." Such functional subdivision may explain the maintenance of related proneural factors during zebrafish evolution. It will now be most important to analyze these factors in more detail and understand their exact contribution to generic as well as cell- or stage-specific aspects of zebrafish neurogenesis.

Our understanding of the zebrafish embryonic and larval brain also remains relatively fragmentary. Comparably fewer molecular and biochemical tools are available than in other vertebrate models to characterize the different brain neuronal types, and their networks of axonal connections remain largely unknown. It is our hope that future selective GFP transgenic lines will permit to extend detailed projection tracing to the unwiring of most brain circuits, as well as to isolate and molecularly characterize most neuronal types.

Table I
Zebrafish Markers and Tools to Assess Proliferation Status and Neuronal Differentiation Stages

Name	Ab/RNA	Type of cells labelled	References	Transgenic line
Proliferation status				
Phosphohistone 3	Polyclonal Ab	Cells in late G2 and M-phase	Wei and Allis (1998)	
Ccnb1	RNA	Cells in late G2 and M-phase	Kassen *et al.* (2008)	*Ccnb1:gfp*
Cdkn1c	RNA		Park *et al.* (2005)	
PCNA	Ab, monoclonal and polyclonal	All proliferating cells (the antigen remains present for about 24 h)	Wullimann and Knipp (2000) Thummel *et al.* (2008)	
MCM5	RNA, Ab	All proliferating cells	Ryu *et al.* (2005)	
Histone H1	Ab	Cells in G1 phase	Huang and Sato (1998) and Tarnowka *et al.* (1978)	
BrdU	mAb	Cells in S-Phase after BrdU treatment		
Histone H2A		Visualization of mitotic figures		*H2A.F/Z:gfp* (Pauls *et al.*, 2001)
Progenitor cells				
Gfap	RNA, Ab	Radial glial cells	Bernardos and Raymond (2006)	Gfap:gfp
Glutamine synthetase				
S100β	Ab	Radial glial cells	Grandel *et al.* (2006)	
BLBP/FABP	RNA, Ab	Radial glial cells	Adolf *et al.* (2006)	
Nestin	RNA	Radial Glial cells	Mahler and Driever (2007)	
Cyp19b (AroB)	RNA, Ab	Radial Glial cells	Pellegrini *et al.* (2007)	arob:gfp
her4			Yeo *et al.* (2007)	her4:gfp; her4:nls-gfp her4::dRFP

(continues)

Table I (*continued*)

Name	Ab/RNA	Type of cells labelled	References	Transgenic line
her5	RNA	Midbrain progenitors	Chapouton *et al.* (2006) and Tallafuss and Bally-Cuif (2003)	her5:gfp
Notch1a	RNA	Proliferative cells during primary and secondary neurogenesis	Mueller and Wullimann (2002a)	UAS:notch1a[ac] (Scheer *et al.*, 2001)
dlA	RNA	Neuronal precursors	Haddon *et al.* (1998) and Mueller and Wullimann (2002a)	
dlD	RNA	Neuronal precursors	Dornseifer *et al.* (1997) and Haddon *et al.* (1998)	deltaD:gfp (Hans and Campos-Ortega, 2002)
Progenitor and differentiating postmitotic neurons				
dlB	RNA	Subpopulation of deltaA and deltaB expressing cells: singled-out primary neurons	Haddon *et al.* (1998)	
Onecut1	RNA	Neuronal precursors, early differentiating neurons, except in the telencephalon	Hong *et al.* (2002)	
Coe2		Neuronal precursors, early postmitotic neurons	Bally-Cuif *et al.* (1998)	
Neurog1 (ngn1)	RNA Ab	Neuronal precursors, early postmitotic neurons	Blader *et al.* (1997), Mueller and Wullimann (2002a), and Thummel *et al.* (2008)	Several neurog1: gfp lines (Blader *et al.*, 2003)
Neurod4 (zath3)	RNA	Neuronal precursors, early postmitotic neurons	Park *et al.* (2003) and Wang *et al.* (2003)	
Ash1a	RNA	Subpopulation of neuronal precursors	Allende and Weinberg (1994)	
Ash1b	RNA	Subpopulation of neuronal precursors	Allende and Weinberg (1994)	
Pou3 (tai-ji)	RNA	Precursor cells	Huang and Sato (1998)	
Pou5f1 (pou2)	RNA	Subpopulation of precursors	Hauptmann and Gerster (1996)	
Sox19	RNA	Subpopulation of precursors	Vriz *et al.* (1996)	

(*continues*)

Table I   (*continued*)

Name	Ab/RNA	Type of cells labelled	References	Transgenic line
Sox31	RNA	Subpopulation of precursors	Girard *et al.* (2001)	
Sox11a	RNA	Subpopulation of precursors	de Martino *et al.* (2000)	
Sox11b	RNA	Subpopulation of precursors	de Martino *et al.* (2000)	
Sox21	RNA	Subpopulation of precursors	Rimini *et al.* (1999)	
Neurod2	RNA	Differentiating neurons	Korzh *et al.* (1998) and Mueller and Wullimann (2002b)	
Dcc	RNA	First neuronal clusters	Hjorth *et al.* (2001)	
β-Thymosin	RNA	Early differentiating neurons	Roth *et al.* (1999)	
Polysialic acid (PSA)	Ab	Differentiating neurons, expression on cell bodies	Marx *et al.* (2001)	
Cntn2 (tag1)	RNA	Outgrowing and migrating neurons	Warren *et al.* (1999)	
L2/HNK1	mAb Zn12	Outgrowing neurons	Metcalfe *et al.* (1990)	
elavl3 (HuC)	RNA	Early differentiating and mature neurons. Start at 1 somite	Kim *et al.* (1996), Mueller and Wullimann (2002a), and Park *et al.* (2000a)	huC:gfp (Park *et al.*, 2000c) huC:kaede (Sato *et al.*, 2006)
elavl4 (HuD)	RNA	Subset of postmitotic neurons. Start at 10 somites	Park *et al.* (2000a)	
tuba1 (α1-Tubulin)	RNA	Early differentiating and regenerating neurons	Hieber *et al.* (1998)	α1-Tubulin:egfp (Goldman *et al.*, 2001; Senut *et al.*, 2004)
Elavl3 + 4 (HuC + D)	mAb 16A11	Early postmitotic and mature neurons	Mueller and Wullimann (2002a)	
β1-Tubulin	RNA	Early differentiating, start at 24 hpf	Oehlmann *et al.* (2004)	
gap43	RNA	Postmitotic neurons in the phase of axonal growth, and regenerating neurons start at 17 hpf	Reinhard *et al.* (1994)	(rat) gap43:gfp (Udvadia *et al.*, 2001) (Fugu) gap43:gfp (Udvadia, 2008)
Mature neurons				
Eno2	RNA		Bai *et al.* (2007)	*eno2:gfp*
α2-Tubulin	RNA	Mature neurons	Hieber *et al.* (1998)	
acetylated Tubulin	Ab	Membrane staining of all differentiated neurons		

(*continues*)

Table I   (*continued*)

Name	Ab/RNA	Type of cells labelled	References	Transgenic line
neurofilament	Ab RMO-44	Mature neurons, reticulospinal neurons (bodies + axons)	Gray *et al.* (2001) and Lee *et al.* (1987)	
Gefiltin (intermediate filament)			Leake *et al.* (1999)	
plasticin (intermediate filament)	Ab RNA	Subset of neurons extending axons	Canger *et al.* (1998)	
Nadl1.1 (L1.1)	RNA	Reticulospinal neurons during axonogenesis	Becker *et al.* (1998)	
E587 (L1 related)	mAb E17	Axons of primary tracts and commissures. start at 17hpf	Weiland *et al.* (1997)	
	CON1	Subset of axons	Bernhardt *et al.* (1990)	
	zn-1	All neurons	Trewarrow *et al.* (1990)	

Table II
Zebrafish Markers and Tools to Assess Neuronal Identity

Identity	Marker	Ab/RNA	References	Transgenic line
GABAergic neurons				
		Antibodies	Extensive list in Marc and Cameron (2001) and Yazulla and Studholme (2001)	
	gad2 (gad65)	RNA	Martin *et al.* (1998)	
	gad1 (gad67)	RNA	Martin *et al.* (1998)	
Monoaminergic neurons				
	Vmat2	RNA	Wen *et al.* (2008)	ET vmat2:gfp
Catecholaminergic neurons (dopamine, noradrenalin, adrenalin)				
Dopaminergic	*TH* (tyrosyne hydrxylase)	RNA Ab(Chemicon)	Guo *et al.* (1999b)	
	Uch-L1	RNA	Son *et al.* (2003)	
	nr4a2a (Nurr1) *Nr4a2b*	RNA	Blin *et al.* (2008), Filippi *et al.* (2007), and Luo *et al.* (2008)	
	slc6a3 (dat)	RNA	Holzschuh *et al.* (2001)	
Noradrenergic and adrenergic neurons	*dbh (dopamine β hydroxylase)*	RNA	Guo *et al.* (1999b)	
	TH (tyrosyne hydrxylase)	RNA Ab(Chemicon)	Guo *et al.* (1999b)	

(*continues*)

Table II   (*continued*)

Identity	Marker	Ab/RNA	References	Transgenic line
Serotonergic neurons				
	5HT (serotonin)	Ab (Chemicon)		
	tph (tphD1)	RNA	Bellipanni *et al.* (2002)	
	tph2 (tphD2)	RNA	Bellipanni *et al.* (2002)	
	tphR	RNA	Teraoka *et al.* (2004)	
	Pet1	RNA	Lillesaar *et al.* (2007, 2009)	*pet1:gfp*
	slc6a4a; slc6a4b	RNA	Norton *et al.* (2008)	
Glutamatergic neurons				
	vglut2.1 vglut2.2	RNA	Higashijima *et al.* (2004a)	
		Antibodies	Extensive list in Marc and Cameron (2001) and Yazulla and Studholme (2001)	
Cholinergic neurons				
		Antibodies	Extensive list in Marc and Cameron (2001) and Yazulla and Studholme (2001)	
Glycinergic neurons				
	glyt2a, glyt2b	RNA	Batista *et al.* (2008) and Higashijima *et al.* (2004a)	
	glra1, glra4a, glrb (glyR subunits)	RNA	Imboden *et al.* (2001)	
Histaminergic neurons				
	l-histidine decarboxylase, HA	RNA, Ab	Kaslin and Panula (2001)	
Orexin/hypocretin neurons				
	preproORX	RNA, antibodies	Kaslin (2004) #482	
Motoneurons				
Primary and secondary motoneurons	HB9		Flanagan-Steet *et al.* (2005)	hb9:gfp hb9:mGFP
Primary motoneurons, early processes	znp-1	mAb	Melancon *et al.* (1997) and Trewarrow *et al.* (1990)	
Secondary motoneurons during axonal growth	Alcam (DM-GRASP/ neurolin)	Ab zn-5	Fashena and Westerfield (1999) and Ott *et al.* (2001)	
Primary (RoP, MiP, VaP, CaP) and secondary motoneurons	*lhx3 (lim3)*	RNA pAb	Appel *et al.* (1995) and Glasgow *et al.* (1997)	

(*continues*)

Table II (*continued*)

Identity	Marker	Ab/RNA	References	Transgenic line
MiP + RoP, secondary motoneurons and cranial motoneurons	*islet1*	RNA Ab(DSHB, 40.2D6 and 39.5D5)	Inoue *et al.* (1994), Korzh *et al.* (1993), and Segawa *et al.* (2001)	*islet1:gfp* (Higashijima *et al.*, 2000) (only for cranial motoneurons)
CaP and VaP	*islet2*	RNA	Appel *et al.* (1995)	
Primary motoneurons	*Pnx*	RNA	Bae *et al.* (2003)	
Primary motoneurons	*olig2*	RNA	Park *et al.* (2002)	*olig2:egfp* (Shin *et al.*, 2003)
Cranial motoneurons	*tbx20*	RNA	Ahn *et al.* (2000)	
Occulomotor and trochlear motoneurons	*phox2a*	RNA	Guo *et al.* (1999a)	
Spinal motoneurons (transiently) + vagal motor nucleus(nX)	*sst1 (ppss1)* (somatostatin)	RNA	Devos *et al.* (2002)	
	cxcr4b	RNA	Chong *et al.* (2001)	
	appa (app)	RNA	Musa *et al.* (2001)	
Reticulospinal interneurons				
	tlx3a (tlxA)	RNA	Andermann and Weinberg (2001)	
Mauthner neuron	3A10	Ab	Hatta (1992)	
Mauthner neuron	*RMO-44*	mAb	Gray *et al.* (2001) and Lee *et al.* (1987)	
Mauthner neuron	*glra1, glra4a, glrb (glyR* subunits)	RNA	Imboden *et al.* (2001)	
Interneurons of the spinal cord				
Dorsal interneurons	*pax2a (pax2.1)*	RNA Ab	Mikkola *et al.* (1992)	*pax2.1:gfp* (Picker *et al.*, 2002)
P2 progenitors, V2a interneurons	*vsx1*	RNA, Ab	Kimura *et al.* (2008) and Passini *et al.* (1998)	*vsx1:gfp* (Kimura et al., 2008)
V2b interneurons	*Scl*	RNA, Ab	Kimura *et al.* (2008)	
V1 (CiA) interneurons	*Eng1b*	RNA, Ab	Higashijima *et al.* (2004b)	
	Pnx	RNA	Bae *et al.* (2003)	
V1 interneurons	*dbx1a (hlx1)*	RNA	Fjose *et al.* (1994) and Gribble *et al.* (2007)	
KA', VeLD	*Olig2*	RNA	Park *et al.* (2004) and Shin *et al.* (2007)	*olig2:gfp* (Shin *et al.*, 2003)
	cxcr4a	RNA	Chong *et al.* (2001)	
Small number of interneurons	*islet1*	RNA, Ab	Korzh *et al.* (1993)	
Rohon-Beard primary sensory neurons				
	Pnx	RNA	Bae *et al.* (2003)	
	cxcr4b	RNA	Chong *et al.* (2001)	
	islet1	RNA Ab	Korzh *et al.* (1993)	
	islet2	RNA	Segawa *et al.* (2001)	
	tlx3a (tlxA)	RNA	Andermann and Weinberg (2001) and Langenau *et al.* (2002)	

(*continues*)

Table II   (*continued*)

Identity	Marker	Ab/RNA	References	Transgenic line
	tlx3b	RNA	Langenau *et al.* (2002)	
	Cbfb	RNA	Blake *et al.* (2000)	
	runx3	RNA	Kalev-Zylinska *et al.* (2003)	
Subpopulation of RB	*ntrk3a (trkC)*		Williams *et al.* (2000)	
Cerebellar granule cells				
	atoh1 (ath1)	RNA	Koster and Fraser (2001)	
Purkinje cells				
	zebrin2	mAb	Jaszai *et al.* (2003)	
	M1	mAb	Miyamura and Nakayasu (2001)	
Retinal cells				
	See extensive reviews	Antibodies	Marc and Cameron (2001) and Yazulla and Studholme (2001)	
	Brn3a		Sato *et al.* (2007)	*Brn3ahsp70:gfp*
Amacrine cells	*pax6a (pax6.1)*	RNA	Kay *et al.* (2001), Raymond *et al.* (2006), and Thummel *et al.* (2008)	*pax6:gfp* (Kay *et al.*, 2001)
Ganglion cells (RGC)	*atoh7 (ath5)*	RNA	Kay *et al.* (2001) and Masai *et al.* (2000)	
RGC and INL	*islet 1*	RNA, Ab	Masai *et al.* (2000) and Korzh *et al.* (1993)	
	islet3		Kikuchi *et al.* (1997)	
RGC and INL	*lhx3 (lim3)*	RNA, Ab	Glasgow *et al.* (1997) and Masai *et al.* (2000)	
RGC	*cxcr4b*	RNA	Chong *et al.* (2001)	
RGC	*Dacha*	RNA	Hammond *et al.* (2002)	
RGC	zn-5	mAb	Kawahara *et al.* (2002)	
Epiphyseal/habenula neurons				
	brn3a		Aizawa *et al.* (2007)	*brn3a-hsp70:gfp*
	Serotonin-N-acetyl-transferase-2 (aanat2)	RNA	Gothilf *et al.* (2002)	*zfAANAT-2:gfp* (Gothilf *et al.*, 2002)
	islet1	RNA, Ab	Korzh *et al.* (1993)	
	lhx3 (lim3)	RNA, pAb	Glasgow *et al.* (1997)	
	tph (tphD1)	RNA	Bellipanni *et al.* (2002)	
	Cxcrb	RNA	Chong *et al.* (2001)	
Pituitary gland neurons				
	tbx20	RNA	Ahn *et al.* (2000)	
	lhx3 (lim3)	RNA, pAb	Glasgow *et al.* (1997)	

(*continues*)

Table II (*continued*)

Identity	Marker	Ab/RNA	References	Transgenic line
Hypothalamic clusters				
	Histamine	Ab	Kaslin and Panula (2001)	
	Neurog3 (ngn2, ngn3), expression start at 24hpf	RNA	Wang *et al.* (2003)	
	Cxcrb	RNA	Chong *et al.* (2001)	
	sst1 (ppss1), start at 5dpf	RNA	Devos *et al.* (2002)	
	tph (tphD1)	RNA	Bellipanni *et al.* (2002)	
	Dacha	RNA	Hammond *et al.* (2002)	

Table III
Zebrafish Glial Markers

Identity	Marker	Ab/RNA	References	Transgenic lines
Radial glial cells				
	Gfap	RNA, Ab	Bernardos and Raymond (2006)	*Gfap:gfp*
	S100β	Ab	Grandel *et al.* (2006)	
	BLBP/FABP	RNA, Ab	Adolf *et al.* (2006)	
	Nestin	RNA	Mahler and Driever (2007)	
	Cyp19b (AroB)	RNA, Ab	Pellegrini *et al.* (2007)	
	zrf-1	mAb	Trewarrow *et al.* (1990)	
	zrf-3	mAb	Trewarrow *et al.* (1990)	
	gfap GFAP	RNA Ab	Marcus and Easter (1995) and Nielsen and Jorgensen (2003)	
	FABP (fatty acid-binding protein)/BLBP	RNA Ab	Adolf *et al.* (2006), Denovan-Wright *et al.* (2000), and Liu *et al.* (2003)	
	zrf-1 = zns-2	mAb	Wullimann and Rink (2002) and Trewarrow *et al.* (1990)	
	sox11a	RNA	de Martino *et al.* (2000)	
	sox11b	RNA	de Martino *et al.* (2000)	
	C-4 subset of gfap-expressing cells, tested in adult	mAb	Tomizawa *et al.* (2000b)	
Mueller glial cells	HNK1	mAb zn12	Peterson *et al.* (2001)	
Mueller glial cells	Glutamin synthetase	mAb	Kay *et al.* (2001) and Peterson *et al.* (2001)	
Mueller glial cells	Carbonic anhydrase		Peterson *et al.* (2001)	

(*continues*)

Table III   (*continued*)

Identity	Marker	Ab/RNA	References	Transgenic lines
Oligodendrocytes				
	olig2	RNA	Park *et al.* (2002)	*Olig2:gfp* (Shin et al., 2003)
	mpz (myelin protein zero)	RNA	Brosamle and Halpern (2002)	
	plp1a (DM20)	RNA	Brosamle and Halpern (2002)	
	Myelin basic protein (MBP)	RNA Ab	Brosamle and Halpern (2002) and Tomizawa *et al.* (2000a)	
	34 kDa protein band	mAb	Tomizawa *et al.* (2000a)	
	A-20	mAb	Arata and Nakayasu (2003)	
	Quaking 6	Ab	Chapouton *et al.* (2006)	
Astrocytes				
Eticular astrocytes (restricted to the optic nerve)	Cytokeratin	Ab	Macdonald *et al.* (1997) and Maggs and Scholes (1990)	
Reticular astrocytes (restricted to the optic nerve)	*pax2a (pax2.1)*	RNA Ab	Macdonald *et al.* (1997) and Mikkola *et al.* (1992)	
Subpopulation of astrocytes (tested only in adults)	A-22	mAb	Kawai *et al.* (2001)	

Table IV
Zebrafish Mutants of Lateral Inhibition or Neuronal Identity

Name	Abbreviation	Gene mutated	Phenotype	References
Neurogenesis mutants				
narrowminded	nrd		Complete loss of Rohon-Beard neurons + reduction of early neural crest cells	Artinger *et al.* (1999)
mind bomb (or *white tail*)	*mib (wit)*	Ubiquitin ligase (required for notch signaling)	Neurons produced in excessive number within the neural plate. Increase in primary neurons and reduction of secondary motor neurons and eye and hindbrain radial glial cells	Haddon *et al.* (1998), Itoh *et al.* (2003), Jiang *et al.* (1996), and Schier *et al.* (1996)
after eight	aei	dlD	Increase in primary sensory neurons	Holley *et al.* (2000)
deadly seven	des	notch1a	Increase in reticulospinal neurons and slight increase in motoneurons	Gray *et al.* (2001) and Liu *et al.* (2003)

(*continues*)

Table IV   (*continued*)

Name	Abbreviation	Gene mutated	Phenotype	References
heart and soul	has	prkci (aPKCλ)	Defects in cell polarity leading to failure of brain ventricles to inflate, defects in retina (+ other defects in endodermal organs)	Horne-Badovinac *et al.* (2001) and Schier *et al.* (1996)
*dlA^dx2^*	dx2	dlA (hypermorph)	Excess of RB neurons	Appel *et al.* (1999)
Neuronal specification mutants				
Motionless	mot		Less dopamiergic hypothalamic neurons, lacking brain ventricles, cell death in telencephalon and lens by 50 hpf	Guo *et al.* (1999b)
Foggy	fog	supt5h (spt5)	Lack of dopaminergic neurons in hypothalamus, telencephalon and retina, and lack of noradrenergic neurons in the locus Coeruleus (+cardiovascular defects)	Guo *et al.* (1999b, 2000)
Too few	tof	Fezl	Reduction of hypothalamic dopaminergic neurons	Guo *et al.* (1999b)
Soulless	sll	phox2a	Loss of locus coeruleus and arch associated catecholaminergic neurons	Guo *et al.* (1999a)
Lakritz	lak	atoh7 (ath5)	Loss of retinal ganglion cells	Kay *et al.* (2001)
Young	yng	Smarca4	Blocked final differentiation of retinal cells	Gregg *et al.* (2003) and Link *et al.* (2000)
Perplexed	plx		Cell death of retinal cells before exiting the cell cycle	Link *et al.* (2001)
Confused	cfs		Cell death in a subset of retinal postmitotic cells	Link *et al.* (2001)

Table V
Gene and Enhancer–Trap Screens

Insertion	Lab	References
Enhancer trap, viral YFP insertion	T. S. Becker	Ellingsen *et al.* (2005) and Kikuta *et al.* (2007)
Gene trap, tol2-mediated GFP insertion	S. Ekker	Sivasubbu *et al.* (2006)
Gene trap, tol2-mediated GFP insertion	K. Kawakami	Kawakami *et al.* (2004)
Gene trap, tol2-mediated GFP insertion	V. Korzh	Choo *et al.* (2006)
Enhancer trap, tol2-mediated Gal4 insertion	H. Baier	Scott *et al.* (2007)

Acknowledgments

We are grateful to Jovica Ninkovic for Fig. 3A and to all lab members for regular discussions. Many thanks to Drs. J. A. Campos-Ortega, M. Hammerschmidt, and U. Straehle for their input on some aspects of this review. Work in L. B.-C's laboratory, L. B. -C., M. C., and P. C. are funded by the VolkswagenStiftung, with additional funds from the European Commission as part of the *ZF-MOD-ELS* Integrated Project in the 6th Framework Programme (Contract No. LSHG-CT-2003-;503496), the Life Science Stiftung (grant GSF 2005/01), and the Center for Protein Science Munich (CIPSM).

References

Adolf, B., Chapouton, P., Lam, C. S., Topp, S., Tannhauser, B., Strahle, U., Gotz, M., and Bally-Cuif, L. (2006). Conserved and acquired features of adult neurogenesis in the zebrafish telencephalon. *Dev. Biol.* **295,** 278–293.

Ahn, D., Ruvinsky, I., Oates, A., Silver, L., and Ho, R. (2000). tbx20 a new vertebrate T-box gene expressed in the cranial motor neurons and developing cardiovascular structures in zebrafish. *Mech. Dev.* **95,** 253–258.

Aizawa, H., Goto, M., Sato, T., and Okamoto, H. (2007). Temporally regulated asymmetric neurogenesis causes left-right difference in the zebrafish habenular structures. *Dev. Cell* **12,** 87–98.

Akimenko, M. A., Ekker, M., Wegner, J., Lin, W., and Westerfield, M. (1994). Combinatorial expression of three zebrafish genes related to distal-less: Part of a homeobox gene code for the head. *J. Neurosci.* **14,** 3475–3486.

Allende, M. L., and Weinberg, E. S. (1994). The expression pattern of two zebrafish achaete-scute homolog (ash) genes is altered in the embryonic brain of the cyclops mutant. *Dev. Biol.* **166,** 509–530.

Alvarez-Buylla, A., and Garcia-Verdugo, M. (2002). Neurogenesis in adult subventricular zone. *J. Neurosci.* **22,** 629–634.

Amoyel, M., Cheng, Y. C., Jiang, Y. J., and Wilkinson, D. G. (2005). Wnt1 regulates neurogenesis and mediates lateral inhibition of boundary cell specification in the zebrafish hindbrain. *Development* **132,** 775–785.

Andermann, P., and Weinberg, E. (2001). Expression of zTlxA, a Hox11-like gene, in early differentiating embryonic neurons and cranial sensory ganglia of the zebrafish embryo. *Dev. Dyn.* **222,** 595–610.

Andreazzoli, M., Gestri, G., Cremisi, F., Casarosa, S., Dawid, I. B., and Barsacchi, G. (2003). Xrx1 controls proliferation and neurogenesis in Xenopus anterior neural plate. *Development* **130,** 5143–5154.

Appel, B., and Chitnis, A. (2002). Neurogenesis and specification of neuronal identity. *Results Probl. Cell Differ.* **40,** 237–251.

Appel, B., and Eisen, J. S. (1998). Regulation of neuronal specification in the zebrafish spinal cord by Delta function. *Development* **125,** 371–380.

Appel, B., Fritz, A., Westerfield, M., Grunwald, D. J., Eisen, J. S., and Riley, B. B. (1999). Delta-mediated specification of midline cell fates in zebrafish embryos. *Curr. Biol.* **9,** 247–256.

Appel, B., Givan, L. A., and Eisen, J. S. (2001). Delta-Notch signaling and lateral inhibition in zebrafish spinal cord development. *BMC Dev. Biol.* **1,** 13.

Appel, B., Korzh, V., Glasgow, E., Thor, S., Edlund, T., Dawid, I. B., and Eisen, J. S. (1995). Motoneuron fate specification revealed by patterned LIM homeobox gene expression in embryonic zebrafish. *Development* **121,** 4117–4125.

Aquilina-Beck, A., Ilagan, K., Liu, Q., and Liang, J. O. (2007). Nodal signaling is required for closure of the anterior neural tube in zebrafish. *BMC Dev. Biol.* **7,** 126.

Arata, N., and Nakayasu, H. (2003). A periaxonal net in the zebrafish central nervous system. *Brain Res. Dev. Brain Res.* **961,** 179–189.

Artinger, K., Chitnis, A., Mercola, M., and Driever, W. (1999). Zebrafish narrowminded suggests a genetic link between formation of neural crest and primary sensory neurons. *Development* **126,** 3969–3979.

Bae, Y. K., Shimizu, T., and Hibi, M. (2005). Patterning of proneuronal and inter-proneuronal domains by hairy- and enhancer of split-related genes in zebrafish neuroectoderm. *Development* **132,** 1375–1385.

Bae, Y. K., Shimizu, T., Yabe, T., Kim, C. H., Hirata, T., Nojima, H., Muraoka, O., Hirano, T., and Hibi, M. (2003). A homeobox gene, pnx, is involved in the formation of posterior neurons in zebrafish. *Development* **130,** 1853–1865.

Bai, Q., Garver, J. A., Hukriede, N. A., and Burton, E. A. (2007). Generation of a transgenic zebrafish model of Tauopathy using a novel promoter element derived from the zebrafish eno2 gene. *Nucleic Acids Res.* **35,** 6501–6516.

Bainter, J., Boos, A., and Kroll, K. L. (2001). Neural Induction takes a transcriptionlal twist. *Dev. Dyn.* **222,** 315–327.

Bally-Cuif, L., Dubois, L., and Vincent, A. (1998). Molecular cloning of Zcoe2, the zebrafish homolog of Xenopus Xcoe2 and mouse EBF-2, and its expression during primary neurogenesis. *Mech. Dev.* **77,** 85–90.

Bally-Cuif, L., and Hammerschmidt, M. (2003). Induction and patterning of neuronal development, and its connection to cell cycle control. *Curr. Opin. Neurobiol.* **13,** 16–25.

Barth, K., Kishimoto, Y., Rohr, K., Seydler, C., Schulte-Merker, S., and Wilson, S. (1999). Bmp activity establishes a gradient of positional information throughout the entire neural plate. *Development* **126,** 4997–4987.

Batista, M. F., Jacobstein, J., and Lewis, K. E. (2008). Zebrafish V2 cells develop into excitatory CiD and Notch signalling dependent inhibitory VeLD interneurons. *Dev. Biol.* **322,** 263–275.

Baye, L. M., and Link, B. A. (2007). Interkinetic nuclear migration and the selection of neurogenic cell divisions during vertebrate retinogenesis. *J. Neurosci.* **27,** 10143–10152.

Becker, T., Bernhardt, R. R., Reinhard, E., Wullimann, M. F., Tongiorgi, E., and Schachner, M. (1998). Readiness of zebrafish brain neurons to regenerate a spinal axon correlates with differential expression of specific cell recognition molecules. *J. Neurosci.* **18,** 5789–5803.

Bellefroid, E. J., Kobbe, A., Gruss, P., Pieler, T., Gurdon, J. B., and Papalopulu, N. (1998). Xiro3 encodes a Xenopus homolog of the Drosophila Iroquois genes and functions in neural specification. *EMBO J.* **17,** 191–203.

Bellipanni, G., Rink, E., and Bally-Cuif, L. (2002). Cloning of two tryptophane hydroxylase genes expressed in the diencephalon of the developing zebrafish brain. *Mech. Dev.* **119S,** S215–S220.

Bellipanni, G., Varga, M., Maegawa, S., Imai, Y., Kelly, C., Myers, A. P., Chu, F., Talbot, W. S., and Weinberg, E. S. (2006). Essential and opposing roles of zebrafish beta-catenins in the formation of dorsal axial structures and neurectoderm. *Development* **133,** 1299–1309.

Belting, H. G., Hauptmann, G., Meyer, D., Abdelilah-Seyfried, S., Chitnis, A., Eschbach, C., Soll, I., Thisse, C., Thisse, B., Artinger, K. B., Lunde, K., and Driever, W. (2001). spiel ohne grenzen/pou2 is required during establishment of the zebrafish midbrain-hindbrain boundary organizer. *Development* **128,** 4165–4176.

Bernardos, R. L., and Raymond, P. A. (2006). GFAP transgenic zebrafish. *Gene Expr. Patterns* **6,** 1007–1013.

Bernhardt, R. R., Chitnis, A. B., Lindamer, L., and Kuwada, J. Y. (1990). Identification of spinal neurons in the embryonic and larval zebrafish. *J. Comp. Neurol.* **302,** 603–616.

Blader, P., Fischer, N., Gradwohl, G., Guillemont, F., and Strähle, U. (1997). The activity of neurogenin1 is controlled by local cues in the zebrafish embryo. *Development* **124,** 4557–4569.

Blader, P., Lam, C. S., Rastegar, S., Scardigli, R., Nicod, J. C., Simplicio, N., Plessy, C., Fischer, N., Schuurmans, C., Guillemot, F., and Strahle, U. (2004). Conserved and acquired features of neurogenin1 regulation. *Development* **131,** 5627–5637.

Blader, P., Plessy, C., and Strahle, U. (2003). Multiple regulatory elements with spatially and temporally distinct activities control neurogenin1 expression in primary neurons of the zebrafish embryo. *Mech. Dev.* **120,** 211–218.

Blake, T., Adya, N., Kim, C. H., Oates, A. C., Zon, L., Chitnis, A., Weinstein, B. M., and Liu, P. P. (2000). Zebrafish homolog of the leukemia gene CBFB: Its expression during embryogenesis and its relationship to scl and gata-1 in hematopoiesis. *Blood* **96,** 4178–4184.

Blin, M., Norton, W., Bally-Cuif, L., and Vernier, P. (2008). NR4A2 controls the differentiation of selective dopaminergic nuclei in the zebrafish brain. *Mol. Cell. Neurosci.*

Bray, S., and Furriols, M. (2001). Notch pathway: Making sense of suppressor of hairless. *Curr. Biol.* **11,** R217–R221.

Brewster, R., Lee, J., and Ruiz i Altaba, A. (1998). Gli/Zic factors pattern the neural plate by defining domains of cell differentiation. *Nature* **393,** 579–583.

Brosamle, C., and Halpern, M. (2002). Characterization of myelination in the developing zebrafish. *Glia* **39,** 47–57.

Campos-Ortega, J. A. (1993). Mechanisms of early neurogenesis in Drosophila melanogaster. *J. Neurobiol.* **10,** 1305–1327.

Canger, A. K., Passini, M. A., Asch, W. S., Leake, D., Zafonte, B. T., Glasgow, E., and Schechter, N. (1998). Restricted expression of the neuronal intermediate filament protein plasticin during zebrafish development. *J. Comp. Neurol.* **399,** 561–572.

Cau, E., Quillien, A., and Blader, P. (2008). Notch resolves mixed neural identities in the zebrafish epiphysis. *Development* **135,** 2391–2401.

Cau, E., and Wilson, S. W. (2003). Ash1a and Neurogenin1 function downstream of floating head to regulate epiphysial neurogenesis. *Development* **130,** 2455–2466.

Chapouton, P., Adolf, B., Leucht, C., Tannhauser, B., Ryu, S., Driever, W., and Bally-Cuif, L. (2006). her5 expression reveals a pool of neural stem cells in the adult zebrafish midbrain. *Development* **133,** 4293–4303.

Chapouton, P., Jagasia, R., and Bally-Cuif, L. (2007). Adult neurogenesis in non-mammalian vertebrates. *Bioessays* **29,** 745–757.

Cheng, Y. C., Amoyel, M., Qiu, X., Jiang, Y. J., Xu, Q., and Wilkinson, D. G. (2004). Notch activation regulates the segregation and differentiation of rhombomere boundary cells in the zebrafish hindbrain. *Dev. Cell* **6,** 539–550.

Chitnis, A., and Kuwada, J. Y. (1990). Axonogenesis in the brain of zebrafish embryos. *J. Neurosci.* **10,** 1892–1905.

Chong, S., Emelyanov, A., Gong, Z., and Korzh, V. (2001). Expression pattern of two zebrafish genes, cxcr4a and cxcr4b. *Mech. Dev.* **109,** 347–354.

Choo, B., Kondricjin, I., Parinov, S., Emelyanov, A., Go, W., Toh, W., and Korzh, V. (2006). Zebrafish transgenic enhancer TRAP line database (ZETRAP). *BMC Dev. Biol.* **6,** doi:10.1186/1471-213X-6-5.

Ciruna, B., Jenny, A., Lee, D., Mlodzik, M., and Schier, A. F. (2006). Planar cell polarity signalling couples cell division and morphogenesis during neurulation. *Nature* **439,** 220–224.

Cornell, R. A., and Eisen, J. S. (2000). Delta signaling mediates segregation of neural crest and spinal sensory neurons from zebrafish lateral neural plate. *Development* **127,** 2873–2882.

Cornell, R. A., and Eisen, J. S. (2002). Delta/Notch signaling promotes formation of zebrafish neural crest by repressing Neurogenin1 function. *Development* **129,** 2639–2648.

Cornell, R. A., and Eisen, J. S. (2005). Notch in the pathway: The roles of Notch signaling in neural crest development. *Semin. Cell Dev. Biol.* **16,** 663–672.

Cunliffe, V. T. (2004). Histone deacetylase 1 is required to repress Notch target gene expression during zebrafish neurogenesis and to maintain the production of motoneurones in response to hedgehog signalling. *Development* **131,** 2983–2995.

de Martino, S., Yan, Y. L., Jowett, T., Postlethwait, J. H., Varga, Z. M., Ashworth, A., and Austin, C. A. (2000). Expression of sox11 gene duplicates in zebrafish suggests the reciprocal loss of ancestral gene expression patterns in development. *Dev. Dyn.* **217,** 279–292.

Dee, C. T., Hirst, C. S., Shih, Y. H., Tripathi, V. B., Patient, R. K., and Scotting, P. J. (2008). Sox3 regulates both neural fate and differentiation in the zebrafish ectoderm. *Dev. Biol.* **320,** 289–301.

Del Bene, F., Wehman, A. M., Link, B. A., and Baier, H. (2008). Regulation of neurogenesis by interkinetic nuclear migration through an apical-basal notch gradient. *Cell* **134,** 1055–1065.

Denovan-Wright, E., Pierce, M., and Wright, J. (2000). Nucleotide sequence of cDNA clones coding for a brain-type fatty acid binding protein and its tissue-specific expression in adult zebrafish (Danio rerio). *BBA Gene Struct. Expr.* **1492,** 221–226.

Dessaud, E., McMahon, A. P., and Briscoe, J. (2008). Pattern formation in the vertebrate neural tube: A sonic hedgehog morphogen-regulated transcriptional network. *Development* **135**, 2489–2503.

Devos, N., Deflorian, G., Biemar, F., Bortolussi, M., Martial, J., Peers, B., and Argenton, F. (2002). Differential expression of two somatostatin genes during zebrafish embryonic development. *Mech. Dev.* **115**, 133–137.

Diks, S. H., Bink, R. J., van de Water, S., Joore, J., van Rooijen, C., Verbeek, F. J., den Hertog, J., Peppelenbosch, M. P., and Zivkovic, D. (2006). The novel gene asb11: A regulator of the size of the neural progenitor compartment. *J. Cell. Biol.* **174**, 581–592.

Diks, S. H., Sartori da Silva, M. A., Hillebrands, J. L., Bink, R. J., Versteeg, H. H., van Rooijen, C., Brouwers, A., Chitnis, A. B., Peppelenbosch, M. P., and Zivkovic, D. (2008). d-Asb11 is an essential mediator of canonical Delta-Notch signalling. *Nat. Cell Biol.* **10**, 1190–1198.

Dornseifer, P., Takke, C., and Campos-Ortega, J. A. (1997). Overexpression of a zebrafish homologue of the Drosophila neurogenic gene Delta perturbs differentiation of primary neurons and somite development. *Mech. Dev.* **63**, 159–171.

Dorsky, R. I., Moon, R. T., and Raible, D. W. (1998). Control of neural crest cell fate by the Wnt signalling pathway. *Nature* **396**, 370–373.

Dutton, K. A., Pauliny, A., Lopes, S. S., Elworthy, S., Carney, T. J., Rauch, J., Geisler, R., Haffter, P., and Kelsh, R. N. (2001). Zebrafish colourless encodes sox10 and specifies non-ectomesenchymal neural crest fates. *Development* **128**, 4113–4125.

Easter, S. S., Jr., Burrill, J., Marcus, R. C., Ross, L., Taylor, J. S. H., and Wilson, S. W. (1994). Initial tract formation in the vertebrate brain. *Prog. Brain Res.* **102**, 79–93.

Eisen, J. S., Pike, S. H., and Romancier, B. (1990). An identified motoneuron with variable fates in embryonic zebrafish. *J. Neurosci.* **10**, 34–43.

Ellingsen, S., Laplante, M., Konig, M., Furmanek, T., Kikuta, H., Hoivik, E., and Becker, T. (2005). Large-scale enhancer detection in the zebrafish genome. *Development* **132**, 3799–3811.

Elsen, G. E., Choi, L. Y., Millen, K. J., Grinblat, Y., and Prince, V. E. (2008). Zic1 and Zic4 regulate zebrafish roof plate specification and hindbrain ventricle morphogenesis. *Dev. Biol.* **314**, 376–392.

Erter, C. E., Wilm, T. P., Basler, N., Wright, C. V., and Solnica-Krezel, L. (2001). Wnt8 is required in lateral mesendodermal precursors for neural posteriorization *in vivo*. *Development* **128**, 3571–3583.

Fame, R. M., Brajon, C., and Ghysen, A. (2006). Second-order projection from the posterior lateral line in the early zebrafish brain. *Neural Dev.* **1**, 4.

Fashena, D., and Westerfield, M. (1999). Secondary motoneuron axons localize DM-GRASP on their fasciculated segments. *J. Comp. Neurol.* **406**, 415–424.

Fausett, B., and Goldman, D. (2006). The proneural basic helix-loop-helix gene ascl1a is required for retina regeneration. *J. Neurosci.* **28**, 1109–1117.

Fekany, K., Yamanaka, Y., Leung, T., Sirotkin, H. I., Topczewski, J., Gates, M. A., Hibi, M., Renucci, A., Stemple, D., Radbill, A., Schier, A. F., Driever, W., *et al.* (1999). The zebrafish bozozok locus encodes Dharma, a homeodomain protein essential for induction of gastrula organizer and dorsoanterior embryonic structures. *Development* **126**, 1427–1438.

Filippi, A., Durr, K., Ryu, S., Willaredt, M., Holzschuh, J., and Driever, W. (2007). Expression and function of nr4a2, lmx1b, and pitx3 in zebrafish dopaminergic and noradrenergic neuronal development. *BMC Dev. Biol.* **7**, 135.

Filippi, A., Tiso, N., Deflorian, G., Zecchin, E., Bortolussi, M., and Argenton, F. (2005). The basic helix-loop-helix olig3 establishes the neural plate boundary of the trunk and is necessary for development of the dorsal spinal cord. *Proc. Natl. Acad. Sci. USA* **102**, 4377–4382.

Fjose, A., Izpisua-Belmonte, J. C., Fromental-Ramain, C., and Duboule, D. (1994). Expression of the zebrafish gene hlx-1 in the prechordal plate and during CNS development. *Development* **120**, 71–81.

Flanagan-Steet, H., Fox, M. A., Meyer, D., and Sanes, J. R. (2005). Neuromuscular synapses can form *in vivo* by incorporation of initially aneural postsynaptic specializations. *Development* **132**, 4471–4481.

Furthauer, M., Thisse, C., and Thisse, B. (1997). A role for FGF-8 in the dorsoventral patterning of the zebrafish gastrula. *Development* **124**, 4253–4264.

Gahtan, E., Tanger, P., and Baier, H. (2005). Visual prey capture in larval zebrafish is controlled by identified reticulospinal neurons downstream of the tectum. *J. Neurosci.* **25,** 9294–9303.

Geldmacher-Voss, B., Reugels, A. M., Pauls, S., and Campos-Ortega, J. A. (2003). A 90-degree rotation of the mitotic spindle changes the orientation of mitoses of zebrafish neuroepithelial cells. *Development* **130,** 3767–3780.

Geling, A., Itoh, M., Tallafuss, A., Chapouton, P., Tannhäuser, B., Kuwada, J. Y., Chitnis, A. B., and Bally-Cuif, L. (2003). bHLH transcription factor Her5 links patterning to regional inhibition of neurogenesis at the midbrain-hindbrain boundary. *Development* **130,** 1591–1604.

Geling, A., Plessy, C., Rastegar, S., Straehle, U., and Bally-Cuif, L. (2004). Her5 acts as a prepattern factor that blocks neurogenin1 and coe2 expression upstream of Notch to inhibit neurogenesis at the midbrain-hindbrain boundary. *Development* **131,** 1993–2006.

Girard, F., Cremazy, F., Berta, P., and Renucci, A. (2001). Expression pattern of the Sox31 gene during Zebrafish embryonic development. *Mech. Dev.* **100,** 71–73.

Glasgow, E., Karavanov, A., and Dawid, I. (1997). Neuronal and neuroendocrine expression of LIM3, a LIM class homeobox gene, is altered in mutant zebrafish with axial signaling defects. *Dev. Biol.* **192,** 405–419.

Goldman, D., Hankin, M., Li, Z., Dai, X., and Ding, J. (2001). Transgenic zebrafish for studying nervous system development and regeneration. *Transgenic Res.* **10,** 21–33.

Golling, G., Amsterdam, A., Sun, Z., Antonelli, M., Maldonado, E., Chen, W., Burgess, S., Haldi, M., Artzt, K., Farrington, S., Lin, S.-Y., Nissen, R. M., *et al.*. (2002). Insertional mutagenesis in zebrafish rapidly identifies genes essential for early vertebrate development. *Nat. Genet.* **31,** 135–140.

Gothilf, Y., Toyama, R., Coon, S., Du, S., Dawid, I., and Klein, D. (2002). Pineal-specific expression of green fluorescent protein under the control of the serotonin-N-acetyltransferase gene regulatory regions in transgenic zebrafish. *Dev. Dyn.* **225,** 241–249.

Gotz, M., Hartfuss, E., and Malatesta, P. (2002). Radial glial cells as neuronal precursors: A new perspective on the correlation of morphology and lineage restriction in the developing cerebral cortex of mice. *Brain Res. Bull.* **57,** 777–788.

Gotz, M., and Huttner, W. B. (2005). The cell biology of neurogenesis. *Nat. Rev. Mol. Cell. Biol.* **6,** 777–788.

Grandel, H., Kaslin, J., Ganz, J., Wenzel, I., and Brand, M. (2006). Neural stem cells and neurogenesis in the adult zebrafish brain: Origin, proliferation dynamics, migration and cell fate. *Dev. Biol.* **295,** 263–277.

Gray, M., Moens, C. B., Amacher, S. L., Eisen, J. S., and Beattie, C. E. (2001). Zebrafish deadly seven functions in neurogenesis. *Dev. Biol.* **237,** 306–323.

Gregg, R. G., Willer, G. B., Fadool, J. M., Dowling, J. E., and Link, B. A. (2003). Positional cloning of the young mutation identifies an essential role for the Brahma chromatin remodeling complex in mediating retinal cell differentiation. *Proc. Natl. Acad. Sci. USA* **100,** 6535–6540.

Gribble, S. L., Nikolaus, O. B., and Dorsky, R. I. (2007). Regulation and function of Dbx genes in the zebrafish spinal cord. *Dev. Dyn.* **236,** 3472–3483.

Grinblat, Y., Gamse, J., Patel, M., and Sive, H. (1998). Determination of the zebrafish forebrain: Induction and patterning. *Development* **125,** 4403–4416.

Grinblat, Y., and Sive, H. (2001). zic Gene expression marks anteroposterior pattern in the presumptive neurectoderm of the zebrafish gastrula. *Dev. Dyn.* **222,** 688–693.

Gritsman, K., Zhang, J., Cheng, S., Heckscher, E., Talbot, W. S., and Schier, A. F. (1999). The EGF-CFC protein one-eyed pinhead is essential for nodal signaling. *Cell* **97,** 121–132.

Grunwald, D. J., Kimmel, C. B., Westerfield, M., Walker, C., and Streisinger, G. (1988). A neural degeneration mutation that spares primary neurons in the zebrafish. *Dev. Biol.* **126,** 115–128.

Guo, S., Brush, J., Teraoka, H., Goddard, A., Wilson, S. W., Mullins, M. C., and Rosenthal, A. (1999a). Development of noradrenergic neurons in the zebrafish hindbrain requires BMP, FGF8, and the homeodomain protein soulless/Phox2a. *Neuron* **24,** 555–566.

Guo, S., Wilson, S. W., Cooke, S., Chitnis, A. B., Driever, W., and Rosenthal, A. (1999b). Mutations in the zebrafish unmask shared regulatory pathways controlling the development of catecholaminergic neurons. *Dev. Biol.* **208**, 473–487.

Guo, S., Yamaguchi, Y., Schilbach, S., Wada, T., Lee, J., Goddard, A., French, D., Handa, H., and Rosenthal, A. (2000). A regulator of transcriptional elongation controls vertebrate neuronal development. *Nature* **408**, 366–369.

Haddon, C., Smithers, L., Schneider-Maunoury, S., Coche, T., Henrique, D. and Lewis, J. (1998). Multiple delta genes and lateral inhibition in zebrafish primary neurogenesis. *Development* **125**, 359–370.

Hale, M., Ritter, D., and Fetcho, J. (2001). A confocal study of spinal interneurons in living larval zebrafish. *J. Comp. Neurol.* **437**, 1–16.

Hammerschmidt, M., and Mullins, M. C. (2002). Dorsoventral patterning in the zebrafish: Bone morphogenetic proteins and beyond. *Results Probl. Cell Differ.* **40**, 72–95.

Hammond, K. L., Hill, R. E., Whitfield, T. T., and Currie, P. D. (2002). Isolation of three zebrafish dachshund homologues and their expression in sensory organs, the central nervous system and pectoral fin buds. *Mech. Dev.* **112**, 183–189.

Hanneman, E., and Westerfield, M. (1989). Early expression of acetylcholinesterase activity in functionally distinct neurons of the zebrafish. *J. Comp. Neurol.* **284**, 350–361.

Hans, S., and Campos-Ortega, J. A. (2002). On the organisation of the regulatory region of the zebrafish deltaD gene. *Development* **129**, 4773–4784.

Hans, S., Scheer, N., Riedl, I., v Weizsacker, E., Blader, P., and Campos-Ortega, J. A. (2004). her3, a zebrafish member of the hairy-E(spl) family, is repressed by Notch signalling. *Development* **131**, 2957–2969.

Hardcastle, Z., and Papalopulu, N. (2000). Distinct effects of XBF-1 in regulating the cell cycle inhibitor p27(XIC1) and imparting a neural fate. *Development* **127**, 1303–1314.

Hatta, K. (1992). Role of the floor plate in axonal patterning in the zebrafish CNS. *Neuron* **9**, 629–642.

Hauptmann, G., and Gerster, T. (1995). Pou-2-a zebrafish gene active during cleavage stages and in the early hindbrain. *Mech. Dev.* **51**, 127–138.

Hauptmann, G., and Gerster, T. (1996). Complex expression of the zp-50 pou gene in the embryonic zebrafish brain is altered by overexpression of sonic hedgehog. *Development* **122**, 1769–1780.

Helde, K., Wilson, E., Cretekos, C., and Grunwald, D. (1994). Contribution of early cells to the fate map of the zebrafish gastrula. *Science* **265**, 517–520.

Hemmati-Brivanlou, A., and Melton, D. (1997). Vertebrate embryonic cells will become nerve cells unless told otherwise. *Cell* **88**, 13–17.

Hernandez-Lagunas, L., Choi, I. F., Kaji, T., Simpson, P., Hershey, C., Zhou, Y., Zon, L., Mercola, M., and Artinger, K. B. (2005). Zebrafish narrowminded disrupts the transcription factor prdm1 and is required for neural crest and sensory neuron specification. *Dev. Biol.* **278**, 347–357.

Hieber, V., Dai, X., Foreman, M., and Goldman, D. (1998). Induction of alpha-1-tubulin gene expression during development and regeneration of the fish central nervous system. *J. Neurobiol.* **37**, 429–440.

Higashijima, S., Hotta, Y., and Okamoto, H. (2000). Visualization of cranial motor neurons in live transgenic zebrafish expressing green fluorescent protein under the control of the islet-1 promoter/enhancer. *J. Neurosci.* **20**, 206–218.

Higashijima, S., Mandel, G., and Fetcho, J. R. (2004a). Distribution of prospective glutamatergic, glycinergic, and GABAergic neurons in embryonic and larval zebrafish. *J. Comp. Neurol.* **480**, 1–18.

Higashijima, S., Masino, M. A., Mandel, G., and Fetcho, J. R. (2004b). Engrailed-1 expression marks a primitive class of inhibitory spinal interneuron. *J. Neurosci.* **24**, 5827–5839.

Hjorth, J. T., Gad, J., Cooper, H., and Key, B. (2001). A zebrafish homologue of deleted in colorectal cancer (zdcc) is expressed in the first neuronal clusters of the developing brain. *Mech. Dev.* **109**, 105–109.

Holley, S. A., Geisler, R., and Nusslein-Volhard, C. (2000). Control of her1 expression during zebrafish somitogenesis by a delta-dependent oscillator and an independent wave-front activity. *Genes Dev.* **14**, 1678–1690.

Holley, S. A., Julich, D., Rauch, G. J., Geisler, R., and Nusslein-Volhard, C. (2002). her1 and the notch pathway function within the oscillator mechanism that regulates zebrafish somitogenesis. *Development* **129**, 1175–1183.

Holzschuh, J., Ryu, S., Aberger, F., and Driever, W. (2001). Dopamine transporter expression distinguishes dopaminergic neurons from other catecholaminergic neurons in the developing zebrafish embryo. *Mech. Dev.* **101**, 237–243.

Hong, E., and Brewster, R. (2006). N-cadherin is required for the polarized cell behaviors that drive neurulation in the zebrafish. *Development* **133**, 3895–3905.

Hong, S., Kim, C., Yoo, K., Kim, H., Kudoh, T., Dawid, I., and Huh, T. (2002). Isolation and expression of a novel neuron-specific onecut homeobox gene in zebrafish. *Mech. Dev.* **112**, 199–202.

Horne-Badovinac, S., Lin, D., Waldron, S., Schwarz, M., Mbamalu, G., Pawson, T., Jan, Y., Stainier, D., and Abdelilah-Seyfried, S. (2001). Positional cloning of heart and soul reveals multiple roles for PKC lambda in zebrafish organogenesis. *Curr. Biol.* **11**, 1492–1502.

Houart, C., Caneparo, L., Heisenberg, C., Barth, K., Take-Uchi, M., and Wilson, S. (2002). Establishment of the telencephalon during gastrulation by local antagonism of Wnt signaling. *Neuron* **35**, 255–265.

Houart, C., Westerfield, M., and Wilson, S. W. (1998). A small population of anterior cells patterns the forebrain during zebrafish gastrulation. *Nature* **391**, 788–792.

Huang, S., and Sato, S. (1998). Progenitor cells in the adult zebrafish nervous system express a Brn-1-related Pou gene, Tai-ji. *Mech. Dev.* **71**, 23–35.

Imboden, M., Devignot, V., Korn, H., and Goblet, C. (2001). Regional distribution of glycine receptor messenger RNA in the central nervous system of zebrafish. *Neuroscience* **103**, 811–830.

Inoue, A., Takahashi, M., Hatta, K., Hotta, Y., and Okamoto, H. (1994). Developmental regulation of islet-1 mRNA expression during neuronal differentiation in embryonic zebrafish. *Dev. Dyn.* **199**, 1–11.

Ishibashi, M., and McMahon, A. P. (2002). A sonic hedgehog-dependent signaling relay regulates growth of diencephalic and mesencephalic primordia in the early mouse embryo. *Development* **129**, 4807–4819.

Itoh, M., Kim, C. H., Palardy, G., Oda, T., Jiang, Y. J., Maust, D., Yeo, S. Y., Lorick, K., Wright, G. J., Ariza-McNaughton, L., Weissman, A. M., Lewis, J., *et al.* (2003). Mind bomb is a ubiquitin ligase that is essential for efficient activation of Notch signaling by Delta. *Dev. Cell* **4**, 67–82.

Itoh, M., Kudoh, T., Dedekian, M., Kim, C. H., and Chitnis, A. B. (2002). A role for iro1 and iro7 in the establishment of an anteroposterior compartment of the ectoderm adjacent to the midbrain-hindbrain boundary. *Development* **129**, 2317–2327.

Jaszai, J., Reifers, F., Picker, A., Langenberg, T., and Brand, M. (2003). Isthmus-to-midbrain transformation in the absence of midbrain-hindbrain organizer activity. *Development* **130**, 6611–6623.

Jiang, Y. J., Brand, M., Heisenberg, C. P., Beuchle, D., Furutani-Seiki, M., Kelsh, R. N., Warga, R. M., Granato, M., Haffter, P., Hammerschmidt, M., Kane, D. A., Mullins, M. C., *et al.* (1996). Mutations affecting neurogenesis and brain morphology in the zebrafish, Danio rerio. *Development* **123**, 205–216.

Julich, D., Hwee Lim, C., Round, J., Nicolaije, C., Schroeder, J., Davies, A., Geisler, R., Lewis, J., Jiang, Y. J., and Holley, S. A. (2005). beamter/deltaC and the role of Notch ligands in the zebrafish somite segmentation, hindbrain neurogenesis and hypochord differentiation. *Dev. Biol.* **286**, 391–404.

Kalev-Zylinska, M. L., Horsfield, J. A., Flores, M. V., Postlethwait, J. H., Chau, J. Y., Cattin, P. M., Vitas, M. R., Crosier, P. S., and Crosier, K. E. (2003). Runx3 is required for hematopoietic development in zebrafish. *Dev. Dyn.* **228**, 323–336.

Kaslin, J., and Panula, P. (2001). Comparative anatomy of the histaminergic and other aminergic systems in zebrafish (Danio rerio). *J. Comp. Neurol.* **440**, 342–377.

Kaslin, J., Nystedt, J. M., Ostergård, M., Peitsaro, N., and Panula, P. (2004). The orexin/hypocretin system in zebrafish is connected to the aminergic and cholinergic systems. *J. Neurosci.* **24**, 2678–2689.

Kassen, S. C., Ramanan, V., Montgomery, J. E., Burket, C. T., Liu, C. G., Vihtelic, T. S., and Hyde, D. R. (2007). Time course analysis of gene expression during light-induced photoreceptor cell death and regeneration in albino zebrafish. *Dev. Neurobiol.* **67**, 1009–1031.

Kassen S. C., Thummel R., Burket, C. T., Campochiaro, L. A., Harding, M. J., and Hyde D. R. (2008). The Tg(ccnb1:EGFP) transgenic zebrafish line labels proliferating cells during retinal development and regeneration. *Mol. Vis.* **19**, 951–963.

Kawahara, A., Chien, C. B., and Dawid, I. B. (2002). The homeobox gene mbx is involved in eye and tectum development. *Dev. Biol.* **248**, 107–117.

Kawai, H., Arata, N., and Nakayasu, H. (2001). Three-dimensional distribution of astrocytes in zebrafish spinal cord. *Glia* **36**, 406–413.

Kawakami, K., Takeda, H., Kawakami, N., Kobayashi, M., Matsuda, N., and Mishina, M. (2004). A transposon-mediated gene trap approach identifies developmentally regulated genes in zebrafish. *Dev. Cell* **7**, 133–144.

Kay, J., Finger-Baier, K., Roeser, T., Staub, W., and Baier, H. (2001). Retinal ganglion cell genesis requires lakritz, a zebrafish atonal homolog. *Neuron* **30**, 725–736.

Ke, Z., Kondrichin, I., Gong, Z., and Korzh, V. (2008). Combined activity of the two Gli2 genes of zebrafish play a major role in Hedgehog signaling during zebrafish neurodevelopment. *Mol. Cell. Neurosci.* **37**, 388–401.

Kelly, G. M., Greenstein, P., Erezyilmaz, D. F., and Moon, R. T. (1995). Zebrafish wnt8 and wnt8b share a common activity but are involved in distinct developmental pathways. *Development* **121**, 1787–1799.

Kelly, G. M., and Moon, R. T. (1995). Involvement of wnt1 and pax2 in the formation of the midbrain-hindbrain boundary in the zebrafish gastrula. *Dev. Genet.* **17**, 129–140.

Kikuchi, Y., Segawa, H., Tokumoto, M., Tsubokawa, T., Hotta, Y., Uyemura, K., and Okamoto, H. (1997). Ocular and cerebellar defects in zebrafish induced by overexpression of the LIM domains of the islet-3 LIM/homeodomain protein. *Neuron* **18**, 369–382.

Kikuta, H., Laplante, M., Navratilova, P., Komisarczuk, A. Z., Engstrom, P. G., Fredman, D., Akalin, A., Caccamo, M., Sealy, I., Howe, K., Ghislain, J., Pezeron, G., *et al.* (2007). Genomic regulatory blocks encompass multiple neighboring genes and maintain conserved synteny in vertebrates. *Genome Res.* **17**, 545–555.

Kim, C. H., Ueshima, E., Muraoka, O., Tanaka, H., Yeo, S. Y., Huh, T. L., and Miki, N. (1996). Zebrafish elav/HuC homologue as a very early neuronal marker. *Neurosci. Lett.* **216**, 109–112.

Kim, H., Shin, J., Kim, S., Poling, J., Park, H. C., and Appel, B. (2008). Notch-regulated oligodendrocyte specification from radial glia in the spinal cord of zebrafish embryos. *Dev. Dyn.* **237**, 2081–2089.

Kimmel, C., and Warga, R. (1986). Tissue specific cell lineages originate in the gastrula of the zebrafish. *Science* **231**, 356–368.

Kimmel, C., and Westerfield, M. (1990). Primray neurons of the zebrafish. *In* "Signal and Sense" (G. M. Edelman, W. E. Gall, and M. W. Cowan, eds.), pp. 561–588. Wiley-Liss, New York.

Kimmel, C. B. (1993). Patterning the brain of the zebrafish embryo. *Annu. Rev. Neurosci.* **16**, 707–732.

Kimmel, C. B., Powell, S. L., and Metcalfe, W. K. (1982). Brain neurons which project to the spinal cord in young larvae of the zebrafish. *J. Comp. Neurol.* **205**, 112–127.

Kimmel, C. B., Sessions, S. K., and Kimmel, R. J. (1981). Morphogenesis and synaptogenesis of the zebrafish Mauthner neuron. *J. Comp. Neurol.* **198**, 101–120.

Kimura, Y., Satou, C., and Higashijima, S. (2008). V2a and V2b neurons are generated by the final divisions of pair-producing progenitors in the zebrafish spinal cord. *Development* **135**, 3001–3005.

Koos, D. S., and Ho, R. K. (1999). The nieuwkoid/dharma homeobox gene is essential for bmp2b repression in the zebrafish pregastrula. *Dev. Biol.* **215**, 190–207.

Korzh, V., Edlund, T., and Thor, S. (1993). Zebrafish primary neurons initiate expression of the LIM homeodomain protein Isl-1 at the end of gastrulation. *Development* **118**, 417–425.

Korzh, V., Sleptsova, I., Liao, J., He, J., and Gong, Z. (1998). Expression of zebrafish bHLH genes ngn1 and nrd defines distinct stages of Neural differentiation. *Dev. Dyn.* **213**, 92–104.

Koshida, S., Shinya, M., Mizuno, T., Kuroiwa, A., and Takeda, H. (1998). Initial anteroposterior pattern of the zebrafish central nervous system is determined by differential competence of the epiblast. *Development* **125,** 1957–1966.

Koster, R. W., and Fraser, S. E. (2001). Direct imaging of in vivo neuronal migration in the developing cerebellum. *Curr. Biol.* **11,** 1858–1863.

Kudoh, T., Wilson, S. W., and Dawid, I. B. (2002). Distinct roles for Fgf, Wnt and retinoic acid in posteriorizing the neural ectoderm. *Development* **129,** 4335–4346.

Kuwada, J. Y., Bernhardt, R. R., and Nguyen, N. (1990). Development of spinal neurons and tracts in the zebrafish embryo. *J. Comp. Neurol.* **302,** 617–628.

Langenau, D., Palomero, T., Kanki, J., Ferrando, A., Zhou, Y., Zon, L., and Look, A. (2002). Molecular cloning and developmental expression of Tlx (Hox11) genes in zebrafish (Danio rerio). *Mech. Dev.* **117,** 243–248.

Leake, D., Asch, W., Canger, A., and Schechter, N. (1999). Gefiltin in zebrafish embryos: Sequential gene expression of two neurofilament proteins in retinal ganglion cells. *Differentiation* **65,** 181–189.

Lecaudey, V., Anselme, I., Dildrop, R., Ruther, U., and Schneider-Maunoury, S. (2005). Expression of the zebrafish Iroquois genes during early nervous system formation and patterning. *J. Comp. Neurol.* **492,** 289–302.

Lecaudey, V., Anselme, I., Rosa, F., and Schneider-Maunoury, S. (2004). The zebrafish Iroquois gene iro7 positions the r4/r5 boundary and controls neurogenesis in the rostral hindbrain. *Development* **131,** 3121–3131.

Lecaudey, V., Thisse, C., Thisse, B., and Schneider-Maunoury, S. (2001). Sequence and expression pattern of ziro7, a novel, divergent zebrafish iroquois homeobox gene. *Mech. Dev.* **109,** 383–388.

Lee, S. A., Shen, E. L., Fiser, A., Sali, A., and Guo, S. (2003). The zebrafish forkhead transcription factor Foxi1 specifies epibranchial placode-derived sensory neurons. *Development* **130,** 2669–2679.

Lee, V. M., Carden, M. J., Schlaepfer, W. W., and Trojanowski, J. Q. (1987). Monoclonal antibodies distinguish several differently phosphorylated states of the two largest rat neurofilament subunits (NF-H and NF-M) and demonstrate their existence in the normal nervous system of adult rats. *J. Neurosci.* **7,** 3474–3488.

Lekven, A. C., Buckles, G. R., Kostakis, N., and Moon, R. T. (2003). Wnt1 and wnt10b function redundantly at the zebrafish midbrain-hindbrain boundary. *Dev. Biol.* **254,** 172–187.

Lekven, A. C., Thorpe, C. J., Waxman, J. S., and Moon, R. T. (2001). Zebrafish wnt8 encodes two wnt8 proteins on a bicistronic transcript and is required for mesoderm and neurectoderm patterning. *Dev. Cell* **1,** 103–114.

Leung, T., Bischof, J., Soll, I., Niessing, D., Zhang, D., Ma, J., Jackle, H., and Driever, W. (2003). bozozok directly represses bmp2b transcription and mediates the earliest dorsoventral asymmetry of bmp2b expression in zebrafish. *Development* **130,** 3639–3649.

Lewis, J. (1998). Notch signling and the control of cell fate choices in vertebrates. *Semin. Cell Dev. Biol.* **9,** 583–589.

Lewis, K. E., Bates, J., and Eisen, J. S. (2005). Regulation of iro3 expression in the zebrafish spinal cord. *Dev. Dyn.* **232,** 140–148.

Li, M., Zhao, C., Wang, Y., Zhao, Z., and Meng, A. (2002). Zebrafish sox9b is an early neural crest marker. *Dev. Genes Evol.* **212,** 203–206.

Li, Y., Allende, M. L., Finkelstein, R., and Weinberg, E. S. (1994). Expression of two zebrafish orthodenticle-related genes in the embryonic brain. *Mech. Dev.* **48,** 229–244.

Liao, J., He, J., Yan, T., Korzh, V., and Gong, Z. (1999a). A class of neuroD-related basic helix-loop-helix transcription factors expressed in developing central nervous system in zebrafish. *DNA Cell Biol.* **18,** 333–344.

Liao, J., He, J., Yan, T., Korzh, V., and Gong, Z. (1999b). A class of neuroD-related basic helix-loop-helix transcription factors expressed in developing central nervous system in zebrafish. *DNA Cell Biol.* **18,** 333–344.

Liedtke, D., and Winkler, C. (2008). Midkine-b regulates cell specification at the neural plate border in zebrafish. *Dev. Dyn.* **237,** 62–74.

Lillesaar, C., Tannhauser, B., Stigloher, C., Kremmer, E., and Bally-Cuif, L. (2007). The serotonergic phenotype is acquired by converging genetic mechanisms within the zebrafish central nervous system. *Dev. Dyn.* **236**, 1072–1084.

Lillesaar, C., Stigloher, C., Tannhäuser, B., Wullimann, M. F., and Bally-Cuif, L. (2009). Axonal projections originating from raphe serotonergic neurons in the developing and adult zebrafish, Danio rerio, using transgenics to visualize raphe-specific pet1 expression. *J. Comp. Neurol.* **512**, 158–182.

Link, B., Fadool, J., Malicki, J., and Dowling, J. (2000). The zebrafish young mutation acts non-cell-autonomously to uncouple differentiation from specification for all retinal cells. *Development* **127**, 2177–2188.

Link, B., Kainz, P., Ryou, T., and Dowling, J. (2001). The perplexed and confused mutations affect distinct stages durig the transition from proliferating to post-mitotic cells within the zebrafish retina. *Dev. Biol.* **236**, 436–453.

Liu, K. S., Gray, M., Otto, S. J., Fetcho, J. R., and Beattie, C. E. (2003). Mutations in deadly seven/notch1a reveal developmental plasticity in the escape response circuit. *J. Neurosci.* **23**, 8159–8166.

Luo, G. R., Chen, Y., Li, X. P., Liu, T. X., and Le, W. D. (2008). Nr4a2 is essential for the differentiation of dopaminergic neurons during zebrafish embryogenesis. *Mol. Cell. Neurosci.* **39**, 202–210.

Lyons, D. A., Guy, A. T., and Clarke, J. D. (2003). Monitoring neural progenitor fate through multiple rounds of division in an intact vertebrate brain. *Development* **130**, 3427–3436.

Macdonald, R., Scholes, J., Strahle, U., Brennan, C., Holder, N., Brand, M., and Wilson, S. W. (1997). The Pax protein Noi is required for commissural axon pathway formation in the rostral forebrain. *Development* **124**, 2397–2408.

Maggs, A., and Scholes, J. (1990). Reticular astrocytes in the fish optic nerve: Macroglia with epithelial characteristics form an axially repeated lacework pattern, to which nodes of Ranvier are apposed. *J. Neurosci.* **10**, 1600–1614.

Mahler, J., and Driever, W. (2007). Expression of the zebrafish intermediate neurofilament Nestin in the developing nervous system and in neural proliferation zones at postembryonic stages. *BMC Dev. Biol.* **7**, 89.

Marc, R., and Cameron, D. (2001). A molecular phenotype atlas of the zebrafish retina. *J. Neurocytol.* **30**, 593–654.

Marcus, R., and Easter, S. J. (1995). Expression of glial fibrillary acidic protein and its relation to tract formation in embryonic zebrafish (Danio rerio). *J. Comp. Neurol.* **359**, 365–381.

Martin, S., Heinrich, G., and JH., S. (1998). Sequence and expression of glutamic acid decarboxylase isoforms in the developing zebrafish. *J. Comp. Neurol.* **396**, 253–266.

Martinez, S. (2001). The isthmic organizer and brain regionalization. *Int. J. Dev. Biol.* **45**, 367–371.

Marx, M., Rutishauser, U., and Bastmeyer, M. (2001). Dual function of polysialic acid during zebrafish central nervous system development. *Development* **128**, 4949–4958.

Masai, I., Stemple, D., Okamoto, H., and Wilson, S. (2000). Midline signals regulate retinal neurogenesis in zebrafish. *Neuron* **27**, 251–263.

McLean, D. L., Fan, J., Higashijima, S., Hale, M. E., and Fetcho, J. R. (2007). A topographic map of recruitment in spinal cord. *Nature* **446**, 71–75.

Megason, S. G., and McMahon, A. P. (2002). A mitogen gradient of dorsal midline Wnts organizes growth in the CNS. *Development* **129**, 2087–2098.

Melancon, E., Liu, D. W., Westerfield, M., and Eisen, J. S. (1997). Pathfinding by identified zebrafish motoneurons in the absence of muscle pioneers. *J. Neurosci.* **17**, 7796–7804.

Mendelson, B. (1986a). Development of reticulospinal neurons of the zebrafish. I. Time of origin. *J. Comp. Neurol.* **251**, 160–171.

Mendelson, B. (1986b). Development of reticulospinal neurons of the zebrafish. II. Early axonal outgrowth and cell body position. *J. Comp. Neurol.* **251**, 172–184.

Metcalfe, W., Myers, P., Trevarrow, B., Bass, M., and Kimmel, C. (1990). Primary neurons that express the L2/HNK-1 carbohydrate during early development in the zebrafish. *Development* **110**, 491–504.

Mikkola, I., Fjose, A., Kuwada, J. Y., Wilson, S., Guddal, P. H., and Krauss, S. (1992). The paired domain-containing nuclear factor pax[b] is expressed in specific commissural interneurons in zebrafish embryos. *J. Neurobiol.* **23**, 933–946.

Miyamura, Y., and Nakayasu, H. (2001). Zonal distribution of Purkinje cells in the zebrafish cerebellum: Analysis by means of a specific monoclonal antibody. *Cell Tissue Res.* **305**, 299–305.

Mueller, T., and Wullimann, M. F. (2002a). BrdU-, neuroD (nrd)- and Hu-studies reveal unusual non-ventricular neurogenesis in the postembryonic zebrafish forebrain. *Mech. Dev.* **117**, 123–135.

Mueller, T., and Wullimann, M. F. (2002b). Expression domains of neuroD (nrd) in the early postembryonic zebrafish brain. *Brain Res. Bull.* **57**, 377–379.

Mueller, T., and Wullimann, M. F. (2003). Anatomy of neurogenesis in the early zebrafish brain. *Brain Res. Dev. Brain Res.* **140**, 137–155.

Mueller, T., and Wullimann, M. F. (2005). "Atlas of early zebrafish brain development, a tool for molecular neurogenetics." Elsevier B.V., The Netherlands.

Müller, M., v. Weizsäcker, E., and Campos-Ortega, J. A. (1996). Transcription of a zebrafish gene of the hairy-enhancer of split family delineates the midbrain anlage in the neural plate. *Dev. Genes Evol.* **206**, 153–160.

Mumm, J., and Kopan, R. (2000). Notch signaling: From the outside in. *Dev. Biol.* **228**, 151–165.

Musa, A., Lehrach, H., and Russo, V. (2001). Distinct expression patterns of two zebrafish homologues of the human APP gene during embryonic development. *Dev. Genes Evol.* **211**, 563–567.

Myers, P. Z., Eisen, J. S., and Westerfield, M. (1986). Development and axonal outgrowth of identified motoneurons in the zebrafish. *J. Neurosci.* **6**, 2278–2289.

Nakayama, H., and Oda, Y. (2004). Common sensory inputs and differential excitability of segmentally homologous reticulospinal neurons in the hindbrain. *J. Neurosci.* **24**, 3199–3209.

Nguyen, V., Trout, J., Connors, S., Andermann, P., Weinberg, E. S., and Mullins, M. (2000). Dorsal and intermediate neuronla cell types of the spinal cord are established by a BMP signaling pathway. *Development* **127**, 1209–1220.

Nielsen, A., and Jorgensen, A. (2003). Structural and functional characterization of the zebrafish gene for glial fibrillary acidic protein, GFAP. *Gene* **3310**, 123–132.

Ninkovic, J., Stigloher, C., Lillesaar, C., and Bally-Cuif, L. (2008). Gsk3beta/PKA and Gli1 regulate the maintenance of neural progenitors at the midbrain-hindbrain boundary in concert with E(Spl) factor activity. *Development* **135**, 3137–3148.

Ninkovic, J., Tallafuss, A., Leucht, C., Topczewski, J., Tannhauser, B., Solnica-Krezel, L., and Bally-Cuif, L. (2005). Inhibition of neurogenesis at the zebrafish midbrain-hindbrain boundary by the combined and dose-dependent activity of a new hairy/E(spl) gene pair. *Development* **132**, 75–88.

Nissen, R. M., Yan, J., Amsterdam, A., Hopkins, N., and Burgess, S. M. (2003). Zebrafish foxi one modulates cellular responses to Fgf signaling required for the integrity of ear and jaw patterning. *Development* **130**, 2543–2554.

Norton, W. H., Folchert, A., and Bally-Cuif, L. (2008). Comparative analysis of serotonin receptor (HTR1A/HTR1B families) and transporter (slc6a4a/b) gene expression in the zebrafish brain. *J. Comp. Neurol.* **511**, 521–542.

O'Malley, D. M., Kao, Y. H., and Fetcho, J. R. (1996). Imaging the functional organization of zebrafish hindbrain segments during escape behaviors. *Neuron* **17**, 1145–1155.

Odenthal, J., and Nusslein-Volhard, C. (1998). fork head domain genes in zebrafish. *Dev. Genes Evol.* **208**, 245–258.

Oehlmann, V., Berger, S., Sterner, C., and Korsching, S. (2004). Zebrafish beta tubulin expression is limited to the nervous system throughout development, and in the adult brain is restricted to a subset of proliferative regions. *Gene. Expr. Patterns* **4**, 191–198.

Ohnuma, S., Hopper, S., Wang, K. C., Philpott, A., and Harris, W. A. (2002). Co-ordinating retinal histogenesis: Early cell cycle exit enhances early cell fate determination in the Xenopus retina. *Development* **129**, 2435–2446.

Ohnuma, S., Philpott, A., and Harris, W. A. (2001). Cell cycle and cell fate in the nervous system. *Curr. Opin. Neurobiol.* **11**, 66–73.

Ott, H., Diekmann, H., Stuermer, C., and Bastmeyer, M. (2001). Function of neurolin (DM-GRASP/SC-1) in guidance of motor axons during zebrafish development. *Dev. Biol.* **235**, 86–97.

Panchision, D. M., Pickel, J. M., Studer, L., Lee, S. H., Turner, P. A., Hazel, T. G., and McKay, R. D. (2001). Sequential actions of BMP receptors control neural precursor cell production and fate. *Genes Dev.* **15**, 2094–2110.

Panhuysen, M., Vogt-Weisenhorn, D., Blanquet, V., Brodski, C., Heinzmann, U., Beister, W., and Wurst, W. (2004). Effects of Wnt signaling on proliferation in the developing mid-hindbrain region. *Mol. Cell. Neurosci.* **26**, 101–111.

Papan, C., and Campos-Ortega, J. A. (1994). On the formation of the neural keel and neural tube in the zebrafish Danio (Brachydanio) rerio. *Roux's Arch. Dev. Biol.* **203**, 178–186.

Papan, C., and Campos-Ortega, J. A. (1997). A clonal analysis of spinal cord development in the zebrafish. *Dev. Genes Evol.* **207**, 71–81.

Park, H., and Appel, B. (2003). Delta-Notch signaling regulates oligodendrocyte specification. *Development* **130**, 3747–3755.

Park, H., Hong, S., Kim, H., Kim, S., Yoon, E., Kim, C., Miki, N., and Huh, T. (2000a). Structural comparison of zebrafish Elav/Hu and their differential expressions during neurogenesis. *Neurosci. Lett.* **279**, 81–84.

Park, H., Shin, J., Roberts, R., and Appel, B. (2007). An olig2 reporter gene marks oligodendrocyte precursors in the postembryonic spinal cord of zebrafish. *Dev. Dyn.* **236**, 3402–3407.

Park, H. C., Boyce, J., Shin, J., and Appel, B. (2005). Oligodendrocyte specification in zebrafish requires notch-regulated cyclin-dependent kinase inhibitor function. *J. Neurosci.* **25**, 6836–6844.

Park, H. C., Hong, S. K., Kim, H. S., Kim, S. H., Yoon, E. J., Kim, C. H., Miki, N., and Huh, T. L. (2000b). Structural comparison of zebrafish Elav/Hu and their differential expressions during neurogenesis. *Neurosci. Lett.* **279**, 81–84.

Park, H. C., Kim, C. H., Bae, Y. K., Yeo, S. Y., Kim, S. H., Hong, S. K., Shin, J., Yoo, K. W., Hibi, M., Hirano, T., Miki, N., Chitnis, A. B., *et al.*. (2000c). Analysis of upstream elements in the HuC promoter leads to the establishment of transgenic zebrafish with fluorescent neurons. *Dev. Biol.* **227**, 279–293.

Park, H. C., Mehta, A., Richardson, J. S., and Appel, B. (2002). olig2 is required for zebrafish primary motor neuron and oligodendrocyte development. *Dev. Biol.* **248**, 356–368.

Park, H. C., Shin, J., and Appel, B. (2004). Spatial and temporal regulation of ventral spinal cord precursor specification by Hedgehog signaling. *Development* **131**, 5959–5969.

Park, S. H., Yeo, S. Y., Yoo, K. W., Hong, S. K., Lee, S., Rhee, M., Chitnis, A. B., and Kim, C. H. (2003). Zath3, a neural basic helix-loop-helix gene, regulates early neurogenesis in the zebrafish. *Biochem. Biophys. Res. Commun.* **308**, 184–190.

Parvin, M. S., Okuyama, N., Inoue, F., Islam, M. E., Kawakami, A., Takeda, H., and Yamasu, K. (2008). Autoregulatory loop and retinoic acid repression regulate pou2/pou5f1 gene expression in the zebrafish embryonic brain. *Dev. Dyn.* **237**, 1373–1388.

Passini, M. A., Kurtzman, A. L., Canger, A. K., Asch, W. S., Wray, G. A., Raymond, P. A., and Schechter, N. (1998). Cloning of zebrafish vsx1: Expression of a paired-like homeobox gene during CNS development. *Dev. Genet.* **23**, 128–141.

Pauls, S., Geldmacher-Voss, B., and Campos-Ortega, J. A. (2001). A zebrafish histone variant H2A.F/Z and a transgenic H2A.F/Z:GFP fusion protein for in vivo studies of embryonic development. *Dev. Genes Evol.* **211**, 603–610.

Pellegrini, E., Mouriec, K., Anglade, I., Menuet, A., Le Page, Y., Gueguen, M. M., Marmignon, M. H., Brion, F., Pakdel, F., and Kah, O. (2007). Identification of aromatase-positive radial glial cells as progenitor cells in the ventricular layer of the forebrain in zebrafish. *J. Comp. Neurol.* **501**, 150–167.

Peterson, R., Fadool, J., McClintock, J., and Linser, P. (2001). Muller cell differentiation in the zebrafish neural retina: Evidence of distinct early and late stages in cell maturation. *J. Comp. Neurol.* **429**, 530–540.

Picker, A., Scholpp, S., Bohli, H., Takeda, H., and Brand, M. (2002). A novel positive transcriptional feedback loop in midbrain-hindbrain boundary development is revealed through analysis of the zebrafish pax2.1 promoter in transgenic lines. *Development* **129**, 3227–3239.

Prince, V. E., Moens, C. B., Kimmel, C. B., and Ho, R. K. (1998). Zebrafish hox genes: Expression in the hindbrain region of wild-type and mutants of the segmentation gene, valentino. *Development* **125**, 393–406.

Quint, E., Zerucha, T., and Ekker, M. (2000). Differential expression of orthologous Dlx genes in zebrafish and mice: Implications for the evolution of the Dlx homeobox gene family. *J. Exp. Zool.* **288**, 235–241.

Raymond, P. A., Barthel, L. K., Bernardos, R. L., and Perkowski, J. J. (2006). Molecular characterization of retinal stem cells and their niches in adult zebrafish. *BMC Dev. Biol.* **6**, 36.

Reifers, F., Adams, J., Mason, I. J., Schulte-Merker, S., and Brand, M. (2000). Overlapping and distinct functions provided by fgf17, a new zebrafish member of the Fgf8/17/18 subgroup of Fgfs. *Mech. Dev.* **99**, 39–49.

Reifers, F., Bohli, H., Walsh, E. C., Crossley, P. H., Stainier, D. Y., and Brand, M. (1998). Fgf8 is mutated in zebrafish acerebellar (ace) mutants and is required for maintenance of midbrain-hindbrain boundary development and somitogenesis. *Development* **125**, 2381–2395.

Reimer, M. M., Sorensen, I., Kuscha, V., Frank, R. E., Liu, C., Becker, C. G., and Becker, T. (2008). Motor neuron regeneration in adult zebrafish. *J. Neurosci.* **28**, 8510–8516.

Reinhard, E., Nedivi, E., Wegner, J., Skene, J. H., and Westerfield, M. (1994). Neural selective activation and temporal regulation of a mammalian GAP-43 promoter in zebrafish. *Development* **120**, 1767–1775.

Rentzsch, F., Bakkers, J., Kramer, C., and Hammerschmidt, M. (2004). Fgf signaling induces posterior neuroectoderm independently of Bmp signaling inhibition. *Dev. Dyn.* **231**, 750–757.

Reugels, A. M., Boggetti, B., Scheer, N., and Campos-Ortega, J. A. (2006). Asymmetric localization of Numb:EGFP in dividing neuroepithelial cells during neurulation in Danio rerio. *Dev. Dyn.* **235**, 934–948.

Reyes, R., Haendel, M., Grant, D., Melancon, E., and Eisen, J. S. (2004). Slow degeneration of zebrafish Rohon-Beard neurons during programmed cell death. *Dev. Dyn.* **229**, 30–41.

Rhinn, M., and Brand, M. (2001). The midbrain-hindbrain boundary organizer. *Curr. Opin. Neurobiol.* **11**, 34–42.

Rimini, R., Beltrame, M., Argenton, F., Szymczak, D., Cotelli, F., and Bianchi, M. E. (1999). Expression patterns of zebrafish sox11A, sox11B and sox21. *Mech. Dev.* **89**, 167–171.

Ritter, D., Bhatt, D., and Fetcho, J. (2001). *In vivo* imaging of zebrafish reveals diffeences in the spinal networks for escape and swimming movements. *J. Neurosci.* **21**, 8956–8965.

Ross, L., Parrett, T., and Easter, S. J. (1992). Axonogenesis and morphogenesis in the embryonic zebrafish brain. *J. Neurosci.* **12**, 467–482.

Roth, L., Bormann, P., Bonnet, A., and Reinhard, E. (1999). beta-thymosin is required for axonal tract formation in developing zebrafish brain. *Development* **126**, 1365–1374.

Ruiz i Altaba, A., Sanchez, P., and Dahmane, N. (2002). Gli and hedgehog in cancer: Tumours, embryos and stem cells. *Nat. Rev. Cancer.* **2**, 361–372.

Ruiz, I. A. A., Palma, V., and Dahmane, N. (2002). Hedgehog-Gli signalling and the growth of the brain. *Nat. Rev. Neurosci.* **3**, 24–33.

Ryu, S., Holzschuh, J., Erhardt, S., Ettl, A. K., and Driever, W. (2005). Depletion of minichromosome maintenance protein 5 in the zebrafish retina causes cell-cycle defect and apoptosis. *Proc. Natl. Acad. Sci. USA* **102**, 18467–18472.

Sato, T., Hamaoka, T., Aizawa, H., Hosoya, T., and Okamoto, H. (2007). Genetic single-cell mosaic analysis implicates ephrinB2 reverse signaling in projections from the posterior tectum to the hindbrain in zebrafish. *J. Neurosci.* **27**, 5271–5279.

Sato, T., Takahoko, M., and Okamoto, H. (2006). HuC:Kaede, a useful tool to label neural morphologies in networks in vivo. *Genesis* **44**, 136–142.

Scheer, N., Groth, A., Hans, S., and Campos-Ortega, J. A. (2001). An instructive function for Notch in promoting gliogenesis in the zebrafish retina. *Development* **128**, 1099–1107.

Schier, A. F., Neuhauss, S. C., Harvey, M., Malicki, J., Solnica-Krezel, L., Stainier, D. Y., Zwartkruis, F., Abdelilah, S., Stemple, D. L., Rangini, Z., Yang, H., and Driever, W. (1996). Mutations affecting the development of the embryonic zebrafish brain. *Development* **123**, 165–178.

Schmid, B., Furthauer, M., Connors, S. A., Trout, J., Thisse, B., Thisse, C., and Mullins, M. C. (2000). Equivalent genetic roles for bmp7/snailhouse and bmp2b/swirl in dorsoventral pattern formation. *Development* **127**, 957–967.

Schmitz, B., Papan, C., and Campos-Ortega, J. A. (1993). Neurulation in the anterior trunk region of the zebrafusg Brachydanio rerio. *Roux's Arch. Dev. Biol.* **202**, 250–259.

Scott, E. K., Mason, L., Arrenberg, A. B., Ziv, L., Gosse, N. J., Xiao, T., Chi, N. C., Asakawa, K., Kawakami, K., and Baier, H. (2007). Targeting neural circuitry in zebrafish using GAL4 enhancer trapping. *Nat. Methods* **4**, 323–326.

Segawa, H., Miyashita, T., Hirate, Y., Higashijima, S., Chino, N., Uyemura, K., Kikuchi, Y., and Okamoto, H. (2001). Functional repression of Islet-2 by disruption of complex with Ldb impairs peripheral axonal outgrowth in embryonic zebrafish. *Neuron* **30**, 423–436.

Senut, M. C., Gulati-Leekha, A., and Goldman, D. (2004). An element in the alpha1-tubulin promoter is necessary for retinal expression during optic nerve regeneration but not after eye injury in the adult zebrafish. *J. Neurosci.* **24**, 7663–7673.

Sheng, G., dos Reis, M., and Stern, C. D. (2003). Churchill, a zinc finger transcriptional activator, regulates the transition between gastrulation and neurulation. *Cell* **115**, 603–613.

Shin, J., Park, H. C., Topczewska, J. M., Mawdsley, D. J., and Appel, B. (2003). Neural cell fate analysis using olig2 BAC transgenics. *Methods Cell Sci.* **25**, 7–14.

Shin, J., Poling, J., Park, H. C., and Appel, B. (2007). Notch signaling regulates neural precursor allocation and binary neuronal fate decisions in zebrafish. *Development* **134**, 1911–1920.

Shiomi, K., Uchida, H., Keino-Masu, K., and Masu, M. (2003). Ccd1, a novel protein with a DIX domain, is a positive regulator in the Wnt signaling during zebrafish neural patterning. *Curr. Biol.* **13**, 73–77.

Simpson, P. (1997). Notch signaling in development: On equivalence groups and asymmetric developmental potential. *Curr. Opin. Genet. Dev.* **7**, 537–542.

Sivasubbu, S., Balciunas, D., Davidson, A. E., Pickart, M. A., Hermanson, S. B., Wangensteen, K. J., Wolbrink, D. C., and Ekker, S. C. (2006). Gene-breaking transposon mutagenesis reveals an essential role for histone H2afza in zebrafish larval development. *Mech. Dev.* **123**, 513–529.

Solnica-Krezel, L., and Cooper, M. S. (2002). Cellular and genetic mechanisms of convergence and extension. *Results Probl. Cell Differ.* **40**, 136–165.

Solomon, K. S., Kudoh, T., Dawid, I. B., and Fritz, A. (2003). Zebrafish foxi1 mediates otic placode formation and jaw development. *Development* **130**, 929–940.

Son, O., Kim, H., Ji, M., Yoo, K., Rhee, M., and Kim, C. (2003). Cloning and expression analysis of a Parkinson's disease gene, uch-L1, and its promoter in zebrafish. *BBRC* **312**, 601–607.

Stern, C. (2002). Induction and initial patterning of the nervous system - the chick embryo enters the scene. *Curr. Opin. Genet. Dev.* **12**, 447–451.

Streit, A., Berliner, A., Papnayotou, C., Sirulnik, A., and Stern, C. (2000). Initiation of neural induction by FGF signaling before gastrulation. *Nature* **406**, 74–78.

Svoboda, K. R., Linares, A. E., and Ribera, A. B. (2001). Activity regulates programmed cell death of zebrafish Rohon-Beard neurons. *Development* **128**, 3511–3520.

Takamiya, M., and Campos-Ortega, J. A. (2006). Hedgehog signalling controls zebrafish neural keel morphogenesis via its level-dependent effects on neurogenesis. *Dev. Dyn.* **235**, 978–997.

Takke, C., Dornseifer, P., v Weizsäcker, E., and Campos-Ortega, J. A. (1999). her4, a zebrafish homologue of the Drosophila neurogenic gene E(spl), is a target of NOTCH signalling. *Development* **126**, 1811–1821.

Tallafuss, A., and Bally-Cuif, L. (2003). Tracing of her5 progeny in zebrafish transgenics reveals the dynamics of midbrain-hindbrain neurogenesis and maintenance. *Development* **130**, 4307–4323.

Tarnowka, M. A., Baglioni, C., and Basilico, C. (1978). Synthesis of H1 histones by BHK cells in G1. *Cell* **15**, 163–171.

Taupin, P., and Gage, F. (2002). Adult neurogenesis and neural stem cells of the central nervous system in mammals. *J. Neurosci. Res.* **69**, 745–749.

Tawk, M., Araya, C., Lyons, D. A., Reugels, A. M., Girdler, G. C., Bayley, P. R., Hyde, D. R., Tada, M., and Clarke, J. D. (2007). A mirror-symmetric cell division that orchestrates neuroepithelial morphogenesis. *Nature* **446**, 797–800.

Teraoka, H., Russell, C., Regan, J., Chandrasekhar, A., Concha, M., Yokoyama, R., Higashi, K., Take-Uchi, M., Dong, W., Hiraga, T., Holder, N., and Wilson, S. (2004). Hedgehog and Fgf signaling pathways regulate the development of tphR-expressing serotonergic raphe neurons in zebrafish embryos. *J. Neurobiol.* **60**, 275–288.

Thisse, C., Thisse, B., and Postlethwait, J. H. (1995). Expression of snail2, a second member of the zebrafish snail family, in cephalic mesendoderm and presumptive neural crest of wild-type and spadetail mutant embryos. *Dev. Biol.* **172**, 86–99.

Thummel, R., Kassen, S. C., Enright, J. M., Nelson, C. M., Montgomery, J. E., and Hyde, D. R. (2008). Characterization of Muller glia and neuronal progenitors during adult zebrafish retinal regeneration. *Exp. Eye. Res.* **87**, 433–444.

Tomizawa, K., Inoue, Y., Doi, S., and Nakayasu, H. (2000a). Monoclonal antibody stains oligodendrocytes and Schwann cells in zebrafish (Danio rerio). *Anat. Embryol.* **201**, 399–406.

Tomizawa, K., Inoue, Y., and Nakayasu, H. (2000b). A monoclonal antibody stains radial glia in the adult zebrafish (Danio rerio) CNS. *J. Neurocytol.* **29**, 119–128.

Topp, S., Stigloher, C., Komisarczuk, A. Z., Adolf, B., Becker, T. S., and Bally-Cuif, L. (2008). Fgf signaling in the zebrafish adult brain: Association of Fgf activity with ventricular zones but not cell proliferation. *J. Comp. Neurol.* **510**, 422–439.

Trewarrow, B., Marks, D., and Kimmel, C. B. (1990). Organization of hindbrain segments in the zebrafish embryos. *Neuron* **4**, 669–679.

Udvadia, A. J. (2008). 3.6 kb genomic sequence from Takifugu capable of promoting axon growth-associated gene expression in developing and regenerating zebrafish neurons. *Gene. Expr. Patterns* **8**, 382–388.

Udvadia, A. J., Koster, R. W., and Skene, J. H. (2001). GAP-43 promoter elements in transgenic zebrafish reveal a difference in signals for axon growth during CNS development and regeneration. *Development* **128**, 1175–1182.

van Eeden, F. J., Granato, M., Schach, U., Brand, M., Furutani-Seiki, M., Haffter, P., Hammerschmidt, M., Heisenberg, C. P., Jiang, Y. J., Kane, D. A., Kelsh, R. N., Mullins, M. C., *et al.* (1996). Mutations affecting somite formation and patterning in the zebrafish, Danio rerio. *Development* **123**, 153–164.

Varga, Z. M., Wegner, J., and Westerfield, M. (1999). Anterior movement of ventral diencephalic precursors separates the primordial eye field in the neural plate and requires cyclops. *Development* **126**, 5533–5546.

von Trotha, J. W., Campos-Ortega, J. A., and Reugels, A. M. (2006). Apical localization of ASIP/PAR-3:EGFP in zebrafish neuroepithelial cells involves the oligomerization domain CR1, the PDZ domains, and the C-terminal portion of the protein. *Dev. Dyn.* **235**, 967–977.

Vriz, S., Joly, C., Boulekbache, H., and Condamine, H. (1996). Zygotic expression of the zebrafish Sox-19, an HMG box-containing gene, suggests an involvement in central nervous system development. *Brain Res. Mol. Brain Res.* **40**, 221–228.

Walshe, J., and Mason, I. (2003). Unique and combinatorial functions of Fgf3 and Fgf8 during zebrafish forebrain development. *Development* **130**, 4337–4349.

Wang, X., Chu, L. T., He, J., Emelyanov, A., Korzh, V., and Gong, Z. (2001). A novel zebrafish bHLH gene, neurogenin3, is expressed in the hypothalamus. *Gene* **275**, 47–55.

Wang, X., Emelyanov, A., Korzh, V., and Gong, Z. (2003). Zebrafish atonal homologue zath3 is expressed during neurogenesis in embryonic development. *Dev. Dyn.* **227**, 587–592.

Warren, J. T., Jr., Chandrasekhar, A., Kanki, J. P., Rangarajan, R., Furley, A. J., and Kuwada, J. Y. (1999). Molecular cloning and developmental expression of a zebrafish axonal glycoprotein similar to TAG-1. *Mech. Dev.* **80,** 197–201.

Wei, Y., and Allis, C. (1998). Pictures in cell biology. *Trends Cell Biol.* **8,** 266.

Weiland, U. M., Ott, H., Bastmeyer, M., Schaden, H., Giordano, S., and Stuermer, C. A. (1997). Expression of an L1-related cell adhesion molecule on developing CNS fiber tracts in zebrafish and its functional contribution to axon fasciculation. *Mol. Cell. Neurosci.* **9,** 77–89.

Wen, L., Wei, W., Gu, W., Huang, P., Ren, X., Zhang, Z., Zhu, Z., Lin, S., and Zhang, B. (2008). Visualization of monoaminergic neurons and neurotoxicity of MPTP in live transgenic zebrafish. *Dev. Biol.* **314,** 84–92.

Westerfield, M., McMurray, J. V., and Eisen, J. S. (1986). Identified motoneurons and their innervation of axial muscles in the zebrafish. *J. Neurosci.* **6,** 2267–2277.

Westin, J., and Lardelli, M. (1997). Three novel Notch genes in zebrafish: Implications for vertebrate Notch gene evolution and function. *Dev. Genes Evol.* **207,** 51–63.

Williams, J. A., Barrios, A., Gatchalian, C., Rubin, L., Wilson, S. W., and Holder, N. (2000). Programmed cell death in zebrafish rohon beard neurons is influenced by TrkC1/NT-3 signaling. *Dev. Biol.* **226,** 220–230.

Wilson, E., Cretekos, C., and Helde, K. (1995). Cell mixing during early epiboly in the zebrafish embryo. *Dev. Genet.* **17,** 6–15.

Wilson, S., and Edlund, T. (2001). Neural induction: Toward a unifying mechanism. *Nat. Neurosci.* **4**(Suppl.), 1161–1168.

Wilson, S., Graziano, E., Harland, R. M., Jessell, T., and Edlund, T. (2000). An early requirement for FGF signaling in the acquisition of neural cell fate in the chick embryo. *Curr. Biol.* **10,** 421–429.

Wilson, S., Ross, L., Parrett, T., and Easter, S. (1990). The development of a simple scaffold of axon tracts in the brain o the embryonic zebrafish Brachydanio rerio. *Development* **108,** 121–145.

Wilson, S. W., Brand, M., and Eisen, J. S. (2002). Patterning the zebrafish central nervous system. *Results Probl. Cell Differ.* **40,** 181–215.

Wilson, S. W., and Easter, S. S., Jr. (1991a). A pioneering growth cone in the embryonic zebrafish brain. *Proc. Natl. Acad. Sci. USA* **88,** 2293–2296.

Wilson, S. W., and Easter, S. S., Jr. (1991b). Stereotyped pathway selection by growth cones of early epiphysial neurons in the embryonic zebrafish. *Development* **112,** 723–746.

Woo, K., and Fraser, S. E. (1995). Order and coherence in the fate map of the zebrafish nervous system. *Development* **121,** 2595–2609.

Woo, K., and Fraser, S. E. (1997). Specification of the zebrafish nervous system by nonaxial signals. *Science* **277,** 254–257.

Woo, K., and Fraser, S. E. (1998). Specification of the hindbrain fate in the zebrafish. *Dev. Biol.* **197,** 283–296.

Wright, G. J., Leslie, J. D., Ariza-McNaughton, L., and Lewis, J. (2004). Delta proteins and MAGI proteins: An interaction of Notch ligands with intracellular scaffolding molecules and its significance for zebrafish development. *Development* **131,** 5659–5669.

Wullimann, M., and Knipp, S. (2000). Proliferation pattern changes in the zebrafish brain from embryonic through early postembryonic stages. *Anat. Embryol.* **202,** 385–400.

Wullimann, M., and Rink, E. (2002). The teleostean forebrain: A comparative and developmental view based on early proliferation, Pax6 activity and catecholaminergic organization. *Brain Res. Bull.* **57,** 363–370.

Wullimann, M. F., and Mueller, T. (2002). Expression of Zash-1a in the postembryonic zebrafish brain allows comparison to mouse Mash1 domains. *Brain Res. Gene Expr. Patterns* **1,** 187–192.

Wurst, W., and Bally-Cuif, L. (2001). Neural plate patterning: Upstream and downstream of the isthmic organizer. *Nat. Rev. Neurosci.* **2,** 99–108.

Xu, J., Liu, Z., and Ornitz, D. M. (2000). Temporal and spatial gradients of Fgf8 and Fgf17 regulate proliferation and differentiation of midline cerebellar structures. *Development* **127,** 1833–1843.

Yamaguchi, M., Tonou-Fujimori, N., Komori, A., Maeda, R., Nojima, Y., Li, H., Okamoto, H., and Masai, I. (2005). Histone deacetylase 1 regulates retinal neurogenesis in zebrafish by suppressing Wnt and Notch signaling pathways. *Development* **132,** 3027–3043.

Yamanaka, Y., Mizuno, T., Sasai, Y., Kishi, M., Takeda, H., Kim, C. H., Hibi, M., and Hirano, T. (1998). A novel homeobox gene, dharma, can induce the organizer in a non-cell autonomous manner. *Genes Dev.* **12,** 2345–2353.

Yazulla, S., and Studholme, K. (2001). Neurochemical anatomy of the zebrafish retina as determined by immunocytochemistry. *J. Neurocytol.* **30,** 551–592.

Yeo, S. Y., and Chitnis, A. B. (2007). Jagged-mediated Notch signaling maintains proliferating neural progenitors and regulates cell diversity in the ventral spinal cord. *Proc. Natl. Acad. Sci. USA* **104,** 5913–5918.

Yeo, S. Y., Kim, M., Kim, H. S., Huh, T. L., and Chitnis, A. B. (2007). Fluorescent protein expression driven by her4 regulatory elements reveals the spatiotemporal pattern of Notch signaling in the nervous system of zebrafish embryos. *Dev. Biol.* **301,** 555–567.

Zecchin, E., Conigliaro, A., Tiso, N., Argenton, F., and Bortolussi, M. (2005). Expression analysis of jagged genes in zebrafish embryos. *Dev. Dyn.* **233,** 638–645.

Zhang, C., Li, Q., and Jiang, Y. J. (2007a). Zebrafish Mib and Mib2 are mutual E3 ubiquitin ligases with common and specific delta substrates. *J. Mol. Biol.* **366,** 1115–1128.

Zhang, C., Li, Q., Lim, C. H., Qiu, X., and Jiang, Y. J. (2007b). The characterization of zebrafish antimorphic mib alleles reveals that Mib and Mind bomb-2 (Mib2) function redundantly. *Dev. Biol.* **305,** 14–27.

CHAPTER 14

Time-Lapse Microscopy of Brain Development

Reinhard W. Köster[*] and Scott E. Fraser[†]

[*]Zebrafish Neuroimaging Group
Helmholtz Zentrum München
Institute of Developmental Genetics
85764 München-Neuherberg
Germany

[†]Biological Imaging Center
Beckman Institute (139-74)
California Institute of Technology
Pasadena, California 91125

I. Introduction: Why and When to Use Intravital Imaging

During embryogenesis the vertebrate brain undergoes dramatic morphological changes. On the cellular level neural cells have to undergo fate decisions, control proliferation and phenotypic differentiation, sense and interpret positional signals, establish polarity along their cell body, regulate motility and adherence, communicate with homotypic and heterotypic cell types, and form axonal connections throughout the brain. Thus the behavior of neural cells is highly dynamic and requires dynamic analytical methods to be fully understood. Moreover, a functional brain is unlikely to result from the independent execution of these developmental processes; instead, they must be orchestrated carefully in a proper spatial and temporal order. All these complex cellular interactions can hardly be recapitulated in isolated environments of cellular cultures. Noninvasive observation by means of intravital imaging of neural cells and their behavior within their natural environment, the embryonic brain, can thus offer a wealth of information about the interactions and decisions that lead to the formation of a vertebrate brain.

Zebrafish as a see-through vertebrate model organism for embryogenesis with its external development and rapid brain formation represents an ideal model organism to tackle these demanding experimental questions of continuous *in vivo* imaging. Furthermore, the accessibility of zebrafish for genetic methods, transient gene knockdown (Nasevicius and Ekker, 2000) and ectopic gene expression as well as the possibility to establish mutant (Doyon *et al.*, 2008; Driever *et al.*, 1996; Haffter *et al.*, 1996; Meng *et al.*, 2008) and stable transgenic lines (Korzh, 2007; Udvadia and Linney, 2003) allow patterning and signal transduction events to be linked spatially and temporally with their morphological consequences on the tissue, cellular, and molecular level *in vivo*.

Initially, dyes were used to mark cells and follow their developmental time course. The advent of the green fluorescent protein (GFP) (Chalfie *et al.*, 1994) and its derived and related variants has propelled intravital imaging into new dimensions. These intravital, genetically encoded dyes are regulated and can thus be controlled by cellular transcription and translation; they can be fused to endogenous proteins as fluorescent tags and even be functionalized to report intracellular molecular events (Miyawaki, 2005).

Recent advances in both optics and genetics have allowed for increasing the temporal and spatial resolution beyond the observation of cellular dynamics. Fast scanning methods now allow optical sectioning in the millisecond range. Genetic coexpression systems have been established to specifically mark and observe several subcellular structures and organelles simultaneously to address their interdependence (Kwan *et al.*, 2007; Provost *et al.*, 2007). Therefore, high-speed time-lapse imaging in zebrafish offers the possibility to merge the two research fields of developmental genetics and cell biology to challenge and readdress in the context of a developing vertebrate organism long-standing models derived from cell culture. Finally, novel fluorescent imaging technologies using photoacoustic detection methods provide access to *in vivo* monitoring of nontransparent adult zebrafish at nearly cellular resolution deep inside the brain (Razansky *et al.*, 2007; Razansky *et al.*, unpublished data). Soon, combining time-lapse imaging of zebrafish embryos with adult imaging approaches will therefore enable researchers to relate embryonic cellular events to the organization and function of the mature brain.

This chapter introduces some of the techniques that can be used for intravital noninvasive time-lapse confocal microscopy of zebrafish embryonic brain development. There is no single experimental approach that covers all the questions of brain development. But the different approaches have in common that they all follow the basic line of labeling the embryo, embedding it, followed by data recording and data analysis. Thus the following sections will keep this order, introducing and weighing different known techniques for each experimental step.

II. Techniques for Vital Staining of the Nervous System

Labeling of the specimen is the most important and often the technically most challenging part of intravital imaging. In most experiments efficient labeling will decide about success or failure of the project. That is why some careful thinking should be dedicated to the labeling approach, the question that is asked and the specificity of the labeling that is desired before the experiment is started. Different answers can be obtained from different types of labeling.

A. Overview of Techniques

1. Ubiquitous Labeling

Labeling all or almost all cells throughout the embryo is helpful to obtain information about cellular morphologies, for exmaple cell polarization, to address morphological changes of tissues, to distinguish between static and motile areas within a tissue of interest or just as histological counter-stain. The easiest labeling method is soaking a dye into zebrafish embryos by adding it to the rearing medium.

Most of these vital dyes however are nonfixable for further immunohistochemical applications. In addition, dyes can be pressure-injected into zebrafish embryos at the single-cell stage to get distributed to all daughter cells during the subsequent cell divisions. Alternatively, mRNA encoding for variants of GFP can be injected as it gets evenly distributed among dividing cells. Injecting dyes has the advantage that successful labeling can be scored and recorded right away. Injecting GFP-mRNA causes a delay in labeling due to required time for translation and protein maturation within the cells, but the label is amplified manifold through rounds of translation and virtually any subcellular structure can be targeted with proper GFP-fusion proteins.

2. Labeling Cell Clusters Randomly

In contrast to ubiquitous labeling a mosaic labeling is favorable if individual cellular behavior is to be studied. The very good contrast between the labeled cell or small cell clusters and the unlabeled cellular environment allows even membrane protrusions to be visualized (Köster and Fraser, 2001a). Such a labeling can be achieved by pressure injection of dyes or mRNA encoding GFP into individual blastomeres during early cleavage stages. Only descendants of the injected cell will be labeled, as distinct cellular fates are not yet established during these early stages of embryogenesis the labeled cells will be distributed randomly.

Injected dyes get diluted through cell divisions and mRNA is degraded over time thus these labeling approaches are mostly limited to the first 30 h of embryonic development. In contrast, injected GFP-DNA expression vectors are stable over weeks. But even when injected at the single-cell stage they get distributed in a random mosaic manner and thus result in a mosaic labeling usually marking fewer cells than obtained by dye or mRNA injection. It has been reported though that a higher degree of labeling can be achieved by either enhancing the integration frequency of the delivered DNA during early-blastula stages by using the meganuclease I-SceI (Thermes et al., 2002), by flanking the enhancer-transgene unit with insulating sequences (Hsiao et al., 2001), by promoting transgene integration with the help of transposable elements (Kawakami, 2007; Korzh, 2007) or by internal amplification through strong transcriptional activators (Köster and Fraser, 2001b).

3. Targeted Labeling of Specific Cell Clusters

If the lineage of a distinct cell population or the behavior of a distinct cell type of interest is to be studied a targeted labeling procedure has to be chosen. Using dyes or GFP-mRNA one can inject embryos at the one-cell stage and later transplant labeled cells from injected donor embryos into gastrula-stage embryos (Chen and Schier, 2001) into regions of known fate (for fate map of the nervous system see Woo and Fraser, 1995). Alternatively, photoactivatable dyes such as caged fluorescein-dextran (Concha et al., 2003) or photoconvertible fluorescent proteins (Ando et al., 2002; Patterson and Lippincott-Schwartz, 2002; Tsutsui et al., 2005;

Wiedenmann *et al.*, 2004) or photoactivatable caged GFP-mRNA (Ando *et al.*, 2001) can be injected at the single-cell stage and subsequently converted into active fluorophores by focused UV-light onto the cells of interest. With respect to photoconvertible fluorescent proteins it has to be kept in mind though that the majority of the existing proteins form homodimers or tetramers and are often not useful to generate fusion proteins targeted to specific subcellular structures. This can be circumvented by the generation of tandem dimers that act as a pseudomonomer (Wiedenmann *et al.*, 2004); recently the identification of a monomeric photoconvertible fluorescent protein has been reported (Chudakov *et al.*, 2007).

Plasmid-DNA can be injected at the one-cell stage when the expression construct contains a suitable enhancer driving expression reliably inside the neural cells of interest, but expression will still be mosaic within this cell population. Alternatively, plasmid-DNA as well as mRNA can be electroporated right into the tissue or cells of interest (Concha *et al.*, 2003; Teh *et al.*, 2003).

4. Targeted Single Cell Labeling

Fate mapping requires that single individual cells be labeled. This can be achieved via single-cell iontophoresis during which a dye is directly applied to cell by an electrical field (Woo and Fraser, 1995). This requires that the cells are accessible for capillary injection and limits this technique mostly to surface cells. Single cell labeling deeper inside tissues of interest is manageable by uncaging photoactivatable dyes with excitation light focused down to sizes smaller than one cell diameter. Similarly, it has been shown that expression vectors driven by heat-shock inducible promoters can be activated in individual cells of interest by local warming using focused laser beams (Halloran *et al.*, 2000) or a soldering iron (Hardy *et al.*, 2007). Using two-photon excitation to convert the emission profile of photo-convertible fluorescent proteins to higher wavelengths, even single organelle regions, for example, inside the mitochondrial network can be marked (Ivanchenko *et al.*, 2007).

5. Targeted Labeling of Entire Cell Populations

When the mosaic labeling achieved with most transient transgenic methods is to overcome, cell-type specific enhancers can be used to establish stable transgenic GFP-expressing lines. Usually established imaging conditions such as excitation intensity can be transferred from one labeled specimen to the next, as the labeling intensity remains relatively unchanged. Although most laborious, this nonrandom, targeted, cell-type specific, genetic labeling offers a broad range of applications such as mutant analysis (Neumann and Nüsslein-Volhard, 2000), screening for mutants and visualization of regeneration processes (Udvadia *et al.*, 2001). A detailed overview over strategies to generate stable transgenic zebrafish lines containing a list of currently established transgenic strains has been reported recently (Udvadia and Linney, 2003).

With the discovery and optimization of transposable systems in the past years the generation of stable transgenic zebrafish lines has been significantly facilitated. Several systems have been established in parallel such as Sleeping Beauty, Tol, or the Ds element from maize (Balciunas and Ekker, 2005; Emelyanov *et al.*, 2006; Kawakami, 2007). Nevertheless, the generation of a stable transgenic zebrafish strain with tissue-specific transgene expression requires that a regulatory element of interest is at hand and the identification of such an element can be very laborious. Here, due to their high efficiency of mediating transgene integration, the MLV retrovirus or again transposable elements can be of help in enhancer detection screens. Several of such enhancer trap screens have reported the isolation of a large collection of transgenic zebrafish strains with cell-type specific fluorescent protein expression in many tissues of interest including almost all regions of the zebrafish central nervous system (Balciunas and Ekker, 2005; Choo *et al.*, 2006; Ellingsen *et al.*, 2005; Nagayoshi *et al.*, 2008).

B. Details of Techniques

1. Soaking Embryos in Dyes for Ubiquitous Labeling

The easiest way of vital staining is soaking the zebrafish embryo in fluorescent dyes. The most versatile dye for cytoplasmic labeling is the quite photostable Bodipy, its lipophilic derivative Bodipy-Ceramide is more specific in labeling only the cytoplasmic membrane, the Golgi apparatus and the interstitial fluid (Cooper *et al.*, 1999; Dynes and Ngai, 1998). As these Bodipy dyes leave the cellular nucleus unlabeled, individual cells can easily be identified. Bodipy comes in different colors (see Table I) and can thus be used in combination with other labeling methods such as GFP, DsRed, or mRFP expressing cells (Cooper *et al.*, 1999; Dynes and Ngai, 1998; Köster and Fraser, 2001b). Depending on the attempted depth of image soaking for optimal labeling can vary between 30 min and 24 h. Bodipy and its derivatives are nonfixable dyes thus labeling is lost during histological staining procedures such as fluorescent immunostaining. As these dyes work almost as specific on fixed tissue as on living samples Bodipy-dyes can be reapplied to immunostained samples to recover the labeling.

a. Protocol

1. Prepare stock solution by dissolving Bodipy in DMSO at a concentration of 2.5 $\mu g/\mu L$ (store at $-20\ ^\circ C$).
2. Place dechorionated embryo in small container (e.g., Nunc produces four-well dishes each of the wells holds 1 mL).
3. Replace embryo medium with 1:1000 dilution of Bodipy (2.5 $\mu g/mL$) in 30% Danieau and incubate embryo at 28 $^\circ C$ for desired time.
4. Shortly rinse off excessive dye on the embryo's skin and image specimen.

Table I

Characteristics of Dyes for Vital Stainings in Developing Zebrafish Embryos

Dye (Exc_{max}/Em_{max})	Delivery	Range	Labeled structure
Bodipy (505/515)	Soaking	Ubiquitous	Cytoplasm, interstitial fluid
Bodipy (548/578)			
Bodipy Ceramide (505/511)	Soaking	Ubiquitous	Cytoplasmic membrane, Golgi apparatus, interstitial fluid
Bodipy CeramideTR (589/617)			
Hoechst-33342 (350/461)	Soaking	Ubiquitous	Nucleus
YO-PRO (491/509)	Soaking	Apoptotic cells	Nucleus
TO-PRO (515/531)	Soaking	Apoptotic cells	Nucleus
FM 4–64 (515/640)	Soaking	Endocytotic cells	Membrane, synaptic vesicles, endosome
Fluorescein (494/521)	Pressure injection	Stays within injected cells	Entire cell
Rhodamine (555/580)	Pressure injection	Stays within injected cells	Entire cell
DiO (484/510)	Pressure injection into cytoplasm/extracellular region or body cavities	Stays within injected cell/is carried along by motile cells	Cytoplasm and cellular membrane/cellular membrane
DiI (549/565)			
DiD (644/665)			

Note: A detailed overview over many dyes and their characteristics for vital stainings can be obtained from the "The Handbook: A Guide to Fluorescent Probes and Labeling Technologies" available free of charge under: http://probes.invitrogen.com/handbook/.

The following article provides a good introduction into different labeling strategies, their strengths and limitations (Giepmans *et al.*, 2006).

Vital nuclear staining can be achieved using Hoechst-33342 (1:1000 dilution of 10 μg/mL-stock in 30% Danieau). Detection of the fluorescence of this dye like many other DNA-binding dyes requires excitation with UV-light (350 nm). The recent incorporation of UV-diode lasers in confocal microscopy systems as well as multiphoton excitation microscopes, although still expensive equipment, will broaden the applicability of this vital nuclear stain. Other nuclear dyes such as YO-PRO and TO-PRO (see Table I) and their related DNA-binding dyes (1:300 dilution of 1 μg/mL DMSO-stock in 30% Danieau) share the advantage that they absorb and emit in the visible spectrum. In contrast to Hoechst they cannot penetrate the cytoplasmic membrane of living cells, with the exception of mechanosensory hair cells, and stain dying or dead cells only. They can thus be used as fluorescent cell death markers.

To counterstain EGFP-labeled cells FM 4–64 (1 mM stock in ethanol, 10 μM in 30% Danieau) can be used; this dye enters cells via endocytosis and thus labels the cytoplasmic membrane and endosomes of endocytotically active cells.

2. Pressure Injection of Dyes for Labeling Groups of Cells

To achieve a more cell specific label fluorescent dyes can be injected into groups of cells or individual cells. This is commonly used for cell tracing in mapping approaches, to address cell autonomy questions or to analyze projection patterns of neurons. Two different types of dyes are usually being used for this approach, hydrophilic fluorescent dye-coupled dextrans (fluorescein- or rhodamine-coupled) or lipophilic carbocyanines (DiO, DiI, or DiD). Aqueous solutions of fluorescein- or rhodamine-coupled dextrans are filled into glass capillaries and are injected through mild air pressure pulses into the cytoplasm of the cells of interest. Here the dye diffuses quickly and labels the entire cytoplasm and nucleus resulting in brightly marked cells, but subcellular structures cannot be visualized. The size of the dextran determines whether the dye is allowed to cross gap junctions (MW < 1 kDa) or remains within the labeled cell (usually dextrans around 10,000 and 15,000 kDa are used). Furthermore, fluorescein can also be acquired as photo-caged compound for light inducible labeling approaches (*Note*: injection and subsequent incubation and handling of embryos have to occur in the dark to prevent premature fluorophore uncaging (Concha *et al.*, 2003)). It is important to note though that fluorescein and rhodamine are bleached easily, thus extensive excitation prior to image recording should be avoided. Due to their dextran residue, these dyes can be fixed in place after live imaging procedures and the analyzed embryo can be subjected to immunohistochemical double labeling techniques.

a. Protocol

1. Dissolve fluorescein/rhodamine dextran at 100 mg/mL.
2. Apply solution to Micro-Centricon ultrafiltration tube (Amicon, threshold of membrane: 3 kDa) and spin at maximum speed until most of the liquid has drained.
3. Wash several times with distilled water to remove synthetic byproducts.
4. Finally dissolve remaining supernatant in water at 100 mg/mL, aliquot into 100 μL volumes and store at $-20\,^{\circ}$C until needed for injection.

As DiO, DiI, or DiD is lipophilic dyes, stock solutions are usually prepared in 70% or 100% ethanol at a concentration of 0.5% w/v. Depending on the way of administration different types of labeling can be achieved. Similar to the dextran-coupled dyes these carbocyanines can be pressure-injected into cells of interest with glass capillaries resulting in individually labeled cells. Due to immediate precipitation in the aqueous environment these dyes stay within the injected cell and its descendants; but the insolubility of DiI-derivatives in water makes vital labeling of zebrafish embryonic cells challenging. Cells labeled by injection can be monitored with regard to their behavior immediately after being labeled. In contrast, focal injections into the extracellular space of tissue regions containing motile cells (e.g., migrating neuronal precursors in the hindbrain) result in a bright depository of DiI or its derivatives. Cell labeling and subsequent observation relies on the fact that precipitating dye-crystals attach to the cell membranes. When the motile cells leave the injection site they carry the dye crystals with them resulting in bright cellular membrane labeling in an unlabeled environment. Alternatively, large amounts of DiI and its derivatives can be injected as isotonic sucrose solutions into body cavities such as the lumen of the neural tube. Again, cells migrating away from the ventricular surface carry the dye crystals as bright fluorescent label with them. While the ventricle is labeled too brightly to be imaged, the cells leaving the ventricular region enter a nonfluorescent environment that provides a good contrast. Once attached to a cell the DiI crystals are not passed onto other cells during the course of cellular migration ensuring that the same individual cells can be observed over their developmental time course. The introduction of Celltracker, CM-DiI, which binds the dye through a thiol-reactive moiety to cellular compounds, retains the label within the cell throughout fixation and permeabilization and thus allows immunostainings to be performed on top of DiI-labeling.

b. Protocol

1. Dissolve DiD, DiI, DiO in 100% ethanol at a concentration of 0.5% w/v.
2a. Inject 1:10 dilution (in 100% ethanol) into cells of interest with self-pulled glass capillaries.
2b. Focally inject 1:10 dilution (in 100% ethanol) in tissue region of interest.
2c. Inject 1:10 dilution (in isotonic sucrose) into lumen of neural tube or other body cavities.
3. Record individually labeled cells or motile cells leaving the brightly labeled injection site carrying DiI-crystals with them attached to their membranes.

3. Iontophoretic Labeling of Single Cells

Labeling of individual cells can be achieved by single-cell injection using iontophoresis at later embryonic stages. Instead of pressure injection, the charged dye (dye-coupled dextrans or DiI-derivatives) is transferred to the pierced cell through directed migration in an electrical field. Intracellular injection using electrophysiology equipment permits single cells to be labeled with certainty by recording the change in potential that accompanies the penetration of the cell membrane (Woo and Fraser, 1995). After injection, the presence of a single labeled cell can be confirmed by brief inspection of the fluorescence signal. Care must be taken to minimize exposure of the newly injected cells to the exciting wavelengths for the dye, as the bleaching that would otherwise result will create toxic by-products. The needed components for intracellular iontophoretic labeling and the key features of successful injections are presented elsewhere (Fraser, 1996).

If single cell labeling is not absolutely required, a much simpler and inexpensive approach can be employed. Cell groups as small as single cells (often doublets or small clusters of cells) can be labeled with lipophilic dyes such as DiI, using visual guidance and a simple battery power supply. The carbocyanine dye, diluted in an organic solvent, is ejected from the very sharp pipet tip by current, allowing much greater control than possible with pressure injections. A simple experimental setup and procedure has been described in detail (Fraser, 1996).

4. Quantum Dots

Recently quantum dots have captured much attention as new biological labeling agents. In principal they behave like other dyes with extreme photostability. They consist of semiconductor material such as cadmiumselenide with only the crystal size determining the excitation and emission characteristics. Thus with the same chemistry a wide range of colors can be generated (Jaiswal et al., 2003). Furthermore, their small size allows high-fluorophore concentrations to be achieved. When coated by micelles (Dubertret et al., 2002), zinc sulfide (Mattoussi et al., 2001), or silica (Bruchez et al., 1998) (Chan and Nie, 1998) these crystals are nontoxic to the organism and can be further functionalized through added adapter molecules (Lidke et al., 2004). In zebrafish quantum dots can be used for cell tracing, live tissue labeling combined with subsequent immunohistochemistry, or microangiography to visualize the vasculature even in deep brain areas (Rieger et al., 2005). Thus quantum dots represent a very photostable alternative for long-term time-lapse imaging of dye-labeled cells or tissues.

5. Genetic Labeling

Dye labeling techniques have in common that the dye dilutes with growth and mitosis and is often lost by photobleaching. In contrast, genetically encoded dyes such as GFP, DsRed, or mRFP can be produced continuously by the cell through ongoing transcription and translation, thereby replacing photodamaged fluorophores. A battery of different GFP-variants has been produced and a list, certainly incomplete, of useful GFP-variants for intravital zebrafish imaging is provided in Table II.

When used in stable transgenic lines GFP can label distinct cell types. Some useful strains for imaging brain development are listed in Table III.

These transgenic strains are powerful tools to analyze the dynamics and progression of phenotypes when crossed into mutant strains. But for many genes mutant strains have not been identified. In these cases interference strategies with the gene product of interest have to be established. For the first 24–30 h of embryogenesis RNA-based strategies are commonly used with the injection of antisense-morpholinos for loss-of-function (Nasevicius and Ekker, 2000) and capped mRNA for gain-of-function studies. When injected into stable transgenic GFP-lines or dye-soaked embryos intravital imaging approaches can be used as dynamic analytical method. With progressing embryogenesis, however, injected RNA becomes degraded making the injection of plasmid-DNA, which is stable

Table II

Different Colors, Subcellular Localizations and Functional Properties of Fluorescent Proteins that Are Useful for Intravital Imaging

GFP-variant (Exc_{max}/Em_{max})	Reference
Different colors	
EBFP (383 nm/447 nm)	Finley *et al.* (2001)
ECFP (439 nm/476 nm)	Heim and Tsien (1996)
EGFP (484 nm/510 nm)	Cormack *et al.* (1996)
EYFP (512 nm/529 nm)	Heim and Tsien (1996)
DsRed (558 nm/583 nm)	Matz *et al.* (1999)
mRFP (584 nm/607 nm)	Campbell *et al.* (2002)
Targeted fusions	
NLS-GFP (nuclear)	Linney *et al.* (1999)
H2B-GFP (nuclear)	Kanda *et al.* (1998)
lyn-GFP (membrane)	Teruel *et al.* (1999)
unc-GFP (neurites)	Dynes and Ngai (1998)
gap43-GFP (neuritis)	Moriyoshi *et al.* (1996)
mito-GFP (mitochondria)	Rizzuto *et al.* (1995)
actin-GFP (actin skeleton)	Westphal *et al.* (1997)
tubulin-GFP (microtubules)	Clontech (1998)
Miscellaneous variants	
Venus (fast folding YFP)	Nagai *et al.* (2002)
Keima (family of fluorescent proteins with large Stokes shift)	Kogure *et al.* (2006)
Timer (time-dependent green-to-red conversion)	Verkhusha *et al.* (2001)
Kaede (UV-induced green-to-red conversion)	Ando *et al.* (2002)
PA-GFP (photoconvertible)	Patterson and Lippincott-Schwartz (2002)
mKikGR (photoconvertible)	Tsutsui *et al.* (2005)
EosFP (photoconvertible)	Wiedenmann *et al.* (2004)
Dendra (photoconvertible)	Chudakov *et al.*, 2007)
Flash-pericam (Ca^{2+}-indicator)	Nagai *et al.* (2001)
Inverse pericam (Ca^{2+}-indicator)	Nagai *et al.* (2001)
Ratiometric pericam (Ca^{2+}-indicator)	Nagai *et al.* (2001)
Protein tyrosine kinase activity reporters	Ting *et al.* (2001)
Cell-cycle reporter (red: G1, green: S, G2, M phase)	Sakaue-Sawano *et al.* (2008)

Note: Given the rapid development of this field and the fast appearance of novel fluorescent proteins with unique properties, a list of fluorescent proteins is quickly out of date. A few informative references are Shaner *et al.* (2004), Chudakov *et al.* (2005), and Miyawaki (2005); in addition the following websites of companies offering expression vectors encoding fluorescent proteins are usually detailed and allow one to stay informed.

Fluorescent protein providers:

Clontech: http://www.clontech.com/

Evrogen: http://www.evrogen.com/

Invitrogen: http://www.invitrogen.com/

over several weeks, the interference method of choice (Hammerschmidt *et al.*, 1999). As injected plasmid-DNA leads to a mosaic distribution of vector-containing and expressing cells the challenge is to label the transgene-expressing cells.

Table III
Stable Transgenic Zebrafish Lines with GFP-Expression Inside the Developing Nervous System that Are Particularly Useful for *In Vivo* Imaging Approaches

Transgenic line	Labeled tissue	Reference
HuC:gfp	Postmitotic neuronal precursors and mature neurons	Park *et al.* (2000)
Gap43:gfp	Neuronal precursors	Udvadia *et al.* (2001)
α-tubulin:gfp	Neuronal precursors	Hieber *et al.* (1998)
netrin:gfp	Floorplate, hypochord	Rastegar *et al.* (2002)
Shh:gfp	Retinal ganglion cells	Neumann and Nüsslein-Volhard (2000)
Her5:gfp	Neuroectodermal cells at mid-hindbrain boundary	Tallafuss and Bally-Cuif (2003)
Pax2.1:gfp	Dorsoposterior telencephalon, anteroposterior diencephalon, posterior tectum, tegmentum, anterior cerebellum, hombomere 3 and 5, otic vesicle, and spinal cord interneurons	Picker *et al.* (2002)
islet1:gfp	Cranial motor neurons	Higashijima *et al.* (2000)
gata2:gfp	Neuronal precursors of vrc in the diencephalon	Bak and Fraser (2003) and Meng *et al.* (1997)
omp:gfp	Neuronal precursors in the olfactory placode	Yoshida *et al.* (2002)
Flh:egpf	Neuronal precursors of epithalamus in the telencephalon	Concha *et al.* (2003)
Ngn1:gfp, ngn1:rfp	Neuronal precursors throughout neural plate (except spinal cord interneurons)	Blader *et al.* (2003)
apoE:GFP	Precursor and mature microglia	Peri and Nüsslein-Volhard (2008)
NBT:DsRed	Postmitotic neuronal precursor and mature neurons	Peri and Nüsslein-Volhard, 2008)

Note: The number of transgenic zebrafish strains expressing fluorescent proteins in a cell-type specific manner has skyrocketed since the availability of efficient retroviral insertion and transposable elements for generating stable transgenic strains. Many of these strains have been introduced into stock centers from where they are publicly available. The reader may best look at the webpages of these stock centers or laboratories for a description of the available fluorescent protein expressing strains.

British Zebrafish Research at UCL (UCL, London): http://www.ucl.ac.uk/zebrafish-group/fishfacility/index.php

Medaka Stockcenter (Nagoya): http://biol1.bio.nagoya-u.ac.jp:8000/

Tübingen Zebrafish Stockcenter (MPI, Tübingen): http://www.eb.tuebingen.mpg.de/core-facilities/zebrafish-stockcenter/tubingen-zebrafish-stockcenter/

Zebrafish International Resource Center (Eugene, Oregon): http://zfin.org

This can be achieved by creating GFP-fusion proteins but their functional integrity has to be ensured. Alternatively simultaneous coexpression of at least two transgenes with one of them being a marker (GFP) allows the transgene expressing cells to be identified and imaged. Three different strategies can be employed. One is the use of internal ribosomal entry sites (IRES) where an expression vector gives rise to a single bicistronic mRNA that gets translated into two different proteins, the protein of interest and GFP. The efficiency at which the IRES-dependent transgene is being translated varies strongly with the IRES that is being used but in general the IRES-dependent cistron is expressed at lower levels than the cap-dependent cistron (Fahrenkrug *et al.*, 1999; Köster *et al.*, 1996; Kwan *et al.*, 2007). A second approach makes use of short viral 2A-peptide sequences. These sequences when cloned in frame between two transgenes mediate a break in the peptide strand during translation via a mechanism called ribosome slippage. With this method both proteins are produced in an equimolar ratio. It has to be kept in mind though that both proteins carry a short additional peptide sequence at the C- or N-terminus, respectively (Provost *et al.*, 2007). A third type of vectors relies on the indirect expression of two cistrons by an enhancer/promoter through the diffusible transcriptional activator Gal4Vp16. Once being transcribed and translated Gal4Vp16 binds back to the expression vector to several Gal4-specific binding-sites (UAS) thereby activating the expression of the Gal4-dependent transgene of interest and Gal4-dependent GFP (Köster and Fraser, 2001b). Due to their long-lasting strong fluorescence these vectors are of use when long-term time-lapse imaging or observation of small size subcellular structures are attempted (Niell *et al.*, 2004). Recently, many stable transgenic zebrafish strains with tissue specific expression of the Gal4 activator have been generated allowing one to use multicistron UAS-vectors with cell-type specific control (Asakawa *et al.*, 2008; Davison *et al.*, 2007; Sott *et al.*, 2007).

III. Preparation of the Zebrafish Specimen

A. Imaging Chambers

For imaging the labeled embryo has to be mounted in an imaging chamber that provides the embryo with the necessary aqueous medium and oxygen but also establishes optimal conditions for the light path. Usually zebrafish embryos are being raised in plastic dishes but most objectives are corrected for glass cover slips. In addition imaging through plastic strongly scatters the light reducing the picture quality. A simple but very efficient imaging chamber can be produced by drilling a hole into the bottom of a cell culture dish (Fig. 1A). A cover slip is subsequently "glued" onto the dish bottom using silicon grease to seal the dish again providing the desired optics of glass for nonimmersive imaging (Fig. 1B and C). When an inverted confocal microscope is used simply covering the imaging dish with the plastic dish cover efficiently restricts evaporation (Fig. 1E). In case an upright microscope is used plastic rings obtained from culture dishes can be used in a similar manner as imaging chamber. After sealing one side of the plastic ring with a cover slip the embryo is embedded on this cover slip, the dish is filled with 30% Danieau and sealed on the other side with a second cover slip. Finally the entire dish can be flipped over and mounted on the scope (Distel and Köster, 2007).

Immersion objectives can be dipped right into the embryonic rearing medium just above the tissue that is to be imaged. As these objectives are corrected for aqueous solutions their use does not require specific imaging dishes. However, in case long-term time-lapse recording experiments are being performed excessive evaporation from this open dish has to be prevented. Sealing the space between the objective and the imaging chamber can be achieved by plastic wrapping of the

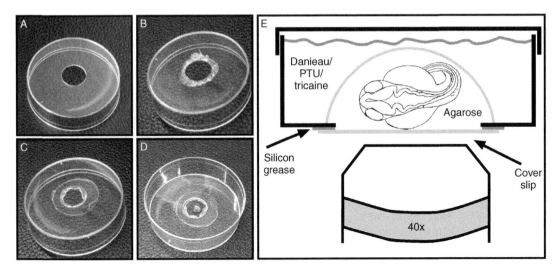

Fig. 1 Agarose embedding for zebrafish time-lapse recordings. To ensure a light path that accounts for the cover slip correction of most objectives and ensures optimal working distances embryos are embedded in special imaging chambers. (A) These chambers are custom-made by drilling a hole into the bottom of a round 5 cm Petri dish. (B) Silicon grease is applied to the rim of this hole from the bottom of the dish. (C) This allows a glass cover slip to be glued underneath the dish sealing the hole and providing glass optics for imaging. (D) Subsequently, a zebrafish embryo is transferred within a drop of ultra low gelling agarose to the cover slip and mounted in the proper position prior to cooling and thus solidifying the agarose. (E) Finally, the embedded embryo is overlaid with 30% Danieau/PTU/Tricaine, the dish is covered to avoid evaporation and the entire imaging chamber is mounted on the stage of an inverted confocal laser scanning microscope.

space between the rim of the dish and the objective (using a condom is one easy approach to reduce evaporation (Potter, 2000)).

Time-lapse imaging, especially at high magnifications, requires that the embryo is lying still, fixed in a desired position as any embryonic movement would move the imaged cells out of focus and field of vision. Thus embryos are mounted in a matrix that allows them to further develop but prevents or restricts movements to a minimum. Several mounting techniques varying mostly in the matrix media have been found useful:

B. Stabilizing the Embryo

1. Methylcelluose-Embedding

Methylcelluose in Danieau provides a very viscous matrix into which the dechorionated embryo can be transferred and oriented using manipulation needles or forceps (Westerfield, 1995). As the matrix does not become solid repositioning of the embryo in the desired orientation can be performed. The matrix does not confer constraints onto the embryo; it is thus a favorable embedding method for young, fragile embryos during gastrulation or neurulation stages with extensive axis growth. As the matrix is not solid, movements of the embryo cannot be fully prevented. Thus methylcellulose embedding is usually chosen for shorter time-lapse sessions lasting a couple of hours.

a. Protocol

1. Prepare a 3% methylcellulose solution in 30% Danieau by stirring for at least 24 h.
2. Pipet a small droplet of methylcellulose solution into imaging chamber.
3. Transfer labeled and dechorionated embryo into methylcellulose droplet and orient it using micromanipulation needles, forceps or a hairloop.
4. Overlay droplet and fill imaging chamber with 30% Danieau.

2. Agarose-Embedding

High-magnification time-lapse imaging and long-term imaging approaches lasting over several days require that the embryo is restricted from moving. Ultra-low gelling agarose provides a matrix that solidifies and thus keeps the embryo still in any desired position (Fig. 1E). It is porous enough to provide the embryo with oxygen also allowing diffusion of added chemicals such as phenylthiourea (PTU) to prevent pigmentation of the imaged embryo. As the agarose gels at 17 °C, mounting can be performed at temperatures that do not harm the embryo. To some extent axial growth is still allowed by the agarose. If extensive axial growth of the embryo is expected, the agarose matrix can be removed along the trunk and tail after it has solidified holding down the head only.

a. Protocol

1. Dissolve 1.2% of ultra low gelling agarose in 30% Danieau by boiling and subsequent cooling to 28 °C (a stock of melted agarose/30% Danieau can be kept inside the embryo incubator).
2. Pipet labeled and dechorionated embryo into the agarose solution.
3. Transfer drop of agarose containing the embryo onto the glass cover slip of the imaging chamber.
4. Place the imaging chamber onto cooled surface (e.g., cell culture dish on ice water) and orient embryo with micromanipulation needles into desired position until agarose solidifies.
5. Remove imaging chamber from cooled surface and overlay embryo with 30% Danieau containing 0.01–0.005% Tricaine (MS22, 3-aminobenzonic acid ethylester) for sedation, PTU at a 0.75 mM concentration should be added to the Danieau solution if pigmentation interferes with imaging.

3. Plasma Clot-Embedding

The embedding techniques described above do not require further manipulation of the labeled embryo. Proper orientation of the specimen prior to imaging provides convenient access to dorsal and lateral sides of the brain. The ventral side instead is hard to access and requires immense imaging depth. For such imaging approaches an embedding method has been described that allows the embryo to be freed from the yolk and interfering tissue to be dissected away from the ventral side of the brain. Mechanical forces exerted by healing processes that would extensively reshape the dissected brain are inhibited by the paralyzing agent AMP-PNP. Subsequently the dissected brain is held in position by a plasma clot (Langenberg *et al.*, 2003). Although not strictly live imaging of an intact embryo, still the entire brain or head is being cultured providing the cells with their complex natural environment and giving access to regions of the brain that cannot be obtained by other embedding methods.

a. Protocol

1. Drop ca. 20 μL of reconstituted bovine plasma onto cover slip and let dry
2. wet plasma covered area with thrombin (100 U/mL)
3. cover plasma clot with L15 amphibian culture medium containing penicillin/streptomycin/antimycotics (at final concentration of 100 U/mL penicillin)
4. inject 8 nL of 40 mM AMP-PNP into yolk cell of labeled embryo
5. remove yolk with microneedle and dissect head or brain
6. place tissue on cover slip and mount in desired position, further stabilization by addition of reconstituted plasma is optional, for detailed description of this technique please refer to the original manuscript (Langenberg *et al.*, 2003).

IV. The Microscopic System

A. Heating Chamber

To guarantee proper temporal development of the embryo during the image-recording period the environmental temperature of the microscopic stage needs to be adjusted to 29/30 °C. Microscope companies often provide incubation chambers for their microscopic setup. Self-made chambers can avoid these additional costs. Cardboard covered with insulating foil can be used to design custom-fit chambers around the microscopic stage still leaving enough space to move the stage for focusing purposes (Fig. 2A and B). Strips of Velcro are helpful to hold the individual pieces of the chamber together and to mount the chamber on the confocal microscope. For image quality the heating chamber should not cover the photomultipliers as their signal to noise ratio correlates inversely with the temperature (Distel and Köster, 2007; Kulesa and Kasemeier-Kulesa, 2007).

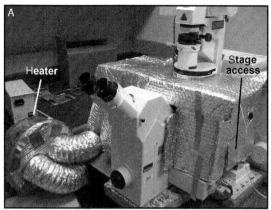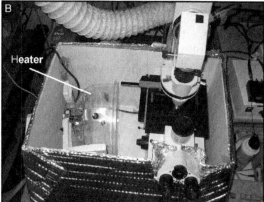

Fig. 2 Heating chambers during time-lapse recording sessions provide the desired environmental temperature for proper temporal development of the imaged specimen. Efficient heating chambers can be built from cardboard covered with insulating foil (A, B). Individual chamber parts can be held together with tape or Velcro. By sealing the microscopic stage (A) they serve both stabilizing the interior temperature and preventing environmental light from passing through the objective to the detectors. Thus, they provide the imaged object with its ideal developmental temperature, minimize evaporation and condensation, and improve the signal to noise ratio respectively. Depending on the available space the heater can feed the incubation chamber from outside by flexible tubes (A) or the heater can be included into the chamber (B) with the restriction that it should not blow the heated air directly across the microscopic stage. Note that the detectors should not be included into the chamber as their signal to noise performance decreases with increasing temperatures.

B. Heater

Depending on the available space the heater can either be positioned outside (Fig. 2A) or included into the heating chamber (Fig. 2B). When mounted outside, the heated air can be fed into the chamber by flexible plastic or aluminum tubes that can be purchased at almost any hardware store. The heated air should not be blown directly across the microscopic stage but rather heat the environmental space surrounding the stage. This prevents temperature differences building up within the imaging chamber and avoids movements of the medium inside the dish. Thus heaters with a weak fan are preferable. Also heaters with low heating power but sensitive temperature sensors should be chosen. Although requiring some preparatory time for heating the interior of the chamber to the desired temperature, overshooting and extensive fluctuation around the aimed temperature is avoided. This is important as temperature changes during the image recording period will lead to shrinkage or expansion of the mounting medium holding the specimen and thus result in a change of the focal plane. We found that small incubators used for breeding of chicken eggs can fulfill these requirements within affordable costs. In order to keep the temperature inside the heating chamber stable the incubator can be controlled by a thermostat with its sensor placed inside the heating chamber next to the microscope stage.

===== V. Data Recording

A. Preparations

Before starting to record actual images the specimen should be mounted and the imaging chamber placed on the microscope for equilibration purposes for at least an hour up to several hours ahead. Thus time-lapse experiments require proper planning to ensure that the actual developmental process that is to be observed does not commence until stable image recording is achieved. Warming of the microscopic stage, the objectives, and the optical parts inside the confocal microscope extensively influence the position of the optical plane.

Three-dimensional imaging over time is performed by recording a stack of images for a field of interest along its z-axis with repetition of this procedure after a certain time interval for the duration of the entire session. Every experimental time-lapse approach requires its specific imaging conditions that have to be established by test experiments. Imaging depth, magnification, laser power, size of pinhole, distance of individual sections along the z-axis, recording period and recording interval have to be balanced. These careful adjustments are necessary to avoid heating of the observed tissue and cells by laser light causing phototoxicity effects, to minimize bleaching of the label and to choose the proper mounting setup (e.g., methylcellulose vs agarose embedding). The following suggestions and thumb rules might help to find the proper conditions faster.

B. Mounting

Long time-lapse recording periods over more than 24 h require stable mounting of the embryo usually provided by agarose embedding. To prevent the agarose from floating off the cover slip during long imaging periods a plastic mesh can be "glued" onto the cover slip with silicon grease prior to applying the embryo within the liquid agarose. Incorporation of the mesh into the solidifying agarose will hold down the mounted specimen for several days.

C. Choice of Specimen

A brightly labeled cell will require less laser power to return a bright signal than a moderately labeled one, but in the long run the brightly labeled cell tends to be more sensitive to heating and phototoxicity. Thus the brightest embryo is not necessarily the best for time-lapse imaging.

D. Saturation

To ensure that the best resolution of the recorded images is achieved their histogram should be checked regularly. For best results the number of used gray levels should be maximized avoiding underexposure or over saturation of the sample as much as possible by adjusting the gain and black level of the detector. Usually these conditions do not vary extensively during a time-lapse recording session but they may need to be adjusted regularly as bleaching of the label (or probably developmental increase in GFP-expression levels), growth and movement of the tissue (change in imaging depth) influence these parameters.

E. Pinhole

The size of the pinhole is tightly correlated to the distance at which individual confocal z-sections need to be recorded. The best axial resolution will be achieved if the pinhole is closed down to one airy unit, which itself is dependent on the objective that is used and its numerical aperture. At this setting, photons from outside the focal plane are hindered to reach the detector, while all the photons from within the focal plane are allowed to pass. Closing the pinhole below one airy unit does not yield much further resolution along the z-axis but severely decreases the number of photons from the focal plane and thus the brightness and contrast of the image. In most cases however, closing the pinhole down to one airy unit will require too many z-sections to cover the entire image depth without running into phototoxicity problems. Thus, a compromise between laser power, number of z-sections for a given imaging depth, and gain of the detector has to be found. Keeping the cells alive is better than recording the perfect image. A proper three-dimensional reconstruction can still be achieved if an object (e.g., cell, nucleus, growth cone) appears in at least two consecutive z-sections.

F. Time Interval

Choosing the proper time interval after which image recording along the z-axis is repeated to generate time-lapse movies strongly depends on the developmental or cellular process that is being analyzed and on the experimental question that is being asked. For example, fate-mapping approaches require the continuous identification of the same individual cell over time. To unambiguously trace individual cells from images of one time point to the next they should not have moved by more than half of their cellular diameter. Thus the maximal time interval that can be chosen between two recording time points depends on the speed with which the observed cell moves. Table IV gives some approximate time intervals as orientation that can be chosen as starting conditions for different cellular processes when time-lapse imaging brain development:

G. Localizing Cells

Ideally a transmitted light picture of the imaged tissue is taken at every recording time point to locate the imaged cells within the context of the surrounding tissue. At least at the beginning, the end, and several times during the image-recording period such a transmitted light picture should be taken for orientation purposes.

Table IV
Dependence of the Approximate Recording Interval on the Observed
Cellular or Subcellular Process when Recording Images for Generat-
ing Time-Lapse Movies

Process	Time interval
Brain morphogenesis	15–20 min
Neuronal migration	10–15 min
Axonal pathfinding	1–5 min
Nuclear translocation in migrating cells	1–5 min
Calcium currents	20 s
Mitochondrial translocations	3–5 s
Actin dynamics	1–5 s

Alternatively, a fluorescent counterstain with different excitation/emission properties than the actual label (e.g., GFP-labeling and Bodipy548/578-counterstain
(Dynes and Ngai, 1998)) can be used.

H. Refocusing

Long time-lapse recording sessions require occasional refocusing. Sometimes the
imaged specimen slowly moves out of the field of vision due to continued growth
and ongoing morphogenetic processes. This requires moving the embryo back into
the field of vision and capturing of a new transmitted light image/counterstain
image prior continuing the time-lapse recording. Image realignment during data
analysis procedures can correct for the sudden "jump" of the imaged cells that
have been caused by moving the microscopic stage.

VI. Data Analysis

A single time-lapse recording session will usually yield a large amount of data in
the range of several hundred megabytes or even gigabytes. To get a first impression
of the quality of the data, the developmental processes that it covers and the
cellular behavior that it reveals an image analysis and rendering software is needed
that is not too time-consuming to obtain first movie animations.

A. NIH-Image

In our hands, NIH-image (rsb.info.nih.gov/nih-image/) has been found to be
very suitable for this purpose. Developed by Wayne Rasband (Rasband and
Bright, 1995) at the NIH, it is a public domain program and thus free of costs
(which is not to underestimate as commercial image analysis products can cost
several 10,000 USD). Originally developed for Macintosh, a Windows version was

introduced called Scion Image (www.scioncorp.com), due to the success of the original program. Meanwhile, a Java version, ImageJ (rsb.info.nih.gov/ij/), that runs on any Java-supporting platform has been released. Besides allowing basic image rendering, measuring and analysis operations the strength of NIH-Image and its updated versions is that user-developed macros can be applied. This enables almost any image rendering and analysis step to be automated speeding up the initial analysis conveniently. Furthermore, a NIH-Image user group has been set up that allows a huge community of NIH-Image users to be reached and addressed with specific image recording and analysis questions (see web site: https://list.nih.gov/). Besides writing to the community for help, all the messages distributed in the past have been archived which can be searched for possible solutions. Thus, this community represents a very valuable pool for any kind of macro that is needed to automate the initial data analysis. On this end another option for getting help with image recording difficulties using laser-scanning microscopy is provided by a user group called "Confocal Microscopy List" that also contains a searchable archive (http://listserv.buffalo.edu/cgi-bin/wa?A0 = CONFOCAL).

B. Projections

When working with data from three-dimensional time-lapse recordings a first quick analysis can be obtained by projecting the individual image stacks of each recorded time-point into a single plain (Fig. 3A). This can be done either by a mean value projection or a brightest point projection. A mean value projection will calculate an average brightness from all pixels with the same xy-coordinates within a given z-stack and project it to the single image formed from the z-stack. A brightest point projection will instead use the brightest level found along the z-axis for given xy-coordinates (Fig. 3A). Both types of projections can be useful depending on the dynamic process that is to be observed and should be tried. This flattening of the three-dimensional data into two dimensions results in a single image for each recorded time-point (Fig. 3C), and these images can be animated right away into a movie. If the embryo had to be moved during the image recording session (see above) a sudden "jump" of labeled structures will appear in the movie. This movement can be corrected by using image rendering programs such as Adobe Photoshop to apply a movement in the opposite direction to the images that were recorded after the microscopic stage had been moved. This however prerequisites that either the labeled structures only move minutely during two subsequent time points (which has to be kept in mind when choosing the recording interval) or that static points are available in the pictures. For example, auto-fluorescing nonmotile skin cells could be used as orientation for realigning pictures. Using such programs as NIH-Image and Photoshop for the described movie rendering process has the advantage that all rendering steps can be automated giving a relatively quick access to a first movie.

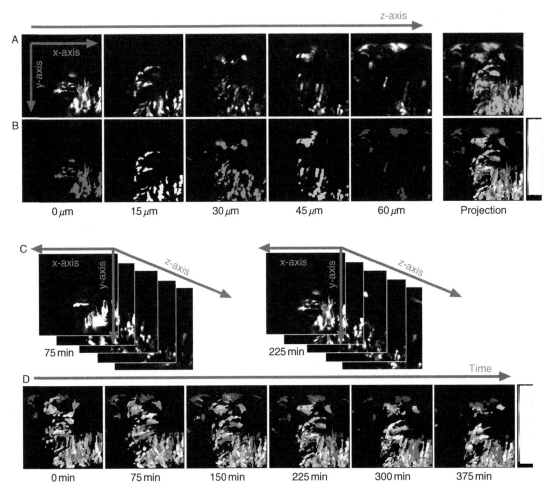

Fig. 3 Analysis of three-dimensional time-lapse recordings. (A) A fast and efficient approach to analyze z-stacks of images that have been recorded over time is to project the individual images of every z-stack/time-point into a single plane by either mean value or brightest point projection (shown here). (B) If the three-dimensional nature of the data is to be retained the recorded signal can be coded with pseudocolors reflecting the individual z-values within each z-stack. (C) Once every z-stack of each recorded time-point is projected into a single plain these projections can be animated into a movie to allow individual cells to be followed. (D) In the case of color-coded z-levels movements of cells along the z-axis can be observed by cells changing colors while migrating. (See Plate no. 8 in the Color Plate Section.)

C. LSM Software

Recently, the laser scanning microscopy manufacturers have started to provide their own image rendering software packages that are quite powerful (Leica: LCS-Leica Confocal Software, Nikon: GUI-Graphical User Interface, Olympus:

FluoView-software, Zeiss: AxioVision). An advantage of using these software tools is that the compatibility of the recorded data with the accepted format of the rendering and analysis software is not an issue. Furthermore, many useful data such as the applied imaging conditions are being stored together with the recorded images facilitating measurements and scaling. In addition these software packages offer some advanced data analysis features that go beyond the capabilities of NIH-Image and can only be found otherwise in software packages that are specialized in image rendering such as Amira (Indeed-Visual Concepts GmbH), Metamorph (Universal Imaging Corporation), Imaris (Bitplane AG), Slidebook (Intelligent Imaging Innovations), Volocity (Improvision), or VoxelView (Vital Images, Inc.). For these specialized software, a comparison of their different properties, advantages, and disadvantages has been reported (Megason and Fraser, 2003). For some of these software packages demo versions can be downloaded and tested free of charge

Amira: http://www.amiravis.com/trial.html

Imaris: http://www.bitplane.com/go/download

Volocity: http://www.improvision.com/downloads

D. Three-Dimensional Renderings Over Time

In order to keep the three-dimensional nature of the recorded image stacks the obtained data can be visualized when using these advanced softwares by creating stereo-optic pairs of projections from each recorded z-stack of pictures for every individual time point. An alternative is represented by red-green rendering of the projections creating a three-dimensional impression to the viewer when wearing red-green glasses. As helpful as these movies are they have the disadvantage of being difficult to publish and being inaccessible to people with red-green blindness that affects about 10% of all humans. A further option that allows presenting the three-dimensional data in two dimensions can be obtained by color-coding the different z-values (Fig. 3B and D). Thus, structures such as cells, axons, mitochondria that move along the z-axis will change their color over time. This allows one, for example, to address whether cells cross each other's pathways during migration, migrate toward each other, or disperse. To obtain information about absolute changes of imaged structures along the z-axis, like cells moving deeper into the brain or toward its surface, one has to make sure that the embedded embryo itself is not moving along the z-axis. Fix-points such as nonmotile skin cells have to stay in the same z-level (thus they do not change their color) within an analyzed stack over time.

E. Cell-Tracking

A challenge for most software is the automated tracking of cells or other labeled structures in three-dimensional time-lapse recordings within a living embryo. For tracking in two dimensions over time using projections of 3D-image stacks several software packages exist with the DIAS software (http://www.solltechnologies. com) being capable of analyzing movies in Quick Time format. From such tracking data a wealth of information can be obtained such as the directionality of movements or changes in migratory speeds (Glickman *et al.*, 2003; Kulesa and Fraser, 1998, 2000). However, most software appear to fail in reliably tracking individual cells over time automatically under these *in vivo* conditions and the tracking has to be done manually image by image which can be quite laborious. Improved labeling techniques and multicolor labeling of different subcellular structures might help to solve this problem by providing the tracking software with more individual information for every single cell.

VII. Pitfalls to Avoid

Time-lapse imaging though not difficult *per se* requires some practice to identify general pitfalls and pitfalls specific for each approach depending on the labeling procedure and quality, the age of the embryo, the tissue and the process that is to be observed. Thus an exhaustive list of potential pitfalls cannot be provided and will not be attempted here. But some of the following general thumb rules and suggestions might be helpful.

A. Technical Pitfalls

1. DNA-Purification

Successful mosaic labeling by the injection of plasmid-DNA is very dependent on the purity of the DNA and the embryonic stage at which the DNA is being injected. In our hands further cleaning of maxi prep DNA with the Geneclean Turbo Kit from Qbiogene prior to injection worked best. In addition, the plasmid-DNA should be injected as early as possible at the one-cell stage just when the blastodisc of the fertilized egg becomes visible.

2. Fixation of Mounted Embryo

Time-lapse recordings that last over several days can be demanding on the individual embedding techniques. Agarose mounted on a cover slip can come loose during these long recordings with the embedded embryo floating out of focus. This can be prevented by incorporating a small size plastic mesh into the agarose by gluing the mesh with silicon grease onto the cover slip of the imaging chamber prior to applying the embryo within the liquid agarose drop. The mesh will hold down the agarose and prevent it from floating away.

3. Colocalization of Fluorophores

For colocalization of two different fluorophores the excitation and detection settings have to separate both emissions properly. Ideally, controls within the same scanning plane should be attempted to ensure that overlapping signal is obtained from the presence of both fluorescent probes instead of one fluorophore being detected by both detection channels. Also, it has to be kept in mind that the pinhole has to be adjusted differently for both dyes in order to record data from an optical section of identical thickness as the thickness of an optical slice for a given pinhole depends on the wavelength of the recorded light.

4. Storage During Data Recording

Data obtained from time-lapse recordings is often stored in the temporary memory of the computer. Once the temporary memory is filled, the confocal microscope will stop recording and tends to crash. Thus, if long time-lapse recording sessions are attempted one should make sure that the recorded images are being saved directly to the hard drive. This option is usually available in the settings menu of the confocal microscope software. Also, ensure yourself that the hard drive contains enough empty space for saving the expected amount of data.

5. Image Resolution

This point leads immediately to the resolution at which images are being recorded. Modern confocal microscopes allow data to be recorded not only at a 512×512 pixel resolution but also at 1024×1024 and 2048×2048 pixels. This means that an individual grayscale image increases in size from 256 KB to 1 MB to 4 MB. A 24 h 3D time-lapse recording at 2048×2048 pixel resolution will lead to large data files for which enough space on the hard drive has to be available. In addition, later handling of these files might be difficult and imaging software is often unable to perform 4D-time-lapse rendering processes with these large amounts of data. Thus, it is important to test the resolution that is required to sufficiently answer the experimental question with short time-lapse recordings to make sure that one does not run into later analysis problems with the obtained data.

B. Analytical Pitfalls

1. Annotation of Recorded Data

Three-dimensional time-lapse data that has been projected into a single plane needs to be reanalyzed in three dimensions if answers about cell-cell contacts are to be obtained. Interacting cells need to be localized within the same plane instead of moving above and underneath each other. Often it is helpful to generate rocking or rotating projections of z-stacked pictured to look at the three-dimensional volume from different angles.

2. Reference Points

Depending on the angle at which a specimen is being optically sectioned different results can be obtained. Two different cell types can appear within the same plane, while in the next specimen they do not share the same z-value. Similarly distances between structures can vary from specimen to specimen as the position at which the embryo is embedded and thus the angle of sectioning changes every time. To ensure that meaningful data is obtained independent reference points to correlate the data with are needed.

3. Identifying Individual Cells

In tracing experiments it is required that individual cells can be doubtlessly followed from one time point to the next. As a thumb rule cells should not move further than half of their cell diameter during the time interval of two subsequent recordings to allow an unambiguous correlation of cells.

VIII. Summary

Zebrafish embryos represent an ideal vertebrate model organism for noninvasive intravital imaging due to their optical clarity, external embryogenesis, and fast development. Many different labeling techniques have been adopted from other model organisms or newly developed to address a wealth of different developmental questions directly inside the living organism. The parallel advancements in the field of optical imaging let us now observe dynamic processes at the cellular and subcellular resolution. Zebrafish as model organism can therefore serve to merge the large research areas of developmental genetics and cell biology in the context of a living organism. Combined with the repertoire of available surgical and genetic manipulations zebrafish embryos provide the powerful and almost unique possibility to observe the interplay of molecular signals with cellular, morphological, and behavioral changes directly within a living and developing vertebrate organism. A bright future for zebrafish is yet to come, let there be light.

Acknowledgments

We are grateful to Dr. Laure Bally-Cuif for the schematic drawing in Fig. 1. R. W. K. is supported by a BioFuture-Award (0311889) of the German Ministry of Education and Research (BMBF) and a grant of the Deutsche Forschungsgemeinschaft (DFG, KO 1949/3-1). S. E. F. is supported by the Biological Imaging Center of the Beckman Institute, Caltech and the NIH.

References

Ando, H., Furuta, T., Tsien, R. Y., and Okamoto, H. (2001). Photo-mediated gene activation using caged RNA/DNA in zebrafish embryos. *Nat. Genet.* **28**, 317–325.

Ando, R., Hama, H., Yamamoto-Hino, M., Mizuno, H., and Miyawaki, A. (2002). An optical marker based on the UV-induced green-to-red photoconversion of a fluorescent protein. *Proc. Natl. Acad. Sci. USA* **99**, 12651–12656.

Asakawa, K., Suster, M. L., Mizusawa, K., Nagayoshi, S., Kotani, T., Urasaki, A., Kishimoto, Y., Hibi, M., and Kawakami, K. (2008). Genetic dissection of neural circuits by Tol2 transposon-mediated Gal4 gene and enhancer trapping in zebrafish. *Proc. Natl. Acad. Sci. USA* **105**, 1255–1260.

Bak, M., and Fraser, S. E. (2003). Axon fasciculation and differences in midline kinetics between pioneer and follower axons within commissural fascicles. *Development* **130**, 4999–5008.

Balciunas, D., and Ekker, S. C. (2005). Traping fish genes with transposons. *Zebrafish* **1**, 335–341.

Blader, P., Plessy, C., and Strähle, U. (2003). Multiple regulatory elements with spatially and temporally distinct activities control neurogenin1 expression in primary neurons of the zebrafish embryo. *Mech. Dev.* **120**, 211–218.

Bruchez, M. J., Moronne, M., Gin, P., Weiss, S., and Alivisatos, A. P. (1998). Semiconductor nano-crystals as fluorescent biological labels. *Science* **281**, 2013–2016.

Campbell, R. E., Tour, O., Palmer, A. E., Steinbach, P. A., Baird, G. S., Zacharias, D. A., and Tsien, R. Y. (2002). A monomeric red fluorescent protein. *Proc. Natl. Acad. Sci. USA* **99**, 7877–7882.

Chalfie, M., Tu, Y., Euskirchen, G., Ward, W. W., and Prasher, D. C. (1994). Green fluorescent protein as a marker for gene expression. *Science* **263**, 802–805.

Chan, W. C., and Nie, S. (1998). Quantum dot bioconjugates for ultrasensitive nonisotopic detection. *Science* **281**, 2016–2018.

Chen, Y., and Schier, A. F. (2001). The zebrafish nodal signal squint functions as a morphogen. *Nature* **411**, 607–610.

Choo, B. G., Kondrichin, I., Parinov, S., Emelyanov, A., Go, W., Toh, W. C., and Korzh, V. (2006). Zebrafish transgenic enhancer trap line database (ZETRAP). *BMC Dev. Biol.* **6**, 5.

Chudakov, D. M., Lukyanov, S., and Lukyanov, K. A. (2005). Fluorescent proteins as a toolkit for *in vivo* imaging. *Trends Biotechnol.* **23**, 605–613.

Chudakov, D. M., Lukyanov, S., and Lukyanov, K. A. (2007). Tracking intracellular protein movements using photoswitchable fluorescent proteins PS-CFP2 and Dendra2. *Nat. Protoc.* **2**, 2024–2032.

Clontech, A. (1998). Living colors subcellular localization vectors. *CLONTECHniques.* **XIII**, 8–9.

Concha, M. L., Russel, C., Regan, J. C., Tawk, M., Sidi, S., Gilmour, D. T., Kapsimali, M., Sumoy, L., Goldstone, K., Amaya, E., Kimelman, D., Nicolson, T., *et al.* (2003). Local tissue interactions across the dorsal midline of the forebrain establish CNS laterality. *Neuron* **39**, 423–438.

Cooper, M. S., D'Amico, L. A., and Henry, C. A. (1999). Confocal microscopy analysis of morphogenetic movements. *Methods Cell Biol.* **59**, 179–204.

Cormack, B. P., Valdivia, R. H., and Falkow, S. (1996). FACS-optimized mutants of the green fluorescent protein (GFP). *Gene* **173**, 33–38.

Davison, J. M., Akitake, C. M., Goll, M. G., Rhee, J. M., Gosse, N., Baier, H., Halpern, M. E., Leach, S. D., and Parsons, M. J. (2007). Transactivation from Gal4-VP16 transgenic insertions for tissue-specific cell labeling and ablation in zebrafish. *Dev. Biol.* **304**, 811–824.

Distel, M., and Köster, R. W. (2007). *In vivo* time-lapse imaging of zebrafish embryonic development. *CSH Protoc.* **2**, 4816.

Doyon, Y., McCammon, J. M., Miller, J. C., Faraji, F., Ngo, C., Katibah, G. E., Amora, R., Hocking, T. D., Zhang, L., Rebar, E. J., Gregory, P. D., Urnov, F. D., *et al.* (2008). Heritable targeted gene disruption in zebrafish using designed zinc-finger nucleases. *Nat. Biotechnol.* **26**, 702–708.

Driever, W., Solnica-Krezel, L., Schier, A. F., Neuhauss, S. C. F., Malicki, J., Stemple, D. L., Stainier, D. Y. R., Zwartkruis, F., Abdelilah, S., Rangini, Z., Belak, J., and Boggs, C. (1996). A genetic screen for mutations affecting embryogenesis in zebrafish. *Development* **123**, 37–46.

Dubertret, B., Skourides, P., Norris, D. J., Noireaux, V., Brivanlou, A. H., and Libchaber, A. (2002). *In vivo* imaging of quantum dots encapsulated in phospholipid micelles. *Science* **298,** 1759–1762.

Dynes, J. L., and Ngai, J. (1998). Pathfinding of olfactory neuron axons to stereotyped glomerular targets revealed by dynamic imaging in living zebrafish embryos. *Neuron* **20,** 1081–1091.

Ellingsen, S., Laplante, M. A., König, M., Kikuta, H., Furmanek, T., Hoivik, E. A., and Becker, T. S. (2005). Large-scale enhancer detection in the zebrafish genome. *Development* **132,** 3799–3811.

Emelyanov, A., Gao, Y., Naqvi, N. I., and Parinov, S. (2006). Trans-kingdom transposition of the maize dissociation element. *Genetics* **174,** 1095–1104.

Fahrenkrug, S. C., Clark, K. J., Dahlquist, M. O., and Hackett, P. B. (1999). Dicistronic gene expressoin in developing zebrafish. *Mar. Biotechnol.* **1,** 552–561.

Finley, K. R., Davidson, A. E., and Ekker, S. C. (2001). Three-color imaging using fluorescent proteins in living zebrafish embryos. *Biotechniques* **31,** 66–73.

Fraser, S. E. (1996). Iontophoretic dye labeling of embryonic cells. *Methods Cell Biol.* **51,** 147–160.

Giepmans, B. N., Adams, S. R., Ellisman, M. H., and Tsien, R. Y. (2006). The fluorescent toolbox for assessing protein location and function. *Science* **312,** 217–224.

Glickman, N. S., Kimmel, C. B., Jones, M. A., and Adams, R. J. (2003). Shaping the zebrafish notochord. *Development* **130,** 873–887.

Haffter, P., Granato, M., Brand, M., Mullins, M. C., Hammerschmidt, M., Kane, D. A., Odenthal, J., van Eeden, F. J. M., Jiang, Y. J., Heisenberg, C. P., Kelsh, R. N., Furutani-Seiki, M., *et al.* (1996). The identification of genes with unique and essential functions in the development of the zebrafish, Danio rerio. *Development* **123,** 1–36.

Halloran, M. C., Sato-Maeda, M., Warren, J. T., Su, F. Y., Lele, Z., Krone, P. H., Kuwada, J. Y., and Shoji, W. (2000). Laser-induced gene expression in specific cells of transgenic zebrafish. *Development* **127,** 1953–1960.

Hammerschmidt, M., Blader, P., and Strahle, U. (1999). Strategies to perturb zebrafish development. *Methods Cell Biol.* **59,** 87–115.

Hardy, M. E., Ross, L. V., and Chien, C. B. (2007). Focal gene misexpression in zebrafish embryos induced by local heat shock using a modified soldering iron. *Dev. Dyn.* **236,** 3071–3076.

Heim, R., and Tsien, R. Y. (1996). Engineering green fluorescent protein for improved brightness, longer wavelenghts and fluorescence resonance energy transfer. *Curr. Biol.* **6,** 178–182.

Hieber, V., Dai, X. H., Foreman, M., and Goldman, D. (1998). Induction of alpha 1-tubulin gene expression during development and regeneration of the fish central nervous system. *J. Neurobiol.* **37,** 429–440.

Higashijima, S., Hotta, Y., and Okamoto, H. (2000). Visualization of cranial motor neurons in live transgenic zebrafish expressing green fluorescent protein under the control of the islet-1 promoter/enhancer. *J. Neurosci.* **20,** 206–218.

Hsiao, C. D., Hsieh, F. J., and Tsai, H. J. (2001). Enhanced expression and stable transmission of transgenes flanked by inverted terminal repeats from adeno-associated virus in zebrafish. *Dev. Dyn.* **220,** 323–336.

Ivanchenko, S., Glaschik, S., Röcker, C., Oswald, F., Wiedenmann, J., and Nienhaus, G. U. (2007). Two-photon axcitation and photoconversion of EosFP in dual-color-4Pi confocal microscopy. *Biophys. J.* **92,** 4451–4457.

Jaiswal, J. K., Mattoussi, H., Mauro, J. M., and Simon, S. M. (2003). Long-term multiple color imaging of live cells using quantum dot bioconjugate. *Nat. Biotechnol.* **21,** 47–51.

Kanda, T., Sullivan, K. F., and Wahl, G. M. (1998). Histone-GFP fusion enables sensitive analysis of chromosome dynamics in living mammalian cells. *Curr. Biol.* **8,** 377–385.

Kawakami, K. (2007). Tol2: A versatile gene transfer vector in vertebrates. *Genome Biol.* **8**(Suppl. 1), S7.

Kogure, T., Karasawa, S., Araki, T., Saito, K., Kinjo, M., and Miyawaki, A. (2006). A fluorescent variant of a protein from the stony coral Montipora facilitates dual-color-single-laser fluorescence cross-correlation spectroscopy. *Nat. Biotechnol.* **24,** 577–581.

Korzh, V. (2007). Transposons as tools for enhancer trap screens in vertebrates. *Genome Biol.* **8**(Suppl. 1), S8.

Köster, R., Götz, R., Altschmied, J., Sendtner, R., and Schartl, M. (1996). Comparison of monocistronic and bicistronic constructs for neurotrophin transgene and reporter gene expression in fish cells. *Mol. Mar. Biol. Biotechnol.* **5**, 1–8.

Köster, R. W., and Fraser, S. E. (2001a). Direct imaging of *in vivo* neuronal migration in the developing cerebellum. *Curr. Biol.* **11**, 1858–1863.

Köster, R. W., and Fraser, S. E. (2001b). Tracing transgene expression in living zebrafish embryos. *Dev. Biol.* **233**, 329–346.

Kulesa, P. M., and Fraser, S. E. (1998). Neural crest cell dynamics revealed by time-lapse video microscopy of whole embryo chick explant cultures. *Dev. Biol.* **204**, 327–344.

Kulesa, P. M., and Fraser, S. E. (2000). In ovo-time-lapse analysis of chick hindbrain neural crest cell migration shows cell interactions during migration to the branchial arches. *Development* **127**, 1161–1172.

Kulesa, P. M., and Kasemeier-Kulesa, J. C. (2007). Construction of a heated incubation chamber around a microscope stage for time-lapse imaging. *CSH Protoc.* **2**. 4792.

Kwan, K. M., Fujimoto, E., Grabher, C., Mangum, B. D., Hardy, M. E., Campbell, D. S., Parant, J. M., Yost, H. J., Kanki, J. P., and Chien, C. B. (2007). The Tol2kit: A multisite gateway-based construction kit for Tol2 transposon transgenesis constructs. *Dev. Dyn.* **236**, 3088–3099.

Langenberg, T., Brand, M., and Cooper, M. S. (2003). Imaging brain development and organogenesis in zebrafish using immobilized embryonic explants. *Dev. Dyn.* **228**, 464–474.

Lidke, D. S., Nagy, P., Heintzmann, R., Arndt-Jovin, D. J., Post, J. N., Grecco, H. E., Jares-Erijman, J., and Jovin, T. M. (2004). Quantum dot ligands provide new insights into erbB/HER receptro-mediated signal transduction. *Nat. Biotechnol.* **22**, 198–203.

Linney, E., Hardison, N. L., Lonze, B. E., Lyons, S., and DiNapoli, L. (1999). Transgene expression in zebrafish: A comparison of retroviral-vector and DNA-injection approaches. *Dev. Biol.* **213**, 207–216.

Mattoussi, H., Mauro, J. M., Goldman, E. R., Green, T. M., Anderson, G. P., Sundar, V. C., and Bawendi, M. G. (2001). Bioconjugation of highly luminescent colloidal CdSe-ZnS quantum dots with an engineered two-domain recombinant protein. *Physica Status Solidi* **224**, 277–283.

Matz, M. V., Fradkov, A. F., Labas, Y. A., Savitsky, A. P., Zaraisky, A. G., Markelov, M. L., and Lukyanov, S. A. (1999). Fluorescent proteins from nonbioluminescent Anthozoa species. *Nat. Biotechnol.* **17**, 969–973.

Megason, S. G., and Fraser, S. E. (2003). Digitizing life at the level of the cell: High-performance laser-scanning microscopy and image analysis for in toto imaging of development. *Mech. Dev.* **120**, 1407–1420.

Meng, A., Tang, H., Ong, B. A., Farrell, M. J., and Lin, S. (1997). Promoter analysis in living zebrafish embryos identifies a cis-acting motif required for neuronal expression of GATA-2. *Proc. Natl. Acad. Sci. USA* **94**, 6267–6272.

Meng, X., Noyes, M. B., Zhu, L. J., Lawson, N. D., and Wolfe, S. A. (2008). Targeted gene inactivation in zebrafish using engineered zinc-finger nucleases. *Nat. Biotechnol.* **26**, 695–701.

Miyawaki, A. (2005). Innovations in the imaging of brain functions using fluorescent proteins. *Neuron* **48**, 189–199.

Moriyoshi, K., Richards, L. J., Akazawa, C., O'Leary, D. D. M., and Nakanishi, S. (1996). Labeling neural cells using adenoviral gene transfer of membrane-targeted GFP. *Neuron* **16**, 255–260.

Nagai, T., Sawano, A., Park, E. S., and Miyawaki, A. (2002). A variant of yellow fluorescent protein with fast and efficient maturation for cell-biological applications. *Nat. Biotechnol.* **20**, 87–90.

Nagai, T., Iabata, K., Park, E. S., Kubota, M., Mikoshiba, K., and Miyawaki, A. (2001). Circularly permuted green fluorescent proteins engineered to sense Ca^{2+}. *Proc. Natl. Acad. Sci. USA* **98**, 3197–3202.

Nagayoshi, S., Hayashi, E., Abe, G., Osato, N., Asakawa, K., Urasaki, A., Horikawa, K., Ikeo, K., Takeda, H., and Kawakami, K. (2008). Insertional mutagenesis by the Tol2 transposon-mediated enhancer trap approach generated mutations in tow developmental genes: tcf7 and synembryn-like. *Development* **135**, 159–169.

Nasevicius, A., and Ekker, S. C. (2000). Effective targeted gene "knockdown" in zebrafish. *Nat. Genet.* **26,** 216–220.

Neumann, C. J., and Nüsslein-Volhard, C. (2000). Patterning of the zebrafish retina by a wave of sonic hedgehog activity. *Science* **289,** 2137–2139.

Niell, C. M., Meyer, M. P., and Smith, S. J. (2004). *In vivo* imaging of synapse formation on a growing dendritic arbor. *Nat. Neurosci.* **7,** 254–260.

Park, H. C., Kim, C. H., Bae, Y. K., Yeo, S. Y., Kim, S. H., Hong, K. S., Shin, J., Yoo, K. W., Hibi, M., Hirano, T., Miki, N., Chitnis, A. B., *et al.* (2000). Analysis of upstream elements in the HuC promoter leads to the establishment of transgenic zebrafish with fluorescent neurons. *Dev. Biol.* **227,** 279–293.

Patterson, G. H., and Lippincott-Schwartz, J. (2002). A photoactivatable GFP for selective photolabeling of proteins and cells. *Science* **297,** 1873–1877.

Peri, F., and Nüsslein-Volhard, C. (2008). Live imaging of neuronal degradation by microglia reveals a role for v0-ATPase a1 in phagosomal fusion *in vivo*. *Cell* **135,** 916–927.

Picker, A., Scholpp, S., Böhli, H., Takeda, H., and Brand, M. (2002). A novel positive transcriptional feedback loop in midbrain-hindbrain boundary development is revealed through analysis of the zebrafish pax2.1 promoter in transgenic lines. *Development* **129,** 3227–3239.

Potter, S. M. (2000). Two-photon microscopy for 4D imaging of living neurons. *In* "Imaging Neurons: A Laboratory Manual" (L. A. K. Yuste, ed.), pp. 20.1–20.16. Cold Spring Harbor Laboratory Press, Cold Spring Harbor.

Provost, E., Rhee, J., and Leach, S. D. (2007). Viral 2A peptides allow expression of multiple proteins from a single ORF in transgenic zebrafish embryos. *Genesis* **45,** 625–629.

Rasband, W. S., and Bright, D. S. (1995). NIH Image: A public domain image processing program for the Macintosh. *Microbeam Anal. Soc. J.* **4,** 137–149.

Rastegar, S., Albert, S., Le Roux, I., Fischer, N., Blader, P., Müller, F., and Strähle, U. (2002). A floor plate enhancer of the zebrafish netrin1 gene requires Cyclops (Nodal) signalling and the winged helix transcription factor FoxA2. *Dev. Biol.* **252,** 1–14.

Razansky, D., Vinegoni, C., and Ntziachristos, V. (2007). Multispectral photoacoustic imaging of fluorochromes in small animals. *Opt. Lett.* **32,** 2891–2893.

Rieger, S., Volkmann, K., and Köster, R. W. (2005). Quantum dots are powerful multipurpose vital labeling agents in zebrafish embryos. *Dev. Dyn.* **234,** 670–681.

Rizzuto, R., Brini, M., Pizzo, P., Murgia, M., and Pozzan, T. (1995). Chimeric green fluorescent protein as a tool for visualizing subcellular organelles in living cells. *Curr. Biol.* **5,** 635–642.

Sakaue-Sawano, A., Kurokawa, H., Morimura, T., Hanyu, A., Hama, H., Osawa, H., Kashiwagi, S., Fukami, K., Miyata, T., Miyoshi, H., Imamura, T., Ogawa, M., *et al.* (2008). Visualizing spatiotemporal dynamics of multicellular cell-cycle progression. *Cell* **132,** 487–498.

Shaner, N. C., Campbell, R. E., Steinbach, P. A., Giepmans, B. N., Palmer, A. E., and Tsien, R. Y. (2004). Improved monomeric red, orange and yellow fluorescent proteins derived from *Discosoma* sp. Red fluorescent protein. *Nat. Biotechnol.* **22,** 1567–1572.

Sott, E. K., Mason, L., Arrenberg, A. B., Ziv, L., Gosse, N. J., Xiao, T., Chi, N. C., Asakawa, K., Kawakami, K., and Baier, H. (2007). Targeting neural circuitry in zebrafish using Gal4 enhancer trapping. *Nat. Methods* **4,** 323–326.

Tallafuss, A., and Bally-Cuif, L. (2003). Tracing of her5 progeny in zebrafish transgenics reveals the dynamics of midbrain-hindbrain neurogenesis and maintenance. *Development* **130,** 4307–4323.

Teh, C., Chong, S. W., and Korzh, V. (2003). DNA delivery into anterior neural tube of zebrafish embryos by electroporation. *Biotechniques* **35,** 950–954.

Teruel, M. N., Blanpied, T. A., Shen, K., Augustine, G. J., and Meyer, T. (1999). A versatile microporation technique for the transfection of cultured CNS neurons. *J. Neurosci. Methods* **93,** 37–48.

Thermes, V., Grabher, C., Ristoratore, F., Bourrat, F., Choulika, A., Wittbrodt, J., and Joly, J. S. (2002). I-*Sce*I meganuclease mediates highly efficient transgenesis in fish. *Mech. Dev.* **118,** 91–98.

Ting, A. Y., Kain, K. H., Klemke, R. L., and Tsien, R. Y. (2001). Genetically encoded fluorescent reporters of protein tyrosine kinase activities in living cells. *Proc. Natl. Acad. Sci. USA* **98,** 15003–15008.

Tsutsui, H., Karasawa, S., Shimizu, H., Nukina, N., and Miyawaki, A. (2005). Semi-rational engineering of a coral fluorescent protein into an efficient higlighter. *EMBO Rep.* **6,** 233–238.

Udvadia, A. J., Koster, R. W., and Skene, J. H. P. (2001). GAP-43 promoter elements in transgenic zebrafish reveal a difference in signals for axon growth during CNS development and regeneration. *Development* **128,** 1175–1182.

Udvadia, A. J., and Linney, E. (2003). Windows into development: Historic, current, and future perspectives on transgenic zebrafish. *Dev. Biol.* **256,** 1–17.

Verkhusha, V. V., Otsuna, H., Awasaki, T., Oda, H., Tsukita, S., and Ito, K. (2001). An enhanced mutant of red fluorescent protein DsRed for double labeling and developmental timer of neural fiber bundle formation. *J. Biol. Chem.* **276,** 29621–29624.

Westerfield, M. (1995). "The Zebrafish Book." University of Orgeon Press, Eugene, OR.

Westphal, M., Jungbluth, A., Heidecker, M., Muhlbauer, B., Heizer, C., Schwartz, J. M., Marriott, G., and Gerisch, G. (1997). Microfilament dynamics during cell movement and chemotaxis monitored using a GFP-actin fusion protein. *Curr. Biol.* **7,** 176–183.

Wiedenmann, J., Ivanchenko, S., Oswald, F., Schmitt, F., Röcker, C., Salih, A., Spindler, K. D., and Nienhaus, G. U. (2004). EosFP, a fluorescnet marker protein with UV-inducible green-to-red fluorescence conversion. *Proc. Natl. Acad. Sci. USA* **101,** 15905–15910.

Woo, K., and Fraser, S. E. (1995). Order and coherence in the fate map of zebrafish nervous system. *Development* **121,** 2595–2609.

Yoshida, T., Ito, A., Matsuda, N., and Mishina, M. (2002). Regulation by protein kinase A switching of axonal pathfinding of zebrafish olfactory sensory neurons through the olfactory placoe-olfactory bulb boundary. *J. Neurosci.* **22,** 4964–4972.

Further Reading

Pawley, J. B. (1995). "Handbook of Biological Confocal Microscopy," 2nd edn., Plenum Press, New York.

Mason, W. T. (1999). "Fluorescent and Luminescent Probes for Biological Activity," 2nd edn., Academic Press, London.

Inoué, S., and Spring, K. R. (1997). "Video Microscopy," 2nd edn., Plenum Press, New York.

Neher, R., and Neher, E. (2003). Optimizing imaging parameters for the separation of multiple labels in a fluorescence image. *J. Microsc.* **213,** 46–62.

CHAPTER 15

Development of the Peripheral Sympathetic Nervous System in Zebrafish

Rodney A. Stewart,* A. Thomas Look,* John P. Kanki,* and Paul D. Henion†

*Department of Pediatric Oncology
Dana-Farber Cancer Institute
Boston, Massachusetts 02115

†Center for Molecular Neurobiology and Department of Neuroscience
Ohio State University
Columbus, Ohio 43210

I. Introduction

The combined experimental attributes of the zebrafish model system accommo-date cellular, molecular, and genetic approaches, making it particularly well suited for studying mechanisms underlying normal developmental processes, as well as disease states, such as cancer. The ability to analyze developing tissues in the optically clear embryo, combined with the unbiased nature of forward genetic screens, allows one to identify previously elusive genes and to analyze their *in vivo* function. In this chapter, we describe the advantages of using the zebrafish for

ESSENTIAL ZEBRAFISH METHODS: CELL AND DEVELOPMENTAL BIOLOGY
Reprinted from *Methods in Cell Biology*, Volume 76 (Elsevier Inc., 2004).
DOI: 10.1016/B978-0-12-374599-6.00015-7

identifying genes and their functional regulation of the developing peripheral sympathetic nervous system (PSNS). We provide a brief overview of the genetic pathways regulating vertebrate PSNS development, the rationale for developing a zebrafish model, and our current understanding of zebrafish PSNS development. We also include examples that illustrate the potential of mutant analysis in zebrafish PSNS research. Finally, we explore the potential of the zebrafish system for discovering genes that are disrupted in neuroblastoma, a highly malignant cancer of the PSNS.

II. The Peripheral Autonomic Nervous System

A. Overview

The internal organs, smooth muscles, skin, and exocrine glands of the vertebrate body are innervated by the peripheral autonomic nervous system (ANS), which comprises the PSNS and the parasympathetic (PAS) and enteric nervous systems (ENS). These three components are structurally and functionally distinguished by differences in the characteristic locations of their cell bodies, the targets they innervate, the neurotransmitters they utilize, and the molecular pathways controlling their development (Brading, 1999). In the sympathetic nervous system, the cell bodies of the preganglionic neurons are generally found in the thoracic and lumbar areas of the spinal cord. Their axons exit ventrally and innervate peripheral ganglia lying near the spinal cord where they synapse with postganglionic neurons, which in turn, innervate target organs. Sympathetic neurons are predominantly adrenergic, producing the neurotransmitter noradrenalin along with one or more neuropeptides. In contrast, the cell bodies of the central component of the PAS lie in the brain and sacral regions of the spinal cord, and synapse with peripheral cell bodies located within the immediate vicinity or in the target organs. PAS neurons are cholinergic, producing acetylcholine along with other neuromodulators (Thexton, 2001). These structural and functional differences allow the sympathetic and parasympathetic systems to function largely in opposition to each other to maintain homeostasis, such as vascular tone, and to generate the fight-or-flight response. Although the ANS normally consists of central preganglionic and peripheral ganglionic neurons to regulate the function of a target organ, an exception can be found in the sympathetic nervous system where chromaffin cells in the adrenal medulla are directly innervated by central preganglionic neurons, but do not innervate a target organ. Chromaffin cells are neural crest-derived endocrine cells closely related lineally and neurochemically to postganglionic neurons of the PSNS. However, rather than functioning in neural transmission, chromaffin cells exhibit an endocrine function secreting hormones, such as adrenalin (Gabella, 2001).

The function of the enteric nervous system is relatively independent of the CNS and other components of the ANS. In most vertebrates, the enteric neurons form two layers of ganglionic plexuses located along the entire length of the gastrointestinal tract, consisting of a microscopic meshwork of ganglia connected to each other by short nerve trunks. The inner myenteric plexus, situated between the longitudinal and circular muscle layers, is mainly responsible for muscle contraction. The submucosal plexus controls motility, secretion, and microcirculation processes in the gut. The functions of the enteric nervous system are complex, and 17 different neuronal types have been identified that produce a variety of different neurotransmitters including, nitric oxide, adenosine triphosphate (ATP), and 5-hyroxytryptamine. Enteric ganglia function locally to integrate sensory and reflex activities for the coordination of the sequential relaxation and contractions occurring during peristalsis (Hansen, 2003).

B. Molecular Pathways Underlying PSNS Development

The early development of the PSNS can be divided into four overlapping stages, based on both morphologic and molecular criteria:

1. Formation and fate specification of neural crest cells that will develop into sympathoadrenal (SA) progenitors.
2. Bilateral migration of SA cells and their coalescence in regions adjacent to the dorsal aorta.
3. Neuronal and noradrenergic differentiation of SA progenitors.
4. Maintenance of PSNS neurons in fully developed ganglia and the establishment of their efferent synaptic connections.

Considerable progress has been made elucidating the cellular and molecular mechanisms underlying PSNS development, which represents one of the best described genetic pathways establishing vertebrate neuronal and neurotransmitter identity (Anderson, 1993; Francis and Landis, 1999; Goridis and Rohrer, 2002; Fig. 1). Briefly, neural crest progenitors form at the border between the neural and nonneural ectoderm through a process regulated by bone morphogenetic proteins (BMPs) and Wnt signaling (Knecht and Bronner-Fraser, 2002). These neural crest progenitor cells express genes such as *slug, snail, tfap2α,* and *foxd3* that appear to play roles in their induction or early development (reviewed in Knecht and Bronner-Fraser (2002)). Subsequent to the morphogenetic movements that result in neural tube closure, neural crest progenitors are localized to the most dorsal aspect of the neural tube (Aybar *et al.,* 2003; Bronner-Fraser, 2002; Kos *et al.,* 2001). The premigratory neural crest progenitors then undergo an epithelial-mesenchymal transition and begin to migrate away from the neural tube. Neural crest cells first migrate out ventromedially and later, others follow a dorsolateral pathway (Goridis and Rohrer, 2002; LeDouarin and Kalcheim, 1999; Fig. 1A). During migration, precursors of the SA lineage are exposed to signaling factors from the neural tube and notochord, such as sonic hedgehog (Shh) and Neuregulin-1, an epidermal growth factor (EGF)-like growth factor (Crone and Lee, 2002; Krauss *et al.,* 1993; Patten and Placzek, 2000; Williams *et al.,* 2000b). Neuregulin-1 expression is associated with the origin, migration, and target site of SA

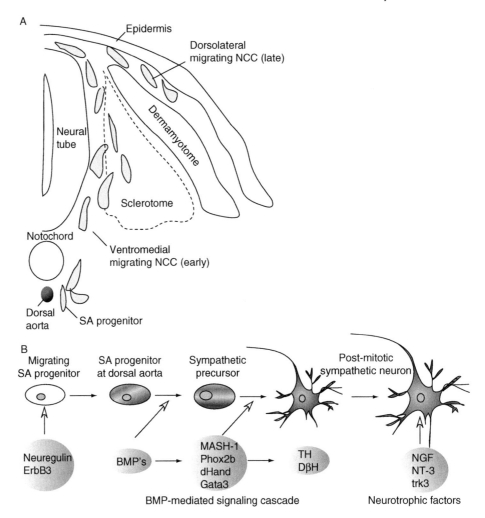

Fig. 1 Neural crest-derived SA progenitors migrate along a ventromedial pathway to bilateral regions adjacent to the dorsal aorta. (A) Schematic diagram of a transverse section through the trunk of a vertebrate embryo (embryonic day 10.5 in the mouse, 2.5 in the chick, and approximately 28 hpf in the zebrafish embryo). In avian and rodent embryos, presumptive SA progenitor cells derived from the neural crest migrate ventromedially within the sclerotome region of the somites and ultimately cease migration in the region of the dorsal aorta. In zebrafish, neural crest-derived SA precursors migrate ventromedially between the neural tube and somites to the dorsal aorta region. (B) Molecular pathways governing sympathetic neuron development. During migration, signaling via the neuregulin-1 growth factor is required for the development of at least some SA progenitors. Once the SA progenitor cells arrive at the dorsal aorta, BMP signaling activates the transcriptional regulators MASH-1 and Phox2b that ultimately lead to the expression of the transcription factors Phox2a, GATA-3, and dHand. Together, these factors are responsible for differentiation of SA progenitors into noradrenergic neurons. Fully differentiated neurons express biosynthetic enzymes responsible for the synthesis of noradrenalin, such as tyrosine hydroxylase and dopamine-β-hydroxylase. Survival of the differentiated sympathetic neurons is governed by a number of neurotrophic factors, such as NGF and NT-3.

progenitors. Mice lacking components of the neuregulin-1 pathway, such as the ErbB3 receptor, exhibit severe hypoplasia of the primary sympathetic ganglion chain (Britsch *et al.*, 1998). Further analysis of these mice indicates the requirement for neuregulin signaling for the normal migration of some SA progenitors, rather than their subsequent differentiation into sympathetic neurons (Britsch *et al.*, 1998; Crone and Lee, 2002; Murphy *et al.*, 2002). As the SA precursors aggregate in the vicinity of the dorsal aorta (around embryonic day 10 in the mouse and E2.5 in the chick), a molecular signaling cascade is initiated in response to BMPs that are secreted by dorsal aorta cells (Fig. 1B) (Reissmann *et al.*, 1996; Schneider *et al.*, 1999; Shah *et al.*, 1996; Varley *et al.*, 1995). BMP signaling induces the expression of the proneural gene *MASH-1*, an *achaete-scute* homolog, and the homeodomain transcription factor *phox2b* by sympathetic neuroblasts (Ernsberger *et al.*, 1995; Groves *et al.*, 1995; Guillemot *et al.*, 1993; Hirsch *et al.*, 1998). Several other critical transcription factors are then activated, including the homeobox proteins Phox2a, the basic Helix-Loop-Helix (bHLH) transcription factor dHand, and the zinc-finger protein GATA-3 (Hirsch *et al.*, 1998; Howard *et al.*, 2000; Lim *et al.*, 2000; Lo *et al.*, 1998; Pattyn *et al.*, 1997, 1999; Schneider *et al.*, 1999). Together, these regulatory factors drive SA differentiation further by activating the expression of pan-neural genes and genes encoding enzymes for the synthesis of catecholaminergic neurotransmitters, such as dopamine-β-hydroxylase (DβH) and tyrosine hydroxylase (TH) (Ernsberger *et al.*, 2000; Kim *et al.*, 2001; Seo *et al.*, 2002). The specification of the neuronal and noradrenergic phenotype of sympathetic neuron precursors by these transcription factors remains incompletely understood, involving a complex interaction of their regulatory pathways, rather than a strictly linear developmental progression (see Goridis and Rohrer (2002)).

A later stage of PSNS development consists of modeling of the sympathetic ganglia through the regulation of cell proliferation and maintenance. The neurotrophic factors, nerve growth factor (NGF) and NT-3, have been shown to control sympathetic neuron survival and the maintenance of their synaptic connections (Birren *et al.*, 1993; DiCicco-Bloom *et al.*, 1993; Francis and Landis, 1999). In the embryo, NGF is secreted from sympathetic target tissues when sympathetic neurons arrive (Chun and Patterson, 1977; Heumann *et al.*, 1984; Korsching and Thoenen, 1983; Shelton and Reichardt, 1984). Analysis of NGF and its high affinity receptor, tyrosine kinase A (TrkA), in mouse mutants confirmed their requirement for the *in vivo* survival of sympathetic neurons (Fagan *et al.*, 1996; Smeyne *et al.*, 1994). In their absence, sympathetic neuron development proceeds normally, but is then followed by neuronal cell death. A similar phenotype is observed in NT-3 mouse mutants, although unlike mutants with NGF-signaling loss, neuronal death occurs at later embryonic stages (Ernfors *et al.*, 1994; Farinas *et al.*, 1994; Francis and Landis, 1999; Wyatt *et al.*, 1997). Therefore, in chick and rodents the action of neurotrophic factors, such as NT-3 and NGF, are largely responsible for establishing and maintaining mature ganglion neuronal cell numbers during embryonic or early postnatal development.

Although many of the inductive signaling pathways affecting different stages of neural crest development have been identified, the regulatory mechanisms controlling these pathways remains poorly understood. While substantial evidence links BMP signaling with the induction of SA progenitor cell development, the genetic control of SA cell responsiveness to BMP signals and their specification remain unclear. While sympathetic precursors are competent to express MASH1 and *phox2b* in response to BMP signaling near the dorsal aorta, they do not respond to BMPs present in the overlying ectoderm during earlier premigratory stages. Furthermore, the molecular mechanisms regulating the interactions of transcription factors and downstream pathways specifying neuronal and noradrenergic differentiation are incompletely understood. Other signaling pathways, such as cAMP (Lo *et al.*, 1999), may also contribute to this process. Finally, proliferation of sympathoblasts and differentiated sympathetic neurons occurs throughout embryogenesis (Birren *et al.*, 1993; Marusich *et al.*, 1994; Rohrer and Thoenen, 1987; Rothman *et al.*, 1978). However, little is known about the genetic pathways that actively promote or inhibit SA progenitor cell proliferation. An inability to control cell proliferation in sympathetic ganglia is of particular medical interest, since it can lead to neuroblastoma, the most common human cancer in infants younger than 1 year of age. While the study of PSNS development in tetrapods will continue to contribute to our knowledge of PSNS development, a goal of this chapter is to demonstrate the power of the zebrafish, *Danio rerio*, exploiting its forward genetic potential and advantages as an embryologic system, for making significant contributions to PSNS research. We also propose that the study of zebrafish PSNS development will contribute to our understanding of both normal and abnormal PSNS development and may ultimately provide a genetic model for human neuroblastoma.

III. The Zebrafish as a Model System for Studying PSNS Development

A. Overview

One of the most powerful attributes of the zebrafish system is its capacity for large-scale genetic screens. The unbiased nature of phenotype-based genetic screens enables new genes to be identified without prior knowledge of their function or expression in the tissue of interest. This approach is particularly attractive for study of the PSNS, as many signaling components involved in determining sympathetic fate are either unknown or incompletely understood. Also, most of our current understanding of PSNS development has relied on functional assays on isolated sympathetic cells in culture or analyses of genes that have been expressed incorrectly (Francis and Landis, 1999; Goridis and Rohrer 2002). While these studies can determine whether certain genes are sufficient to direct sympathetic development, they do not address whether those genes are normally required for PSNS development. Murine gene knockout models have been used in loss-of-function studies to confirm the *in vivo* requirement for particular genes in PSNS development (Guillemot *et al.*, 1993; Lim *et al.*, 2000; Morin *et al.*, 1997; Pattyn *et al.*, 1999). For example, although both Phox2a and Phox2b can induce a sympathetic phenotype when expressed incorrectly in the chick, only the selective knockout of *Phox2b* appears to be necessary for sympathetic development *in vivo*, as the PSNS in *Phox2a*$^{-/-}$ mutant mice appears relatively normal (Morin *et al.*, 1997; Pattyn *et al.*, 1999).

Together, these studies provide valuable insights into the regulatory pathways directing sympathetic neuron development and emphasize the advantages of using mutants to dissect genetic pathways *in vivo*. The capacity for experimental mutagenesis and manipulation of gene expression in the developing embryo is a major strength of the zebrafish system. Forward genetic zebrafish screens can be especially valuable for identifying genes affecting complex signaling pathways that rely on interactions between the developing PSNS and surrounding tissues, which may be impossible to address using *in vitro* assays. Critical roles for extrinsic factors are particularly evident in PSNS development and SA progenitors migrate past a number of tissues expressing different signaling molecules, such as the neural tube, notochord, and floor-plate. In addition, zebrafish mutant embryos can often survive for a longer period of time during embryogenesis than knockout mice lacking orthologous genes. This may be attributed to the external development of the zebrafish embryo and can allow the analysis of the PSNS to extend through later stages of sympathetic neuron differentiation and maintenance. Finally, the zebrafish system offers the most amenable vertebrate model for performing large-scale mutagenesis screens to identify novel genes affecting all aspects of PSNS development, as the time, space, and expense associated with mutagenesis techniques in mice can be prohibitive.

Establishing the zebrafish as a useful vertebrate model for identifying new genes important for PSNS development will require (1) an analysis of zebrafish sympathetic neuron development and its comparison with other vertebrates, (2) an analysis of the genetic programs regulating zebrafish PSNS development and their conservation in other vertebrates, and (3) the generation of efficient mutagenesis protocols and screening assays for the isolation of PSNS mutants. Each of these areas is addressed in the next sections.

B. Development of the PSNS in Zebrafish

1. Neural Crest Development and Migration

The different stages of PSNS development in the zebrafish have been recently analyzed (An *et al.*, 2002; Raible and Eisen, 1994). The findings show that the morphogenesis and differentiation of sympathetic neurons in zebrafish is qualitatively very similar to other vertebrates. The migration and cell fate specification of trunk sympathetic precursors in zebrafish was analyzed by labeling single neural crest cells with vital dyes and following their subsequent development (Raible and Eisen, 1994). In the trunk, neural crest migration begins around 16 hpf, at the level of somite 7, and sympathetic neurons are only derived from neural crest cells undergoing early migration along the ventromedial pathway. Hence, the ventro-medial migration of SA precursor cells is conserved in zebrafish. These studies also demonstrated the existence of both multipotent and fate-restricted neural crest precursors that generate a limited number of neural crest derivatives, such as sympathetic neurons, before or during the initial stages of neural crest migration (Raible and Eisen, 1994). Although little is known about the molecular mechanisms underlying such fate decisions, the ability to analyze the fate restriction of sympathetic neurons in zebrafish, using different neural crest mutants, affords a powerful method to dissect the genetic pathways underlying this process.

2. Gene Expression in Migrating SA Progenitors

Many of the genes capable of inducing the development SA progenitors in birds and rodents have been identified in zebrafish (Table I). Furthermore, the expression of some of these genes in dorsal aorta cells and in neural crest-derived cells in its vicinity, where noradrenergic neurons form, is consistent with their role in fish PSNS development. A number of *BMP* homologs have been identified in zebrafish and some have been shown to be expressed by the dorsal aorta (Dick *et al.*, 2000; Nguyen *et al.*, 1998). Several zebrafish mutants exhibit midline defects affecting structures that may be responsible for BMP signaling. In *flh* mutant embryos, the notochord and dorsal aorta fail to develop (Fouquet *et al.*, 1997; Talbot *et al.*, 1995) and *bmp4* expression is absent. The loss of the local source of BMP signaling corresponds with a

Table I
Expression of Conserved PSNS Genes in Zebrafish

Genes	Expression during PSNS development	References
ErbB3	Not available	Lo *et al.* (2003)
BMP4	Dorsal aorta	Fig. 2A; Dick *et al.* (2000) and Nguyen *et al.* (1998)
Crestin	Migrating SA, nascent SCG	Luo *et al.* (2001)
Zash1a	SCG	Allende and Weinberg (1994) and Stewart (unpublished data)
Phox2a	SCG	Guo *et al.* (1999a) and Holzschuh *et al.* (2003)
Phox2b	Migrating SA, SCG	Guo (unpublished data) and Stewart (unpublished data)
HuC/D	SCG, trunk sympathetic ganglia	An *et al.* (2002)
GATA3	ND	Neave *et al.* (1995)
dHand	SCG	Fig. 2C and D; Yelon *et al.* (2000)
TH	SCG, trunk sympathetic ganglia	An *et al.* (2002), Guo *et al.* (1999b) and Holzschuh *et al.* (2001)
DβH	SCG, trunk sympathetic ganglia	An *et al.* (2002) and Holzschuh *et al.* (2003)
PNMT	SCG	Stewart (unpublished data)
Trk receptors	ND	Martin *et al.* (1995) and Williams *et al.* (2000a)

failure of sympathetic neurons (Hu$^+$/TH$^+$) to form (Henion, unpublished data, 2002). Interestingly, neural crest-derived cells (crestin$^+$) continue to populate this region where sympathetic neurons would normally develop. These observations suggest that, like other vertebrates, dorsal aorta-derived BMPs are required for SA development in zebrafish (Henion, unpublished data). In contrast, floor-plate cells, which are lacking in the *cyclops* mutant, do not appear to be required for dorsal aorta development, BMP expression, or SA development. All these functions appear to be normal in these mutant embryos (Henion, unpublished data). Interestingly, SA development appears normal in *no tail* mutants (Fig. 2A and B) even though dorsal aorta development is impaired (Fouquet *et al.*, 1997) and BMP expression is reduced. It is possible that weak BMP persists in *ntl* because of the continued presence of notochord precursor cells that fail to differentiate properly in this mutant (Melby *et al.*, 1997). However, whether the notochord is directly responsible for BMP expression and dorsal aorta development is unclear.

Most of the described transcription factors known to direct the development of the sympathetic precursors in other species are present in the zebrafish and exhibit appropriate gene expression patterns. The zebrafish *Zash1a* gene, a homolog of *MASH-1*, is transiently expressed in cells near the dorsal aorta by 48 hpf (Allende and Weinberg, 1994; Stewart, unpublished data). Preliminary gene knockdown experiments using antisense morpholinos to specifically target the *Zash1a* gene, resulted in the loss of *TH*-expressing noradrenergic neurons in the developing PSNS (Yang and Stewart, unpublished data). The *Phox2a*, *Phox2b*, *GATA-3*, and *dHand* genes have also been cloned in zebrafish. The *dHand* and both *Phox2* genes are also expressed in the first PSNS neurons that develop by 2 dpf (Fig. 2C and D; Guo *et al.*, 1999a; Holzschuh *et al.*, 2003; Neave *et al.*, 1995; Yelon *et al.*, 2000). Therefore, although the expression of *GATA-3* in the developing PSNS has not yet been characterized, many of these early markers of SA precursors appear to be conserved in zebrafish. The future analysis of *Zash1a, Phox2b, Phox2a, GATA-3*, and *dHand* expression, together with the examination of *TH* and *DβH* expression in SA precursors, will provide insight into the functions of these genes with respect to sympathetic neuron differentiation (discussed later). Importantly, analysis of compound mutants utilizing existing zebrafish mutants such as *soulless* (*phox2a*) (Guo *et al.*, 1999a) and *hands down* (*dHand*) (Miller *et al.*, 2000), together with other mutants described in this chapter, will contribute to our understanding of the functional roles of these genes in sympathetic neuron development. The combination of transient gene knockdown techniques using morpholinos and the identification of new zebrafish mutants affecting other regulators of sympathetic neuron development, should also contribute to novel insights relevant to PSNS development across vertebrate species.

3. Neuronal Dierentiation and Coalescence into Sympathetic Ganglia

The timing of overt neuronal differentiation of sympathetic precursors and their transition to fully differentiated noradrenalin (NA)-producing neurons has been described in detail in zebrafish (An *et al.*, 2002). The pan-neuronal antibody 16A11

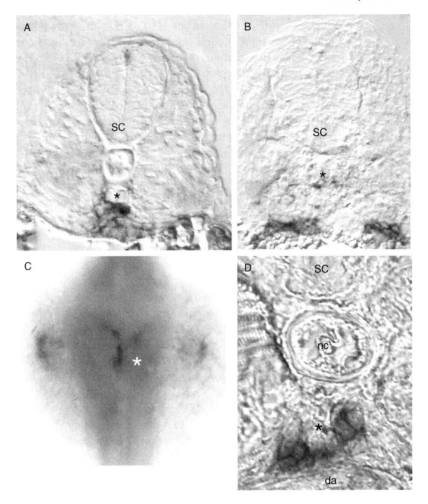

Fig. 2 Expression of *bmp* by the dorsal aorta and dHand by sympathetic neurons. (A) Expression of *bmp4* in transverse sections at the mid-trunk level in wild type (*top*) and *ntl* mutant embryos at 48 hpf. In *bmp4*, expression is evident in cells of the dorsal aorta (*asterisk*) of wild-type embryos. (B) In *ntl* mutants, the dorsal aorta does not form completely (*asterisk*), although expression of *bmp4* is present albeit to a reduced extent. (C) Expression of zebrafish dHand at 58 hpf in cervical sympathetic neurons (*asterisk*). (D) Coexpression of dHand and TH in sympathetic neurons (*asterisk*) in a transverse section in the anterior trunk region of a 3 dpf embryo. Sc, spinal cord; nc,notochord; da, dorsal aorta. Top is dorsal in (A, B, and D) and anterior in (C).

recognizes members of the Hu family of ribonucleic acid (RNA)-binding proteins (Marusich *et al.*, 1994) and labels sympathetic precursors located ventrolateral to the notochord and adjacent to the dorsal aorta (Fig. 3; An *et al.*, 2002). Sympathetic neurons were found to differentiate at different times in the zebrafish embryo

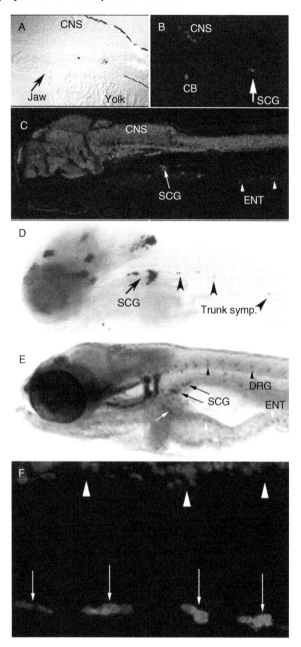

Fig. 3 Development of the peripheral sympathetic nervous system in zebrafish embryos. (A–C) Parasagittal section of 3.5-dpf embryo. (A) High magnification DIC and (B) fluorescence of the same field, showing TH-IR (*red*) in the SCG (*arrow*), carotid body (CB), and a group of anterior cells in the midbrain (CNS). (C) Low magnification view of 3.5-dpf embryo labeled with anti-Hu to reveal all neurons. A subset of cervical sympathetic neurons is indicated by the arrow and enteric neurons (ENT)

and two populations of sympathetic ganglion neurons were defined. The most rostral population develops at 2 dpf and comprises the superior cervical ganglion (SCG) complex that consists of two separate ganglia arranged in an hourglass shape. Several days later, more caudal trunk sympathetic neurons develop as irregular, bilateral rows of single neurons adjacent to the dorsal aorta, presumably analogous to the primary sympathetic chain in other vertebrates. These neurons differentiate in an anterior to posterior temporal progression, extending caudally as far as the level of the anus, and eventually form regular arrays of segmentally distributed sympathetic ganglia (An *et al.*, 2002).

The reason for the delay in the differentiation of the caudal sympathetic neurons is not known, since the formation of the dorsal aorta (Fouquet *et al.*, 1997) and its expression of *bmps* (see Fig. 2A; Martinez-Barbera *et al.*, 1997), occur well before the differentiation of sympathetic neurons is observed. Importantly, ventrally migrating neural crest-derived cells populate the region adjacent to the dorsal aorta between 24 and 36 hpf. Therefore, the delay in caudal PSNS development may be attributed to a delay in neural crest-derived cells becoming competent to respond to BMP signaling. Since the expression of some zebrafish BMPs have not been examined in the dorsal aorta, it remains possible that different types of BMPs may be selectively expressed by dorsal aorta cells or that SA progenitors exhibit differential responsiveness to different BMPs.

4. Dierentiation of Noradrenergic Neurons

One of the key events in PSNS differentiation is the acquisition of the NA-neurotransmitter phenotype, indicated by the expression of noradrenalin and the genes, such as TH and DβH, which are required for its synthesis through the enzymatic conversion of the amino acid L-tyrosine. In zebrafish, *TH* expression has been used as the principal marker for the presence and formation of fully differentiated sympathetic neurons, although it is also expressed by other catecholaminergic neurons in the central nervous system (CNS) (Fig. 3A-D; An *et al.*, 2002; Guo *et al.*, 1999b; Holzschuh *et al.*, 2001). Both the TH protein and mRNA are readily detectable in the SCG complex beginning at 48 hpf. Consistent with the expression of Hu proteins, most sympathetic neurons located posterior to the SCG complex do not begin to express *TH* mRNA until approximately 5 days of

by arrowheads. (D) Lateral view of whole-mount *TH* RNA *in situ* preparation at 5 dpf. *TH* RNA is strongly expressed in the SCG (*arrow*) at this stage and is beginning to be expressed in the trunk sympathetic chain (*arrowheads*). A description of *TH* RNA expression in the head is described in Guo *et al.* (1999a, b). (E) Whole-mount antibody preparation of a 7-dpf larva labeled with anti-Hu to reveal neurons. Black arrows indicate SCG, black arrowheads indicate dorsal root ganglion (DRG) sensory neurons, and white arrow and white arrowheads indicate enteric neurons (ENT). (F) Parasagittal section in the mid-trunk region of a 17-dpf embryo labeled with anti-Hu. Ventral spinal cord neurons are evident at the top (*arrowheads*). Four segmental sympathetic ganglia (*arrows*) are located ventral to the notochord adjacent to the dorsal aorta. (See Plate no. 9 in the Color Plate Section.)

development, in a few of the more rostral trunk segments (Fig. 3D; An *et al.*, 2002). By 10 dpf, all of the sympathetic ganglia contain neurons expressing *TH*, although some neurons within the nascent sympathetic ganglia do not express TH protein. However, by 28 dpf, all of the neurons uniformly express TH, suggesting the complete maturation of sympathetic ganglia by this time. The expression of DβH protein and mRNA has also been used as markers of PSNS differentiation (Fig. 4A; An *et al.*, 2002; Holzschuh *et al.*, 2003). Because of its requirement for the conversion of dopamine to noradrenaline, *DβH* is expressed in sympathetic neurons of the PSNS, and a subset of catecholaminergic neurons in the CNS (An *et al.*, 2002; Holzschuh *et al.*, 2003). The expression of DβH is generally observed slightly later than TH in differentiating sympathetic neurons (An *et al.*, 2002). However, DβH is expressed along with TH, as early as 2 dpf in the SCG complex (Holzschuh *et al.*, 2003).

Another element of PSNS function is the regulated release of adrenalin and noradrenalin by chromaffin cells of the adrenal gland, which form in or around the developing kidney. Chromaffin cells represent a specialized component of the PSNS and express in the catecholaminergic pathway an additional enzyme, called phenylethanolamine N-methyltransferase (PNMT), which converts noradrenaline into adrenaline (Kalcheim *et al.*, 2002; Schober *et al.*, 2000). In mammals, chromaffin cells are localized to the adrenal medulla, which is located within the cortex of the adrenal glands overlying the kidneys; whereas in zebrafish, both adrenocortical and chromaffin cells are interspersed within the anterior region of the kidney, referred to as the interrenal gland (Hsu *et al.*, 2003; Liu *et al.*, 2003). As in other species, both noradrenergic and adrenergic chromaffin cells have been described in zebrafish. Noradrenergic cells have heterogeneous vesicles containing asymmetrically localized electron-dense granules, while adrenergic cells form smaller vesicles with homogenous electron-lucent granules (Hsu *et al.*, 2003). Initial observations indicated that nonneuronal (16A11$^-$), TH, and DβH-positive cells are present in the SCG complex at 2 dpf, which continue to migrate ventrally to the anterior pronephros (also referred to as the head kidney; Fig. 4; An *et al.*, 2002). Preliminary mRNA *in situ* hybridization assays using *PNMT* expression as a marker for chromaffin cells, support these observations and suggest that these nonneuronal SCG cells represent chromaffin cells (Stewart, unpublished data). Further analysis of chromaffin cell development, using *PNMT* as a marker, should provide a more complete understanding of how SA progenitors form chromaffin cells in the zebrafish.

5. Modeling of Sympathetic Ganglia

In rodents and birds, neurotrophic factors, such as NGF and NT-3, control sympathetic neuron cell numbers through the regulation of their survival and continued maintenance of their synaptic connections (Francis and Landis, 1999; Schober and Unsicker, 2001). In zebrafish, the ability of these factors to control the survival of sympathetic neurons remains unknown. However, NT-3 has been

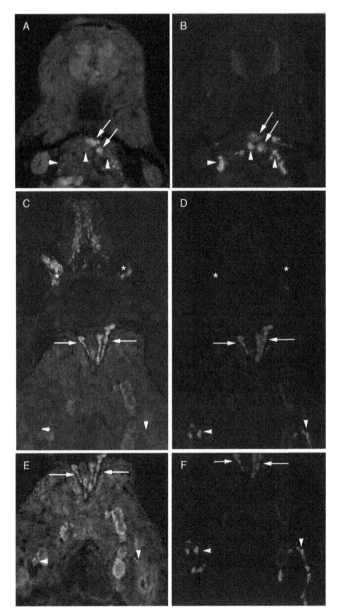

Fig. 4 Sympathoadrenal derivatives in embryonic and juvenile zebrafish. (A and B) Transverse section of a 3.5-dpf embryo double-labeled with anti-Hu (*green*) and DβH (*red*). Arrows indicate Hu$^+$/DβH$^+$ sympathetic neurons of the cervical ganglion. Arrowheads indicate Hu$^-$/DβH$^+$ presumptive chromaffin cells. (C, D) Transverse section through the mid-trunk region at 28 dpf, double-labeled with anti-Hu (*green*) and anti-TH (*red*). (E, F) Higher magnification of C and D, including a slightly more ventral region. Arrows indicate sympathetic neurons, arrowheads indicate chromaffin cells, and asterisks denote dorsal root ganglia. (See Plate no. 10 in the Color Plate Section.)

shown to act as a neurotrophic factor regulating cell death in Rohon-Beard sensory neurons (Williams *et al.*, 2000a). In teleost sympathetic ganglia, the proliferation of cells occurs during early development and may possibly continue throughout adult life (An *et al.*, 2002; Weis, 1968). Analysis of 5-bromo-2-deoxyuridine (BrdU) incorporation and the expression of phospho-histone H3 indicate that cells proliferate within the developing sympathetic ganglia (An *et al.*, 2002). Interestingly, some of these cells also expressed the pan-neuronal marker, 16A11, suggesting that preexisting neuronal cells proliferated within the ganglia, a process that has also been observed in both chick and mouse PSNS (Birren *et al.*, 1993; Cohen, 1974; DiCicco-Bloom *et al.*, 1993; Marusich *et al.*, 1994; Rohrer and Thoenen, 1987; Rothman *et al.*, 1978). However, in chick and rodents, sympathetic neurons become postmitotic during embryonic development, while they may remain competent to divide throughout life in the zebrafish (Weis, 1968).

C. Mutations Affecting PSNS Development

1. Introduction

PSNS development, from the induction of neural crest through the overt differentiation of sympathetic ganglia, can be readily observed within the first 5 days of zebrafish development in the SCG (see Fig. 3D; An *et al.*, 2002). During this time, dynamic changes in both the numbers and distribution of sympathetic cells within the SCG can be easily visualized by *TH* mRNA whole-mount *in situ* hybridization. At 2 dpf, bilateral rows containing approximately 5 TH-positive cells are ventrally located near the dorsal aorta. By 5 dpf, the number of *TH*-positive cells have increased fivefold and coalesced into a V-shaped ganglia, including some appearing to migrate ventrally toward the kidney, which may represent putative adrenal chromaffin cells (An *et al.*, 2002). Therefore, the evaluation of the SCG at 5 dpf represents an excellent assay for early PSNS development that can be used in genetic screens to detect mutations affecting any stage of PSNS development. Assaying the SCG would also be able to confirm mutations found to affect very early neural crest development. Such mutagenesis screens are currently being performed and some examples of the different mutant classes that have been isolated thus far are discussed in the next sections.

2. Mutations Affecting Early PSNS Development

Mutations affecting early PSNS development can fail to form either neural crest precursors or SA progenitor cells. They may also disrupt the specification of these cells within the premigratory crest. One example is *sympathetic mutant 1* (*sym1*), which was discovered in a diploid gynogenetic screen designed to identify mutations disrupting *TH* expression in the SCG complex at 5 dpf (Fig. 5A and C). The *sym1* mutation causes a severe reduction or absence of *TH*- and *PNMT*-expressing cells in the SCG complex of the PSNS, but *TH* expression is

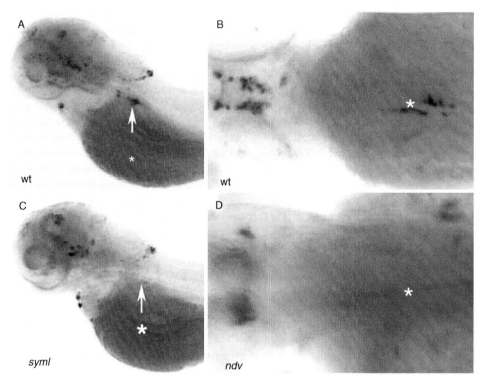

Fig. 5 Isolation of PSNS mutants in zebrafish. Whole-mount *TH in situ* preparation showing expression of *TH* mRNA in (A and B) wild type, (C) *sym1*, and (D) *nosedive* mutant embryos. (A and C) Lateral view of *TH* expression at 5 dpf. (C) The *sym1* mutant phenotype was identified in an *in situ* screen at 5 dpf for mutations that specifically lack *TH* expression in the SCG (C, *asterisk*), but leave others areas of TH expression in the CNS and carotid body unaffected. (B and D) Dorsal view of *TH* expression in the SCG region at 3 dpf in wild-type embryos (B, *asterisk*). Expression of *TH* is absent in the SCG in *nosedive* mutants (D, *asterisk*). Analysis of Hu immunoreactivity revealed that the lack of *TH* expression in the cervical region in *nosedive* mutant embryos is attributed to the absence of sympathetic neurons (data not shown).

not affected in other regions of the CNS. Analysis of the *sym1* phenotype indicates that early neural crest development is disrupted, and the cells exhibit a reduction in the expression of early neural crest markers, such as *crestin* (*cnt*) (Luo *et al.*, 2001). Another example of a mutant within this class is *colourless* (*cls*), which was isolated in screens for neural crest mutants affecting pigment cell development (Kelsh and Eisen, 2000). The zebrafish *cls* mutation disrupts the *sox10* gene, which is required for the development of most nonectomesenchymal neural crest lineages, including the PSNS (Dutton *et al.*, 2001; Kelsh and Eisen, 2000). The *sox10* gene is expressed in early premigratory neural crest cells and then undergoes rapid down-regulation. Analysis of the *cls* phenotype revealed that premigratory neural crest cells fail to migrate and subsequently undergo apoptosis (Dutton *et al.*, 2001).

Interestingly, the *sym1* and *cls* mutations differentially affect complementary sets of neural crest derivatives. For example, unlike *cls*, *sym1* mutants have severe defects in craniofacial cartilage development, while unlike *sym1*, the *cls* mutant lacks pigment cells (Kelsh and Eisen, 2000). The genetic mechanisms specifying the fate restriction of subsets of premigratory neural crest are unknown, although evidence suggests that they exist in the zebrafish (Raible and Eisen, 1994). It is possible that the genes disrupted by the *sym1* and *cls* mutations may contribute to the generation of fate-restricted subsets of neural crest precursors, such as SA progenitors.

In a screen for mutations affecting neural crest derivatives (Henion *et al.*, 1996), three mutants were identified based on the absence of neural crest-derived dorsal root ganglion (DSG) sensory neurons, based on the expression of 16A11 immunoreactivity. Subsequently, it has been found that these mutants also completely lack sympathetic and enteric neurons based on the expression of TH mRNA and 16A11, respectively (Fig. 5B and D; Henion, unpublished data, 2002). In contrast, the development of trunk neural crest-derived chromatophores and glia appears normal in all three mutants. These observations suggest that the genes disrupted by these mutations do not affect all neural crest derivatives, but selectively function during the development of crest-derived neurons. It is tempting to speculate that the phenotypes of these mutants further indicate an early lineage segregation of neurogenic precursors or suggest the function of these genes in the development of multiple neural crest-derived neuronal subtypes.

3. Mutations Affecting Later Stages of PSNS Development

A number of genes are expressed in SA precursors once they have migrated to the region adjacent to the dorsal aorta, indicating further differentiation toward mature sympathetic neurons (see Fig. 1). Therefore, mutations that disrupt later stages of neuronal differentiation may result in normal neural crest migration and expression of early neural crest markers, but exhibit a failure of sympathetic neuron differentiation. The *lockjaw* or *mount blanc* mutant (*tfap2a*) is an example of this mutant class. It shows a lack of *TH-* and *DβH-positive neurons in the region of the SCG complex* (Holzschuh *et al.*, 2003; Knight *et al.*, 2003). The *tfap2a* gene is normally expressed both in the premigratory neural crest and in the region of the developing SCG. Analysis of neurogenesis, using the 16A11 antibody, revealed normal migration and neural differentiation of Sa progenitors at the SCG in *tfap2a* mutants (Holzschuh *et al.*, 2003; Knight *et al.*, 2003). Furthermore, the expression of *Phox2a* in cells in the region of the SCG in these mutants indicates that the signaling cascade required to induce the initial stages of noradrenergic differentiation at the dorsal aorta is intact in these mutants (Holzschuh *et al.*, 2003). The failure of *tfap2a* mutants to express the noradrenergic differentiation markers *TH* and *DβH*, is likely attributed to the requirement for Tfap2a to activate these genes, as Tfap2a has conserved DNA binding regions in both *TH* and *DβH* promoters (Holzschuh *et al.*, 2003; Seo *et al.*, 2002). Retinoic acid (RA) signaling pathways may possibly function upstream of *tfap2a* in the differentiation of noradrenergic neurons because incubation of wild-type zebrafish embryos in RA induces ectopic *TH*-positive cells in the region of the SCG. This effect is blocked in *tfap2a* mutants (Holzschuh *et al.*, 2003). In addition, mutations in *neckless/rald2*, which disrupt the biosynthesis of RA from vitamin A, have fewer *TH*-expressing cells in the SCG (Holzschuh *et al.*, 2003). These studies demonstrate mutants affecting late stages of sympathetic neuron differentiation and further highlight the advantages of zebrafish mutant analysis for identifying novel genes and signaling pathways affecting PSNS development. Studies in other organisms have not previously evaluated the *in vivo* role of either *tfap2a* or RA signaling in PSNS development.

4. Mutations Affecting Sympathetic Ganglia Modeling

Mutations of this class affect the maintenance or survival of the PSNS. These mutants display abnormal *TH* expression in the SCG because of a disruption of genes affecting the number, morphology, or survival of *TH*-positive cells. Many dynamic processes occur during modeling of the SCG. In addition, mutations affecting the coalescence of differentiated sympathetic neurons into the discrete ganglia, or cell proliferation within the ganglia, can be identified in the zebrafish embryo by 4-5 dpf. Another PSNS mutant isolated from a diploid gynogenetic screen, is the *sympathetic2* (*sym2*) mutant, which displays abnormal *TH* expression in the SCG (Stewart, unpublished data). The earlier stages of PSNS development and differentiation proceed normally in *sym2* mutants. These include neural crest induction, migration, aggregation, and noradrenergic differentiation. Although the numbers of sympathetic neurons appear normal in this mutant, changes in the size and shape of the SCG cells become evident at 3 dpf, which worsen until the embryos die at 6 dpf. Preliminary data suggest that *sym2* plays a role in the coalescence of sympathetic neurons into discrete ganglia once they reach the region of the dorsal aorta. Furthermore, the analysis of pigment cells in *sym2* shows that the cells are capable of migration and differentiate normally, but fail to aggregate into discrete lateral or ventral stripes in the developing embryo. Therefore, *sym2* may possibly function in the formation of ganglia from differentiated sympathetic neurons, a critical process that has yet to be addressed at the molecular level.

IV. Zebrafish as a Novel Model for Studying Neuroblastoma

Neuroblastoma (NB) is an embryonic tumor of the PSNS and the most common extracranial solid tumor of children, often arising in the adrenal medulla (40% of cases; Maris and Matthay, 1999). NB affects 650 children in the United States each year and is the leading cause of cancer deaths in children 1–4 years of age (Goodman, 1999). Clinically, NB manifests diverse behavior and is one of the few cancers that can spontaneously regress. Often in infants younger than 1 year of age, NB tumors may regress or differentiate without receiving treatment. However, older children with advanced disease account for 70% of all NB patients, and their poor long-term survival rate has not risen above 20–30% (Maris and Matthay, 1999). Despite intensive research efforts during the mid-1970s, -1980s, -1990s, and early in the millennium, many of the genetic pathways disrupted in this cancer remain unknown, posing a major obstacle to understanding its molecular pathology and to the development of effective therapies for this devastating disease.

Analysis of NB by cytogenetic criteria has identified a number of chromosomal regions that are consistently deleted in NB cells (Maris and Matthay, 1999; Westermann and Schwab, 2002). These regions include allelic losses of 1p36.1, 2q, 3p, 4p, 5q, 9p, 11q23, 14q23-qter, 16p12-13, and 18q-, which have been identified in 15–44% of primary neuroblastomas (Brodeur, 2003; Brodeur et al., 1997; Maris and Matthay, 1999). Although the critical target genes within these chromosomal regions remain to be identified, the findings suggest that multiple tumor suppressor genes may function during different stages of NB pathogenesis. Such genes may include those that normally suppress cell proliferation, induce developmentally regulated apoptosis, or promote cell differentiation, therefore preventing the uncontrolled proliferation and survival of sympathetic neuroblasts in NB. Of the tumor suppressor genes found in all types of human tumors, we hypothesize that those contributing to NB are among the most likely to also play critical roles in the normal embryologic development of the target tissue, in this case the developing PSNS. Population-based, neonatal screening of infants with elevated levels of catecholamines demonstrated NB with allelic losses in eight chromosomal regions, indicating that loss of heterozygosity of at least one allele of key tumor suppressor genes, occurred within sympathetic neuroblasts during embryologic development (Takita et al., 1995). For these reasons, we postulate that the identification of genes affecting normal embryonic PSNS development in zebrafish, through large-scale genetic screens, should be particularly relevant to NB and provide important insights into the molecular pathways that are dysfunctional in this disease. Such novel approaches are clearly needed to foster the development of therapies to successfully target this deadly cancer.

V. Conclusion and Future Directions

Impressive advances have been made in understanding the genetic mechanisms that regulate PSNS development through studies in birds and rodents. Zebrafish studies that further define the anatomic and morphologic aspects of the developing PSNS should accelerate the pace of discovery in this field. Exploiting the forward genetics of the zebrafish system can contribute significantly to the identification of new genes and pathways that regulate PSNS development. These mutants also provide the means to genetically dissect PSNS developmental processes in vivo. Finally, we postulate that knowledge of the genes responsible for normal PSNS development in the zebrafish will help identify the molecular pathways that are affected in neuroblastoma. Ultimately, second generation suppressor screens based on established PSNS mutants exhibiting proliferative abnormalities, typical of NB, can be used to identify potential genes and pathways that may be relevant to the development of effective therapies for this disease.

Acknowledgments

We thank Hermann Rohrer for comments on the manuscript and Su Guo and Hong Wei Yang for unpublished data. R.A.S. was supported by the Hope Street Kids foundation. This work was supported in part by NIH grant CA 104605 to A.T.L., and NIH grant NS38115 and NSF grant IBN0315765 to P.D.H.

References

Allende, M. L., and Weinberg, E. S. (1994). The expression pattern of two zebrafish achaete-scute homolog (ash) genes is altered in the embryonic brain of the cyclops mutant. *Dev. Biol.* **166,** 509–530.

An, M., Luo, R., and Henion, P. D. (2002). Differentiation and maturation of zebrafish dorsal root and sympathetic ganglion neurons. *J. Comp. Neurol.* **446,** 267–275.

Anderson, D. J. (1993). Cell fate determination in the peripheral nervous system: The sympathoadrenal progenitor. *J. Neurobiol.* **24,** 185–198.

Aybar, M. J., Nieto, M. A., and Mayor, R. (2003). Snail precedes slug in the genetic cascade required for the specification and migration of the Xenopus neural crest. *Development* **130,** 483–494.

Birren, S. J., Lo, L., and Anderson, D. J. (1993). Sympathetic neuroblasts undergo a developmental switch in trophic dependence. *Development* **119,** 597–610.

Brading, A. (1999). "The Autonomic Nervous System and its Effectors". Blackwell Science, Malden, MA.

Britsch, S., Li, L., Kirchhoff, S., Theuring, F., Brinkmann, V., Birchmeier, C., and Riethmacher, D. (1998). The ErbB2 and ErbB3 receptors and their ligand, neuregulin-1, are essential for development of the sympathetic nervous system. *Genes Dev.* **12,** 1825–1836.

Brodeur, G. M. (2003). Neuroblastoma: Biological insights into a clinical enigma. *Nat. Rev. Cancer* **3,** 203–216.

Brodeur, G. M., Maris, J. M., Yamashiro, D. J., Hogarty, M. D., and White, P. S. (1997). Biology and genetics of human neuroblastomas. *J. Pediatr. Hematol. Oncol.* **19,** 93–101.

Bronner-Fraser, M. (2002). Molecular analysis of neural crest formation. *J. Physiol. Paris* **96,** 3–8.

Chun, L. L., and Patterson, P. H. (1977). Role of nerve growth factor in the development of rat sympathetic neurons *in vitro*. I. Survival, growth, and differentiation of catecholamine production. *J. Cell Biol.* **75,** 694–704.

Cohen, A. M. (1974). DNA synthesis and cell division in differentiating avian adrenergic neuroblasts. *In* "Wenner-Gren Center International Symposium Series: Dynamics of Degeneration and Growth in Neurons" (K. Fuxe, L. Olson, and Y. Zotterman, eds.), pp. 359–370. Pergamon, Oxford.

Crone, S. A., and Lee, K. F. (2002). Gene targeting reveals multiple essential functions of the neuregulin signaling system during development of the neuroendocrine and nervous systems. *Ann. N. Y. Acad. Sci.* **971,** 547–553.

DiCicco-Bloom, E., Friedman, W. J., and Black, I. B. (1993). NT-3 stimulates sympathetic neuroblast proliferation by promoting precursor survival. *Neuron* **11,** 1101–1111.

Dick, A., Hild, M., Bauer, H., Imai, Y., Maifeld, H., Schier, A. F., Talbot, W. S., Bouwmeester, T., and Hammerschmidt, M. (2000). Essential role of Bmp7 (snailhouse) and its prodomain in dorsoventral patterning of the zebrafish embryo. *Development* **127,** 343–354.

Dutton, K. A., Pauliny, A., Lopes, S. S., Elworthy, S., Carney, T. J., Rauch, J., Geisler, R., Haffter, P., and Kelsh, R. N. (2001). Zebrafish colourless encodes sox10 and specifies non-ectomesenchymal neural crest fates. *Development* **128,** 4113–4125.

Ernfors, P., Lee, K. F., Kucera, J., and Jaenisch, R. (1994). Lack of neurotrophin-3 leads to deficiencies in the peripheral nervous system and loss of limb proprioceptive afferents. *Cell* **77,** 503–512.

Ernsberger, U., Patzke, H., Tissier-Seta, J. P., Reh, T., Goridis, C., and Rohrer, H. (1995). The expression of tyrosine hydroxylase and the transcription factors cPhox-2 and Cash-1: Evidence for distinct inductive steps in the differentiation of chick sympathetic precursor cells. *Mech. Dev.* **52**, 125–136.

Ernsberger, U., Reissmann, E., Mason, I., and Rohrer, H. (2000). The expression of dopamine beta-hydroxylase, tyrosine hydroxylase, and Phox2 transcription factors in sympathetic neurons: Evidence for common regulation during noradrenergic induction and diverging regulation later in development. *Mech. Dev.* **92**, 169–177.

Fagan, A. M., Zhang, H., Landis, S., Smeyne, R. J., Silos-Santiago, I., and Barbacid, M. (1996). TrkA, but not TrkC, receptors are essential for survival of sympathetic neurons *in vivo. J. Neurosci.* **16**, 6208–6218.

Farinas, I., Jones, K. R., Backus, C., Wang, X. Y., and Reichardt, L. F. (1994). Severe sensory and sympathetic deficits in mice lacking neurotrophin-3. *Nature* **369**, 658–661.

Fouquet, B., Weinstein, B. M., Serluca, F. C., and Fishman, M. C. (1997). Vessel patterning in the embryo of the zebrafish: Guidance by notochord. *Dev. Biol.* **183**, 37–48.

Francis, N. J., and Landis, S. C. (1999). Cellular and molecular determinants of sympathetic neuron development. *Annu. Rev. Neurosci.* **22**, 541–566.

Gabella, G. (2001). "Autonomic Nervous System, vol. 2004." Nature Publishing Group, New York.

Goodman, N. W. (1999). An open letter to the Director General of the Cancer Research Campaign. *J. R. Coll. Physicians Lond.* **33**, 93.

Goridis, C., and Rohrer, H. (2002). Specification of catecholaminergic and serotonergic neurons. *Nat. Rev. Neurosci.* **3**, 531–541.

Groves, A. K., George, K. M., Tissier-Seta, J. P., Engel, J. D., Brunet, J. F., and Anderson, D. J. (1995). Differential regulation of transcription factor gene expression and phenotypic markers in developing sympathetic neurons. *Development* **121**, 887–901.

Guillemot, F., Lo, L. C., Johnson, J. E., Auerbach, A., Anderson, D. J., and Joyner, A. L. (1993). Mammalian achaete-scute homolog 1 is required for the early development of olfactory and autonomic neurons. *Cell* **75**, 463–476.

Guo, S., Brush, J., Teraoka, H., Goddard, A., Wilson, S. W., Mullins, M. C., and Rosenthal, A. (1999a). Development of noradrenergic neurons in the zebrafish hindbrain requires BMP, FGF8, and the homeodomain protein soulless/Phox2a. *Neuron* **24**, 555–566.

Guo, S., Wilson, S. W., Cooke, S., Chitnis, A. B., Driever, W., and Rosenthal, A. (1999b). Mutations in the zebrafish unmask shared regulatory pathways controlling the development of catecholaminergic neurons. *Dev. Biol.* **208**, 473–487.

Hansen, M. B. (2003). The enteric nervous system I: Organisation and classification. *Pharmacol. Toxicol.* **92**, 105–113.

Henion, P. D., Raible, D. W., Beattie, C. E., Stoesser, K. L., Weston, J. A., and Eisen, J. S. (1996). Screen for mutations affecting development of Zebrafish neural crest. *Dev. Genet.* **18**, 11–17.

Heumann, R., Korsching, S., Scott, J., and Thoenen, H. (1984). Relationship between levels of nerve growth factor (NGF) and its messenger RNA in sympathetic ganglia and peripheral target tissues. *EMBO J.* **3**, 3183–3189.

Hirsch, M. R., Tiveron, M. C., Guillemot, F., Brunet, J. F., and Goridis, C. (1998). Control of noradrenergic differentiation and Phox2a expression by MASH1 in the central and peripheral nervous system. *Development* **125**, 599–608.

Holzschuh, J., Barrallo-Gimeno, A., Ettl, A. K., Durr, K., Knapik, E. W., and Driever, W. (2003). Noradrenergic neurons in the zebrafish hindbrain are induced by retinoic acid and require tfap2a for expression of the neurotransmitter phenotype. *Development* **130**, 5741–5754.

Holzschuh, J., Ryu, S., Aberger, F., and Driever, W. (2001). Dopamine transporter expression distinguishes dopaminergic neurons from other catecholaminergic neurons in the developing zebrafish embryo. *Mech. Dev.* **101**, 237–243.

Howard, M. J., Stanke, M., Schneider, C., Wu, X., and Rohrer, H. (2000). The transcription factor dHAND is a downstream effector of BMPs in sympathetic neuron specification. *Development* **127**, 4073–4081.

Hsu, H. J., Lin, G., and Chung, B. C. (2003). Parallel early development of zebrafish interrenal glands and pronephros: Differential control by wt1 and ff1b. *Development* **130**, 2107–2116.

Kalcheim, C., Langley, K., and Unsicker, K. (2002). From the neural crest to chromaffin cells: Introduction to a session on chromaffin cell development. *Ann. N. Y. Acad. Sci.* **971**, 544–546.

Kelsh, R. N., and Eisen, J. S. (2000). The zebrafish colourless gene regulates development of non-ectomesenchymal neural crest derivatives. *Development* **127**, 515–525.

Kim, H. S., Hong, S. J., LeDoux, M. S., and Kim, K. S. (2001). Regulation of the tyrosine hydroxylase and dopamine beta-hydroxylase genes by the transcription factor AP-2. *J. Neurochem.* **76**, 280–294.

Knecht, A. K., and Bronner-Fraser, M. (2002). Induction of the neural crest: A multigene process. *Nat. Rev. Genet.* **3**, 453–461.

Knight, R. D., Nair, S., Nelson, S. S., Afshar, A., Javidan, Y., Geisler, R., Rauch, G. J., and Schilling, T. F. (2003). Lockjaw encodes a zebrafish tfap2a required for early neural crest development. *Development* **130**, 5755–5768.

Korsching, S., and Thoenen, H. (1983). Nerve growth factor in sympathetic ganglia and corresponding target organs of the rat: Correlation with density of sympathetic innervation. *Proc. Natl. Acad. Sci. USA* **80**, 3513–3516.

Kos, R., Reedy, M. V., Johnson, R. L., and Erickson, C. A. (2001). The winged-helix transcription factor FoxD3 is important for establishing the neural crest lineage and repressing melanogenesis in avian embryos. *Development* **128**, 1467–1479.

Krauss, S., Concordet, J. P., and Ingham, P. W. (1993). A functionally conserved homolog of the Drosophila segment polarity gene hh is expressed in tissues with polarizing activity in zebrafish embryos. *Cell* **75**, 1431–1444.

LeDouarin, N., and Kalcheim, C. (1999). The Neural Crest. Cambridge University Press, New York.

Lim, K. C., Lakshmanan, G., Crawford, S. E., Gu, Y., Grosveld, F., and Engel, J. D. (2000). Gata3 loss leads to embryonic lethality due to noradrenaline deficiency of the sympathetic nervous system. *Nat. Genet.* **25**, 209–212.

Liu, Y. W., Gao, W., Teh, H. L., Tan, J. H., and Chan, W. K. (2003). Prox1 is a novel coregulator of Ff1b and is involved in the embryonic development of the zebra fish interrenal primordium. *Mol. Cell. Biol.* **23**, 7243–7255.

Lo, J., Lee, S., Xu, M., Liu, F., Ruan, H., Eun, A., He, Y., Ma, W., Wang, W., Wen, Z., and Peng, J. (2003). 15000 unique zebrafish EST clusters and their future use in microarray for profiling gene expression patterns during embryogenesis. *Genome Res.* **13**, 455–466.

Lo, L., Morin, X., Brunet, J. F., and Anderson, D. J. (1999). Specification of neurotransmitter identity by Phox2 proteins in neural crest stem cells. *Neuron* **22**, 693–705.

Lo, L., Tiveron, M. C., and Anderson, D. J. (1998). MASH1 activates expression of the paired homeodomain transcription factor Phox2a, and couples pan-neuronal and subtype-specific components of autonomic neuronal identity. *Development* **125**, 609–620.

Luo, R., An, M., Arduini, B. L., and Henion, P. D. (2001). Specific pan-neural crest expression of zebrafish Crestin throughout embryonic development. *Dev. Dyn.* **220**, 169–174.

Maris, J. M., and Matthay, K. K. (1999). Molecular biology of neuroblastoma. *J. Clin. Oncol.* **17**, 2264–2279.

Martin, S. C., Marazzi, G., Sandell, J. H., and Heinrich, G. (1995). Five Trk receptors in the zebrafish. *Dev. Biol.* **169**, 745–758.

Martinez-Barbera, J. P., Toresson, H., Da Rocha, S., and Krauss, S. (1997). Cloning and expression of three members of the zebrafish Bmp family: Bmp2a, Bmp2b and Bmp4. *Gene* **198**, 53–59.

Marusich, M. F., Furneaux, H. M., Henion, P. D., and Weston, J. A. (1994). Hu neuronal proteins are expressed in proliferating neurogenic cells. *J. Neurobiol.* **25**, 143–155.

Melby, A. E., Kimelman, D., and Kimmel, C. B. (1997). Spatial regulation of floating head expression in the developing notochord. *Dev. Dyn.* **209**, 156–165.

Miller, C. T., Schilling, T. F., Lee, K., Parker, J., and Kimmel, C. B. (2000). Sucker encodes a zebrafish Endothelin-1 required for ventral pharyngeal arch development. *Development* **127**, 3815–3828.

Morin, X., Cremer, H., Hirsch, M. R., Kapur, R. P., Goridis, C., and Brunet, J. F. (1997). Defects in sensory and autonomic ganglia and absence of locus coeruleus in mice deficient for the homeobox gene Phox2a. *Neuron* **18,** 411–423.

Murphy, S., Krainock, R., and Tham, M. (2002). Neuregulin signaling via erbB receptor assemblies in the nervous system. *Mol. Neurobiol.* **25,** 67–77.

Neave, B., Rodaway, A., Wilson, S. W., Patient, R., and Holder, N. (1995). Expression of zebrafish GATA 3 (gta3) during gastrulation and neurulation suggests a role in the specification of cell fate. *Mech. Dev.* **51,** 169–182.

Nguyen, V. H., Schmid, B., Trout, J., Connors, S. A., Ekker, M., and Mullins, M. C. (1998). Ventral and lateral regions of the zebrafish gastrula, including the neural crest progenitors, are established by a bmp2b/swirl pathway of genes. *Dev. Biol.* **199,** 93–110.

Patten, I., and Placzek, M. (2000). The role of Sonic hedgehog in neural tube patterning. *Cell Mol. Life Sci.* **57,** 1695–1708.

Pattyn, A., Morin, X., Cremer, H., Goridis, C., and Brunet, J. F. (1997). Expression and interactions of the two closely related homeobox genes Phox2a and Phox2b during neurogenesis. *Development* **124,** 4065–4075.

Pattyn, A., Morin, X., Cremer, H., Goridis, C., and Brunet, J. F. (1999). The homeobox gene Phox2b is essential for the development of autonomic neural crest derivatives. *Nature* **399,** 366–370.

Raible, D. W., and Eisen, J. S. (1994). Restriction of neural crest cell fate in the trunk of the embryonic zebrafish. *Development* **120,** 495–503.

Reissmann, E., Ernsberger, U., Francis-West, P. H., Rueger, D., Brickell, P. M., and Rohrer, H. (1996). Involvement of bone morphogenetic protein-4 and bone morphogenetic protein-7 in the differentiation of the adrenergic phenotype in developing sympathetic neurons. *Development* **122,** 2079–2088.

Rohrer, H., and Thoenen, H. (1987). Relationship between differentiation and terminal mitosis: Chick sensory and ciliary neurons differentiate after terminal mitosis of precursor cells, whereas sympathetic neurons continue to divide after differentiation. *J. Neurosci.* **7,** 3739–3748.

Rothman, T. P., Gershon, M. D., and Holtzer, H. (1978). The relationship of cell division to the acquisition of adrenergic characteristics by developing sympathetic ganglion cell precursors. *Dev. Biol.* **65,** 322–341.

Schneider, C., Wicht, H., Enderich, J., Wegner, M., and Rohrer, H. (1999). Bone morphogenetic proteins are required *in vivo* for the generation of sympathetic neurons. *Neuron* **24,** 861–870.

Schober, A., Krieglstein, K., and Unsicker, K. (2000). Molecular cues for the development of adrenal chromaffin cells and their preganglionic innervation. *Eur. J. Clin. Invest.* **30,** 87–90.

Schober, A., and Unsicker, K. (2001). Growth and neurotrophic factors regulating development and maintenance of sympathetic preganglionic neurons. *Int. Rev. Cytol.* **205,** 37–76.

Seo, H., Hong, S. J., Guo, S., Kim, H. S., Kim, C. H., Hwang, D. Y., Isacson, O., Rosenthal, A., and Kim, K. S. (2002). A direct role of the homeodomain proteins Phox2a/2b in noradrenaline neurotransmitter identity determination. *J. Neurochem.* **80,** 905–916.

Shah, N. M., Groves, A. K., and Anderson, D. J. (1996). Alternative neural crest cell fates are instructively promoted by TGFbeta superfamily members. *Cell* **85,** 331–343.

Shelton, D. L., and Reichardt, L. F. (1984). Expression of the beta-nerve growth factor gene correlates with the density of sympathetic innervation in effector organs. *Proc. Natl. Acad. Sci. USA* **81,** 7951–7955.

Smeyne, R. J., Klein, R., Schnapp, A., Long, L. K., Bryant, S., Lewin, A., Lira, S. A., and Barbacid, M. (1994). Severe sensory and sympathetic neuropathies in mice carrying a disrupted Trk/NGF receptor gene. *Nature* **368,** 246–249.

Takita, J., Hayashi, Y., Kohno, T., Shiseki, M., Yamaguchi, N., Hanada, R., Yamamoto, K., and Yokota, J. (1995). Allelotype of neuroblastoma. *Oncogene* **11,** 1829–1834.

Talbot, W. S., Trevarrow, B., Halpern, M. E., Melby, A. E., Farr, G., Postlethwait, J. H., Jowett, T., Kimmel, C. B., and Kimelman, D. (1995).A homeobox gene essential for zebrafish notochord development. *Nature* **378,** 150–157.

Thexton, A. (2001). "Vertebrate Peripheral Nervous System, Vol. 2001." Nature Publishing Group, New York.

Varley, J. E., Wehby, R. G., Rueger, D. C., and Maxwell, G. D. (1995). Number of adrenergic and islet-1 immunoreactive cells is increased in avian trunk neural crest cultures in the presence of human recombinant osteogenic protein-1. *Dev. Dyn.* **203,** 434–447.

Weis, J. S. (1968). Analysis of the development of nervous system of the zebrafish, Brachydanio rerio. I. The normal morphology and development of the spinal cord and ganglia of the zebrafish. *J. Embryol. Exp. Morphol.* **19,** 109–119.

Westermann, F., and Schwab, M. (2002). Genetic parameters of neuroblastomas. *Cancer Lett.* **184,** 127–147.

Williams, J. A., Barrios, A., Gatchalian, C., Rubin, L., Wilson, S. W., and Holder, N. (2000a). Programmed cell death in zebrafish rohon beard neurons is influenced by TrkC1/NT-3 signaling. *Dev. Biol.* **226,** 220–230.

Williams, Z., Tse, V., Hou, L., Xu, L., and Silverberg, G. D. (2000b). Sonic hedgehog promotes proliferation and tyrosine hydroxylase induction of postnatal sympathetic cells *in vitro. Neuroreport* **11,** 3315–3319.

Wyatt, S., Pinon, L. G., Ernfors, P., and Davies, A. M. (1997). Sympathetic neuron survival and TrkA expression in NT3-deficient mouse embryos. *EMBO J.* **16,** 3115–3123.

Yelon, D., Ticho, B., Halpern, M. E., Ruvinsky, I., Ho, R. K., Silver, L. M., and Stainier, D. Y. (2000). The bHLH transcription factor hand2 plays parallel roles in zebrafish heart and pectoral fin development. *Development* **127,** 2573–2582.

CHAPTER 16

Approaches to Study the Zebrafish Retina

Jarema Malicki and Andrei Avanesov

Department of Ophthalmology
Harvard Medical School/MEEI
Boston, Massachusetts 02114

I. Introduction

Over the last decade or so, the zebrafish has become one the leading models for studies of vertebrate eye development and function. In the most recent past, particularly impressive advances have been made in the genetic analysis as well as imaging of its embryonic development: two areas traditionally thought of as

ESSENTIAL ZEBRAFISH METHODS:
CELL AND DEVELOPMENTAL BIOLOGY
Copyright 2009, Elsevier Inc. All rights reserved.

DOI: 10.1016/B978-0-12-374599-6.00016-9

strengths of this model system. A growing variety of fluorescent marker proteins and their derivatives facilitates the imaging of increasingly more complex cellular features in the retina, frequently in multiple cell types at the same time. These advances are further enhanced by new genetic approaches that allow one to efficiently assemble complex expression constructs and use them stably or transiently in embryos and larvae. The use of transposons as mutagenic agents is another important development, which greatly enhances the ability to use forward genetics to dissect developmental and physiological pathways in zebrafish. These advances, combined with promising tools of reverse genetic analysis, such as engineered zinc finger nucleases, will assure that the zebrafish model continues to contribute important advances for years to come.

The vertebrate central nervous system is enormously complex. The human cerebral cortex alone is estimated to contain in excess of 10^9 neurons (Jacobson, 1991). Each neuron is characterized by the morphology of its soma and processes, its synaptic connections with other cells, the receptors expressed on its surface, the neurotransmitters it releases, and numerous other molecular and cellular properties. Together these characteristics define cell identity. To understand the development of the central nervous system, the multiple steps involved in generation of the numerous cell identities must be determined. One way to approach this enormously complicated task is to choose a region of the central nervous system characterized by relative simplicity.

The retina is such a region. Several characteristics make the retina more approachable than most other areas of the central nervous system. Cajal noted that the separation of different cells into distinct layers, the small size of dendritic fields, and the presence of layers consisting almost exclusively of neuronal projections are fortuitous characteristics of the retina (Cajal, 1893). The retina contains a relatively small number of neuronal cell classes characterized by stereotypical positions and distinctive morphologies. Even in very crude histological preparations, the identity of individual cells can be frequently and correctly determined based on their location. In addition, the eye becomes isolated from other parts of the central nervous system early in embryogenesis. Cell migrations into the retina are limited to the optic nerve and the optic chiasm (Burrill and Easter, 1994; Watanabe and Raff, 1988). This relative isolation facilitates the interpretation of developmental events within the retina. Taken together, all these qualities make the retina an excellent model system for the studies of vertebrate neuronal development.

Teleost retinae have been studied for over a century (Cajal, 1893; Dowling, 1987; Muller, 1857; Rodieck, 1973). The eyes of teleosts in general and zebrafish in particular are large and their neuroanatomy well characterized. An important advantage of the zebrafish retina for genetic and developmental research is that it is formed and becomes functional very early in development. Neurogenesis in the central retina of the zebrafish eye is essentially complete by 60 hpf (Nawrocki, 1985) and, as judged by the startle and optokinetic responses, the zebrafish eye detects light stimuli surprisingly early, starting between 2.5 and 3.5 dpf (Clark, 1981; Easter and Nicola, 1996). Studies of the zebrafish retina benefit from many general qualities of the system: high fecundity, transparency and external development of embryos, the ease of maintenance in large numbers, the short length of the life cycle, the ability to study haploid development, and most recently, progress in zebrafish genomics, including the genome sequencing project.

The vertebrate retina has been remarkably conserved in evolution. Early investigators noted that even retinae of divergent vertebrate phyla have similar organization (Cajal, 1893; Muller, 1857). Gross morphological and histological features of mammalian and teleost retinae display few differences. Accordingly, human and zebrafish retinae contain the same major cell classes organized in the same layered pattern. Similarities extend beyond histology and morphology. Pax-2/noi and Chx10/Vsx-2 expression patterns, for example, are very similar in mouse and zebrafish eyes (Liu *et al.*, 1994; Macdonald and Wilson, 1997; Nornes *et al.*, 1990; Passini *et al.*, 1997). Likewise, several genetic defects of the zebrafish retina are reminiscent of human disorders. For example, mutations affecting photoreceptor cells (Doerre and Malicki, 2002; Malicki *et al.*, 1996; Neuhauss *et al.*, 1999) very much resemble retinitis pigmentosa and cone-rod dystrophies (Dryja and Li, 1995; Merin, 1991; Yagasaki and Jacobson, 1989) and mutations in the Pax-2/noi gene produce abnormal optic nerve development in zebrafish as well as in humans (Macdonald and Wilson, 1997; Sanyanusin *et al.*, 1995, 1996). This similarity between human and zebrafish retinae allows us to use the zebrafish as a model of retinal disorders, a fortuitous circumstance given that vision is the major sense used by humans in their interactions with the environment. Throughout the world, diseases of the retina affect millions. In the United States alone, retinitis pigmentosa is estimated to affect 50,000-100,000 people (Dryja and Li, 1995), age-related macular degeneration over 15 million people (Seddon, 1994), and in some populations the prevalence of glaucoma can reach over 4% in people above 65 (Cedrone *et al.*, 1997). Thus, in addition to being an excellent model for studies of vertebrate neurogenesis, the zebrafish retina is likely to provide us with insights into the nature of human disorders.

II. Development of the Zebrafish Retina

A. Early Morphogenetic Events

Fate mapping studies indicate that the retina originates from a single field of cells positioned roughly between the telencephalic and the diencephalic precursor fields during early gastrulation (Woo and Fraser, 1995). During late gastrulation, the anterior and lateral migration of diencephalic precursors is thought to subdivide the retinal field into two separate primordia (Rembold *et al.*, 2006; Varga *et al.*, 1999). Neurulation in teleosts proceeds somewhat differently than in higher vertebrates. The primordium of the central nervous system does not take the form of a tube (the neural tube), and instead is shaped in the form of a solid rod called the neural keel (Fig. 1B and C) (Kimmel *et al.*, 1995; Lowery and Sive, 2004;

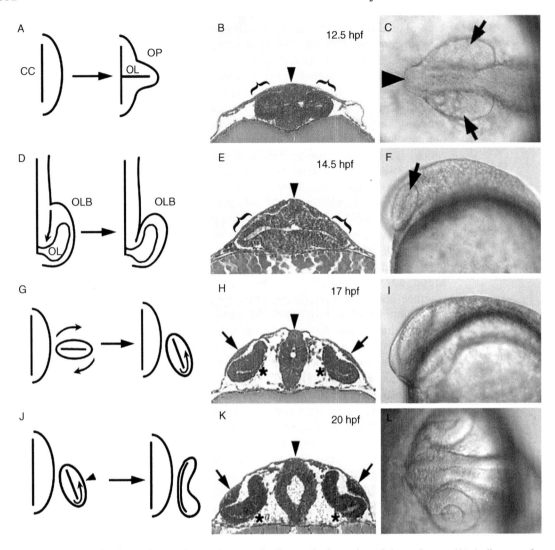

Fig. 1 Early morphogenetic events leading to the formation of the optic cup. (A) A diagram of a transverse section through anterior neural keel illustrating morphogenetic transformation that leads to the formation of optic lobes. Solid horizontal line represents the ventricular lumen (OL) of the optic lobe. (B) A transverse plastic section through the anterior portion of the neural keel and optic lobes (brackets). (C) Dorsal view of anterior neural keel and optic lobes (arrows) at 12.5 hpf. (D) A schematic representation of anterior neural tube (dorsal view, anterior down). Wing-shaped optic primordia gradually detach from the neural tube starting posteriorly (arrow). (E) A transverse plastic section through anterior neural keel and optic lobes (brackets) at 14.5 hpf. (F) Lateral view of anterior neural keel and optic lobe (arrow) at the same stage. (G) A diagram of dorso-ventral reorientation of the optic lobe. (H) A transverse plastic section through neural keel and optic lobes during the reorientation at ca. 17 hpf. At about the same time, the lens rudiments start to form (arrows) and the medial layer of the optic lobe becomes thinner as it begins to differentiate into the pigmented epithelium (asterisks). The lateral surface of the optic lobe starts to invaginate. (I) A lateral view of anterior neural keel during optic

Schmitz *et al.*, 1993). Consistent with that, optic vesicles are not present, and the equivalent structures are called optic lobes. These first become evident as bilateral thickenings of the anterior neural tube at about 11.5 hpf and gradually become more and more prominent (Fig. 1A–C) (Schmitt and Dowling, 1994). They are initially flattened and protrude laterally on both sides of the brain (Fig. 1B and C, arrows). At approximately 13 hpf, the posterior portions of optic lobes start to separate from the brain. The anterior portions, on the other hand, remain attached (Fig. 1D). This attachment will persist later in development as the optic stalk. In parallel, the optic lobe turns around its antero-posterior axis so that its lower surface becomes directed toward the brain and the upper surface toward the outside environment (Fig. 1G). Later in development, this outside surface will form the neural retina. Fate mapping studies suggest that starting at ca. 15 hpf, cells migrate from the medial to lateral epithelial layer of the optic cup (Fig. 1G) (Li *et al.*, 2000b). The medial layer becomes thinner and subsequently differentiates as the retinal pigmented epithelium (RPE) (Fig. 1H and K, asterisks). At about the same time, an invagination forms on the lateral (upper, before turning) surface of the optic lobe (Schmitt and Dowling, 1994). This is accompanied by the appearance of a thickening in the epithelium overlying the optic lobe, the lens rudiment (Fig. 1H, arrows). Subsequently, over a period of several hours both the invagination and the lens placode become increasingly more prominent, transforming the optic lobe into the optic cup (Fig. 1J–L). The choroid fissure forms in the rim of the optic cup next to the optic stalk. The lens placode continues to grow and by 24 hpf it is detached from the epidermis. At the beginning of day 2, the optic cup consists of two closely connected sheets of cells: the pseudostratified columnar neuroepithelium (rne) and the cuboidal pigmented epithelium (pe) (Fig. 2A). Starting at about 24 hpf, melanin granules appear in the cells of the pigmented epithelium. In the first half of day 2, concomitant to the expansion of the ventral diencephalon, the eye rotates so that the choroid fissure, which at 24 hpf was pointing above the yolk sack, is now directed toward the heart (Kimmel *et al.*, 1995; Schmitt and Dowling, 1994). Throughout this period, the optic stalk gradually becomes less prominent. In the first half of day 2, as ganglion cells begin to differentiate, the optic stalk provides support for their axons. Later in development, it is no longer present as a distinct structure and its cells may contribute to the optic nerve (Macdonald *et al.*, 1997).

---

cup formation. (J) A schematic representation of morphogenetic movements that accompany optic cup formation. Cells migrate (arrow) from the medial to the lateral cell layer around the ventral edge of the lobe. Simultaneously, the initially flat lobe invaginates (arrowhead) to become the concave eye cup. (K) A transverse plastic section through the anterior neural tube during optic cup formation at 20 hpf. Lens rudiments are quite prominent by this stage (arrows). Most of the medial cell layer already displays a flattened morphology, except for the ventral-most regions, which still retain columnar appearance (asterisks). (L) A dorsal view of anterior neural keel and optic lobes at 20 hpf. Vertical arrowheads in B, E, H, and K indicate the midline. CC, central canal; OL, optic lumen; OP, optic primordium; OLB, optic lobe; hpf, hours postfertilization. Except D, in all panels dorsal is up. Panels A, D, G, and J are based on Easter and Malicki (2002). The remaining panels reprinted from Pujic and Malicki (2001) with permission from Elsevier.

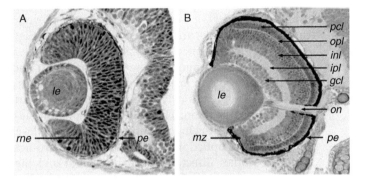

Fig. 2 Histology of the zebrafish retina. (A) A section through the zebrafish eye during early stages of neurogenesis at approximately 36 hpf. At this stage, the retina mostly consists of two epithelial layers: the pigmented epithelium and the retinal neuroepithelium. Although some retinal cells are already postmitotic at this stage, they are not numerous enough to form distinct layers. (B) A section through the zebrafish eye at 72 hpf. With the exception of the marginal zone, where cell proliferation will continue throughout the lifetime of the animal, retinal neurogenesis is mostly completed. The major nuclear and plexiform layers, as well as the optic nerve and the pigmented epithelium, are well differentiated. gcl, ganglion cell layer; inl, inner nuclear layer; ipl, inner plexiform layer; le, lens; mz, marginal zone; on, optic nerve; opl, outer plexiform layer; pcl, photoreceptor cell layer; pe, pigmented epithelium; rne, retinal neuroepithelium.

Rotation of the optic cup around its medio-lateral axis is the last major transformation in zebrafish eye development (Schmitt and Dowling, 1994).

B. Neurogenesis

At the beginning of the second day of development, the zebrafish neural retina still consists of a single sheet of pseudostratified neuroepithelium. Similar to other neural tube epithelia, the retinal neuroepithelium is a highly polarized tissue, characterized by the presence of apico-basal nuclear movements, which correlate with cell-cycle phase (Baye and Link, 2007; Das *et al.*, 2003; Hinds and Hinds, 1974). Nuclei of cells that are about to divide translocate to the apical surface of the neuroepithelium, where both nuclear division and cytokinesis take place. Although it has been assumed for a long time that dividing cells lose their contact with the basal surface of the neuroepithelium (Hinds and Hinds, 1974), more recent two-photon imaging studies in zebrafish show that this view is most likely incorrect, because a tenuous cytoplasmic process extends toward the basal surface during nuclear division of the neuroepithelial cell (Das *et al.*, 2003). Interestingly, in the brain neuroepithelium, but possibly also the retina, this process splits into two or more branches prior to the cleavage of the cell body, and the daughter processes are inherited either symmetrically or asymmetrically by the daughter cells (Kosodo *et al.*, 2008).

Between mitotic divisions, cell nuclei of neuroepithelial cells translocate basally. The depth of this translocation is very heterogeneous, ranging from 10% to 90% of neuroepithelial thickness. Interestingly, deeper nuclear migration correlates with divisions that generate postmitotic cells (Baye and Link, 2007). Mitotic divisions are observed nearly exclusively at the apical surface of the neuroepithelium until about 1.5 dpf. Following that, between 40 and 50 hpf, approximately 50% of mitoses occur in the INL (Godinho *et al.*, 2005). Very few mitotic divisions are observed in the central retina at later stages.

Despite its uniform morphological appearance, the retinal neuroepithelium is the site of many transformations, apparent in the changes of cell-cycle length and in dynamic characteristics of gene expression patterns. After a period of very slow cell-cycle progression during early stages of optic cup morphogenesis, the cell-cycle shortens to ca. 10 h by 24 hpf, and later its duration appears even shorter (Baye and Link, 2007; Hu and Easter, 1999; Li *et al.*, 2000a; Nawrocki, 1985). Imaging of individual neuroepithelial cells between 24 and 40 hpf revealed that their cell cycle varies greatly in length from about 4 to 11 h during this period, averaging approximatley 6.5 h (Baye and Link, 2007). The significance of changes in the length of the cell-cycle or the genetic mechanisms that regulate them are not understood. In parallel to the fluctuations of cell-cycle length, the expression patterns of numerous loci display dramatic changes in the retinal neuroepithelium during this time. While the transcription of some early expressed genes, such as *rx3* or *six3*, is downregulated, other loci become active. The zebrafish *atonal 5* homolog, *lakritz*, is one interesting example of an important genetic regulator characterized by a dynamic expression pattern. The *lakritz* gene becomes transcriptionally active in a small group of cells in the ventral retina by 25 hpf, and from there its expression spreads into the nasal, dorsal, and finally temporal eye (Masai *et al.*, 2000). This gradual advance of expression around the retinal surface is noteworthy because it characterizes many other developmental regulators and neuronal differentiation markers (reviewed in Pujic and Malicki, 2004).

Another noteworthy feature of neuroepithelial cells is the orientation of their mitotic spindles. The mitotic spindle position and its role in cell-fate determination is an interesting, albeit contentious issue. It has been proposed that in some species the vertical (apico-basal) reorientation of the mitotic spindle characterizes asymmetric cell divisions, which produce cells of different identities, a progenitor cell and a postmitotic neuron for example (Cayouette and Raff, 2003; Cayouette *et al.*, 2001). As such divisions first appear in the neuroepithelium at the onset of neurogenesis, so should vertically oriented mitotic spindles. The analysis of zebrafish neuroepithelial cells found, however, little support for the presence of vertically oriented mitotic spindles: the majority, if not all, of zebrafish neuroepithelial cells divide horizontally (Das *et al.*, 2003).

As the morphogenetic movements that shape and orient the optic cup come to completion, the first retinal cells become postmitotic and differentiate. Gross morphological characteristics of the major retinal cell classes are very well conserved in all vertebrates. Six major classes of neurons arise during neurogenesis:

ganglion, amacrine, bipolar, horizontal, interplexiform, and photoreceptor cells. The Muller glia are also generated in the same period. Ganglion cell precursors are the first to become postmitotic in a small patch of ventrally located cells between 27 and 28 hpf (Hu and Easter, 1999; Nawrocki, 1985). The early onset of ganglion cell differentiation is again conserved in many vertebrate phyla (Altshuler *et al.*, 1991). Similar to expression patterns that characterize the genetic regulators of retinal neurogenesis, differentiated ganglion cells first appear in the ventral retina, nasal to the optic nerve (Burrill and Easter, 1995; Schmitt and Dowling, 1996). The rudiments of the ganglion cell layer are recognizable in histological sections by 36 hpf. Approximately 10 hours after the first ganglion neuron progenitors exit the cell cycle, cells that contribute to the inner nuclear layer also become postmitotic. Again, this first happens in a small ventral group of cells (Hu and Easter, 1999). By 34–36 hpf, and possibly even earlier, terminal divisions of retinal progenitor cells give rise to daughter pairs containing a ganglion and most often a photoreceptor cell, indicating that these two cell classes are generated in overlapping windows of time (Poggi *et al.*, 2005).

By 60 hpf, over 90% of neurons in the central retina are postmitotic, and the major neuronal layers are distinguishable by morphological criteria. Cells of different layers become postmitotic in largely nonoverlapping windows of time. This is particularly obvious for ganglion cell precursors, most of which, if not all, are postmitotic before the first inner nuclear layer cells exit the cell cycle (Hu and Easter, 1999). This is different from *Xenopus*, where the times of cell-cycle exit for different cell classes overlap extensively (Holt *et al.*, 1988). In contrast to mammals, neurogenesis in teleosts and larval amphibians continues at the retinal margin throughout the lifetime of the organism (Marcus *et al.*, 1999). In adult zebrafish, as well as in other teleosts, neurons are also added in the outer nuclear layer. In contrast to the marginal zone, where many cell types are generated, only rods are added in the outer nuclear layer (Mack and Fernald, 1995; Marcus *et al.*, 1999).

Photoreceptor morphogenesis starts shortly after the exit of photoreceptor precursor cells from the cell cycle (reviewed in Tsujikawa and Malicki, 2004a). The photoreceptor cell layer can be distinguished in histological sections by 48 hpf. Photoreceptor outer segments first appear in the ventral patch by 60 hpf, and ribbon synapses of photoreceptor synaptic termini are detectable by 62 hpf (Branchek and Bremiller, 1984; Schmitt and Dowling, 1999). Rods are the first to express opsin, shortly followed by blue and red cones, and somewhat later by short-single cones (Raymond *et al.*, 1995; Robinson *et al.*, 1995; Takechi *et al.*, 2003). The photoreceptor cell layer of the zebrafish retina contains five types of photoreceptor cells: rods, short-single cones, long single cones, and short and long members of double cone pairs. The differentiation of morphologically distinct photoreceptor types becomes apparent by 4 dpf, and by 12 dpf all zebrafish photoreceptor classes can be distinguished on the basis of their morphology (Branchek and Bremiller, 1984). The photoreceptor cells of the zebrafish retina are organized in a regular pattern, referred to as the "photoreceptor mosaic." In the adult, cones form regular rows. The spaces between these rows are occupied by rods, which do not display any obvious pattern. Within a single row of cones, double cones are separated from each other by alternating long and short-single cones. Adjacent rows of cones are staggered relative to each other so that short-single cones of one row are flanked on either side by long single cones of the two neighboring rows (Fadool, 2003; Larison and Bremiller, 1990). In addition to morphology, individual types of photoreceptors are uniquely characterized by spectral sensitivities and visual pigment expressions. Long single cones express blue light-sensitive opsin; short-single cones, UV-sensitive opsin; double cones, red sensitive, and green-sensitive opsins; whereas rods express rod opsin (Hisatomi *et al.*, 1996; Raymond *et al.*, 1993). The number of opsin genes exceeds the number of photoreceptor types; two and four independent loci encode red and green opsins, respectively (Chinen *et al.*, 2003). Each green and red opsin gene is expressed in a different subpopulation of double cones. Of the two red opsin genes, LWS-2 is expressed in the central retina, while LWS-1 in the retinal periphery (Takechi and Kawamura, 2005). Similarly, the expression domains of green opsin genes, RH2-1 and RH2-2 occupy largely overlapping areas in the central retina while RH2-3 and RH3-4 are expressed at the retinal circumference in what appear to be nonoverlapping regions (Takechi and Kawamura, 2005).

C. Development of Retinotectal Projections

As this aspect of retinal development is discussed at length in an accompanying Chapter 9, we comment here on some of the most basic observations only. The neuronal network of the retina is largely self-contained. The only retinal neurons that send their projections outside are the ganglion cells. Their axons navigate through the midline of the ventral diencephalon into the dorsal part of the midbrain, the optic tectum. The ganglion cells produce axonal processes shortly after the final mitosis, while they are already migrating toward the vitreal surface (Bodick and Levinthal, 1980). The projections proceed toward the inner surface of the retina and subsequently along the inner limiting membrane toward the optic nerve head. In zebrafish, the first ganglion cell axons exit the eye between 34 and 36 hpf and navigate along the optic stalk and through the ventral region of the brain toward the midline (Burrill and Easter, 1995; Macdonald and Wilson, 1997). At about 2 dpf, the zebrafish optic nerve contains ca. 1800 axons at the exit point from the retina (Bodick and Levinthal, 1980). Cross sections near the nerve head reveal a crescent-shaped optic nerve. Axons of centrally located ganglion cells occupy the outside (dorsal) surface of the crescent whereas the axons of more peripheral (younger) cells localize to the inside (ventral) surface. With the exception of the axonal trajectories of cells separated by the choroid fissure, axons of neighboring ganglion cells travel together in the optic nerve (Bodick and Levinthal, 1980). In addition to ganglion cell axons, the optic nerve contains retinopetal projections. These appear after 5 dpf and originate in the nucleus olfactoretinalis of the rostral telencephalon (Burrill and Easter, 1994). After crossing the midline, the axonal projections of the ganglion cells split into the dorsal and ventral branches of the optic tract. The ventral branch contains mostly axons of the dorsal retinal ganglion cells, the dorsal branch mostly of the ventral cells (Baier *et al.*, 1996). The growth cones of the retinal ganglion cells first enter the optic tectum between 46 and 48 hpf. In addition to the optic tectum, the retinal axons innervate nine other, much smaller targets in the zebrafish brain (Burrill and Easter, 1994).

Spatial relationships between individual ganglion cells in the retina are precisely reproduced by their projections in the tectum. The exactitude of this pattern has long fascinated biologists and has been a subject of intensive research in many vertebrate species (Drescher *et al.*, 1997; Fraser, 1992; Sanes, 1993). The spatial coordinates of the retina and the tectum are reversed. The ventral-nasal ganglion cells of the zebrafish retina project to the dorsal-posterior optic tectum whereas the dorsal-temporal cells innervate the ventral-anterior tectum (Stuermer, 1988; Karlstrom *et al.*, 1996; Trowe *et al.*, 1996). By 72 hpf, axons from all quadrants of the retina are in contact with their target territories in the optic tectum.

In summary, development of the zebrafish retina proceeds at a rapid pace. By the end of day 3, all major retinal cell classes have been generated and are organized in distinct layers (Fig. 2B), the photoreceptor cells have developed outer segments, and the ganglion cell axons have innervated the optic tectum. It is also about this time that the zebrafish visual system becomes functional (Clark, 1981; Easter and Nicola, 1996). The brevity of eye morphogenesis and retinal neurogenesis is a major advantage offered by the zebrafish eye as a model system.

D. Nonneuronal Tissues

In many vertebrates, the retina is intimately associated with the vascular system (Wise *et al.*, 1971). The mature zebrafish retina features two vessel systems, the choroidal and retinal vasculatures. The first of these tightly surrounds the retinal pigment epithelium, whereas the second differentiates on the inner surface of the retina (Alvarez *et al.*, 2007; Kitambi *et al.*, 2009). The development of the eye vasculature can be efficiently visualized using transgenic lines. Carriers of the fli-GFP and flk-GFP transgenes are suitable for this purpose (Choi *et al.*, 2007; Lawson and Weinstein, 2002). In these strains, GFP-positive cells first appear in the retinal choroid fissure and the retina toward the end of the first 24 h of embryogenesis (Kitambi *et al.*, 2009). By 48 hpf, a vascular bed forms on the medial surface of the lens (Alvarez *et al.*, 2007; Kitambi *et al.*, 2009). Initially, retinal blood vessels appear to adhere tightly to the lens. As the organism matures, however, vasculature appears to progressively lose contact with the lens and starts to adhere to the vitreal surface of the retina (Alvarez *et al.*, 2007). In contrast to many mammals, including primates, blood vessels do not penetrate the neural retina in zebrafish (Alvarez *et al.*, 2007). In addition to the vasculature, several other nonneuronal ocular tissues, such as the cornea, the iris, the ciliary body, and the lens, have been characterized in the zebrafish in detail (Dahm *et al.*, 2007; Gray *et al.*, 2008; Soules and Link, 2005; Zhao *et al.*, 2006).

═══════════   ## III. Analysis of Wild-Type and Mutant Visual System

Diverse research approaches have been used to study the zebrafish retina. This chapter provides an overview of the available methods. While some techniques are described in detail, the majority are discussed only briefly due to space constraints and references to sources of more comprehensive protocols are provided. Where applicable, other chapters of this volume are referenced as the source of more complete information. Table I lists some of the most important techniques currently available for the analysis of the zebrafish retina.

Observations of retinal development in the zebrafish embryo after 30 hpf are hampered by pigmentation of the retinal pigmented epithelium. In immunohistochemical experiments, for example, the staining pattern is not accessible to visual inspection in whole embryos unless they are sectioned or their pigmentation is inhibited. To inhibit pigmentation, developing zebrafish embryos are raised in media containing 1-phenyl-2-thiourea (PTU). Concentrations ranging from 75 to 200 μm are recommended (Karlsson *et al.*, 2001; Westerfield, 2000). In the presence of PTU, however, zebrafish embryonic development does not proceed entirely normally. Starting between 2 and 3 dpf, embryogenesis is somewhat delayed, hatching is inhibited, and pectoral fins are abnormal (Karlsson *et al.*, 2001). Appropriate controls have to be included to account for these deviations from normal embryogenesis. An additional disadvantage of using PTU is that it does not inhibit the differentiation of iridophores, which are present on the surface of the eye by 42 hpf, and by 4 dpf are dense enough to impair the visualization of retinal cells with fluorescent probes. An alternative to using PTU is to conduct experiments on pigmentation-deficient animals. The *albino; roy* double mutant line is the most useful for this purpose because it lacks both RPE pigmentation and iridophores (Ren *et al.*, 2002). As crossing a mutation of interest into a pigmentation-deficient background takes two generations, this approach is, however, time consuming.

A. Histological Analysis

A major goal of future eye research in zebrafish will be to characterize phenotypes obtained in the course of new generations of forward and reverse genetic experiments. Even in the most comprehensive of the genetic screens performed in zebrafish so far, the number of multiple hits per locus was low, indicating that many more genes will be discovered before saturation is achieved (Driever *et al.*, 1996; Golling *et al.*, 2002; Haffter *et al.*, 1996). Thus, almost certainly, future genetic screens will enrich the already impressive collection of the zebrafish eye mutants even further. Following morphological description, the first and simplest step in the analysis of a phenotype is histological analysis. It allows one to evaluate how a mutation influences the major cell classes in the retina. Because of the exquisitely precise organization of the zebrafish retinal neurons, histological

Table I
Techniques Available to Study the Zebrafish Retina and Their Sources/Examples of Use

Protocol	Goal	Sources/examples of use
Histological analysis		
Electron microscopy	Evaluation of phenotype on a subcellular level	Allwardt et al. (2001); Doerre and Malicki (2002); Kimmel et al. (1981)
Light microscopy	Evaluation of phenotype on a cellular level	Malicki et al. (1996); Schmitt and Dowling (1994)
Molecular marker analysis		
Antibody staining (whole mount)	Determination of expression pattern on protein level	Schmitt and Dowling (1996)
Antibody staining (sections)	Determination of expression pattern on protein level	Pujic and Malicki (2001); Wei and Malicki (2002)
In situ hybridization—double labeling	Parallel determination of two expression patterns on transcript level	Hauptmann and Gerster (1994); Jowett (2001), Jowett and Lettice (1994); Strahle et al. (1994)
In situ hybridization—frozen sections	Determination of expression pattern on transcript level	Barthel and Raymond (1993); Hisatomi et al. (1996)
In situ hybridization—whole mount	Determination of expression pattern on transcript level	Oxtoby and Jowett (1993); Thisse et al. (2004)
Gene function analysis		
Implantation	Test of function for a factor (most often diffusible) via the implantation of a bead saturated with this substance	Hyatt et al. (1996); Martinez-Morales et al. (2005)
Morpholino knockdown	Test of gene function based on antisense inhibition of its activity	Eisen and Smith (2008); Nasevicius and Ekker (2000); Tsujikawa and Malicki (2004)
Overexpression (DNA injections)	Test of gene function based on enhancement of its activity through DNA injections	Koster and Fraser (2001); Mumm et al. (2006)
Overexpression (light-mediated RNA/DNA uncaging)	Identification of gene function through enhancement of its activity in selected tissues at specific developmental stages	Ando et al. (2001); Ando and Okamoto (2003)
Overexpression (RNA injections)	Test of gene function based on enhancement of its activity through RNA injections	Macdonald et al. (1995); reviewed in Malicki et al. (2002)
Overexpression (UAS-GAL4 system)	Test of gene function through enhancement of its activity in selected tissues using stable transgenic lines	Del Bene et al. (2008); Scheer and Campos-Ortega (1999)
TILLING (Targeting Induced Local LESIONS IN Genomes)	Identification of chemically induced mutant alleles in a specific genetic locus	Colbert et al. (2001); Wienholds et al. (2002)
Embryological techniques		
Cell labeling (caged fluorophore)	Fate determination for a specific group of cells	Take-uchi et al. (2003)
Cell labeling (iontophoretic)	Determination of morphogenetic movements or cell lineage relationships	Li et al. (2000); Varga et al. (1999); Woo and Fraser (1995)
Cell labeling (lipophilic tracers)	Analysis of ganglion cell development (e.g., retinotectal projection)	Baier et al. (1996); Malicki and Driever (1999); Mangrum et al. (2002)
Cell labeling (fluorescent protein transgenes)	Determination of cell-fate and fine differentiation features in living animals	Hatta et al. (2006); Mumm et al. (2006); Neumann and Nuesslein-Volhard (2000)

Method	Description	References
Mitotic activity detection (BrdU)	Identification of mitotically active cell populations; birthdating	Hu and Easter (1999); Larison and Bremiller (1990)
Mitotic activity detection (tritiated thymidine)	Identification of mitotically active cell populations; birthdating	Nawrocki (1985)
Tissue ablation	Functional test for a field of cells via their removal by surgical means	Masai et al. (2000)
Transplantation (whole eye)	Test whether a defect (in axonal navigation, for example) originates within or outside the eye	Fricke et al. (2001)
Transplantation (fragment of tissue)	Functional test for a field of cells via their transplantation to an ectopic position by surgical means	Masai et al. (2000)
Transplantation (blastomere)	Test of cell autonomy of a mutant phenotype by generating a genetically mosaic embryo	Ho and Kane (1990); Jensen et al. (2001); Malicki and Driever (1999)
Behavioral tests		
Optokinetic response	Test of vision based on eye movements; allows for evaluation of visual acuity	Brockerhoff et al. (1995); Clark (1981); Neuhauss et al. (1999)
Optomotor response	Test of vision based on swimming behavior	Clark (1981); Neuhauss et al. (1999)
Startle response	Simple test of vision based on swimming behavior	Easter and Nicola (1996)
Electrophysiological tests		
ERG	Test of retinal function based on the detection of electrical activity of retinal neurons and glia	Avanesov et al. (2005); Brockerhoff et al. (1995)
Biochemical approaches		
Coimmunoprecipitation from embryo extracts	Identification of direct and indirect protein binding partners	Insinna et al. (2008); Krock and Perkins (2008)
Tandem affinity purification from embryo extracts	Identification of direct and indirect protein binding partners	Omori et al. (2008)
Chemical screens		
Screens of small molecule libraries	Identification of chemicals that affect a developmental of process	Kitambi et al. (2008)
Screening approaches		
Behavioral	Detection of mutant phenotypes by behavioral tests	Muto et al. (2005); Neuhauss et al. (1999)
Histological	Detection of mutant phenotypes via histological analysis of sections	Mohideen et al. (2003)
Marker/tracer labeling	Detection of mutant phenotypes via staining with antibodies, rnaprobes, or lipophilic tracers	Baier et al. (1996); Guo et al. (1999)
Morphological	Detection of mutant phenotypes by morphological criteria	Malicki et al. (1996)
Transgene guided	Detection of mutant phenotypes in transgenic lines expressing fluorescent proteins in specific cell populations	Xiao et al. (2005)

We primarily cite experiments performed on the retina. Only where references to work on the eye are not available, we refer to studies of other organs. Most forward genetic approaches such as mutagenesis, mapping, and positional cloning methods do not contain visual system-specific features and thus are not listed. These approaches are discussed in-depth in other sections of this volume. Within each section of the table, entries are listed alphabetically.

analysis is frequently informative. Plastic sections offer very good tissue preservation for histological analysis. Both epoxy (epon, araldite) and methacrylate (JB4) resins are available for tissue embedding (Polysciences Inc.). Epoxy resins can be used both for light and electron microscopy. Several fixation methods suitable for plastic sections are routinely used (Li *et al.*, 2000b; Malicki *et al.*, 1996). For light microscopy, plastic sections are frequently prepared at 1–8 μm thickness and stained with an aqueous solution of 1% methylene blue and 1% azure II (Humphrey and Pittman, 1974; Malicki *et al.*, 1996; Schmitt and Dowling, 1999).

Following transmitted light microscopy histological analysis of mutant phenotypes can be performed at a higher resolution using electron microscopy. This allows one to inspect morphological details of subcellular structures, such as the photoreceptor outer segments, cell junctions, cilia, synaptic ribbons, mitochondria and many other organelles (Schmitt and Dowling, 1999). These cellular elements frequently offer insight into the nature of a mutant phenotype. Electron microscopy can be used in combination with diaminobenzidine (DAB) labeling of specific cell populations. Oxidation of DAB results in the formation of polymers which are chelated with osmium tetroxide and subsequently observed in the electron microscope (Hanker, 1979). Prior to microscopic analysis, specific cells can be selectively DAB labeled using several approaches: photoconversion, (Burrill and Easter, 1995) antibody staining combined with peroxidase detection, (Metcalfe *et al.*, 1990) or retrograde labeling with horse radish peroxidase (HRP) (Metcalfe, 1985).

B. The Use of Molecular Markers

A variety of molecular markers are used to study the zebrafish retina before, during, and after neurogenesis. Endogenous transcripts and proteins are among the most frequently used markers, although smaller molecules, such as neurotransmitters, and neuropeptides can also be used. During early embryogenesis, the analysis of marker distribution allows one to determine whether the eye field is specified correctly. Several RNA probes are available to visualize the optic lobe during embryogenesis (Table II). Some of them label all cells of the optic lobe uniformly, while others can be used to monitor the optic stalk area (Table II). After the completion of neurogenesis, cell class-specific markers are used to determine whether particular cell populations are specified and occupy correct positions. Some of these markers are also listed in Table II. Many transcript and protein detection methods have been described. Detailed protocols for most of these are available and we reference many of them in Table I. Below we discuss in detail the main types of molecular probes used to study the zebrafish visual system.

1. Antibodies

Antibody staining experiments can be performed in a several ways. Staining of whole embryos is the easiest. Many antibodies produce high background in whole-mount experiments, however, and the eye pigmentation needs to be eliminated

Table II
Selected Molecular Markers Available to Study the Zebrafish Retina

Name	Type	Expression pattern	References[a]/sources
Optic lobe, optic stalk markers			
pax2a (pax 2)	RNA probe and Ab (poly)	Nasal retina, optic stalk (≤24 hpf); ON (2 dpf)	Kikuchi et al. (1997); Macdonald et al. (1997); Covance PRB-276P
rx1 (zrx1)	RNA probe	Anterior neural keel, optic primordia (≤11 hpf)	Chuang et al. (1999); Pujic and Malicki (2001)
rx2 (zrx2)	RNA probe	Anterior neural keel, optic primordia (≤11 hpf)	Chuang et al. (1999); Pujic and Malicki (2001)
rx3 (zrx3)	RNA probe	Anterior neural plate (≤9 hpf); optic primordia (≤12 hpf)	Chuang et al. (1999); Pujic and Malicki (2001)
six3a (six3)	RNA probe	Neural keel, optic primordia (≤11 hpf)	Pujic and Malicki (2001); Seo et al. (1998)
six3b (six6)	RNA probe	Anterior neural keel, optic primordia (≤11 hpf)	Pujic and Malicki (2001); Seo et al. (1998)
vax2	RNA probe	Optic stalk (≤15 hpf); optic stalk, ventral retina (≤18 hpf)	Take-uchi et al. (2003)
Ganglion cell markers			
alcam[b] (Neurolin)	RNA probe, Ab (mono and poly)	Ganglion cells (28 hpf, RNA; ≤32 hpf protein)	Fashena and Westerfield (1999); Laessing et al. (1994); Laessing and Stuermer (1996); Zn-5/Zn-8 DSHB and ZIRC
cxcr4b	RNA probe	Ganglion cells (30 hpf)	Pujic et al. (2006)
gc34	RNA probe	Ganglion cells (≤36 hpf)	Pujic et al. (2006)
L3	RNA probe	Ganglion cells (30 hpf)	Brennan et al. (1997)
Tg (ath5:GFP)	Transgene	Ganglion cells (25 hpf)	Masai et al. (2003); Masai et al. (2005)
Tg (brn3c: gap43-GFP)	Transgene	Ganglion cells (42 hpf)	Xiao et al. (2005)
Amacrine cell markers			
Ap2α	RNA probe	Amacrine cells (1.5-2 dpf)	Pujic et al. (2006)
Ap2β	RNA prove	Amacrine cells (≤36 hpf)	Pujic et al. (2006)
Choline Acetyltransferase	Ab (poly)	Subset in INL and GCL, IPL (≤5 dpf)	Avanesov et al. (2005); Millipore, cat# AB144P
GABA	Ab (poly)	Subset in INL and GCL, IPL (2.5 dpf); ON (2 dpf)	Sandell et al. (1994); Millipore, cat# AB131; Sigma, cat# A2052
GAD67	Ab (poly)	Subset in INL and few in GCL, IPL (≤7 dpf)	Connaughton et al. (1999); Kay et al. (2001); Millipore, cat# AB9706
Hu C/D	Ab (mono)	INL and GCL (≤3 dpf)	Kay et al. (2001); Link et al. (2000); Invitrogen, cat# A21271
Neuro-peptide Y	Ab (poly)	Subset in INL, IPL (≤4 dpf)	Avanesov et al. (2005); ImmunoStar, cat# 22940

(continues)

Table II (*continued*)

Name	Type	Expression pattern	References[a]/sources
Parvalbumin	Ab (mono)	Subset in INL and GCL, IPL (≤ 3 dpf)	Malicki *et al.* (2003); *Millipore, cat# MAB1572*
pax6a (pax6.1)	RNA probe Ab (poly)	Neuroepithelium (12-34 hpf); INL and GCL (2 dpf); INL (5 dpf)	Hitchcock *et al.* (1996); Macdonald and Wilson (1997)
Serotonin	Ab (poly)	Subset in INL (≤ 5 dpf)	Avanesov *et al.* (2005); *Sigma, cat# S5545*
Somatostatin	Ab (poly)	Subset in INL (≤ 5 dpf)	Malicki lab (unpublished data); *ImmunoStar, cat# 20067*
Substance P	Ab (mono)	Subset in INL (≤ 5 dpf)	Malicki lab (unpublished data); *AbCam, cat# AB6338*
Tyrosine hydroxylase	Ab (mono)	Subset in INL (3–3.5 dpf)	Biehlmaier *et al.* (2003); Pujic and Malicki (2001); *ImmunoStar, cat# 22941; Millipore, cat# MAB318*
Bipolar cell markers			
vsx1	RNA probe	Neuroepithelium (31 hpf); outer INL (50 hpf)	Passini *et al.* (1997)
vsx2	RNA probe	Neuroepithelium (24 hpf); primarily or exclusively in the bipolar cells (50 hpf)	Passini *et al.* (1997)
Protein kinase C $\beta 1$	Ab (poly)	IPL, OPL (2.5 dpf); bipolar cells somata (4 dpf)	Biehlmaier *et al.* (2003); Kay *et al.* (2001); *Santa Cruz, cat# sc-209*
Tg(nyx::Gal4VP16; UAS:: MYFP)	Transgene	ON bipolar cells (2.5 dpf)	Schroeter *et al.* (2006)
Horizontal cell markers			
Cx 52.6	Ab (poly)	Horizontal cells (≤ 7 dpf)	Shields *et al.* (2007)
Cx 55.5	Ab (poly)	Horizontal cells (≤ 7 dpf)	Shields *et al.* (2007)
Horizin	RNA probe	Horizontal cells, weak staining in GCL and inner INL (≤ 60 hpf)	Pujic *et al.* (2006)
Photoreceptor markers			
Blue opsin	Ab (poly)	Blue cones (≤ 3 dpf)	Doerre and Malicki (2001); Vihtelic *et al.* (1999)
Blue opsin[c]	RNA probe	Blue cones (52 hpf)	Chinen *et al.* (2003); Raymond *et al.* (1995); Vihtelic *et al.* (1999)
Green opsin	Ab (poly)	Green cones (≤ 3 dpf)	Doerre and Malicki (2001); Vihtelic *et al.* (1999)
Green opsins (four genes)	RNA probes	Green cones (40–45 hpf)	Chinen *et al.* (2003); Takechi and Kawamura (2005); Vihtelic *et al.* (1999)
NDRG1	RNA probe	Photoreceptors (36–48 hpf)	Pujic *et al.* (2006)

Marker	Type	Expression onset	References
Red opsin	Ab (poly)	Red cones (≤3 dpf)	Doerre and Malicki (2001); Vihtelic et al. (1999)
Red opsins (two genes)	RNA probes	Red cones (40–45 hpf)	Chinen et al. (2003); Raymond et al. (1995); Takechi and Kawamura (2005)
Rod opsin	Ab (poly)	Rods (≤3 dpf)	Doerre and Malicki (2001); Vihtelic et al. (1999)
Rod opsin [c]	RNA probe	Rods (50 hpf)	Chinen et al. (2003); Raymond et al. (1995)
Tg (opn1sw1: EGFP)	Transgene	UV cones (≤56 hpf)	Takechi et al. (2003)
Tg (xops: GFP)	Transgene	Rods	Fadool (2003)
UV opsin	RNA probe	UV cones (56 hpf)	Hisatomi et al. (1996); Takechi et al. (2003)
UV opsin	Ab (poly)	UV cones (≤3 dpf)	Doerre and Malicki (2001); Vihtelic et al. (1999)
Zpr1 (Fret 43)	Ab (mono)	Double cones in larvae (48 hpf); double cones and bipolar cell subpopulation in the adult	Larison and Bremiller (1990); ZIRC
Zpr3 (Fret11)	Ab (mono)	Rods (50 hpf)	Schmitt and Dowling (1996); ZIRC
Zs-4	Ab (mono)	Rod inner segments (adult), onset unknown	Vihtelic and Hyde (2000); ZIRC
Muller glia markers			
cahz (carbonic anhydrase)	RNA probe Ab (poly)	Mueller glia (≤4 dpf)	Peterson et al. (1997, 2001)
GFAP	Ab (poly)	Mueller glia (5 dpf)	Malicki Lab (unpublished data); DAKO, cat# Z0334
Glutamine synthetase	Ab (poly)	Mueller glia (60 hpf)	Peterson et al. (2001)
Tg (gfap: GFP)	Transgene	Muller glia (48 hpf)	Bernardos and Raymond (2006)
Plexiform layer markers			
Phalloidin	Fungal toxin	IPL, OPL, ON (≤60 hpf)	Malicki et al. (2003); Invitrogen, cat# A-12379
Snap-25	Ab (poly)	OPL, IPL (≤2.5 dpf)	Biehlmaier et al. (2003); StressGen, cat# VAP-SV002
SV2	Ab (mono)	IPL, OPL (≤2.5 dpf)	Biehlmaier et al. (2003); DSHB
Syntaxin-3	Ab (poly)	OPL (2.5 dpf); faint IPL (5 dpf)	Biehlmaier et al. (2003); Alamone labs, cat# ANR-005

Approximate time of the expression onset is indicated in parenthesis. Sources of commercially available reagents are listed, including catalog numbers where appropriate. Names of markers are listed alphabetically within each section.

DSHB, Developmental Studies Hybridoma Bank (http://www.uiowa.edu/dshbwww/); ZIRC, Zebrafish International Resource Center (http://zfin.org/zirc/home/guide.php); dpf, days postfertilization; hpf, hours postfertilization; GCL, ganglion cell layer; INL, inner nuclear layer; IPL, inner plexiform layer; OPL, outer plexiform layer; OL, optic lobe; ON, optic nerve.

[a] When references to work performed on zebrafish are not available, experiments on related fish species are cited.

[b] Zn5 and Zn8 antibodies both recognize neurolin (Kawahara et al., 2002).

[c] Transcript expression onset was estimated by using goldfish probes (Raymond et al., 1995).

after 30 hpf as described above. At later stages of development, tissue penetration may become an additional problem. This can be alleviated by increasing detergent concentration above the standard level of 0.5% (2.5% Triton in both blocking and staining solution works well for anti-Pax-2 antibody, see Riley *et al.*, 1999) or by enzymatic digestion of embryos (for example collagenase treatment, see Doerre and Malicki, 2002). When background or tissue penetration is a problem, a useful alternative to whole mounts is either frozen or paraffin sections. Confocal microscopy applied to the analysis of sections allows one to reduce background even further.

For cryosectioning, embryos should be fixed as appropriate for a particular antigen and infiltrated in sucrose for cryoprotection. While for many antigens simple overnight fixation in 4% paraformaldehyde (PFA) at 4 °C is sufficient, some others require special treatments. Anti-GABA staining of amacrine cells, for example, requires fixation in both glutaraldehyde and paraformaldehyde (2% each, see Sandell *et al.*, 1994) (Fig. 3F). Glyoxal-based fixatives (such as Prefer fix supplied by Anatech) may also be useful when testing new antibodies (Dapson, 2007; Pathak *et al.*, 2007). Proper orientation of fixed specimen can be accomplished in molds prepared from eppendorf tubes by cutting them transversely into ca. 3–4 mm wide rings. These are then placed flat on a glass slide and filled with embedding medium (Richard-Allan Scientific Inc.). Embryos are placed in the medium, oriented with a probe, and transferred into a cryostat chamber that is cooled to −20 °C. Once the medium solidifies, plastic rings are removed with a razor blade.

Antibody staining can be efficiently performed on 15–30 μm sections and analyzed by confocal microscopy. For conventional microscopy, thinner sections may be desired. Upon the application of modified infiltration and embedding protocols, 3 μm sections of the zebrafish embryos can be prepared and analyzed using a conventional microscope equipped with UV illumination (Barthel and Raymond, 1990). Some antigens require the application of additional steps during staining protocols. Antigen retrieval procedure is necessary, for example, in the case of antiserotonin or anticholine acetlytransferase staining (Fig. 3G and H) (Avanesov *et al.*, 2005). While staining for the presence of these antigens, sections are immersed in near-boiling solution of 10 mM sodium citrate for 10 min prior to the application of blocking solution. Acetone treatment is required for some anti-γ-tubulin antibodies (Pujic and Malicki, 2001).

In some cases, such as the detection of GABA, antibody staining can be performed on plastic sections. Both epoxy (Epon-812, Electron Microscopy Sciences Inc.) and methacrylate (JB-4, Polysciences Inc.) resins can be used as the embedding medium. Plastic sections preserve tissue morphology very well and consequently are more informative than frozen sections. In the GABA protocol, primary antibody can be detected using avidin-HRP conjugate (Vector Laboratories Inc.) or a fluorophore-conjugated secondary antibody (Fig. 2F and Malicki and Driever, 1999; Sandell *et al.*, 1994). An extensive collection of antibodies that can be used to visualize features of the retina in the adult zebrafish has been characterized by Yazulla and Studholme (2001).

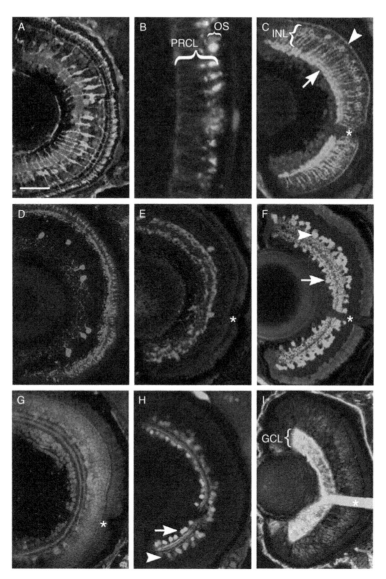

Fig. 3 Transverse sections through the center of the zebrafish eye reveal several major retinal cell classes and their subpopulations. (A) Antirod opsin antibody detects rod photoreceptor outer segments (red), which are fairly uniformly distributed throughout the outer perimeter of the retina by 5 dpf. On the same section, an antibody to carbonic anhydrase labels cell bodies of Mueller glia in the INL as well as their radially oriented processes. (B) A higher magnification of the photoreceptor cell layer shows the distribution of rod opsin (red signal) and UV opsin (green signal) in the outer segments (OS) of rods and short-single cones, respectively. (C) A subpopulation of bipolar cells is detected using antibody directed to protein kinase C β (PKC). While cell bodies of PKC-positive bipolar neurons are situated in the central region of the INL, their processes travel radially into the inner (arrow) and outer (arrowhead) plexiform layers, where they make synaptic connections. (D) Tyrosine hydroxylase-positive

2. mRNA Probes

In situ hybridization with most RNA probes is conveniently done on whole embryos according to standard protocols (Oxtoby and Jowett, 1993). Following hybridization, embryos are dehydrated in ethanol and embedded in plastic as described above (Pujic and Malicki, 2001). Expression patterns are subsequently analyzed on 1–5 μm sections. Several *in situ* protocols can be used to monitor the expression of two genes simultaneously (Jowett, 2001, and references in Table II; Jowett and Lettice, 1994). In the experiment shown in Fig. 4B, expression patterns of two opsins are detected simultaneously using two different chromogenic substrates of alkaline phosphatase (AP) (Hauptmann and Gerster, 1994). *In situ* hybridization can also be combined with antibody staining (Novak and Ribera, 2003; Prince *et al.*, 1998). In embryos older than 5 dpf, *in situ* reagents sometimes do not penetrate to the center of the retina. In such cases, hybridization procedures can be performed more successfully on sections (Hisatomi *et al.*, 1996). Given the small size of zebrafish embryos, *in situ* hybridization experiments can be performed in a high-throughput fashion using hundreds or even thousands of probes to screen for genes expressed in specific organs, tissues, or even-specific cell types (Thisse *et al.*, 2004). This approach was also applied to the retina and lead to the identification of numerous transcripts expressed in subpopulations of retinal cells (Pujic *et al.*, 2006). Some of these transcripts can be used as markers of specific retinal cell classes.

3. Lipophilic Tracers

Details of cell morphology can also be studied using lipophilic carbocyanine dyes, which label cell membranes: DiI and DiO (Honig and Hume, 1986, 1989). In the retina, these are especially useful in the analysis of ganglion cells. Carbocyanine dyes can be used as anterograde as well as retrograde tracers. When applied to the retina, DiI and DiO allow one to trace the retinotectal projections

---

interplexiform cells are relatively sparse in the larval retina. (E) Similarly, the distribution of neuropeptide Y is limited to only a few cells per section. (F) The distribution of GABA, a major inhibitory neurotransitter. GABA is largely found in amacrine neurons in the INL (arrowhead), although some GABA-positive cells are also found in the GCL (arrow). (G) Choline acetyltransferase, an enzyme of acetylcholine biosynthetic pathway, is restricted to a relatively small amacrine cell subpopulation. (H) Antibodies directed to a calcium-binding protein, parvalbumin, recognize another fairly large subpopulation of amacrine cells in the INL (green, arrowhead). Some parvalbumin-positive cells localize also to the GCL and most likely represent displaced amacrine neurons (arrow). By contrast, serotonin-positive neurons (red) are exclusively found in the INL. (I) Ganglion cells stain with the Zn-8 antibody directed to neurolin, a cell-surface antigen (Fashena and Westerfield, 1999). In addition to neuronal somata, strong Zn8 staining exists in the optic nerve (asterisk). In all panels lens is left, dorsal is up. (A–H) show the retina at 5 dpf, (I) shows a 3 dpf retina. Asterisks indicate the optic nerve. Scale bar equals 50 μm in (A), and (C–I) and 10 μm in (B). dpf, days postfertilization; GCL, ganglion cell layer; INL, inner nuclear layer; OS, outer segments; PRCL, photoreceptor cell layer. Panels D, G, and H are reprinted from Pujic and Malicki (2004) with permission from Elsevier. (See Plate no. 11 in the Color Plate Section.)

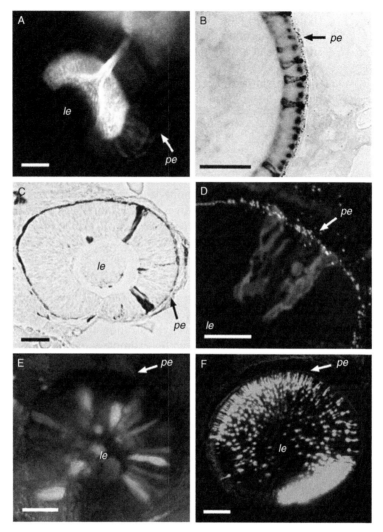

Fig. 4 Selected techniques available to study neurogenesis in the zebrafish retina. (A) DiI incorporation into the optic tectum retrogradely labels the optic nerve and ganglion cell somata. (B) A transverse plastic section through the zebrafish retina at 3 dpf. *In situ* mRNA hybridization using two probes, each targeted to a different opsin transcript and detected using a different enzymatic reaction, visualizes two types of photoreceptor cells. (C) A plastic section through a genetically mosaic retina at ca. 30 hpf. Biotinylated dextran labeled donor-derived cells incorporate into retinal neuroepithelial sheet of a host embryo and can be detected using HRP staining (brown precipitate). (D) A transverse cryosection through a genetically mosaic zebrafish eye at 36 hpf. In this case, donor-derived clones of neuroepithelial cells are detected with fluorophore-conjugated avidin (red). The apical surface of the neuroepithelial sheet is visualized with anti-γ-tubulin antibody, which stains centrosomes (green). (E) GPF expression in the eye of a zebrafish embryo following injection of a DNA construct containing the GFP gene under the control of a heat-shock promoter. The transgene is expressed in only a small subpopulation of cells. (F) A confocal z-series through the eye of a living transgenic zebrafish, carrying a GFP transgene under the control of a rod opsin promoter (Fadool, 2003).Bright expression is present in rod photoreceptor cells (ca. 3 dpf). Scale bar, 50 μm. pe, pigmented epithelium; le, lens. Panel E reprinted from Malicki *et al.* (2002) with permission from Elsevier. (See Plate no. 12 in the Color Plate Section.)

(Baier *et al.*, 1996). When applied to the optic tectum or the optic tract, they can be used to determine the position of ganglion cell pericarya, and even to study the stratification and branching of ganglion cell dendrites (Burrill and Easter, 1995; Malicki and Driever, 1999; Mangrum *et al.*, 2002). Since DiI and DiO have different emission spectra, they can be used simultaneously to label two different populations of cells (Baier *et al.*, 1996).

4. Fluorescent Proteins

Fluorescent proteins (FPs), frequently fused to other polypeptides, are a very rich source of markers to visualize tissues, cells, and even subcellular structures. These can be expressed in the embryos either transiently or from stably integrated transgenes. Numerous derivatives of two fluorescent proteins, GFP (green fluorescent protein from jellyfish, *Aequorea victoria*) and RFP (red fluorescent protein from coral species), are currently available (reviewed in Shaner *et al.*, 2007) and differ in brightness as well as emission spectra. Many of them have been applied in zebrafish. The uses of fluorescent proteins can be grouped in at least three categories:

1. *Visualization of gene activity.* The purpose of these experiments is to determine where and when a gene of interest is transcribed. Although the same goal can be accomplished via *in situ* hybridization, the use of FP fusions may result in higher sensitivity of detection (see for example a *sonic hedgehog* study by Neumann and Nuesslein-Volhard, 2000) and allows one to monitor, and even video record, gene expression continuously in living organisms. The biggest challenge in this type of study is to include all regulatory elements in a transgene in order to faithfully reproduce wild-type expression pattern. The best way to accomplish this is to insert a fluorescent protein coding sequence into the open reading frame of a gene in an artificial chromosome (PAC or BAC). For example, to study the expression of zebrafish green opsin genes, a modified PAC clone of approximatley 85 kb was used to generate transgenic lines. To visualize expression, the first exon after the initiation codon was replaced with GFP sequence in each gene (Tsujimura *et al.*, 2007). The use of artificial chromosomes is frequently necessary because distant regulatory elements are likely to affect the expression. Nevertheless, one has to remember that even artificial chromosomes do not ensure that all relevant regulatory elements will be included in a transgene.

In some experiments, when temporal characteristics of expression need to be faithfully reproduced, excessive stability of FP may pose a problem. FPs tend to be stable and may persist much longer than the transcript and the protein product of the gene being studied, making it difficult to determine when the gene of interest turns off. To circumvent this difficulty, fluorescent proteins characterized by reduced stability, such as dRFP (destabilized RFP) or shGFP (short half life GFP), are available (Yeo *et al.*, 2007; Yu *et al.*, 2007). dRFP was used, for example, to study Notch pathway activity in the zebrafish retinal neuroepithelium (Del Bene *et al.*, 2008).

2. *Visualization of subcellular localization.* In this type of experiment, it is not necessary to recapitulate the tissue distribution of the protein being studied and thus expression can be driven ubiquitously. Consequently, transient expression methods based on mRNA or DNA injection are preferred. Although they usually do not allow for the targeting of expression to particular tissues, they are much less time consuming, compared to generating stable transgenic lines. The expression of FP fusions is especially valuable when antibodies are difficult to generate, as has been the case for the Elipsa protein (Omori *et al.*, 2008). This procedure is not without drawbacks, however. First, adding GFP polypeptide to a protein may change its binding properties, and thus cause aberrant localization in the cell. Second, because FP fusions are frequently expressed at a higher level compared to their wild-type counterparts, they may display nonspecific binding. Finally, fusion proteins may be toxic to cells. These problems can be largely, although not entirely, eliminated by placing FP tags in multiple locations and testing whether the resulting fusion proteins rescue mutant/morphant phenotypes.

3. *Monitoring of fate, differentiation, and cell physiology.* In these studies, FP fusions are used solely as markers of cells or subcellular structures. In the simplest case, the differentiation and survival of cells can be monitored. In more sophisticated uses of this approach, one monitors cell division patterns, migration trajectories, or specific aspects of cell morphology, such as the shape of dendritic processes, subcellular distribution of organelles, or intracellular transport. Zebrafish FP transgenic lines have been generated to monitor the differentiation of fine morphological features of several retinal cell classes, including bipolar interneurons (Schroeter *et al.*, 2006), horizontal interneurons (Shields *et al.*, 2007), amacrine interneurons (Godinho *et al.*, 2005; Kay *et al.*, 2004), ganglion cells (Xiao *et al.*, 2005), and Muller glia (Bernardos and Raymond, 2006) (Table II). These transgenic lines allow for continuous observation of fine features of cells, including the entire trajectory of the retinotectal projection, or the phylopodia of differentiating bipolar cell axon terminals. In most studies conducted so far, FP fusions were expressed from stably integrated transgenes, although in some cases the GAL4-VP16-based system (Koster and Fraser, 2001, see below) is used to drive transient expression in retinal interneurons (Mumm *et al.*, 2006; Shields *et al.*, 2007). While generating stable transgenic lines, the injection of the same construct can produce very different expression patterns in different lines, most likely due to position-specific effects. For example, depending on the integration site, a hexamer of the DF4 regulatory element of the Pax6 gene can drive expression either throughout the retina, or in subsets of amacrine cells (Godinho *et al.*, 2005; Kay *et al.*, 2004).

In some experiments, it is useful to monitor the behavior of organelles. This is accomplished by generating FPs fused to subcellular localization signals or to entire proteins that display a desirable subcellular localization. The H2A-GFP transgene, for example, allows one not only to visualize cell nuclei but also to reveal when they are undergoing mitosis, and even to determine the orientation of mitotic spindles in the retinal neuroepithelium (Cui *et al.*, 2007; Pauls *et al.*, 2001).

Similarly, GFP-centrin can be used to monitor the position of the centrosome in differentiating ganglion cells (Zolessi *et al.*, 2006), and GFP fused to a mitochondrial localization sequence can be applied to observe the distribution of mitochondria (Kim *et al.*, 2008). GFP fused to the 44 C-terminal amino acids of rod opsin is targeted to the photoreceptor outer segment and can be used as a specific marker of this structure (Perkins *et al.*, 2002). FPs can also be applied to mark specific cell membrane domains: PAR-3/EGFP fusion, for example, labels the apical surface of retinal neuroepithelial cells (Zolessi *et al.*, 2006).

Photoconvertible FPs are yet another class of markers that can be used to visualize cell morphology. Kaede and Dronpa have been used most frequently in the zebrafish so far (Aramaki and Hatta, 2006; Hatta *et al.*, 2006; Sato *et al.*, 2006). Kaede is irreversibly converted from green to red fluorescence using UV irradiation, whereas Dronpa green fluorescence can be reversibly activated and deactivated multiple times by irradiating it with blue and UV light, respectively. The advantage of these FPs is that they can be used to reveal morphology of single neurons by selective photoconversion in the cell soma (anterograde labeling) or in cell processes (retrograde labeling). This is particularly useful when appropriate regulatory elements are not available to drive FP expression in specific cell populations.

The number of different FPs and the variety of their applications in zebrafish have been growing at a breathtaking pace. Given the multitude of available promoter sequences, the diversity of spectral variants, and the variety of methods for protein expression in the zebrafish embryo, one is frequently confronted with the task of generating multiple combinations of regulatory elements and FP tags. This can be made easier by recombination cloning approaches (Kwan *et al.*, 2007; Villefranc *et al.*, 2007, see comments below). The use of FPs to monitor the divisions, movements, and differentiation of cells has been one of the fastest growing approaches in the studies of zebrafish embryogenesis.

C. Analysis of Cell Movements and Lineage Relationships

Apart from transgenic methods, one of the best-established and most versatile approaches to cell labeling in living zebrafish embryos is iontophoresis. This technique was applied in numerous zebrafish cell-fate studies (Collazo *et al.*, 1994; Devoto *et al.*, 1996; Raible *et al.*, 1992). In the context of visual system development, iontophoretic cell labeling was used to determine the developmental origins of the optic primordium (Woo and Fraser, 1995) and later to study cell rearrangements that accompany optic cup morphogenesis (Li *et al.*, 2000b). Iontophoretic cell labeling has been applied to study cell lineage relationships in the developing retina of *Xenopus laevis* (Holt *et al.*, 1988; Wetts and Fraser, 1988). Lineage analysis has been performed in the zebrafish retina to a very limited extent only, perhaps due to the perception that it would be unlikely to add much to the results previously obtained in higher vertebrates (Holt *et al.*, 1988; Turner and Cepko, 1987; Turner *et al.*, 1990). One study of lineage relationships in the zebrafish eye took advantage of a transgenic line that expresses GFP in retinal progenitor cells (Poggi *et al.*, 2005). A potentially very informative variant of cell-fate analysis is to perform it in the retinae of mutant animals (Poggi *et al.*, 2005; Varga *et al.*, 1999). An alternative to iontophoresis is the activation of caged fluorophores using a laser beam. Caged flourescein (Molecular Probes, Inc.) is particularly popular in this type of experiment, and was applied to study cell-fate changes caused by a double knockdown of *vax1* and *vax2* gene function (Take-uchi *et al.*, 2003).

D. Analysis of Cell and Tissue Interactions

Transplantation techniques are used to determine cell or tissue interactions. The size of a transplant varies from a small group of cells or even a single cell to the entire organ. In the case of mutations that affect retinotectal projections, it is important to determine whether defects originate in the eye or in brain tissues. This can be accomplished by transplanting the entire optic lobe at 12 hpf and allowing the animals to develop until later stages (Fricke *et al.*, 2001). Smaller size fragments of tissue can be transplanted to document cell-cell signaling events within the optic cup. This approach has been applied to demonstrate inductive properties of the optic stalk tissue, and to test the presence of cell-cell interactions within the optic cup (Kay *et al.*, 2005; Masai *et al.*, 2000). Transplantation can also be used to study lens mutants. Lens transplantation in zebrafish can be performed following a procedure similar to that developed for *Astyanax mexicanus* embryos (Yamamoto and Jeffery, 2002; Zhang *et al.*, 2009).

Mosaic analysis is a widely used approach that combines genetic and embryological manipulations (Ho and Kane, 1990). The goal of such experiments is to determine in which cell or tissue a mutant gene is active. The site of gene activity does not necessarily correlate with the mutant phenotype. Cell-autonomous phenotypes are caused by gene function defects within the affected cells, while cell-nonautonomous phenotypes are caused by defects in other (frequently neighboring) cells. In contrast to *Drosophila*, zebrafish genetic mosaics are generated by embryological means: blastomere transplantations (Ho and Kane, 1990; Westerfield, 2000). As this technique has been widely used in zebrafish, also in the context of eye development, we provide a more extensive description of how it is applied. In the first step, the donor embryos are labeled at the 1–8-cell stage with a tracer. Dextrans conjugated with biotin or a fluorophore are the most commonly used tracers, and frequently a mix of these two tracers is used. Prior to a transplantation experiment, it is desirable to purify dextran solution by filtering it through a spin column (Microcon YM-3, Millipore Inc.). This procedure removes small molecular weight contaminants and increases the survival of donor cells. Within a few minutes after injection into the yolk, tracers diffuse throughout the embryo, labeling all blastomeres. Subsequently, starting at about 3 hpf, blastomeres are transplanted using a glass needle from tracer-labeled donor embryos to unlabeled host embryos. The number of transplanted blastomeres usually varies from a few to hundreds, depending on the experimental context. One donor embryo is frequently sufficient to supply blastomeres for several hosts. The transplanted blastomeres become incorporated into the host embryo and randomly contribute to various tissues including those of experimental interest. To increase the frequency of donor-derived cells in the retina, blastomeres should be transplanted into the animal pole of a host embryo (Moens and Fritz, 1999). Cells in that region will later contribute to eye and brain structures (Woo and Fraser, 1995). Embryos that contain descendants of the donor blastomeres in the eye are identified using UV illumination between 24 and 30 hpf, when the retina is only weakly pigmented and contains large radially oriented neuroepithelial cells (see Fig. 4C and D). An elegant control for the cell autonomy test can be generated by transplanting cells from two donor embryos—one wild type, one mutant—into a single host (Ho and Kane, 1990). In such a case, each of the donors has to be labeled with a different tracer. A relatively demanding step of the blastomere transplantation procedure is the preparation of a transplantation needle with an appropriate opening diameter and a sharp, preferentially beveled tip, a feature that helps to penetrate the embryo. A needle preparation method and other technical details of blastomere transplantation protocol have been described (Westerfield, 2000).

Analysis of donor-derived cells in mosaic embryos can proceed in several ways. In the simplest case, the donor-derived cells are labeled with a fluorescent tracer only and directly analyzed in whole embryos (Ho and Kane, 1990). Confocal microscopy can be used to achieve this goal. Such analysis is sufficient to provide information about the position and sometimes the morphology of donor-derived cells. When more detailed analysis is necessary, the donor-derived cells can be further analyzed on sections. In such cases, the donor blastomeres are usually labeled with both fluorophore- and biotin-conjugated dextrans. The fluorophore-conjugated tracer is used to distinguish which embryos contain donor-derived cells in the desired tissue as described above. The biotin-conjugated dextran, on the other hand, is used in detailed analysis at later developmental stages. The HRP-conjugated streptavidin version of the ABC kit (Vector Laboratories Inc.) or fluorophore-conjugated avidin (Jackson ImmunoResearch Inc., Molecular Probes Inc.) can be used to detect biotinylated dextran (Fig. 4C and D, respectively). HRP detection can be performed in whole mounts and analyzed on plastic sections, as described above for histological analysis. In contrast to that, fluorophore-conjugated tracers are preferably used after sectioning of frozen tissue, due to degradation of some flurophores during plastic embedding. In these experiments, cryosections are prepared as described for antibody staining above. In some experiments, it is desirable to analyze the donor-derived cells for the expression of molecular markers (for example, see Fig. 4D). On frozen sections, avidin detection of the donor-derived cells can be combined with antibody staining. Another way to reach this goal is to combine HRP detection of donor-derived cells with *in situ* hybridization or antibody staining (Halpern *et al.*, 1993; Schier *et al.*, 1997). When HRP is used for the detection of donor-derived cells, the resulting reaction product inhibits the detection of the *in situ* probe with alkaline phosphatase (AP) (Schier *et al.*, 1997). Because of this, the opposite sequence of enzymatic detection reactions is preferred: *in situ* probe detection first, HRP staining second.

When mosaic analysis is performed in the zebrafish retina at 3 dpf or later, the dilution of a donor-cell tracer can make the interpretation of the results difficult. This is because the descendants of a single transplanted blastomere divide a variable number of times. Thus in the donor-derived cells which undergo the highest number of divisions, the label may be diluted so much that it is no longer detectable. In mosaic animals, such a situation can lead to the appearance of a mutant phenotype or to the rescue of a mutant phenotype in places seemingly not associated with the presence of donor cells and complicate the interpretation of experimental results. Increasing the concentration of the tracer or, in the case of whole-mount experiments, improving the penetration of staining reagents, can sometimes alleviate this problem. Similar to antibody staining, collagenase can be used to improve reagent penetration during the detection of donor cells (Doerre and Malicki, 2001). The amount of injected dextran should be increased carefully as excessively high concentrations are lethal for labeled cells.

An excellent alternative to dextran tracers is the use of transgenes. An ideal transgene to mark donor cells in mosaic analysis experiments would drive the expression of membrane-bound FP at a high level in all cells throughout development. In the context of the retina, the Q01 line, which expresses mCFP, meets most of these requirements, although its expression becomes somewhat dimmer as development advances (Godinho *et al.*, 2005). This line has been used, for example, to study photoreceptor and glia defects in *ale oko* mutant retinae (Malicki lab, unpublished). The use of transgenic lines has two advantages. First, it eliminates the need to inject tracer. Second, because a transgene is continuously expressed by donor-derived cells, it is not diluted during the growth of the organism. A disadvantage of transgene use in this context is that it takes one generation to introduce a transgene into a mutant line. Mosaic analysis is an important approach that has been used to analyze numerous retinal mutants in zebrafish (Doerre and Malicki, 2001, 2002; Goldsmith *et al.*, 2003; Jensen *et al.*, 2001; Krock *et al.*, 2007; Link *et al.*, 2000; Malicki and Driever, 1999; Malicki *et al.*, 2003; Pujic and Malicki, 2001; Wei and Malicki, 2002).

E. Analysis of Cell Proliferation

Several techniques are available to study cell proliferation in the retina. The amount of cell proliferation, the timing of cell-cycle exit (birth date), and cell-cycle length can be evaluated by H^3-thymidine labeling (Nawrocki, 1985) or via BrdU injections into the embryo (Hu and Easter, 1999). Such studies can be very informative in mutant animals (Kay *et al.*, 2001; Link *et al.*, 2000; Yamaguchi *et al.*, 2008). To identify the population of cells that exit the cell cycle in a particular window of time, BrdU labeling can be combined with IdU (Del Bene *et al.*, 2008). Finally, a useful technique that can be used to test for cell-cycle defects in mutant strains is FACS sorting of dissociated retinal cells (Plaster *et al.*, 2006; Yamaguchi *et al.*, 2008).

F. Behavioral Studies

Several vision-dependent behavioral responses have been described in zebrafish larvae and adults: the optomotor response (Clark, 1981), the optokinetic response (Clark, 1981; Easter and Nicola, 1996), the startle response (Easter and Nicola, 1996), the phototaxis (Brockerhoff *et al.*, 1995), the escape response (Li and Dowling, 1997), and the dorsal light reflex (Nicolson *et al.*, 1998). Not surprisingly, larval feeding efficiency also depends on vision (Clark, 1981). While some of these behaviors are already present by 72 hpf, others have been described in adult fish only (for a recent review see Neuhauss, 2003). The vision-dependant behaviors of zebrafish proved to be very useful in genetic screening (see genetic screens below). The optokinetic response appears to be the most robust and versatile. It is useful both in quick tests of vision and in quantitative estimates of visual acuity. In addition to genetic screens, behavioral tests have been used to study the function of the zebrafish optic tectum (Roeser and Baier, 2003). An in-depth discussion of behavioral approaches is provided elsewhere in this volume.

G. Electrophysiological Analysis of Retinal Function

In addition to behavioral tests, measurements of electrical activity in the eye are another more precise way to evaluate retinal function. Electrical responses of the zebrafish retina can be evaluated by electroretinography (ERG) by 4 dpf (for example, Avanesov *et al.*, 2005). Similar to other vertebrates, the zebrafish ERG response contains two main waves: a small negative a-wave, originating from the photoreceptor cells, and a large positive b-wave, which reflects the function of the inner nuclear layer (Dowling, 1987; Makhankov *et al.*, 2004). The goal of an ERG study in zebrafish is no different from that of a similar procedure performed on the human eye. ERG can be used to evaluate the site of retinal defects in mutant animals. Ganglion cell defects do not affect the ERG response (Gnuegge *et al.*, 2001), whereas the absence of the a-wave or the b-wave suggests a defect in the photoreceptor or in the inner nuclear layers, respectively. The a-wave is relatively small in ERG measurements due to an overlap with the b-wave. To measure the a-wave amplitude, the b-wave can be blocked pharmacologically (Kainz *et al.*, 2003). An additional ERG wave, the d-wave, is produced when longer (ca. 1 s) flashes of light are used. Referred to as the OFF response, the d-wave is thought to reflect the activity of OFF-bipolar cells and photoreceptors (Kainz *et al.*, 2003; Makhankov *et al.*, 2004).

Retinal responses are usually elicited using a series of light stimuli that vary by several orders of magnitude in intensity (Allwardt *et al.*, 2001; Kainz *et al.*, 2003). This allows the evaluation of the visual response threshold, a parameter that is sometimes abnormal in mutant animals (Li and Dowling, 1997). Another important variable in ERG measurements is the level of background illumination. ERG measurements can be performed on light-adapted retinae using background illu-mination of a constant intensity, or in dark-adapted retinae, which are maintained in total darkness for at least 20 min prior to measurements (Kainz *et al.*, 2003). Most frequently recordings are performed on intact anesthetized animals (Makhankov *et al.*, 2004). Alternatively, eyes are gently removed from larvae and bathed in an oxygenated buffer solution. This ensures the oxygen supply to the retina in the absence of blood circulation (Kainz *et al.*, 2003). ERG recordings have become a standard assay while evaluating zebrafish eye mutants (Allwardt *et al.*, 2001; Avanesov *et al.*, 2005; Biehlmaier *et al.*, 2007; Brockerhoff *et al.*, 1998; Kainz *et al.*, 2003; Makhankov *et al.*, 2004; Morris *et al.*, 2005).

In addition to ERG, other more sophisticated electrophysiological measurements can be used to evaluate zebrafish (mutant) retinae. The ganglion cell function, for example, can be evaluated by recording action potentials from the optic nerve (Emran *et al.*, 2007). Such measurements revealed ganglion cell defects in the retinae of *nbb* and *mao* mutants (Gnuegge *et al.*, 2001; Li and Dowling, 2000). Similarly, photoreceptor function has been evaluated by measuring outer segment currents in isolated cells (Brockerhoff *et al.*, 2003).

H. Biochemical Approaches

Genetic experiments in animal model systems are frequently supplemented with studies of protein-protein interactions. Although this type of analysis has not traditionally been a strength of the zebrafish model, zebrafish embryos can be used to analyze protein-protein binding interactions. In the context of the visual system, biochemical analysis has been largely applied to study the intraflagellar transport in photoreceptor outer segment formation. As this process occurs in many tissues, it can be studied via coimmunoprecipitation from embryonic or larval extracts (Krock and Perkins, 2008). Alternatively, extracts from the retinae of adult animals can be used (Insinna *et al.*, 2008). An advantage of using larvae is that one can apply biochemical methods to analyze mutant phenotypes. As most zebrafish mutants are lethal at embryonic or larval stages, adult retinae are not suitable for this purpose. In addition to immunoprecipitation experiments, a more sophisticated but also more laborious and technically demanding approach is tandem affinity purification (TAP) (reviewed in Collins and Choudhary, 2008). The TAP tag procedure involves attaching a peptide tag to the protein of interest, and expressing it in zebrafish embryos. Following the preparation of embryonic extract, the peptide tag is used to purify the target protein along with its binding partners using appropriate affinity columns. The identities of the binding partners are established using mass spectrometry. The TAP tag approach was applied in the zebrafish to identify the binding partners of Elipsa, a determinant of outer segment differentiation (Omori *et al.*, 2008). It is a relatively demanding technique, because it requires the expression of the bait protein in thousands of embryos. As more efficient affinity purification tags are engineered (Burckstummer *et al.*, 2006), TAP is likely to become easier to apply in the zebrafish.

I. Chemical Screens

Another approach that is gaining popularity in zebrafish is the screening of chemical libraries for compounds that affect developmental processes. The same characteristics that make the zebrafish suitable for genetic experiments: small size, rapid development, and transparency, also make it exceptionally useful for small molecule screening (Peterson *et al.*, 2000; Tran *et al.*, 2007; Zon and Peterson, 2005). In this type of experiment, hundreds or even thousands of small batches of embryos are each exposed to a different chemical compound, and analyzed for developmental changes. Such an approach can be applied either to wild-type embryos or to carriers of genetic defects. In the latter case, compounds that rescue a mutant phenotype can be searched for. When mutations that resemble human abnormalities are used, this can be a powerful way to identify chemicals of potential therapeutic importance (Hong *et al.*, 2006).

Chemical libraries ranging in size from hundreds to tens of thousands of molecules are commercially available. Phenotype detection methods in a small molecule screen are potentially as varied as in a genetic screen (see below). Gross evaluation of morphological features is the simplest option. Transgenic lines that express fluorescent proteins in target tissues make it possible to detect subtle phenotypes. In a recent experiment, for example, a flk-GFP transgenic line was used to screen approximately 2000 small molecules for their effects on retinal vasculature (Kitambi *et al.*, 2009). Although little precedent exists at this time for small molecule screens focusing on retinal development, this approach has been successful in the analysis of other zebrafish organs (Hong *et al.*, 2006; North *et al.*, 2007; Sachidanandan *et al.*, 2008), and thus is also likely to find its way into the studies of the visual system.

IV. Analysis of Gene Function in the Zebrafish Retina

A. Reverse Genetic Approaches

A series of mutant alleles of varying severity is arguably the most informative tool of gene function analysis. Although a great variety of mutant lines have been identified in forward genetic screens (see below), for many loci chemically induced mutant alleles are not yet available. In these cases, other approaches must be applied to study gene function. In this section we briefly discuss advantages and disadvantages of different loss-of-function and gain-of-function approaches in the context of the zebrafish visual system, and we provide references to more comprehensive discussions of each.

1. Loss–of-function Analysis

In the absence of chemically induced loss-of-function alleles, antisense-based interference is by far the most common way to obtain information about gene function in the zebrafish embryo (Nasevicius and Ekker, 2000). The reasons for this popularity are low-cost and low-labor expense involved in their use. Although morpholino-modified oligonucleotides have been shown to reproduce chemically induced mutant phenotypes quite well, their use suffers from two main disadvantages. First, they become progressively less effective as development proceeds, presumably due to degradation. Second, some morpholinos produce nonspecific toxicity, which must be distinguished from specific features of a morpholino-induced phenotype. Morpholino oligos can be used to interfere either with translation initiation or with splicing. Importantly, the efficiency of splice-site morpholinos can be monitored by RT-PCR (Draper et al., 2001; Tsujikawa and Malicki, 2004b). In general, splice-site morpholinos reduce wild-type transcript expression below the level of RT-PCR detection throughout the first 36 h of development, although some have been reported to remain active until 3 or even 5 dpf (Tsujikawa and Malicki, 2004b). Most morpholinos are thus sufficient to interfere with genetic pathways involved in retinal neurogenesis but not to study later differentiation events or retinal function. Some help designing morpholinos can also be obtained from their manufacturer (Gene Tools LLC). Detailed protocols for the use of morpholinos, including their target-site homology requirements, injection protocols, and methods to control for specificity are available in literature (reviewed by Eisen and Smith, 2008; Malicki et al., 2002).

A powerful alternative to the use of morpholinos is targeting induced local leisions IN genomes (TILLING) (Colbert et al., 2001; Wienholds et al., 2002). This approach, essentially a combination of chemical mutagenesis and PCR-based mutation detection protocol, produces a series of variable strength alleles for the target locus. Its main disadvantage is the vast amount of preparation that needs to be done to initiate these experiments. One particularly labor-intensive step is the collection of thousands of sperm and DNA samples from mutagenized males. Because of this limitation, TILLING experiments need to be performed by core facilities, which serve a group of laboratories, or the entire research community. A detailed discussion of this approach is provided elsewhere in this volume.

A recent addition to mutagenesis approaches in zebrafish is the use of zinc finger nucleases (ZFNs) to induce lesions in specific genes. ZFNs consist of a DNA recognition module, essentially a tandem array of 2-4 zinc finger-type DNA binding domains, and a catalytic module, which mediates DNA cleavage (reviewed in Porteus and Carroll, 2005). The catalytic module is usually derived from the FokI restriction endonuclease. ZFNs are designed to introduce lesions in a preselected gene, and so their DNA binding specificity is critical. The ability to manipulate DNA binding of ZFNs is based on several findings: individual zinc fingers primarily interact with a single triplet of the DNA sequence; this interaction displays a significant degree of sequence specificity; and multiple zinc fingers can be assembled together to recognize longer target sequences (reviewed in Porteus and Carroll, 2005). In zebrafish, pilot studies confirmed that ZFNs can be used to induce mutations in desired genes with good efficiency and specificity (Doyon et al., 2008; Meng et al., 2008). Nonetheless, the engineering of zinc finger binding domains of predetermined specificities remains laborious because it requires lengthy *in vitro* and/or *in vivo* selection procedures

(Doyon *et al.*, 2008; Meng *et al.*, 2008). Once these steps are streamlined, ZNFs may become the mutagenesis tool of choice in zebrafish.

2. Approaches to Gene Overexpression

To obtain a comprehensive understanding of gene function, one often needs to supplement loss-of-function analysis with overexpression data. In the simplest scenario, this can be accomplished in zebrafish by RNA or DNA injections into the embryo. Several variants of this procedure exist, each with unique advantages and drawbacks (reviewed in Malicki *et al.*, 2002). A main disadvantage of injecting RNA into embryos is its limited stability. The injection of DNA constructs, on the other hand, produces expression for a much longer period of time but only in a small number of cells. The fraction of cells that express a gene of interest following the injection of a DNA construct into the embryo can be increased by placing the gene to be studied under the control of (UAS) upstream activating sequence sites (multiple sites are used in tandem) and driving its expression with GAL4-VP16 fusion protein expressed from either a ubiquitous or a tissue-specific promoter (Koster and Fraser, 2001). Alternatively, expression efficiency (estimated as the fraction of cells that express DNA construct) can be greatly improved by using Tol2 transposon-based vectors (Kawakami, 2004). These are injected into 1-4-cell embryo along with transposase mRNA (Kawakami, 2004; Kwan *et al.*, 2007). The integration of these constructs into the genome relies on terminal transposon sequences, including the terminal inverted repeats (TIRs). The Tol2-derived terminal sequences can be as short as 150-200 bp, but tend to be longer in older vectors, such as T2KXIG (Kawakami, 2004; Urasaki *et al.*, 2006). In addition to transposon terminal sequences, these vectors contain an FP marker that helps to follow the pattern of transgene inheritance in embryonic tissues. Genes of interest can also be placed in these vectors under the control of appropriate regulatory elements. The heat-shock promoter has been used, for example, to drive the expression of a *crumbs* gene from a Tol2-based vector in the zebrafish retinal neuroepithelium. This approach produced expression in approximately half of neuroepithelial cells (Omori *et al.*, 2008).

Overexpression phenotypes can also be studied in stable transgenic lines, provided that the resulting dominant phenotype is viable or can be conditionally induced. Several efficient methods for generating transgenic zebrafish are available (recently reviewed in Soroldoni *et al.*, 2009; Suster *et al.*, 2009; Yang *et al.*, 2009). To develop a good understanding of its function, a gene under investigation may have to be overexpressed under the control of several regulatory elements and/or as a fusion with more than one tag (FP tags with different spectral characteristics and/ or a myc tag, for example). As generating appropriate expression constructs using traditional cloning approaches is laborious, recombination cloning-based strategies can greatly help to reach this goal (Kwan *et al.*, 2007; Villefranc *et al.*, 2007). These methods utilize a set of bacteriophage λ recombination enzymes to transfer DNA fragments from so-called entry vectors into so-called destination vectors, and are referred to as Gateway cloning (Hartley *et al.*, 2000). One of the most obvious advantages of the Gateway system is that it allows one to combine several different DNA elements relatively efficiently in a single enzymatic reaction. In one example of how this method can be applied, three entry clones were assembled in the correct configuration into a Tol2-based zebrafish destination vector (Kwan *et al.*, 2007). The use of the Gateway system requires some preparatory work. Recombination sites need to be added to generate entry vectors, and, similarly, the destination vectors have to be prepared by inserting recombination sites and selection markers. These procedures are nonetheless straightforward, and most standard laboratory vectors can be fairly easily converted into destination vectors. To make this approach even more attractive several destination vectors are already available for use in the zebrafish (Kwan *et al.*, 2007; Villefranc *et al.*, 2007).

A frequent limitation of overexpression studies is the pleiotropy of mutant phenotypes: for many loci, early embryonic phenotypes are so severe that they preclude the analysis of late developmental processes, such as retinal neurogenesis. Several experimental tools are available to overcome this problem, including the use of heat-shock promoters, the GAL4-UAS overexpression system, and caged nucleic acids. Similar to invertebrate model systems, the use of heat-shock induced expression in zebrafish relies on the hsp70 promoter (Halloran *et al.*, 2000). An interesting variant of this protocol involves the activation of a heat-shock promoter-driven transgene in a small group of cells in a living embryo by heating them gently with a laser beam, which provides both temporal and spatial control of overexpression pattern (Halloran *et al.*, 2000).

GAL4-UAS system provides another method to achieve spatial control of gene expression. Modeled after *Drosophila* (Brand and Perrimon, 1993), the GAL4-UAS overexpression approach takes advantage of two transgenic strains. The activator strain expresses the GAL4 transcriptional activator in a desired subset of tissues, while the effector strain carries the gene of interest under the control of a GAL4 responsive promoter. The effector transgene is activated by crossing its carrier strain to a line that carries the activator transgene (Scheer *et al.*, 2002). One variant of this system involves a fusion of the Gal4 DNA binding domain to the viral VP16 activation domain and uses a multimer of 14 UAS sites in the reporter construct (Koster and Fraser, 2001, see also comments above). The GAL4-UAS system was initially used in the zebrafish eye to study *notch* function (Scheer *et al.*, 2001), and since then has gained popularity (Del Bene *et al.*, 2008; Godinho *et al.*, 2005; Mumm *et al.*, 2006; Yeo *et al.*, 2007). Enhancer trap screens have generated hundreds of transgenic strains that express the Gal4 activator in a variety of patterns, and can be used to drive expression of UAS effector transgenes in many organs, including the eye (Asakawa and Kawakami, 2008; Scott *et al.*, 2007). Finally, an interesting method to control gene overexpression patterns takes advantage of Bhc-caged nucleic acids (Ando *et al.*, 2001). In this approach, embryos are injected with an inactive form of an overexpression construct, which is then later activated in a selected tissue using UV illumination. Both RNA and DNA templates can be used to produce overexpression in this approach (Ando and Okamoto, 2003).

B. Forward Genetics

The use of zebrafish in genetic studies offers several obvious advantages. The most important of these is the possibility of performing efficient forward genetic screens. Genetic screening is feasible because adult zebrafish are highly fecund and are easily maintained in large numbers in a fairly small laboratory space. Screens performed in the zebrafish so far identified hundreds of visual system mutants (Baier *et al.*, 1996; Fadool *et al.*, 1997; Malicki *et al.*, 1996; Muto *et al.*, 2005; Neuhauss *et al.*, 1999). While designing a genetic screen, one has to consider three important variables: the type of mutagen, the design of breeding scheme, and mutant defect recognition criteria. Each of these is discussed below.

1. Mutagenesis Approaches

The majority of screens performed in zebrafish so far involved the use of *N*-ethyl-*N*-nitrosourea (ENU) (Mullins *et al.*, 1994; Solnica-Krezel *et al.*, 1994). This mutagenesis approach is very effective as evidenced by the fact that the vast majority of mutations isolated so far are ENU-induced. A powerful alternative to chemical mutagenesis is insertional retroviral mutagenesis. Although the efficiency of this mutagenesis approach is still lower than that of chemical methods, an obvious advantage of a retroviral mutagen is that it provides means for very rapid identification of mutant genes (Amsterdam *et al.*, 1999; Golling *et al.*, 2002). Retroviral mutagenesis has also been applied on a large scale to identify hundreds of mutant strains (Golling *et al.*, 2002). The photoreceptor mutant *nrf* is an example of a retinal defect induced using this approach (Becker *et al.*, 1998). More recently, a rescreen of 250 retrovirus-induced mutants led to the identification of defects in several aspects of eye development (Gross *et al.*, 2005).

In addition to chemical mutagens and retroviral vectors, transposons provide a basis for another effective mutagenesis approach. Transposable elements of the *Tc-1/mariner* (*Sleeping beauty*) and *hAT* (*Tol2*) families integrate into the zebrafish genome in a transposase-dependant manner (Fadool *et al.*, 1998; Kawakami *et al.*, 2000; Raz *et al.*, 1998). Although initial efforts to induce mutations using transposon-based vectors were unsuccessful (Balciunas *et al.*, 2004; Kawakami *et al.*, 2004), recent experiments that relay on improved vector design generated mutants with high efficiency (Nagayoshi *et al.*, 2008; Sivasubbu *et al.*, 2006). Both *Tol2*- and *Sleeping beauty*-based constructs were used in these efforts. Transposon-based mutagenesis is an attractive alternative to retrovirus-mediated one because transposon-based vectors efficiently integrate into the zebrafish genome, and their mutagenicity (measured as the fraction of genome insertion events that lead to mutant phenotypes in homozygous animals) already exceeds that of retroviral mutagenesis (Nagayoshi *et al.*, 2008; Sivasubbu *et al.*, 2006). The use of transposons does not require technically difficult packaging of DNA into viral particles, and appears to pose fewer safety concerns. As the efficiency of transposon-mediated mutagenesis is gradually improving, future genetic screens are likely to be performed with the help of transposons.

Transposon-mediated mutagenesis is usually performed using enhancer or gene trap vectors, which carry FP reporter genes (reviewed in Balciunas *et al.*, 2004). These are expressed following trasposon integration in the vicinity of genes. Such a design is important for several reasons. First, it allows one to visually detect integration events in the vicinity of genes. These are much more likely to produce phenotypic defects, compared to insertions into nontranscribed regions of the genome. Second, as different integration events tend to produce different expression patterns of a reporter gene, at least in some cases one can distinguish them from each other via simple inspection of living embryos. Consequently, potentially mutagenic insertions can be driven to homozygocity already in the F2 generation of a screen (Nagayoshi *et al.*, 2008). Moreover, as gene/enhancer trap expression patterns suggest the function of genes in which insertions have occurred, they may allow one to focus a genetic screen on a specific developmental or physiological process. Finally, trap-induced mutant alleles are easier to maintain as their presence can be tracked in heterozygotes based on reporter gene expression pattern. Although retroviral mutagenesis vectors can also be engineered to function as traps (Ellingsen *et al.*, 2005), mutants generated using retroviral trap vectors have not been reported so far in zebrafish. Insertional mutagenesis strategies are discussed in detail in other chapters of this volume.

2. Breeding Schemes

The second important consideration is the type of breeding scheme that will carry genetic defects from mutagenized animals (G0) to the generation in which the screening for mutant phenotypes is performed. The most straightforward option, but also the most space- and time-consuming one, is screening for recessive defects in F3 generation embryos. This procedure was used in large-scale genetic screens that have been performed to date (Amsterdam *et al.*, 1999; Driever *et al.*, 1996; Haffter *et al.*, 1996). Its main disadvantage is that it requires a very large number of tanks to raise the F2 generation to adulthood. As the majority of laboratories do not have access to several thousands of fish tanks, more space-efficient procedures are frequently required. In this regard, the zebrafish offers some possibilities not available in other genetically studied vertebrates—haploid and early pressure screens (for a review, see Malicki, 2000). The major asset of these screening strategies is that one generation of animals is omitted and consequently time and the amount of laboratory space required is dramatically reduced. Although there are obvious advantages, these two screening strategies also suffer from some limitations. The most significant disadvantage of using haploids is that their development does not proceed in the same way as wild-type embryogenesis. Haploid embryos do not survive beyond 5 dpf, and even at earlier stages of development they display obvious defects. Although the eyes of haploid zebrafish appear fairly normal at least until 3 dpf, the architecture of their retinae tends to be disorganized. By 5 dpf, haploid embryos are markedly smaller than the wild type and display numerous abnormalities. In the context of the visual system, haploid screens appear useful to search for early patterning defects prior to the onset of neurogenesis.

Screening of embryos generated via the application of early pressure (Streisinger *et al.*, 1981) is another strategy that can be used to save both time and space. Similar to haploidization, this technique also allows one to screen for recessive defects in F2 generation embryos. The early pressure technique also involves some shortcomings. Embryos produced via this method display a high background of developmental abnormalities, which complicate the detection of mutant phenotypes, especially at early developmental stages. Another limitation of early pressure screens is that the fraction of homozygous mutant animals in a clutch of early pressure-generated embryos depends on the distance of a mutant locus from the centromere. For centromeric loci, the fraction of mutant embryos approaches 50%, whereas for telomeric genes it decreases below 10% (Streisinger *et al.*, 1986). In other types of screens, mutant phenotypes can be distinguished from nongenetic developmental abnormalities based on their frequencies (25% in the case of screens on F3 embryos). Clearly, this criterion cannot be used in early pressure screens. Despite these limitations, early pressure screens are a very useful approach, especially in small-scale endeavors. The experimental techniques involved in haploid and early pressure screens have been previously reviewed in depth (Beattie *et al.*, 1999; Walker, 1999).

While the approaches discussed above are used to identify recessive mutant phenotypes, an entirely different breeding scheme is used in searches for dominant defects. These can already be detected in embryos, larvae, or adults of the F1 generation. Although this category of screens requires just a single generation and consequently a very small amount of tank space, few experiments focusing on dominant defects have been performed in zebrafish so far (van Eeden *et al.*, 1999). An example of a search for dominant defects of the visual system is provided by a small behavioral screen of adult animals for defects of visual perception, which identified a late-onset photoreceptor degeneration phenotype (Li and Dowling, 1997).

3. Phenotype Detection Methods

The third important consideration while designing a genetic screen is the choice of mutant phenotype detection method. This aspect of screening allows for substantial creativity. Phenotype detection criteria range from very simple to very sophisticated. Ideally, the mutant phenotype recognition strategy should fulfill the following requirements: (1). Involve minimal effort. (2). Detect gross abnormalities as well as subtle changes. (3). Exclude phenotypes irrelevant to the targeted process. One class of irrelevant phenotypes are nonspecific defects. In large-scale mutagenesis screens performed so far, more than two thirds of all phenotypes were classified as nonspecific (Driever *et al.*, 1996; Golling *et al.*, 2002; Haffter *et al.*, 1996). The most frequent nonspecific phenotypes in zebrafish are early degeneration spreading across the entire embryo, and developmental retardation affecting brain, eyes, fins, and jaw. The latter class of mutants affects tissues that display robust proliferation between 3 and 5 dpf. Nonspecific phenotypes are not necessarily without value, but are usually considered uninteresting because they are likely to be produced by defects in a broad range of housekeeping mechanisms (such as metabolic pathways or DNA replication machinery, see, for example, Allende *et al.*, 1996; Plaster *et al.*, 2006). Another category of irrelevant phenotypes includes specific defects of no interest to the investigators performing the screen. Such phenotypes are isolated when a screening procedure detects mutations affecting multiple organs, only one of which is of interest. A good example of such a situation is provided by behavioral screens involving the optomotor response. Lack of the optomotor response may be due to defects of photoreceptor neurons or skeletal muscles. These two cell types are rarely of interest to the same group of investigators. It is one of the virtues of a well-designed screen that irrelevant phenotypes are efficiently eliminated.

The simplest way to screen for mutant phenotypes is by visual inspection. The most significant disadvantage of this method is that it detects changes only in structures easily recognizable using a microscope (preferably a dissecting scope). Thus visual inspection screens are suitable to search for defects in zebrafish blood vessels (which are easy to see in larvae), but would not detect a loss of a small population of neurons hidden in the depths of the brain. Visual inspection criteria work well when the aim of a screen is the detection of gross morphological changes. Within the eye, such changes may reflect specific defects in a single neuronal lamina. In several mutants, the changes of eye size are caused by a degeneration of photoreceptor cells (Malicki *et al.*, 1996). In this case, the affected cell population is numerous enough to cause a major change of morphology. Most likely, a morphological screen would not detect abnormalities in a less numerous cell class.

Changes confined to small populations of cells cannot usually be identified in a visual inspection screen. To detect these changes, the target cell population must somehow be made accessible to inspection. Several options exist in this regard: analysis of histological sections, whole-mount antibody staining, *in situ* hybridization, retrograde or anterograde labeling of neurons, and cell class-specific transgenes. One technically simple but rather laborious approach is to embed zebrafish larvae in paraffin and prepare histological sections. This approach was used to screen more than 2000 individuals from ca. 50 clutches of F2 early pressure-generated mutagenized larvae and led to the identification of two photoreceptor mutants (Mohideen *et al.*, 2003). In addition to histological analysis, individual cell populations can be visualized in mutgenized animals using antibody staining or *in situ* hybridization. In one screening endeavor, staining of 700 early pressure-generated egg clutches with antityrosine hydroxylase antibody led to the isolation of two retinal mutants (Guo *et al.*, 1999).

An excellent example of a genetic screen that involves labeling of a specific neuronal population has been performed to uncover defects of the retinotectal projection (Baier *et al.*, 1996; Karlstrom *et al.*, 1996; Trowe *et al.*, 1996). In this screen, two subpopulations of retinal ganglion cells were labeled with the carbocyanine tracers, DiI and DiO. Labeling procedures usually make screening much more laborious. To reduce the workload in this screen, DiI and DiO labeling were highly automated. For tracer injection, fish larvae were mounted in a standardized fashion in a temperature-controlled mounting apparatus. After filling the apparatus with liquid agarose and mounting the larvae, the temperature was lowered allowing the agarose to solidify. Subsequently, the blocks of agarose containing mounted larvae were transferred into the injection setup. Upon injection, the larvae were stored overnight at room temperature to allow for the diffusion of the injected tracer, and then transferred to a microscope stage for phenotypic analysis. The authors of this experiment estimate that using this highly automated screening procedure allowed them to inspect over 2000 larvae per day and to reduce the time spent on the analysis of a single individual to less than 1 min (Baier *et al.*, 1996). Other labeling procedures can also be scaled up to process many clutches of embryos in a single experiment. Antibody or *in situ* protocols, for example, involve multiple changes of staining and washing solutions. To perform these protocols on many embryos in parallel, one can use multiwell staining dishes with stainless steel mesh at the bottom. Such staining dishes can be quickly transferred from one solution to another. Since many labeling procedures are time consuming, it is essential that during a screen they are performed in parallel on many embryos.

Recent advances provide an additional way to label specific cell populations in a much less labor-intensive way by using GFP transgenes, such as the ones described earlier in this chapter. Transgenic GFP lines can be either directly mutagenized or crossed to mutagenized males and the resulting progeny is used to search for defects in fine features of retinal cell populations. In contrast to other cell labeling procedures, the use of GFP transgenes requires very little additional effort, compared to simple morphological observations of the external phenotype.

Behavioral tests are yet another screening alternative. Several screens based on behavioral criteria have been performed in recent years, leading to the isolation of interesting developmental defects (Brockerhoff *et al.*, 1997, 2003; Li and Dowling, 1997; Muto *et al.*, 2005; Neuhauss *et al.*, 1999). Behavioral screens allow one to detect subtle defects of function, which might evade other search criteria. They can be used to search for both recessive and dominant defects in larvae as well as in adult fish (Li and Dowling, 1997). Similar to many labeling procedures, however, behavioral screens tend to be laborious. In one instance of a screen involving the optokinetic response, the authors estimate that screening of a single zebrafish larva took, on average, 1 min (Brockerhoff *et al.*, 1995). Since optomotor tests can be performed on populations of animals, they tend to be less time consuming, compared to optokinetic response tests. They do, however, produce more false-positve hits (Muto *et al.*, 2005). In addition, since behavioral responses usually involve the cooperation of many cell classes, screens of this type tend to detect a wide range of defects. The optokinetic response screens, for example, may lead to the isolation of defects in the differentiation of lens cells, the specification of the retinal neurons or glia, the formation of synaptic connections, the mechanisms of neurotransmitter release, or the development of ocular muscles. Additional tests are usually necessary to assure that the isolated mutants belong to the desired category. To be useful for screening, the behavioral response should be robust and reproducible, and should involve the simplest possible neuronal circuitry. In light of these criteria, the optokinetic response appears to be superior to other behaviors; both optomotor and startle responses require functional optic tecta while the optokinetic response does not (Clark, 1981; Easter and Nicola, 1996). The optokinetic response also appears to be more robust than the optomotor response and phototaxis (Brockerhoff *et al.*, 1995; Clark, 1981). The most extensive visual behavior-based screen conducted so far relied on two tests conducted in parallel: for optokinetic and for optomotor responses (Muto *et al.*, 2005). Although the results of this experiment are quite informative, they also illustrate problems associated with the use of behavioral tests as a screening tool. First, the initial round of screening was characterized by a very high false positive rate (>90% for the optomotor test). Second, surprisingly, the two behavioral tests used in this study uncovered largely nonoverlapping sets of mutants. Following retests, it turned out, however, that all mutants display both optomotor and optokinetic defects to varying degrees. Finally, as pointed out above, a broad range of phenotypic abnormalities in different cell classes were found in this experiment.

4. Positional and Candidate Cloning

Molecular characterization of defective loci is usually a crucial step that follows the isolation of mutant lines. The development of positional and candidate gene cloning strategies is one of the most significant advances in the field of zebrafish genetics within the last decade. These approaches are currently well established and have played a key role in many important contributions to the understanding of eye development and function. The positional cloning strategy involves a standard set of steps, such as mapping, chromosomal walking, transcript identification, and the delivery of a proof that the correct gene has been cloned. These steps are largely the same, regardless of the nature of a mutant phenotype, and are discussed in-depth in other chapters of this volume. An example of a positional cloning strategy, laborious but eventually successful, is the cloning of the *nagie oko* locus (Wei and Malicki, 2002).

5. Mutant Strains Available

Large and small mutagenesis screens identified numerous genetic defects of retinal development in zebrafish. Mutant phenotypes affect a broad range of developmental stages, starting with the specification of the eye primordia, through optic lobe morphogenesis, the specification of neuronal identities, and including the final steps of differentiation, such as outer segment development in photoreceptor cells. Lists of mutant lines, excluding these that produce nonspecific degeneration of the entire retina have been provided previously (Avanesov and Malicki, 2004; Malicki, 1999). Although these are still useful, many new mutants have been generated in recent years. The descriptions of these are available in the Zebrafish Model Organism Database (ZFIN, http://zfin.org).

V. Summary

Similar to other vertebrate species, the zebrafish retina is simpler than other regions of the central nervous system. Relative simplicity, rapid development, and accessibility to genetic analysis make the zebrafish retina an excellent model system for the studies of neurogenesis in the vertebrate CNS. Numerous genetic screens have led to the isolation of an impressive collection of mutants affecting the retina and the retinotectal projection in zebrafish. Mutant phenotypes are being studied using a rich variety of markers: antibodies, RNA probes, retrograde and antero-grade tracers, as well as transgenic lines. A particularly impressive progress has been made in the characterization of the zebrafish genome. Consequently, positional and candidate cloning of mutant loci are now fairly easy to accomplish. Many mutant zebrafish genes have been cloned, and their analysis has provided insights into genetic circuitries that regulate retinal pattern formation, and later the differentiation of neurons and glia. Genetic screens for visual system defects will continue in the future, and progressively more sophisticated screening approaches will make it possible to detect an increasingly broad and varied assortment of mutant phenotypes. The remarkable evolutionary conservation of the vertebrate eye provides the basis for the use of the zebrafish retina as a model of human inherited eye defects. As new techniques are being introduced and improved at a rapid pace, the zebrafish will continue to be an important model organism for the studies of the vertebrate visual system.

Acknowledgments

The authors are grateful to Brian Perkins, Ichiro Masai, Brian Link, and Kyle McCulloch for critical reading of the manuscript and helpful comments. The authors' research on the retina is supported by grants from the National Eye Institute and the Glaucoma Foundation.

References

Allende, M. L., Amsterdam, A., Becker, T., Kawakami, K., Gaiano, N., and Hopkins, N. (1996). Insertional mutagenesis in zebrafish identifies two novel genes, pescadillo and dead eye, essential for embryonic development. *Genes Dev.* **10**, 3141–3155.

Allwardt, B. A., Lall, A. B., Brockerhoff, S. E., and Dowling, J. E. (2001). Synapse formation is arrested in retinal photoreceptors of the zebrafish nrc mutant. *J. Neurosci.* **21**, 2330–2342.

Altshuler, D., Turner, D., and Cepko, C. (1991). Specification of cell type in the vertebrate retina. *In* "Development of the Visual System" (D. Lam and C. Shatz, eds.), pp. 37–58. MIT Press, Cambridge, MA.

Alvarez, Y., Cederlund, M. L., Cottell, D. C., Bill, B. R., Ekker, S. C., Torres-Vazquez, J., Weinstein, B. M., Hyde, D. R., Vihtelic, T. S., and Kennedy, B. N. (2007). Genetic determinants of hyaloid and retinal vasculature in zebrafish. *BMC Dev. Biol.* **7**, 114.

Amsterdam, A., Burgess, S., Golling, G., Chen, W., Sun, Z., Townsend, K., Farrington, S., Haldi, M., and Hopkins, N. (1999). A large-scale insertional mutagenesis screen in zebrafish. *Genes Dev.* **13**, 2713–2724.

Ando, H., Furuta, T., Tsien, R. Y., and Okamoto, H. (2001). Photo-mediated gene activation using caged RNA/DNA in zebrafish embryos. *Nat. Genet.* **28**, 317–325.

Ando, H., and Okamoto, H. (2003). Practical procedures for ectopic induction of gene expression in zebrafish embryos using Bhc-diazo-caged mRNA. *Methods Cell. Sci.* **25**, 25–31.

Aramaki, S., and Hatta, K. (2006). Visualizing neurons one-by-one *in vivo*: Optical dissection and reconstruction of neural networks with reversible fluorescent proteins. *Dev. Dyn.* **235**, 2192–2199.

Asakawa, K., and Kawakami, K. (2008). Targeted gene expression by the Gal4-UAS system in zebrafish. *Dev. Growth Differ.* **50**, 391–399.

Avanesov, A., Dahm, R., Sewell, W. F., and Malicki, J. J. (2005). Mutations that affect the survival of selected amacrine cell subpopulations define a new class of genetic defects in the vertebrate retina. *Dev. Biol.* **285**, 138–155.

Avanesov, A., and Malicki, J. (2004). Approaches to study neurogenesis in the zebrafish retina. *Methods Cell. Biol.* **76**, 333–384.

Baier, H., Klostermann, S., Trowe, T., Karlstrom, R. O., Nusslein-Volhard, C., and Bonhoeffer, F. (1996). Genetic dissection of the retinotectal projection. *Development* **123**, 415–425.

Balciunas, D., Davidson, A. E., Sivasubbu, S., Hermanson, S. B., Welle, Z., and Ekker, S. C. (2004). Enhancer trapping in zebrafish using the Sleeping Beauty transposon. *BMC Genomics.* **5**, 62.

Barthel, L. K., and Raymond, P. A. (1990). Improved method for obtaining 3-microns cryosections for immunocytochemistry. *J. Histochem. Cytochem.* **38**, 1383–1388.

Barthel, L. K., and Raymond, P. A. (1993). Subcellular localization of alpha-tubulin and opsin mRNA in the goldfish retina using digoxigenin-labeled cRNA probes detected by alkaline phosphatase and HRP histochemistry. *J. Neurosci. Methods* **50**, 145–152.

Baye, L. M., and Link, B. A. (2007). Interkinetic nuclear migration and the selection of neurogenic cell divisions during vertebrate retinogenesis. *J. Neurosci.* **27**, 10143–10152.

Beattie, C. E., Raible, D. W., Henion, P. D., and Eisen, J. S. (1999). Early pressure screens. *Methods Cell. Biol.* **60**, 71–86.

Becker, T. S., Burgess, S. M., Amsterdam, A. H., Allende, M. L., and Hopkins, N. (1998). Not really finished is crucial for development of the zebrafish outer retina and encodes a transcription factor highly homologous to human Nuclear Respiratory Factor-1 and avian Initiation Binding Repressor. *Development* **125**, 4369–4378.

Bernardos, R. L., and Raymond, P. A. (2006). GFAP transgenic zebrafish. *Gene. Expr. Patterns* **6**, 1007–1013.

Biehlmaier, O., Makhankov, Y., and Neuhauss, S. C. (2007). Impaired retinal differentiation and maintenance in zebrafish laminin mutants. *Invest. Ophthalmol. Vis. Sci.* **48**, 2887–2894.

Biehlmaier, O., Neuhauss, S. C., and Kohler, K. (2003). Synaptic plasticity and functionality at the cone terminal of the developing zebrafish retina. *J. Neurobiol.* **56**, 222–236.

Bodick, N., and Levinthal, C. (1980). Growing optic nerve fibers follow neighbors during embryogenesis. *Proc. Natl. Acad. Sci. USA* **77**, 4374–4378.

Branchek, T., and Bremiller, R. (1984). The development of photoreceptors in the zebrafish, Brachydanio rerio. I. Structure. *J. Comp. Neurol.* **224**, 107–115.

Brand, A. H., and Perrimon, N. (1993). Targeted gene expression as a means of altering cell fates and generating dominant phenotypes. *Development* **118**, 401–415.

Brennan, C., Monschau, B., Lindberg, R., Guthrie, B., Drescher, U., Bonhoeffer, F., and Holder, N. (1997). Two Eph receptor tyrosine kinase ligands control axon growth and may be involved in the creation of the retinotectal map in the zebrafish. *Development* **124**, 655–664.

Brockerhoff, S. E., Dowling, J. E., and Hurley, J. B. (1998). Zebrafish retinal mutants. *Vis. Res.* **38**, 1335–1339.

Brockerhoff, S. E., Hurley, J. B., Janssen-Bienhold, U., Neuhauss, S. C., Driever, W., and Dowling, J. E. (1995). A behavioral screen for isolating zebrafish mutants with visual system defects. *Proc. Natl. Acad. Sci. USA* **92**, 10545–10549.

Brockerhoff, S. E., Hurley, J. B., Niemi, G. A., and Dowling, J. E. (1997). A new form of inherited red-blindness identified in zebrafish. *J. Neurosci.* **17,** 4236–4242.

Brockerhoff, S. E., Rieke, F., Matthews, H. R., Taylor, M. R., Kennedy, B., Ankoudinova, I., Niemi, G. A., Tucker, C. L., Xiao, M., Cilluffo, M. C., *et al.* (2003). Light stimulates a transducin-independent increase of cytoplasmic Ca2+ and suppression of current in cones from the zebrafish mutant nof. *J. Neurosci.* **23,** 470–480.

Burckstummer, T., Bennett, K. L., Preradovic, A., Schutze, G., Hantschel, O., Superti-Furga, G., and Bauch, A. (2006). An efficient tandem affinity purification procedure for interaction proteomics in mammalian cells. *Nat. Methods* **3,** 1013–1019.

Burrill, J. D., and Easter, S. S., Jr. (1994). Development of the retinofugal projections in the embryonic and larval zebrafish (Brachydanio rerio). *J. Comp. Neurol.* **346,** 583–600.

Burrill, J., and Easter, S. (1995). The first retinal axons and their microenvironment in zebrafish cryptic pioneers and the pretract. *J. Neurosci.* **15,** 2935–2947.

Cayouette, M., and Raff, M. (2003). The orientation of cell division influences cell-fate choice in the developing mammalian retina. *Development* **130,** 2329–2339.

Cayouette, M., Whitmore, A. V., Jeffery, G., and Raff, M. (2001). Asymmetric segregation of Numb in retinal development and the influence of the pigmented epithelium. *J. Neurosci.* **21,** 5643–5651.

Cedrone, C., Culasso, F., Cesareo, M., Zapelloni, A., Cedrone, P., and Cerulli, L. (1997). Prevalence of glaucoma in Ponza, Italy: A comparison with other studies. *Ophthal. Epidemiol.* **4,** 59–72.

Chinen, A., Hamaoka, T., Yamada, Y., and Kawamura, S. (2003). Gene duplication and spectral diversification of cone visual pigments of zebrafish. *Genetics* **163,** 663–675.

Choi, J., Dong, L., Ahn, J., Dao, D., Hammerschmidt, M., and Chen, J. N. (2007). FoxH1 negatively modulates flk1 gene expression and vascular formation in zebrafish. *Dev. Biol.* **304,** 735–744.

Chuang, J. C., Mathers, P. H., and Raymond, P. A. (1999). Expression of three Rx homeobox genes in embryonic and adult zebrafish. *Mech. Dev.* **84,** 195–198.

Clark, T. (1981). Visual responses in developing zebrafish (Brachydanio rerio). (Eugene, ed.). University of Oregon Thesis, Eugene, Oregon.

Colbert, T., Till, B. J., Tompa, R., Reynolds, S., Steine, M. N., Yeung, A. T., McCallum, C. M., Comai, L., and Henikoff, S. (2001). High-throughput screening for induced point mutations. *Plant Physiol.* **126,** 480–484.

Collazo, A., Fraser, S. E., and Mabee, P. M. (1994). A dual embryonic origin for vertebrate mechan-oreceptors. *Science* **264,** 426–430.

Collins, M. O., and Choudhary, J. S. (2008). Mapping multiprotein complexes by affinity purification and mass spectrometry. *Curr. Opin. Biotechnol.* **19,** 324–330.

Connaughton, V. P., Behar, T. N., Liu, W. L., and Massey, S. C. (1999). Immunocytochemical localization of excitatory and inhibitory neurotransmitters in the zebrafish retina [In Process Citation]. *Vis. Neurosci.* **16,** 483–490.

Cui, S., Otten, C., Rohr, S., Abdelilah-Seyfried, S., and Link, B. A. (2007). Analysis of aPKClambda and aPKCzeta reveals multiple and redundant functions during vertebrate retinogenesis. *Mol. Cell. Neurosci.* **34,** 431–444.

Dahm, R., Schonthaler, H. B., Soehn, A. S., van Marle, J., and Vrensen, G. F. (2007). Development and adult morphology of the eye lens in the zebrafish. *Exp. Eye Res.* **85,** 74–89.

Dapson, R. W. (2007). Glyoxal fixation: How it works and why it only occasionally needs antigen retrieval. *Biotech. Histochem.* **82,** 161–166.

Das, T., Payer, B., Cayouette, M., and Harris, W. A. (2003). *In vivo* time-lapse imaging of cell divisions during neurogenesis in the developing zebrafish retina. *Neuron* **37,** 597–609.

Del Bene, F., Wehman, A. M., Link, B. A., and Baier, H. (2008). Regulation of neurogenesis by interkinetic nuclear migration through an apical-basal notch gradient. *Cell* **134,** 1055–1065.

Devoto, S. H., Melancon, E., Eisen, J. S., and Westerfield, M. (1996). Identification of separate slow and fast muscle precursor cells *in vivo*, prior to somite formation. *Development* **122,** 3371–3380.

Doerre, G., and Malicki, J. (2001). A mutation of early photoreceptor development, *mikre oko*, reveals cell-cell interactions involved in the survival and differentiation of zebrafish photoreceptors. *J. Neurosci.* **21,** 6745–6757.

Doerre, G., and Malicki, J. (2002). Genetic analysis of photoreceptor cell development in the zebrafish retina. *Mech. Dev.* **110**, 125–138.

Dowling, J. (1987). "The Retina." Harvard University Press, Cambridge, MA.

Doyon, Y., McCammon, J. M., Miller, J. C., Faraji, F., Ngo, C., Katibah, G. E., Amora, R., Hocking, T. D., Zhang, L., Rebar, E. J., *et al.* (2008). Heritable targeted gene disruption in zebrafish using designed zinc-finger nucleases. *Nat. Biotechnol.* **26**, 702–708.

Draper, B. W., Morcos, P. A., and Kimmel, C. B. (2001). Inhibition of zebrafish fgf8 pre-mRNA splicing with morpholino oligos: A quantifiable method for gene knockdown. *Genesis* **30**, 154–156.

Drescher, U., Bonhoeffer, F., and Muller, B. K. (1997). The Eph family in retinal axon guidance. *Curr. Opin. Neurobiol.* **7**, 75–80.

Driever, W., Solnica-Krezel, L., Schier, A. F., Neuhauss, S. C., Malicki, J., Stemple, D. L., Stainier, D. Y., Zwartkruis, F., Abdelilah, S., Rangini, Z., *et al.* (1996). A genetic screen for mutations affecting embryogenesis in zebrafish. *Development* **123**, 37–46.

Dryja, T., and Li, T. (1995). Molecular genetics of retinitis pigmentosa. *Hum. Mol. Genet.* **4**, 1739–1743.

Easter, S., and Nicola, G. (1996). The development of vision in the zebrafish (Danio rerio). *Dev. Biol.* **180**, 646–663.

Eisen, J. S., and Smith, J. C. (2008). Controlling morpholino experiments: Don't stop making antisense. *Development* **135**, 1735–1743.

Ellingsen, S., Laplante, M. A., Konig, M., Kikuta, H., Furmanek, T., Hoivik, E. A., and Becker, T. S. (2005). Large-scale enhancer detection in the zebrafish genome. *Development* **132**, 3799–3811.

Emran, F., Rihel, J., Adolph, A. R., Wong, K. Y., Kraves, S., and Dowling, J. E. (2007). OFF ganglion cells cannot drive the optokinetic reflex in zebrafish. *Proc. Natl. Acad. Sci. USA* **104**, 19126–19131.

Fadool, J. M. (2003). Development of a rod photoreceptor mosaic revealed in transgenic zebrafish. *Dev. Biol.* **258**, 277–290.

Fadool, J. M., Brockerhoff, S. E., Hyatt, G. A., and Dowling, J. E. (1997). Mutations affecting eye morphology in the developing zebrafish (Danio rerio). *Dev. Genet.* **20**, 288–295.

Fadool, J. M., Hartl, D. L., and Dowling, J. E. (1998). Transposition of the mariner element from Drosophila mauritiana in zebrafish. *Proc. Natl. Acad. Sci. USA* **95**, 5182–5186.

Fashena, D., and Westerfield, M. (1999). Secondary motoneuron axons localize DM-GRASP on their fasciculated segments. *J. Comp. Neurol.* **406**, 415–424.

Fraser, S. (1992). Patterning of retinotectal connections in the vertebrate visual system. *Curr. Opin Neurobiol.* **2**, 83–87.

Fricke, C., Lee, J. S., Geiger-Rudolph, S., Bonhoeffer, F., and Chien, C. B. (2001). Astray, a zebrafish roundabout homolog required for retinal axon guidance. *Science* **292**, 507–510.

Gnuegge, L., Schmid, S., and Neuhauss, S. C. (2001). Analysis of the activity-deprived zebrafish mutant macho reveals an essential requirement of neuronal activity for the development of a fine-grained visuotopic map. *J. Neurosci.* **21**, 3542–3548.

Godinho, L., Mumm, J. S., Williams, P. R., Schroeter, E. H., Koerber, A., Park, S. W., Leach, S. D., and Wong, R. O. (2005). Targeting of amacrine cell neurites to appropriate synaptic laminae in the developing zebrafish retina. *Development* **132**, 5069–5079.

Goldsmith, P., Baier, H., and Harris, W. A. (2003). Two zebrafish mutants, ebony and ivory, uncover benefits of neighborhood on photoreceptor survival. *J. Neurobiol.* **57**, 235–245.

Golling, G., Amsterdam, A., Sun, Z., Antonelli, M., Maldonado, E., Chen, W., Burgess, S., Haldi, M., Artzt, K., Farrington, S., *et al.* (2002). Insertional mutagenesis in zebrafish rapidly identifies genes essential for early vertebrate development. *Nat. Genet.* **31**, 135–140.

Gray, M. P., Smith, R. S., Soules, K. A., John, S. W., and Link, B. A. (2008). The aqueous humor outflow pathway of zebrafish. *Invest. Ophthalmol. Vis. Sci.* **50**, 1515–1521.

Gross, J. M., Perkins, B. D., Amsterdam, A., Egana, A., Darland, T., Matsui, J. I., Sciascia, S., Hopkins, N., and Dowling, J. E. (2005). Identification of zebrafish insertional mutants with defects in visual system development and function. *Genetics* **170**, 245–261.

Guo, S., Wilson, S. W., Cooke, S., Chitnis, A. B., Driever, W., and Rosenthal, A. (1999). Mutations in the zebrafish unmask shared regulatory pathways controlling the development of catecholaminergic neurons. *Dev. Biol.* **208**, 473–487.

Haffter, P., Granato, M., Brand, M., Mullins, M. C., Hammerschmidt, M., Kane, D. A., Odenthal, J., van Eeden, F. J., Jiang, Y. J., Heisenberg, C. P., *et al.* (1996). The identification of genes with unique and essential functions in the development of the zebrafish, Danio rerio. *Development* **123,** 1–36.

Halloran, M. C., Sato-Maeda, M., Warren, J. T., Su, F., Lele, Z., Krone, P. H., Kuwada, J. Y., and Shoji, W. (2000). Laser-induced gene expression in specific cells of transgenic zebrafish. *Development* **127,** 1953–1960.

Halpern, M., Ho, R., Walker, C., and Kimmel, C. (1993). Induction of muscle pioneers and floor plate is distinguished by the zebrafish no tail mutation. *Cell* **75,** 99–111.

Hanker, J. S. (1979). Osmiophilic reagents in electronmicroscopic histocytochemistry. *Prog. Histochem. Cytochem.* **12,** 1–85.

Hartley, J. L., Temple, G. F., and Brasch, M. A. (2000). DNA cloning using *in vitro* site-specific recombination. *Genome Res.* **10,** 1788–1795.

Hatta, K., Tsujii, H., and Omura, T. (2006). Cell tracking using a photoconvertible fluorescent protein. *Nat. Protoc.* **1,** 960–967.

Hauptmann, G., and Gerster, T. (1994). Two-color whole-mount *in situ* hybridization to vertebrate and Drosophila embryos. *Trends Genet.* **10,** 266.

Hinds, J., and Hinds, P. (1974). Early ganglion cell differentiation in the mouse retina: An electron microscopic analysis utilizing serial sections. *Dev. Biol.* **37,** 381–416.

Hisatomi, O., Satoh, T., Barthel, L. K., Stenkamp, D. L., Raymond, P. A., and Tokunaga, F. (1996). Molecular cloning and characterization of the putative ultraviolet-sensitive visual pigment of gold-fish. *Vis. Res.* **36,** 933–939.

Hitchcock, P. F., Macdonald, R. E., VanDeRyt, J. T., and Wilson, S. W. (1996). Antibodies against Pax6 immunostain amacrine and ganglion cells and neuronal progenitors, but not rod precursors, in the normal and regenerating retina of the goldfish. *J. Neurobiol.* **29,** 399–413.

Ho, R. K., and Kane, D. A. (1990). Cell-autonomous action of zebrafish spt-1 mutation in specific mesodermal precursors. *Nature* **348,** 728–730.

Holt, C., Bertsch, T., Ellis, H., and Harris, W. (1988). Cellular determination in the Xenopus retina is independent of lineage and birth date. *Neuron* **1,** 15–26.

Hong, C. C., Peterson, Q. P., Hong, J. Y., and Peterson, R. T. (2006). Artery/vein specification is governed by opposing phosphatidylinositol-3 kinase and MAP kinase/ERK signaling. *Curr. Biol.* **16,** 1366–1372.

Honig, M. G., and Hume, R. I. (1986). Fluorescent carbocyanine dyes allow living neurons of identified origin to be studied in long-term cultures. *J. Cell. Biol.* **103,** 171–187.

Honig, M. G., and Hume, R. I. (1989). DiI and diO: Versatile fluorescent dyes for neuronal labelling and pathway tracing. *Trends Neurosci.* **12,** 333–335, 340–341.

Hu, M., and Easter, S. S. (1999). Retinal neurogenesis: The formation of the initial central patch of postmitotic cells. *Dev. Biol.* **207,** 309–321.

Humphrey, C., and Pittman, F. (1974). A simple methylene blue-azure II-basic fuchsin stain for epoxy-embedded tissue sections. *Stain Technol.* **49,** 9–14.

Hyatt, G. A., Schmitt, E. A., Marsh-Armstrong, N., McCaffery, P., Drager, U. C., and Dowling, J. E. (1996). Retinoic acid establishes ventral retinal characteristics. *Development* **122,** 195–204.

Insinna, C., Pathak, N., Perkins, B., Drummond, I., and Besharse, J. C. (2008). The homodimeric kinesin, Kif17, is essential for vertebrate photoreceptor sensory outer segment development. *Dev. Biol.* **316,** 160–170.

Jacobson, M. (1991). "Developmental Neurobiology." Plenum Press, New York.

Jensen, A. M., Walker, C., and Westerfield, M. (2001). Mosaic eyes: A zebrafish gene required in pigmented epithelium for apical localization of retinal cell division and lamination. *Development* **128,** 95–105.

Jowett, T. (2001). Double *in situ* hybridization techniques in zebrafish. *Methods* **23,** 345–358.

Jowett, T., and Lettice, L. (1994). Whole-mount un situ hybridizations on zebrafish embryos using a mixture of digoxigenin- and fluorescein-labelled probes. *TIG* **10,** 73–74.

Kainz, P. M., Adolph, A. R., Wong, K. Y., and Dowling, J. E. (2003). Lazy eyes zebrafish mutation affects Muller glial cells, compromising photoreceptor function and causing partial blindness. *J. Comp. Neurol.* **463,** 265–280.

Karlsson, J., von Hofsten, J., and Olsson, P. E. (2001). Generating transparent zebrafish: A refined method to improve detection of gene expression during embryonic development. *Mar. Biotechnol. (NY)* **3,** 522–527.

Karlstrom, R. O., Trowe, T., Klostermann, S., Baier, H., Brand, M., Crawford, A. D., Grunewald, B., Haffter, P., Hoffmann, H., Meyer, S. U., *et al.* (1996). Zebrafish mutations affecting retinotectal axon pathfinding. *Development* **123,** 427–438.

Kawahara, A., Chien, C. B., and Dawid, I. B. (2002). The homeobox gene mbx is involved in eye and tectum development. *Dev. Biol.* **248,** 107–117.

Kawakami, K. (2004). Transgenesis and gene trap methods in zebrafish by using the Tol2 transposable element. *Methods Cell Biol.* **77,** 201–222.

Kawakami, K., Shima, A., and Kawakami, N. (2000). Identification of a functional transposase of the Tol2 element, an Ac-like element from the Japanese medaka fish, and its transposition in the zebrafish germ lineage. *Proc. Natl. Acad. Sci. USA* **97,** 11403–11408.

Kawakami, K., Takeda, H., Kawakami, N., Kobayashi, M., Matsuda, N., and Mishina, M. (2004). A transposon-mediated gene trap approach identifies developmentally regulated genes in zebrafish. *Dev. Cell.* **7,** 133–144.

Kay, J. N., Finger-Baier, K. C., Roeser, T., Staub, W., and Baier, H. (2001). Retinal ganglion cell genesis requires lakritz, a Zebrafish atonal Homolog. *Neuron* **30,** 725–736.

Kay, J. N., Link, B. A., and Baier, H. (2005). Staggered cell-intrinsic timing of ath5 expression underlies the wave of ganglion cell neurogenesis in the zebrafish retina. *Development* **132,** 2573–2585.

Kay, J. N., Roeser, T., Mumm, J. S., Godinho, L., Mrejeru, A., Wong, R. O., and Baier, H. (2004). Transient requirement for ganglion cells during assembly of retinal synaptic layers. *Development* **131,** 1331–1342.

Kikuchi, Y., Segawa, H., Tokumoto, M., Tsubokawa, T., Hotta, Y., Uyemura, K., and Okamoto, H. (1997). Ocular and cerebellar defects in zebrafish induced by overexpression of the LIM domains of the islet-3 LIM/homeodomain protein. *Neuron* **18,** 369–382.

Kim, M. J., Kang, K. H., Kim, C. H., and Choi, S. Y. (2008). Real-time imaging of mitochondria in transgenic zebrafish expressing mitochondrially targeted GFP. *Biotechniques* **45,** 331–334.

Kimmel, C. B., Ballard, W. W., Kimmel, S. R., Ullmann, B., and Schilling, T. F. (1995). Stages of embryonic development of the zebrafish. *Dev. Dyn.* **203,** 253–310.

Kimmel, C. B., Sessions, S. K., and Kimmel, R. J. (1981). Morphogenesis and synaptogenesis of the zebrafish Mauthner neuron. *J. Comp. Neurol.* **198,** 101–120.

Kitambi, S. S., McCulloch, K. J., Peterson, R. T., and Malicki, J. J. (2009). Small molecule screen for compounds that affect vascular development in the zebrafish retina. *Mech. Dev.* **126,** 464–477.

Kosodo, Y., Toida, K., Dubreuil, V., Alexandre, P., Schenk, J., Kiyokage, E., Attardo, A., Mora-Bermudez, F., Arii, T., Clarke, J. D., *et al.* (2008). Cytokinesis of neuroepithelial cells can divide their basal process before anaphase. *EMBO J.* **27,** 3151–3163.

Koster, R. W., and Fraser, S. E. (2001). Tracing transgene expression in living zebrafish embryos. *Dev. Biol.* **233,** 329–346.

Krock, B. L., Bilotta, J., and Perkins, B. D. (2007). Noncell-autonomous photoreceptor degeneration in a zebrafish model of choroideremia. *Proc. Natl. Acad. Sci. USA* **104,** 4600–4605.

Krock, B. L., and Perkins, B. D. (2008). The intraflagellar transport protein IFT57 is required for cilia maintenance and regulates IFT-particle-kinesin-II dissociation in vertebrate photoreceptors. *J. Cell Sci.* **121,** 1907–1915.

Kwan, K. M., Fujimoto, E., Grabher, C., Mangum, B. D., Hardy, M. E., Campbell, D. S., Parant, J. M., Yost, H. J., Kanki, J. P., and Chien, C. B. (2007). The Tol2kit: A multisite gateway-based construction kit for Tol2 transposon transgenesis constructs. *Dev. Dyn.* **236,** 3088–3099.

Laessing, U., Giordano, S., Stecher, B., Lottspeich, F., and Stuermer, C. A. (1994). Molecular charac-terization of fish neurolin: A growth-associated cell surface protein and member of the immunoglob-ulin superfamily in the fish retinotectal system with similarities to chick protein DM-GRASP/SC-1/BEN. *Differentiation* **56,** 21–29.

Laessing, U., and Stuermer, C. A. (1996). Spatiotemporal pattern of retinal ganglion cell differentiation revealed by the expression of neurolin in embryonic zebrafish. *J. Neurobiol.* **29,** 65–74.

Larison, K., and Bremiller, R. (1990). Early onset of phenotype and cell patterning in the embryonic zebrafish retina. *Development* **109,** 567–576.

Lawson, N. D., and Weinstein, B. M. (2002). *In vivo* imaging of embryonic vascular development using transgenic zebrafish. *Dev. Biol.* **248,** 307–318.

Li, L., and Dowling, J. E. (1997). A dominant form of inherited retinal degeneration caused by a non-photoreceptor cell-specific mutation. *Proc. Natl. Acad. Sci. USA* **94,** 11645–11650.

Li, L., and Dowling, J. E. (2000). Disruption of the olfactoretinal centrifugal pathway may relate to the visual system defect in night blindness b mutant zebrafish. *J. Neurosci.* **20,** 1883–1892.

Li, Z., Hu, M., Ochocinska, M. J., Joseph, N. M., and Easter, S. S., Jr. (2000a). Modulation of cell proliferation in the embryonic retina of zebrafish (Danio rerio). *Dev. Dyn.* **219,** 391–401.

Li, Z., Joseph, N. M., and Easter, S. S., Jr. (2000). The morphogenesis of the zebrafish eye, including a fate map of the optic vesicle. *Dev. Dyn.* **218,** 175–188.

Li, Z., Joseph, N. M., and Easter, S. S., Jr. (2000b). The morphogenesis of the zebrafish eye, including a fate map of the optic vesicle. *Dev. Dyn.* **218,** 175–188.

Link, B. A., Fadool, J. M., Malicki, J., and Dowling, J. E. (2000). The zebrafish young mutation acts non-cell-autonomously to uncouple differentiation from specification for all retinal cells. *Development* **127,** 2177–2188.

Liu, I. S., Chen, J. D., Ploder, L., Vidgen, D., van der Kooy, D., Kalnins, V. I., and McInnes, R. R. (1994). Developmental expression of a novel murine homeobox gene (Chx10): Evidence for roles in determination of the neuroretina and inner nuclear layer. *Neuron* **13,** 377–393.

Lowery, L. A., and Sive, H. (2004). Strategies of vertebrate neurulation and a re-evaluation of teleost neural tube formation. *Mech. Dev.* **121,** 1189–1197.

Macdonald, R., Barth, K. A., Xu, Q., Holder, N., Mikkola, I., and Wilson, S. W. (1995). Midline signalling is required for Pax gene regulation and patterning of the eyes. *Development* **121,** 3267–3278.

Macdonald, R., Scholes, J., Strahle, U., Brennan, C., Holder, N., Brand, M., and Wilson, S. W. (1997). The Pax protein Noi is required for commissural axon pathway formation in the rostral forebrain. *Development* **124,** 2397–2408.

Macdonald, R., and Wilson, S. (1997). Distribution of Pax6 protein during eye development suggests discrete roles in proliferative and differentiated visual cells. *Dev. Genes Evol.* **206,** 363–369.

Mack, A. F., and Fernald, R. D. (1995). New rods move before differentiating in adult teleost retina. *Dev. Biol.* **170,** 136–141.

Makhankov, Y. V., Rinner, O., and Neuhauss, S. C. (2004). An inexpensive device for non-invasive electroretinography in small aquatic vertebrates. *J. Neurosci. Methods* **135,** 205–210.

Malicki, J. (1999). Development of the retina. *Methods Cell. Biol.* **59,** 273–299.

Malicki, J. (2000). Harnessing the power of forward genetics—analysis of neuronal diversity and patterning in the zebrafish retina. *Trends Neurosci.* **23,** 531–541.

Malicki, J., and Driever, W. (1999). *oko meduzy* mutations affect neuronal patterning in the zebrafish retina and reveal cell-cell interactions of the retinal neuroepithelial sheet. *Development* **126,** 1235–1246.

Malicki, J., Jo, H., and Pujic, Z. (2003). Zebrafish N-cadherin, encoded by the *glass onion* locus, plays an essential role in retinal patterning. *Dev. Biol.* **259,** 95–108.

Malicki, J., Jo, H., Wei, X., Hsiung, M., and Pujic, Z. (2002). Analysis of gene function in the zebrafish retina. *Methods* **28,** 427–438.

Malicki, J., Neuhauss, S. C., Schier, A. F., Solnica-Krezel, L., Stemple, D. L., Stainier, D. Y., Abdelilah, S., Zwartkruis, F., Rangini, Z., and Driever, W. (1996). Mutations affecting development of the zebrafish retina. *Development* **123,** 263–273.

Mangrum, W. I., Dowling, J. E., and Cohen, E. D. (2002). A morphological classification of ganglion cells in the zebrafish retina. *Vis. Neurosci.* **19,** 767–779.

Marcus, R. C., Delaney, C. L., and Easter, S. S., Jr. (1999). Neurogenesis in the visual system of embryonic and adult zebrafish (Danio rerio). *Vis. Neurosci.* **16,** 417–424.

Martinez-Morales, J. R., Del Bene, F., Nica, G., Hammerschmidt, M., Bovolenta, P., and Wittbrodt, J. (2005). Differentiation of the vertebrate retina is coordinated by an FGF signaling center. *Dev. Cell.* **8,** 565–574.

Masai, I., Lele, Z., Yamaguchi, M., Komori, A., Nakata, A., Nishiwaki, Y., Wada, H., Tanaka, H., Nojima, Y., Hammerschmidt, M., *et al.* (2003). N-cadherin mediates retinal lamination, maintenance of forebrain compartments and patterning of retinal neurites. *Development* **130,** 2479–2494.

Masai, I., Stemple, D. L., Okamoto, H., and Wilson, S. W. (2000). Midline signals regulate retinal neurogenesis in zebrafish. *Neuron* **27,** 251–263.

Masai, I., Yamaguchi, M., Tonou-Fujimori, N., Komori, A., and Okamoto, H. (2005). The hedgehog-PKA pathway regulates two distinct steps of the differentiation of retinal ganglion cells: The cell-cycle exit of retinoblasts and their neuronal maturation. *Development* **132,** 1539–1553.

Meng, X., Noyes, M. B., Zhu, L. J., Lawson, N. D., and Wolfe, S. A. (2008). Targeted gene inactivation in zebrafish using engineered zinc-finger nucleases. *Nat. Biotechnol.* **26,** 695–701.

Merin, S. (1991). "Inherited Eye Disease." Marcel Dekker, Inc., New York.

Metcalfe, W., Myers, P., Trevarrow, B., Bass, M., and Kimmel, C. (1990). Primary neurons that express the L2/HNK-1 carbohydrate during early development in the zebrafish. *Development* **110,** 491–504.

Metcalfe, W. K. (1985). Sensory neuron growth cones comigrate with posterior lateral line primordial cells in zebrafish. *J. Comp. Neurol.* **238,** 218–224.

Moens, C. B., and Fritz, A. (1999). Techniques in neural development. *Methods Cell Biol.* **59,** 253–272.

Mohideen, M. A., Beckwith, L. G., Tsao-Wu, G. S., Moore, J. L., Wong, A. C., Chinoy, M. R., and Cheng, K. C. (2003). Histology-based screen for zebrafish mutants with abnormal cell differentiation. *Dev. Dyn.* **228,** 414–423.

Morris, A. C., Schroeter, E. H., Bilotta, J., Wong, R. O., and Fadool, J. M. (2005). Cone survival despite rod degeneration in XOPS-mCFP transgenic zebrafish. *Invest. Ophthalmol. Vis. Sci.* **46,** 4762–4771.

Muller, H. (1857). Anatomisch-physiologische untersuchungen uber die Retina bei Menschen und Wirbelthieren. *Z. Wiss. Zool.* **8,** 1–122.

Mullins, M. C., Hammerschmidt, M., Haffter, P., and Nusslein-Volhard, C. (1994). Large-scale mutagenesis in the zebrafish: In search of genes controlling development in a vertebrate. *Curr. Biol.* **4,** 189–202.

Mumm, J. S., Williams, P. R., Godinho, L., Koerber, A., Pittman, A. J., Roeser, T., Chien, C. B., Baier, H., and Wong, R. O. (2006). *In vivo* imaging reveals dendritic targeting of laminated afferents by zebrafish retinal ganglion cells. *Neuron* **52,** 609–621.

Muto, A., Orger, M. B., Wehman, A. M., Smear, M. C., Kay, J. N., Page-McCaw, P. S., Gahtan, E., Xiao, T., Nevin, L. M., Gosse, N. J., *et al.* (2005). Forward genetic analysis of visual behavior in zebrafish. *PLoS Genet.* **1,** e66.

Nagayoshi, S., Hayashi, E., Abe, G., Osato, N., Asakawa, K., Urasaki, A., Horikawa, K., Ikeo, K., Takeda, H., and Kawakami, K. (2008). Insertional mutagenesis by the Tol2 transposon-mediated enhancer trap approach generated mutations in two developmental genes: tcf7 and synembryn-like. *Development* **135,** 159–169.

Nasevicius, A., and Ekker, S. C. (2000). Effective targeted gene "knockdown" in zebrafish. *Nat. Genet.* **26,** 216–220.

Nawrocki, W. (1985). Development of the neural retina in the zebrafish, Brachydanio rerio., (Eugine, ed.), University of Oregon Thesis, Eugene, Oregon.

Neuhauss, S. C. (2003). Behavioral genetic approaches to visual system development and function in zebrafish. *J. Neurobiol.* **54,** 148–160.

Neuhauss, S. C., Biehlmaier, O., Seeliger, M. W., Das, T., Kohler, K., Harris, W. A., and Baier, H. (1999). Genetic disorders of vision revealed by a behavioral screen of 400 essential loci in zebrafish. *J. Neurosci.* **19**, 8603–8615.

Neumann, C. J., and Nuesslein-Volhard, C. (2000). Patterning of the zebrafish retina by a wave of sonic hedgehog activity. *Science* **289**, 2137–2139.

Nicolson, T., Rusch, A., Friedrich, R. W., Granato, M., Ruppersberg, J. P., and Nusslein-Volhard, C. (1998). Genetic analysis of vertebrate sensory hair cell mechanosensation: The zebrafish circler mutants. *Neuron* **20**, 271–283.

Nornes, H. O., Dressler, G. R., Knapik, E. W., Deutsch, U., and Gruss, P. (1990). Spatially and temporally restricted expression of Pax2 during murine neurogenesis. *Development* **109**, 797–809.

North, T. E., Goessling, W., Walkley, C. R., Lengerke, C., Kopani, K. R., Lord, A. M., Weber, G. J., Bowman, T. V., Jang, I. H., Grosser, T., *et al.* (2007). Prostaglandin E2 regulates vertebrate haematopoietic stem cell homeostasis. *Nature* **447**, 1007–1011.

Novak, A. E., and Ribera, A. B. (2003). Immunocytochemistry as a tool for zebrafish developmental neurobiology. *Methods Cell Sci.* **25**, 79–83.

Omori, Y., Zhao, C., Saras, A., Mukhopadhyay, S., Kim, W., Furukawa, T., Sengupta, P., Veraksa, A., and Malicki, J. (2008). Elipsa is an early determinant of ciliogenesis that links the IFT particle to membrane-associated small GTPase Rab8. *Nat. Cell Biol.* **10**, 437–444.

Oxtoby, E., and Jowett, T. (1993). Cloning of the zebrafish krox-20 gene (krx-20) and its expression during hindbrain development. *NAR* **21**, 1087–1095.

Passini, M. A., Levine, E. M., Canger, A. K., Raymond, P. A., and Schechter, N. (1997). Vsx-1 and Vsx-2: Differential expression of two paired-like homeobox genes during zebrafish and goldfish retinogenesis. *J. Comp. Neurol.* **388**, 495–505.

Pathak, N., Obara, T., Mangos, S., Liu, Y., and Drummond, I. A. (2007). The zebrafish fleer gene encodes an essential regulator of cilia tubulin polyglutamylation. *Mol. Biol. Cell* **18**, 4353–4364.

Pauls, S., Geldmacher-Voss, B., and Campos-Ortega, J. A. (2001). A zebrafish histone variant H2A.F/Z and a transgenic H2A.F/Z:GFP fusion protein for *in vivo* studies of embryonic development. *Dev. Genes Evol.* **211**, 603–610.

Perkins, B. D., Kainz, P. M., O'Malley, D. M., and Dowling, J. E. (2002). Transgenic expression of a GFP-rhodopsin COOH-terminal fusion protein in zebrafish rod photoreceptors. *Vis. Neurosci.* **19**, 257–264.

Peterson, R. E., Fadool, J. M., McClintock, J., and Linser, P. J. (2001). Muller cell differentiation in the zebrafish neural retina: Evidence of distinct early and late stages in cell maturation. *J. Comp. Neurol.* **429**, 530–540.

Peterson, R. T., Link, B. A., Dowling, J. E., and Schreiber, S. L. (2000). Small molecule developmental screens reveal the logic and timing of vertebrate development. *Proc. Natl. Acad. Sci. USA* **97**, 12965–12969.

Peterson, R. E., Tu, C., and Linser, P. J. (1997). Isolation and characterization of a carbonic anhydrase homologue from the zebrafish (Danio rerio). *J. Mol. Evol.* **44**, 432–439.

Plaster, N., Sonntag, C., Busse, C. E., and Hammerschmidt, M. (2006). p53 deficiency rescues apoptosis and differentiation of multiple cell types in zebrafish flathead mutants deficient for zygotic DNA polymerase delta1. *Cell. Death Differ.* **13**, 223–235.

Poggi, L., Vitorino, M., Masai, I., and Harris, W. A. (2005). Influences on neural lineage and mode of division in the zebrafish retina *in vivo. J. Cell. Biol.* **171**, 991–999.

Porteus, M. H., and Carroll, D. (2005). Gene targeting using zinc finger nucleases. *Nat. Biotechnol.* **23**, 967–973.

Prince, V. E., Joly, L., Ekker, M., and Ho, R. K. (1998). Zebrafish hox genes: Genomic organization and modified colinear expression patterns in the trunk. *Development* **125**, 407–420.

Pujic, Z., and Malicki, J. (2001). Mutation of the zebrafish glass onion locus causes early cell-nonautonomous loss of neuroepithelial integrity followed by severe neuronal patterning defects in the retina. *Dev. Biol.* **234**, 454–469.

Pujic, Z., and Malicki, J. (2004). Retinal pattern and the genetic basis of its formation in zebrafish. *Semin. Cell Dev. Biol.* **15**, 105–114.

Pujic, Z., Omori, Y., Tsujikawa, M., Thisse, B., Thisse, C., and Malicki, J. (2006). Reverse genetic analysis of neurogenesis in the zebrafish retina. *Dev. Biol.* **293**, 330–347.

Raible, D. W., Wood, A., Hodsdon, W., Henion, P. D., Weston, J. A., and Eisen, J. S. (1992). Segregation and early dispersal of neural crest cells in the embryonic zebrafish. *Dev. Dyn.* **195**, 29–42.

Raymond, P., Barthel, L., and Curran, G. (1995). Developmental patterning of rod and cone photo-receptors in embryonic zebrafish. *J. Comp. Neurol.* **359**, 537–550.

Raymond, P., Barthel, L., Rounsifer, M., Sullivan, S., and Knight, J. (1993). Expression of rod and cone visual pigments in godfish and zebrafish: A rhodopsin-like gene is expressed in cones. *Neuron* **10**, 1161–1174.

Raz, E., van Luenen, H. G., Schaerringer, B., Plasterk, R. H., and Driever, W. (1998). Transposition of the nematode Caenorhabditis elegans Tc3 element in the zebrafish Danio rerio. *Curr. Biol.* **8**, 82–88.

Ramón y Cajal, S. (1892). La retine des vertebres. *La Cellule* **9**, 121–255.

Rembold, M., Loosli, F., Adams, R. J., and Wittbrodt, J. (2006). Individual cell migration serves as the driving force for optic vesicle evagination. *Science* **313**, 1130–1134.

Ren, J. Q., McCarthy, W. R., Zhang, H., Adolph, A. R., and Li, L. (2002). Behavioral visual responses of wild-type and hypopigmented zebrafish. *Vis. Res.* **42**, 293–299.

Riley, B. B., Chiang, M., Farmer, L., and Heck, R. (1999). The deltaA gene of zebrafish mediates lateral inhibition of hair cells in the inner ear and is regulated by pax2.1. *Development* **126**, 5669–5678.

Robinson, J., Schmitt, E., and Dowling, J. (1995). Temporal and spatial patterns of opsin gene expression in zebrafish (Danio rerio). *Vis. Neurosci.* **12**, 895–906.

Rodieck, R. W. (1973). "The Vertebrate Retina. Principles of Structure and Function." W. H. Freeman & Co, San Francisco, CA.

Roeser, T., and Baier, H. (2003). Visuomotor behaviors in larval zebrafish after GFP-guided laser ablation of the optic tectum. *J. Neurosci.* **23**, 3726–3734.

Sachidanandan, C., Yeh, J., Peteerson, Q., and Peteerson, R. (2008). Identification of a novel retinoid by small molecule screening with zebrafish embryos. *PLoS ONE* **3**, 1–9.

Sandell, J., Martin, S., and Heinrich, G. (1994). The development of GABA immunoreactivity in the retina of the zebrafish. *J. Comp. Neurol.* **345**, 596–601.

Sanes, J. R. (1993). Topographic maps and molecular gradients. *Curr. Opin. Neurobiol.* **3**, 67–74.

Sanyanusin, P., Schimmenti, L. A., McNoe, L. A., Ward, T. A., Pierpont, M. E., Sullivan, M. J., Dobyns, W. B., and Eccles, M. R. (1995). Mutation of the PAX2 gene in a family with optic nerve colobomas, renal anomalies and vesicoureteral reflux *Nat. Genet.* **9**, 358–364.

Sanyanusin, P., Schimmenti, L. A., McNoe, T. A., Ward, T. A., Pierpont, M. E., Sullivan, M. J., Dobyns, W. B., and Eccles, M. R. (1996). Mutation of the gene in a family with optic nerve colobomas, renal anomolies and vesicoureteral reflux. *Nat. Genet.* **13**, 129.

Sato, T., Takahoko, M., and Okamoto, H. (2006). HuC:Kaede, a useful tool to label neural morphol-ogies in networks *in vivo. Genesis* **44**, 136–142.

Scheer, N., and Campos-Ortega, J. A. (1999). Use of the Gal4-UAS technique for targeted gene expression in the zebrafish. *Mech. Dev.* **80**, 153–158.

Scheer, N., Groth, A., Hans, S., and Campos-Ortega, J. A. (2001). An instructive function for Notch in promoting gliogenesis in the zebrafish retina. *Development* **128**, 1099–1107.

Scheer, N., Riedl, I., Warren, J. T., Kuwada, J. Y., and Campos-Ortega, J. A. (2002). A quantitative analysis of the kinetics of Gal4 activator and effector gene expression in the zebrafish. *Mech. Dev.* **112**, 9–14.

Schier, A. F., Neuhauss, S. C., Helde, K. A., Talbot, W. S., and Driever, W. (1997). The one-eyed pinhead gene functions in mesoderm and endoderm formation in zebrafish and interacts with no tail. *Development* **124**, 327–342.

Schmitt, E., and Dowling, J. (1994). Early eye morphogenesis in the Zebrafish, Brachydanio rerio. *J. Comp. Neurol.* **344**, 532–542.

Schmitt, E. A., and Dowling, J. E. (1996). Comparison of topographical patterns of ganglion and photoreceptor cell differentiation in the retina of the zebrafish, Danio rerio. *J. Comp. Neurol.* **371**, 222–234.

Schmitt, E. A., and Dowling, J. E. (1999). Early retinal development in the zebrafish, Danio rerio: Light and electron microscopic analyses. *J. Comp. Neurol.* **404**, 515–536.

Schmitz, B., Papan, C., and Campos-Ortega, J. (1993). Neurulation in the anterior trunk of the zebrafish Brachydanio rerio. *Roux''s Arch. Dev. Biol.* **202**, 250–259.

Schroeter, E. H., Wong, R. O., and Gregg, R. G. (2006). *In vivo* development of retinal ON-bipolar cell axonal terminals visualized in nyx::MYFP transgenic zebrafish. *Vis. Neurosci.* **23**, 833–843.

Scott, E. K., Mason, L., Arrenberg, A. B., Ziv, L., Gosse, N. J., Xiao, T., Chi, N. C., Asakawa, K., Kawakami, K., and Baier, H. (2007). Targeting neural circuitry in zebrafish using GAL4 enhancer trapping. *Nat. Methods* **4**, 323–326.

Seddon, J. (1994). Age-related macular degeneration: Epidemiology. *In* "Principles and Practice of Ophthalmology" (B. Albert and F. Jakobiec, eds.), pp. 1266–1274. Saunders, W. B, Philadelphia.

Seo, H. C., Drivenes, E. S., and Fjose, A. (1998). Expression of two zebrafish homologues of the murine Six3 gene demarcates the initial eye primordia. *Mech. Dev.* **73**, 45–57.

Shaner, N. C., Patterson, G. H., and Davidson, M. W. (2007). Advances in fluorescent protein technology. *J. Cell. Sci.* **120**, 4247–4260.

Shields, C. R., Klooster, J., Claassen, Y., Ul-Hussain, M., Zoidl, G., Dermietzel, R., and Kamermans, M. (2007). Retinal horizontal cell-specific promoter activity and protein expression of zebrafish connexin 52.6 and connexin 55.5. *J. Comp. Neurol.* **501**, 765–779.

Sivasubbu, S., Balciunas, D., Davidson, A. E., Pickart, M. A., Hermanson, S. B., Wangensteen, K. J., Wolbrink, D. C., and Ekker, S. C. (2006). Gene-breaking transposon mutagenesis reveals an essential role for histone H2afza in zebrafish larval development. *Mech. Dev.* **123**, 513–529.

Solnica-Krezel, L., Schier, A., and Driever, W. (1994). Efficient recovery of ENU-induced mutations from the zebrafish germline. *Genetics* **136**, 1–20.

Soroldoni, D., Hogan, B. M., and Oates, A. C. (2009). Simple and efficient transgenesis with meganuclease constructs in zebrafish. *Methods Mol. Biol.* **546**, 117–130.

Soules, K. A., and Link, B. A. (2005). Morphogenesis of the anterior segment in the zebrafish eye. *BMC Dev. Biol.* **5**, 12.

Strahle, U., Blader, P., Adam, J., and Ingham, P. W. (1994). A simple and efficient procedure for non-isotopic in situ hybridization to sectioned material. *Trends Genet.* **10**, 75–76.

Streisinger, G., Singer, F., Walker, C., Knauber, D., and Dower, N. (1986). Segregation analyses and gene-centromere distances in zebrafish. *Genetics* **112**, 311–319.

Streisinger, G., Walker, C., Dower, N., Knauber, D., and Singer, F. (1981). Production of clones of homozygous diploid zebra fish (Brachydanio rerio). *Nature* **291**, 293–296.

Suster, M. L., Kikuta, H., Urasaki, A., Asakawa, K., and Kawakami, K. (2009). Transgenesis in zebrafish with the tol2 transposon system. *Methods Mol. Biol.* **561**, 41–63.

Take-uchi, M., Clarke, J. D., and Wilson, S. W. (2003). Hedgehog signalling maintains the optic stalk-retinal interface through the regulation of Vax gene activity. *Development* **130**, 955–968.

Takechi, M., Hamaoka, T., and Kawamura, S. (2003). Fluorescence visualization of ultraviolet-sensitive cone photoreceptor development in living zebrafish. *FEBS Lett.* **553**, 90–94.

Takechi, M., and Kawamura, S. (2005). Temporal and spatial changes in the expression pattern of multiple red and green subtype opsin genes during zebrafish development. *J. Exp. Biol.* **208**, 1337–1345.

Thisse, B., Heyer, V., Lux, A., Alunni, V., Degrave, A., Seiliez, I., Kirchner, J., Parkhill, J. P., and Thisse, C. (2004). Spatial and temporal expression of the zebrafish genome by large-scale *in situ* hybridization screening. *Methods Cell Biol.* **77**, 505–519.

Tran, T. C., Sneed, B., Haider, J., Blavo, D., White, A., Aiyejorun, T., Baranowski, T. C., Rubinstein, A. L., Doan, T. N., Dingledine, R., *et al.* (2007). Automated, quantitative screening assay for antiangiogenic compounds using transgenic zebrafish. *Cancer Res.* **67**, 11386–11392.

Trowe, T., Klostermann, S., Baier, H., Granato, M., Crawford, A. D., Grunewald, B., Hoffmann, H., Karlstrom, R. O., Meyer, S. U., Muller, B., *et al.* (1996). Mutations disrupting the ordering and topographic mapping of axons in the retinotectal projection of the zebrafish, Danio rerio. *Development* **123**, 439–450.

Tsujikawa, M., and Malicki, J. (2004). Intraflagellar transport genes are essential for differentiation and survival of vertebrate sensory neurons. *Neuron* **42**, 703–716.

Tsujikawa, M., and Malicki, J. (2004a). Genetics of photoreceptor development and function in zebrafish. *Int. J. Dev. Biol.* **48**, 925–934.

Tsujikawa, M., and Malicki, J. (2004b). Intraflagellar transport genes are essential for differentiation and survival of vertebrate sensory neurons. *Neuron* **42**, 703–716.

Tsujimura, T., Chinen, A., and Kawamura, S. (2007). Identification of a locus control region for quadruplicated green-sensitive opsin genes in zebrafish. *Proc. Natl. Acad. Sci. USA* **104**, 12813–12818.

Turner, D., and Cepko, C. (1987). A common progenitor for neurons and glia persists in rat retina late in development. *Nature* **328**, 131–136.

Turner, D., Snyder, E., and Cepko, C. (1990). Lineage-Independent Determination of Cell Type in the Embryonic Mouse Retina. *Neuron* **4**, 833–845.

Urasaki, A., Morvan, G., and Kawakami, K. (2006). Functional dissection of the Tol2 transposable element identified the minimal cis-sequence and a highly repetitive sequence in the subterminal region essential for transposition. *Genetics* **174**, 639–649.

van Eeden, F. J., Granato, M., Odenthal, J., and Haffter, P. (1999). Developmental mutant screens in the zebrafish. *Methods Cell Biol.* **60**, 21–41.

Varga, Z. M., Wegner, J., and Westerfield, M. (1999). Anterior movement of ventral diencephalic precursors separates the primordial eye field in the neural plate and requires cyclops. *Development* **126**, 5533–5546.

Vihtelic, T. S., Doro, C. J., and Hyde, D. R. (1999). Cloning and characterization of six zebrafish photoreceptor opsin cDNAs and immunolocalization of their corresponding proteins. *Vis. Neurosci.* **16**, 571–585.

Vihtelic, T. S., and Hyde, D. R. (2000). Light-induced rod and cone cell death and regeneration in the adult albino zebrafish (Danio rerio) retina. *J. Neurobiol.* **44**, 289–307.

Villefranc, J. A., Amigo, J., and Lawson, N. D. (2007). Gateway compatible vectors for analysis of gene function in the zebrafish. *Dev. Dyn.* **236**, 3077–3087.

Walker, C. (1999). Haploid screens and gamma-ray mutagenesis. *Methods Cell Biol.* **60**, 43–70.

Watanabe, T., and Raff, M. (1988). Retinal astrocytes are immigrants from the optic nerve. *Nature* **332**, 834–837.

Wei, X., and Malicki, J. (2002). nagie oko, encoding a MAGUK-family protein, is essential for cellular patterning of the retina. *Nat. Genet.* **31**, 150–157.

Westerfield, M. (2000). "The Zebrafish Book. A Guide for the Laboratory Use of Zebrafish (*Danio rerio*)." University of Oregon Press, Eugene, Oregon.

Wetts, R., and Fraser, S. (1988). Multipotent precursors can give rise to all major cell types of the frog retina. *Science* **239**, 1142–1145.

Wienholds, E., Schulte-Merker, S., Walderich, B., and Plasterk, R. H. (2002). Target-selected inactivation of the zebrafish rag1 gene. *Science* **297**, 99–102.

Wise, G., Dollery, C., and Henkind, P. (1971). "The Retinal Circulation." Harper & Row, New York.

Woo, K., and Fraser, S. E. (1995). Order and coherence in the fate map of the zebrafish nervous system. *Development* **121**, 2595–2609.

Xiao, T., Roeser, T., Staub, W., and Baier, H. (2005). A GFP-based genetic screen reveals mutations that disrupt the architecture of the zebrafish retinotectal projection. *Development* **132**, 2955–2967.

Yang, Z., Jiang, H., and Lin, S. (2009). Bacterial artificial chromosome transgenesis for zebrafish. *Methods Mol. Biol.* **546**, 103–116.

Yagasaki, K., and Jacobson, S. G. (1989). Cone-rod dystrophy. Phenotypic diversity by retinal function testing. *Arch. Ophthalmol.* **107**, 701–708.

Yamaguchi, M., Fujimori-Tonou, N., Yoshimura, Y., Kishi, T., Okamoto, H., and Masai, I. (2008). Mutation of DNA primase causes extensive apoptosis of retinal neurons through the activation of DNA damage checkpoint and tumor suppressor p53. *Development* **135**, 1247–1257.

Yamamoto, Y., and Jeffery, W. R. (2002). Probing teleost eye development by lens transplantation. *Methods* **28**, 420–426.

Yazulla, S., and Studholme, K. M. (2001). Neurochemical anatomy of the zebrafish retina as determined by immunocytochemistry. *J. Neurocytol.* **30,** 551–592.

Yeo, S. Y., Kim, M., Kim, H. S., Huh, T. L., and Chitnis, A. B. (2007). Fluorescent protein expression driven by her4 regulatory elements reveals the spatiotemporal pattern of Notch signaling in the nervous system of zebrafish embryos. *Dev. Biol.* **301,** 555–567.

Yu, C. J., Gao, Y., Willis, C. L., Li, P., Tiano, J. P., Nakamura, P. A., Hyde, D. R., and Li, L. (2007). Mitogen-associated protein kinase- and protein kinase A-dependent regulation of rhodopsin promoter expression in zebrafish rod photoreceptor cells. *J. Neurosci. Res.* **85,** 488–496.

Zhang, Y., Mcculloch, K., and J, M. (2009). Lens transplantation in zebrafish and its application in the analysis of eye mutants. *J. Vis. Exp.* **28.**

Zhao, X. C., Yee, R. W., Norcom, E., Burgess, H., Avanesov, A. S., Barrish, J. P., and Malicki, J. (2006). The zebrafish cornea: Structure and development. *Invest. Ophthalmol. Vis. Sci.* **47,** 4341–4348.

Zolessi, F. R., Poggi, L., Wilkinson, C. J., Chien, C. B., and Harris, W. A. (2006). Polarization and orientation of retinal ganglion cells *in vivo. Neural Dev.* **1,** 2.

Zon, L. I., and Peterson, R. T. (2005). *In vivo* drug discovery in the zebrafish. *Nat. Rev. Drug. Discov.* **4,** 35–44.

CHAPTER 17

Instrumentation for Measuring Oculomotor Performance and Plasticity in Larval Organisms

James C. Beck,★ Edwin Gilland,† Robert Baker,★ and David W. Tank‡,§

★Department of Physiology and Neuroscience
New York University School of Medicine
New York 10016

†Department of Anatomy
Howard University School of Medicine
Washington, District of Columbia 20059

‡Department of Physics
Princeton University
Princeton, New Jersey 08544

§Department of Molecular Biology
Princeton University
Princeton, New Jersey 08544

ESSENTIAL ZEBRAFISH METHODS:
CELL AND DEVELOPMENTAL BIOLOGY
Copyright 2009, Elsevier Inc. All rights reserved.

401

DOI: 10.1016/B978-0-12-374599-6.00017-0

I. Introduction

The next stage for genomics in neurobiology is to discern how genetic networks operate to regulate neural function and control behavior (Bate, 1998; Gerlai, 2003). A direct investigation of the genetic basis of motor behavior can only be achieved when both the motor dynamics and underlying neural circuit dynamics are measurable; thus, the selection of which motor behavior to study from an animal's unique repertoire is critical. The vertebrate oculomotor system is an ideal candidate because both the stimuli, visual and vestibular motion, and the evoked eye movements are well defined, reproducible, and quantifiable (Robinson, 1968). Measuring oculomotor performance requires not only control of retinal visual slip and head acceleration (both linear and angular); but it also necessitates a means of animal restraint, facilitating uniform presentation of sensory stimuli while accurately measuring eye movements. Available oculomotor instrumentation has been optimized for studies in large, adult animals; nevertheless, these systems provide a basis for designing a purpose-built apparatus for investigations of oculomotor behavior and plasticity in small genetic model organisms.

The quest for understanding the genetic basis of behavior has prompted several initial studies of oculomotor behavior in larval zebrafish. However, these investigations employed relatively simple recording and stimulating technologies and none explored the oculomotor repertoire with the degree of sophistication that has become routine in studies of adult goldfish (Marsh and Baker, 1997; Pastor *et al.*, 1992), cat (Godaux *et al.*, 1983; Robinson, 1976; Shinoda and Yoshida, 1974), rabbit (Barmack, 1981; Collewijn, 1969; Ito *et al.*, 1979), primate (Fernandez and Goldberg, 1971; Lisberger *et al.*, 1981; Melvill Jones, 1977; Skavenski and Robinson, 1973), or mouse (Boyden and Raymond, 2003; van Alphen *et al.*, 2001).

The only investigation of the angular vestibuloocular reflex (VOR) in larval zebrafish used hand driven, rotation to provide the acceleration stimulus (Easter and Nicola, 1997). With no defined control or measure of table velocity, only a qualitative response was reported. In contrast, previous studies of the zebrafish optokinetic reflex (OKR) have employed motor driven (Brockerhoff *et al.*, 1995; Carvalho *et al.*, 2002; Easter and Nicola, 1996, 1997; Rick *et al.*, 2000) or video projected (Roeser and Baier, 2003) striped drums to provide visual stimuli; however, a constant drum velocity was used in all experiments. While useful for genetic screens, this simple stimulus paradigm fails to distinguish visual from oculomotor performance and to quantitatively assess visuomotor maturation (Easter and Nicola, 1997, 1996). In order to evaluate the developing dynamic and integrative functions of central neurons responsible for the oculomotor behaviors, such as eye position (Robinson, 1975, 1989) and velocity integration (Cohen et al., 1977; Raphan et al., 1979), it is essential to employ dynamic stimuli in the form of changing stimulus frequency and/or amplitude combined with sinusoids and steps. The motor designs in these previous investigations lacked both the position and velocity feedback necessary to control drum kinetics for the accurate reproduction of complex motion.

Since genetic manipulations of the neural circuits underlying motor behavior may yield subtle differences in oculomotor performance, it is important to begin with an accurate measurement of eye movements. Prior work relied on two methods of measuring oculomotor output in larval animals: either direct observation (Brockerhoff et al., 1995, 1997; Carvalho et al., 2002) or video recordings (Easter and Nicola, 1997; Moorman et al., 1999; Neuhauss et al., 1999; Rick et al., 2000; Riley and Moorman, 2000; Roeser and Baier, 2003). While suitable for large-scale screens of mutant animals, direct observation yielded only a binary answer—movement or no movement. This simple result does not provide sufficient quantitative resolution to link genetic expression and motor behavior. Video recordings of eye movements can permit a more quantitative approach to behavioral analysis. Still, most reports have analyzed the behavior manually, frame by frame (e.g., Easter and Nicola, 1996, 1997; Moorman et al., 1999; Neuhauss et al., 1999; Riley and Moorman, 2000). This approach was laborious, of low resolution, and practical only for short periods of data. More recent efforts have begun to utilize computer-aided video analysis but have been restricted to off-line processing at low frame rates (<5 Hz) (Bahadori et al., 2003; Rick et al., 2000; Roeser and Baier, 2003) and limited accuracy ($\pm 2°$, Roeser and Baier, 2003).

Another major compromise in larval eye measurement recordings has been the necessity to perform experiments in the light. While not a concern for visually driven behaviors, isolation, and measurement of vestibular reflexes requires the removal of visual feedback, which can be achieved by measuring VOR with infrared illumination (Wilson and Melvill Jones, 1979). However, investigations into both linear (Moorman, 2001; Moorman et al., 1999; Riley and Moorman, 2000) and angular (Easter and Nicola, 1997) VOR did not use available technology to measure eye movements in the dark. As an alternative, these studies were performed with a featureless drum (ganzfeld) to minimize visual feedback during vestibular stimulation. The ganzfeld employed by Easter and Nicola (1997) did not eliminate all visual motion cues since their observation of adult-like angular VOR in 5 dpf zebrafish was inconsistent with measurements made using infrared video (Beck et al., 2004).

Overall what is needed in assessing larval oculomotor performance is a means of accurately measuring eye movements in the dark or light during dynamic visual and/or vestibular stimulation. Measurements available in real time would, in addition, allow the use of behavioral feedback learning paradigms to explore the ontogeny of oculomotor plasticity (Major et al., 2004; Mensh et al., 2004) and to potentially address the molecular mechanisms underlying memory and learning, that is long-term potentiation and depression (Boyden and Raymond, 2003; Hartell, 2002; Ito, 2002).

To satisfy the prerequisites for measuring oculomotor performance in larval animals, a vertical axis vestibular turntable and optokinetic drum were constructed. This apparatus accommodates animals of small sizes (0–16 mm) and operates over the wide frequency range (0.01–10 Hz) necessary for both optokinetic and vestibular stimulation (Wilson and Melvill Jones, 1979). The table and drum were independent, each driven by a feedback servo-controller to provide accurate angular acceleration and visual motion stimulation, required to test visuo-vestibular interactions. In real time (60 Hz), CCD video images of the subject's head and trunk, illuminated from below with infrared LEDs, were captured to computer, and analyzed using vision processing algorithms to determine the relative position of the eye with respect to body axis with high fidelity ($\pm 0.1°$). Eye positions could be used to control drum kinetics in real time for visuomotor plasticity paradigms. Larval and juvenile zebrafish, goldfish, *Xenopus*, and medaka, were compared under both light and dark conditions using classic oculomotor testing paradigms, demonstrating the powerful utility and flexibility of this instrumentation. Part of this work has appeared elsewhere in abstract form (Beck *et al.*, 2002).

The methods and tools described have been successfully employed to study a variety of larval animals that now include zebrafish, medaka, goldfish, flatfish, *Xenopus*, midshipmen, and cichlids. A prolonged delay in the acquisition of angular vestibuloocular reflexes to the optokinetic and gravitointerial systems due to semicircular canal size maturation limited oculomotor investigations in the popular zebrafish model (Beck and Baker, 2005; Beck *et al.*, 2004; Lambert *et al.*, 2008). Consequently, a direct, quantifiable measure of higher order processing involving cerebellar function, for example, as well as VOR learning and memory, is not addressable until after 28 dpf in zebrafish. This time point is too late for the majority of mutants, for the use of up/down expression vectors, or for the use of morpholinos. Cerebellar function may also be addressed by measuring oculomotor integrator plasticity after sensory visual feedback training which is another technique this methodology permits (Major *et al.*, 2004). However, an overlap with the developmental time course underlying integrator stability (from 5 to 21 dpf) also hampers the use of feedback training.

Another model animal that has gained great attention in the scientific community and circumvents these sensory/motor developmental constraints is the African cichlid *Astatotilapia burtoni*. Unlike the cyprinids (zebrafish, goldfish), the semicircular canals of larval *A. burtoni* are functional by 9 dpf and the angular VOR exhibits robust quantifiable plasticity as early as 11 dpf. Much like in adult animals, larval cichlids can rapidly modify any motor performance from gaze (eye and body position) through locomotion (Bassett *et al.*, 2008). Thus the highly derived status and explosive radiation of cichlid species (>3000 extant) combined with the push in determining their genomics makes the cichlid model a very compelling system for future study (Kocher, 2004).

Note to reader: To aid in understanding and visualization of this system in operation, supplemental movies are provided in QuickTime 6.0 format, playable on both Mac and PC computers (http://www.apple.com/quicktime/download/standalone/). Movies are available on the Web at the homepage for this volume at the Methods in Cell Biology website.

II. Methods

A. Vestibular Turntable and Drum (Supplemental Movie 1)

The vertical axis turntable comprised an aluminum base, housing the drive motor assembly, and an aluminum top connected to the base via a hollow, stainless steel shaft (Fig. 1A, Movie 1, http://www.elsevierdirect.com/companion.jsp?ISBN=9780123745996). The surface of the table contained a compact, three-axis micromanipulator for positioning the specimen, a motorized optokinetic drum, and a CCD camera centered on the table's axis. Table top rotation was belt driven by a compact and powerful 16 W motor (Faulhaber, Clearwater, FL; model 3042W024C001G) with 40 lb-in, planetary gear head (Faulhaber, model 30/1) inline coupled with the motor shaft. Motor velocity was monitored via an inline tachometer (Faulhaber, model 0436C). Separately connected to the table top was a precision position potentiometer that provided a voltage proportional to the angular position of the table (ETI Systems, Carlsbad, CA; model SP40B). The table motor operated with a servo-controller (Western Servo Design, Hayward, CA; model LDH-S1) that utilized both of these velocity and position feedback signals (see Movie 1).

A small drum centered on the table top provided visual stimuli to induce optokinetic reflexes (OKR, Fig. 1B). Clear, acrylic drums of different sizes, 3.5–8 cm, nested concentrically around the specimen holder, provided stationary, patterned, or *ganzfeld* (featureless visual surround) stimuli for both monocular and binocular testing. The moving drum was illuminated from below, through the table shaft, by visible light LEDs and was belt driven by a smaller servomotor than that used in the table base (Faulhaber, model 2251U012S1.5G). Gearing (Faulhaber, model 23/1) and tachometer (Faulhaber, model 0431) arrangement were similar. The drum was servo-controlled (Western Servo Design; model LDH-A1) independently from the table and also employed both position (ETI Systems; model SP22GS) and velocity feedback (see Movie 1). Stimulus waveforms controlling table and drum motion were created with two Agilent (Palo Alto, CA) 33120A phase linked, function generators.

The table and drum were located within a removable, light-tight wooden enclosure with a black felt curtain on the front to facilitate access to the specimen. As needed, the enclosure was maintained between 18 and 30 °C with a YSI temperature controller (Yellow Springs, OH) and a screw base radiant heater (McMaster-Carr, Dayton, NJ).

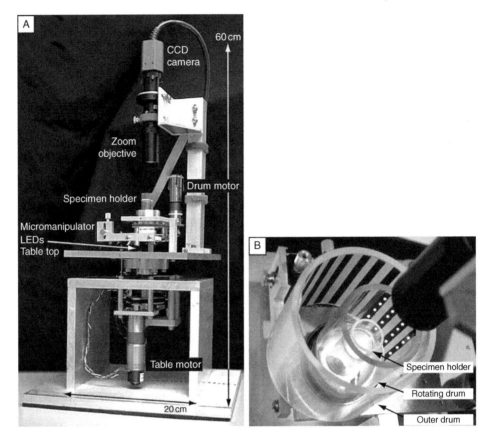

Fig. 1 Vertical axis vestibular turntable and optokinetic drum scaled to accommodate larval and juvenile animals. (A) Side view of vestibular turntable showing table motor in the base with the optokinetic drum motor, CCD camera with zoom objective, three-axis micromanipulator, and specimen holder on the table top. The drum has been removed to reveal the specimen holder, lying directly below are the visible light and infrared LEDs. (B) Top view of specimen holder with surrounding drums depicting potential visual stimuli arrangements that could be used for *en bloc* or, here, monocular stimulation. The outer drum provides stationary stimuli to one eye while the inner drum rotates (white dots added for illustration). Also see Supplemental Movie 1.

Movie 1 Operation of motorized vestibular turntable and optokinetic drum. A brief identification of the parts of the apparatus followed by a demonstration of vestibular, optokinetic, and visuo-vestibular stimuli. Vestibular stimuli in order shown are sinusoids at 0.125 Hz, ±20 °/s and 0.5 Hz, ±60 °/s as well as velocity steps at 1.0 Hz, ±60 °/s and 10.0 Hz, ±10 °/s. Optokinetic stimuli are sinusoids at 0.125 Hz, ±30 °/s, ±30 °/s velocity steps at 0.1 Hz, and high-frequency sinusoids (0.5 and 2.0 Hz, ±10 °/s). The movie concludes with the turntable and drum producing stimuli for VOR in the light (gain 1), gain up training (gain 2x), and gain down training (gain 0 or cancellation). Length: 2:43; size: 5.7 MB.

B. Specimen Holder and Mounting (Supplemental Movie 2)

A transparent glass specimen holder contained the subject in 2.5 mL of water. The specimen holder comprised two pieces: (1) a round, 19 mm diameter, glass chamber that was 12 mm in height having a floor and roof of 18 mm diameter, round coverslips (Fisher Scientific, Pittsburgh, PA) and (2) a round, 19 mm diameter, glass base of 35 mm in height (Figs. 1 and 2B; Movie 2, http://www.elsevierdirect.com/companion. jsp?ISBN=9780123745996). The specimen holder base and chamber, as well as the coverslip floor, were held together with clear epoxy (81190; Loctite Corp, Avon, OH). The roof of the specimen chamber was held in place by surface tension of the water and could be left ajar to permit oxygen exchange. The specimen holder was placed into a clear, acrylic mounting stage attached to a three-axis micromanipulator, allowing the head to be centered on the table and to focus the animal for the camera.

Animals were held in the specimen chamber embedded in a drop of low-melting temperature agarose (2.0%; Sigma; see Movie 2). The head region, rostral to the swim bladder in fish, was freed of agarose with a scalpel, permitting unrestricted movement of the eyes and respiration (Fig. 2A, Movie 2, http://www.elsevierdirect. com/companion.jsp?ISBN=9780123745996). Larger and older animals were capable of aspirating the molten agarose, occasionally clogging the pharynx once solidified. This was mitigated by covering the head/pharynx with 6% methylcellu-lose just prior to enveloping with agarose. In addition, large animals could also extricate themselves from the agarose once their head was freed, which was reduced by flexing the tail into a "J" configuration before the agarose solidified. Alternatively, an electrolytically sharpened tungsten or platinum 10 μm wire, coated with a 10% benzocaine/ethanol solution, inserted perpendicular to the body through the epithelial ridge in the caudal tail was sufficient to prevent escape (Fig. 2B(3)). The solidified agarose block was affixed with several 100 μm minutien pins to a 1 mm thick, clear disk (Sylgard 184; Dow Corning, Midland, MI) that snuggly fit within the bottom of the specimen holder (Fig. 2B, Movie 2, http://www.elsevierdirect.com/companion.jsp?ISBN=9780123745996). Animals that were removed from the agarose after <1 h swam normally and could be reared to adult stages.

C. Measuring Eye Movements (Supplemental Movie 3)

A visible light blocking filter (Wratten 88A; Kodak, Rochester, NY) in the base of the stage provided a dark area below the animal, while permitting the transmis-sion of infrared light. Transillumination of the animal with an infrared (880 nm) LED (1) increased the contrast between the eyes and the head and (2) allowed video recording of eye movements in either the light or dark (as required for measuring either VOR or spontaneous fixations in open loop conditions). The zoom objective (VZM 300i; Edmund Industrial Optics, Barrington, NJ), fitted to a monochrome CCD camera (6412; COHU, San Diego, CA), was adjusted to maximize the head and rostral trunk of the animal within the field of view. Video output was digitized with a National Instruments frame-grabber (PCI-1409;

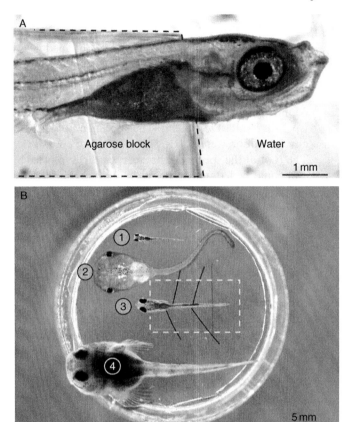

Fig. 2 Larval animals restrained in agarose with head free. (A) Medaka embedded in a block of low-melting temperature agarose (dashed line), with the area rostral to the swim bladder freed to allow for unrestricted eye movements and respiration. The upturned mouth of medaka is characteristic of a surface feeding fish. (B) Agarose block (dashed line) pinned to a clear Sylgard disk (1 mm thick) and placed in a clear glass specimen holder (19 mm diameter). Four larval animals that were tested are illustrated, scaled to size: (1) zebrafish, (2) *X. laevis*, (3) goldfish, and (4) midshipman. Animals 1, 2, and 4 were digitally superimposed upon the specimen chamber containing the goldfish (3). Also see Supplemental Movie 2 and behavioral movies (5–10).

Movie 2 A step-by-step demonstration of embedding a 5 dpf zebrafish in agarose and preparing the animal for eye movement recordings. Although all steps are shown in their entirety, events not pertaining directly to the process have been edited for time (e.g., changing focus). Length: 4:28; size: 6.0 MB.

Austin, TX) and analyzed with software developed in National Instruments' Lab-VIEW and IMAQ Vision (version 6 for both). The video analysis program was executed on a dual processor PC running Microsoft Windows XP Professional with two 2.1 GHz Athlon 2400+ MP processors (see Movie 3).

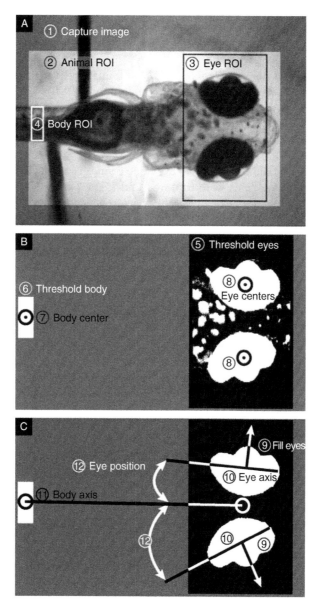

Fig. 3 Eye tracking algorithm. (A) An initial still video image was captured (1) and a region of interest (ROI) drawn around the entire animal (2, dark border). ROIs were drawn around the eyes (3, black rectangle) and around a fixed point along the body axis (4, white rectangle). (B) As video images were captured in real time, the ROIs around the eyes (5) and body (6) were thresholded and inverted; and the centers of the body (7) and the two eyes (8) were determined. (C) From the center of the eye, similar pixel values were filled (9); and the body axis (11) subtracted from the major axis of each eye (10) in order to determine absolute eye position (12). Also see Supplemental Movie 3.

Movie 3 Step-by-step demonstration of the eye movement algorithm in operation, from initial acquisition to adjusting parameters for optimal eye tracking. Length: 0:58; size: 11.0 MB.

Prior to video analysis, a still image was captured (Fig. 3A(1), Movie 3, http:// www.elsevierdirect.com/companion.jsp?ISBN=9780123745996) and a region of interest (ROI) was drawn around the body to reduce overall image size used in computation (Fig. 3A(2)). In addition, two ROIs were selected, one around the animal's eyes (Fig. 3A(3)) and another around a reference on the body axis (Fig. 3A(4)). Two video fields, odd and even, each scanned at 60 Hz are interlaced to provide a standard 30 Hz NTSC video frame. To achieve a 60 Hz sampling rate, individual video fields were captured and treated as independent images. This resulted in a reduction in height of the processed video image from 640 pixels in a video frame to 320 pixels in a video field. This change in pixel aspect ratio was taken into account within the software.

Once the real time video processing began, the two ROIs selected around the eyes and along the body axis were binary thresholded and inverted, producing white eyes (Fig. 3B(5)) and body reference (Fig. 3B(6), Movie 3, http://www. elsevierdirect.com/companion.jsp?ISBN=9780123745996). The center of mass of the body reference was then calculated (Fig. 3B(7)). To select only the eyes within the thresholded ROI (Fig. 3B(5)) and to reject pigments (melanophores) in the head around the eyes, the size of each object in the eye ROI was computed. The two largest objects in the eye ROI (Fig. 3A(3)) were the eyes and the center of mass of each was then determined (Fig. 3B(8)). The center of each eye served as the starting point for a pixel seeding operation ("magic wand") that smoothly filled in the entire shape of each eye by selecting neighboring pixels that were of similar intensity (Fig. 3C(9), Movie 3, http://www.elsevierdirect.com/companion.jsp? ISBN=9780123745996). The equivalent ellipse major and minor axes of each filled eye (Fig. 3C(10)) and the body axis were then determined (Fig. 3C(11)). The body axis connected the midpoint between the eyes to the center of the thresholded body reference (Fig. 3B(6)). Eye positions (Fig. 3C(12)) were then calculated as the angle of the major axis of each eye (Fig. 3C(10)) relative to the body axis (Fig. 3C(11)). Left and right directions were established by evaluating the vector cross product of the eye and body axis. Computed eye positions were converted to voltages (16 bit) using D/A converters on a National Instruments computer card (PCI-6052E). Copies of this LabVIEW code will be made available upon request to either J. C. B. or D. W. T.

D. Data Acquisition and Analysis

Data was digitized with an Axon Instruments (Union City, CA) Digidata 1200B and Axoscope 9.0 software executed on a computer separate from the one used to calculate eye positions. Position potentiometer and tachometer outputs for both the table and drum, as well as eye positions, and the visible light LED voltage were digitized to hard disk at 200 Hz. Eye position records were subsequently imported into MATLAB and differentiated to produce eye velocity. Both the position and resulting velocity traces were filtered with a sliding-average window of 50 ms (10 sample points at 200 Hz). Drum and turntable tachometer and position traces were filtered with a sliding-average window of 10–30 ms.

E. Experimental Animals

Animals were used in accordance with the *Guide for the Care and Use of Laboratory Animals* (http://www.nap.edu/readingroom/books/labrats/). Specific protocols were approved by the NYU School of Medicine Institutional Animal Care and Use Committee. Larval species tested included established physiological model animals: goldfish (*Carassius auratus*) and midshipmen (*Porichthys notatus*, not illustrated), as well as established developmental and genetic vertebrate models: zebrafish (*Danio rerio*), medaka (*Oryzias latipes*) and larval frogs (*Xenopus laevis* and *Xenopus tropicalis*). Goldfish eggs were obtained from Hunting Creek Fisheries (Thurmond, MD); zebrafish and medaka adults were obtained from Aquatic Research Organisms, Inc. (Hampton, NH) and larvae spawned in lab; and *X. laevis* (Movie 5, http://www.elsevierdirect.com/companion.jsp?ISBN=9780123745996) and *X. tropicalis* larvae were obtained from NASCO (Fort Atkinson, WI). Midshipmen were collected in Bodega Bay, CA and provided by Dr. Andrew Bass at Cornell University.

F. Updates on the methodology

• For zebrafish <21 dpf, build a translation (tilt) rather than rotational vestibular table—as otoliths are functional by 4 days (Beck *et al.*, 2004). By contrast, cichlids have angular and linear VOR reflexes by 9 dpf that exhibit adaptive plasticity that can be precisely quantified.

• For immobilizing larger larvae, >6 mm, very brief doses of MS-222 at high concentrations (0.84–1.68 mg/mL) have been found to minimize aspiration of agarose and thus maintain animal's respiration and behavior. The most important step is to immediately begin the mounting procedures once gilling slows/stops, as speed is always essential to minimize anoxia.

• Analog video cameras with CCD sensors generally have the best IR sensitivity; however, consider the newer CMOS video cameras that easily exceed 120 fps and use either firewire or Ethernet connections. The latest Intel and AMD processors allow the real-time processing of high frame rates (>120 fps).

• High camera frame rates mean reduced light reaching the imaging sensor because of the increased shutter speed. Use a high-power IR LED to compensate (≥880 nm).

• Wale Apparatus (http://www.waleapparatus.com/) is a good source of glass to make the specimen chambers. Glass can be cut by a glass blower most frequently located in Chemistry departments.

========= III. Results and Discussion

A. Table and Drum (Supplemental Movie 1)

Current technology used to investigate oculomotor behavior in large animals generates angular vestibular stimulation by rotating an adult animal on a vestibular turntable with visual motion stimuli frequently produced by a rotating, striped optokinetic drum (Collewijn, 1991; Wilson and Melvill Jones, 1979). The drum is sufficiently large to physically surround the vestibular platform and thus occupies a large area. In our instrument, the dimensions of the vestibular turntable and optokinetic drum were significantly reduced such that the pair operated as a bench top device (Fig. 1, Movie 1). The compact size facilitated the ease of recording from millimeter sized animals.

Powerful motors are required to overcome the inertial mass of the table and drum to faithfully reproduce the command stimulus required for experimental paradigms that cover a wide frequency and amplitude range. Despite their size, the miniature motors chosen to drive the table and drum demonstrated the capability of producing a range of stimuli on par with systems used for oculomotor investigations in much larger animals (Fig. 4). When driven with a waveform generator, the motorized turntable produced sinusoids of up to 10 Hz as well as position triangles and steps with an average latency of 104 ms (Fig. 4) and accelerations over $1000°/s^2$ at 0.5Hz, ±10 °/s. Like the table, the motorized drum produced sinusoids of up to 10 Hz as well as position steps with latencies of <55 ms (Fig. 4). Because of their independent control, the inertial differences between the table and drum could be compensated for by adjusting the drum phase relative to the table, enabling visuo-vestibular conflict training paradigms. In sum, established behavioral and stimulation paradigms commonly used in the study of the adult oculomotor system could be applied directly to larval and juvenile animals.

B. Animal Immobilization (Supplemental Movie 2)

Another issue that required equal consideration for these investigations was how to properly restrain small animals. A primary concern was that the head remained centered relative to the stimulus; otherwise, off-axis angular vestibular stimulation would activate both the semicircular canals and the otoliths, making it impossible to distinguish between the linear and angular acceleration systems (Wilson and Melvill Jones, 1979). As reported in behavioral studies of the developing reticulospinal network, larval animals were successfully embedded in low-melting temperature agarose, with the tail subsequently freed to study swimming (Higashijima *et al.*, 2003; Ritter *et al.*, 2001). Unfortunately, locking the head in agarose also locks the eyes in place, defeating the ability to study eye movements. Alternatively, small animals were also frequently immobilized by immersing them in a highly viscous aqueous mixture of methylcellulose (Brockerhoff *et al.*, 1995; Neuhauss *et al.*, 1999; Rick *et al.*, 2000; Roeser and Baier, 2003) or low-concentration

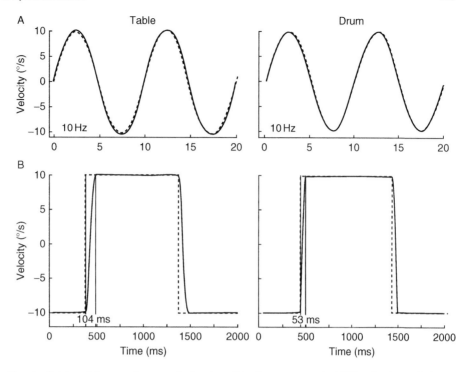

Fig. 4 Table and drum performance during sinusoid and step commands. (A) Despite the small sizes of the drive motors, both the table and drum servo-controllers were able to faithfully reproduce the command signal of 10 Hz sinusoids at ±10 °/s. (B) Response to a 0.5 Hz ±10 °/s step command demonstrates short latencies for both the table and drum, 104 and 53 ms, respectively, and accurate signal reproduction. Also see Supplemental Movie 1. Solid line: table, drum velocity. Dashed line: command signal.

agarose (Clark, 1981); but these methods have several draw backs as animals immersed in thick liquids cannot respire normally and must exchange oxygen through the skin, which is only effective in the very youngest (~7 dpf) cases (Rombough, 2002). In addition, larger animals can move freely through viscous solutions also limiting its use to early ontogenetic studies (Clark, 1981). More germane, it is likely that the use of a highly viscous mixture to restrain the larval animal would impact oculomotor performance by creating significant drag to eye movement, a possibility tested by placing 5 dpf zebrafish in a 3% methylcellulose solution (see below, Fig. 8).

Embedding the animal in a block of low-melting temperature agarose, as done for swimming (Higashijima *et al.*, 2003; Ritter *et al.*, 2001) but with the head and gills exposed, represented an ideal solution to the problems outlined above (Fig. 2, Movie 2). After initial encasement, the agarose covering the head and gills could be quickly removed; and the block trimmed to an appropriate size to fit in the specimen holder. While encased in agarose, the specimen was easily manipulated and could be transferred from Petri dish to holder without injury. Importantly, respiration and eye movements were unencumbered (Fig. 2A). The agarose also acted as a support for minutien (100 um) pins, allowing the block to be fixed in place (Fig. 2B, Movie 2). If need be, the animal could be safely removed from the agarose holder and reared to adulthood, providing the opportunity for longitudinal observations and the preservation of isolated mutant animals. Thus, agarose provides a means of gentle, yet effective immobilization with little to no biological perturbation.

By combining a specimen holder constructed entirely of glass with the transparency of the agarose, there was an almost unobstructed visual field encircling the preparation (Fig. 2B). The 19 mm diameter of the specimen holder easily accommodated a wide range of species, including zebrafish, goldfish, medaka, *Xenopus*, and the midshipmen *P. notatus* (Fig. 2B). These animals ranged in size from under 4 mm to greater than 16 mm in length. The largest animals were accommodated by flexing the tail to match the curvature of the holder. By enlarging the diameter of the specimen holder, the oculomotor behavior of even juvenile animals could be studied.

C. Eye Position Measurements (Supplemental Movies 2–4)

The *sine qua non* of any investigation of motor performance is the ability to measure behavioral output. The current standard of eye position measurement is the use of a scleral search coil arrangement (Robinson, 1963). Comprised of a few turns of wire, a coil is attached directly to the eye. Within an alternating magnetic field, coil voltage is proportional to eye position and can be measured with an accuracy of 0.025° at 1000 Hz (Fuchs and Robinson, 1966). Because this approach is clearly intractable for larval animals, video recordings of eye movements have been the preferred alternative (Easter and Nicola, 1997; Neuhauss *et al.*, 1999; Rick *et al.*, 2000; Roeser and Baier, 2003). Video imaging provides a noninvasive yet quantitative approach for measuring eye position and exploits the transparency found in many young animals. Like the scleral search coil system, standard video recordings have a comparatively high-temporal resolution that could be employed for real-time measurements. In previous investigations, the limited spatial resolution (±2°, Roeser and Baier, 2003) and low-sampling rate used (<5 Hz, Bahadori *et al.*, 2003; Rick *et al.*, 2000; Roeser and Baier, 2003) failed to capitalize on this potential.

In contrast, the algorithm presented here (see Fig. 3, Movie 3) operated at 60 Hz by individually capturing the odd and even video fields provided by a standard NTSC CCD camera. The resulting video images were scaled to 320 × 240 pixels to insure a square aspect ratio for eye position measurement, which minimized the complexity in calculating eye position. The efficiency of the algorithm, combined with the computational speed of the dual processor computer, generated eye position records in real time with minimal delay (see below).

The eye position measurement algorithm was tested for accuracy and resolution by measuring the rotation angle of computer generated eyes (Fig. 5, Movie 4, http://www.elsevierdirect.com/companion.jsp?ISBN=9780123745996). To generate the virtual eye, an ellipse of approximately the same size and location as a larval zebrafish eye was linearly rotated from −40° to +40° in 0.01° increments

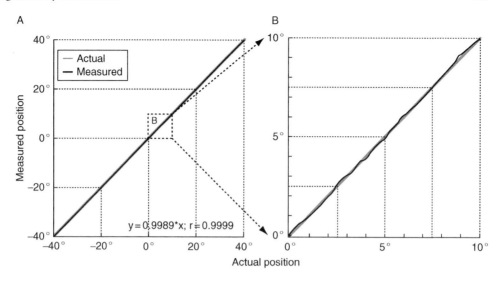

Fig. 5 Accuracy of video eye position measurement system. (A) A computer generated virtual eye (ellipse) was rotated in 0.01° increments over ±40° (actual position, thick solid gray line), and measurements of angular rotation determined using the video measurement algorithm, slightly filtered with a five-point sliding window (measured position, black line). Linear regression of the measured positions was well fit ($r = 0.9999$). (B) Magnification of a smaller range (0°–10°) highlights the correlation between measured and actual rotation. Also see Supplemental Movie 4.

Movie 4 An example of the eye tracking algorithm in action, measuring the position of a computer generated virtual eye (blue) that was rotated linearly ±40° in 0.1° increments. The green trace superimposed upon the measured position record is the actual position record. Length: 0:34; size: 3.5 MB.

(Adobe Illustrator), adequately encompassing the normal ocular range of ±20°. The computer generated images were processed using the software algorithm (Fig. 5, Movie 4), and the measured position was compared to the actual position.

Eye movement measurements produced from video systems are typically noisier than those obtained using the well-refined scleral search coil technique. Nevertheless, the algorithm described here demonstrated a remarkable degree of resolution and accuracy throughout the entire range of motion. Linear regression analysis of the measured versus actual position of the virtual eye yielded an excellent linear fit with a slope of 0.9989 ($r = 0.9999$, see Fig. 5B). Because the actual change in virtual eye position was a straight line, the average difference between each step in the measured eye position determined the measurement resolution or mean resolvable step, which was 0.00997°, close to the actual step value of 0.01°. The accuracy was quantified by calculating the standard deviation of the difference between the measured position and the actual position. Measurement accuracy for this virtual eye was determined to be ±0.13°, an order of magnitude improvement over the off-line measurement of $\pm 2°$ reported (Roeser and Baier, 2003) and closer to that of the search coil (Fuchs and Robinson, 1966).

It is important to note that the measurement accuracy was calculated for only a 320×240 image size, representing a trade-off required to achieve a high-frame rate with a standard CCD camera. Indeed, accuracy and image size are closely related. By doubling pixel dimensions (640×480), one could improve measurement accuracy to approximately ± 0.05 (data not shown). This improvement could be achieved by utilizing a nonstandard, progressive scan camera capturing whole frames (640×480) at 60 Hz. Alternatively, accuracy could also be increased by filling the video image with the eyes and a caudal point on the head firmly embedded in agarose.

Accurate timing relationships between the calculated eye position and the table or drum output signals are critical for studying latency during step stimulation or phase during Bode analysis. Digitizing all of the signals through the same apparatus was a simple solution to solve the majority of issues surrounding signal synchrony. Even though the video analysis system operated in real time, delays were introduced into the eye position records. To determine the average latency, two ellipses (the eyes) and a square (body) were printed on a piece of clear acetate and directly attached to the optokinetic drum. The position of one ellipse was measured during sinusoidal rotations of the drum (0.065–4 Hz). If a processing delay was not produced by the video system, the phase lag between eye velocity and drum velocity would be $0°$ throughout the frequency range. However, a noticeable phase lag was observed that increased with frequency (Fig. 6). Processing lag was calculated to be 38.1 ms (± 1.1), which was then used to shift the eye position records relative to the table and drum output by eight sample points (40 ms). Correcting for the phase lag in this manner yielded an average delay of 2.8 ms (± 1.2), below the 5 ms sampling resolution for a 200 Hz digitization rate. For higher frequencies, more accurate phase correction was achieved by directly subtracting the processing lag from any calculated timing values.

The eye position measurements presented here were limited by the 60 Hz scan rate of the camera. A comparison of how sample rate affects the capture of a high-frequency (2 Hz) signal can be seen in Fig. 6C. Sampling at 5 or 10 Hz (as in Rick *et al.*, 2000; Roeser and Baier, 2003), produced waveforms that were highly distorted and only vaguely resembling sinusoids. A 20 Hz sampling rate was a large improvement; but higher sampling rates (60 and 200 Hz) were necessary to capture the signal with fidelity. While a 60 Hz sampling rate was found to be adequate to measure many aspects of oculomotor performance, that is, slow phase, the temporal resolution of 17 ms limited utility in saccade and step analysis. Advances in imaging technology have made widely available infrared sensitive cameras that are both high in speed and in resolution yet small in size. For example, the recently introduced Pulnix PC-640CL is a digital CMOS camera that can achieve >180 Hz at nearly full (640×480) resolution. Therefore, combined with the availability of ever increasing computational power, it is reasonable to expect newer systems utilizing this measurement algorithm to operate in excess of 180 Hz with a computational accuracy approaching $\pm 0.05°$.

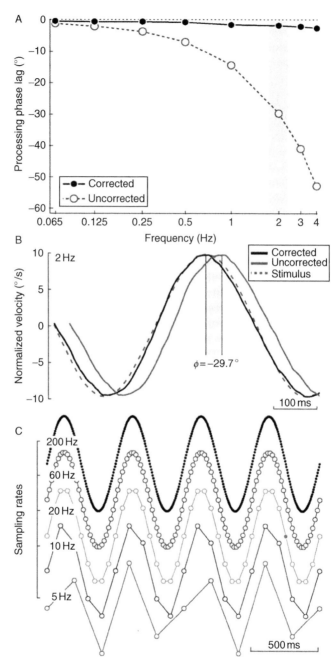

Fig. 6 Processing of phase lag and of sampling rate on signal acquisition. (A) Computing eye positions from video images introduced an average time delay of 38.1 ms (±1.1) into the eye position records that caused a marked phase shift as command frequency increased (open circles). (B) At 2 Hz (dark band in A), the phase lag was 29.7°. Correcting for the time delay reduced the lag to only a few degrees even at 4 Hz (A, filled circles). (C). Sampling rate has a large impact on the ability to accurately capture high-frequency signals, here 2 Hz.

D. Behavioral Recordings (Supplemental Movies 6–10)

1. Optokinetic Measurements

Compensatory eye movements are the result of both vestibular and optokinetic pathways that work together to minimize retinal motion. Each sensory subsystem can be isolated for study by using either vestibular stimuli in the absence of visual input or visual stimuli in the absence of head movement, respectively (Wilson and Melvill Jones, 1979). Significantly, oculomotor performance cannot be fully assessed using only simple constant velocity stimuli (Carvalho *et al.*, 2002; Easter and Nicola, 1997, 1996; Rick *et al.*, 2000; Roeser and Baier, 2003). Instead, more dynamic stimuli are required to determine latencies in pathways as well as to distinguish positional and velocity components of neuronal signaling. The more common method to dynamically assess the performance of each pathway, independent of the other, is through the use of linear systems frequency analysis (Fernandez and Goldberg, 1971; Melvill Jones and Milsum, 1971). This technique (called Bode analysis) often employs sinusoidal stimuli over a wide frequency range at a constant velocity amplitude. The Bode plot generated by this analysis provides a measure of eye velocity gain—the ratio of eye velocity to stimulus velocity—and phase relative to the stimulus (also see Fig. 9, Beck *et al.*, 2004). (Beck *et al.*, 2004)

An example of OKR testing is illustrated with a low-frequency visual stimulus (0.0325 Hz) in a 35 dpf zebrafish (Fig. 7A, Movie 6, http://www.elsevierdirect.com/companion.jsp?ISBN=9780123745996). OKR is a combination of slow phase eye movements (Fig. 7A, top trace, solid lines) and fast, resetting phases (Fig. 7A, top trace, vertical dashed lines). Because stimuli for Bode analysis normally maintain a constant velocity amplitude with increasing frequency, the position amplitude of the stimulus actually *decreases* proportionately. Consequently, stimulation with high-frequency sinusoids produces small rotation excursions and often elicits even smaller excursions of eye position. This represents a real test of the accuracy and resolution of the eye tracking system. For example, maintaining a drum velocity amplitude of ± 10 °/s, as in Fig. 7A, but at 2 Hz reduced the drum position amplitude to approximately $\pm 1°$. As illustrated in Fig. 7B (Movie 7, http://www.elsevierdirect.com/companion.jsp?ISBN=9780123745996), a 20 dpf zebrafish was visually stimulated with sinusoids at 2 Hz ± 10 °/s. Despite the small change in eye position ($\pm 1°$), the eye tracking system exhibited sufficient resolution (Fig. 7B, top traces, inset) to generate velocity traces for gain and phase analysis with little noise (Fig. 7B, bottom traces). The average gain in this case was 0.88, with a $-58°$ phase lag.

Previous vertebrate studies divided the visuomotor system into direct and indirect components for analysis (Cohen *et al.*, 1977; Marsh and Baker, 1997). The only means of observing these aspects of optokinetic behavior is through the use of abrupt high-acceleration stimuli, for exmple, a velocity step. Step performance is determined, in part, by resetting saccades that are of sufficient frequency and amplitude to maintain a compensatory eye velocity; and, thus, velocity steps are also a tool to indirectly assess saccadic performance. In addition, prolonged step stimuli are required to measure the time constant of velocity storage integration.

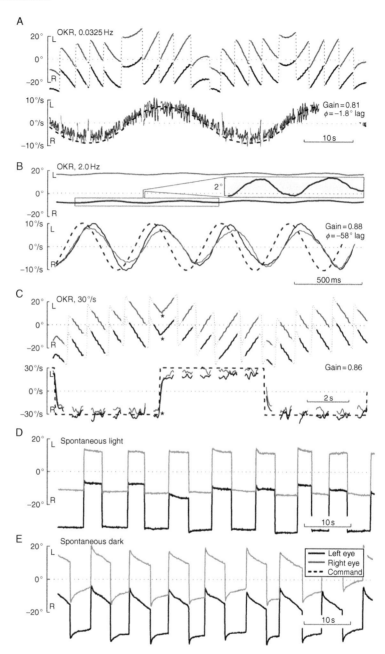

Fig. 7 Behavioral recordings of optokinetic stimulation. (A) Low-frequency optokinetic stimulation (0.032 Hz, ±10 °/s) showing slow and fast phases (upper trace; solid and dashed lines, respectively) in a 35 dpf zebrafish. Eye velocity (bottom) follows stimulus velocity (Movie 6). (B) At 2 Hz ±10 °/s, drum rotation and eye movements were approximately ±1° in a 20 dpf zebrafish (also see Movie 7). (C) OKR

Movie 5 Optokinetic response in *Xenopus laevis*. Wild-type *X. laevis* (35 dpf, 13.4 mm total length, stage 48–50) was stimulated with a drum velocity of ±10°/s at 0.25 Hz, demonstrating the utility of the apparatus for testing other species. Eye velocity was similar to stimulus velocity (mean gain = 0.82); however, eye movements were jerky and not as smooth as in larval fish. Superimposed on the video image of the animal (top) are lines depicting the body axis (green) and direction of eye rotation (blue: left eye; red: right eye). Below the video image are measurements of eye position as well as eye velocity and stimulus velocity (green). Measurement of time is along the abscissa in seconds. To permit reduced movie file size, sampling rate was reduced to 24 Hz. Eye velocity traces were smoothed with a Gaussian filter offline. Stimulus waveforms were based on original recording. Length: 0:42; size: 9.3 MB.

Movie 6 Measurement of OKR in a 25 dpf zebrafish at a low frequency (0.03125 Hz). Traces are same as in Movie 5. The abrupt changes in the eye position traces (middle) correspond to resetting fast phases in eye velocity. Eye velocity closely matched stimulus velocity and phase (mean: gain = 0.84, phase = 0.4° lag). Movie plays at 2x normal speed. Length: 0:45; size: 9.1 MB.

Movie 7 High-frequency OKR in a 40 dpf goldfish. Drum rotation was 2.0 Hz ±10 °/s, producing modest changes in drum position and small changes in eye position. Eye velocity nearly matched drum velocity (gain = 0.67). Due to processing delays in the visuomotor system, the eye velocity is out of phase with stimulus velocity and lags by 53°. Length: 0:42; size: 12.3 MB.

Movie 8 OKR drum velocity steps of ±30 °/s at 0.1 Hz in a 30 dpf medaka, used to assess both fixed and dynamic neural components of the visuomotor system. A slight buildup in eye velocity during the course of the step was observed. Length: 0:45; size: 10.0 MB.

To illustrate, optokinetic drum velocity steps of ±30 °/s at 0.1 Hz were employed to induce an OKR in a 15 dpf medaka (Fig. 7C, Movie 8, http://www.elsevierdirect.com/companion.jsp?ISBN=9780123745996). The high-speed processing of the eye position tracking algorithm was able to capture both the sharp change in eye position during the drum turnaround (Fig. 7C, top traces, asterisks), as well as the fast phases, in sufficient detail (Fig. 7C, top traces, dashed lines). The fact that eye velocity, and not eye position, closely correlated with drum velocity (average gain = 0.86) is evidence of the importance to present and analyze data in the velocity domain (cf. Fig. 7A).

Measurements of spontaneous activity often require several minutes of recordings to accumulate enough data for analysis can be easily acquired in either the light or dark. Spontaneous activity in the light provides a measure of how well the oculomotor system utilizes visual feedback to maintain eye position, with position

---

stimulation using drum velocity steps of ±30 °/s at 0.1 Hz in a 15 dpf medaka (Movie 8). Change in direction of the slow phase (asterisks) during reversal of drum velocity (bottom trace, dashed line). Recordings of spontaneous activity in a 10 dpf zebrafish both in the light (D) and dark (E). Most animals (A–E) exhibited a 10° nasal bias in eye position; zero eye position was the eye direction perpendicular to the body axis. Left eye, black traces; right eye, gray traces; command velocity, black dashed line.

drifts indicating abnormal function. Here, a 10 dpf zebrafish exhibited stable eye positions with no visible drift with time (Fig. 7D). Having established a properly functioning visuomotor system, spontaneous activity in the dark can then be assessed. Measurements in the dark remove visual feedback (open loop) and permit an ontogenetic assessment of oculomotor neural integration. In adults, maintaining eye position in the dark with minimal drift is consistent with a properly functioning and tuned oculomotor integrator (Aksay *et al.*, 2000; Major *et al.*, 2004; Mensh *et al.*, 2004; Pastor *et al.*, 1994; Seung, 1996). The same, young zebrafish (Fig. 7E) exhibited a centripetal drift of eye position, which is part of the orderly development of position in holding larval fish (Beck *et al.*, 2003).

2. Negative Effect of Methylcellulose on Optokinetic Performance

Methylcellulose has been extensively used in the oculomotor testing of larval fish (Brockerhoff *et al.*, 1995; Easter and Nicola, 1997, 1996; Moorman, 2001; Moorman *et al.*, 1999; Neuhauss *et al.*, 1999; Rick *et al.*, 2000; Riley and Moorman, 2000; Roeser and Baier, 2003) since, at low concentrations, it is excellent at restricting movement. Thus, it was logical to assume that methylcellulose might restrict movement of the eyes as well. To test this, 5 dpf larval zebrafish ($n = 5$) were mounted in agarose blocks as described and OKR was measured from 0.065 to 2.0 Hz $\pm 10°/s$ (Fig. 8). Water surrounding the animals was then replaced with a 3% methylcellulose solution, selecting an average of concentrations previously used. When OKR was measured across the same frequency range, oculomotor performance was decreased. With a 0.065 Hz stimulus, control OKR gain was 0.78 (Fig. 8A) and, in the presence of methylcellulose (Fig. 8B), eye velocity gain dropped to 0.29, returning to control levels upon washout (Fig. 8C); median drop for all frequencies was 63% (range 37–351%). A summary of the results (Fig. 8D) demonstrates that methylcellulose impacted performance at all frequencies, showing the largest drop at the highest.

3. Summary of Optokinetic and Vestibular Performance in Zebrafish

A quantitative, ontogenetic analysis of oculomotor performance in larval zebrafish, medaka, and goldfish has been recently performed (Beck *et al.*, 2004). The results were in contrast to previous investigations that concluded the oculomotor system of larval zebrafish was adult-like by 4 dpf (Easter and Nicola, 1997, 1996). Optokinetic performance measured from 5 to 35 dpf (Fig. 9) showed that, while larval zebrafish of all ages have a wide OKR frequency response (Fig. 9A), their ability to follow stimuli of increasing velocity amplitude was initially poor and improved with maturity (Fig. 9B and C). Saccade performance (velocity, amplitude, and frequency) also improved with maturity and significantly impacted eye velocity at higher stimulus velocities (Beck *et al.*, 2004). Although it is tempting to utilize saccade frequency as a measure of OKR performance (Rick *et al.*, 2000), it should be avoided since optokinetic gain is only loosely correlated with saccade frequency (Beck *et al.*, 2004).

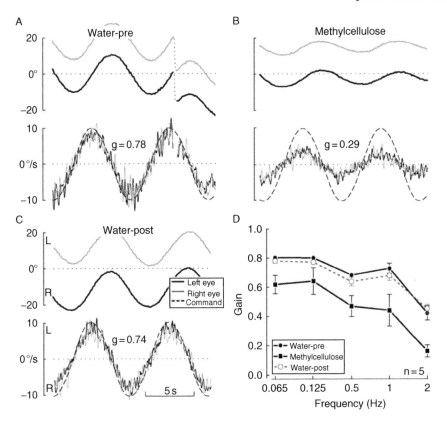

Fig. 8 Effect of methylcellulose on oculomotor performance. OKR eye velocity in a 5 dpf zebrafish at 0.065 Hz ±10 °/s (gain = 0.78) in water (A), after replacement with a 3% methylcellulose solution (B, gain = 0.29), and after washout of methylcellulose (C, gain = 0.74). Black trace, left eye; gray trace, right eye; dashed line, drum velocity. (D) OKR Bode frequency plot before, during, and after 3% methylcellulose treatment in 5 dpf zebrafish. Bars are standard error, *n* = 5.

Also in contrast to other investigations (Easter and Nicola, 1997), a vestibuloocular reflex was never observed in young larval zebrafish (Movie 9, http://www.elsevierdirect. com/companion.jsp?ISBN=9780123745996) and only consistently appeared in older animals (Fig. 10, Beck *et al.*, 2004). After 2 weeks of age, a VOR could be observed but was of very low amplitude even with high accelerations (Beck *et al.*, 2004). For example, a typical zebrafish exhibited almost no VOR at low-stimulus intensities (Fig. 10A, 0.5 Hz, ±60°/s) but demonstrated a slightly stronger response with increased frequency (Fig. 10B) as accelerations became higher (here ±1000°/s^2). As zebrafish grew in size, their VOR became more sensitive to lower accelerations (Fig. 10C, 0.5 Hz, ±60°/s, Movie 10, http://www.elsevierdirect.com/companion. jsp?ISBN=9780123745996). For comparison, at 0.5 Hz, the average OKR gain in a 5 dpf animal was 0.8 while the VOR gain in a 35 dpf zebrafish barely reaches a gain of 0.2 (Fig. 10D, Beck *et al.*, 2004). By adulthood, VOR gain in zebrafish was close

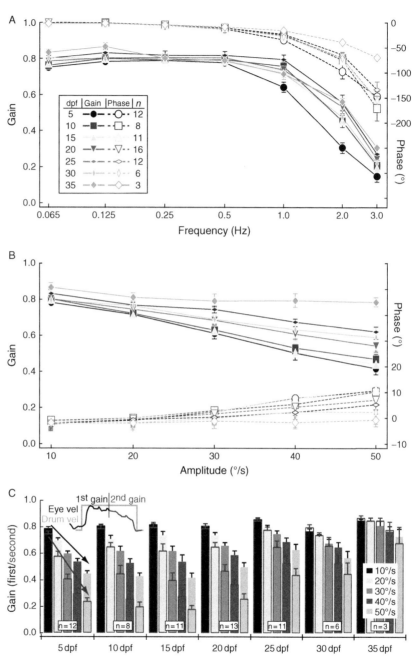

Fig. 9 Development of zebrafish OKR. (A) Bode analysis of zebrafish (5–35 dpf) from 0.065 to 3.0 Hz at ±10 °/s. (B) Phase and gain plot of zebrafish OKR in response to sinusoids (0.125 Hz) of increasing velocity. Traces same as legend in A. (C) Eye velocity gains in response to bi-directional velocity steps (0.1 Hz). Gains were calculated at each drum velocity for (1) the first 2.5 s of the 5 s step (solid bar) and (2) the remaining 2.5 s (open, red bars), also see inset. Arrows follow trend in decreasing performance with increasing amplitude between first (black) and second (red) step halves. All data averaged over several cycles and offset slightly (A–B) for readability. (See Plate no. 13 in the Color Plate Section.)

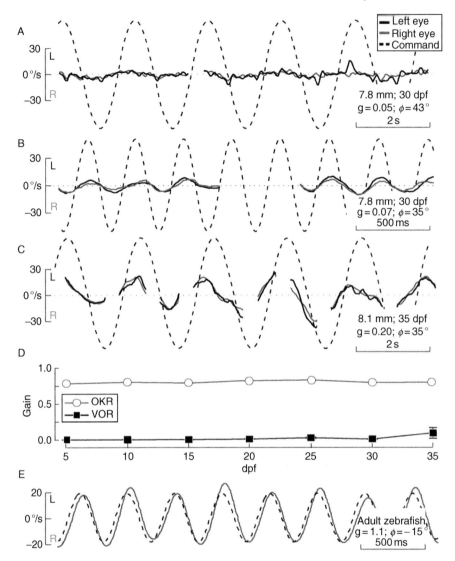

Fig. 10 Zebrafish VOR. VOR was barely detected (A) in a 30 dpf zebrafish at 0.5 Hz, ±60 °/s. At 3.0 Hz, ±45 °/s (B), acceleration forces are higher (±1000 °/s²) and a slightly larger response is observed. (C) Older and larger animals exhibit increased VOR sensitivity at lower table amplitudes (0.5 Hz, ±60 °/s). (D) Age comparison of OKR and VOR gain at 0.5 Hz; *n* same as legend in Fig. 9A. (E) Adult zebrafish show an almost perfect compensatory eye velocity with a table stimulus of 3.0 Hz, ±20 °/s. Table velocity (A–E) was inverted to facilitate comparison with eye velocity; spaces in traces are where fast phases have been removed. See Movies 9 and 10.

Movie 9 Angular VOR was absent in 4 dpf larvae. A small, young zebrafish larvae (here, 3.8 mm; 4 dpf) was rotated sinusoidally at 0.5 Hz, $\pm 30°$/s. A motion tracer within the water column partially reflects the angular acceleration forces the zebrafish larva is experiencing. The inset within the larval video image is the table activity during this stimulus paradigm. Length: 0:28; size: 11 MB.

Movie 10 Measurement of the angular VOR in a 25 dpf medaka during sinusoidal rotation in the dark at 1 Hz, $\pm 60°$/s. The angular VOR was robust as eye velocity was nearly the same as the table velocity (gain $= 0.9$, phase $= 9°$ lead). Table velocity trace was inverted to facilitate comparison with eye velocity. Length: 0:38; size: 7.4 MB.

to the ideal gain value of 1.0, especially with increasing frequency (Fig. 10E, Suwa *et al.*, 1998). Based on theoretical work relating semicircular canal morphology (i.e., size) and sensitivity (Muller, 1994, 1999; Rabbitt, 1999; Rabbitt *et al.*, 2003), we propose that the lack of an effective angular VOR in larval fish is the result of the small canal lumen diameter, with VOR sensitivity increasing proportional to animal size. Recently, this hypothesis has been validated in larval *Xenopus* by demonstrating that a minimum canal size must be reached prior to the onset of vestibular function (Lambert *et al.*, 2008). Therefore, contrary to the conclusions of Easter and Niccola (1997), who found no difference in the ontogenetic timing of OKR and VOR behaviors, our findings strongly suggest that natural selection has favored the evolution of a system detecting visual over angular motion for early survival in freely swimming larva.

IV. Conclusion

The miniaturized vestibular turntable and optokinetic drum, combined with the novel approach to larval animal immobilization and real-time eye position measurement, represent significant advances in the ability to study eye movements in several species of larval and juvenile animals. Classical physiological methods, combined with imaging approaches, allow structural changes to neuronal architecture as well as genetic misexpression to be effectively and quantitatively linked with functional changes in oculomotor behavior.

Acknowledgments

The authors thank Alfred Benedek for constructing the vestibular turntable and drum, Ray Stepnoski (Lucent Technology) for assistance with the construction of an early prototype, and Dr. Sebastian Seung (M.I.T.) for initial advice with eye detection software. This work was supported by grants from the National Institutes of Health (D. W. T and R. B.) as well as a National Research Service Award from the National Eye Institute (J. C. B.).

References

Aksay, E., Baker, R., Seung, H. S., and Tank, D. W. (2000). Anatomy and discharge properties of pre-motor neurons in the goldfish medulla that have eye-position signals during fixations. *J. Neurophysiol.* **84,** 1035–1049.

Bahadori, R., Huber, M., Rinner, O., Seeliger, M. W., Geiger-Rudolph, S., Geisler, R., and Neuhauss, S. C. (2003). Retinal function and morphology in two zebrafish models of oculo-renal syndromes. *Eur. J. Neurosci.* **18,** 1377–1386.

Barmack, N. H. (1981). A comparison of the horizontal and vertical vestibulo-ocular reflexes of the rabbit. *J. Physiol.* **314,** 547–564.

Bassett, J. P., Baker R., and Beck J. C. (2008) Larval oculomotor behaviors in the mouth-brooding cichlid, *Astatotilapia burtoni.* *In* "2008 Abstract Viewer/Itinerary Planner," p. 29.29. Society for Neuroscience, Washington, DC.

Bate, M. (1998). Making sense of behavior. *Int. J. Dev. Biol.* **42,** 507–509.

Beck, J. C., and Baker, R. (2005). Static and dynamic measurement of otolithic VOR during the absence of canal function in larval zebrafish. *In* "2005 Abstract Viewer/Itinerary Planner," p. 391.316. Society for Neuroscience, Washington, DC.

Beck, J. C., Gilland E., Tank D. W., and Baker R. (2004). Quantifying the ontogeny of optokinetic and vestibuloocular behaviors in zebrafish, medaka, and goldfish. *J. Neurophysiol.* **92,** 3546–3561.

Beck, J. C., Tank, D. W., and Baker, R. (2003). Ontogeny of persistent neural activity: Maturation of gaze holding in zebrafish. *In* "2003 Abstract Viewer/Itinerary Planner." Society for Neuroscience, Washington, DC.

Beck, J. C., Tank, D. W., Gilland, E., and Baker, R. (2002). Instrumentation for measuring oculomotor performance and learning in larval and juvenile fish. *In* "2002 Abstract Viewer/Itinerary Planner." Society for Neuroscience, Washington, DC.

Boyden, E. S., and Raymond, J. L. (2003). Active reversal of motor memories reveals rules governing memory encoding. *Neuron* **39,** 1031–1042.

Brockerhoff, S. E., Hurley, J. B., Janssen-Bienhold, U., Neuhauss, S. C., Driever, W., and Dowling, J. E. (1995). A behavioral screen for isolating zebrafish mutants with visual system defects. *Proc. Natl. Acad. Sci. USA* **92,** 10545–10549.

Brockerhoff, S. E., Hurley, J. B., Niemi, G. A., and Dowling, J. E. (1997). A new form of inherited red-blindness identified in zebrafish. *J. Neurosci.* **17,** 4236–4242.

Carvalho, P. S. M., Noltie, D. B., and Tillitt, D. E. (2002). Ontogenetic improvement of visual function in the medaka Oryzias latipes based on an optomotor testing system for larval and adult fish. *Anim. Behav.* **64,** 1–10.

Clark, D. T. (1981). Visual responses in developing zebrafish (*Brachydanio rerio*). p. xi, 138 leaves.

Cohen, B., Matuso, V., and Raphan, T. (1977). Quantitative analysis of the velocity characteristics of the optokinetic nystagmus and optokinetic after-nystagmus. *J. Physiol.* **270,** 321–344.

Collewijn, H. (1969). Optokinetic eye movements in the rabbit: input-output relations. *Vision. Res.* **9,** 117–132.

Collewijn, H. (1991). The optokinetic contribution. *In* "Eye Movements" (R. H. S. Carpenter, ed.), pp. 45–70. CRC Press, Boca Raton.

Easter, S. S., Jr., and Nicola, G. N. (1996). The development of vision in the zebrafish (*Danio rerio*). *Dev. Biol.* **180,** 646–663.

Easter, S. S., Jr., and Nicola, G. N. (1997). The development of eye movements in the zebrafish (*Danio rerio*). *Dev. Psychobiol.* **31,** 267–276.

Fernandez, C., and Goldberg, J. M. (1971). Physiology of peripheral neurons innervating semicircular canals of the squirrel monkey. II. Response to sinusoidal stimulation and dynamics of peripheral vestibular system. *J. Neurophysiol.* **34,** 661–675.

Fuchs, A. F., and Robinson, D. A. (1966). A method for measuring horizontal and vertical eye movement chronically in the monkey. *J. Appl. Physiol.* **21,** 1068–1070.

Gerlai, R. (2003). Zebra fish: an uncharted behavior genetic model. *Behav. Genet.* **33,** 461–468.

Godaux, E., Gobert, C., and Halleux, J. (1983). Vestibuloocular reflex, optokinetic response, and their interactions in the alert cat. *Exp. Neurol.* **80**, 42–54.

Hartell, N. A. (2002). Parallel fiber plasticity. *Cerebellum* **1**, 3–18.

Higashijima, S. i., Masino, M. A., Mandel, G., and Fetcho, J. R. (2003). Imaging neuronal activity during zebrafish behavior with a genetically encoded calcium indicator. *J. Neurophysiol.* **90**, 3986–3997.

Ito, M. (2002). The molecular organization of cerebellar long-term depression. *Nat. Rev. Neurosci.* **3**, 896–902.

Ito, M., Jastreboff, P. J., and Miyashita, Y. (1979). Adaptive modification of the rabbit's horizontal vestibulo-ocular reflex during sustained vestibular and optokinetic stimulation. *Exp. Brain. Res.* **37**, 17–30.

Kocher, T. D. (2004). Adaptive evolution and explosive speciation: the cichlid fish model. *Nat. Rev. Genet.* **5**, 288–298.

Lambert, F. M., Beck J. C., Baker R., and Straka H. (2008). Semicircular canal size determines the developmental onset of angular vestibuloocular reflexes in larval *Xenopus. J. Neurosci.* **28**, 8086–8095.

Lisberger, S. G., Miles, F. A., Optican, L. M., and Eighmy, B. B. (1981). Optokinetic response in monkey: underlying mechanisms and their sensitivity to long-term adaptive changes in vestibulo-ocular reflex. *J. Neurophysiol.* **45**, 869–890.

Major, G., Baker R., Aksay E., Mensh B., Seung H. S., and Tank D. W. (2004). Plasticity and tuning by visual feedback of the stability of a neural integrator. *Proc. Natl. Acad. Sci. USA* **101**, 7739–7744.

Marsh, E., and Baker, R. (1997). Normal and adapted visuooculomotor reflexes in goldfish. *J. Neurophysiol.* **77**, 1099–1118.

Melvill Jones, G. (1977). Plasticity in the adult vestibulo-ocular reflex arc. *Philos. Trans. R. Soc. Lond., B, Biol. Sci.* **278**, 319–334.

Melvill Jones, G., and Milsum, J. H. (1971). Frequency-response analysis of central vestibular unit activity resulting from rotational stimulation of the semicircular canals. *J. Physiol.* **219**, 191–215.

Mensh, B. D., Aksay, E., Lee, D. D., Seung, H. S., and Tank, D. W. (2004). Spontaneous eye movements in goldfish: Oculomotor integrator performance, plasticity, and dependence on visual feedback. *Vision. Res.* **44**, 711–726.

Moorman, S. J. (2001). Development of sensory systems in zebrafish (*Danio rerio*). *ILAR J.* **42**, 292–298.

Moorman, S. J., Burress, C., Cordova, R., and Slater, J. (1999). Stimulus dependence of the development of the zebrafish (Danio rerio) vestibular system. *J. Neurobiol.* **38**, 247–258.

Muller, M. (1994). Semicircular duct dimensions and sensitivity of the vertebrate vestibular system. *J. Theor. Biol.* **167**, 239–256.

Muller, M. (1999). Size limitations in semicircular duct systems. *J. Theor. Biol.* **198**, 405–437.

Neuhauss, S. C., Biehlmaier, O., Seeliger, M. W., Das, T., Kohler, K., Harris, W. A., and Baier, H. (1999). Genetic disorders of vision revealed by a behavioral screen of 400 essential loci in zebrafish. *J. Neurosci.* **19**, 8603–8615.

Pastor, A. M., de la Cruz, R. R., and Baker, R. (1992). Characterization and adaptive modification of the goldfish vestibuloocular reflex by sinusoidal and velocity step vestibular stimulation. *J. Neurophysiol.* **68**, 2003–2015.

Pastor, A. M., De la Cruz, R. R., and Baker, R. (1994). Eye position and eye velocity integrators reside in separate brainstem nuclei. *Proc. Natl. Acad. Sci. USA* **91**, 807–811.

Rabbitt, R. D. (1999). Directional coding of three-dimensional movements by the vestibular semicircular canals. *Biol. Cybern.* **80**, 417–431.

Rabbitt, R. D., Damiano, E. R., and Grant, J. W. (2003). Biomechanics of the vestibular semicircular canals and otolith organs. *In* "The Vestibular System" (S. M.Highstein, R. R.Fay, and A. N. Popper, eds.), pp. 153–201. Springer-Verlag.

Raphan, T., Matsuo, V., and Cohen, B. (1979). Velocity storage in the vestibulo-ocular reflex arc (VOR). *Exp. Brain. Res.* **35**, 229–248.

Rick, J. M., Horschke, I., and Neuhauss, S. C. (2000). Optokinetic behavior is reversed in achiasmatic mutant zebrafish larvae. *Curr. Biol.* **10**, 595–598.

Riley, B. B., and Moorman, S. J. (2000). Development of utricular otoliths, but not saccular otoliths, is necessary for vestibular function and survival in zebrafish. *J. Neurobiol.* **43,** 329–337.

Ritter, D. A., Bhatt, D. H., and Fetcho, J. R. (2001). *In vivo* imaging of zebrafish reveals differences in the spinal networks for escape and swimming movements. *J. Neurosci.* **21,** 8956–8965.

Robinson, D. A. (1963). A method of measuring eye movement using a scleral search coil in a magnetic field. *IEEE. Trans. Biomed. Electrng.* **10,** 137–145.

Robinson, D. A. (1968). Eye movement control in primates. The oculomotor system contains specialized subsystems for acquiring and tracking visual targets. *Science* **161,** 1219–1224.

Robinson, D. A. (1975). Oculomotor signal controls. *In* "Basic Mechanisms of Ocular Motility and Their Clinical Implications" (G. Lennerstrand and P.Bach-y-Rita), pp. 337–374. Pergamon, New York.

Robinson, D. A. (1976). Adaptive gain control of vestibuloocular reflex by the cerebellum. *J. Neurophysiol.* **39,** 954–969.

Robinson, D. A. (1989). Integrating with neurons. *Annu. Rev. of. Neurosci.* **12,** 33–45.

Roeser, T., and Baier, H. (2003). Visuomotor behaviors in larval zebrafish after GFP-guided laser ablation of the optic tectum. *J. Neurosci.* **23,** 3726–3734.

Rombough, P. (2002). Gills are needed for ionoregulation before they are needed for O_2 uptake in developing zebrafish, Danio rerio. *J. Exp. Biol.* **205,** 1787–1794.

Seung, H. S. (1996). How the brain keeps the eyes still. *Proc. Natl. Acad. Sci. USA* **93,** 13339–13344.

Shinoda, Y., and Yoshida, K. (1974). Dynamic characteristics of responses to horizontal head angular acceleration in vestibuloocular pathway in the cat. *J. Neurophysiol.* **37,** 653–673.

Skavenski, A. A., and Robinson, D. A. (1973). Role of abducens neurons in vestibuloocular reflex. *J. Neurophysiol.* **36,** 724–738.

Suwa, H., Straka, H., and Baker, R. (1998). Functional organization of the vestibulo-oculomotor system in zebrafish. *In* "Zebrafish Development & Genetics," p. 163. Cold Spring Harbor Laboratory, New York.

van Alphen, A. M., Stahl, J. S., and De Zeeuw, C. I. (2001). The dynamic characteristics of the mouse horizontal vestibulo-ocular and optokinetic response. *Brain. Res.* **890,** 296–305.

Wilson, V. J., and Melvill Jones, G. (1979). "Mammalian Vestibular Physiology." Plenum Press, New York, p. xi, 365.

Supplemental Movie Descriptions

The supplemental movies provided here are referred to throughout the text and are specifically noted at each section heading. The movies were saved in a QuickTime 6.0 format and are playable on both Mac and PC computers with freely available software (http://www.apple.com/quicktime/download/). Movies available may be found at the Elsevier website (http://www.elsevierdirect.com/companion.jsp?ISBN=9780123745996).

CHAPTER 18

Development of Cartilage and Bone

Yashar Javidan, Courtney Alexander, and Thomas F. Schilling

Department of Developmental and Cell Biology
University of California, Irvine
California 92697-2300

I. Introduction

A. The Zebrafish Model

The skeletons of vertebrates are remarkably similar. Centuries of studies of bone morphology and the fact that bones are preserved as fossils have revealed that identical bones form much of the skull, vertebrae and appendicular skeleton in fish, amphibians, reptiles, and mammals (De Beer, 1937; Goodrich, 1930).

Clear homologies between bones can be drawn based not only on fossils but also on commonalities in their articulations with one another and their patterns of development. Genetic studies in humans, mice, and most recently zebrafish are now revealing that these similarities reflect similar developmental mechanisms underlying skeletal morphogenesis.

Zebrafish develop a simple pattern of early larval cartilages and bones, much of which is highly conserved among all vertebrates. The last common ancestor of humans and zebrafish existed approximately 420 million years ago, yet both form the same skeletal cell types, including endochondral and dermal bones as well as cartilages that persist in the adult (Hall and Hanken, 1985). Mutant screens in zebrafish have identified many genes required for early cartilage development, and a number of these mutations are also associated with human syndromes that disrupt the skeleton.

Here we summarize current methods for visualizing cartilage and bone that make zebrafish an excellent genetic model for dissecting the molecular basis of skeletal development. Further, we describe methods of screening for mutants with anomalous craniofacial and other skeletal phenotypes.

B. Zebrafish Skeletal Anatomy

By simply fixing and staining whole zebrafish larvae in Alcian dyes, the entire pattern of cartilage is revealed in whole-mounted specimens (Fig. 1A and B). This pattern emerges between 45 and 72 hours postfertilization (hpf), and includes both the pharyngeal skeleton (Fig. 1A) of the jaw and gills, as well as the neurocranial skeleton, which houses the brain (Fig. 1B; Schilling and Kimmel, 1997). Figure 1C illustrates the simple pattern of neurocranial cartilages, and Fig. 1D shows some of the homologs (e.g., auditory capsule, trabeculae) in the developing human skull (adapted from De Beer, 1937).

In the neurocranium (Fig. 1B and C), the first chondrifications lie parallel to the anterior end of the notochord at 45–48 hpf, and these form the parachordals. The trabecular cartilages chondrify further anteriorly, between the eyes, and eventually fuse in the midline to form the trabeculae communis (~52 hpf) and flare outward laterally as the ethmoid plate. At approximately this stage, cartilage of the parachordals fuses anteriorly with the trabeculae to form the basal plate. Early chondrification patterns in mammals are remarkably similar (compare Fig. 1C and D).

Meanwhile the pharyngeal arch skeleton begins to chondrify (Fig. 1A). Like most teleost fishes, the zebrafish forms a series of seven pharyngeal arches, a mandibular (jaw), hyoid, and five branchials that support the gills. Each pharyngeal arch segment consists of a similar set of serially homologous cartilages. These form a basic pattern of five elements from dorsal to ventral (pharyngobranchial, epibranchial, ceratobranchial, hypobranchial, basibranchial) that is conserved in all vertebrates (reviewed in Schilling, 1997). Only a subset of these is visible in the larval zebrafish (Figs. 1A and 2E). Ceratobranchial elements of the first two arches, Meckel's cartilage and the ceratohyal, are the first to chondrify at

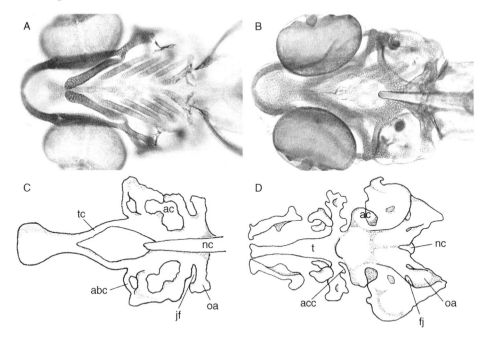

Fig. 1 The simple pattern of early cartilages in the zebrafish larva (A–C) and homologous cartilages in the neurocranium of the human (D). Anterior is to the left in all panels. (A, B) Larvae (~4.0 mm, 5 dpf) are stained with Alcian blue and shown in ventral view (A) to display the pharyngeal arches and dorsal view (B) to show the neurocranium. (C, D) Diagrams illustrating the neurocranium in dorsal view of zebrafish (C; ~4.0 mm, 5 dpf) and human (D; ~20 mm, adapted from De Beer, 1937). Abbreviations: abc, anterior basicapsular commissure; ac, auditory capsule; acc, alicochlear commissure; fj, foramen jugularis; jf, jugular foramen; nc, notochord; oa, occipital arch; tc, trabecula cranii (trabecular plate).

approximately 54 hpf and these are followed by the appearance of five more posterior ceratobranchials between 60 and 72 hpf. Here there are some major differences between fish and mammals in adult morphology. For example, in the first two arches, the dorsal elements in fish (palatoquadrate and hyomandibulae) are involved in jaw articulation, but in mammals form middle ear bones (incus, malleus, and stapes).

As in all bony fishes, zebrafish form several different types of cartilage and bone and their development resembles that of the mammalian skeleton (Hall and Hanken, 1985). The development and chondrification sequence of larval cartilage has been described in the greatest detail (Kimmel *et al.*, 1998; Schilling and Kimmel, 1997). The adult skeletal anatomy and ossification sequence of craniofacial bones in zebrafish has been described by Cubbage and Mabee (1996), and the axial skeleton by Bird and Mabee (2003).

Many bones develop by replacement of cartilage precursors by osteocytes, and in zebrafish these include bones of the neurocranium (perichondral), and

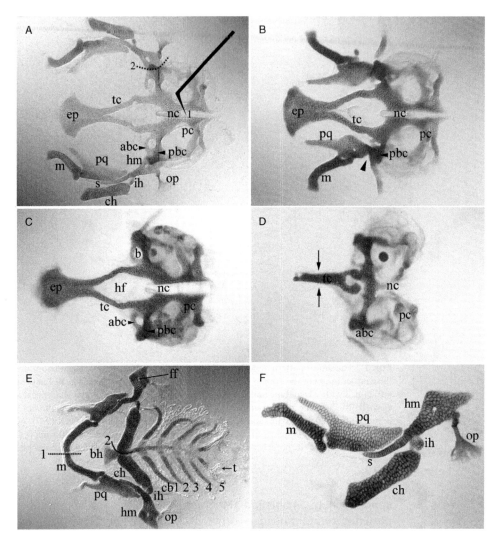

Fig. 2 Flat-mounted preparations of larval cartilages. Early larvae (~3.5 mm, 5 dpf) are stained with Alcian blue, dissected and shown flattened (see Protocol 2) for detailed analysis of morphology. Anterior is to the left. (A) Butterfly preparation of the wild-type neurocranium with the mandibular and hyoid arches attached. Numbers indicate important dissection sites: 1, securing needle; 2, joint between anterior basicranial commissure of the neurocranium and hyomandibular. (B) Butterfly preparation of the *lockjaw* (*tfap2α* mutant) skeleton. Large arrowhead indicates the fusion point between the neurocranium and the mandibular arch. (C) Neurocranial preparation of a wild type. (D) Similar preparation of the *sonic you* (*shh* mutant) neurocranium, showing fusion of skeletal elements across the midline (arrows). (E) Seven arch preparation of wild-type mandibular, hyoid, and branchial arches. Numbers indicate important dissection sites for the lateral jaw preparation shown in F: 1, midline fusion point of Meckel's cartilages; 2, midline attachment between ceratohyal and basihyal cartilages. (F) Lateral jaw preparation of wild-type mandibular and hyoid arches. Abbreviations: abc, anterior basicapsular commissure; b, basicapsular fenestra; bh, basihyal cartilage; cb, ceratobranchial; ch, ceratohyal; ep, ethmoid plate; ff, facial nerve foramen (VII); hf, hypophyseal fenestra; hm, hyomandibular; ih, interhyal; m, Meckel's; nc, notochord; op, opercle; pbc, posterior basicapsular commissure; pc, parachordal cartilage; pq, palatoquadrate; s, symplectic; tc, trabeculae.

pharyngeal arches (endochondral). While most neurocranial bones ossify relatively late, many endochondral arch bones begin to ossify by 6–7 days postfertilization (dpf). The first dermal bones (membrane bone) ossify even earlier. These bones form directly from a mesenchymal condensation, without a cartilage intermediate. Examples include the opercle (3 dpf), parasphenoid (4 dpf), and branchiostegal rays (4–5 dpf). Dermal plates that cover the skull (parietal, frontal, etc.) develop later, when fish are approximately 8.0 mm in length (3–4 weeks) and these have clear homologs in humans. Most teleosts, including ostariophysans (zebrafish), are reported to have enclosed osteocytes like other vertebrates, while certain higher orders have only acellular bone (Parenti, 1986; Moss and Posner, 1960).

C. Mutations that Disrupt the Skeleton

Mutations provide entry points into a molecular analysis of developmental mechanisms controlling skeletal development, via cloning of the mutated genes. The zebrafish is the only animal in which large-scale screens have been performed to identify mutations that disrupt the skeleton. To date, such screens have largely focused on finding mutations that disrupt early mouth and pharyngeal morphology during the first 5–6 days of development. These have yielded a large collection of mutants that disrupt larval craniofacial development (Neuhauss *et al.*, 1996; Piotrowski *et al.*, 1996; Schilling *et al.*, 1996), and for an increasing number, the underlying genetic causes are known (Table I). Figure 2B is an example of the pharyngeal defects in *lockjaw*, a mutant in *tfap2a*, which lacks hyoid cartilage and instead develops fusions of mandibular cartilage to the neurocranium.

Several classes of mutations with similar phenotypes have been characterized, including those that predominantly disrupt the neurocranium, pharyngeal arches, or more generally alter chondrocyte differentiation (Table I). In the neurocranium, mutations that disrupt midline development and either Nodal (*cyclops, one-eyed pinhead*) or Hedgehog signaling (*sonic you, slow muscle omitted*) also disrupt trabecular formation and chondrification of the ethmoid plate (Eberhart *et al.*, 2006; Kimmel *et al.*, 1998; Schilling, 1997; Wada *et al.*, 2005). Figure 2D is an example of the neurocranial defects in *sonic you*, a mutation in *sonic hedgehog*, showing characteristic fusion across the midline of structures such as the trabeculae. In contrast, loss of signaling through platelet-derived growth factor receptor alpha (*Pdgfra*) leads to ethmoid loss and a split between trabeculae, or palatal "clefting" (Eberhart *et al.*, 2008).

Specific classes of mutations that disrupt pharyngeal arches include *sucker*, a mutation in *endothelin-1 (edn1)*, which leads to defective formation of the lower jaw and jaw joint (Miller *et al.*, 2000). *sucker* resembles several other "anterior" arch mutants, the cloning of which has identified novel components of the Edn1 signaling pathway involved in craniofacial development, such as *phospholipase C beta 3 (plcβ3), myocyte enhancing factor (mef2c), and furinA* (Miller *et al.*, 2004, 2007; Piotrowski *et al.*, 1996; Walker *et al.*, 2006, 2007). Mutations that disrupt endodermal development, such as *casanova* (Dickmeis *et al.*, 2001; Kikuchi *et al.*, 2001) and *van gogh* (Piotrowski *et al.*, 2003), also show pharyngeal cartilage

Table I
Cloned and Published Zebrafish Mutations that Disrupt the Skeleton

Mutant name [(alleles)]	Gene	Protein type (gene family)	Skeletal phenotype	Associated human syndrome	Reference
Neurocranial and midline defects					
cyclops (cyc[tf291])	nodal related factor (ndr2)	Secreted ligand (TGFβ)	Absence of anterior and lateral neurocranium		Rebagliati et al. (1998)
one-eye pinhead (oep[m134] and [tz57])	EGF-CFC	Cofactor activin type II receptor	Severely reduced pharyngeal and neurocranial cartilage		Zhang et al. (1998)
sonic you (syu[tq252])	sonic hedgehog (shh)	Secreted ligand (Hedgehog)	Fused anterior neurocranium	Holoprosencephaly (HPE) and cleft palate	Eberhart et al. (2006), Schauerte et al. (1998), and Wada et al. (2005)
slow muscle omitted (smu[b577] and [b641])	smoothened (smu)	Transmembrane signaling protein (Hedgehog family)	Severely reduced pharyngeal and neurocranial cartilage	Basal cell carcinomas	Barresi et al. (2000)
dolphin (dol[t1230g] and [t2495])	cyp26b1	Cytochrome p450 (retinoic acid degrading enzyme)	Bone overgrowth, midline fusion of ceratohyals, loss of basibranchial and ethmoid		Laue et al. (2008b), Spoorendonk et al. (2008)
platelet-derived growth factor receptor alpha (pdgfra[b1059])	platelet-derived growth factor receptor alpha (pdgfra)	Platelet-derived growth factor receptor	Neurocranium split at the midline		Eberhart et al. (2008)
Pharyngeal cartilage defects					
lockjaw (low[ts213]); mont blanc (mob[m610])	tfap2a	Transcription Factor (Ap2)	Hyoid arch defects, including reduced hyomandibular and ceratohyal cartilages	Deafness; branchio-oculo-facial syndrome	Barallo-Gimeno et al. (2004) and Knight et al. (2003)
sucker (suc[tf216b])	endothelin-1 (edn1)	Secreted ligand (Edn)	Loss of ventral mandibular and hyoid cartilages		Miller et al. (2000)
sturgeon (stu[tg419, td204e, and b963])	furinA	Endoprotease (proprotein convertase)	Reduction or loss of ventral mandibular and hyoid cartilages		Walker et al. (2006)
schmerle (she[tg203e and th210])	phospholipase C beta3 (plcβ3)	Enzyme (phospholipase)	Fusion of mandibular and hyoid joints, reduced ventral cartilages in arches 1–4		Walker et al. (2007)
hoover (hoo[tn213])	monocyte enhancing factor (mef2ca)	Transcription factor (Mef)	Fusion of mandibular and hyoid joints, reduced ventral cartilages in arches 1–4		Miller et al. (2007)
van gogh (vgo[tm208] and [tu285])	tbx1	Transcription Factor (T-Box)	Reduced pharyngeal cartilage	DiGeorge deletion (DGS)	Piotrowski et al. (2003)
integrinα5[b926]	integrinα5 (itga5)	Cell adhesion molecule	Hyoid arch defects		Crump et al. (2004)

Gene	Molecular identity	Function	Phenotype		Reference
valentino (*val^b337*)	*mafβ*	Transcription Factor (bZIP)	Ectopic cartilage in 3rd branchial arch		Prince et al. (1998)
lazarus (*lzr^b557*)	*pbx 4*	Hox cofactor (Pbx)	Fusion of mandibular and hyoid arch cartilages		Popperl et al. (2000)
foxi one (*foo*)	*foxi 1*	Transcription Factor (Forkhead)	Hyoid arch defects		Nissen et al. (2003)
casanova (*cas^ta56 and s4*)	*sox32*	Transcription Factor (Sox)	Loss of pharyngeal cartilage		Dickmeis (2001) and Kikuchi (2001)
monocytic leukemia zinc finger (*moz^b719*)	*moz*	Histone acetyltransferase (Hat)	Hyoid arch defects (partial transformation to mandibular identity)		Miller et al. (2004)
brpf1^t20002, b943 and t25114	*bromodomain and PHD finger*	Subunit of *moz* histone acetyltransferase complex	Like *moz* mutant, partial transformation of hyoid and arches 3–6		Laue et al. (2008a)
pescadillo (*pes^hi2*)	*pescadillo* (*pes*)	Regulator of cell cycle and ribosome biogenesis	Loss of branchial arches, midline fusion of Meckels cartilage, reduced hyosymplectic and pectoral fin		Allende et al. (1996)
mother superior (*mos^m188*)	*foxd3*	Transcription factor (Fox)	Meckels cartilages point ventrally; variable hyoid defects Reduced branchial arches		Montero-Balguer et al. (2006)
babyface (*bab^tb210, tw220c*)	*RERE/atrophin-2* (*atr2*)	Transcription factor	Reduced pharyngeal cartilages, loss of joints		Plaster et al. (2007)
flathead (*fla^ta53c*)	*polymerase delta 1* (*polδ1*)	DNA polymerase	Reduced neurocranium and arches 1–3; loss of arches 4–7		Plaster et al. (2006)
pinball eye (*piy^rw255*)	*primase polypeptide 1* (*prim*)	Subunit of DNA primase (Prim1)	Similar to *fla* mutant; loss of arches 3–7, reduced Meckels	Osteosarcoma	Yamaguchi et al. (2008)
dirty south (*dys^hi3812*)	*WD repeat domain 68* (*wdr68*)	Unknown—WD40 repeat-containing protein	Variable asymmetric cartilage defects; reduced mandibular and ceratohyal cartilages		Nissen et al. (2006)
nil per os (*npo^fw07g*)	*RNA binding motif protein 19* (*rbm19*)	RNA binding protein	Reduced jaw and arches		Mayer and Fishman (2003)
perplexed^n52	*Cad*	Enzyme (pyrimidine synthesis)	Reduced cartilage and fins		Willer et al. (2005)
lessen (*lsn^w24*)	*mediator complex subunit 24* (*med24*)	Component of TRAP/mediator transcriptional regulation complex	Loss of ceratobranchials 4–5		Pietsch et al. (2005)

(continues)

Table I (continued)

Mutant name [(alleles)]	Gene	Protein type (gene family)	Skeletal phenotype	Associated human syndrome	Reference
trapped (*trp*[25870]); *kohtalo* (*kto*[y81] *and* [y82])	*mediator complex subunit 12* (*med12*)	Component of TRAP/mediator transcriptional regulation complex	Complete loss of cartilage and pectoral fins	Opotiz-Kaveggia syndrome	Hong *et al.* (2005), Rau *et al.* (2006), and Risheg *et al.* (2007)
WDR55[hi2786B]	*WD repeat domain 55*	Unknown–contains a tryptophan-aspertate-repeat motif	Short and abnormally shaped		Iwanami *et al.* (2008)
Cartilage differentiation defects other					
jellyfish (*jef*[hi1134 and tw37])	*sox 9a*	Transcription factor (HMG)	Lack of neurocranial, pharyngeal, and pectoral girdle cartilages	Campomelic dysplasia	Yan *et al.* (2002)
sox9b[b971]	*sox9b*	Transcription factor (HMG)	Severely reduced mandibular, hyoid, and neurocranial cartilages	Campomelic dysplasia	Yan *et al.* (2005)
jekyll (*jek*[m151 and m310])	*ugdh*	Enzyme (Ugdh)	Defect in cartilage differentiation (no alcian staining)		Walsh and Stainier (2001)
earthd[i23e1]	*glutamine: fructose-6-phosphate amino-transfera se 1* (*gfpt1*)	Enzyme	Shortened cartilages, loss of chondrocyte organization and stacking		Yang *et al.* (2007)
dackel (*dak*[to273b])	*exostoses 2* (*ext2*)	Glycosyltransferase	Short thick cartilage, loss of chondrocyte organization/stacking	Hereditary multiple exostoses (HME)	Clement *et al.* (2003) and Lee *et al.* (2004)

Mutant	Gene	Protein function	Phenotype	Human disease	Reference
boxer (*box^{tw24} and m70g*)	*exostoses 3* (*ext3*)	Glycosyltransferase	Short thick cartilage, loss of chondrocyte organization/stacking		Lee *et al.* (2004)
pinscher (*pic^{to14mx}*)	*papst1*	Sulfate transporter	Short thick cartilage, loss of chondrocyte organization/stacking, stain poorly with Alcian	Hereditary multiple exostoses (HME)	Clement *et al.* (2008)
crusher (*cru^{m299}*)	*sec23 homolog A* (*sec23a*)	COPII complex component—protein transport	Reduced arches and neurocranial cartilage	Craniolenticulosutural dysplasia (CLSD)	Lang *et al.* (2006)
sufu (*sufu^{m146d}*)	*suppressor of fused homolog* (*Su(fu)*)	Negative regulator of shh	Branchial arch cartilages less branched and not stacked, chondrocytes are prehypertrophic, immature	Medulloblastoma	Koudijs *et al.*, 2005
acerebellar (*ace^{ti282a}*)	*fgf8*	Secreted ligand (Fgf)	Hyoid arch defects		Reifers *et al.* (1998)
hands-off (*han^{s6} and c99*)	*hand 2*	Transcription Factor (bHLH)	Loss of pectoral fins		Yelon *et al.* (2000)
chihuahua (*chi^{dc124}*)	*collagen1* (*col1*)	Extracellular matrix structural protein	Adult skeletal dysplasia, short thick bones	Osteogenesis imperfecta	Fisher *et al.* (2003)
nackt (*nkt^{dt35243x}*)	*ectodysplasin* (*eda*)	Ligand	Loss of scales in adults; reduced teeth and gill rakers	Hypohidrotic ectodermal dysplasia (HED)	Harris *et al.* (2008)
finless (*fls^{te370f}*)	*ectodysplasin receptor* (*edar*)	Transmembrane protein	Loss of scales and fins in adults; reduced teeth and gill rakers	Hypohidrotic ectodermal dysplasia (HED)	Harris *et al.* (2008)
patched 1 (*ptc1^{hu1602}*)	*patched 1*	Transmembrane receptor-Shh receptor	Increased size of pectoral fins	Nevoid basal-cell carcinoma syndrome (NBCCS)	Koudijs *et al.* (2005)
chordino (*chd*)	*chordin*	Secreted BMP antagonist	Defective axial skeleton and pectoral fins		Fisher and Halpern (1999)

defects, which are secondary consequences of a loss of chondrogenic signals from the endoderm (David *et al.*, 2002).

jellyfish is one of a large class of mutants originally called "hammerheads" that disrupt chondrocyte differentiation. Cartilages and the facial structures that they support, do not extend in these mutants, causing the larval eyes to protrude laterally. The *jellyfish* mutation disrupts *sox9a*, a relative of the mammalian HMG transcription factor, Sox9, required for condensation and differentiation of cartilage precursors (Yan *et al.*, 2002). In zebrafish, however, *sox9a* functions together with a fish duplicate, *sox9b*, in cartilage, and mutant analyses have revealed a subfunctionalization in Sox9 genes in teleost evolution (Yan *et al.*, 2005). The *jekyll* mutation disrupts a *uridine 5′ diphosphate -glucose dehydrogenase (udgh)* required for production of much of the proteoglycan component of cartilage extracellular matrix (Walsh and Stainier, 2001). Recent identification of mutants in components of the secretory pathway, such as Sec23, in which chondrocytes remain small and have swollen endoplasmic reticulum, further suggest the importance of secretion of large matrix molecules in chondrocyte development and morphogenesis (Lang *et al.*, 2006).

Zebrafish form a much more complex skeleton at later stages than the larval structures that we have considered so far, and a few recent screens have begun to identify mutants that survive larval development and disrupt cartilage, bone and in some cases the integument (teeth, scales, fins) of adults (Harris *et al.*, 2008). Several additional mutations with skeletal defects have been generated by insertional mutagenesis, such as the *hands-off* mutant that lacks pectoral fins (Yelon *et al.*, 2000), and *chihuahua*, which exhibits skeletal dysplasia, caused by a mutation in type I collagen (Fisher *et al.*, 2003). *chi* was detected through a radiography screen, which is another method by which zebrafish adults can be rapidly screened for new phenotypes (see below).

II. Cartilage Visualization Techniques

A. Alcian Blue Staining

Alcian blue and green are dyes that stain the proteoglycan components of the extracellular matrix associated with chondrocytes and that are used to visualize cartilage patterns in larvae and in adults (Fig. 1A). Animals are fixed in formalin and stained for several hours to days in the dye, which will label different types of proteoglycans depending on the pH and aqueous content of the staining solution. Here we describe a standard method for staining in 80% ethanol at pH ~5.7 and without $MgCl_2$. Under these conditions, Alcian dye labels most anionic tissue components, mucosubstances with carboxyl and sulfate groups, including hyaluronate and chondroitin sulfate. At higher $MgCl_2$ concentrations only highly sulfated proteoglycans such as heparin sulfate are stained (Wardi and Allen, 1972; Whiteman, 1973).

Alcian Blue first labels the earliest differentiating chondrocytes at 54 hpf (Schilling and Kimmel, 1997), but the full pattern of early cranial cartilages is best labeled after

72 hpf. Mouth extension has proven to be a useful staging tool for this phase of development (Schilling and Kimmel, 1997).

1. Protocol 1: Alcian Blue Staining of Cartilage

1. Anesthetize larvae in 5–10% methylethane sulfonate ("tricaine").
2. Fix in 3.7% neutral buffered formaldehyde (10% formalin in phosphate buffered saline (PBS)) at room temperature for several hours to overnight. Care should be taken not to overfix the preparation.
3. Wash larvae in PBT (PBT: PBS, 0.1% Tween-20) 3–5 times for at least 5 min per wash.
4. Transfer into a 0.1% solution of Alcian blue (Sigma) dissolved in 80% ethanol, 20% glacial acetic acid for at least 6–8 h to overnight (do not over-stain). Alcian solution should be syringe filtered (0.22 μm) before use.
5. Rinse larvae in ethanol and rehydrate gradually into PBS. Use ethanol, PBT solutions of 70–30%, 50–50%, and 30–70%, respectively.
6. If fish have not been raised in phenyl-thiourea (PTU), remove pigmentation by bleaching in 3% hydrogen peroxide/1% potassium hydroxide for 1–3 hours. This reaction should be monitored carefully and stopped once eye pigmentation disappears. Bubbles may form within the embryo during bleaching but will disappear in subsequent steps. Leave tube caps open during bleaching.
7. Clear tissue in 0.05% trypsin dissolved in PBS for 1–2 hours. A 0.01% solution may be use for overnight trypsinization; however, this can be performed at room temperature if clearing is monitored closely. Stop the trypsin digestion once the brain tissue has cleared.
8. Wash larvae in PBS 2–3 times for 5 min per wash.

   • Alizarin red staining for bone visualization should be performed at this stage if desired (see Protocol 4a).

9. Transfer embryos into glycerol gradually through a series of 30% and 50% dilutions in PBS and store in 70% glycerol.

B. Microdissection of Larval Craniofacial Cartilage

Cartilage dissection and flat mounting allows for close observation of the cartilage pattern that is not possible in whole-mounted preparations. Flat mounts allow the specimen to be photographed in one focal plane (Fig. 2), producing very informative depictions of each cartilage. However, flat mounting disturbs three-dimensional relationships and should be used in combination with whole-mounted specimens to develop the most accurate configuration of skeletal structures. Dissection of the larval skeleton is a bit intricate; however, with some patience the following methods allow detailed analyses of cartilage pattern.

1. Protocol 2: Microdissection and Mounting of Stained Cartilage

1. Stain larval skeleton with Alcian Blue (see Protocol 1) or acid-free double stain (see Protocol 4b) and store in 70% glycerol.

2. Place stained larvae in depression slides under a dissecting stereomicroscope submerged in a large droplet (~100 μl) of 100% glycerol. Use 40× magnification for sorting and securing embryos and 80–100× for dissection.

3. Use two tungsten needles to remove eyes and discard. Gently remove brain tissue by holding the larva with one needle and scraping dorsal to the skeleton and notochord. Hold the larva by placing one needle directly on the notochord, which is the strongest tissue at this stage. Carefully cut the head free from the trunk and tail at the level of the pectoral fins (severing the notochord can be tough). Use very thin tungsten needles (0.1 or 0.2 mm in diameter), which we find work better than glass needles. Secure needles to the ends of glass pipettes by melting glass or with superglue. Sharpen needles either by flaming the tips briefly over a Bunsen burner or with an electrolytic cell containing NaOH using the needle as the anode. Use tweezers to bend the tips of the needles (~1 mm from the end) at a 90° angle to provide maximum versatility.

4. Use one needle to secure the larva in place and the second to dissect cartilage with your dominant hand. We prefer to secure the specimen with a needle attached to a micromanipulator in order to minimize motion. Place the tip of the securing needle at the base of the notochord and skull, near the posterior end of the head. (Fig. 2A, place needle at position 1).

5. Depending on the desired preparation (see below), skeletal structures may be dissected by placing the bent tip at a joint between two cartilage elements and pulling away very gently. The pharyngeal arches (mandibular, hyoid, and the 5 branchial arches) can be separated from the neurocranium and displayed intact (Fig. 2E). To retain neurocranial and arch attachments, we use a "butterfly" preparation (Fig. 2A). We used this type of preparation to display the fusion between the skull and the arches in the *lockjaw* mutant (Knight *et al.*, 2003; Fig. 2B). Lateral preparations of the jaw and hyoid are dissected free from the neurocranium to show detail at higher magnification (Fig. 2F).

a. Suggested Preparations

a. Neurocranial preparation (Fig. 2C)

1. Scrape away brain tissue and separate head from trunk and tail.

2. After positioning the securing needle (Fig. 2A), place dissecting needle at the joint between the tip of the palatoquadrate (pq) and the ethmoid plate (ep) and gently pull to detach. Repeat for the contralateral joint.

3. Place dissecting needle above the hyomandibular (hm) cartilage, at the joint between hm and the neurocranium. Now gently pull to detach at the joint. Repeat for the contralateral joint (Fig. 2A, cut along line 2).

4. Gently glide the dissecting needle horizontally, along the plane of the neurocranium and dorsal to the arches to remove any remaining soft tissue. Repeat until the jaw and arches separate from the neurocranium (save arches for preparation).

b. Seven arch preparation (jaw and posterior arches—Fig. 2E)

1. Follow procedures for the neurocranial preparation but save the arches.

2. Be careful when mounting to orient the pq cartilage such that the tip is positioned laterally relative to Meckels cartilage (m). The pq cartilage will naturally fold inward and obscure the joint with hm. To achieve the best configuration, flip pq around and remove any soft tissue attached. Once the arches are in the proper orientation, carefully place the coverslip over the specimen and adjust gently, as needed.

c. Butterfly preparation (neurocranium, hyoid, and mandibular arches— Fig. 2A)

1. Scrape away brain tissue and separate head from trunk and tail.

2. After positioning the securing needle, place dissecting needle at the midline of m and press down with a cutting motion to separate at the fusion point in the midline (Fig. 2E, cut along line 1).

3. Place dissecting needle at the joint between the tip of pq and ep and gently pull to detach.

4. Place dissecting needle at the midline between the two ceratohyal (ch) cartilages. Gently pull to separate (Fig. 2E, cut along line 2) and then remove the more posterior ceratobranchial (cb) cartilages.

5. For mounting this preparation, position the mandibular and hyoid arches to the sides as displayed in Fig. 2A before securing the cover slip.

d. Lateral jaw preparation (hyoid and mandibular arches—Fig. 2F)

1. Remove eyes and brain by scraping, and separate the head from trunk and tail.

2. Place dissecting needle at the midline of m and press down with a cutting motion to detach at the midline (Fig. 2E, cut along line 1).

3. Place dissecting needle at the joint between the tip of pq and ep and gently pull to detach.

4. Place dissecting needle at the midline between the two ch cartilages. Gently detach and split the posterior arches at the midline (Fig. 2E, cut along line 2).

5. Place needle on hs, near the joint between hs and the neurocranium and gently pull away (Fig. 2A, cut along line 2).

6. Flat mount desired structures by transferring them (using a glass pipette) on to a clean glass slide containing a small droplet (~50–80 μl) of 100% glycerol. Manipulate the specimen with needles under the microscope to achieve desired orientation before adding the cover slip. Coverslips should be elevated slightly to prevent crushing the specimen. Place small amounts of Vaseline or modeling clay at four spots around the specimen, in approximately a 1 cm square array. Use Vaseline for flat structures (i.e., neurocranium) and clay for thicker (i.e., arches) structures. Place coverslip on Vaseline/clay bridges and gently slide the cover slip to manipulate specimen and obtain the ideal orientation.

C. BrdU Labeling

Bromodeoxyuridine (BrdU) is incorporated into newly synthesized DNA and can be used as a label for dividing chondroblasts and osteoblasts in the skeleton (Fig. 3A). One can label animals with BrdU at any stage, but for incorporation into skeletal condensations one should inject/incubate animals after 54 hpf. It is important to dissect and flat-mount cartilages free of other tissues, which will also be BrdU labeled. We describe an incubation time in BrdU of 24 hours, which gives strong labeling, but shorter pulsed treatments are possible. Both injection of BrdU (Larison and Bremiller, 1990) and immersion (Kimmel *et al.*, 1998) work at early larval stages.

1. Protocol 3: BrdU Labeling of Dividing Cartilage Cells Counter Stained with Alcian Blue

1. Incubate larvae in 1 mM BrdU (Vector Laboratories Inc.) in embryo medium for 24 h (change medium every 6 h to increase BrdU incorporation).
2. Fix larvae in 4% paraformaldehyde (PFA) overnight at 4 °C (can store in MeOH and rehydrate through a MeOH/PBS series if required).
3. Rinse 3 times for 5 min each in PBT (PBT: PBS, 0.1% Tween-20).
4. Rinse 5 min in dH$_2$O.
5. Place larvae in 0.1% trypsin for < 2 min (or in acetone at −20 °C for 7 min) to increase antibody penetration.
6. Rinse once in dH$_2$O and 3 times in PBT for 5 min per rinse.
7. Place in blocking solution for up to 1 h (blocking solution: 1× PBS, 1% DMSO, 10% serum, 0.1% Tween-20).
8. Detect BrdU by adding anti-BrdU primary antibody 1:200 (antibody G3G4 developed by S. J. Kaufman and obtained from Developmental Hybridoma Bank, University of Iowa) in blocking solution and incubate overnight at 4 °C with mild agitation on a nutator or shaker.
9. Rinse 5 times for 10–15 min each in PBT + 1% DMSO.
10. Incubate in biotinylated secondary antibody, diluted 1:500 in blocking solution overnight at 4 °C.
11. Rinse 5 times for 10–15 min each in PBT (perform step 12 while waiting).

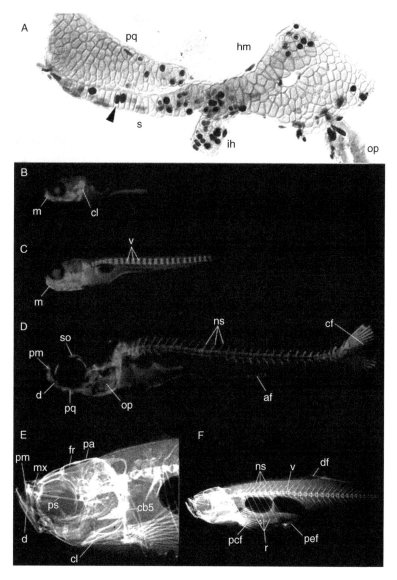

Fig. 3 Techniques for visualizing the zebrafish skeleton. Anterior is to the left in all panels. (A) Partial lateral jaw preparation, showing BrdU-labeled cartilages of the hyoid arch. Arrowhead indicates one of many labeled nuclei that have taken up the BrdU and are labeled with an anti-BrdU antibody. (B–D) Whole larvae (5, 11, 21 dpf respectively) in lateral view, showing calcein stained cartilages and bones. (E, F) Radiographs of adult zebrafish. Abbreviations: af, anal fin; cb5, ceratobranchial 5; cf, caudal fin; cl, cleithrum; d, dentary; ds, dorsal fin; fr, frontal; hm, hyomandibular; ih, interhyal; m, Meckel's; mx, maxilla; ns, neural (dorsal) spine; op, opercle pa, parietal; pcf, pectoral fin; pef, pelvic fin; pm, premaxilla; pq, palatoquadrate; ps, parasphenoid r, ribs; s, symplectic; so, supraorbital; v, vertebrae.

12. Prepare "AB" complex: 1 ml of blocking solution, 20 μl vectastain "A," 20 μl vectastain "B" and incubate for 30 min before applying to specimens.

13. Incubate specimens for 1 h in the "AB" complex at room temperature.

14. Rinse 5 times for 15 min each in PBT.

15. Rinse 3 times for 10 min each in 1× PBS to remove Tween-20 (Tween inhibits subsequent enzymatic reactions).

16. Preincubate in diaminobenzidine (DAB) solution:1 ml PBS, 1 ml H_2O, 20 μl DMSO, 25 μl DAB for 15–20 min (0.5 ml per tube).

17. Add 5 μl of 3% H_2O_2 to each tube (add 5 μl of 3% H_2O_2 for every 500 μl of DAB).

18. Monitor development from 1 to 30 min.

19. Rinse in PBT for at least three washes to stop staining.

20. Counterstain preparations with Alcian Blue (Protocal 1.1) and dissect and flat-mount cartilages (Protocol 2) as previously described.

III. Bone Visualization Techniques

A. Alizarin Red Staining

Alizarin red stains calcified matrix associated with bone, and can be used to distinguish cartilage and bone by counter-staining with Alcian blue (Cubbage and Mabee, 1996). For single alizarin red staining of bone without additional cartilage staining, follow Protocol 4a. Traditional double bone and cartilage staining methods can negatively affect the quality of the bone stain, since the acidity of the Alcian blue solution can decalcify the bone matrix. Controlling magnesium concentration rather than pH of Alcian blue solution is a method to avoid matrix degradation. Protocol 4b details the acid-free double stain method adapted from Walker and Kimmel (2007). The acid-free double stain may also be used in place of the Alcian blue stain in Protocol 1 for similar results.

Alizarin red first labels the earliest differentiating bones at 3–4 days of development, and bone development continues for many weeks before achieving the adult pattern. Body length, rather than age, as well as changes in pigmentation and fin morphology are useful staging tools at these ages (Schilling, 2002).

1. Protocol 4a: Alizarin Red Staining of Bone

1. Fix larvae in 4% PFA at room temperature for several hours to overnight.

2. Wash embryos in PBT 3–5 times for at least 5 min per wash.

3. Transfer larvae into a 0.05% solution of Alizarin red for several hours.

4. Clear embryos in glycerol, gradually through a series of 30% and 50% dilutions in PBS and store in 70% glycerol.

2. Protocol 4b: Acid–Free Double Cartilage and Bone Stain

1. Fix larvae in 4% PFA in PBS for 2 h at room temperature, with agitation. Do not overfix as this can significantly reduce bone staining.

2. Wash with 50% ethanol for 10 min at room temperature.

3. Add 1 ml of acid-free double stain solution. This is made by mixing the two separate solutions just prior to staining. Solution A includes 0.02% Alcian blue dissolved in 70% ethanol/20–60 mM $MgCl_2$. Higher or lower concentrations of MgCl2 can result in reduced cartilage staining. Solution B contains 0.5% Alizarin red S powder dissolved in water. Combine 10 μl of solution B with 1 ml of solution A to prepare the final staining solution. Incubate overnight at room temperature with agitation.

4. If necessary, remove pigmentation by bleaching. Wash stain from larvae with 1 ml water and add 1 ml bleach solution (equal volumes of 3% H_2O_2 and 2% KOH). Leave larvae in bleaching solution for 20 min with lids open.

5. Tissues are further cleared with successive washes of glycerol and KOH solutions. Add 1 ml of a 20% glycerol and 0.25% KOH solution and agitate 30 min to overnight at room temperature. Transfer larvae to a solution of 50% glycerol and 0.25% KOH and agitate 2 h to overnight.

6. Transfer larvae to a solution of 50% glycerol and 0.1% KOH for storage.

3. Protocol 4c: Live Alizarin Red Staining

1. Alizarin will fluoresce bright red under ultraviolet light. Larval zebrafish can be reared in a 0.05% solution of Alizarin red in embryo medium (Flemming et al., 2004) or stained overnight in a 4 μg/ml solution (Kimmel et al., 2003).

B. Adult/Larval Calcein Staining

The fluorescent chromophore, calcein ($C_{30}H_{26}N_2O_{13}$), specifically binds to calcium, fluorescently staining the calcified skeletal structures in living zebrafish (Fig. 3B–D). This allows analysis of bone with high sensitivity, though it also labels some cartilages at larval stages. Calcein staining can be used to follow the development of skeletal structures from 2 to 21 dpf and is potentially an effective screening tool for identifying skeletal mutants (Du et al., 2001b).

1. Protocol 5: Calcein Staining

1. Immerse living animals (not anesthetized) in 0.2% calcein solution in Petri dishes from 3 to 10 min, depending on the size of the larvae or juveniles. Prepare solution by dissolving 2 g of calcein powder (Sigma Chemical, St. Louis, MO) in 1 l

of deionized water. To counter calcein's strong acidifying affects, add an appropriate amount of NaOH (0.5 N) to the solution to restore neutral pH (~7.0–7.5).

2. Rinse 3–4 times in dH$_2$O and then allow to soak for 10 min so that excess, unbound calcein diffuses out of the tissues.

3. Euthanize or anesthetize as desired in 10–20% tricaine.

4. Mount on glass depression slides with 3% methyl-cellulose.

5. View using a microscope with a filter set for GFP or FITC (excitation wavelength is around 495 nm and the emission wavelength is around 525 nm).

C. Radiographic Visualization of Adult Zebrafish Skeleton

Radiographic analysis is an excellent technique for visualizing skeletal anatomy and bone morphology (Fig. 3E) and has been previously described by Fisher *et al.* (2003). This technique, in comparison to traditional histological methods such as Alcian blue/Alizarin red staining, is a quick and efficient way to detect subtle skeletal abnormalities in adult zebrafish. This is useful since much of the skeleton matures relatively late in zebrafish development, when the animals are large and difficult to stain and analyze in large numbers. Radiography has been used with zebrafish anywhere between 5 dpf and adulthood, and shown to be a powerful method for genetic screens to identify subtle phenotypes, such as *chihuahua* (Fisher *et al.*, 2003).

1. Protocol 6: Radiography of Adult Zebrafish

1. Anesthetize larval or adult fish with 5–10% tricaine.

2. Lay animals flat on X-ray platform.

3. Use a small specimen radiography X-ray machine for higher resolution, such as a Faxitron MX-20 cabinet.

4. Expose 3–4 s at 17–20 kV (use high resolution Min-R 2000 film and intensifying screens).

5. To assign bones, use stained and dissected skeletal preparations as a reference for identification and comparison.

D. Osteoblast and Osteoclast Histology

One can observe bone cell types at a much more detailed level in sections, particularly beyond early larval stages. Here we describe a tartrate-resistant acid phosphatase (TRAP)-staining method modified from Witten and Villwock (1997) and Witten *et al.* (2001). Osteoclasts associated with bone resorption can be identified in light microscopic sections, stained with toluidine blue, and labeled by the TRAP method. For identification of early osteoblasts, alkaline phosphatase (ALP) is visualized (Miyake *et al.*, 1997). While only mononucleate osteoclasts are present at larval stages, multinucleate osteoclasts appear after approximately 40 dpf. To label osteoclasts the sections must be decalcified.

1. Protocol 7: TRAP Staining to Visualize Osteoclasts

1. Fix specimens for 1 h at 4 °C in 10% formaldehyde (methanol-free) in 50 mM Tris buffer, pH 7.2.

2. Rinse in tap water for 1 h and decalcify for 48 h in Tris buffer (100 mM, pH 7.2) containing 10% EDTA.

3. Dehydrate for 1.5 h in an acetone series (30%, 50%, 70%, 90%, 100%).

4. Soak specimens in a glycol methacrylate monomer solution for 60 min to impregnate bone: 80 ml (2-hydroxyethyl)-methacrylate, 200 ppm *p*-methoxyphenol, 12 ml ethylene glycol monobutyl ether, 270 mg benzoyl peroxide.

5. Change monomer solution and soak for 24 h to achieve further impregnation.

6. Add 2% catalyst (1 ml *N,N*-dimethylaniline, 10 ml poly-ethyleneglycol-200) to the monomer solution just prior to embedding.

7. Embed in polyethylene jars with tight lids.

8. Polymerize at 4 °C for 24 h and another 24 h at room temperature. Tissue blocks can be stored at 4 °C.

9. Section specimens (5 μm thickness), float on demineralized water (25 °C), mount on uncoated slides, and dry at 25 °C.

10. For TRAP-labeling reincubate sections for 30 min at 20 °C in 0.1 M acetate buffer 50 mM disodium tartrate dihydrate, pH 5.5. Incubate in TRAP solution for 30 min. This solution is prepared with naphthol AS-TR phosphate (N-AS-TR-P) as substrate, and hexazotized pararosaniline (PRS): 1 g PRS dissolved in 19.25 ml. Demineralized water, 5.75 ml 32% HCl, heated, and stored in the dark at 4 °C. For hexazotiation, add 2 ml 4% $NaNO_2$ (0.58 M) to 1 ml of prepared PRS solution at 20 °C.

11. Incubate for 30 minutes at 20 °C. Prepare final TRAP incubation solution by adding 1 ml hexazotized PRS, 600 ml 2% MgCl solution, 2 ml enzyme substrate solution (2 mg N-ASTR-P dissolved in 2 ml *N,N*-dimethylformamide), and 100 mM disodium tartrate dihydrate to 30 ml of 0.1 M acetate buffer at pH 5.5.

12. Rinse with demineralized water.

13. Counterstain sections with Mayers hematoxylin for 10 min, rinse in running tap water for 15 min, flush with demineralized water, dry at 40 °C, and mount with DPX.

14. Possible controls for staining specificity include: (1) heating at 90 °C for 10 min prior to incubation, (2) incubation without substrate, (3) incubation without tartrate, and (4) adding NaF (10 mmol/l) to the incubation solution.

2. Protocol 8: Alkaline Phosphatase Staining to Visualize Osteoblasts

1. Fix in acetone at 4 °C and perform all subsequent preparation steps (embedding, cutting, mounting) as described above but do not decalcify.
2. Preincubate in Tris buffer (50 mmol/l, pH 9.5) for 1 h. Incubate in 30 ml Tris buffer (50 mmol/l, pH 9.5) for 2×1 h at 20 °C; include hexazotized PRS (see above) and 60 mg napthol-AS-BI-phosphate dissolved in 0.2 ml *N,N*-dimethylformamide in the incubation solution.
3. Visualize staining based on the fluorescence of the reaction product at 568 nm, following the protocol of Sakakura *et al.* (1998).
4. Possible controls include: (1) heating at 90 °C for 10 min prior to incubation, (2) incubation without substrate, (3) adding 10 mmol levamisole to the incubation solution, and (4) adding EDTA (10 mg/ml) to the incubation solution.

IV. Molecular Markers and Transgenic Lines

A. Molecular Markers of Progenitor Populations

The skeleton derives from multiple embryonic origins. For example, the appendicular skeletons of the limbs derive from lateral plate mesoderm, while much of the craniofacial skeleton derives from migratory cells of the neural crest. Thus, to understand the earliest events in skeletogenesis, it is important to define early molecular markers for these progenitors. A large number of transcription factors are expressed throughout the premigratory cranial neural crest in zebrafish, including *crestin* (Luo *et al.*, 2001), *foxd3* (Odenthal and Nusslein-Volhard, 1998), *sna2* (Thisse *et al.*, 1995), *sox9a* (Yan *et al.*,1995), *sox10* (Dutton *et al.*, 2001), and *tfap2a* (Knight *et al.*, 2003). Others are more segmentally restricted, such as *hox* group 1–3 genes (Amores *et al.*, 1998; Prince *et al.*, 1998) and *krox20* (Oxtoby and Jowett, 1993). Genes encoding cell surface proteins of the Eph family are also expressed in segment-specific domains within the premigratory neural crest (Xu *et al.*, 1996). For most of these early neural crest markers, expression persists during at least the early stages of neural crest migration into the periphery.

B. Molecular Markers of Cartilage and Bone Lineages

A second set of transcription factors that mark putative skeletogenic neural crest are expressed after migration. These include the homeobox-containing genes *dlx2-7* (Akimenko *et al.*, 1994), *msx* (Ekker *et al.*, 1997), and the ETS-domain containing genes *fli1* (Brown *et al.*, 2000) and *pea3* (Brown *et al.*, 1998). Ventral crest cells within each pharyngeal arch express *dHand* (Yelon *et al.*, 2000). Many of these markers continue to be expressed up to the stage of cartilage differentiation in the larvae. By this stage expression of several markers becomes restricted to skeletogenic populations, including *sox9a*, and new markers become expressed, such as *chondromodulin* and *runx* (Fisher *et al.*, 2003).

C. Transgenics

Traditional methods for visualizing gene expression patterns (i.e., *in situ* hybridization, immunohistochemistry) have the drawback that they require tissue fixation and only reveal expression at one time point in development. Transgenic technologies tagging genes with fluorescent proteins derived from other species, such as green fluorescent protein (GFP), allow one to monitor gene expression in living cells, for which the transparent embryos of zebrafish are ideally suited. Recent improvements in techniques for transgenesis, such as a modified Tol2 transposon (Kawakami, 2005), have dramatically increased the efficiency of obtaining stable integration of gene-specific regulatory elements driving GFP expression into the zebrafish genome. Several such lines are now available that label skeletal progenitors, particularly the craniofacial skeleton, with many more anticipated in the near future that label other skeletal lineages.

The cranial skeleton derives largely from neural crest, and several transgenic lines mark these cells as they migrate and/or differentiate as cartilage. Sox10 is a member of the sox transcription factor family required for the formation of many different neural crest-derived lineages. A *sox10:gfp* transgenic line in which 7.2 kb of the *sox10* promoter drives *egfp* labels both premigratory and migrating cranial neural crest cells (Wada *et al.*, 2005; Carney *et al.*, 2006). Importantly, *sox10:gfp* expression is strong in cartilage progenitors and differentiated cartilage in the pharyngeal arches and neurocranium at later stages (unlike the endogenous *sox10* gene), allowing for easy visualization of cartilage morphology in living embryos and larvae.

The Dlx (distal-less related) genes are homeodomain containing transcription factors that play critical roles in craniofacial development in mice. Zebrafish *dlx2b* expression is a specific marker of tooth forming (odontogenic) epithelia on pharyngeal arch #7 (zebrafish and its close relatives have pharyngeal rather than oral teeth). A *dlx2b:gfp* transgenic line fluorescently labels developing pharyngeal tooth progenitors beginning at 60 hpf and expression persists until the stage of tooth attachment (Jackman and Stock, 2006). In addition to being a useful marker for tooth development, *dlx2b:gfp* expression also labels the developing median fin fold and pectoral fins.

Expression of the ETS-domain containing transcription factor Fli1 marks endothelial cells in developing blood vessels but in zebrafish is also found in migrating cranial neural crest cells and in cartilage of the jaw. A *fli1:gfp* transgene fluorescently marks migrating neural crest cells and cartilage of the jaw up to 7 days postfertilization (Lawson and Weinstein, 2002).

Fibroblast growth factors (Fgfs) are important signaling molecules in skeletal development, in general, and in early embryonic development of craniofacial structures and limbs (David *et al.*, 2002). Dual specificity phosphatase 6 (dusp6) is an enzyme involved in dephosphorylating extracellular signal-regulated kinases

(ERKs) that is directly regulated by Fgf signaling. A *dusp6:gfp* transgene, when stably integrated into the zebrafish genome, is only expressed in cells responding to Fgf signaling and therefore acts as a reliable FGF reporter in living animals (Molina *et al.*, 2007). In 24 hpf embryos, *dusp6:gfp* is expressed in the pharyngeal arches and by 50 hpf in the pectoral fins.

Similarly, several members of the Hedgehog family of signaling molecules (see section on mutations and Table 1), are crucial for skeletal differentiation, as well as patterning the craniofacial and limb skeletons. Transgenics in which the promoters of two relatives of Sonic hedgehog (Shh) drive GFP (*shha:gfp* and *shhb:gfp*) have been generated and while both mark the ventral neural tube, somites and eyes, *shhb:gfp* (shhb previously known as tiggy-winkle hedgehog) also marks the branchial arches and pectoral fin buds (Du and Dienhart, 2001a). Several *shha:gfp* lines have been constructed using different portions of the upstream regulatory regions of *shha* and these differentially mark pharyngeal endoderm, distal tips of growing fin rays, and pectoral fin bud mesenchyme (Hadzhiev *et al.*, 2007). One line, 2.2shh:gfp:ABC#15, also labels caudal fin fold mesenchyme and the adult caudal fin, though this likely reflects a position effect of the insertion point and does not accurately represent endogenous *shha* expression.

V. Strategy and Potential of Future Screens for Skeletal Mutants

In zebrafish, one can screen large numbers of animals for mutations that disrupt most any aspect of skeletal development. Synchronous development of progeny from single crosses allows efficient recognition of morphological defects by simple comparison between siblings. Previous screens have yielded a large collection of larval cartilage mutants. Simple protocols for staining cartilage and bone and imaging the living skeleton with X-rays can be used effectively in mutant screens to find more subtle or later phenotypes in the skeleton.

Screens for mutations that disrupt skeletal features after the first week of development, however, require much more careful control of larval health and staging. A larval zebrafish must fill its air bladder by 5–6 dpf to survive, and a majority of the recessive lethal mutants found in previous screens fail to fill their swim bladders. In addition, the larva begins to feed, having consumed yolk up to this stage, and underfeeding can lead to delayed development and a failure in skeletal maturation. Mouth extension is an accurate method for staging between 54 and 72 hpf (Schilling and Kimmel, 1997), but for larvae, juveniles, and adults, staging requires examining other body features (Schilling, 2002).

Skeletal screens also require long-term maintenance of isolated fish during the analysis of their offspring. At minimum, to examine early larval cartilage formation in a typical screen for recessive lethal mutations, each pair of F2 fish must be kept in a separate container while their progeny are raised for 72 h in addition to time required for staining (Protocol 1) and analysis. Extending the screen to early stages of bone development requires an additional day or two for the larvae to mature as well as an additional day for Alizarin red staining, calcein staining, or X-ray (Protocols 4–6), and this means keeping their parents in separate containers for at least a week. The following is one example of steps involved in a combined screen for cartilage and bone defects at 6 dpf.

1. Mutagenize male zebrafish. Raise founders (2–4 months).

2. Intercross F1 founders to create F2 families of mutagenized animals. Raise F2 families (2–4 months).

3. Intercross F2 family members and raise embryos in embryo medium to at least 3 dpf for cartilage, and 5 days or more for bone. For raising to these stages embryos should not be overcrowded (approximately 50 embryos per 30 ml Petri dish), and media changed at least once after hatching (2–3 days).

4. Anesthetize at least 25 larvae from each of at least six crosses between family members. If possible, first examine living larvae for defects in air bladder formation, head shape, jaw elongation, gill, and pectoral fin formation, all of which can be hallmarks of problems with skeletal development.

5. Fix at least 25 larvae in 3.7% neutral buffered formalin and perform Alcian staining (Protocol 1), followed by Alizarin staining (Protocol 4) if desired.

6. Examine stained specimens for cartilage in at least two steps. First in ventral view, examine the large ventral elements (ceratobranchials) of each of the seven pharyngeal arches. These develop in an anterior to posterior sequence between 60 and 72 hpf, and screeners should be aware that the posteriormost arches are often reduced or lost if animals are developmentally retarded. Then focus dorsally to examine cartilage of the trabeculae and parachordal cartilages of the braincase.

7. As for earlier phenotypes, criteria for keeping mutants should include uniformity of phenotype and Mendelian segregation in more than one cross from the family. Potential mutant carriers should also be rescreened immediately.

Acknowledgments

We thank members of the Schilling lab for helpful comments on the manuscript. Support was provided by the NIH (NS-41353, DE-13828) to T. S.

References

Akimenko, M. A., Ekker, M., Wegner, J., Lin, W., and Westerfield, M. (1994). Combinatorial expression of three zebrafish genes related to distal-less: Part of a homeobox gene code for the head. *J. Neurosci.* **14,** 3475–3486.

Allende, M. L., Amsertdam, A., Becker, T., Kawakami, K., Gaiano, N., and Hopkins, N. (1996). Insertional mutagenesis in zebrafish identifies two novel genes, pescadillo and dead eye, essential for embryonic development. *Genes Dev.* **10,** 3141–3155.

Amores, A., Force, A., Yan, Y. L., Joly, L., Amemiya, C., Fritz, A., Ho, R. K., Langeland, J., Prince, V., Wang, Y. L., Westerfield, M., Ekker, M., *et al.* (1998). Zebrafish hox clusters and vertebrate genome evolution. *Science* **282**, 1711–1714.

Barrallo-Gimeno, A., Holzschuh, J., Driever, W., and Knapik, E. W. (2004). Neural crest survival and differentiation in zebrafish depends on mont blanc/tfap2a gene function. *Development* **131**, 1463–1477.

Barresi, M. J., Stickney, H. L., and Devoto, S. H. (2000). The zebrafish slow-muscle-omitted gene product is required for Hedgehog signal transduction and the development of slow muscle identity. *Development* **127**, 2189–2199.

Bird, N. C., and Mabee, P. M. (2003). Developmental morphology of the axial skeleton of the zebrafish, *Danio rerio* (Ostariophysi: Cyprinidae). *Dev. Dyn.* **228**, 337–357.

Brown, L. A., Amores, A., Schilling, T. F., Jowett, T., Baert, J. L., de Launoit, Y., and Sharrocks, A. D. (1998). Molecular characterization of the zebrafish PEA3 ETS-domain transcription factor. *Oncogene* **17**, 93–104.

Brown, L. A., Rodaway, A. R., Schilling, T. F., Jowett, T., Ingham, P. W., Patient, R. K., and Sharrocks, A. D. (2000). Insights into early vasculogenesis revealed by expression of the ETS-domain transcription factor Fli-1 in wild-type and mutant zebrafish embryos. *Mech. Dev.* **90**, 237–252.

Carney, T. J., Dutton, K. A., Greenhill, E., Delfino-Machin, M., Dufourcq, P., Blader, P. and Kelsh, R. N. (2006). A direct role for Sox10 in specification of neural crest-derived sensory neurons. *Development* **133**, 4619–4630.

Clement, A., Wiweger, M., von der Hardt, S., Rusch, M. A., Selleck, S. B., Chien, C. B., and Roehl, H. H. (2008). Regulation of zebrafish skeletogenesis by ext2/dackel and papst1/pinscher. *PLoS Genet.* **4**, e1000136.

Crump, J. G., Swartz, M. E., and Kimmel, C. B. (2004). An integrin-dependent role of pouch endoderm in hyoid cartilage development. *PLoS Biol.* **2**, E244.

Cubbage, C. C., and Mabee, P. M. (1996). Development of the cranium and paired fins in the zebrafish, *Danio rerio* (Ostariophysi, Cyprinidae). *J. Morphol.* **229**, 121–160.

David, N., Saint-Etienne, L., Schilling, T. F., and Rosa, F. (2002). Critical role for endoderm and FGF signaling in ventral head skeleton induction. *Development* **129**, 4457–4468.

De Beer, G. R. (1937). "The Development of the Vertebrate Skull." Oxford University Press, Oxford [Reprinted 1985, Chicago University Press, Chicago].

Dickmeis, T., Mourrain, P., Saint-Etienne, L., Fischer, N., Aanstad, P., Clark, M., Strahle, U., and Rosa, F. (2001). A crucial component of the endoderm formation pathway, CASANOVA, is encoded by a novel sox-related gene. *Genes Dev.* **15**, 1487–1492.

Du, S. J., and Dienhart, M. (2001a). Zebrafish tiggy-winkle hedgehog promoter directs notochord and floor plate green fluorescence protein expression in transgenic zebrafish embryos. *Dev. Dyn.* **222**, 655–666.

Du, S. J., Frenkel, V., Kindschi, G., and Zohar, Y. (2001b). Visualizing normal and defective bone development in zebrafish embryos using the fluorescent chromophore calcein. *Dev. Biol.* **238**, 239–246.

Dutton, K. A., Pauliny, A., Lopes, S. S., Elworthy, S., Carney, T. J., Rauch, J., Geisler, R., Haffter, P., and Kelsh, R. N. (2001). Zebrafish colourless encodes sox10 and specifies non-ectomesenchymal neural crest fates. *Development* **128**, 4113–4125.

Eberhart, J. K., He, X., Swartz, M. E., Yan, Y. L., Song, H., Boling, T. C., Kunerth, A. K., Walker, M. B., Kimmel, C. B., and Postlethwait, J. H. (2008). MicroRNA Mirn140 modulates Pdgf signaling during palatogenesis. *Nat. Genet.* **40**, 290–298.

Eberhart, J. K., Swartz, M. E., Crump, J. G., and Kimmel, C. B. (2006). Early Hedgehog signaling from neural to oral epithelium organizes anterior craniofacial development. *Development* **133**, 1069–1077.

Ekker, M., Akimenko, M. A., Allende, M. L., Smith, R., Drouin, G., Langille, R. M., Weinberg, E. S., and Westerfield, M. (1997). Relationships among msx gene structure and function in zebrafish and other vertebrates. *Mol. Biol. Evol.* **14**, 1008–1022.

Fisher, S., and Halpern, M. E. (1999). Patterning the zebrafish axial skeleton requires early chordin function. *Nat. Genet.* **23**, 442–446.

Fisher, S., Jagadeeswaran, P., and Halpern, M. E. (2003). Radiographic analysis of zebrafish skeletal defects. *Dev. Biol.* **264**, 64–76.

Fleming, A., Keynes, R., and Tannahill, D. (2004). A central role for the notochord in vertebral patterning. *Development* **131**, 873–880.

Goodrich, E. S. (1930). "Studies on the Structure and Development of Vertebrates." University of Chicago Press, Chicago.

Hadzhiev, Y., Lele, Z., Schindler, S., Wilson, S. W., Ahlberg, P., Strahle, U., and Muller, F. (2007). Hedgehog signaling patterns the outgrowth of unpaired skeletal appendages in zebrafish. *BMC Dev. Biol.* **7**, 75.

Hall, B. K., and Hanken, J. (1985). "Foreword to Reissue of *The Development of the Vertebrate Skull*, by G.N. deBeer." University of Chicago Press, Chicago.

Harris, M. P., Rohner, N., Schwarz, H., Perathoner, S., Konstantinidis, P., and Nusslein-Volhard, C. (2008). Zebrafish eda and edar mutants reveal conserved and ancestral roles of ectodysplasin signaling in vertebrates. *PLoS Genet.* **4**, e1000206.

Hong, S. K., Haldin, C. E., Lawson, N. D., Weinstein, B. M., Dawid, I. B., and Hukriede, N. A. (2005). The zebrafish kohtalo/trap230 gene is required for the development of the brain, neural crest, and pronephric kidney. *Proc. Natl. Acad. Sci. USA* **102**, 18473–18478.

Iwanami, N., Higuchi, T., Sasano, Y., Fujiwara, T., Hoa, Q., Okada, M., Talukder, S. R., Kunimatsu, S., Li, J., Saito, F., Bhattacharya, C., Matin, A., *et al.* (2008). WDR55 is a nucleolar modulator of ribosomal RNA synthesis, cell cycle progression, and teleost organ development. *PLos Genet.* **4**, e1000171.

Jackman, W. R., and Stock, D. W. (2006). Transgenic analysis of Dlx regulation in fish tooth development reveals evolutionary retention of enhancer function despite organ loss. *Proc. Natl. Acad. Sci. USA* **103**, 19390–19395.

Kawakami, K. (2005). Transposon tools and methods in zebrafish. *Dev. Dyn.* **234**, 244–254.

Kikuchi, Y., Agathon, A., Alexander, J., Thisse, C., Waldron, S., Yelon, D., Thisse, B., and Stainier, D. Y. (2001). casanova encodes a novel Sox-related protein necessary and sufficient for early endoderm formation in zebrafish. *Genes Dev.* **15**, 1493–1505.

Kimmel, C. B., Miller, C. T., Kruze, G., Ullmann, B., BreMiller, R. A., Larison, K. D., and Snyder, H. C. (1998). The shaping of pharyngeal cartilages during early development of the zebrafish. *Dev. Biol.* **203**, 246–263.

Kimmel, C. B., Ullmann, B., Walker, M., Miller, C. T., and Crump, J. G. (2003). Endothelin 1-mediated regulation of pharyngeal bone development in zebrafish. *Development* **130**, 1339–1351.

Knight, R. D., Nair, S., Nelson, S., Afshar, A., Javidan, Y., Geisler, R., Rauch, G. J., and Schilling, T. F. (2003). *lockjaw* encodes a zebrafish tfap2a required for early neural crest development. *Development* **130**, 5755–5768.

Koudijs, M. J., den Broeder, M. J., Keijser, A., Wienholds, E., Houwing, S., van Rooijen, E. M., Geisler, R., and van Eeden, F. J. (2005). The zebrafish mutants dre, uki, and lep encode negative regulators of the hedgehog signaling pathway. *PLos Genet.* **1**, e19.

Lang, M. R., Lapierre, L. A., Frotshcer, M., Goldenring, J. R., and Knapik, E. W. (2006). Secretory COPII coat component Sec23a is essential for craniofacial chondrocyte maturation. *Nat. Genet.* **38**, 1198–1203.

Larison, K. D. and Bremiller, R. (1990). Early onset of phenotype and cell patterning in the embryonic zebrafish retina. *Development* **109**, 567–576.

Laue, K., Daujat, S., Crump, J. G., Plaster, N., Roehl, H. H., Tübingen 2000 Screen Consortium, Kimmel, C. B., Schneider, R., and Hammerschmidt, M. (2008a). The multidomain protein Brpf1 binds histones and is required for Hox gene expression and segmental identity. *Development* **135**, 1935–1946.

Laue, K., Janicke, M., Plaster, N., Sonntag, C., and Hammerschmidt, M. (2008b). Restriction of retinoic acid activity by Cyp26b1 is required for proper timing and patterning of osteogenesis during zebrafish development. *Development* **135**, 3775–3787.

Lawson, N. D., and Weinstein, B. M. (2002). *In vivo* imaging of embryonic vascular development using transgenic zebrafish. *Dev. Biol.* **248**, 307–318.

Lee, J. S., von der Hardt, S., Rusch, M. A., Stringer, S. E., Stickney, H. L., Talbot, W. S., Geisler, R., Nusslein-Volhard, C., Selleck, S. B., Chien, C. B., and Roehl, H. (2004). Axon sorting in the optic tract requires HSPG synthesis by ext2 (dackel) and extl3 (boxer). *Neuron* **44**, 947–960.

Luo, R., An, M., Arduini, B. L., and Henion, P. D. (2001). Specific pan-neural crest expression of zebrafish Crestin throughout embryonic development. *Dev. Dyn.* **220**, 169–174.

Mayer, A. N., and Fishman, M. C. (2003). Nil per os encodes a conserved RNA recognition motif protein required for morphogenesis and cytodifferentiation of digestive organs in zebrafish. *Development* **130**, 3917–3928.

Miller, C. T., Maves, L., and Kimmel, C. B. (2004). moz regulates Hox expression and pharyngeal segmental identity in zebrafish. *Development* **131**, 2443–2461.

Miller, C. T., Schilling, T. F., Lee, K. H., Parker, J., and Kimmel, C. B. (2000). *Sucker* encodes a zebrafish Endothelin-1 required for ventral pharyngeal arch development. *Development* **127**, 3815–3828.

Miller, C. T., Swartz, M. E., Khuu, P. A., Walker, M. B., Eberhart, J., and Kimmel, C. B. (2007). Mef2ca is required for cranial neural crest to effect Endothelin1 signaling in zebrafish. *Dev. Biol.* **308**, 144–157.

Miyake, T., Cameron, A. M., and Hall, B. K. (1997). Stage-specific expression patterns of alkaline phosphatase during development of the first arch skeleton in inbred C57BL/6 mouse embryos. *J. Anat.* **190**, 239–260.

Molina, G. A., Watkins, S. C., and Tsang, M. (2007). Generation of FGF reporter transgenic zebrafish and their utility in chemical screens. *BMC Dev. Biol.* **7**, 62.

Montero-Balaguer, M., Lang, M. R., Sachdev, S. W., Knappmeyer, C., Stewart, R. A., De La Guardia, A., Hatzopoulos, A. K., and Knapik, E. W. (2006). The *mother superior* mutation ablates foxd3 activity in neural crest progenitor cells and depletes neural crest derivatives in zebrafish. *Dev. Dyn.* **235**, 3199–3212.

Moss, M. L., and Posner, A. S. (1960). X-ray diffraction of acellular telost bone. *Nature* **188**, 1037–1038.

Neuhauss, S. C., Solnica-Krezel, L., Schier, A. F., Zwartkruis, F., Stemple, D. L., Malicki, J., Abdelilah, S., Stainier, D. Y., and Driever, W. (1996). Mutations affecting craniofacial development in zebrafish. *Development* **123**, 357–367.

Nissen, R. M., Amsterdam, A., and Hopkins, N. (2006). A zebrafish screen for craniofacial mutants identifies wdr68 as a highly conserved gene required for endothelin-1 expression. *BMC Dev. Biol.* **6**, 28.

Nissen, R. M., Yan, J., Amsterdam, A., Hopkins, N., and Burgess, S. M. (2003). Zebrafish foxi one modulates cellular responses to Fgf signaling required for the integrity of ear and jaw patterning. *Development* **130**, 2543–2554.

Odenthal, J., and Nusslein-Volhard, C. (1998). fork head domain genes in zebrafish. *Dev. Genes Evol.* **208**, 245–258.

Oxtoby, E., and Jowett, T. (1993). Cloning of the zebrafish krox-20 gene (krx-20) and its expression during hindbrain development. *Nucleic Acids Res.* **21**, 1087–1095.

Parenti, L. R. (1986). The phylogenetic significance of bone types in euteleost fishes. *Zool. J. Linn. Soc.* **87**, 37–51.

Pietsch, J., Delalande, J. M., Jakaitis, B., Stensby, J. D., Dohle, S., Talbot, W. S., Raible, D. W., and Shepherd, I. T. (2005). Lessen encodes a zebrafish Trap100 required for enteric nervous system development. *Development* **133**, 395–406.

Piotrowski, T., Ahn, D. G., Schilling, T. F., Nair, S., Ruvinsky, I., Geisler, R., Rauch, G. J., Haffter, P., Zon, L. I., Zhou, Y., Foott, H., Dawid, I. B., *et al.* (2003). The zebrafish *van gogh* mutation disrupts tbx1, which is involved in the DiGeorge deletion syndrome in humans. *Development* **130**, 5043–5052.

Piotrowski, T., Schilling, T. F., Brand, M., Jiang, Y. J., Heisenberg, C. P., Beuchle, D., Grandel, H., Van Eeden, F. J. M., Furutani-Seiki, M., Granato, M., Haffter, P., Hammerschmidt, M., *et al.* (1996). Jaw and branchial arch mutants in zebrafish. II: Anterior arches and cartilage differentiation. *Development* **123**, 345–356.

Plaster, N., Sonntag, C., Busse, C. E., and Hammerschmidt, M. (2006). p53 deficiency rescues apoptosis and differentiation of multiple cell types in zebrafish flathead mutants deficient for zygotic DNA polymerase delta1. *Cell Death Differ.* **13**, 223–235.

Plaster, N., Sonntag, G., Schilling, T. F., and Hammerschmidt, M. (2007). Atrophin-2 interacts with Fgf8 signaling in a histone-acetylase dependent manner. *Dev. Dyn.* **236**, 1891.

Popperl, H., Rikhof, H., Chang, H., Haffter, P., Kimmel, C. B., and Moens, C. B. (2000). *lazarus* is a novel *pbx* gene that globally mediates *hox* gene function in zebrafish. *Mol. Cell.* **6**, 255–267.

Prince, V. E., Moens, C. B., Kimmel, C. B., and Ho, R. K. (1998). Zebrafish *hox* genes: Expression in the hindbrain region of the wild-type and mutants of the segmentation gene, *valentino*. *Development* **125**, 393–406.

Rau, M. J., Fischer, S., and Neumann, C. J. (2006). Zebrafish Trap230/Med12 is required as a coactivator for Sox9-dependent neural crest, cartilage and ear development. *Dev. Biol.* **296**, 83–93.

Rebagliati, M. R., Toyama, R., Haffter, P., and Dawid, I. B. (1998). cyclops encodes a nodal-related factor involved in midline signaling. *Proc. Natl. Acad. Sci. USA* **95**, 9932–9937.

Reifers, F., Bohli, H., Walsh, E. C., Crossley, P. H., Stainier, D. Y., and Brand, M. (1998). Fgf8 is mutated in zebrafish acerebellar (ace) mutants and is required for maintenance of midbrain-hindbrain boundary development and somitogenesis. *Development* **125**, 2381–2395.

Risheg, H., Graham, J. M., Clark, R. D., Rogers, R. C., Opitz, J. M., Moeschler, J. B., Peiffer, A. P., May, M., Joseph, S. M., Jones, J. R., Stevenson, R. E., Schwartz, C. E., *et al.* (2007). A recurrent mutation in MED12 leading to R961W causes Opotiz-Kaveggia syndrome. *Nat. Genet.* **39**, 451–453.

Sakakura, Y., Yajima, T. and Tsuruga, E. (1998). Confocal laser scanning microscopic study [corrected] of tartrate-resistant acid phosphatase-positive cells in the dental follicle during early morphogenesis of mouse embryonic molar teeth. *Arch. Oral Biol.* **43**, 353–360.

Schauerte, H. E., van Eeden, F. J., Fricke, C., Odenthal, J., Strahle, U., and Haffter, P. (1998). Sonic hedgehog is not required for the induction of medial floor plate cells in the zebrafish. *Development* **125**, 2983–2993.

Schilling, T. F. (1997). Genetic analysis of craniofacial development in the vertebrate embryo. *BioEssays* **19**, 459–468.

Schilling, T. F. (2002). The morphology of larval and adult zebrafish. *In* "Zebrafish: A Practical Approach" (R. Dahm and C. Nusslein-Volhard, eds.), pp. 59–94. Oxford University Press, Oxford.

Schilling, T. F., and Kimmel, C. B. (1997). Musculoskeletal patterning in the pharyngeal segments of the zebrafish embryo. *Development* **124**, 2945–2960.

Schilling, T. F., Piotrowski, T., Grandel, H., Brand, M., Heisenberg, C. P., Jiang, Y. J., Beuchle, D., Hammerschmidt, M., Kane, D. A., Mullins, M. C., van Eeden, F. J. M., Kelsh, R. N., *et al.* (1996). Jaw and branchial arch mutants in zebrafish. I: Branchial arches. *Development* **123**, 329–344.

Spoorendonk, K. M., Peterson-Maduro, J., Renn, J., Trowe, T., Kranenbarg, S., Winkler, C., and Schulte-Merker, S. (2008). Retinoic acid and Cyp26b1 are critical regulators of osteogenesis in the axial skeleton. *Development* **135**, 3765–3774.

Thisse, C., Thisse, B., and Postlethwait, J. H. (1995). Expression of snail2, a second member of the zebrafish snail family, in cephalic mesendoderm and presumptive neural crest of wild-type and spadetail mutant embryos. *Dev. Biol.* **172**, 86–99.

Wada, N., Javidan, Y., Nelson, S., Carney, T. J., Kelsh, R. N., and Schilling, T. F. (2005). Hedgehog signaling is required for cranial neural crest morphogenesis and chondrogenesis at the midline in the zebrafish skull. *Development* **132**, 3977–3988.

Walker, M. B., Miller, C. T., Coffin Talbot, J., Stock, D. W., and Kimmel, C. B. (2006). Zebrafish furin mutants reveal intricacies in regulating Endothelin1 signaling in craniofacial patterning. *Dev. Biol.* **295**, 194–205.

Walker, M. B., Miller, C. T., Swartz, M. E., Eberhart, J. K., and Kimmel, C. B. (2007). Phospholipase C, beta 3 is required for Endothelin1 regulation of pharyngeal arch patterning in zebrafish. *Dev. Biol.* **304**, 194–207.

Walker, M. B., and Kimmel, C. B. (2007). A two-color acid-free cartilage and bone stain for zebrafish larvae. *Biotech. Histochem.* **82**, 23–28.

Walsh, E. C., and Stainier, D. Y. (2001). UDP-glucose dehydrogenase required for cardiac valve formation in zebrafish. *Science* **293,** 1670–1673.

Wardi, A. H., and Allen, W. S. (1972). Alcian blue staining of glycoproteins. *Anal. Biochem.* **48,** 621–623.

Whiteman, P. (1973). The quantitative measurement of Alcian Blue-glycosaminoglycan complexes. *Biochem. J.* **131,** 343–350.

Willer, G. B., Lee, V. M., Gregg, R. G., and Link, B. A. (2005). Analysis of the Zebrafish perplexed mutation reveals tissue-specific roles for de novo pyrimidine synthesis during development. *Genetics* **170,** 1827–1837.

Witten, P. E., Hansen, A., and Hall, B. K. (2001). Features of mono-and multinucleated bone resorbing cells of the zebrafish Danio rerio and their contribution to skeletal development, remodeling, and growth. *J. Morphol.* **250,** 197–207.

Witten, P. E., and Villwock, W. (1997). Growth requires bone resorption at particular skeletal elements in a teleost fish with acellular bone (*Oreochromis niloticus*, Cichlidae). *J. Appl. Ichtyol.* **13,** 149–158.

Xu, Q., Alldus, G., Macdonald, R., Wilkinson, D. G., and Holder, N. (1996). Function of the Ephrelated kinase rtk1 in patterning of the zebrafish forebrain. *Nature* **381,** 319–322.

Yamaguchi, M., Fujimori-Tonou, N., Yoshimura, Y., Kishi, T., Okamoto, H., and Masai, I. (2008). Mutation of DNA primase causes extensive apoptosis of retinal neurons through the activation of DNA damage checkpoint and tumor suppressor p53. *Development* **135,** 1247–1257.

Yan, Y. L., Hatta, K., Riggleman, B., and Postlethwait, J. H. (1995). Expression of a type-II collagen gene in the zebrafish embryonic axis. *Dev. Dyn.* **203,** 363–376.

Yan, Y. L., Miller, C. T., Nissen, R. M., Singer, A., Liu, D., Kirn, A., Draper, B., Willoughby, J., Morcos, P. A., Amsterdam, A., Chung, B. C., Westerfield, M., *et al.* (2002). A zebrafish sox9 gene required for cartilage morphogenesis. *Development* **129,** 5065–5079.

Yan, Y. L., Willoughby, J., Liu, D., Crump, J. G., Wilson, C., Miller, C. T., Singer, A., Kimmel, C., Westerfield, M., and Postlethwait, J. H. (2005). A pair of Sox: Distinct and overlapping functions of zebrafish sox9 co-orthologs in craniofacial and pectoral fin development. *Development* **132,** 1069–1083.

Yang, C. T., Hindes, A. E., Hultman, K. A., and Johnson, S. L. (2007). Mutations in gfpt1 and skiv2l2 cause distinct stage-specific defects in larval melanocyte regeneration in zebrafish. *PLoS Genet.* **3,** e88.

Yelon, D., Ticho, B., Halpern, M. E., Ruvinsky, I., Ho, R. K., Silver, L. M., and Stainier, D. Y. (2000). The bHLH transcription factor hand2 plays parallel roles in zebrafish heart and pectoral fin development. *Development* **127,** 2573–2582.

Zhang, J., Talbot, W. S., and Schier, A. F. (1998). Positional cloning identifies zebrafish *one-eyed pinhead* as a permissive EGF-related ligand required during gastrulation. *Cell* **92,** 241–251.

CHAPTER 19

Morphogenesis of the Jaw: Development Beyond the Embryo

R. Craig Albertson* and Pamela C. Yelick[†]

*Department of Biology
Syracuse University
Syracuse, New York 13244

[†]Department of Oral and Maxillofacial Pathology
Tufts University
Boston, Massachusetts 02111

I. Larval Zebrafish Craniofacial Cartilage Development

Since the mid-1990s, the zebrafish model has been used to provide significant insight into the molecular/genetic pathways guiding craniofacial development. Analyses of wild-type zebrafish and the identification and characterization of classes of phenotypically related zebrafish craniofacial mutants have revealed important signaling cascades and tissue interactions that are required for the proper placement, patterning, and differentiation of craniofacial cartilages (Kimmel *et al.*, 1998; Neuhauss *et al.*, 1996; Piotrowski *et al.*, 1996; Schilling *et al.*, 1996; Yelick and Schilling, 2002). As discussed in more detail by Javidan and Schilling in Chapter 18 of this volume, the recent molecular characterization of

DOI: 10.1016/B978-0-12-374599-6.00019-4

five previously identified zebrafish craniofacial mutants has revealed the importance of signals derived from the endothelium (and of certain transcription factors
mediating these signals) for proper craniofacial development. Correlation of the
phenotypes of the *suc/et1* zebrafish mutant, caused by a point mutation in the
endothelin-1 gene (Miller *et al.*, 2000), with the mouse mutants *piebald (s)*, which
encode a B-type endothelin receptor (Hosoda *et al.*, 1994), and *lethal spotting (ls)*,
which encodes the endothelin 3 ligand (Baynash *et al.*, 1994), has demonstrated the
importance of the endodermally derived endothelin signaling for proper dorsoventral patterning and differentiation of neural crest cells (NCCs) that give rise to
craniofacial cartilage structures. The *suc/et1* phenotype closely resembles human
DiGeorge (DGS) and velocardiofacial syndrome phenotypes, thereby providing a
useful model for analyses of these human diseases (Kurihara *et al.*, 1995; Paylor
et al., 2001). Roles for endothelin signaling pathways in the later developmental
events of mineralized craniofacial skeletal elements were also revealed, as discussed
later.

Analyses of the craniofacial transcription factor mutants *colorless (cls)/sox10,
van Gough (vgo)*, and *lockjaw (low)/tfap2a* have revealed potential genetic interactions of these genes in dorsoventral patterning of the craniofacial complex.
The combined pigment and enteric neuronal defects exhibited by *cls/sox10* mutants
resemble those of the already mentioned mouse mutants *piebald (s)* and
lethal spotting (ls) and the *Dominant megacolon (Dom)* mutant, which encodes
Sox10—the functional homolog of *cls* (Herbarth *et al.*, 1998; Southard-Smith
et al., 1998). The *cls/sox10* phenotype also bears a striking resemblance to the
human Waardenburg-Shah syndrome and Hirschsprung's disease (Hassinger
et al., 1980).

The ventral arch mutant *van Gough (vgo)*, caused by a mutation in the T-Box
family member, *tbx1* (Piotrowski *et al.*, 2003), exhibits severe craniofacial skeletal
segmentation defects where cartilages of adjacent arches often fuse. Mutations in
the *tbx1* gene are thought to be major contributors to the cardiovascular defects in
human DGS patients, a disease affecting several NCC derivatives of the pharyngeal arches (Kurihara *et al.*, 1995). Analysis of the *vgo/tbx1* zebrafish mutant
demonstrated that *tbx1* acts cell autonomously in the pharyngeal mesendoderm
to secondarily influence the development of NCC-derived cartilages. This study
also identified regulatory interactions between *vgo/tbx1, ednl*, and *hand2*, genes
also implicated in DGS.

An additional mutant, *lockjaw(low)*, caused by a mutation in the *ap2a* gene,
exhibits NCC-derived skeletal and pigment defects (Knight *et al.*, 2003). Studies of
low suggest a model where *tfap2a* functions independently in the specification of
subpopulations of NCC-derived pigment cells and in patterning of the pharyngeal
skeleton through the regulation of Hox genes.

Together, the successful application of a forward genetic approach to identify
and characterize genes essential for early zebrafish craniofacial development has

proven extremely beneficial to increasing our understanding of the molecular genetic signals regulating normal human craniofacial development, and in elucidation of molecular defects in human craniofacial syndromes including DGS/VCs, Waardenburg-Shah, and Hirschsprung's disease. In Chapter 18 of this volume, many of the techniques that have been perfected to facilitate analyses of early zebrafish craniofacial cartilage growth and patterning up to approximately 5 days of development are described in detail. In this chapter, we describe methods used to characterize the molecular/genetic signaling events regulating later stages of craniofacial development, including the growth and differentiation of the craniofacial skeleton, and of replacement tooth development.

II. Analysis of Craniofacial Skeletal and Replacement Tooth Development

Forward genetic mutagenesis screens in zebrafish have provided a detailed knowledge of the molecular signaling cascades regulating *early* embryonic events. The efficiency of this method, however, is simultaneously a limitation. The vast majority of described zebrafish craniofacial mutants exhibit gross qualitative defects and are generally lethal by 5–7 dpf (Neuhauss *et al.*, 1996; Piotrowski *et al.*, 1996; Schilling *et al.*, 1996; Yelick and Schilling, 2002). At this developmental stage, zebrafish normally have begun to mineralize a bilateral series of 10–11 pharyngeal bones and exhibit 1–3 teeth in each bilateral ceratobranchial 5 (cb5) arch. Bones and teeth present in the 7-dpf larval zebrafish are illustrated in Fig. 1, panel C. Two dermal and two cartilage replacement bones are derived from the mandibular arch, three dermal and two endochondral bones arise from the hyoid arch; whereas the only mineralized structure derived from more posterior arches at this stage is cb5. Since these represent only a fraction of the more than 50 bones and 11 bilateral teeth present in adult zebrafish, previously identified homozygous recessive craniofacial mutants are not useful for developmental analyses of the vast majority of skeletal and tooth structures (Cubbage and Mabee, 1996). Therefore, the early lethal phenotypes exhibited by the majority of previously identified craniofacial mutants have resulted in a substantial gap in our knowledge of the genetic control of *later* developmental events including: (1) primary and replacement tooth development, (2) development of the craniofacial skeleton, and (3) growth and remodeling of the craniofacial complex.

In the sections that follow, we discuss experimental approaches that can be used to facilitate analyses of the later developmental events of skeletogenesis and replacement tooth development. To begin, we review the incipient body of literature composing our current understanding of the molecular basis of craniofacial skeletal and pharyngeal tooth development. Next, we propose a new paradigm to use in the characterization and interpretation of the molecular events regulating bone and tooth development in zebrafish.

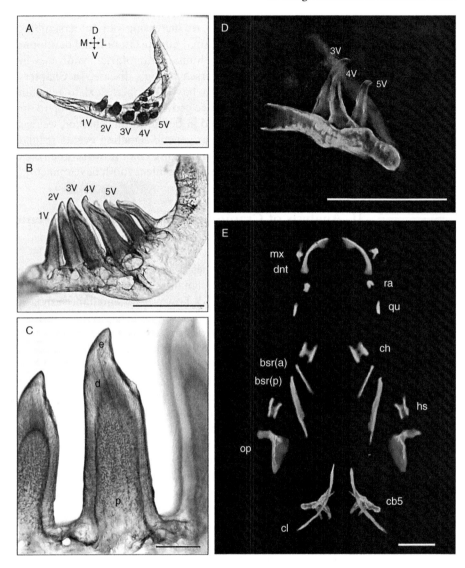

Fig. 1 Adult pharyngeal teeth and larval mineralized skeleton. (A–C) Dissected and Alizarin red-stained cb5 arch and teeth of the adult zebrafish. The large ventral row of pharyngeal teeth is labeled IV–5V. Adult zebrafish also possess discrete medial and dorsal rows of teeth (*not labeled*). Dorsal, ventral, medial, and lateral are designated D, V, M, and L, respectively. (C) High magnification of 2V shows the pulp cavity (p), dentin tubules (d), and enameloid (e). (D) Quercetin-stained cb5 in wild-type 7-dpf larvae with three pharyngeal teeth, 3V–5V. (E) Quercetin-stained mineralized pharyngeal skeleton in 7-dpf larval zebrafish viewed under ultraviolet (UV) illumination. Dermal bones are labeled to the left. Cartilage replacement bones are to the right. Scale bars in A and B represent 0.5 mm. Bars in C–E represent 100 μm. Abbreviations are as follows: mx, maxilla; dnt, dentary; ra, retroarticular; qu, quadrate; ch, ceratohyal; bsr(a), anterior branchiostegal ray; posterior branchiostegal ray; hs, hyosymplectic; op, opercle; cl, cleithrum; cb5, ceratobranchial 5. (See Plate no. 14 in the Color Plate Section.)

A. Zebrafish Pharyngeal Tooth Development

Like other cyprinid fishes, zebrafish lack teeth in their oral jaws but rather exhibit a series of bilaterally symmetric teeth on each cb5 arch (Fig. 1). These pharyngeal teeth exhibit all of the characteristics of teeth present in "higher" vertebrates, including a pulp cavity, dentin, and enameloid surface. Recent morphological and histological studies have described the pattern of tooth development in the larval zebrafish (Perrino and Yelick, 2004; Van der Heyden and Huysseune, 2000; Wautier et al., 2001), and the regulation of adult replacement tooth formation (Huysseune and Sire, 2003; Perrino and Yelick, 2004; Van der Heyden et al., 2001, 2000). Adult zebrafish teeth closely resemble mammalian teeth, exhibiting distinct pulp cavity, dentin, and enameloid tissues (Fig. 1, panel B). Zebrafish teeth arise in a synchronized fashion to form three distinct dorsoventral rows (Fig. 1, panel A). The development of each tooth consists of distinct morphological stages similar to those of mammalian teeth, including initiation, morphogenesis, and cytodifferentiation stages (Thesleff, 2003; Thesleff et al., 1991).

Distinct from mammalian tooth development, zebrafish pharyngeal teeth undergo continuous tooth replacement, characterized by coordinated shedding and attachment stages. Where an older functional tooth is shed, the socket becomes remodeled, and a newly formed replacement tooth is subsequently secured to the cb5 skeletal element. Zebrafish teeth are continuously replaced throughout the life of the zebrafish, providing an opportune model to explore the molecular/genetic signaling cascades regulating both primary *and* replacement tooth development.

Although brief descriptions of tooth defects were included in analyses of zebrafish pharyngeal arch mutants described in the original two large-scale zebrafish mutagenesis screens (Neuhauss et al., 1996; Piotrowski et al., 1996; Schilling et al., 1996), we are unaware of the existence of any tooth-specific mutant. In mammals, the major gene families involved in mammalian tooth specification—bone morphogenetic protein (Bmp), Wnt, Hedgehog (Shh), fibroblast growth factor (Fgf), and TNF-alpha families—are reiteratively expressed throughout successive stages of tooth development. Many of these genes are also expressed in the tooth-bearing arches of the zebrafish (for review, see Yelick and Schilling (2002)). Unfortunately, functional characterizations of genes regulating tooth development are less straightforward. In zebrafish, as in all other vertebrates, many of the growth factors that regulate tooth development are also required in very early patterning events of the embryo. Thus, mutations in these genes often result in early lethal phenotypes, resulting in the death of the developing embryo before the initiation or during very early stages, of tooth development. The identification of tooth-specific zebrafish mutants, an ongoing effort in our laboratory, would greatly enhance the understanding of the molecular pathways particularly involved in tooth specification and morphogenesis.

B. Bone Development

Skeletal development in the zebrafish is regulated in much the same way as in other vertebrates, by balanced activities of bone formation by osteoblasts and bone resorption by osteoclasts (Witten *et al.*, 2001). Descriptions of the development of the bony skeleton of zebrafish have been reported (Bird and Mabee, 2003; Cubbage and Mabee, 1996). However, the molecular characterization of this process is virtually nonexistent at this time. Since the mid-1990s, we have seen admirable advances in our understanding of the molecular mechanisms that contribute to the early development of the zebrafish pharyngeal arch cartilages (Kimmel *et al.*, 2001; Miller *et al.*, 2000, 2003; Piotrowski *et al.*, 1996; Schilling *et al.*, 1996). Unfortunately, the relatively late formation of the bony skeleton has precluded its characterization in most analyses of zebrafish homozygous recessive craniofacial mutants. This notable lack of molecular characterization of mineralized tissue formation in the zebrafish has prompted a number of laboratories, including ours, to focus on the elucidation of the molecular/genetic signaling pathways regulating these processes.

Reports of bone development in zebrafish provide insight into certain molecular signals regulating zebrafish skeletal patterning and growth. One study took advantage of the observation that a percentage of *chordin (chd)* homozygous recessive mutants survived to become fertile adults (Fisher and Halpern, 1999). Chordin is a bone morphogenic protein (Bmp) antagonist that plays an important role in the development of the vertebrate gastrula (Miller-Bertoglio *et al.*, 1997; Thomsen, 1997). Mutants deficient in *chordin* exhibit expanded *bmp4* expression in the gastrula, resulting in ventralized phenotypes (Hammerschmidt *et al.*, 1996). While most *chd* mutants exhibited early lethal phenotypes, a small percentage developed into fertile adults (Fisher and Halpern, 1999). Between 4 and 50% of *chd* mutants survive to sexual maturity depending on the allele, providing a unique opportunity to investigate the effects of *chordin* on the later development of bone (Fisher and Halpern, 1999). It was discovered that *chordin*-deficient mutants exhibited defects in axial and caudal skeletal patterning, including the absence, branching, or fusion of the bony processes of the vertebrae and fins (Fisher and Halpern, 1999). These phenotypes were correlated with the ectopic and expanded expression of *bmp4* and its downstream target *msxC*, suggesting a role for *chordin*-mediated BMP activity in skeletal patterning. Interestingly, the craniofacial skeleton was reported as having no noticeable defects.

Another study explored the role of the secreted peptide Endothelin 1 (*edn1*) in regulating pharyngeal bone development in the zebrafish (Kimmel *et al.*, 2003). As already mentioned, the *N*-ethyl-*N*-nitrosourea (ENU)-induced mutant *sucker (suc)*, a point mutation in the *endothelin 1 (edn1)* gene, was identified in a genetic screen for pharyngeal cartilage development (Miller *et al.*, 2000; Piotrowski *et al.*, 1996). Zebrafish *edn1* is expressed in a central core of arch paraxial mesoderm, in both surface ectoderm and pharyngeal endodermal epithelia, and not in skeletogenic neural crest. However, zebrafish *edn1* is thought to act directly on postmigratory NCCs (Kimmel *et al.*, 2003; Miller *et al.*, 2000). Defects exhibited by *suc* mutants suggest *edn1* is required for dorsal-ventral patterning of the first and second pharyngeal arch cartilages (Miller and Kimmel, 2001; Miller *et al.*, 2000; Piotrowski *et al.*, 1996). *Suc* mutants are characterized by fusion of the dorsal and ventral cartilages in both the first and second pharyngeal arches (Miller *et al.*, 2000). The anterior-posterior polarity of the ventral first arch Meckel's cartilage is also reversed, in that it is directed backward (Miller *et al.*, 2000). Similar defects are observed in the dermal bones of *suc* larvae (Kimmel *et al.*, 2003). In the first arch, the bones of the upper and lower jaw are fused and reversed in polarity. The effects of *edn1* deficiency on the dermal bones of the second arch are more dynamic, including bone loss, gain, and fusion, in a manner that is dependent on Edn1 levels (Kimmel *et al.*, 2003). When Edn1 was severely reduced (e.g., using antisense morpholino oligomer targeted depletion strategies) second arch bones were generally reduced or missing. Alternatively, when Edn1 was only mildly reduced, second arch bones were more likely to develop, and the dorsal element (opercle) was often increased in size (Kimmel *et al.*, 2003). Mild reduction of Edn1 also led to a fusion of the dorsal and ventral bones of the hyoid arch, revealing defects in dorsoventral patterning. The range of bone defects observed in *suc* mutants suggests that Edn1 patterns the pharyngeal skeleton through morphogenic gradients, and that pharyngeal bones are acutely sensitive to Edn1 protein levels (Kimmel *et al.*, 2003). Unfortunately, because of the severe nature of their defects, *suc* mutant larvae do not survive much beyond 1 week postfertilization. Thus, the role of Edn1 in regulating bone remodeling and growth remains unknown.

A large-scale screen for skeletal dysplasias in adult ENU-mutagenized zebrafish has been performed (Fisher *et al.*, 2003). Radiography was used to screen living F_1 adults for dominant skeletal mutations. Out of 2000 F_1 fish, only one skeletal mutant was recovered, the *chihuahua*^{dc124} *(chi)* mutant. *Chi* was identified as a dominant mutation that transmitted the mutant phenotype to approximately half of its progeny. Adult (*chi/+*) fish appear shorter than their wild-type siblings, but are otherwise morphologically indistinguishable from a gross perspective. In contrast, examination of the skeletal anatomy of adult *(chi/+)* fish using radiographic methods revealed extensive skeletal dysplasia characterized by irregular bone growth, uneven mineralization, and bone weakness as revealed by the presence of multiple bone fractures (Fisher *et al.*, 2003). Many of the defining characteristics of *chi* mutants are similar to individuals exhibiting osteogenesis imperfecta (OI), a dominantly inherited skeletal dysplasia typically caused by mutations in one of two type I collagen genes, COL1A1 or COL1A2 (Benusiene and Kucinskas, 2003). It was therefore not surprising that *chi* mapped to the same interval on LG3 as *col1a1* (Fisher *et al.*, 2003). Although *chi/chi* homozygous mutants failed to develop swim bladders or feed after 1 week, they were reported to be indistinguishable from age-matched *chi/+* siblings (Fisher *et al.*, 2003). This study by Fisher *et al.* (2003) demonstrates the utility of radiography as a high-throughput screening method for skeletal defects. However, the relatively low frequency that bone-specific mutants were recovered (i.e., 1/2000 or 0.05%) is less encouraging. It will be interesting to further define the effects of the *col1a1* mutation, as present in the *chi* mutant, on bone growth and remodeling.

To date, only one report has focused specifically on characterizing bone growth and remodeling in the zebrafish. Witten *et al.* (2001) endeavored to characterize the manner by which osteoclast activity contributes to skeletal growth and development in zebrafish. Bone remodeling is intimately associated with bone growth (Olsen *et al.*, 2000). Since the zebrafish skeleton continues to grow throughout the life of the fish, osteoclast activity in developing bone was studied from larval to juvenile stages of development to provide an understanding of the mechanisms that regulate this process (Witten *et al.*, 2001). In zebrafish, bone remodeling is initiated between 2 and 3 weeks postfertilization via activation of mononucleated osteoclasts located at the surface of bones poised to undergo the process of remodeling (Witten *et al.*, 2001). As development and growth proceeds, both mononucleated and multinucleated osteoclasts contribute to the process of bone resorption. In general, multinucleated cells were found in association with more robust bones and deep absorption pits: whereas mononucleated osteoclasts were found in association with thin bones and shallow resorption pits (Witten *et al.*, 2001). While the pattern of bone remodeling in zebrafish appears to be consistent with that in mammals, the regulation of osteoclast activity in zebrafish remains largely unknown and may differ from that in mammals (Witten *et al.*, 2001). Zebrafish, for example, have three parathyroid hormone (PTH) receptors (zPTH1R, zPTH2R, and zPTH3R), and at least two ligands (zPTH1 and zPTH2), suggesting a potentially more complex PTH regulatory system in fishes than in mammals (Gensure *et al.*, 2004). Understanding the fundamental processes of osteoclast- and osteoclast-regulated bone remodeling in teleosts is central to understanding the molecular determinants of bone growth and shape in zebrafish.

C. Paradigm Shift: Analyses of Postlarval Tooth and Bone Development

Since the mid-1990s, we have seen significant advances in defining molecular mechanisms that pattern the zebrafish embryo. Significantly less is known of zebrafish development beyond early embryonic stages. This is not for lack of interest, but rather is because many of the tools and methods used to analyze early embryonic processes exhibit limited utility for that of later developmental stages. We argue that to better understand the molecular interactions contributing to development beyond the embryo, traditional techniques must be bolstered by the introduction of new methodologies. Specifically, studies in zebrafish aimed at characterizing the development and growth of bones and teeth would be greatly facilitated by the following: (1) genetic screens targeted for later developmental processes (e.g., tooth development, regeneration, and replacement and bone formation, growth, and remodeling); (2) a shift from *qualitative* to *quantitative* characterizations of phenotypes; and (3) transition away from assessment of defects at isolated developmental stages, to the continuous monitoring of mutant phenotypes throughout extended developmental periods.

An elegant example of the first approach was described in the study by Fisher *et al.* (2003). Similarly, we are currently conducting an ENU-mutagenesis screen for bone and tooth phenotypes in larval and juvenile zebrafish. Our screen begins at 7 dpf, when bones first appear, and continues through 1 month postfertilization. We are screening F_3 families for homozygous recessive and heterozygous mutations, using both an *in vivo* Quercetin fluorescent stain (Sigma, St. Louis, MO), and Alizarin red/Alcian blue staining methods (modified from Potthoff (1984)). The Quercetin and Alizarin red stains both bind free Ca^{++}, and therefore label mineralized tissues (e.g., bone and teeth). At 7 dpf, we screen for the appearance of bone and the development of primary teeth. At 1 month, we examine bone growth, as well as the development and patterning of replacement teeth.

Our screening method is simple, robust, and noninvasive. At each developmental stage, F_3 families are placed in a 0.1 mg/ml solution of Quercetin in system water for 1 to 3 h. The fish are then anesthetized and screened under a Zeiss M2-Bio dissecting microscope fitted with fluorescence. For long-term storage, zebrafish are fixed in buffered 4% paraformaldehyde (PFA), enzymatically cleared with trypsin, stained with Alizarin red in 0.5% KOH, and stored in Glycerol. Using this screening method, we have begun to recover mutants with cartilage, bone, and tooth phenotypes. Some of the newly identified homozygous recessive mutants exhibit severe defects in pharyngeal cartilage development similar to those described in previous mutagenesis screens (Neuhauss *et al.*, 1996; Piotrowski *et al.*, 1996; Schilling *et al.*, 1996; Yelick and Schilling, 2002), making defects in bone and tooth development difficult to interpret. In contrast, analysis of heterozygous mutants has revealed haploinsufficiency bone and tooth phenotypes that appear quite interesting. We have also recovered mutants that exhibit more specific and subtle defects. One such example is shown in Fig. 2. This putative mutant survives well into larval development (i.e., more than 7 dpf), and is distinguished from wild-type siblings by a shortened head and misshapen eyes (Fig. 2, panels A and D). Examination of dermal bones at 8 dpf revealed defects that appear largely restricted to anterior pharyngeal skeletal elements (Fig. 2, panels B and E). First-arch bones appear shorter than those of wild-type siblings, while there is no noticeable difference in the length of second-arch bones. Similar defects are observed in pharyngeal arch cartilages (Fig. 2, panels C and F). In fact, there is a quantifiable difference in the length of Meckel's cartilage (Fig. 2, panel G), but not of the ceratohyal cartilage (Fig. 2, panel H). Thus, this mutation appears to specifically affect the cartilages and dermal bones of the mandibular arch. Subtle defects in skeletal anatomy, such as those exhibited by this mutant, can also be quantified using powerful geometric morphometric (GM) techniques to reveal regional differences in the size and shape of bony structures.

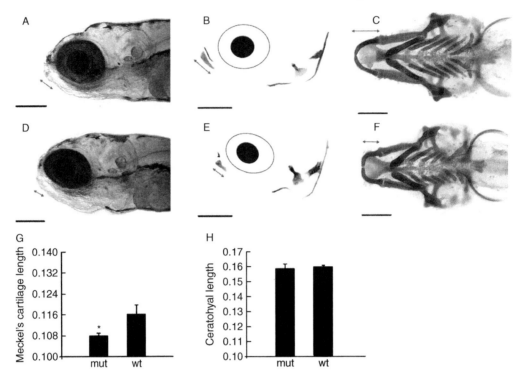

Fig. 2 Characterization of subtle pharyngeal skeletal defects. Eight-dpf wild-type (WT) (A–C) and mutant (D–F) larvae. Mutants are distinguished from WT siblings by a misshaped eye and shortened anterior head structures (A vs D). In the lateral view, first-arch dermal bones appear shorter than those of WT siblings (B vs E). The chondrocranium of mutant zebrafish also exhibits similar defects (C vs F). Meckel's cartilage is significantly shorter in mutant animals (G), while no significant difference in the length of the ceratohyal is detected (H).

D. Use of Geometric Morphometric Tools for Quantification of Skeletal Defects

GM is an exceedingly powerful tool with which to describe shape variation (Rohlf and Marcus, 1993). A geometric approach to shape analysis is based on the establishment of landmark data, which are homologous anatomical points recorded as a Cartesian coordinate system. Instead of reporting shape as lengths, widths, or angles, a landmark-based approach emphasizes the geometry of a given structure, allowing shape variation to be described relative to other structures. In this way, GM is an ideal method with which to quantify regional defects in experimentally manipulated animals (e.g., genetic mutants). Moreover, geometric descriptors of shape can be reported via pictorial representations of the organism/ structure, providing an intuitive and precise biological representation of the observed defects.

The interested reader is encouraged to review Bookstein (1991) and Rohlf and Marcus (1993) for a more thorough description of concepts and methods in GM. In general, geometric methods of shape analysis involve three steps: (1) superimposition of landmark data; (2) decomposition of shape variation into a series of geometric variables; and (3) statistical analysis of those variables. The purpose of the first step is to eliminate variance introduced by orientation, position, and size of the specimens as they are measured. Many superimposition algorithms have been developed. A least-squares approach, which is perhaps the most common, superimposes landmark configurations such that the sum of the squared distances between corresponding landmarks is minimized. Thin-plate spline (TPS) analysis is a geometric technique that quantifies D'Arcy Thompson's concept of Cartesian grid deformations (Thompson, 1917). Deformation grids are constructed from two landmark configurations—the starting form (e.g., mean wild-type shape) and the target form (e.g., mean mutant shape). For morphometrics analysis, the starting form is constrained at some combination of points (i.e., landmarks), but is otherwise free to adopt the target form in a way that minimizes bending energy. This total deformation of the Cartesian grid is then decomposed into a series of geometrically orthogonal variables based on scale (Rohlf and Marcus, 1993). These variables (called partial warps) can be localized to describe precisely what aspects of shape differ between the starting and target forms. When amassing large series of variables, it is often desirable to define major axes of variation. Statistically, partial warps can be treated in the same way as any other variable. Thus, partial warps are perfectly amenable to data reduction analyses such as canonical variate or principal component analysis. The advantage of using geometric variables is that instead of viewing results in terms of graphs and tables, data may be reported as deformation grids. Again, this offers the advantage of direct biological interpretation of the results.

Equipped with these tools and an eye for subtle defects, we can explore cartilage and bone defects in previously identified and in new ENU-induced zebrafish mutants. An example of the application of GM analysis is presented in Fig. 3. Here we have used GM analyses to examine cartilage defects in the larval mutant *acerebellar* (*ace*), which is a mutation in the *fgf8* gene. The *ace/fgf8* mutant was originally classified as a neural mutant based on the lack of midbrain structures (Brand *et al.*, 1996). Since that time, efforts have been made to characterize pharyngeal cartilage defects in *ace* and other *fgf*-deficient mutants (David *et al.*, 2002; Walshe and Mason, 2003). Fig. 3, panel A illustrates the pharyngeal skeleton of wild-type 7-dpf zebrafish larvae in the ventral view. Landmarks that capture variation in first- and second-arch cartilages are shown in red, and are labeled 1-9. Age-matched wild-type sibling and *ace/fgf8* mutant larvae are shown (Fig. 3, panels B and C). Note that the *ace/fgf8* mutant heads apparently shorten (Fig. 3, panel C). Geometric shape analysis was performed on a family derived from an incross of two identified heterozygous *ace/fgf8* adults (Fig. 3, panels D–G). Landmark variation among specimens is illustrated in Fig. 3, panel D. Medially placed landmarks (landmark numbers 1, 5, and 6) exhibit distinct anterior-posterior

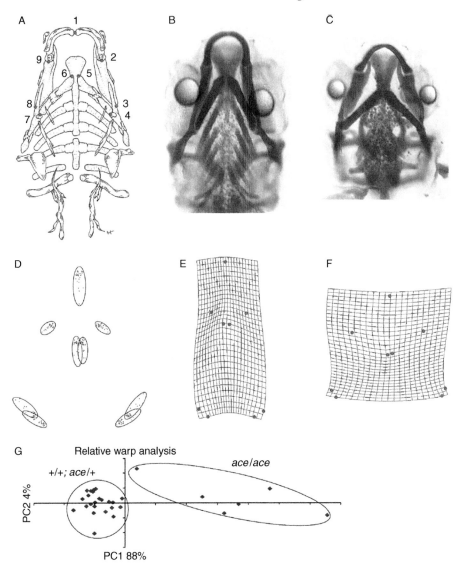

Fig. 3 Shape differences in wild-type (WT) and mutant skeletal elements. Illustration of the 7-dpf WT zebrafish pharyngeal skeleton in the ventral view (A). 7-dpf WT (B) and ace (C) larvae. Note the shortened mutant head. (D–G) Geometric morphometrics analysis of skeletal defects. (D) Variation in landmark configuration after superimposition. (E, F) Deformation of shape along the first principal component axis (PC1). WT shape is depicted in (E); mutant shape in (F). Mutants exhibit greater variance along PC1 (G). (See Plate no. 15 in the Color Plate Section.)

variation; whereas laterally positioned landmark numbers 2, 3, 4, 7, 8, and 9 exhibit distinct variation along the medial-lateral axis. Results of a principal-components analysis on partial warp scores (formally referred to as *relative warp analysis*) are shown (Fig. 3, panels E–G). Deformation grids representing deviation in shape along the first principal component axis explain 88% of the total phenotypic variation (Fig. 3, panels E and F). Mutant animals plot on one side of principal component (PC1), while wild-type siblings plot on the opposite side. As expected, mutants exhibit much greater variation along PC1 than do wild-type sibling larvae (Fig. 3, panel G). Shape variation along PC1 reveals pronounced shortening of both first and second ventral arch cartilages in *ace/fgf8* mutants (Fig. 3, panel F). These elements are also splayed laterally, resulting in a widening of the pharyngeal skeleton. (Note the horizontal displacement of landmarks 2, 3, and 4 and 7, 8, and 9.) As a consequence, first- and second-arch cartilages have become distanced along the anterior-posterior axis. (Note the vertical displacement of landmarks 5 and 6 relative to landmarks 2 and 9.)

Interestingly, shape variation in the wild type versus mutant analysis is similar to that exhibited by a comparison of homozygous *ace/fgf8* larvae at 5 and 7 dpf (Fig. 4). Wild-type and *ace/fgf8* mutants at 7 dpf, and 5-dpf *ace/fgf8* mutant larvae are presented (Fig. 4, panels A-C, respectively). Figure 4, panels D-F represent deformations of shape along the first principal component axis, which explains 94% of the total shape variation (Fig. 4, panel G). Deformation grids represent the mean shape of 7-dpf wild-type larvae (Fig. 4, panel D), 7-dpf *ace/fgf8* mutants (Fig. 4, panel E) and 5-dpf *ace/fgf8* mutants (Fig. 4, panel F). The geometry of shape in this analysis is strikingly similar to that presented in the previous example (Fig. 3), in that both exhibit the displacement of the same sets of landmarks relative to one another. Moreover, only one major axis of shape variation (PC1) exists. That is to say, wild-type animals lie along the same trajectory as the mutants from two developmental stages. Altogether, these observations suggest that defects exhibited by *ace/fgf8* mutants are associated with the growth of the pharyngeal skeleton, consistent with the reported roles for fibroblast growth factor (FGF)-signaling in regulating patterning and growth of the anterior region of the vertebrate craniofacial complex (Abu-Issa *et al.*, 2002; Bachler and Neubuser, 2001; Frank *et al.*, 2002; Mina *et al.*, 2002; Tucker *et al.*, 1999).

The examples described demonstrate the utility of GM in quantifying shape defects in mutant zebrafish. Because this approach is dependent on the accurate placement of homologous landmarks among specimens, GM analyses are not useful when defects are so severe that cartilages are missing or grossly misplaced. However, when defects are subtle, this is a consummate technique with which to describe and quantify regional affects of genetic deficiency. As shown in Fig. 4, GM is also a particularly useful tool to describe growth of the craniofacial complex of wild-type and mutant zebrafish. In contrast to the considerable amount of research on larval pharyngeal cartilage development, to date no published studies describe growth of the zebrafish pharyngeal skeleton. A morphometric characterization of wild-type skeletal growth would significantly extend previous studies by

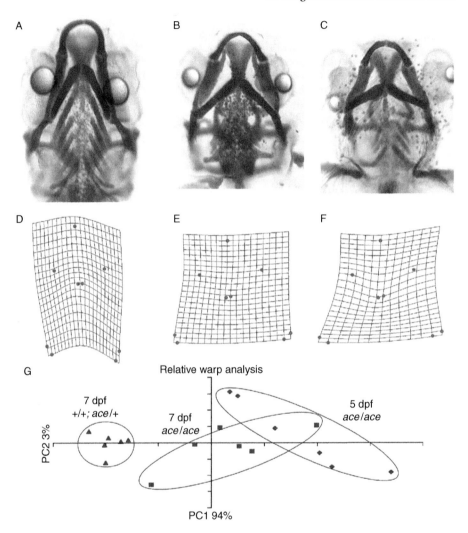

Fig. 4 Geometry and growth of wild-type (WT) and mutant pharyngeal cartilage defects. Representative specimens are shown in (A–C). (A) WT 7-dpf larvae, (B) mutant 7-dpf larvae, and (C) mutant 5-dpf larvae. Shape deformations are illustrated in (D, E), when pharyngeal skeletal defects in both 7-dpf (E) and 5-dpf (F) *ace* are evaluated against 7-dpf WT larvae (D). The similarity in geometry suggests the *ace* mutation effects growth of the pharyngeal skeleton. All specimens plot on the same principal component axis (PC1), which explains 94% of the total shape variation (G). (See Plate no. 16 in the Color Plate Section.)

providing fundamental knowledge of how adult skeletal form is achieved. Moreover, such a study would provide a reference standard for comparison to that of identified mutants. Discrete changes in the slopes of the growth trajectories of mutants as compared with wild-type siblings could be anatomically localized by GM to identify regional defects of individual gene mutations.

As already mentioned, GM can be used to explore the shape and growth of skeletal elements in previously described heterozygous and nonlethal zebrafish mutants. An appropriate mutant for such analysis is the zebrafish mutant *ikarus* (*ika*), which encodes a *fgf24* mutation (Fischer *et al.*, 2003). Homozygous recessive *ika/fgf24* mutants completely lack pectoral fins, but otherwise appear normal and live to reproductive maturity (Fischer *et al.*, 2003). Another suitable candidate is the *somitabun* (*sbn*) mutant, originally identified in a screen for dorsal-ventral defects in the zebrafish embryo (Mullins *et al.*, 1996). *Sbn/Smad5* mutants exhibit a dominant effect with varying penetrance due to a single amino acid substitution in the Smad5 protein (Hild *et al.*, 1999). Homozygous *sbn/Smad5* embryos are severely dorsalized and die before 3 dpf (Mullins *et al.*, 1996). Embryos that possess one *sbn/Smad5* allele survive to adulthood, but have notable posterior axial skeletal defects including short or no caudal fin (Mullins *et al.*, 1996). The *chi* mutation, discussed earlier, is another example of a dominant mutation with effects on skeletal anatomy. Paradoxically, all of the genes encoded by the mutations delineated earlier are expressed in the developing head, yet no craniofacial defects have been reported. It is possible that compensatory gene expression may alleviate craniofacial defects to a certain extent. We suggest that statistical analyses of the craniofacial complex using sensitive GM tools may reveal previously unidentified defects. Quantification of such phenotypes in heterozygous and nonlethal homozygous mutants could eventually facilitate an understanding of the molecules that regulate pharyngeal skeletal shape and growth.

III. Conclusion

Morphogenesis can be defined as *the processes that are responsible for producing the complex shapes of* adults *from the simple ball of cells that derives from division of the fertilized egg* (On-line Medical Dictionary, © 1997-98 Academic Medical Publishing & CancerWEB). By this definition, a large gap exists in our current knowledge of the morphogenesis of the zebrafish craniofacial complex. In stark contrast to an extensive body of literature that analyzes early craniofacial cartilage development, very few studies in zebrafish have examined stages beyond 5 dpf. Since extensive morphogenetic changes occur after 5 dpf, the existing challenge for developmental biologists is to extend the current genetic paradigm beyond the embryo. Recently developed tools, including *in vivo* mineralized tissue stains combined with sensitive GM analyses described earlier, provide the necessary means to elucidate the molecules and pathways that produce shape in the adult craniofacial skeletal complex. It is likely that the application of these methods for the analyses of later-staged zebrafish development will provide significant insight into the regulation of human craniofacial development.

IV. Updates and Recent Advances

In the years following the original publication of this review in 2004 there have been numerous publications that have advanced our understanding of jaw morphogenesis. In keeping with our original theme, here we shall briefly highlight those papers that distinguish themselves from other works by expanding the field of developmental genetics in one of several ways: through the consideration of variance and the application of quantitative methods in experimental studies of morphogenesis; by focusing on development beyond embryonic stages; and through the use of nontraditional model systems to augment work in zebrafish to advance our understanding of craniofacial development.

The importance of variation in developmental biology. Parichy (2005) recently argued that developmental biology would benefit from the incorporation of a population perspective on organismal design, and this assertion is worth reiterating here. The concept of variation is a corner stone of evolutionary biology, whose importance can be overlooked by molecular geneticists, developmental biologists, and physiologists. As a result, we often confer upon induced genetic mutants a status analogous to a holotype, and our tendency toward this typological view can be misleading. Consideration of a "mutant phenotype" as the singular example of the outcome of impaired protein function places precarious emphasis on the "type" and ignores the collective series. This issue is further confounded by the fact that the most extreme phenotype is often presented as the holotype specimen. Alternatively, population considerations are grounded by the idea that organisms are most appropriately described in statistical terms, where the "type," or average, is a statistical abstraction and the variance is real (Mayr, 1979). Evolutionary biology's "modern synthesis" was rooted in population thinking, and led to a considerably enhanced understanding of organismal design and evolution. We argue that just as population thinking and an emphasis on variation provides a more comprehensive understanding of organismal design, so too can it provide a better understanding of the developmental processes that underlie organismal design. Such a shift in thinking will facilitate efforts to better understand the developmental pathways that underlie organismal development, as well as how perturbations in regulatory networks effect the presentation of complex human disease.

A thorough consideration of phenotypic variation requires the application of analytical tools with which to dissect, describe and quantify this metric. The potential benefits of the application of quantitative methods in experimental studies of morphogenesis, as described above and reviewed by Cooper and Albertson (2008), are based on the central thesis that the use of quantitative and statistical methods can improve the accuracy with which the genetic control of development is understood and described in experimental systems. For example, in a recent paper by Parsons *et al.* (2008) investigators used a series of quantitative methods to study the etiology of cleft lip and palate in an inbred strain of mice possessing a mutation in *Wnt9*, which results in varying degrees of clefting.

Quantitative analyses demonstrated that these mutant mice displayed significantly increased levels of shape variation compared to wild-type littermates, as well as a developmental delay in the formation of the maxillary prominence. They concluded that a reduction in mid-facial growth may push mutant mice closer to the threshold for clefting, and that elevated shape variation can explain why some of these mice exhibited clefting while others did not (Parsons *et al.*, 2008). While not an example from zebrafish *per se*, this work beautifully illustrates how quantitative methods can be used to gain a better understanding of the etiology and pathophysiology of complex mutant phenotypes.

Understanding development beyond embryonic stages. The application of quantitative methods in experimental studies of morphogenesis will be particularly important in investigating the subtle anatomical changes that occur during postembryonic stages of development. For example, Albertson and Yelick (2007) recently used quantitative methods to study the effects of the zebrafish *fgf8/ace* mutation on postembryonic craniofacial development. Data from this study support a role for *fgf8* as a negative regulator of bone formation and remodeling during postembryonic craniofacial development in the zebrafish. In another recent study, Elizondo *et al.* (2005) showed that zebrafish *trpm-7/nutria^{j124e2}* mutants displayed significantly accelerated rates of endochondral ossification and correspondingly delayed rates of intramembranous ossification, revealing a complex relationship between different modes of bone formation. These studies provided novel insights into the mechanisms by which the skeletal system develops, grows, and maintains size and shape over time. They also underscore the utility of quantitative methods in terms of both statistical hypothesis testing and characterizing subtle and/or complex patterns of development.

Somewhat surprising insight into genes functioning in craniofacial and skeletal growth was provided by a recent report of the chemically induced zebrafish *crusher* mutant, which provided the first vertebrate model system linking the biology of endoplasmic reticulum to Golgi trafficking with a clinically relevant dysmorphology (Lang *et al.*, 2006). The zebrafish *crusher* variant, caused by a nonsense mutation in the *sec23a* gene, accumulates proteins in a distended endoplasmic reticulum of chondrocytes, resulting in severe reduction of cartilage extracellular matrix (ECM) deposits, including type II collagen, malformed craniofacial skeleton, kinked pectoral fins and a short body length. The crusher mutant is an important model for human cranio-lenticulo-sutural dysplasia syndrome, caused by a lesion in human SEC23A.

Integration of model and nonmodel systems. A comprehensive understanding of jaw morphogenesis will ultimately involve the identification of both the number and mode of action of genes that underlie morphology, as well as the specific signaling pathways that regulate the development of anatomical form. Here quantitative genetic analyses can compliment traditional experimental embryology, with quantitative shape analysis providing the conceptual link between disciplines. This integrative approach is illustrated by Albertson *et al.* (2005), where a role for

Bmp4 in directing the development of mandibular shape was assessed through a combination of quantitative trait loci (QTL) analysis and experimental embryology. Allelic variation in *bmp4* was shown to be associated with quantitative differences in the shape of the lower jaws of cichlid fishes. By artificially manipulating levels of *bmp4* in zebrafish in the lab, investigators were able to recapitulate natural, quantitative variation in jaw shape observed in the cichlid F_2 mapping population. This novel role for *bmp4* in determining jaw shape was revealed using quantitative shape analysis as the means to integrate distinct approaches and developmental models into a coherent picture of how morphology develops and evolves. A similar approach combining experimental studies in zebrafish and genetic mapping in three-spine sticklebacks has been used to generate and test hypotheses regarding natural variation in the shape of the opercular skeleton in fishes (reviewed by Kimmel *et al.* (2007)). These studies demonstrate the utility of a comparative approach in elucidating molecular pathways directing the development of complex morphologies. Just as experimentally manipulated zebrafish have been used to study craniofacial *patterning*, natural fish populations can be used to study the development of craniofacial *shape*. The integration of these data will provide a more robust estimate of the number, type, mode of action, and interactions of genes that regulate jaw morphogenesis, which will ultimately facilitate our efforts to better understand, predict and treat complex human genetic diseases.

References

Abu-Issa, R., Smyth, G., Smoak, I., Yamamura, K., and Meyers, E. N. (2002). *Fgf8* is required for pharyngeal arch and cardiovascular development in the mouse. *Development* **129**, 4613–4625.

Albertson, R. C., Streelman, J. T., Kocher, T. D., and Yelick, P. C. (2005). Integration and evolution of the cichlid mandible: Molecular basis of alternate feeding strategies. *Proc. Natl. Acad. Sci. USA* **102**, 16287–16292.

Albertson, R. C., and Yelick, P. C. (2007). Fgf8 haploinsufficiency results in distinct craniofacial defects in adult zebrafish. *Dev. Biol.* **306**, 505–515.

Bachler, M., and Neubuser, A. (2001). Expression of members of the Fgf family and their receptors during midfacial development. *Mech. Dev.* **100**, 313–316.

Baynash, A. G., Hosoda, K., Giaid, A., Richardson, J. A., Emoto, N., Hammer, R. E., and Yanagisawa, M. (1994). Interaction of *endothelin-3* with *endothelin-B* receptor is essential for development of epidermal melanocytes and enteric neurons. *Cell* **79**, 1277–1285.

Benusiene, E., and Kucinskas, V. (2003). COL1A1 mutation analysis in Lithuanian patients with osteogenesis imperfecta. *J. Appl. Genet.* **44**, 95–102.

Bird, N. C., and Mabee, P. M. (2003). Developmental morphology of the axial skeleton of the zebrafish, *Danio rerio* (Ostariophysi: Cyprinidae). *Dev. Dyn.* **228**, 337–357.

Bookstein, F. L. (1991). "Morphometric Tools for Landmark Data: Geometry and Biology." Cambridge University Press, New York.

Brand, M., Heisenberg, C. P., Jiang, Y. J., Beuchle, D., Lun, K., Furutani-Seiki, M., Granato, M., Haffter, P., Hammerschmidt, M., Kane, D. A., Kelsh, R. N., Mullins, M. C., *et al.* (1996). Mutations in zebrafish genes affecting the formation of the boundary between midbrain and hindbrain. *Development* **123**, 179–190.

Cooper, W.J., and Albertson, R.C. (2008). Quantification and variation in experimental studies of morphogenesis. *Dev. Biol.* **321**(2), 295–302.

Cubbage, C. C., and Mabee, P. M. (1996). Development of the cranium and paired fins in the zebrafish *Danio rerio* (Ostariophysi, Cyprinidae). *J. Morph.* **229**, 121–160.

David, N. B., Saint-Etienne, L., Tsang, M., Schilling, T. F., and Rosa, F. M. (2002). Requirement for endoderm and FGF3 in ventral head skeleton formation. *Development* **129**, 4457–4468.

Elizondo, M. R., Arduini, B. L., Paulsen, J., Macdonald, E. L., Sabel, J. L., Henion, P. D., Cornell, R. A., and Parichy, D. M. (2005). Defective skeletogenesis with kidney stone formation in dwarf zebrafish mutant for *trpm7*. *Curr. Biol.* **15**, 667–671.

Fischer, S., Draper, B. W., and Neumann, C. J. (2003). The zebrafish *fgf24* mutant identifies an additional level of Fgf signaling involved in vertebrate forelimb initiation. *Development* **130**, 3515–3524.

Fisher, S., and Halpern, M. E. (1999). Patterning the zebrafish axial skeleton requires early chordin function. *Nat. Genet.* **23**, 442–446.

Fisher, S., Jagadeeswaran, P., and Halpern, M. E. (2003). Radiographic analysis of zebrafish skeletal defects. *Dev. Biol.* **264**, 64–76.

Frank, D. U., Fotheringham, L. K., Brewer, J. A., Muglia, L. J., Tristani-Firouzi, M., Capecchi, M. R., and Moon, A. M. (2002). An *Fgf8* mouse mutant phenocopies human 22q11 deletion syndrome. *Development* **129**, 4591–4603.

Gensure, R. C., Ponugoti, B., Gunes, Y., Papasani, M. R., Lanske, B., Bastepe, M., Rubin, D. A., and Juppner, H. (2004). Identification and characterization of two PTH-like molecules in zebrafish. *Endocrinology* **145**(4), 1634–1639 (Epub ahead of print).

Hammerschmidt, M., Pelegri, F., Mullins, M. C., Kane, D. A., van Eeden, F. J., Granato, M., Brand, M., Furutani-Seiki, M., Haffter, P., Heisenberg, C. P., Jiang, Y. J., Kelsh, R. N., *et al.* (1996). *dino* and *mercedes*, two genes regulating dorsal development in the zebrafish embryo. *Development* **123**, 95–102.

Hassinger, D. D., Mulvihill, J. J., and Chandler, J. B. (1980). Aarskog's syndrome with Hirschsprung's disease, midgut malrotation, and dental anomalies. *J. Med. Genet.* **17**, 235–237.

Herbarth, B., Pingault, V., Bondurand, N., Kuhlbrodt, K., Hermans-Borgmeyer, I., Puliti, A., Lemort, N., Goossens, M., and Wegner, M. (1998). Mutation of the *Sry*-related *Sox10* gene in Dominant megacolon, a mouse model for human Hirschsprung disease. *Proc. Natl. Acad. Sci. USA* **95**, 5161–5165.

Hild, M., Dick, A., Rauch, G. J., Meier, A., Bouwmeester, T., Haffter, P., and Hammerschmidt, M. (1999). The *smad5* mutation *somitabun* blocks Bmp2b signaling during early dorsoventral patterning of the zebrafish embryo. *Development* **10**, 2149–2159.

Hosoda, K., Hammer, R. E., Richardson, J. A., Baynash, A. G., Cheung, J. C., Giaid, A., and Yanagisawa, M. (1994). Targeted and natural (*piebald*-lethal) mutations of *endothelin-B* receptor gene produce megacolon associated with spotted coat color in mice. *Cell* **79**, 1267–1276.

Huysseune, A., and Sire, J. Y. (2003). The role of epithelial remodeling in tooth eruption in larval zebrafish. *Cell Tissue Res.* **315**, 85–95.

Kimmel, C. B., Miller, C. T., Kruze, G., Ullmann, B., BreMiller, R. A., Larison, K. D., and Snyder, H. C. (1998). The shaping of pharyngeal cartilages during early development of the zebrafish. *Dev. Biol.* **203**, 245–263.

Kimmel, C. B., Miller, C. T., and Moens, C. B. (2001). Specification and morphogenesis of the zebrafish larval head skeleton. *Dev. Biol.* **233**, 239–257.

Kimmel, C. B., Ullmann, B., Walker, M., Miller, C., and Crump, J. G. (2003). Endothelin 1-mediated regulation of pharyngeal bone development in zebrafish. *Development* **130**, 1339–1351.

Kimmel, C. B., Walker, M. B., and Miller, C. T. (2007). Morphing the hyomandibular skeleton in development and evolution. *J. Exp. Zool. B Mol. Dev. Evol.* **308**(5), 609–624.

Knight, R. D., Nair, S., Nelson, S. S., Afshar, A., Javidan, Y., Geisler, R., Rauch, G. J., and Schilling, T. F. (2003). *lockjaw* encodes a zebrafish *tfap2a* required for early neural crest development. *Development* **130**, 5755–5768.

Kurihara, Y., Kurihara, H., Maemura, K., Kuwaki, T., Kumada, M., and Yazaki, Y. (1995). Impaired development of the thyroid and thymus in *endothelin-1* knockout mice. *J. Cardiovasc. Pharmacol.* **3**, S13–S16.

Lang, M. R., Lapierre, L. A., Frotscher, M., Goldenring, J. R., and Knapik, E. W. (2006). Secretory COPII coat component Sec23a is essential for craniofacial chondrocyte maturation. *Nat. Genet.* **38**(10), 1198–1203.

Mayr, E. (1979). Typological versus population thinking. *In* "Evolution and the Diversity of Life: Selected Essays" (S. Sober, ed.). Harvard University Press, Cambridge, MA.

Miller, C. T., and Kimmel, C. B. (2001). Morpholino phenocopies of endothelin 1 (*sucker*) and other anterior arch class mutations. *Genesis* **30**, 186–187.

Miller, C. T., Schilling, T. F., Lee, K., Parker, J., and Kimmel, C. B. (2000). *sucker* encodes a zebrafish Endothelin-1 required for ventral pharyngeal arch development. *Development* **127**, 3815–3828.

Miller, C. T., Yelon, D., Strainier, D. Y. R., and Kimmel, C. B. (2003). Two *endothelin 1* effectors, *hand2* and *bapx1*, pattern ventral pharyngeal cartilage and the jaw joint. *Development* **130**, 1353–1365.

Miller-Bertoglio, V. E., Fisher, S., Sanchez, A., Mullins, M. C., and Halpern, M. E. (1997). Differential regulation of chordin expression domains in mutant zebrafish. *Dev. Biol.* **129**, 537–550.

Mina, M., Wang, Y. H., Ivanisevic, A. M., Upholt, W. B., and Rodgers, B. (2002). Region- and stage-specific effects of FGFs and BMPs in chick mandibular morphogenesis. *Dev. Dyn.* **223**, 333–352.

Mullins, M. C., Hammerschmidt, M., Kane, D. A., Odenthal, J., Brand, M., van Eeden, F. J., Furutani-Seiki, M., Granato, M., Haffter, P., Heisenberg, C. P., Jiang, Y. J., Kelsh, R. N., *et al.* (1996). Genes establishing dorsoventral pattern formation in the zebrafish embryo: The ventral specifying genes. *Development* **123**, 81–93.

Neuhauss, S. C., Solnica-Krezel, L., Schier, A. F., Zwartkruis, F., Stemple, D. L., Malicki, J., Abdelilah, S., Stainier, D. Y., and Driever, W. (1996). Mutations affecting craniofacial development in zebrafish. *Development* **123**, 357–367.

Olsen, B. R., Reginato, A. M., and Wang, W. (2000). Bone development. *Annu. Rev. Cell. Dev. Biol.* **16**, 191–220.

Parichy, D. M. (2005). Variation and developmental biology: Prospects for the future. *In* "Variation: A Hierarchical Examination of a Central Concept in Biology" (B. Hallgrimsson and B. K. Hall, eds.). Academic Press, Burlington, MA.

Parsons, T. E., Kristensen, E., Hornung, L., Diewert, V. M., Boyd, S. K., German, R. Z., and Hallgrimsson, B. (2008). Phenotypic variability and craniofacial dysmorphology: Increased shape variance in a mouse model for cleft lip. *J. Anat.* **212**, 135–143.

Paylor, R., McIlwain, K. L., McAninch, R., Nellis, A., Yuva-Paylor, L. A., Baldini, A., and Lindsay, E. A. (2001). Mice deleted for the DiGeorge/velocardiofacial syndrome region show abnormal sensorimotor gating and learning and memory impairments. *Hum. Mol. Genet.* **10**, 2645–2650.

Perrino, M. A., and Yelick, P. C. (2004). Immunolocalization of *Alk8* during replacement tooth development in zebrafish. *Cells Tissues Organs* **176**, 17–27.

Piotrowski, T., Ahn, D. G., Schilling, T. F., Nair, S., Ruvinsky, I., Geisler, R., Rauch, G. J., Haffter, P., Zon, L. I., Zhou, Y., *et al.* (2003). The zebrafish *van gogh* mutation disrupts *tbx1*, which is involved in the DiGeorge deletion syndrome in humans. *Development* **130**, 5043–5052.

Piotrowski, T., Schilling, T. F., Brand, M., Jiang, Y. J., Heisenberg, C. P., Beuchle, D., Grandel, H., van Eeden, F. J., Furutani-Seiki, M., Granato, M., Haffter, P., Hammerschmidt, M., *et al.* (1996). Jaw and branchial arch mutants in zebrafish II: Anterior arches and cartilage differentiation. *Development* **123**, 345–356.

Potthoff, T. (1984). Clearing and staining technique. *In* Moser, H. G. (ed.) "Ontogeny and Systematics of Fishes. Special Publication No. 1" (H. G. Moser, ed.). American Society of Ichthyologists and Herpetologists, Austin, TX.

Rohlf, F. J., and Marcus, L. F. (1993). A revolution in morphometrics. *Trend. Ecol. Evol.* **8**, 129–132.

Schilling, T. F., Piotrowski, T., Grandel, H., Brand, M., Heisenberg, C. P., Jiang, Y. J., Beuchle, D., Hammerschmidt, M., Kane, D. A., Mullins, M. C., *et al.* (1996). Jaw and branchial arch mutants in zebrafish I: Branchial arches. *Development* **123**, 329–344.

Southard-Smith, E. M., Kos, L., and Pavan, W. J. (1998). *Sox10* mutation disrupts neural crest development in Dom Hirschsprung mouse model. *Nat. Genet.* **18**, 60–64.

Thesleff, I. (2003). Developmental biology and building a tooth. *Quintessence Int.* **34**, 613–620.

Thesleff, I., Partanen, A. M., and Vainio, S. (1991). Epithelial-mesenchymal interactions in tooth morphogenesis: The roles of extracellular matrix, growth factors, and cell surface receptors. *J. Craniofac. Genet. Dev. Biol.* **11,** 229–237.

Thompson, D. A. W. (1917). "On Growth and Form." Cambridge University Press, Cambridge.

Thomsen, G. H. (1997). Antagonism within and around the organizer: BMP inhibitors in vertebrate body patterning. *Trends Genet.* **13,** 209–211.

Tucker, A. S., Yamada, G., Grigoriou, M., Pachnis, V., and Sharpe, P. T. (1999). Fgf-8 determines rostral-caudal polarity in the first branchial arch. *Development* **126**(1), 51–61.

Van der Heyden, C., and Huysseune, A. (2000). Dynamics of tooth formation and replacement in the zebrafish (*Danio rerio*) (Teleostei, Cyprinidae). *Dev. Dyn.* **219,** 486–496.

Van der Heyden, C., Huysseune, A., and Sire, J. Y. (2000). Development and fine structure of pharyngeal replacement teeth in juvenile zebrafish (*Danio rerio*) (Teleostei, Cyprinidae). *Cell Tissue Res.* **302,** 205–219.

Van der Heyden, C., Wautier, K., and Huysseune, A. (2001). Tooth succession in the zebrafish (*Danio rerio*). *Arch. Oral. Biol.* **46,** 1051–1058.

Walshe, J., and Mason, I. (2003). Fgf signaling is required for formation of cartilage in the head. *Dev. Biol.* **264,** 356–522.

Wautier, K., Van der Heyden, C., and Huvsseune, A. (2001). A quantitative analysis of pharyngeal tooth shape in the zebrafish (*Danio rerio,* Teleostei, Cyprinidae). *Arch. Oral. Biol.* **46,** 67–75.

Witten, P. E., Hansen, A., and Hall, B. K. (2001). Features of mono- and multinucleated bone resorbing cells of the zebrafish *Danio rerio* and their contribution to skeletal development, remodeling, and growth. *J. Morph.* **250,** 197–207.

Yelick, P. C., and Schilling, T. F. (2002). Molecular dissection of craniofacial development using zebrafish. *Crit. Rev. Oral. Biol. Med.* **13,** 308–322.

CHAPTER 20

Cardiac Development

Le A. Trinh* and Didier Y. R. Stainier†

*Biology Department
California Institute of Technology
Pasadena, California 91125

†Department of Biochemistry and Biophysics
Programs in Developmental Biology
Genetics and Human Genetics
University of California, San Francisco
San Francisco, California 94143–0448

Update

Our understanding of cardiac development in zebrafish has been pushed forward in recent years by a number of technical advances in high-resolution imaging. These imaging advances in combination with the well-established genetic and embryological tools available in zebrafish have provided a detailed framework of

DOI: 10.1016/B978-0-12-374599-6.00020-0

the genetic and cellular processes that govern the formation of the embryonic heart. The availability of transgenic lines that express multiple fluorescent proteins to mark specific cell-types and subcellular compartments is beginning to provide an unprecedented cellular resolution of the developing heart. The development of new imaging technology and improvements of existing techniques have allowed high-resolution dynamic imaging using these transgenic tools. These advances have shed new light on regulatory mechanisms at play during heart development. In this updated chapter, we have incorporated a discussion of some of these advances with the existing description of cardiac development, highlighting these new areas when applicable.

I. Introduction

The goal of this chapter is to provide a reference guide for the development of the embryonic heart in zebrafish. Here, we provide a description of the steps of heart development, discussing morphogenetic processes at each stage as well as regulatory events and gene expression patterns. Although development of the vascular system is integral to cardiovascular function, a description of vascular development is beyond the scope of this chapter (reviewed in Jin *et al.*, 2002; Lawson and Weinstein, 2002).

As in all vertebrates, the zebrafish heart is the first internal organ to form and function. It is comprised of two layers, an outer muscular layer, the myocardium, and an inner endothelial layer, the endocardium. These two layers are subdivided into two major chambers: the atrium and the ventricle. The zebrafish heart begins to beat at 22 hpf and circulation is initiated by 24 hpf.

Many of the characteristics of the zebrafish have allowed it to emerge as a powerful vertebrate model organism for the study of cardiac development. The external fertilization, rapid development, and optical clarity of the zebrafish provide distinct advantages for the study of organogenesis. The zebrafish heart can be easily observed throughout the stages of its development as it is prominently positioned at the ventral midline of the embryo. In addition, the ability to combine genetics with embryology and cell biology to investigate lineage relationship, cell behavior, and molecular networks has greatly facilitated our understanding of the regulatory processes underlying heart development.

II. Stages of Heart Tube Morphogenesis

Our understanding of heart tube morphogenesis as it occurs in zebrafish comes from a combination of lineage analyses, gene expression studies, and mutations that affect various steps in this process. From these studies, we have divided the stages of heart tube morphogenesis into six discrete phases: heart field formation, migration, heart tube elongation, heart looping, valve formation, and myocardial remodeling. Each phase will be elaborated upon in the following sections by providing a description of the morphological changes that occur at each step followed by a discussion of the regulatory events.

A. Formation of the Heart Fields

Lineage analyses indicate that cardiac progenitors reside in the first four tiers of marginal blastomeres in the early blastula (Keegan *et al.*, 2004; Stainier *et al.*, 1993; Warga and Nüsslein-Volhard, 1999). Within the marginal blastomeres, the cardiac progenitors are located bilaterally around 90°–180° from the dorsal midline (Fig. 1A). This cardiac region encompasses both endocardial and myocardial progenitors with endocardial progenitors positioned more ventrally (Lee *et al.*, 1994).

The endocardial and myocardial cells share a common progenitor prior to gastrulation as labeling a single cell within the cardiac region in either the early or mid-blastula results in progeny contributing to both lineages (Lee *et al.*, 1994). This is not the case with respect to atrial and ventricular lineages (Stainier *et al.*, 1993). Labeling a single cell in the early blastula results in progeny contributing to both the atrium and ventricle, while a single cell labeled in the mid-blastula contributes to either the atrium or the ventricle. These results suggest that the cardiac progenitors have acquired positional information by the mid-blastula-stage embryo. However, it is not known whether this positional information is in the form of a signal that restricts the cells to a chamber specific fate or due to spatial arrangements that limits cell movements during gastrulation.

Recent refined fates maps and mutant analyses have shed light on some of the signaling pathways that are involved in defining the myocardial progenitors in the pregastrula embryo. Fate maps of embryos treated with a retinoic acid receptor antagonist compared to those of control embryos indicate that retinoic acid signaling at the margin restricts the size of the cardiac progenitor pool prior to gastrulation (Keegan *et al.*, 2005). Furthermore, fate maps performed at the 40% epiboly stage show that chamber specific progenitors are spatially distinct by gastrulation stages. At these early stages, the ventricular myocardial progenitors are positioned closer to the margin and dorsal midline than atrial myocardial progenitors (Keegan *et al.*, 2004). At the margin, the ventricular myocardial progenitors are presumably exposed to higher levels of Nodal signal and thus their fate is promoted by Nodal signaling (Keegan *et al.*, 2005).

During gastrulation, cardiac progenitors are among the first mesodermal cells to involute (Warga and Kimmel, 1990). After involuting, they move toward the animal pole and converge dorsally toward the embryonic axis (Stainier and Fishman, 1992). At the end of gastrulation, the cardiac progenitors arrive on either side of the embryonic midline, where they reside as subpopulations of cells within the anterior lateral plate mesoderm (LPM) (Fig. 1B). At the on-set of somitogenesis, the LPM begins to express a number of myocardial differentiation genes. These differentiations factors include the homeodomain gene, *nkx2.5*, and a

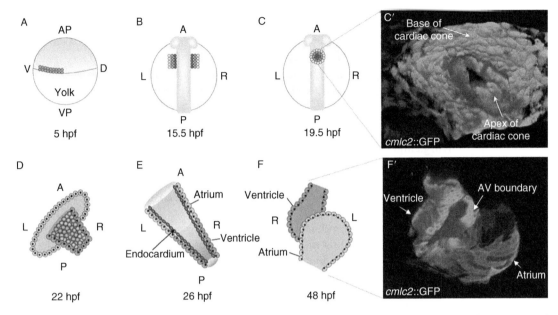

Fig. 1 Zebrafish heart development. (A) In the early blastula, the cardiac progenitors correspond to the ventro-lateral margin of the blastoderm. (B) After gastrulation, the cardiac progenitors arrive on either side of the midline. The three rows of circles represent the relative position of the endocardial precursors (red), ventricular precursors (dark green) and atrial precursors (light green) at this stage. By the 13-somite stage (15.5 hpf), the myocardial precursors are patterned mediolaterally, with ventricular precursors positioned medial to the atrial precursors. (C) At the 21-somite stage (19.5 hpf), the myocardial precursors have migrated to the midline and fused to form the cardiac cone. Within the cardiac cone the endocardial precursors (red) are located in the central lumen, the ventricular precursors (dark green) at the apex, and the atrial precursors (light green) at the base. (C′) Dorsal view of the cardiac cone as seen by projection of confocal optical sections in a transgenic embryo that expresses GFP in the myocardial precursors (*cmlc2*::GFP). (D) Cardiac cone tilting starts with the apex of the cone bending posteriorly and to the right. Subsequently, the base of the cone, consisting of atrial precursors, will gradually coalesce into a tube. (E) By 26 hpf, elongation of the heart tube results in the leftward positioning of the linear heart tube which is composed of two layers, an inner endocardium (red) and an outer myocardium (green). (F) By 48 hpf, gradual bending of the heart tube at the atrioventricular boundary forms an S-shaped loop that positions the ventricle (dark green) to the right of the atrium (light green) as seen in a head-on view. (F′) Projection of confocal optical sections of a 56 hpf embryo with GFP expressed in the myocardium. In the projection, a transverse section through the myocardium shows a thick, compact ventricular wall while the atrial wall appears thinner. (See Plate no.17 in the Color Plate Section.)

number of *gata* genes (4/5/6), which appear to define the cardiac fields within the anterior LPM.

During the early somitogenesis stages (10–16-somite), the cardiac fields appear to regulate in response to injury (Serbedzija *et al.*, 1998). Unilateral laser ablation of the cardiac progenitors at the 10-somite stage results in repopulation of the cardiac fields by LPM cells lateral and anterior to the normal cardiac domain. This ability to regulate the heart field persists to the 16-somite stage.

Additionally, the notochord appears to restrict the posterior extend of the cardiac fields in the LPM (Goldstein and Fishman, 1998). Lineage analysis of the LPM indicates that only LPM anterior to the tip of the notochord contributes to the heart. Laser ablation of the notochord results in the posterior expansion of the myocardial expression domain indicating that the notochord provides an inhibitory signal that limits the extent of the cardiac fields.

B. Migration to the Midline

After the cardiac progenitors arrive on either side of the midline, they undergo a secondary migration toward the midline, where they fuse to form the cardiac cone (Fig. 2A–C). At the 16-somite stage, the myocardial precursors are positioned on either side of the midline, ventral to the anterior endoderm and dorsal to the yolk syncytial layer (YSL) (Stainier et al., 1993). By the 18-somite stage, these bilateral populations have moved toward the midline to make contacts medially (Fig. 2B) (Yelon et al., 1999). The initial contact by the medial myocardial precursors is followed by fusion of the more posterior domain. At the 21-somite stage, the anterior domains fusion to form a cone with a central lumen occupied by endocardial progenitors (Fig. 1C and C′) (Stainier et al., 1993; Yelon et al., 1999).

Time-lapse confocal microscopy of a transgenic line that marks the myocardial precursors during these stages has identified two distinct phases in the migration process (Holtzman et al., 2007). Cell tracking data show that the myocardial precursors initially undergo a medial movement to the midline. This medial movement is followed by a change in direction of anterior and posterior cells to form the circumference of the heart tube. This second movement is characterized by an angular displacement of the cells and is driven by the endocardial precursors, implicating an interplay between the myocardial and endocardial precursors during heart tube formation.

As the myocardial precursors move to the midline, they undergo epithelial maturation (Trinh and Stainier, 2004). At the 16-somite stage, myocardial precursors are cuboidal in shape and show an enrichment of the adherens junction proteins atypical protein kinase (aPKCs) at points of cell-cell contacts (Fig. 2D). As migration proceeds, cell junctional proteins such as aPKCs, zonula occluden-1 (ZO-1), and β-catenin show an increased asymmetric localization in the myocardial precursors (Fig. 2E). The asymmetry in protein localization is accompanied by a change in cell shape, with the medial myocardial precursors becoming columnar. The enrichment of cell junctional proteins in the myocardial precursors throughout the migration stages indicates that these cells are forming tight adhesion to one another and migrate as coherent populations.

A number of genes involved in epithelial polarity have recently been added to the list of essential players involved in heart tube development. *Heart and soul (has)/aPKCλ, nagio oko (nok)*, and *heart and mind (had)/Na⁺, K⁺ ATPase* mutants all exhibit defects in epithelial organization of the myocardial precursors (Cibrian-Uhalte et al., 2007; Rohr et al., 2005). The disruption in epithelial organization in

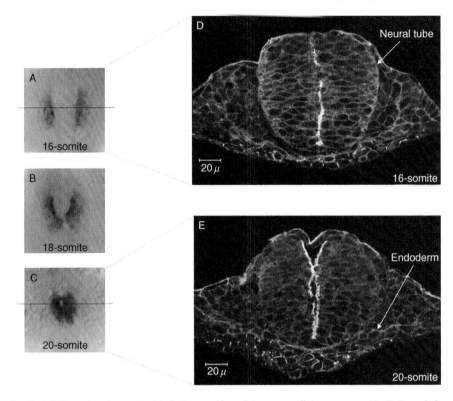

Fig. 2 Midline migration and epithelial maturation of the myocardial precursors. (A–C) Dorsal views of *cmlc2* expression during myocardial migration to the midline. (A) At the 16-somite stage, the myocardial precursors are positioned on either side of the midline. (B) By the 18-somite stage, the medial myocardial precursors make contacts at the midline. (C) At the 20-somite stage, the myocardial precursors are fusing to form the cardiac cone. (D, E) Transverse confocal images of *cmlc2*::GFP (false colored blue) transgenic embryos immunostained with antibodies against β-catenin (red) and aPKCs (green); dorsal to the top. (D) At the 16-somite stage, the cuboidal myocardial precursors show cortical localization of β-catenin and an enrichment of aPKCs on the apical surface and at points of cell-cell contacts. (E) By the 20-somite stage, transverse section through the middle of the cardiac cone shows that the myocardial precursors form two U-shaped structures, with the medial cells appearing columnar in shape, while the lateral cells are cuboidal. aPKCs are restricted to the apicolateral domains, while β-catenin localizes to the basolateral domains of the myocardial precursors. (See Plate no. 18 in the Color Plate section.)

these mutants affects myocardial integrity and heart tube morphogenesis further indicating an essential role for epithelial organization in cardiac development.

Based on expression analyses with the early endothelial progenitor marker, *flk-1*, it is thought that the endocardial progenitors are at the midline by the 18-somite stage (Liao *et al.*, 1997). Similarly, transverse sections of a transgenic line expressing green fluorescent protein (GFP) under the control of the promoter of the *flk-1* gene indicate that the endocardial precursors are at the embryonic midline by the

16-somite stage (Trinh and Stainier, 2004). These results place the endocardial progenitors at the midline prior to myocardial fusion.

Analyses in *cloche* (*clo*) mutants indicate that endocardial-myocardial interaction is required for proper timing of myocardial migration and heart tube formation. In *clo* mutants, which show a complete absence of endocardial cells (Stainier *et al.*, 1995), myocardial migration is delayed (Trinh and Stainier, 2004). Gene expression analyses demonstrate that the endocardial precursors are a source of Fibronectin that is deposited at the midline on the ventral side of the anterior endoderm (Trinh and Stainier, 2004). In *clo* mutants, Fibronectin deposition is absent at the midline, which may provide the substrate for the temporal regulation of myocardial migration. These data suggest that endocardial progenitors play an important role in the timing of myocardial migration.

Recently, dynamic imaging approaches have been applied to describe the morphogenesis of the endocardial precursors. Time-lapse confocal microscopy of a transgenic line marking the endothelial precursors (*Tg(flk1::eGFP)*) confirm molecular expression analyses data suggesting that the endocardial precursors originate from the LPM and migrate to the midline prior to the myocardial precursors (Bussmann *et al.*, 2007). Furthermore, this migration process appears to be dependent on the bHLH transcription factor Scl/Tal1 as *scl/tal1* mutants show aberrant migration (Bussmann *et al.*, 2007).

Large-scale genetic screens in zebrafish have identified eight mutations, *hands off* (*han*), *faust* (*fau*), *casanova* (*cas*), *bonnie and clyde* (*bon*), *one-eye pinhead* (*oep*), *natter* (*nat*), *miles apart* (*mil*), and *two-of-heart* (*toh*), that disrupt the medial migration of the myocardial precursors, resulting in the formation of two separate hearts, a phenotype referred to as cardia bifida (Alexander *et al.*, 1998; Chen *et al.*, 1996; Stainier *et al.*, 1996). Analyses of these mutants have led to the identification of several requirements for the coordinated movement of myocardial precursors to the midline. First, myocardial differentiation appears to be critical for migration as mutations that disrupt myocardial differentiation (e.g., *han*, *fau*, and *oep*) all exhibit migration defects (Reiter *et al.*, 1999, 2001; Schier *et al.*, 1997; Yelon *et al.*, 2000). The *han* mutation encodes the bHLH transcription factor Hand2 and appears to regulate both the number of myocardial precursors and myocardial migration (Trinh *et al.*, 2005; Yelon *et al.*, 2000). *hand2* is expressed exclusively in the LPM during myocardial migration suggesting that an aspect of the migration process is autonomous to the myocardial precursors.

Second, the anterior endoderm appears to be essential for myocardial migration as mutants that lack the anterior endoderm (e.g., *cas*, *bon*, *fau*, and *oep*) display cardia bifida (Alexander *et al.*, 1999; Kikuchi *et al.*, 2000; Reiter *et al.*, 1999; Schier *et al.*, 1997). Additionally, wild-type endoderm when transplanted into a subclass of cardia bifida mutants can rescue myocardial migration (David and Rosa, 2001). Though these studies point to the involvement for the endoderm in myocardial precursor migration, the basis of this requirement remains to be determined.

Third, epithelial organization of the myocardial precursors appears to be critical for the coordinated movement of myocardial precursors (Trinh and Stainier, 2004).

The *nat* mutation disrupts adherens junction clustering in the myocardial epithelia and causes cardia bifida. *nat* encodes Fibronectin which itself is deposited in the basal substratum around the myocardial precursors throughout the migration stages. Additionally, Fibronectin is deposited at the midline between the endoderm and endocardial precursors. In the complete absence of Fibronectin deposition, myocardial migration is disrupted and adherens junctions between the myocardial precursors do not form properly. These findings suggest that the Fibronectin matrix provides a positional cue for establishing cellular asymmetry in the myocardial precursors and that cell-substratum interaction is required for their epithelial organization and migration.

Finally, signaling mediated by the sphingosine 1-phosphate (S1P) receptor, *miles apart* (*mil*), is critical for myocardial migration (Kupperman *et al.*, 2000). S1P is a bioactive lysophospholipid that regulates a wide range of processes including cell proliferation, differentiation, and survival (Hla, 2003; Panetti *et al.*, 2000). Myocardial differentiation appears unaffected in *mil* mutants. Additionally, cell autonomy studies indicate that *mil* is not required in the migrating myocardial precursors. These results have led to a model in which *mil* functions to provide an environment permissive for myocardial migration to the midline.

C. Heart Tube Elongation

Once the myocardial precursors reach the midline and fuse to form the cardiac cone, the cone will extend to form the linear heart tube. Heart tube elongation begins at the 22-somite stage and proceeds until 26 hpf (Yelon *et al.*, 1999). This process begins with the apex of the cardiac cone tilting posteriorly and toward the right side of the embryo, repositioning the cardiac cone from a dorsal-ventral (D-V) axis to an anterior-posterior (A-P) axis (Fig. 1D). The apex of the cone comprises of ventricular cells and establishes the arterial end of the heart tube (Stainier *et al.*, 1993; Yelon *et al.*, 1999). The atrial cells occupy the base of the cone, which coalesce into a tube as the apex tilts (Yelon *et al.*, 1999). By 24 hpf, the ventricular cells are completely repositioned into a leftward slanted tube, while the atrial cells are continuing to telescope into a tube. By 26 hpf, heart tube elongation is completed, resulting in a linear heart tube positioned on the ventral left side of the embryo (Fig. 1E).

The *has* and *had* mutations provide the first molecular insights into our understanding of heart tube elongation. *has* mutants form cardiac cones that fail to tilt and elongate while *had* mutants exhibit a delay in heart tube extension (Horne-Badovinac *et al.*, 2001; Shu *et al.*, 2003). *has* encodes aPKCλ, an adherens junction protein that localizes to the apical domain of polarized epithelia (Horne-Badovinac *et al.*, 2001). *has* mutants exhibit defects in retinal pigmented epithelia and gut tube formation, as well as gut looping, indicating that aPKCλ is critical for the development and morphogenesis of multiple epithelial tissues. Overexpression of aPKCλ specifically in the myocardial precursors of *has* mutant embryos can rescue cardiac function indicating that aPKCλ is acting autonomously in the myocardial precursors to control myocardial morphogenesis (Rohr *et al.*, 2005). Consistent with this result is the finding that *had* encodes the α1 isoform of Na, K-ATPase, an epithelial polarity component (Shu *et al.*, 2003). In addition to transporting Na$^+$ and K$^+$ across the plasma membrane to establish proper chemical and electrical gradients, Na, K-ATPases are required for septate junction formation in polarized epithelia of *Drosophila* (Paul *et al.*, 2003). The analysis of apicobasal polarization of the myocardial precursors in these two mutants further supports an essential role for epithelial organization during heart tube elongation (Cibrian-Uhalte *et al.*, 2007; Rohr *et al.*, 2005).

The orderly fusion of the myocardial precursors in forming the cardiac cone has also been implicated as an important regulatory step for heart tube elongation (Peterson *et al.*, 2001). As previously discussed, the fusion of myocardial precursors at the midline occurs in an orderly fashion with the posterior cells fusing prior to the anterior cells. In embryos treated with the small molecule concentramide and in *has* mutants myocardial fusion is defective and heart cone elongation is blocked (Peterson *et al.*, 2001). However, analyses of cardia bifida mutants indicate that myocardial fusion is not required for heart tube elongation (Yelon *et al.*, 1999). In migration defective mutants, the unfused bifid populations of myocardial precursors undergo tilting and elongation to form tubes as in wild-type embryos indicating that the fusion of the two myocardial primordia is not a necessary step in heart tube elongation (Yelon *et al.*, 1999). Thus, the fusion defects seen in concentramide-treated and *has* mutants may be coincident with a lack of heart tube elongation rather than causal.

Studies of the heart tube elongation process using time-lapse confocal microscopy indicate that the tilting and elongation of the cardiac cone is driven by an asymmetric migration and clockwise rotation of the cardiac cone (Smith *et al.*, 2008). The leftward migration of myocardial precursors correlates with the expression of a number of genes involved in left-right asymmetry, mainly components of the Nodal and BMP signaling pathways (Baker *et al.*, 2008; Rohr *et al.*, 2008; Smith *et al.*, 2008). Functionally, BMP signaling has also been implicated in directing the asymmetric movement and rotation of the cardiac cone as time-lapse microscopy of mutant embryos in the BMP type I receptor gene, *lost-a-fin* (*laf*)/*alk8*, show a reduction in rotation of the cardiac cone and lack of leftward movement (Baker *et al.*, 2008). Furthermore, overexpression of BMP by insertion of beads soaked in BMP can redirect the elongation process implicating BMP signaling as an instructive signal for heart tube elongation (Baker *et al.*, 2008).

D. Heart Looping

Heart looping can be temporally segregated into two distinct steps: the initial leftward placement of the elongated heart tube, followed by a gradual bending at the atrioventricular boundary to form an S-shape loop that positions the ventricle to the right of the atrium (Fig. 1F) (Chen *et al.*, 1997; Chin *et al.*, 2000). The leftward placement of the heart started by heart tube elongation occurs from 22 to

30 hpf and is the first morphological indication of left-right asymmetry in the zebrafish embryo. The subsequent repositioning of the ventricle to the right of the atrium, known as D-looping, occurs between 30 and 48 hpf. Mutant analyses indicate that the initial placement of the heart can be uncoupled from D-looping (Chin *et al.*, 2000). This discordance is seen in mutations that disrupt the formation of the notochord, suggesting that the notochord provides an essential signal either to couple the initial heart position and D-looping or a midline barrier to this coupling signal (Chin *et al.*, 2000; Danos and Yost, 1996).

The cellular processes regulating the morphogenesis of heart looping are poorly understood, however, molecular asymmetries in heart looping have been extensively documented. Prior to the morphological asymmetry in the heart, a number of genes are expressed asymmetrically in the anterior LPM. At 19-20-somite, components of the Nodal signaling pathway, such as *southpaw, cyclop*, and *lefty-1* are expressed in the left LPM (Bisgrove *et al.*, 1999; Long *et al.*, 2003; Rebagliati *et al.*, 1998; Thisse and Thisse, 1999). The homeobox gene *nkx2.5* and the transcription factor *pitx2* are expressed more posteriorly in the left cardiac field than the right at 20-24-somite (Essner *et al.*, 2000; Schilling *et al.*, 1999). *bmp4* is expressed uniformly in the myocardial precursors at the time of cardiac cone fusion, however, as the cardiac cone begins to tilt and elongate, *bmp4* expression accumulates predominantly on the left side of the heart tube (Chen *et al.*, 1997). Mutant analyses have shown that the patterns of asymmetric gene expression correlate with the direction of heart looping, suggesting that left-right asymmetry signaling pathways regulate this process (Bisgrove *et al.*, 2000). While numerous genes and developmental processes have been implicated in establishing the left-right axis, our understanding of how left-right asymmetry signals are translated into cellular mechanisms regulating heart looping remains limited.

E. Valve Formation

Formation of a functioning valve is an essential step to prevent retrograde blood flow through the heart. Valves form from the endocardial layer of the heart tube at three sites along the developing heart: the outflow tract, atrioventricular (AV) boundary, and sinus venosus. While proper valve formation at these three sites is critical for unidirectional flow, much of our understanding of valve formation in the zebrafish comes from studies focusing on the AV boundary. It is unclear if the outflow tract and the sinus venosus form valves in a similar fashion as the AV boundary.

The zebrafish heart begins to beat at 22 hpf and circulation is initiated by 24 hpf (Stainier *et al.*, 1993), however, by 36 hpf, only a single layer of endocardial cells line the lumen of the heart (Fig. 3A) (Beis *et al.*, 2005; Stainier *et al.*, 2002). At these stages, chamber dynamics drive blood flow unidirectionally from the sinus venosus into the atrium, ventricle, and outflow tract. The contraction of the myocardium has been described as starting with peristaltic waves that change to coordinated sequential rhythmic beats of the chambers (Warren *et al.*, 2000). High-speed

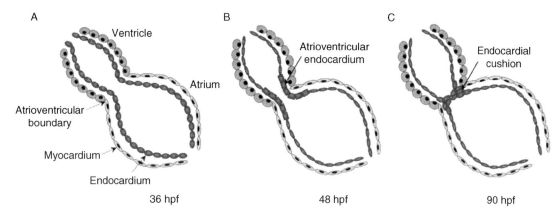

Fig. 3 Atrioventricular (AV) valve formation. (A) At 36 hpf, a single layer of endocardial cells line the lumen of the heart. (B) At 48 hpf, the AV endocardial cells consist of approximately 5–6 cuboidal cells along the A-P axis of the endocardium. (C) By 90 hpf, the AV endocardial cells have undergone an epithelial-to-mesenchymal transition forming endocardial cushions that appear as clusters that are two-cell thick at the AV boundary.

confocal imaging using a laser slit-scanning system and image-processing algorithms that synchronize the variation in heartbeat sequences due to imaging in three-dimensions has found that the zebrafish heart does not act as a peristaltic pump. Instead, the myocardium behaviors appear similar to a suction pump that exhibits elastic wave propagation bidirectionally through the heart tube (Forouhar *et al.*, 2006).

By 48 hpf, the endocardial cells at the AV boundary begin to show cellular characteristics that distinguish them from the rest of the endocardium (Beis *et al.*, 2005; Stainier *et al.*, 2002; Walsh and Stainier; 2001). The AV endocardial cells, consisting of approximately 5–6 cells along the A-P axis of the endocardium, appear cubiodal while non-AV endocardial cells are more squamous (Fig. 3B). In histological sections, cardiac jelly, a type of extracellular matrix, fills the space between the endocardium and myocardium (Hu *et al.*, 2000). Between 60 and 72 hpf, histological sections indicate that the AV endocardial cells appear to undergo an epithelial-to-mesenchyme transition (EMT) (Beis *et al.*, 2005; Stainier *et al.*, 2002). The advent of real-time high-speed imaging using selective plane illumination microscopy (SPIM) indicate that zebrafish AV endocardial cells do not undergo a complete EMT but rather form valve leaflets to prevent retrograde flow (Scherz *et al.*, 2008). By 90 hpf, the AV endocardial cells form endocardial cushions that appear as clusters that are two-cell thick at the AV boundary (Fig. 3C) (Stainier *et al.*, 2002). At 5-days postfertilization, valves at the AV boundary and the outflow tract are visible as two cusps (Hu *et al.*, 2000).

Molecular indications of AV boundary formation occur prior to morphological changes. In the myocardium, *bmp4* and *versician* are initially expressed throughout the heart and become restricted to the AV myocardial cells at 37 hpf (Walsh and Stainier, 2001). In the endocardium, *notch1b* is initially expressed throughout the entire extent of the heart tube and becomes restricted to the AV endocardial cells at 45 hpf (Walsh and Stainier, 2001). A similar restriction of GFP expression in the endocardium is seen in a GFP transgenic line that expresses GFP under the control of the *tie2* promoter (*tie2*::GFP) (Walsh and Stainier, 2001). The restriction of these molecular markers at AV boundary suggests that these signaling pathways may be critical regulators of valve formation.

Molecular insights into valve formation in zebrafish first came from identification of the gene disrupted by the *jekyll* (*jek*) mutation which affects valve formation. *jek* encodes uridine 5′ diphosphate-glucose dehydrogenase (Ugdh), which is also known as Sugarless in *Drosophila*. Ugdh is required for the conversion of UDP-glucose into UDP-glucuronate for the production of heparin sulfate, chondroitin sulfate, and hyaluronic acid (Esko and Selleck, 2002). Proteoglycan biosynthesis has been implicated in a number of signaling pathways including Wnt, BMP, and FGF (Hacker *et al.*, 1997; Jackson *et al.*, 1997; Lin *et al.*, 1999). While the precise substrate for Jek in cardiac valve formation has yet to be identified, the Wnt signaling pathway has been implicated in valve formation via the identification of a mutation in the *adenomatous polyposis coli* (*apc*) gene of zebrafish (Hurlstone *et al.*, 2003). Apc is a component of the Axin-containing complex involved in phosphorylating β-catenin and targeting it for ubiquitination and degradation by the proteasome. Mutations in *apc* lead to the stabilization of β-catenin and constitutive activation of the Wnt signaling pathway (Fodde *et al.*, 2001). Activated canonical Wnt signaling, as seen by nuclear accumulation of β-catenin, is restricted to the AV endocardial cells in wild-type endocardium. In *apc* mutants, canonical Wnt signaling is activated throughout the endocardium and endocardial cushions form outside of the AV boundary.

In *jek* mutants, the AV endocardial cells fail to form endocardial cushions (Walsh and Stainier, 2001). On the molecular level, expression of AV boundary markers such as *bmp4*, *versican*, and *notch1b* fail to restrict to the AV boundary, indicating a defect in boundary formation. Interestingly, *clo* mutants, which lack endocardial cells, also fail to restrict AV myocardial genes (Walsh and Stainier, 2001). These observations suggest that myocardial-endocardial interactions are critical for determining the identity of the AV boundary.

While many aspects of zebrafish development are independent of circulation, blood flow appears to be a critical epigenetic factor in valve formation (Hove *et al.*, 2003). Inhibiting blood flow by insertion of a bead in front of the sinus venosus or the back of the ventricle in 37 hpf embryos leads to lack of heart looping and valve formation defects. Additionally, mutations that affect contractility such as *silent heart* and *cardiofunk* exhibit defects in endocardial cushions formation (Bartman *et al.*, 2004). These results underscore the importance of examining the interplay between genetic and epigenetic factors when analyzing cardiovascular defects.

F. Myocardial Remodeling

Following assembly into a functioning heart tube, considerable differentiation and morphogenesis continue in the embryo to form the mature heart. In addition to heart looping and valve formation, the myocardium undergoes significant maturation in the form of thickening of the ventricular wall and ventricular trabeculation (Hu *et al.*, 2000). At 36 hpf, morphological differences between the atrium and ventricle are visible by normarski optics (Stainier *et al.*, 1996). The ventricle appears thicker than the atrium which may reflect differences in sarcomere arrangement in the two chambers. Similarly, in transverse sections at 48 hpf, the ventricular myocardial cells appear thick and compact in comparison to the atrium (Fig. 1F′) (Berdougo *et al.*, 2003; Hu *et al.*, 2000). By 5-days postfertilization, trabeculae, finger-like projections of the myocardium, are visible in the ventricle (Hu *et al.*, 2000). While cardiac function appears to be important for myocardial maturation as mutations in genes encoding the sarcomere scaffold proteins Titin and Tnnt2, and the alpha subunit of the L-type calcium channel result in defective ventricular growth (Rottbauer *et al.*, 2001; Sehnert *et al.*, 2002; Xu *et al.*, 2002), the cellular and molecular mechanisms responsible for these steps in differentiation remain to be elucidated.

III. Gene Expression

Numerous genes have been reported to be expressed in the embryonic heart in zebrafish. This section is not intended to provide a comprehensive list of the genes expressed in the embryonic heart but rather to focus on the expression of a few genes that mark key stages in the patterning and morphogenesis of the embryonic heart. In addition to genes discussed in these sections, we have also delineated genes involved in heart looping and valve formation in the above sections on heart morphogenesis.

A. Lateral Plate Mesoderm Gene Expression

Both the myocardial and endocardial precursors arise from the anterior LPM, thus, the patterning and differentiation of the LPM is critical for heart tube formation. Morphologically the anterior LPM starts out as narrow bilateral stripes of tissues on either side of the midline. During mid-somitogenesis (10–20-somite stages), these bilateral stripes undergo a spreading process in which they expand along the medial-lateral extent of the embryo. Within the LPM a number of myocardial differentiation genes are expressed in overlapping patterns.

The bHLH transcription factor *hand2* is expressed throughout the A-P extent of the LPM at the on-set of somitogenesis (Yelon *et al.*, 2000). At the 10-somite stage, *hand2* expression segregates between the anterior and posterior domains of the LPM such that a gap forms between the two expression domains. As the anterior LPM spreads, *hand2* expression becomes broader in the anterior expression domain, while the posterior domain remains narrow. The expression pattern of *hand2* is consistent with its role in LPM morphogenesis and myocardial differentiation as mutations in *hand2* result in a lack of LPM spreading and myocardial differentiation defects (Yelon *et al.*, 2000).

Members of the *Gata* family of transcription factor genes (*4/5/6*) are also expressed in the LPM. *Gata* genes are critical regulators of myocardial differentiation and participate in intricate cross-regulatory networks during embryogenesis (Charron *et al.*, 1999; Koutsourakis *et al.*, 1999; Kuo *et al.*, 1997; Molkentin *et al.*, 1997; Reiter *et al.*, 1999). At the 5–10-somite stages, *gata4/5/6* expression domains correspond to different A-P extent of the anterior LPM (Fig. 4). *gata4* expression in the anterior LPM extends to the optic cup and ends posterior to the anterior tip of the notochord (Heicklen-Klein and Evans, 2004; Serbedzija *et al.*, 1998). *gata5* expression extends posterior to the *gata4* expression domain but does not appear to mark the entire LPM when compared to *hand2* expression (Heicklen-Klein and Evans, 2004; Yelon *et al.*, 2000). Although *gata6* expression extends more posteriorly than *gata4*, its expression domain is shorter than that of the *gata5* expression domain (Heicklen-Klein and Evans, 2004). These overlapping expression patterns in the anterior LPM are consistent with the cross-regulation observed among the Gata family of transcription factors. *Gata6* null mice die at E5.5 and fail to express *Gata4* (Koutsourakis *et al.*, 1999). Similarly, *gata5* mutants in zebrafish exhibit a reduction in *gata4* expression in the LPM and overexpression of *gata5* is sufficient to induce ectopic expression of *gata4* and *gata6* (Reiter *et al.*, 1999). Thus, the patterns of overlapping genes may define an overall code for the differentiation of myocardial and endocardial precursors.

The cardiac field, as defined by the expression of the homeodomain gene *nkx2.5*, occupies a subdomain of the anterior LPM (Chen and Fishman, 1996). *nkx2.5*, the homolog of *Drosophila tinman*, is expressed in the myocardial precursors in all vertebrates (Komuro and Izumo, 1993; Lints *et al.*, 1993; Schultheiss *et al.*, 1995; Tonissen *et al.*, 1994;). In *Drosophila*, the *tinman* mutation results in a complete lack of heart formation (Bodmer *et al.*, 1990). In zebrafish, *nkx2.5* expression in the anterior LPM is initially detected at the 5-somite stage (Alexander *et al.*, 1998). At the 10-somite stage, *nkx2.5* expression in the LPM extends posteriorly to the otic vesicle and anteriorly to halfway between the notochord and the eye (Fig. 4) (Serbedzija *et al.*, 1998). The anterior tip of the notochord marks the median of the *nkx2.5* expression domain. The A-P boundary of the *nkx2.5* expression domain is maintained throughout the migration stages. Thereafter, *nkx2.5* is expressed throughout the embryonic myocardium.

Expression of *nkx2.5* anterior to the notochord appears to define the cardiac field. Lineage analyses of the LPM indicate that only cells anterior to the notochord in the *nkx2.5* expression domain contributes progeny to the myocardium (Goldstein and Fishman, 1998). Additionally, when the cardiac progenitors are

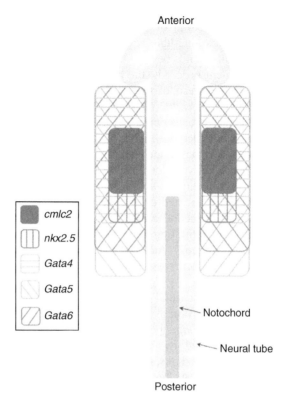

Fig. 4 Overlapping gene expression patterns in the anterior LPM. In the 5–10-somite stages embryo, *gata4/5/6* expression domains correspond to different A-P extent of the anterior LPM. The *gata4* expression domain (green horizontal stripes) in the anterior LPM extends to the optic cup and ends posterior to the tip of the notochord. The *gata5* expression domain (orange diagonal stripes) extends furthest posteriorly, while the *gata6* expression domain (blue diagonal stripes) extends more posteriorly than *gata4*, but anterior to that of the *gata5* expression domain. The *nkx2.5* expression domain (purple vertical stripes) extends posterior to the tip of the notochord and anteriorly to halfway between the notochord and the eye. *cmlc2* expression (solid red) initiates in the 13-somite stage embryo and is limited to the anterior tip of the notochord. (See Pleate no. 19 in the Color Plate Section.)

ablated only LPM cells lateral and anterior to the normal cardiac progenitors can compensate for the ablated cells (Serbedzija *et al.*, 1998). Therefore, the ability to contribute to the heart is limited to the region of the LPM anterior to the notochord and not all *nkx2.5* expressing cells contribute to the myocardium.

B. Myocardial Gene Expression

Myocardial specific gene expression initiates at the 13-somite stage with the expression of distinct myosin genes in the bilateral populations of myocardial precursors (Yelon *et al.*, 1999). *cardiac myosin light chain 2* (*cmlc2*) expression overlaps with that of *nkx2.5* but is more restricted. The posterior boundary of *cmlc2* expression aligns with the anterior boundary of the notochord, while *nkx2.5* expression extends beyond the tip of the notochord (Fig. 4) (Yelon *et al.*, 1999). As development proceeds, the bilateral stripes of *cmlc2*-expressing cells delineate the migration pattern of the myocardial precursors (figure and see   Section II.B). At 18-somite, *cmlc2* expression resembles a butterfly pattern, with the medial *cmlc2*-positive cells making contact at the midline. By 21-somite when myocardial fusion is completed, *cmlc2* expression marks the cardiac cone. Subsequently, *cmlc2* expression is maintained in the myocardium throughout development.

Myocardial ventricular specific gene expression also initiates at the 13-somite stage. At this stage, *ventricular myosin heavy chain* (*vmhc*) expression is restricted to the medially located cells of the bilateral myocardial populations (Yelon *et al.*, 1999). These medial cells compose the leading edge of the migrating myocardial precursors, suggesting that ventricular precursors may play an essential role in the migration process. Additionally, the early expression of *vmhc* indicates that the myocardial precursors exhibit molecular chamber identity prior to morphological differences.

Known myocardial atrial specific gene expression begins at the 19-somite stage with the expression of the *atrial myosin heavy chain* (*amhc*) (Berdougo *et al.*, 2003). At this stage, *amhc*-positive cells mark the outer portion of the forming cardiac cone as defined by *cmlc2* expression. The inner portion of the cardiac cone expresses *vmhc*. Thus, the expression pattern of *amhc* and *vmhc* are complementary to each other. As heart tube morphogenesis proceeds, the two complementary expression patterns are maintained. The segregation of *amhc* and *vmhc* expression at these early stages, indicate that the myocardial precursors are patterned before heart tube formation.

C. Endocardial Gene Expression

Many of the genes expressed in the endocardial cells are expressed throughout the vascular endothelium. The endothelial-specific receptor tyrosine kinase gene *flk-1* is expressed in two bilateral stripes of cells within the LPM at the 5-somite stage (Liao *et al.*, 1997). These bilateral stripes of *flk-1*-expressing cells appear more medial than the *nkx2.5* expression domain (Alexander *et al.*, 1998). As development proceeds, the bilateral stripes of *flk-1* expression extend in both the anterior and posterior directions. By the 15-somite stage, at the level of the cardiac field, *flk-1* expressing cells can be detected medial to the bilateral populations of myocardial precursors. At the 18-somite stage, the medial expression of *flk-1* is seen as a dense cluster and mostly marks the endocardial precursors (Liao *et al.*, 1997). As the heart tube elongates and loops, *flk-1* expression is detected in the inner layer of the tube, as well as in the developing vasculature. Two other receptor tyrosine kinase genes *tie-2* and *flt-4* are expressed in a similar pattern as *flk-1* starting at the 18-somite stage (Liao *et al.*, 1997; Thompson *et al.*, 1998).

The commonality in gene expression between the endocardial and endothelial cells suggests that two cell types share a common progenitor.

One gene expressed in the endocardial precursors, but not the endothelial cells, is the extracellular matrix gene *fibronectin*. *fibronectin* appears to be one of the earliest markers of the endocardial precursors as it is expressed in the endocardial precursors prior to the midline migration of the myocardial precursors (Trinh and Stainier, 2004). At the start of somitogenesis, *fibronectin* is initially expressed in the anterior LPM. However, starting between the 11- and 12-somite stages, *fibronectin* expression can be observed in a few cells medial to the LPM. The medial expression spans the midline by the 15-somite stage and is maintained throughout the stages of myocardial migration to the midline. In transverse sections, the *fibronectin* expressing cells occupy two ventral layers of cells at the midline. This expression pattern is absent in *clo* mutants which lack endocardial cells, indicating the midline *fibronectin* expression corresponds to endocardial expression. The medial progression of the *fibronectin* expression between the 10- and 15-somite stages may reflect the migration pathway of the endocardial precursors to the midline.

IV. Conclusion and Future Directions

In this chapter, we have focused on the events and processes known to govern heart morphogenesis in zebrafish. While we are gaining a better understanding of some of the processes that regulate zebrafish cardiac development, from the initial specification of the cardiac precursors to the subsequent morphogenesis in forming a functional organ, much remains to be done to gain a full understanding of the various genes and morphogenetic processes that control heart development.

The recent development of transgenic zebrafish has provided powerful tools for the analyses of cardiac morphogenesis on a detailed cellular level. In particular, transgenic lines that use cardiac specific promoters to drive expression of fluorescent proteins allow for the visualization of cell behavior throughout development in both wild-type and mutant situations. Several cardiac transgenic lines have been created that have facilitated our understanding of the complex cell behavior and tissue-tissue interactions during different stages in cardiac morphogenesis. In combination with live imaging, these various transgenic lines have provided insights into the cellular processes such as migration, shape changes, and proliferation that are involved in heart tube assembly and remodeling.

Additionally, the identification of additional regulators of cardiac development through the cloning of existing mutations and isolation of new loci will be instrumental to further our understanding of the genetic networks involved in cardiac development. These studies will provide the framework to clarify the relationship between genetic networks that regulate cardiac formation and the morphogenetic processes that drive organogenesis.

References

Alexander, J., Rothenberg, M., Henry, G. L., and Stainier, D. Y. (1999). Casanova plays an early and essential role in endoderm formation in zebrafish. *Dev. Biol.* **215**, 343–357.

Alexander, J., Stainier, D. Y., and Yelon, D. (1998). Screening mosaic F1 females for mutations affecting zebrafish heart induction and patterning. *Dev. Genet.* **22**, 288–299.

Baker, K., Holtzman, N. G., and Burdine, R. D. (2008). Direct and indirect roles of Nodalsignaling in two axis conversions during asymmetric morphogenesis of thezebrafish heart. *PNAS* **105**, 13924–13929.

Bartman, T., Walsh, E. C., Wen, K. K., McKane, M., Ren, J., Alexander, J., Rubenstein, P. A., and Stainier, D. Y. R. (2004). Early myocardial function aVects endocardial cushion development in zebrafish. *PLoS* **2**, 673–681.

Beis, D., Bartman, T., Jin, S. W., Scott, I. C., D'Amico, L. A., Ober, E. A., Verkade, H., Frantsve, J., Field, H. A., Wehman, A., Baier, H., Tallafuss, A., *et al.* (2005). Genetic and cellular analyses of zebrafish atrioventricular cushion and valve development. *Development* **132**, 4193–4204.

Berdougo, E., Coleman, H., Lee, D. H., Stainier, D. Y. R., and Yelon, D. (2003). Mutation of *weak atrium/atrial myosin heavy chain* disrupts atrial function and influences ventricular morphogenesis in zebrafish. *Development* **130**, 6121–6129.

Bisgrove, B. W., Essner, J. J., and Yost, H. J. (1999). Regulation of midline development by antagonism of lefty and nodal signaling. *Development* **126**, 3253–3262.

Bisgrove, B. W., Essner, J. J., and Yost, H. J. (2000). Multiple pathways in the midline regulate concordant brain, heart and gut left-right asymmetry. *Development* **127**, 3567–3579.

Bodmer, R., Jan, L. Y., and Jan, Y. N. (1990). A new homeobox-containing gene, *msh-2* (*tinman*), is transiently expressed during mesoderm formation in Drosophila. *Development* **110**, 661–669.

Bussmann, J., Bakkers, J., and Schulte-Merker, S. (2007). Early endocardial morphogenesis requires scl/tal1. *PLoS* **3**, 1425–1437.

Charron, F., Paradis, P., Bronchain, O., Nemer, G., and Nemer, M. (1999). Cooperative interaction between GATA4 and GATA6 regulates myocardial gene expression. *Mol. Cell. Biol.* **19**, 4355–4356.

Chen, J. N., and Fishman, M. C. (1996). Zebrafish *tinman* homolog demarcates the heart field and initiates myocardial differentiation. *Development* **122**, 3809–3816.

Chen, J. N., HaVter, P., Odenthal, J., Vogelsang, E., Brand, M., van Eedan, F. J. M., Furutani-Seike, M., Granato, M., Hammerschmidt, M., Heisenberg, C. P., *et al.* (1996). Mutations affecting the cardiovascular system and other internal organs in zebrafish. *Development* **123**, 293–302.

Chen, J. N., van Eeden, F. J. M., Warren, K. S., Chin, A., Nüsslein-Volhard, C., HaVter, P., and Fishman, M. C. (1997). Left-right pattern of cardiac bmp4 may drive asymmetry of the heart in zebrafish. *Development* **124**, 4373–4382.

Chin, A. J., Tsang, M., and Weinberg, E. S. (2000). Heart and gut chiralities are controlled independently from initial heart position in the developing zebrafish. *Dev. Biol.* **227**, 403–421.

Cibrian-Uhalte, E., Langenbacher, A., Shu, X., Chen, J. N., and Abdelilah-Seyfried, S. (2007). Involvement of zebrafish Na$^+$, K$^+$ ATPase in myocardial cell junction maintenance. *J. Cell Biol.* **176**, 223–230.

Danos, M. C., and Yost, H. J. (1996). Role of notochord in specification of cardiac left-right orientation in zebrafish and Xenopus. *Dev. Biol.* **177**, 96–103.

David, N. B., and Rosa, F. M. (2001). Cell autonomous commitment to an endodermal fate and behaviour by activation of Nodal signalling. *Development* **128**, 3937–3947.

Esko, J. D., and Selleck, S. B. (2002). Order out of chaos: Assembly of ligand binding sites in heparin sulfate. *Annu. Rev. Biochem.* **71**, 435–471.

Essner, J. J., Branford, W. W., Zhang, J., and Yost, H. J. (2000). Mesentoderm and left-right brain, heart and gut development are diVerentially regulated by pitx2 isoforms. *Development* **127**, 1081–1093.

Fodde, R., Smits, R., and Clevers, H. (2001). APC, signal transduction and genetic instability in colorectal cancer. *Nat. Rev. Cancer* **1**, 55–67.

Forouhar, A. S., Liebling, M., Hickerson, A., Nasiraei-Moghaddam, A., Tsai, H. J., Hove, J. R., Fraser, S. E., Dickinson, M. E., and Gharib, M. (2006). The embryonic vertebrate heart tube is a dynamic suction pump. *Science* **312**, 751–753.

Goldstein, A. M., and Fishman, M. C. (1998). Notochord regulates cardiac lineage in zebrafish embryos. *Dev. Biol.* **201,** 247–252.

Hacker, U., Lin, X., and Perrimon, N. (1997). The *Drosophila sugarless* gene modulates Wingless signaling and encodes an enzyme involved in polysaccharide biosynthesis. *Development* **124,** 3565–3573.

Heicklen-Klein, A., and Evans, T. (2004). T-box binding sites are required for activity of a cardiac GATA4 enhancer. *Dev. Biol.* **267,** 490–504.

Hla, T. (2003). Signaling and biological actions of sphingosine 1-phosphate. *Pharmacol. Res.* **47,** 401–407.

Holtzman, N. G., Schoenebeck, J. J., Tsai, H. J., and Yelon, D. (2007). Endocardium is necessary for cardiomyocyte movement during heart tube assembly. *Development* **134,** 2379–2386.

Horne-Badovinac, S., Lin, D., Waldron, S., Schwarz, M., Mbamalu, G., Pawson, T., Jan, J., Stainier, D. Y. R., and Abdelilah-Seyfried, S. (2001). Positional cloning of heart and soul reveals multiple roles for PKC lambda in zebrafish organogenesis. *Curr. Biol.* **11,** 1492–1502.

Hove, J. R., Koster, R. W., Forouhar, A. S., Acevedo-Bolton, G., Fraser, S. E., and Gharib, M. (2003). Intracardiac fluid forces are an essential epigenetic factor for embryonic cardiogenesis. *Nature* **421,** 172–177.

Hu, N., Sedmera, D., Yost, H. J., and Clark, E. B. (2000). Structure and function of the developing zebrafish heart. *Anat. Rec.* **260,** 148–157.

Hurlstone, A. F. L., Haramis, A. G., Wienholds, E., Begthel, H., Korving, J., van Eeden, F., Cuppen, E., Zivkovic, D., Plasterk, R. H. A., and Clevers, H. (2003). The Wnt/β-catenin pathway regulates cardiac valve formation. *Nature* **425,** 633–637.

Jackson, S. M., Nakato, H., Sugiura, M., Jannuzi, A., Oakes, R., Kaluza, V., Golden, C., and Selleck, S. B. (1997). *Dally,* a *Drosophila* glypican, controls cellular responses to the TGFβ-related morphogen, Dpp. *Development* **124,** 4113–4120.

Jin, S. W., Jungblut, B., and Stainier, D. Y. R. (2002). Angiogenesis in zebrafish. *In* "Genetics of Angiogenesis" (J. Honig, ed.), pp. 101–118. BIOS, London.

Keegan, B. R., Meyer, D., and Yelon, D. (2004). Organization of cardiac chamber progenitors in the zebrafish blastula. *Development* **131,** 3081–3091.

Keegan, B. R., Feldman, J. L., Begemann, G., Ingham, P. W., and Yelon, D. (2005). Retinoic acid signaling restricts the cardiac progenitor pool. *Science* **307,** 247–249.

Kikuchi, Y., Trinh, L. A., Reiter, J. F., Alexander, J., Yelon, D., and Stainier, D. Y. R. (2000). The zebrafish *bonnie and clyde* gene encodes a Mix family homeodomain protein that regulates the generation of endodermal precursors. *Genes Dev.* **14,** 1279–1289.

Komuro, I., and Izumo, S. (1993). *Csx*: A murine homeobox-containing gene specifically expressed in the developing heart. *Proc. Natl. Acad. Sci. USA* **90,** 8145–8149.

Koutsourakis, M., Langeveld, A., Patient, R., Beddington, R., and Grosveld, F. (1999). The transcription factor GATA6 is essential for early extraembryonic development. *Development* **126,** 723–731.

Kuo, C. T., Morrisey, E. E., Anandappa, R., Sigrist, K., Lu, M. M., Parnacek, M. S., Soudais, C., and Leiden, J. M. (1997). GATA4 transcription factor is required for ventral morphogenesis and heart tube formation. *Genes Dev.* **11,** 1048–1060.

Kupperman, E., An, S., Osborne, N., Waldron, S., and Stainier, D. Y. (2000). A sphingosine-1-phosphate receptor regulates cell migration during vertebrate heart development. *Nature* **406,** 192–195.

Lawson, N. D., and Weinstein, B. M. (2002). Arteries and veins: Making a difference with zebrafish. *Nat. Rev. Genet.* **3,** 674–682.

Lee, R. K. K., Stainier, D. Y. R., Weinstein, B. M., and Fishman, M. C. (1994). Cardiovascular development in the zebrafish. II. Endocardial progenitors are sequestered within the heart field. *Development* **120,** 3361–3366.

Liao, W. B., Bisgrove, W., Sawyer, H., Hug, B., Bell, B., Peters, K., Grunwald, D. J., and Stainier, D. Y. R. (1997). The zebrafish gene *cloche* acts upstream of a *flk-1* homologue to regulate endothelial cell differentiation. *Development* **124,** 381–389.

Lin, X., BuV, E. M., Perrimon, N., and Michelson, A. M. (1999). Heparan sulfate proteoglycans are essential for Fgf receptor signaling during *Drosophila* embryonic development. *Development* **126,** 3715–3723.

Lints, T. J., Parsons, L. M., Hartley, L., Lyons, I., and Harvey, R. P. (1993). *Nkx2.5*: A novel murine homeobox gene expressed in early heart progenitor cells and their myogenic descendants. *Development* **119**, 419–431.

Long, S., Ahmad, N., and Rebagliati, M. (2003). The zebrafish nodal-related gene *southpaw* is required for visceral and diencephalic left-right asymmetry. *Development* **130**, 2303–2316.

Molkentin, J. D., Lin, Q., Duncan, S. A., and Olson, E. N. (1997). Requirement of the transcription factor GATA4 for heart tube formation and ventral morphogenesis. *Genes Dev.* **11**, 1061–1072.

Panetti, T. S., Nowlen, J., and Mosher, D. F. (2000). Sphingosine-1-phosphate and lysophosphatidic acid stimulate endothelial cell migration. *Arterioscler. Thromb. Vasc. Biol.* **20**, 1013–1019.

Paul, S. M., Ternet, M., Salvaterra, P. M., and Beitel, G. J. (2003). The Na+/K + ATPase is required for septate junction function and epithelial tube-size control in the Drosophila tracheal system. *Development* **130**, 4963–4974.

Peterson, R. T., Mably, J. D., Chen, J. N., and Fishman, M. C. (2001). Convergence of distinct pathways to heart patterning revealed by the small molecule concentramide and the mutation heart-and-soul. *Curr. Biol.* **11**, 1481–1491.

Rebagliati, M. R., Toyama, R., Fricke, C., HaVter, P., and Dawid, I. B. (1998). Zebrafish nodal-related genes are implicated in axial patterning and establishing left-right asymmetry. *Dev. Biol.* **199**, 261–272.

Reiter, J. F., Alexander, J., Rodaway, A., Yelon, D., Patient, R., Holder, N., and Stainier, D. Y. R. (1999). *Gata5* is required for the development of the heart and endoderm in zebrafish. *Genes Dev.* **13**, 2983–2995.

Reiter, J. F., Verkade, H., and Stainier, D. Y. R. (2001). *Bmp2b* and *Oep* promote early myocardial differentiation through their regulation of *gata5*. *Dev. Biol.* **234**, 330–338.

Rohr, S., Bit-Avragim, N., and Abdelilah-Seyfried, S. (2005). Heart and soul/PRKCi and nagie oko/Mpp5 regulate myocardial coherence and remodeling during cardiac morphogenesis. *Development* **133**, 107–115.

Rohr, S., Otten, C., and Abdelilah-Seyfried, S. (2008). Asymmetric involution of the myocardial field drives heart tube formation in zebrafish. *Circ. Res.* **102**(2), e12–e19.

Rottbauer, W., Baker, K., Wo, Z. G., Mohideen, M. A., Cantiello, H. F., and Fishman, M. C. (2001). Growth and function of the embryonic heart depend upon the cardiac-specific L-type calcium channel alpha1 subunit. *Dev. Cell* **2**, 265–275.

Scherz, P. J., Huisken, J., Sahai-Hernandez, P., and Stainier, D. Y. R. (2008). High-speed imaging of developing heart valves reveals interplay of morphogenesis and function. *Development* **135**, 1179–1187.

Schier, A. F., Neuhauss, S. C., Helde, K. A., Talbot, W. S., and Driever, W. (1997). The *one-eyed-pinhead* gene functions in mesoderm and endoderm formation in zebrafish and interacts with *no tail*. *Development* **124**, 327–342.

Schilling, T. F., Concordet, J. P., and Ingham, P. W. (1999). Regulation of left-right asymmetries in the zebrafish by *shh* and *bmp4*. *Dev. Biol.* **210**, 277–287.

Schultheiss, T. M., Xydas, S., and Lassar, A. B. (1995). Induction of avian cardiac myogenesis by anterior endoderm. *Development* **121**, 4203–4214.

Sehnert, A. J., Huq, A., Weinstein, B. M., Walker, C., Fishman, M., and Stainier, D. Y. (2002). Cardiac troponin T is essential in sarcomere assembly and cardiac contractility. *Nat. Genet.* **31**, 106–110.

Serbedzija, G. N., Chen, J. N., and Fishman, M. C. (1998). Regulation of the heart field in zebrafish. *Development* **125**, 1095–1101.

Shu, S., Cheng, K., Patel, N., Chen, F., Joseph, E., Tsai, H. J., and Chen, J. N. (2003). Na, K-ATPase is essential for embryonic heart development in the zebrafish. *Development* **130**, 6165–6173.

Smith, K. A., Chocron, S., von der Hardt, S., de Pater, E., Soufan, A., Bussmann, J., Schulte-Merker, S., Hammerschmidt, M., and Bakkers, J. (2008). Rotation and asymmetric development of the zebrafish heart requires directed migration of cardiac progenitor cells. *Dev. Cell* **14**, 287–297.

Stainier, D. Y. R., Beis, D., Jungblut, B., and Bartman, T. (2002). Endocardial cushion formation in zebrafish. *Cold Spring Harb Symp. Quant. Biol.* **67**, 49–56.

Stainier, D. Y. R., and Fishman, M. C. (1992). Patterning the zebrafish heart tube: acquisition of anteroposterior polarity. *Dev. Biol.* **153**, 91–101.

Stainier, D. Y., Fouquet, B., Chen, J., Warren, K. S., Weinstein, B. M., Meiler, S. E., Mohideen, M. P. K., Neuhauss, S. C. F., Solnica-Krezel, L., Schier, A. F., Zwartkruis, F., Stemple, D. L., *et al.* (1996). Mutations affecting the formation and function of the cardiovascular system in the zebrafish embryo. *Development* **123**, 285–292.

Stainier, D. Y., Lee, R. K., and Fishman, M. (1993). Cardiovascular development in the zebrafish. I. Myocardial fate map and heart tube formation. *Development* **119**, 31–40.

Stainier, D. Y. R., Weinstein, B. M., Detrich, H. W., Zon, L. I., and Fishman, M. C. (1995). *cloche*, an early acting zebrafish gene, is required by both the endothelial and hematopoietic lineages. *Development* **121**, 3141–3150.

Thisse, C., and Thisse, B. (1999). *Antivin*, a novel and divergent member of the TGF-beta superfamily, negatively regulates mesoderm induction. *Development* **126**, 229–240.

Thompson, M. A., Ransom, D. G., Pratt, S. J., MacLennan, H., Kieran, M. W., Detrich, H. W., Vail, B., Huber, T. L., Paw, B., Brownlie, A. J., Oates, A. C., Fritz, A, *et al.* (1998). The *cloche* and spadetail genes differentially affect hematopoiesis and vasculogenesis. *Dev. Biol.* **197**, 248–269.

Tonissen, K. F., Drysdale, T. A., Lints, T. J., Harvey, R. P., and Krieg, P. A. (1994). *XNkx2.5*, a Xenopus gene related to Nkx2.5 and tinman: Evidence for a conversed role in cardiac development. *Dev. Biol.* **162**, 325–328.

Trinh, L. A., and Stainier, D. Y. R. (2004). Fibronectin regulates epithelial organization During myocardial migration in zebrafish. *Dev. Cell* **3**, 371–382.

Trinh, L. A., Yelon, D., and Stainier, D. Y. R. (2005). Hand2 regulates epithelial formation during myocardial differentiation. *Curr. Biol.* **15**, 441–446.

Walsh, E. C., and Stainier, D. Y. R. (2001). UDP-glucose dehydrogenase required for cardiac valve formation in zebrafish. *Science* **293**, 1670–1673.

Warga, R. M., and Kimmel, C. (1990). Cell movements during epiboly and gastrulation in zebrafish. *Development* **108**, 569–580.

Warga, R. M., and Nüsslein-Volhard, C. (1999). Origin and development of the zebrafish endoderm. *Development* **126**, 827–838.

Warren, K. S., Wu, J. C., Pinet, F., and Fishman, M. C. (2000). The genetic basis of cardiac function: Dissection by zebrafish (*Danio rerio*) screens. *Philos. Trans. R. Soc. Lond. B Biol. Sci.* **355**, 939–944.

Xu, X., Meiler, S. E., Zhong, T. P., Mohideen, M., Crossley, D. A., Burggren, W. W., and Fishman, M. C. (2002). Cardiomyopathy in zebrafish due to mutation in an alternatively spliced exon of *titin*. *Nat. Gen.* **30**, 205–209.

Yelon, D., Horne, S. A., and Stainier, D. Y. R. (1999). Restricted expression of cardiac myosin genes reveals regulated aspects of heart tube assembly in zebrafish. *Dev. Biol.* **214**, 23–37.

Yelon, D., Ticho, B., Halpern, M. E., Ruvinsky, I., Ho, R. K., Silver, L. M., and Stainier, D. Y. R. (2000). The bHLH transcription factor *hand2* plays parallel roles in zebrafish heart and pectoral fin development. *Development* **127**, 2573–2582.

CHAPTER 21

Zebrafish Kidney Development

Iain A. Drummond

Departments of Medicine and Genetics
Harvard Medical School and Nephrology Division
Massachusetts General Hospital
Charlestown, Massachusetts 02129

I. Introduction
II. Pronephric Structure and Function
III. Pronephric Development
    A. Patterning the Intermediate Mesoderm
    B. Development of the Pronephric Nephrons
    C. Pronephric Vascularization
IV. The Zebrafish Pronephros as a Model of Human Disease
    A. Disease of the Glomerular Filtration Apparatus
    B. Disease of the Tubules
V. Conclusions
    References

I. Introduction

Kidney development involves the formation of epithelial tubules from mesenchymal cells followed by interactions of epithelial cells with vascular tissue. The final result is an organ that filters blood and regulates body fluid composition. In the course of vertebrate evolution, three distinct kidneys of increasing complexity have been generated: the pronephros, mesonephros, and metanephros (Saxén, 1987). The pronephros is the first kidney to form during embryogenesis. In vertebrates with free-swimming larvae, including amphibians and teleost fish, the pronephros is the functional kidney of early larval life (Howland, 1921; Tytler, 1988; Tytler *et al.*, 1996; Vize *et al.*, 1997) and is required for proper osmoregulation (Howland, 1921). Later, in juvenile stages of zebrafish development, a mesonephros forms around and along the length of the pronephros and later serves as the kidney of adult fish. Despite some differences in organ morphology between

ESSENTIAL ZEBRAFISH METHODS:
CELL AND DEVELOPMENTAL BIOLOGY
Copyright 2009, Elsevier Inc. All rights reserved.

501

DOI: 10.1016/B978-0-12-374599-6.00021-2

the various kidney forms, many common elements exist at the cellular and molecular level that can be exploited to further our understanding of epithelial and vascular differentiation in particular, and organogenesis in general. To date, the zebrafish pronephros has provided a useful model of nephrogenic mesoderm differentiation, kidney cell type differentiation, nephron patterning, kidney: vasculature interactions, glomerular function, and diseases affecting glomerular filtration and tubule lumen size, that is cystic kidney disease. While much remains to be done, the basic features of zebrafish pronephric development and patterning have emerged from studies using simple histology, cell lineage tracing, gene expression patterns, and analysis of zebrafish mutants affecting this process.

II. Pronephric Structure and Function

The kidney has two principal functions: to remove waste from the blood and to balance ion and metabolite concentrations in the blood within physiological ranges that support proper functioning of all other cells (Vize *et al.*, 2002). Kidney function is achieved largely by first filtering the blood and then recovering useful ions and small molecules by directed epithelial transport. This work is performed by *nephrons*, the functional unit of the kidney (Fig. 1). The nephron is partitioned into distinct segments, each specialized for blood filtration or specific solute transport activities (Fig. 1C and D). The *glomerulus* is the site of blood filtration and primary urine formation. It is composed of specialized epithelial cells called *podocytes* that form a basket like extension of cellular processes around a *capillary tuft*. The basement membrane between podocytes and capillary endothelial cells together with the specialized junctions between the podocyte cell processes (*slit-diaphragms*) function as a blood filtration barrier, allowing passage of small molecules, ions, and blood fluid into the urinary space, while retaining high molecular weight proteins in the vascular system (Fig. 1; see also Fig. 12D). Blood filtrate is collected first in the epithelial capsule surrounding the capillary tuft (the *nephrocoel* or *Bowman's capsule*) and then processed as it travels down the lumen of the kidney tubules and ducts where segment specific transport processes recover useful ions and metabolites (Fig. 1 and Vize *et al.*, 2002).

In general, osmoregulation by the fresh water adult teleost kidney is achieved by a high rate of blood filtration, active recovery of essential salts and other blood solutes, and the excretion of copious amounts of dilute urine (Hickman and Trump, 1969). The pronephros is necessary for larval osmoregulation since most mutants with pronephric defects die of edema. Most texts on fish physiology state that nitrogenous waste removal from the blood is not a major function of the fish kidney and is instead carried out primarily by ammonia transport mechanisms in the gills (Hickman and Trump, 1969). Since newly hatched zebrafish larvae do not yet have gills, it is worth considering that the pronephros may also be the primary site of nitrogenous waste removal in the first days of life.

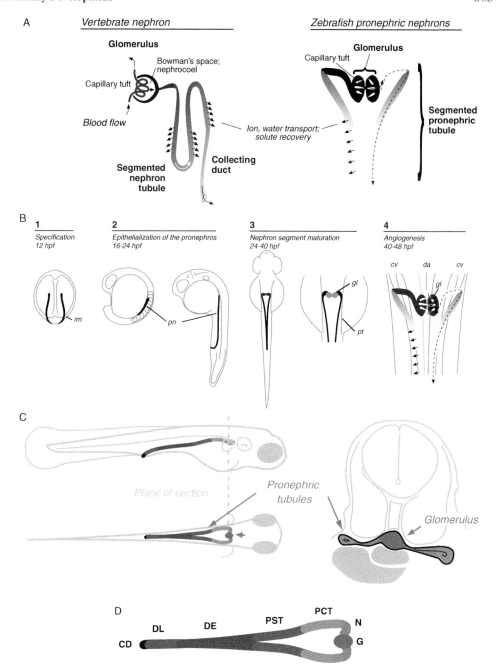

Fig. 1 The zebrafish pronephros. (A) Functional features of the vertebrate nephron and the zebrafish pronephric nephrons. See text for details. (B) Stages in zebrafish pronephric kidney development. (1) Specification of mesoderm to a nephric fate: expression patterns of *pax2.1* and *lim-1* define a posterior region of the intermediate mesoderm (*im*) and suggest that a nephrogenic field is established in early

In zebrafish, and several other teleosts, the larval pronephros consists of only two nephrons with glomeruli fused at the embryo midline just ventral to the dorsal aorta (Fig. 1C; Agarwal and John, 1988; Armstrong, 1932; Balfour, 1880; Drummond, 2000b; Drummond et al., 1998; Goodrich, 1930; Hentschel and Elger, 1996; Marshall and Smith, 1930; Newstead and Ford, 1960; Tytler, 1988; Tytler et al., 1996). Although simple in form, the pronephric glomerulus is composed of cell types typical of higher vertebrates kidneys including fenestrated capillary endothelial cells, podocytes, and polarized tubular epithelial cells (Drummond et al., 1998). Pronephric tubules and ducts complete the nephrons, connecting the glomerulus to the outside world at the cloaca. The zebrafish pronephric nephrons form a closed system of blood filtration, tubular resorption, and fluid excretion.

III. Pronephric Development

Formation of the pronephros occurs during the first two days of zebrafish embryonic development. Pronephric development can be conceptualized in three broad stages: (1) the commitment of undifferentiated mesodermal cells to a nephrogenic fate, (2) the epithelialization and patterning of glomerular and tubule cells of the nephrons, and (3) the tissue interactions between endothelial cells and kidney podocytes and the formation of the glomerular capillary tuft that leads to the onset of blood filtration and nephron function (Fig. 1B). In embryological terms, the first kidney structures to form are the pronephric ducts, so-named based on their anatomical homology with pronephric or Wolffian ducts in amphibians, chickens, and mammals (Vize et al., 2002). Although outwardly similar to the mammalian nephric ducts, the pronephric ducts in zebrafish are now known to form the segmented tubules of pronephric nephrons and are not to be confused with "collecting ducts" in mammalian kidneys (Fig. 1C and D; Wingert et al., 2007). In this review, substitution of the terms "pronephric nephron" or "pronephric tubule" for the term "pronephric duct" reflects a consensus that the pronephros is really a segmented nephron, homologous in many ways to the mammalian nephron. The tubular components of the pronephric nephrons (historically, the pronephric tubules and ducts) are complete and patent to the exterior by 24 hours postfertilization (hpf; Kimmel et al., 1995). At about 30 hpf the

---

development. (2) Epithelialization of the pronephros (pn) follows somitogenesis and is complete by 24 hpf. (3) Patterning of the nephron gives rise to the pronephric glomerulus (gl) and pronephric tubules (pt). (4) Angiogenic sprouts from the dorsal aorta (da) invade the glomerulus and form the capillary loop. The cardinal vein (cv) is apposed to the tubules and receives recovered solutes. (C) Diagram of the mature zebrafish pronephric kidney in 3 days larva. A midline compound glomerulus connects to the segmented pronephric tubules that run laterally. The nephrons are joined at the cloaca where they communicate with the exterior. (D) Patterning of the pronephric nephron generates discrete segments: neck (N), proximal convoluted tubule (PCT), proximal straight tubule (PST), distal early (DE), late distal (DL), and collecting duct (CD).

pronephric tubules start to become convoluted adjacent to the glomerulus and by 48 hpf the glomerulus becomes a functional blood filter (Drummond *et al.*, 1998). This stepwise formation of nephron components echoes the general scheme of vertebrate kidney development.

A. Patterning the Intermediate Mesoderm

1. Origins and Patterning of Nephrogenic Mesoderm

Cell labeling and lineage tracing in zebrafish gastrula-stage embryos has demonstrated that cells destined to form the pronephros lie just dorsal to the heart progenitors in the shield-stage mesodermal germ ring, and appear to overlap somewhat with cells fated to form blood (Fig. 2A; Kimmel *et al.*, 1990). This position in the prospective mesoderm is similar to the region defined in amphibian embryos as the source of kidney tissue (Delarue *et al.*, 1997). These cells emerge shortly after gastrulation as a band of tissue, the intermediate mesoderm, at the ventro-lateral edge of the paraxial mesoderm (Fig. 2B and C). Although currently there is no data on when pronephric cells are specified in zebrafish embryos, it is likely that signaling events occurring at this early stage of development have a major impact on all subsequent differentiation of the pronephros.

The pronephric mesoderm at the blastula and gastrula stages is patterned by morphogen gradients in the early embryo which are dominated by BMP expression on the ventral side of the embryo and β-catenin/BMP inhibitor expression in the dorsal embryonic shield (Fig. 2A; Hammerschmidt *et al.*, 1996b; Neave *et al.*, 1997; Nikaido *et al.*, 1997). The dorsalized mutants *swirl* (*bmp2b*), *snailhouse* (*bmp7*), *somitabun* (*smad5*), and *lost-a-fin* (*alk8*) all lack signals necessary for ventral mesodermal development and show a reduction or elimination of *pax2.1* positive, presumptive kidney intermediate mesoderm (Hild *et al.*, 1999; Kishimoto *et al.*, 1997; Mullins *et al.*, 1996; Nguyen *et al.*, 1998). Conversely, the ventralized mutant *chordino* lacks the dorsal shield determinant Chordin and shows an expansion of the *pax2.1* expression domain at later stages (Hammerschmidt *et al.*, 1996b).

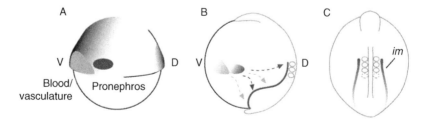

Fig. 2 Origins of the intermediate mesoderm. (A) Approximate positions of cells in a shield stage embryo destined to contribute to the blood/vasculature and pronephric lineages in the ventral (V) germ ring. (D; dorsal shield). (B) migration of cells during gastrulation to populate the intermediate mesoderm (*im*) (C).

The data from these mutants suggests that BMP signaling, in concert with signals from shield region, is required to specify kidney mesoderm.

The steps in pronephric mesoderm patterning subsequent to early BMP signaling are less well characterized. However, it is clear that further development of the lateral posterior mesoderm is critically dependent on two functionally related sets of genes: T-box transcription factors and fibroblast growth factors. The T-box transcription factor genes, *notail (ntl)*, *spadetail (spt)*, and *tbx6* function in concert to pattern the trunk and tail mesoderm (Griffin *et al.*, 1998; Hammerschmidt *et al.*, 1996a; Ho and Kane, 1990). In shield-stage embryos, all three of these genes are expressed in the ventral and lateral germ ring (Griffin *et al.*, 1998). *spadetail* and *tbx6* are later expressed broadly in the hypoblast (4–6-somite stage) out to the lateral edge of the forming mesoderm and are more specifically involved in trunk development (Griffin *et al.*, 1998). Recent work demonstrates that embryos lacking both *notail* and *spadetail* function do not express the pronephric mesoderm marker *pax2.1* (see below) and later show no signs of pronephric tubule development (Amacher *et al.*, 2002). A dramatic loss of all ventral-posterior embryonic structures is observed in embryos expressing a dominant negative FGF receptor (Griffin *et al.*, 1995, 1998). This is most likely due to the requirement for FGF signaling in the initiation or maintenance of *notail*, *spadetail*, and *tbx6* gene expression (Griffin *et al.*, 1995, 1998). In zebrafish, two FGFs, *fgf8* and *fgf24* function redundantly to promote posterior mesoderm development by maintaining *ntl* and *spt* expression (Draper *et al.*, 2003). In these experiments, a combined loss of *fgf8* and either *notail* or *spadetail*, which would be expected to substantially reduce T-box gene activity, did not reduce *pax2.1* expression in the pronephric mesoderm while axial mesoderm was severely affected (Draper *et al.*, 2003). It may be that pronephric mesodermal development requires only low levels of T-box gene function.

Another secreted factor found to influence trunk development and T-box gene expression is Wnt8 (Lekven *et al.*, 2001). The *wnt8* gene is expressed in the germ ring at gastrulation, transiently in the region of the intermediate mesoderm just prior to somitogenesis, and in a more prolonged fashion in the tail bud (Kelly *et al.*, 1995). Deletion mutants of the bicistronic *wnt8* locus show significant reduction in *tbx6* expression and loss of ventral-posterior mesoderm (Lekven *et al.*, 2001). Inhibition of *wnt8* mRNA translation in *ntl* and *spt* mutant embryos using injected morpholino antisense oligonucleotides has synergistic effects resulting in more profound loss of ventral posterior mesoderm (Lekven *et al.*, 2001). Thus trunk development is mediated by at least two secreted signals, *fgf8/24* and Wnt8, that control T-box gene expression and subsequent patterning of trunk mesoderm.

2. Pronephric Nephron Cell Lineage

By the early stages of somite formation, the pronephric intermediate mesoderm is clearly defined by specific pronephric marker gene expression. The Wilms' tumor suppressor, *wt1a*, encodes a zinc finger transcriptional regulator that is required for normal mouse kidney development and, when mutated in humans, gives rise to pediatric kidney tumors. Pax-2, a member of the paired-domain containing homeobox transcriptional regulators, has also been shown to be essential for kidney development in mammals as well as zebrafish (see below). Sim-1 is the vertebrate homolog of the Drosophila transcription factor gene *simple-minded*. Based on the overlapping but distinct expression patterns of the zebrafish homologs *wt1a, pax2.1,* and *sim1* in the zebrafish intermediate mesoderm (IM), Serluca and Fishman (2001) performed lineage studies by uncaging fluorescent dyes in the expression domains of these three genes prior to epithelialization. They showed that the areas of the intermediate mesoderm fated to become glomerulus, and what are now known to be segments of the proximal tubules (previously pronephric tubules and ducts) could be defined as sequential anterior to posterior subdomains of the IM that roughly correspond to the expression domains of *wt1a, pax2.1,* and *sim1* respectively (Fig. 3; Serluca and Fishman, 2001). In terms of gene function, *pax2.1* has been shown to be required for zebrafish pronephric proximal tubule formation (see below). In zebrafish, the mesenchyme to epithelial transition appears to occur *in situ* at all levels of the intermediate mesoderm (Serluca and Fishman, 2001). Ablation of cells within the IM does not result in subsequent absence of kidney tissue, suggesting the existence of a kidney morphogenetic field that can regulate, presumably under the influence of local environmental signals and downstream transcriptional circuits.

3. Gene Expression in the Intermediate Mesoderm

The expression of known signaling molecules and transcription factors in the forming pronephros provides entrance points to pathways that regulate early pronephric development. The zebrafish frizzled gene, *frz8a*, presumably encoding

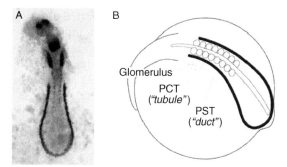

Fig. 3 Derivation of the pronephros from the intermediate mesoderm. (A) The *pax2.1* expression domain in early somitogenesis stage embryos defines a stripe of intermediate mesoderm fated to become the pronephric epithelia. (B) fate map of the nephrogenic intermediate mesoderm derived from fluorescent dye uncaging lineage experiments. Proximal fates previously referred to as "tubule" are now more accurately defined as proximal convoluted tubule (PCT) and more distal fates previously referred to as "duct" are now more accurately termed proximal straight tubule (PST).

a receptor for a locally acting Wnt signal, is expressed throughout the intermediate mesoderm at the 5-6-somite stage (Kim *et al.*, 1998). The timing of expression *frz8a* is significantly later than *wnt8* expression (see above) suggesting that a different Wnt ligand may signal via *frz8a*. *wnt4* is has been reported to be weakly expressed at about this time in bilateral anterior stripes of intermediate mesoderm (Ungar *et al.*, 1995), although we have not been able to confirm this (Liu *et al.*, 2000; Drummond, unpublished observation). *deltaC*, a ligand for notch, is expressed in the intermediate mesoderm adjacent to somites 1–4 at the 7-somite stage, suggesting that notch signaling may play a role in patterning the pronephric primordium (Smithers *et al.*, 2000). This expression domain corresponds to the precursors of the glomerulus and tubules, and extends anterior of the *pax2.1* expression domain (Fig. 3; Serluca and Fishman, 2001; Smithers *et al.*, 2000). The *foxc1a* gene, encoding a member of the forkhead/winged helix transcription factor family, is expressed strongly in forming somites and also in the intermediate mesoderm adjacent to somites 1–4 at the 7-somite stage (Topczewska *et al.*, 2001b). Morpholino inhibition of *foxc1a* expression results in loss of *deltaC* positive anterior kidney mesoderm while more posterior *pax2.1* positive mesoderm is unaffected (Topczewska *et al.*, 2001a). *foxc1a* most likely acts in the anterior pronephric mesoderm and might be expected to play a role in glomerular and/or tubular development, although this has not been experimentally confirmed. Several other transcription factors are also expressed in the presumptive pronephric mesoderm. As noted above, the paired-domain transcription factor *pax2.1* is expressed in a continuous band of intermediate mesoderm from the cloaca up to the posterior boundary of somite 2 (Carroll *et al.*, 1999; Drummond, 2000b; Heller and Brandli, 1999; Krauss *et al.*, 1991; Majumdar *et al.*, 2000; Mauch *et al.*, 2000; Pfeffer *et al.*, 1998; Puschel *et al.*, 1992). A second paired-domain transcription factor, *pax8*, is expressed in a similar pattern. *pax8* expression is initiated in the intermediate mesoderm during gastrulation slightly earlier than *pax2.1* expression (Pfeffer *et al.*, 1998). The *lim1* gene is expressed during early somitogenesis at all A-P levels of the pronephric mesoderm extending from somite 1 to the future cloaca (Toyama and Dawid, 1997). Posterior *lim1* expression is downregulated by the 12-somite stage while anterior expression (adjacent to somites 1 and 2) just rostral the *pax2.1* expression domain persists until 24 hpf (Fig. 4; Toyama and Dawid, 1997). lim-1 deficient mouse embryos completely lack the genitourinary tract (Shawlot and Behringer, 1995). The early and extensive expression of *lim1* in the intermediate mesoderm suggests that it may play a similarly important role in zebrafish kidney development. The Wilms tumor suppressor gene, *wt1a*, is expressed beginning at the 2-3-somite stage and later, at the 8-somite stage, in the intermediate mesoderm extending from somites 1 to 4 (Serluca and Fishman, 2001). In the mature pronephros, *wt1a* is expressed exclusively in podocytes (Drummond *et al.*, 1998; Majumdar and Drummond, 1999b, 2000; Majumdar *et al.*, 2000; Serluca and Fishman, 2001) and is required for glomerular morphogenesis (Perner *et al.*, 2007).

Along with the expression of these identified genes, data from EST *in situ* expression pattern screens is generating a wealth of new intermediate mesoderm

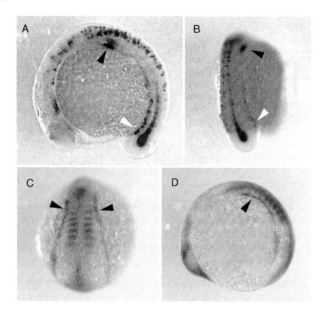

Fig. 4 Gene expression in the intermediate mesoderm. (A, B) *lim1* expression in the anterior pronephric nephron primordium (*black arrows* in A, B) and the posterior pronephros near the forming cloaca (*white arrows* in A, B) at the 15-somite stage. *lim1* is also expressed in the nervous system and the notochord. (C, D). EST ibd2750 expression marks the entire intermediate mesoderm at the 8-somite stage (*arrows*) as well as the forming somites.

markers and potential gene targets for further functional studies (Kudoh *et al.*, 2001; Thisse *et al.*, 2004). Data from two *in situ* screens of this type are an available online and can be searched by organ type for candidate markers/genes (Sprague *et al.*, 2008). As an example, the novel EST ibd2760 (Kudoh *et al.*, 2001) is expressed in the somites and the intermediate mesoderm at the 8-somite stage (Fig. 4).

4. Adjacent Tissues and Kidney Cell Specification

In addition to the pronephros, the intermediate mesoderm also gives rise to the blood and endothelial cells of the major trunk blood vessels (Detrich *et al.*, 1995; Gering *et al.*, 1998; Horsfield *et al.*, 2002; Weinstein *et al.*, 1996). At the 3–5-somite stage the IM expresses the stem cell leukemia gene (*scl*), a basic helix-loop-helix transcription factor essential for blood cell development in both mouse and fish. *scl* is expressed most strongly in the medial IM while kidney markers *pax2.1* and *cadherin17* (*cdh17*) are expressed only in the more lateral IM (Fig. 5; Gering *et al.*, 1998; Horsfield *et al.*, 2002). The formation of distinct kidney and blood lineage cell layers soon after marker gene expression is first detectable suggests that IM cell specification may occur prior to *pax2.1* and *scl* expression (Davidson *et al.*, 2003). Within the vascular lineage alone, cell fate may be determined in the IM well before

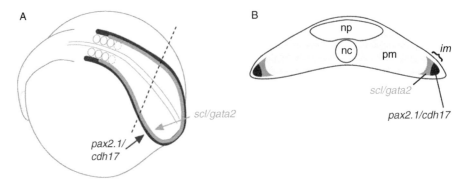

Fig. 5 Blood and kidney progenitor cells in the intermediate mesoderm. (A) During the early stages of somitogenesis, distinct bands of intermediate mesoderm are formed expressing kidney (*pax2.1/cdh17*) and blood/vascular (*scl/gata2*) markers. Cross section diagram (B) of lateral nephrogenic mesoderm and more medial blood/vascular mesoderm. *np*; neural plate, *nc*; notochord, *pm*; presomitic mesoderm.

major vessel formation, since cells labeled in the IM later give rise to the midline vein or aorta but not both (Zhong *et al.*, 2001). On the other hand, cells of the IM also appear to retain some plasticity in cell fate. Overexpression of *scl* causes excess blood cell development at the expense of somite and pronephric cell development (Gering *et al.*, 1998; Horsfield *et al.*, 2002). These studies indicate that forced expression of blood lineage regulators can transfate cells normally destined to become the kidney. It remains to be determined at what stage genes involved in specifying the fates of the intermediate mesoderm act, and when the commitment to kidney fate occurs.

Experiments in the chick and frog have suggested that tissues neighboring the intermediate mesoderm may play a role in inducing or maintaining *pax2* expression in the IM and driving subsequent kidney development (Mauch *et al.*, 2000; Seufert *et al.*, 1999). In zebrafish, expression of *pax2.1* in the intermediate mesoderm and later development of the kidney occurs independently of the notochord and sonic hedgehog signaling. These conclusions are based on studies *of floating head, sonic-you, and you-too* mutant embryos where *pax2.1* expression and/or pronephric morphology, if not perfectly normal, is still observed (Majumdar and Drummond, 2000; Drummond, unpublished observations). *spadetail* embryos lacking proper trunk somite formation still express *pax2.1* in the IM and form the pronephric epithelium (Drummond, unpublished observations). These observations again suggest that specification of the kidney IM in zebrafish may occur relatively early, perhaps as early as gastrulation, and proceed independently of the development of other axial and paraxial mesoderm.

Although pronephric development does not require proper development of the endoderm, overdevelopment of the endoderm can alter pronephric development (Mudumana *et al.*, 2008). Knockdown of the zebrafish *odd-skipped related1* gene (*osr1*) causes expansion of endoderm, which subsequently inhibits epithelialization of the pronephros and enhances angioblast differentiation. Retinoic acid signaling, presumably from paraxial mesoderm expressing *raldh2*, has a major influence on proximo-distal patterning of the pronephros (Wingert *et al.*, 2007). Mutants lacking *raldh2* (*neckless*) or embryos treated with drugs that block RA signaling lack proximal nephron segments (Wingert *et al.*, 2007). Conversely, RA treated embryos have expanded proximal nephron segments (see below).

B. Development of the Pronephric Nephrons

1. Epithelial Tubule Formation

Tubule formation is a central feature of virtually every epithelial organ including the vasculature, the gut, and the liver. Development of the pronephric epithelial tubules is initiated at the boundary of somite two and three on both sides of the embryo and temporally follows the anterior to posterior progression of somitogenesis (Fig. 1B; Kimmel *et al.*, 1995). By 24 hpf the tubules are complete, existing just dorsal the to yolk extension, from somite three to the cloaca (Drummond *et al.*, 1998; Kimmel *et al.*, 1995). In zebrafish this tubular component of the nephron forms primarily through recruitment of intermediate mesoderm at all axial levels followed by *in situ* epithelial differentiation; lineage tracing experiments have not detected any contribution from caudally migrating cells (Serluca and Fishman, 2001). Tubule formation is mediated by a mesenchyme to epithelial transformation; a process central to kidney formation in all vertebrates (Saxén, 1987). By the end of this transition the epithelial cells are polarized with an apical brush border and a basolateral membrane domain containing ion transport proteins that are essential for the osmoregulatory function of the nephron (Fig. 6A; Drummond *et al.*, 1998). Importantly, the pronephric epithelium already shows evidence of segmentation and cell type differentiation by the time polarization is complete

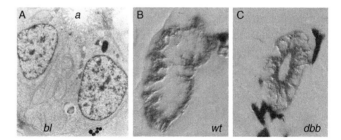

Fig. 6 Epithelial cell polarity in the pronephric tubules. (A) Electron micrograph of 2.5 days pronephric tubule epithelial cells showing apical (*a*) brush border and basolateral cell surfaces and infoldings. Polarized distribution of the NaK ATPase in 2.5 days pronephric tubule epithelial cells visualized by the alpha6F monoclonal antibody. The apical cell surface is devoid of staining while staining is strong on the basolateral cell surface and membrane infoldings (B). Double bubble mutant embryos (C) aberrantly express the NaK ATPase on the apical cytoplasm and cell membrane.

(24 hpf) (see Section III.C; Drummond, 2004; Liu *et al.*, 2007; Nichane *et al.*, 2006; Shmukler *et al.*, 2005; Wingert *et al.*, 2007). In addition, individual cell types, for instance single multiciliated cells are interspersed with transporting epithelial cells along the pronephros (Liu *et al.*, 2007). Thus epithelium formation occurs simultaneously with patterning events that define pronephric nephron segment identity.

Formation of the caudal pronephric opening or cloaca requires BMP signaling (Pyati *et al.*, 2006). The ligand involved here appears to be BMP4 since ENU mutants lacking functional BMP4 exhibit failure to complete the fusion of the pronephric ducts with the epidermis and an absence of a pronephric opening (Stickney *et al.*, 2007), similar to the phenotype seen when a dominant negative BMP receptor is ectopically expressed late in development (Pyati *et al.*, 2006). These studies revealed that cloaca formation is likely to involve developmentally programmed cell death accompanied by cellular rearrangements within the terminus of the pronephric duct and the epidermis. Although zebrafish do not have a urinary bladder, the terminus of the pronephros may be homologous to the end of the common nephric duct in mammals which inserts into the bladder by a mechanism involving programmed cell death (Batourina *et al.*, 2005).

Establishment of cell-cell junctions is an essential step in separating apical and basolateral membrane domains and giving an epithelium its vectorial property. Cadherins are the major proteins of the adherens junction that maintain the integrity of epithelial sheets and separate apical and basolateral membrane domains. *cadherin17* is specifically expressed in the zebrafish intermediate mesoderm and later in the pronephric epithelium (Horsfield *et al.*, 2002). Loss of function studies using antisense morpholino oligonucleotides show that *cadherin17* is essential for tubule and duct morphogenesis. *cadherin17*-null epithelia show a general loss of the normally tight cell-cell adhesion, failure of the bilateral ducts to fuse at the cloaca, and gaps between epithelial cells (Horsfield *et al.*, 2002). Cytoplasmic polarity, in this case the relative placement of nucleus to cytoplasm, is also disrupted. Although it has been thought that cadherins are likely to be essential for mammalian kidney tubule formation, cadherin loss of function studies in the mouse kidney have not yielded as clear a phenotype (Dahl *et al.*, 2002). It may be that fewer members of the cadherin gene family are expressed in the more primative pronephric kidney, potentially avoiding the issue of redundancy in gene function.

Cell polarity and proper targeting of membrane transporters is essential for proper kidney ion transport and function. Several zebrafish mutants have been found to mistarget the NaK ATPase in the tubules from its normal basolateral membrane location to the apical membrane (Fig. 6; Drummond *et al.*, 1998). The activity of the NaK ATPase provides the motive force for many other coupled transport systems (Seldin and Giebisch, 1992); its mislocalization suggests that severe problems in osmoregulation exist in these mutants. In fact, these mutants later develop cysts in the pronephric tubule and the embryos eventually die of edema (Drummond *et al.*, 1998).

2. Nephron Cell Differentiation and Segmentation

Nephron segmentation is a general feature of vertebrate kidneys, where osmoregulatory function is dependent on an ordered array of ion transporters acting sequentially along the length of the nephron (Seldin and Giebisch, 1992). Similar to the mammalian nephron, the zebrafish pronephric nephron is composed of the glomerulus and tubule segments homologous to proximal, distal, and collecting duct segments that are likely to have distinct functional properties (Fig. 1C and D). Wingert et al. have mined the results of large scale in situ hybridization screens to systematize the nomenclature for pronephric nephron segments (Wingert and Davidson, 2008; Wingert et al., 2007). These segments have been termed the neck, proximal convoluted, proximal straight, early distal, late distal, and collecting duct segments (Fig. 1D; Wingert et al., 2007). How nephron segment boundaries are set and maintained in the nephron tubule is not well understood. The linear, anterior-posterior orientation of the zebrafish pronephros makes it an attractive and simplified model system for studies of nephron segmentation and patterning (Wingert and Davidson, 2008).

Examples of segment-specific membrane transporter expression are shown in Fig. 7. The proximal convoluted tubule segment specifically express several transporters associated with acid/base homeostasis: the chloride/bicarbonate anion exchanger AE2, the sodium/bicarbonate cotransporter NBC1, and the sodium/hydrogen exchanger NHE (Fig. 7A), (Nichane et al., 2006; Shmukler et al., 2005; Wingert et al., 2007). Markers of the proximal straight segment include an aspartoacylase homolog (Fig. 7B) and the zebrafish starmaker gene (Fig. 7C; Hukreide et al., personal communication; Sollner et al., 2003; Sprague et al., 2008). The early distal-segment expresses slc12a1 (Na-K-Cl symporter; Fig. 7D), which in mammals is most highly expressed in the distal tubules (loop of Henle, the thick ascending limb, and the distal convoluted tubules). The late distal pronephric nephron segment expresses an EST encoding a putative ABC transporter (Fig. 7G) as well as the zebrafish c-ret homolog ret1 (Marcos-Gutierrez et al., 1997). Some overlap exists between expression of what would be collecting duct markers in mammals (c-ret) and distal tubule markers (ClCK; Wingert et al., 2007) in the late distal segment of the pronephros. Since freshwater vertebrates do not have concentrating kidneys, the lack of a significant segment expressing markers consistent with collecting duct identity is not unusual.

3. Nephron Segment Boundary Patterning

One of the clearest nephron boundaries in the pronephros exists between the glomerulus and the proximal tubule. The paired-domain transcription factor pax2.1 is now known to play a key role in establishing this boundary. pax2.1 is expressed in the early IM and also later in the most proximal tubule and neck segments (Krauss et al., 1991; Majumdar et al., 2000). Fish with mutations in the pax2.1 gene (no isthmus, noi) do not express proximal tubule markers while the

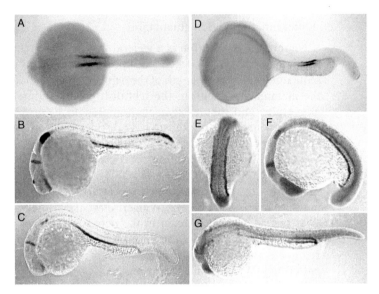

Fig. 7 Ion transporter mRNA expression define pronephric nephron segments. (A) The chloride-bicarbonate anion exchanger (AE2) is expressed in the proximal convoluted tubule. The proximal straight segment specifically express the zebrafish *starmaker* gene (B) and an aspartoacylase homolog (C). Early distal segments express the Na-K-Cl symporter *slc12a1* (D). Expression of a putative ABC transporter (ibd2207) is observed initially throughout most of the forming pronephric tubules at the 15-somite stage (E, F) but becomes restricted primarily to the late distal segment by 24 hpf (G). Embryos in b, c, and e-g are counterstained with *pax2.1* probe in red for reference(a and d: courtesy of Alan Davidson *et al.*; b, c, e-g courtesy of Neil Hukreide).

distal tubules remain relatively unaffected (Majumdar *et al.*, 2000). The expression of glomerular podocyte markers (*wt1a* and *vegf*) on the other hand are expanded into the proximal tubule epithelial cells (Fig. 8). Although these proximal tubule cells maintain an epithelial character (no transdifferentiation to a podocyte morphology is observed), they fail to express normal markers of this segment (NaK ATPase and 3G8, a brush border marker; Majumdar *et al.*, 2000). The data suggest that *pax2.1* is plays an important role in defining the tubule/podocyte boundary by repressing podocyte-specific genes in the proximal tubule. It is conceivable that podocyte transcription factors, perhaps including *wt1a*, play a similar role in repressing tubule-specific genes in podocytes (Yang *et al.*, 1999). Mutually antagonistic signals acting in concert to define a tissue boundary is a prevalent concept in developmental biology (Briscoe *et al.*, 2000; Li and Joyner, 2001; Schwarz *et al.*, 2000) and may also underlie the patterning of the kidney nephron.

As noted above, retinoic acid signaling controls the establishment of nephron boundaries early in development (Wingert *et al.*, 2007). High RA signaling favors proximal while low RA signaling favors distal nephron cell fates. Thus manipulations of RA signaling significantly shift nephron segment boundaries in the pronephros. The position of sources of RA, that is cells expressing

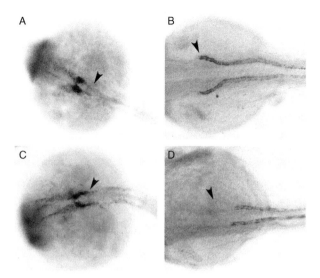

Fig. 8 Formation of the glomerulus-tubule boundary is disrupted in *no isthmus* (*noi; pax2.1*) mutants. Wholemount *in situ* hybridization with *wt1* marks the presumptive podocytes in wild-type embryos (A). *wt1* expression is caudally expanded into the anterior pronephric tubules in *noi*^{tb21} mutant embryos (C) at 24hpf (*arrow* in C). Wholemount antibody staining of wild type (B) and *noi*^{tb21} (D) embryos with mAb alpha6F mAb which recognizes a Na$^+$/K$^+$ ATPase alpha1 subunit. alpha6F marks the pronephric epitheilia in wild type (*arrows* in B) and proximal tubule Na$^+$/K$^+$ ATPase expression is missing in *noi*$^{-/tb21}$ mutant embyros at 2.5 dpf (D).

retinaldehyde dehydrogenase (*raldh2*), are controlled by *cdx* and *hox* genes (Wingert *et al.*, 2007) acting early during somitogenesis. RA may also function later in pronephric development. A retinoic-acid response element/green fluorescent protein reporter transgene shows significant activity in the pronephric tubules at the 18-somite stage (Perz-Edwards *et al.*, 2001) suggesting RA may also maintain nephron cell fates.

   In addition to differences in ion transporter expression, pronephric nephron epithelia can be subdivided into distinct segments identified based on cell morphology. In the mammalian kidney, the proximal tubule is composed of epithelial cells with apical surfaces covered with dense microvillar projections, that is a brush border, while in more distal tubule segments the cells, a dense brush border is absent (Vize *et al.*, 2002). An antibody raised against rabbit smooth muscle actin cross reacts with the zebrafish pronephros brush border, presumably recognizing an isoform of actin (Fig. 9C). The monoclonal antibody 3G8 raised against the frog pronephric tubules, also stains the apical cell surface in the zebrafish pronephros (Fig. 9B; Drummond *et al.*, 1998, unpublished results). Both of these antibodies stain only the proximal tubule of the pronephric nephron. This suggests that (1) distal pronephric nephron segments lack a brush border like the mammalian distal tubules and (2) the proximal pronephric tubules share functional

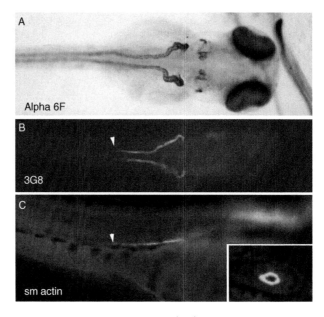

Fig. 9 Segmentation of the pronephros. The Na^+/K^+ ATPase alpha1-subunit (A) is expressed uniformly along the pronephric tubules in a 2.5 dpf embryo as detected by immunohistochemistry with the alpha6F monoclonal antibody. The 3G8 monoclonal antibody stains only the proximal tubule in 2.5 dpf embryos (B). An antibody against rabbit smooth muscle actin (clone 1A4; Sigma) also stains the proximal tubule (C). This monoclonal specifically stains the brush border of pronephric tubule epithelial cells (C, inset: cross section of the pronephric tubule showing intense apical staining). (See Plate no 20 in the Color Plate Section.)

characteristics with the mammalian pronephric tubule. Interestingly, recent work has shown that the receptor for adenovirus (CAR) is required for brush border formation (Raschperger *et al.*, 2008). Functional homology of proximal nephron segments is also supported by the demonstration that the endocytic receptors megalin and cubalin are required for the high rate of endocytic activity in the pronephric proximal segments as they are in the mammalian proximal nephron (Anzenberger *et al.*, 2006).

Current studies of nephron patterning rely primarily on the expression of ion transporter, transcription factor, and signaling receptor genes (Liu *et al.*, 2007; Nichane *et al.*, 2006; Shmukler *et al.*, 2005; Wingert *et al.*, 2007) and much remains to be done to assign physiological transport functions to specific nephron segments. The *nutria* mutant serves as a good example of how zebrafish genetics can link pronephric patterning to kidney function. *nutria* mutants lack a functional *trpM7* gene which is expressed specifically in the proximal straight segment of the pronephric nephron and functions as a calcium channel. *nutria* mutants exhibit kidney stone formation and skeletal defects (Elizondo *et al.*, 2005) indicating an essential role for TrpM7 in calcium uptake from the kidney tubule lumen.

4. Pronephric Tubule Isolation

Historically, the functional aspects of kidney epithelial ion transport have been studied using isolated single epithelial tubules in primary culture. This has not yet been achieved for zebrafish pronephric tubules. However, a useful first step in considering such an approach is larval tissue fractionation and tubule isolation (Fig. 10). Isolation of homogeneous preparations of viable pronephric tubules is also useful for purifying RNA for use in assaying tissue-specific gene expression. The protocol below is useful for the isolation of larval pronephroi as well as spinal cord, individual myotubes, and anything else that can be identified visually, by GFP transgene expression or by other forms of labeling.

Two to three days old zebrafish larvae show a remarkable resistance to collagenase digestion. Since collagenase is one of the least damaging enzymes used for tissue isolation, we have tested various ways to render larvae collagenase-sensitive. We find that a 1 h preincubation in a reducing agent (DTT or N-acetyl-cysteine) allows subsequent incubation in collagenase to be effective. These reducing agents were tested because of their known mucolytic activity which we reasoned would degrade the protective mucous layer of the larval fish skin.

1. Incubate 2–3 days old larvae in 10 mM DTT or N-acetyl-cysteine in egg water for 1 h at room temperature.
2. Wash the larvae 3–4 times with egg water to remove the DTT.
3. Incubate larvae at 28.5° in 5 mg/ml collagenase in tissue culture medium or hanks saline with calcium (Worthington) for 3–6 h. Collagenase requires calcium for activity.
4. Triturate the larvae gently 5 times with a "blue tip" 1000 μl pipet tip. The larvae should disaggregate into chunks of tissue.
5. Disperse the cell/tissue suspension into a 10 cm petri dish containing 10 ml of tissue culture medium or Hanks buffer.

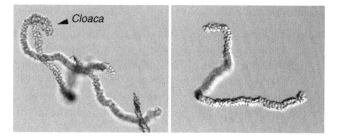

Fig. 10 Pronephric tubules isolated from 48 hpf zebrafish larvae. Two-day-old larvae were treated with 10 mM DTT followed by incubation in 5 mg/ml collagenase for 3 h at 28 °C. Individual pronephric nephrons can be dissected away from trunk tissue, often with the cloaca intact (*arrowhead*), joining the bilateral tubules at the distal segment.

6. Collect pronephric tubules by visual identification under a dissecting microscope.

Further optimizations of tissue isolation and functional assays of zebrafish pronephric tubule transport would be a significant advance in using fish larvae to study the development of mature organ physiology.

C. Pronephric Vascularization

Proper functioning of the kidney requires a structural integration of pronephric epithelia and blood vessels. The glomerulus is a capillary tuft covered in epithelial cells, the podocytes, and specialized for blood filtration (Drummond *et al.*, 1998; Majumdar and Drummond, 1999b; Pavenstadt *et al.*, 2003). Lateral and caudal to the podocytes, the pronephric tubules lie adjacent to the cardinal veins. Both the glomerular podocytes and the epithelial cells of the tubules are separated by only a basement membrane from endothelial cells. While the glomerulus is specialized for filtration of blood fluid into the tubules, the close association of tubule epithelia with veinous blood vessels facilitates the recovery of ions and metabolites from the glomerular filtrate back to the general circulation.

Vascularization of the glomerulus occurs relatively late in development, after pronephric tubule development is complete (Armstrong, 1932; Drummond *et al.*, 1998; Tytler, 1988). In zebrafish, bilateral glomerular primordia coalesce at 36–40 hpf ventral to the notochord, bringing the presumptive podocytes into contact with endothelial cells of the dorsal aorta (Fig. 11; Drummond, 2000a; Drummond *et al.*, 1998; Majumdar and Drummond, 1999b). Between 40 and 48 hpf, new blood vessels sprout from the aorta and podocytes elaborate foot processes to surround the ingrowing endothelial cells (Fig. 11; Drummond *et al.*, 1998). Further capillary growth and invasion by endothelial cells results in extensive convolution of the glomerular basement membrane, capillary formation, and the onset of glomerular filtration (Drummond *et al.*, 1998).

1. The Role of Podocytes

Evidence that podocytes act to organize vessel ingrowth can be found in (1) the expression patterns of genes known to play an important role in angiogenesis and (2) the recruitment of endothelial cells to ectopic clusters of podocytes in mutant embryos that lack the dorsal aorta, the normal blood supply for the pronephric glomerulus. Zebrafish pronephric podocytes express two known mediators of angiogenesis: vascular endothelial growth factor (VEGF) and angiopoietin2 (Carmeliet *et al.*, 1996; Ferrara *et al.*, 1996; Majumdar and Drummond, 1999b, 2000; Pham *et al.*, 2001; Shalaby *et al.*, 1995). In a complementary manner, capillary-forming endothelial cells express *flk1*, a VEGF receptor, and an early marker of the endothelial differentiation program (Majumdar and Drummond, 1999b). In zebrafish embryos at 40 hpf, *flk-1* positive endothelial cells can be

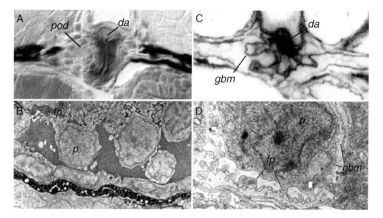

Fig. 11 Interaction of pronephric podocytes with the vasculature. (A) Apposition of nephron primordia at the embryo midline in a 40 hpf zebrafish embryo. Aortic endothelial cells in the cleft separating the nephron primordia are visualized by endogenous alkaline phosphatase activity. Podocytes; *pod,* dorsal aorta; *da.* (B) Ultrastructure of the forming zebrafish glomerulus at 40 hpf. A Longitudinal section shows podocytes (*p*) extending foot processes (*fp*) in a dorsal direction and in close contact with overlying capillary endothelial cells. (C) Rhodamine dextran (10,000 MW) injected embryos show dye in the dorsal aorta (*da*) and in the glomerular basement membrane (*gbm*) shown here graphically inverted from the original fluorescent image. (D) Podocyte foot process formation does not require signals from endothelial cells as evidenced by the appearance of foot processes (*fp*) in *cloche* mutant embryos which lack all vascular structures. Glomerular basement membrane; (*gbm*), podocyte cell body; (*p*).

observed invading the glomerular epithelium. In zebrafish *floating head* mutant embryos, the normal source of glomerular blood vessels, the dorsal aorta, is absent (Fouquet *et al.*, 1997). The *floating head* gene encodes a homeobox transcription factor that is required for notochord formation; the absence of a notochord prevents the normal signaling that induces formation of the aorta. Pronephric nephron primordia in *floating head* embryos do not fuse at the midline, instead remaining at ectopic lateral positions in the embryo (Majumdar and Drummond, 2000). These podocytes in the primordia continue to express *wt1* and *vegf* and appear to recruit *flk-1* positive endothelial cells and go on to form a functional glomerulus (Majumdar and Drummond, 2000). These results support the idea that podocytes, by expressing *vegf*, play a primary role in attracting and assembling the glomerular capillary tuft.

2. The Role of Endothelial Cells

The question of whether complementary signals are emitted from endothelial cells that would stimulate podocyte development has been addressed in studies of the mutant *cloche* (*clo*), which is nearly completely lacking in endothelial cells (Stainier *et al.*, 1995). When assayed by the expression of molecular markers and by an ultrastructural analysis of podocyte morphology, the elaboration of podocyte foot processes in *cloche* embryos occurs normally despite the complete absence of glomerular endothelial cells (Fig. 11; Majumdar and Drummond, 1999a). *cloche* podocytes express *wt1* and *vegf* and form extensive foot processes arranged as pedicels along a glomerular basement membrane (Majumdar and Drummond, 1999a). These findings suggest that once initiated, podocyte differentiation proceeds independently of endothelial cells or endothelial cell derived signals. Other developmental outcomes of signaling between podocytes and endothelial cells remain to be explored. For instance, maturation of the structure and composition of the glomerular basement membrane (Miner and Sanes, 1994) may require a contribution or signal from endothelial cells. Also, proper polarization of podocytes and organization of cell-cell junctions may require the presence of endothelial cells.

3. The Role of Vascular Flow

Zebrafish mutants specifically lacking cardiac function, for instance *silent heart* and *island beat* (*isl*) with mutations in cardiac troponin T and an L-type cardiac calcium channel respectively (Rottbauer *et al.*, 2001; Sehnert *et al.*, 2002), fail to form a proper glomerular capillary tuft, demonstrating that vascular shear force is required to drive capillary formation. In mutants that lack circulation, endothelial cells fail to invade the encapsulated podocytes and the aorta is instead a broadly dilated vessel surrounding the glomerular primordia (Fig. 12; Serluca *et al.*, 2002). The cells themselves appear normal in several ways; for instance expression patterns of *wt1* and *vegf* in podocytes and *flk-1* in endothelial cells remain unaltered and individual cell ultrastructure appears normal (Fig. 6; Serluca *et al.*, 2002). The glomerular morphogenesis defect can be phenocopied by pharmacological and surgical manipulations that disrupt flow through the aorta. This failure in glomerular morphogenesis is likely be related to the expression of matrix metalloproteinase-2 (MMP-2) in endothelial cells since (1) expression of MMP-2 in the trunk vasculature is modulated by vascular flow and (2) inhibition of MMP activity by TIMP1 (tissue inhibitor of metalloproteinase-1) injections into the vasculature result in a similar dilated aorta and failure to form the glomerulus (Serluca *et al.*, 2002). Recent studies of blood flow in the zebrafish heart (Hove *et al.*, 2003) confirm that hydrodynamic forces can have a major influence on tissue morphogenesis.

4. A Simple Assay for Glomerular Filtration

Filtration of blood by the pronephric glomerulus can be detected by injections of fluorescent compounds (10–70 KD rhodamine dextran) into the general circulation and then monitoring the appearance of fluorescent endosomes in the apical cytoplasm of pronephric tubule cells (Fig. 12C; Drummond *et al.*, 1998; Majumdar and Drummond, 2000). From this data, it can be inferred that the fluorescent tracer has passed the glomerular basement membrane and entered the lumen of the

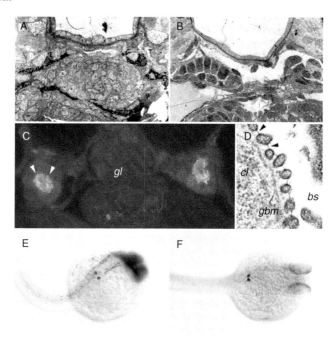

Fig. 12 The glomerular capillary tuft and podocyte slit-diaphragms. (A) An electron micrograph of the forming glomerulus at 2.5 dpf with invading endothelial cells from the dorsal aorta shaded in red and podocytes shaded in blue (image false-colored in Adobe Photoshop). (B) A similar stage glomerulus in the mutant island beat which lacks blood flow due to a mutation in an L-type cardiac specific calcium channel. The endothelial cells and podocytes are present but the aorta has a dilated lumen surrounding the podocytes with no sign of glomerular remodeling and morphogenesis. (C) Rhodamine-dextran filtration and uptake by pronephric epithelial cells. 10 KD lysine-fixable rhodamine dextran injected into the general circulation can be seen as red fluorescence in glomerular capillaries (*gl*), and filtered dye is seen in apical endosomes of pronephric tubule cells (*arrowheads*). Counterstain: FITC wheat germ agglutinin. (D) Electron micrograph of the glomerular basement membrane region in the glomerulus. Individual profiles of podocyte foot processes resting on the glomerular basement membrane (*gbm*) are connected by slit-diaphragms (*arrowheads* at *top*). *cl*; capillary lumen, *bs*; Bowman's space. Whole mount *in situ* hybridization shows expression of zebrafish podocin (E) and nephrin (F) specifically in the forming podocytes. (See Plate no. 21 in the Color Plate Section.)

pronephric tubules where it is actively endocytosed. Using this assay we have established that blood filtration by the zebrafish pronephros begins around 40 hpf (Drummond *et al.*, 1998). The ability to detect filtered dextrans in endocytic vesicles of the tubules has also been adapted to create an assay for disruption of the filtration barrier (i.e., proteinuria or nephrotic syndrome). Kramer-Zucker *et al.* demonstrated that large dextrans (500 KD) do not significantly pass a normal glomerular filter while in gene knockdown embryos affecting human nephrotic syndrome genes, passage of large dextrans could be observed as accumulation in tubule endocytic vesicles (Kramer-Zucker *et al.*, 2005b). This assay has been used subsequently to identify new nephrotic syndrome gene candidates (*mosaic eyes/ epb4l5*) (Kramer-Zucker *et al.*, 2005b). Filtered fluorescent dextrans can also be observed directly as they exit the pronephros at the cloaca in live larvae and used as a qualitative assay of the rate of pronephric fluid output (Kramer-Zucker *et al.*, 2005a).

IV. The Zebrafish Pronephros as a Model of Human Disease

A. Disease of the Glomerular Filtration Apparatus

A major feature of the glomerular blood filter is the podocyte slit-diaphragm, a specialized adherens junction that forms between the finger-like projections of podocytes (podocyte foot processes) (Reiser *et al.*, 2000). Failure of the slit-diaphragm to form results in leakage of high molecular weight proteins into the filtrate, a condition called proteinuria in human patients. Proteinuria is the cardinal feature of several human congenital nephropathies and also a common complication of diabetes (Cooper *et al.*, 2002). Several disease genes known function in the slit-diaphragm have been cloned. Nephrin is a transmembrane protein present in the slit-diaphragm itself and is thought to contribute to the zipper-like extracellular structure between foot processes (Ruotsalainen *et al.*, 1999). Podocin is a podocyte junction-associated protein (Roselli *et al.*, 2002) that resembles stomatin proteins which play a role in regulating mechanosensitive ion channels (Tavernarakis and Driscoll, 1997). Electron microscopy of the zebrafish pronephric glomerulus reveals that like mammalian podocytes, zebrafish podocytes form slit-diaphragms between their foot processes (Fig. 12D). Zebrafish homologs of podocin and nephrin are specifically expressed in podocytes as early as 24 hpf (Fig. 12E and F) and have been shown to be required for proper slit-diaphragm formation in pronephric podocytes (Kramer-Zucker *et al.*, 2005b).

B. Disease of the Tubules

One of the most common human genetic diseases is polycystic kidney disease which affects 1 in 1000 individuals (Calvet and Grantham, 2001). Kidney cysts are the result of grossly expanded kidney tubule lumens and, when present in sufficient size and number, lead to kidney fibrosis and end stage renal failure. A relatively large set of genetic loci associated with cystic pronephroi have been identified in large scale ENU zebrafish mutant screens (Fig. 13). The 15 genetic loci that we originally identified (Drummond *et al.*, 1998) probably underestimates the number of genes involved since the degree of saturation for this phenotype was low (so far, 12 of the 15 loci are represented by one allele). Recently, the results of a large scale retroviral insertional mutagenesis screen have identified 10 zebrafish genes that when mutated cause pronephric cysts (Sun *et al.*, 2004). The requirement for a large number of genes is consistent with the idea that maintenance of lumen size

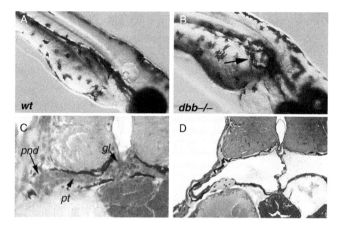

Fig. 13 The zebrafish as a model of polycystic kidney disease. (A, B) Three days larvae showing wild-type (A) and the mutant *double bubble* (*dbb*; B) with a grossly distended pronephric tubule that appears as a bubble (*arrow*) just behind the pectoral fin. (C) Three days wild-type kidney structure showing the pronephric tubule (*pt*) and glomerulus (*gl*). (D) A section of a *dbb* mutant pronephros shows the distended lumen of the pronephric tubule (*asterisk*) and distended glomerulus at the midline.

and epithelial cell shape is a complex process controlled by many cellular proteins or signaling pathways.

A surprising convergence of data from studies of cystic disease, left-right asymmetry, retinal degeneration and flagella formation in Chlamydomonas lead to the idea that defects in the formation or function of cilia may underlie the pathology observed in all of these conditions. Cloning the gene responsible for the oakridge polycystic kidney (*orpk*) mouse was the first link between cilia and kidney cystic disease. The mutant gene, polaris, is a homolog of a Chlamydomonas gene, IFT88, that is required for intraflagellar transport, an essential process in flagellum formation (Pazour *et al.*, 2000; Taulman *et al.*, 2001). Mammalian kidney cells are not flagellated but have a single, nonmotile apical cilium. Until recently the apical cilium had been thought to be a vestigial organelle with uncertain function. The finding that *orpk* mouse kidney epithelial cells have short, malformed apical cilia (Pazour *et al.*, 2000; Taulman *et al.*, 2001; Yoder *et al.*, 2002b) suggested a functional link between cilia and maintenance of epithelial tubule lumen diameter. Subsequent studies revealed that most known cystic mutant genes including poly-cystin1, polycystin2, cystin (*cpk* mouse), polaris, inversin, and the *C. elegans* polycystin homologs lov-1 and pkd2 were, at least in part, localized to cilia (Barr *et al.*, 2001; Morgan *et al.*, 2002; Pazour *et al.*, 2002b; Qin *et al.*, 2001; Yoder *et al.*, 2002a). Cilia function and nodal flow has been implicated in the establishment of left-right asymmetry (Tabin and Vogan, 2003); among the cystic genes, inversin, polaris, and polycystin2 mutations show laterality defects (Morgan *et al.*, 1998; Pennekamp *et al.*, 2002; Taulman *et al.*, 2001). Finally, retinal photoreceptor cells contain a modified cilium that appears to function in membrane transport between the cell body and rod outer membranes. In light of this, the retinal degeneration seen in the orpk/polaris mouse (Pazour et al., 2002a) makes some sense. Interestingly, some zebrafish cyst mutants manifest pleiotropic defects affecting both the kidney and the eye, referred to as "renal-retinal dysplasia" when observed in human patients (Drummond et al., 1998; Godel et al., 1978). Although the exact function of these genes in vertebrate embryos remains to be worked out, recent studies of polycystin1 and polycystin2, the genes responsible for autosomal dominant polycystic kidney disease, in epithelial cells suggests that they act together to mediate calcium entry into cells upon flow-induced cilium deflection (Nauli et al., 2003; Praetorius and Spring, 2001). It may be that the cilium acts as a sensor of tubule lumen mechanics and flow, providing a feedback signal that limits lumen diameter.

Our own observations indicate that cilia in the zebrafish pronephros are motile and have a "9 + 2" microtubule doublet organization that is typical of motile cilia and flagella (Kramer-Zucker et al., 2005a; Liu et al., 2007; Pathak et al., 2007). Motile cilia are often associated with fluid flow and has lead to an alternative hypothesis that cilia act as a fluid pump in the pronephros. When cilia are malformed due to knockdown of IFT57 or IFT88 or immotile due to dynein knockdown, failure to move fluid or particles in the lumen of the pronephric tubules is likely to account for the accumulation of fluid in the proximal segments of the pronephros (Kramer-Zucker et al., 2005a). Other genes linking cilia motility to cystic pronephroi include the leucine rich repeat protein LRRC50 (Sullivan-Brown et al., 2008; van Rooijen et al., 2008) and fleer, a novel TPR protein required for tubulin polyglutamylation (Pathak et al., 2007). Motile cilia are a common feature in amphibian pronephric tubules and ducts and can be observed under pathological conditions in human proximal tubules (Ong and Wagner, 2005).

Several genes have been identified so far in zebrafish that are responsible for tubule cyst formation. Nek8 is a member of the NIMA family of serine/threonine kinases and is mutated in the juvenile cystic kidney (jck) mouse (Liu et al., 2002). Morpholino antisense oligos that disrupt the function of zebrafish nek8 cause a severe cystic distension of the pronephric tubules. In vitro expression of a kinase-dead Nek8 mutant protein or the jck Nek8 allele (a missense mutation in a C-terminal putative protein interaction domain) in cultured cells results in multi-nucleated cells, suggesting a role for nek8 in regulating the cytoskeleton (Liu et al., 2002). Other Nek kinases have links to cytoskeletal functions: Nek2 is localized to centrioles and acts to promote splitting of duplicated centrioles during the cell cycle (Fry, 2002). In Chlamydomonas, FA2 is a Nek kinase necessary for shedding flagella prior to cell division (Mahjoub et al., 2002).

Disruption of the zebrafish homolog of the human cystic disease gene polycystin2 also causes pronephric cyst formation (Obara et al., 2006; Sun et al., 2004) without, however, affecting cilia structure or motility (Obara et al., 2006; Sullivan-Brown et al., 2008). Coinjected human polycystin2 mRNA along with polycystin2 antisense oligos can rescue this phenotype, suggesting that the function of

Polycystin2 has been highly conserved. Laterality defects are also associated with *pkd2* zebrafish mutants (Obara *et al.*, 2006; Schottenfeld *et al.*, 2007), similar to mammals.

Morpholino disruption of the zebrafish *inversin* gene results in both laterality defects and kidney cysts. The human condition nephronophthisis type 2 (NPHP2) is associated with mutations in the human inversin gene. Both inversin and the mammalian NPHP1 gene nephrocystin are found in basal bodies and cilia, and have been shown to interact biochemically. A splice donor morpholino induced C-terminal deletion of the putative nephrocystin binding domain in zebrafish inversin results in severe cyst formation, supporting the idea that NPHP proteins act as a multiprotein complex to regulate the function of basal bodies and/or cilia.

The transcription factor HNF1β is required for normal zebrafish pronephric tubule development (Sun and Hopkins, 2001). Cystic tubules are observed in three different retroviral insertional mutants that harbor a disrupted HNF1β gene. The function of HNF1β affects tissue patterning: *pax2.1* expression is lost from the tubules and *wt1* expression in podocyte progenitors appears expanded (Sun and Hopkins, 2001). Mutations in HNF1β in humans are associated with glomerulocystic disease and maturity onset diabetes of the young, type V (MODY5) in humans (Froguel and Velho, 1999). HNF1β has been shown to regulate a number of genes associated with cystic disease in mammals (Igarashi *et al.*, 2005).

In addition to cystic kidney disease, the kidney of zebrafish and its cyprinid relative the goldfish have been used to model acute renal injury and kidney regeneration (Hentschel *et al.*, 2004; Reimschuessel *et al.*, 1990; Reimschuessel and Williams, 1995). Both larval and adult zebrafish kidney tubules are sensitive to commonly used nephrotoxins like gentamycin. Hentschel *et al.* demonstrated that the rate of glomerular clearance in the context of injury could be quantitated with injections of fluorescent dextrans into the vasculature (Hentschel *et al.*, 2004), creating a new opportunity to study the pathophysiology of acute renal failure. Gentamycin injured adult fish kidneys show a remarkable ability to regenerate from new nephron precursors (Reimschuessel and Williams, 1995; Salice *et al.*, 2001), indicating that zebrafish could be used to uncover new modes of kidney repair following injury.

V. Conclusions

The zebrafish pronephric kidney represents one of the many vertebrate kidney forms that have evolved to solve the problem of blood fluid and electrolyte homeostasis in an osmotically challenging environment. Despite differences in organ morphology between the mammalian and teleost kidneys, many parallels exist at the cellular and molecular levels that can be exploited to further our understanding of kidney cell specification, epithelial tubule formation, and the tissue interactions that drive nephrogenesis. The same genes (for instance *pax2*), and cell types (for instance podocytes, endothelial cells, and tubular epithelial cells) are employed in the development and function of fish, frog, chicken, and mammalian kidneys. Genes mutated in human disease are also essential for the formation and function of the zebrafish pronephros. The zebrafish thus presents a useful and relevant model of vertebrate kidney development: its principal strengths lie in the ease with which it can be genetically manipulated and phenotyped so as to rapidly determine the function of genes and cell-cell interactions that underlie the development of all kidney forms.

Acknowledgments

This work was supported by NIH grants DK53093, DK071041, and DK070263 to I.A.D and by grants from the PKD foundation.

References

Agarwal, S., and John, P. A. (1988). Studies on the development of the kidney of the guppy, *Lebistes reticulatus*. Part 1. The development of the pronephros. *J. Anim. Morphol. Physiol.* **35**, 17–24.

Amacher, S. L., Draper, B. W., Summers, B. R., and Kimmel, C. B. (2002). The zebrafish T-box genes no tail and spadetail are required for development of trunk and tail mesoderm and medial floor plate. *Development* **129**, 3311–3323.

Anzenberger, U., Bit-Avragim, N., Rohr, S., Rudolph, F., Dehmel, B., Willnow, T. E., and Abdelilah-Seyfried, S. (2006). Elucidation of megalin/LRP2-dependent endocytic transport processes in the larval zebrafish pronephros. *J. Cell Sci.* **119**, 2127–2137.

Armstrong, P. B. (1932). The embryonic origin of function in the pronephros through differentiation and parenchyma-vascular association. *Am. J. Anat.* **51**, 157–188.

Balfour, F. M. (1880). "A Treatise on Comparative Embryology." Macmillan and Co., London.

Barr, M. M., DeModena, J., Braun, D., Nguyen, C. Q., Hall, D. H., and Sternberg, P. W. (2001). The *Caenorhabditis elegans* autosomal dominant polycystic kidney disease gene homologs lov-1 and pkd-2 act in the same pathway. *Curr. Biol.* **11**, 1341–1346.

Batourina, E., Tsai, S., Lambert, S., Sprenkle, P., Viana, R., Dutta, S., Hensle, T., Wang, F., Niederreither, K., McMahon, A. P., Carroll, T. J., and Mendelsohn, C. L. (2005). Apoptosis induced by vitamin A signaling is crucial for connecting the ureters to the bladder. *Nat. Genet.* **37**, 1082–1089.

Briscoe, J., Pierani, A., Jessell, T. M., and Ericson, J. (2000). A homeodomain protein code specifies progenitor cell identity and neuronal fate in the ventral neural tube. *Cell* **101**, 435–445.

Calvet, J. P., and Grantham, J. J. (2001). The genetics and physiology of polycystic kidney disease. *Semin. Nephrol.* **21**, 107–123.

Carmeliet, P., Ferreira, V., Breier, G., Pollefeyt, S., Kieckens, L., Gertsenstein, M., Fahrig, M., Vandenhoeck, A., Harpal, K., Eberhardt, C., Declercq, C., Pawling, J., *et al.* (1996). Abnormal blood vessel development and lethality in embryos lacking a single VEGF allele. *Nature* **380**, 435–439.

Carroll, T. J., Wallingford, J. B., and Vize, P. D. (1999). Dynamic patterns of gene expression in the developing pronephros of Xenopus laevis. *Dev. Genet.* **24**, 199–207.

Cooper, M. E., Mundel, P., and Boner, G. (2002). Role of nephrin in renal disease including diabetic nephropathy. *Semin. Nephrol.* **22**, 393–398.

Dahl, U., Sjodin, A., Larue, L., Radice, G. L., Cajander, S., Takeichi, M., Kemler, R., and Semb, H. (2002). Genetic dissection of cadherin function during nephrogenesis. *Mol. Cell Biol.* **22**, 1474–1487.

Davidson, A. J., Ernst, P., Wang, Y., Dekens, M. P., Kingsley, P. D., Palis, J., Korsmeyer, S. J., Daley, G. Q., and Zon, L. I. (2003). cdx4 mutants fail to specify blood progenitors and can be rescued by multiple hox genes. *Nature* **425**, 300–306.

Delarue, M., Saez, F. J., Johnson, K. E., and Boucaut, J. C. (1997). Fates of the blastomeres of the 32-cell stage Pleurodeles waltl embryo. *Dev. Dyn.* **210**, 236–248.

Detrich, H. W., III, Kieran, M. W., Chan, F. Y., Barone, L. M., Yee, K., Rundstadler, J. A., Pratt, S., Ransom, D., and Zon, L. I. (1995). Intraembryonic hematopoietic cell migration during vertebrate development. *Proc. Natl. Acad. Sci. USA* **92**, 10713–10717.

Draper, B. W., Stock, D. W., and Kimmel, C. B. (2003). Zebrafish fgf24 functions with fgf8 to promote posterior mesodermal development. *Development* **130**, 4639–4654.

Drummond, I. A. (2000a). The zebrafish pronephros: A genetic system for studies of kidney development. *Pediatr. Nephrol.* **14**, 428–435.

Drummond, I. A. (2000b). The zebrafish pronephros: A genetic system for studies of kidney development [In Process Citation]. *Pediatr. Nephrol.* **14**, 428–435.

Drummond, I. A. (2004). Zebrafish kidney development. *Methods Cell Biol.* **76**, 501–530.

Drummond, I. A., Majumdar, A., Hentschel, H., Elger, M., Solnica-Krezel, L., Schier, A. F., Neuhauss, S. C., Stemple, D. L., Zwartkruis, F., Rangini, Z., Driever, W., and Fishman, M. C. (1998). Early development of the zebrafish pronephros and analysis of mutations affecting pronephric function. *Development* **125**, 4655–4667.

Elizondo, M. R., Arduini, B. L., Paulsen, J., MacDonald, E. L., Sabel, J. L., Henion, P. D., Cornell, R. A., and Parichy, D. M. (2005). Defective skeletogenesis with kidney stone formation in dwarf zebrafish mutant for trpm7. *Curr. Biol.* **15**, 667–671.

Ferrara, N., Carver-Moore, K., Chen, H., Dowd, M., Lu, L., O'Shea, K. S., Powell-Braxton, L., Hillan, K. J., and Moore, M. W. (1996). Heterozygous embryonic lethality induced by targeted inactivation of the VEGF gene. *Nature* **380**, 439–442.

Fouquet, B., Weinstein, B. M., Serluca, F. C., and Fishman, M. C. (1997). Vessel patterning in the embryo of the zebrafish: Guidance by notochord. *Dev. Biol.* **183**, 37–48.

Froguel, P., and Velho, G. (1999). Molecular genetics of maturity-onset diabetes of the young. *Trends Endocrinol. Metab.* **10**, 142–146.

Fry, A. M. (2002). The Nek2 protein kinase: A novel regulator of centrosome structure. *Oncogene* **21**, 6184–6194.

Gering, M., Rodaway, A. R., Gottgens, B., Patient, R. K., and Green, A. R. (1998). The SCL gene specifies haemangioblast development from early mesoderm. *EMBO J.* **17**, 4029–4045.

Godel, V., Romano, A., Stein, R., Adam, A., and Goodman, R. M. (1978). Primary retinal dysplasia transmitted as X-chromosome-linked recessive disorder. *Am. J. Ophthalmol.* **86**, 221–227.

Goodrich, E. S. (1930). "Studies on the Structure and Development of Vertebrates." Macmillan, London.

Griffin, K., Patient, R., and Holder, N. (1995). Analysis of FGF function in normal and no tail zebrafish embryos reveals separate mechanisms for formation of the trunk and the tail. *Development* **121**, 2983–2994.

Griffin, K. J., Amacher, S. L., Kimmel, C. B., and Kimelman, D. (1998). Molecular identification of spadetail: Regulation of zebrafish trunk and tail mesoderm formation by T-box genes. *Development* **125**, 3379–3388.

Hammerschmidt, M., Pelegri, F., Mullins, M. C., Kane, D. A., Brand, M., van Eeden, F. J., Furutani-Seiki, M., Granato, M., Haffter, P., Heisenberg, C. P., Jiang, Y. J., Kelsh, R. N., et al. (1996a). Mutations affecting morphogenesis during gastrulation and tail formation in the zebrafish, Danio rerio. *Development* **123**, 143–151.

Hammerschmidt, M., Pelegri, F., Mullins, M. C., Kane, D. A., van Eeden, F. J., Granato, M., Brand, M., Furutani-Seiki, M., Haffter, P., Heisenberg, C. P., Jiang, Y. J., Kelsh, R. N., et al. (1996b). dino and mercedes, two genes regulating dorsal development in the zebrafish embryo. *Development* **123**, 95–102.

Heller, N., and Brandli, A. W. (1999). Xenopus Pax-2/5/8 orthologues: Novel insights into Pax gene evolution and identification of Pax-8 as the earliest marker for otic and pronephric cell lineages. *Dev. Genet.* **24**, 208–219.

Hentschel, H., and Elger, M. (1996). Functional morphology of the developing pronephric kidney of zebrafish. *J. Am. Soc. Nephrol.* **7**, 1598.

Hentschel, D. M., Park, K. M., Cilenti, L., Zervos, A. S., Drummond, I., and Bonventre, J. V. (2005). Acute renal failure in zebrafish: A novel system to study a complex disease. *Am. J. Physiol. Renal Physiol.* **288,** F923–F929.

Hickman, C. P., and Trump, B. F. (1969). The kidney. *In* "Fish Physiology" (R. Hoar, ed.), Vol. 1, pp. 91–239. Academic Press, New York.

Hild, M., Dick, A., Rauch, G. J., Meier, A., Bouwmeester, T., Haffter, P., and Hammerschmidt, M. (1999). The smad5 mutation somitabun blocks Bmp2b signaling during early dorsoventral patterning of the zebrafish embryo. *Development* **126,** 2149–2159.

Ho, R. K., and Kane, D. A. (1990). Cell-autonomous action of zebrafish spt-1 mutation in specific mesodermal precursors. *Nature* **348,** 728–730.

Horsfield, J., Ramachandran, A., Reuter, K., LaVallie, E., Collins-Racie, L., Crosier, K., and Crosier, P. (2002). Cadherin-17 is required to maintain pronephric duct integrity during zebrafish development. *Mech. Dev.* **115,** 15–26.

Hove, J. R., Koster, R. W., Forouhar, A. S., Acevedo-Bolton, G., Fraser, S. E., and Gharib, M. (2003). Intracardiac fluid forces are an essential epigenetic factor for embryonic cardiogenesis. *Nature* **421,** 172–177.

Howland, R. B. (1921). Experiments on the effect of the removal of the pronephros of *Ambystoma punctatum. J. Exp. Zool.* **32,** 355–384.

Igarashi, P., Shao, X., McNally, B. T., and Hiesberger, T. (2005). Roles of HNF-1beta in kidney development and congenital cystic diseases. *Kidney Int.* **68,** 1944–1947.

Kelly, G. M., Greenstein, P., Erezyilmaz, D. F., and Moon, R. T. (1995). Zebrafish wnt8 and wnt8b share a common activity but are involved in distinct developmental pathways. *Development* **121,** 1787–1799.

Kim, S. H., Park, H. C., Yeo, S. Y., Hong, S. K., Choi, J. W., Kim, C. H., Weinstein, B. M., and Huh, T. L. (1998). Characterization of two frizzled8 homologues expressed in the embryonic shield and prechordal plate of zebrafish embryos. *Mech. Dev.* **78,** 193–201.

Kimmel, C. B., Warga, R. M., and Schilling, T. F. (1990). Origin and organization of the zebrafish fate map. *Development* **108,** 581–594.

Kimmel, C. B., Ballard, W. W., Kimmel, S. R., Ullmann, B., and Schilling, T. F. (1995). Stages of embryonic development of the zebrafish. *Dev. Dyn.* **203,** 253–310.

Kishimoto, Y., Lee, K. H., Zon, L., Hammerschmidt, M., and Schulte-Merker, S. (1997). The molecular nature of zebrafish swirl: BMP2 function is essential during early dorsoventral patterning. *Development* **124,** 4457–4466.

Kramer-Zucker, A. G., Olale, F., Haycraft, C. J., Yoder, B. K., Schier, A. F., and Drummond, I. A. (2005a). Cilia-driven fluid flow in the zebrafish pronephros, brain and Kupffer's vesicle is required for normal organogenesis. *Development* **132,** 1907–1921.

Kramer-Zucker, A. G., Wiessner, S., Jensen, A. M., and Drummond, I. A. (2005b). Organization of the pronephric filtration apparatus in zebrafish requires Nephrin, Podocin and the FERM domain protein Mosaic eyes. *Dev. Biol.* **285,** 316–329.

Krauss, S., Johansen, T., Korzh, V., and Fjose, A. (1991). Expression of the zebrafish paired box gene pax[zf-b] during early neurogenesis. *Development* **113,** 1193–1206.

Kudoh, T., Tsang, M., Hukriede, N. A., Chen, X., Dedekian, M., Clarke, C. J., Kiang, A., Schultz, S., Epstein, J. A., Toyama, R., and Dawid, I. B. (2001). A gene expression screen in zebrafish embryogenesis. *Genome Res.* **11,** 1979–1987.

Lekven, A. C., thorpe, C. J., Waxman, J. S., Moon, R. T. (2001). Zebrafish wnt8 encodes two wint8 proteins on a bicistronic transcript and is required for mesoderm and neurectoderm patterning. *Dev. Cell* **1,** 103–114.

Li, J. Y., and Joyner, A. L. (2001). Otx2 and Gbx2 are required for refinement and not induction of mid-hindbrain gene expression. *Development* **128,** 4979–4991.

Liu, A., Majumdar, A., Schauerte, H. E., Haffter, P., and Drummond, I. A. (2000). Zebrafish wnt4b expression in the floor plate is altered in sonic hedgehog and gli-2 mutants. *Mech. Dev.* **91,** 409–413.

Liu, S., Lu, W., Obara, T., Kuida, S., Lehoczky, J., Dewar, K., Drummond, I. A., and Beier, D. R. (2002). A defect in a novel Nek-family kinase causes cystic kidney disease in the mouse and in zebrafish. *Development* **129**, 5839–5846.

Liu, Y., Pathak, N., Kramer-Zucker, A., and Drummond, I. A. (2007). Notch signaling controls the differentiation of transporting epithelia and multiciliated cells in the zebrafish pronephros. *Development* **134**, 1111–1122.

Mahjoub, M. R., Montpetit, B., Zhao, L., Finst, R. J., Goh, B., Kim, A. C., and Quarmby, L. M. (2002). The FA2 gene of Chlamydomonas encodes a NIMA family kinase with roles in cell cycle progression and microtubule severing during deflagellation. *J. Cell Sci.* **115**, 1759–1768.

Majumdar, A., and Drummond, I. A. (1999a). Podocyte differentiation in the absence of endothelial cells as revealed in the zebrafish avascular mutant, cloche. *Dev. Genet.* **24**, 220–229.

Majumdar, A., and Drummond, I. A. (1999b). Podocyte differentiation in the absence of endothelial cells as revealed in the zebrafish avascular mutant, cloche [In Process Citation]. *Dev. Genet.* **24**, 220–229.

Majumdar, A., and Drummond, I. A. (2000). The zebrafish floating head mutant demonstrates podocytes play an important role in directing glomerular differentiation. *Dev. Biol.* **222**, 147–157.

Majumdar, A., Lun, K., Brand, M., and Drummond, I. A. (2000). Zebrafish no isthmus reveals a role for pax2.1 in tubule differentiation and patterning events in the pronephric primordia. *Development* **127**, 2089–2098.

Marcos-Gutierrez, C. V., Wilson, S. W., Holder, N., and Pachnis, V. (1997). The zebrafish homologue of the ret receptor and its pattern of expression during embryogenesis. *Oncogene* **14**, 879–889.

Marshall, E. K., and Smith, H. W. (1930). The glomerular development of the vertebrate kidney in relation to habitat. *Biol. Bull.* **59**, 135–153.

Mauch, T. J., Yang, G., Wright, M., Smith, D., and Schoenwolf, G. C. (2000). Signals from trunk paraxial mesoderm induce pronephros formation in chick intermediate mesoderm. *Dev. Biol.* **220**, 62–75.

Miner, J. H., and Sanes, J. R. (1994). Collagen IV alpha 3, alpha 4, and alpha 5 chains in rodent basal laminae: Sequence, distribution, association with laminins, and developmental switches. *J. Cell Biol.* **127**, 879–891.

Morgan, D., Turnpenny, L., Goodship, J., Dai, W., Majumder, K., Matthews, L., Gardner, A., Schuster, G., Vien, L., Harrison, W., Elder, F. F., Penman-Splitt, M., *et al.* (1998). Inversin, a novel gene in the vertebrate left-right axis pathway, is partially deleted in the inv mouse. *Nat. Genet.* **20**, 149–156.

Morgan, D., Eley, L., Sayer, J., Strachan, T., Yates, L. M., Craighead, A. S., and Goodship, J. A. (2002). Expression analyses and interaction with the anaphase promoting complex protein Apc2 suggest a role for inversin in primary cilia and involvement in the cell cycle. *Hum. Mol. Genet.* **11**, 3345–3350.

Mudumana, S. P., Hentschel, D., Liu, Y., Vasilyev, A., and Drummond, I. A. (2008). odd skipped related1 reveals a novel role for endoderm in regulating kidney versus vascular cell fate. *Development* **135**, 3355–3367.

Mullins, M. C., Hammerschmidt, M., Kane, D. A., Odenthal, J., Brand, M., van Eeden, F. J., Furutani-Seiki, M., Granato, M., Haffter, P., Heisenberg, C. P., Jiang, Y. J., Kelsh, R. N., *et al.* (1996). Genes establishing dorsoventral pattern formation in the zebrafish embryo: The ventral specifying genes. *Development* **123**, 81–93.

Nauli, S. M., Alenghat, F. J., Luo, Y., Williams, E., Vassilev, P., Li, X., Elia, A. E., Lu, W., Brown, E. M., Quinn, S. J., Ingber, D. E., and Zhou, J. (2003). Polycystins 1 and 2 mediate mechanosensation in the primary cilium of kidney cells. *Nat. Genet.* **33**, 129–137.

Neave, B., Holder, N., and Patient, R. (1997). A graded response to BMP-4 spatially coordinates patterning of the mesoderm and ectoderm in the zebrafish. *Mech. Dev.* **62**, 183–195.

Newstead, J. D., and Ford, P. (1960). Studies on the development of the kidney of the Pacific Salmon, *Oncorhynchus forbuscha* (Walbaum). 1. The development of the pronephros. *Can. J. Zool.* **36**, 15–21.

Nguyen, V. H., Schmid, B., Trout, J., Connors, S. A., Ekker, M., and Mullins, M. C. (1998). Ventral and lateral regions of the zebrafish gastrula, including the neural crest progenitors, are established by a bmp2b/swirl pathway of genes. *Dev. Biol.* **199,** 93–110.

Nichane, M., Van Campenhout, C., Pendeville, H., Voz, M. L., and Bellefroid, E. J. (2006). The Na+/PO4 cotransporter SLC20A1 gene labels distinct restricted subdomains of the developing pronephros in Xenopus and zebrafish embryos. *Gene Expr. Patterns* **6,** 667–672.

Nikaido, M., Tada, M., Saji, T., and Ueno, N. (1997). Conservation of BMP signaling in zebrafish mesoderm patterning. *Mech. Dev.* **61,** 75–88.

Obara, T., Mangos, S., Liu, Y., Zhao, J., Wiessner, S., Kramer-Zucker, A. G., Olale, F., Schier, A. F., and Drummond, I. A. (2006). Polycystin-2 immunolocalization and function in zebrafish. *J. Am. Soc. Nephrol.* **17,** 2706–2718.

Ong, A. C., and Wagner, B. (2005). Detection of proximal tubular motile cilia in a patient with renal sarcoidosis associated with hypercalcemia. *Am. J. Kidney Dis.* **45,** 1096–1099.

Pathak, N., Obara, T., Mangos, S., Liu, Y., and Drummond, I. A. (2007). The zebrafish fleer gene encodes an essential regulator of cilia tubulin polyglutamylation. *Mol. Biol. Cell* **18,** 4353–4364.

Pavenstadt, H., Kriz, W., and Kretzler, M. (2003). Cell biology of the glomerular podocyte. *Physiol. Rev.* **83,** 253–307.

Pazour, G. J., Dickert, B. L., Vucica, Y., Seeley, E. S., Rosenbaum, J. L., Witman, G. B., and Cole, D. G. (2000). Chlamydomonas IFT88 and its mouse homologue, polycystic kidney disease gene tg737, are required for assembly of cilia and flagella. *J. Cell Biol.* **151,** 709–718.

Pazour, G. J., Baker, S. A., Deane, J. A., Cole, D. G., Dickert, B. L., Rosenbaum, J. L., Witman, G. B., and Besharse, J. C. (2002a). The intraflagellar transport protein, IFT88, is essential for vertebrate photoreceptor assembly and maintenance. *J. Cell Biol.* **157,** 103–113.

Pazour, G. J., San Agustin, J. T., Follit, J. A., Rosenbaum, J. L., and Witman, G. B. (2002b). Polycystin-2 localizes to kidney cilia and the ciliary level is elevated in orpk mice with polycystic kidney disease. *Curr. Biol.* **12,** R378–R380.

Pennekamp, P., Karcher, C., Fischer, A., Schweickert, A., Skryabin, B., Horst, J., Blum, M., and Dworniczak, B. (2002). The ion channel polycystin-2 is required for left-right axis determination in mice. *Curr. Biol.* **12,** 938–943.

Perner, B., Englert, C., and Bollig, F. (2007). The Wilms tumor genes wt1a and wt1b control different steps during formation of the zebrafish pronephros. *Dev. Biol.* **309,** 87–96.

Perz-Edwards, A., Hardison, N. L., and Linney, E. (2001). Retinoic acid-mediated gene expression in transgenic reporter zebrafish. *Dev. Biol.* **229,** 89–101.

Pfeffer, P. L., Gerster, T., Lun, K., Brand, M., and Busslinger, M. (1998). Characterization of three novel members of the zebrafish Pax2/5/8 family: Dependency of Pax5 and Pax8 expression on the Pax2.1 (noi) function. *Development* **125,** 3063–3074.

Pham, V. N., Roman, B. L., and Weinstein, B. M. (2001). Isolation and expression analysis of three zebrafish angiopoietin genes. *Dev. Dyn.* **221,** 470–474.

Praetorius, H. A., and Spring, K. R. (2001). Bending the MDCK cell primary cilium increases intracellular calcium. *J. Membr. Biol.* **184,** 71–79.

Puschel, A. W., Westerfield, M., and Dressler, G. R. (1992). Comparative analysis of Pax-2 protein distributions during neurulation in mice and zebrafish. *Mech. Dev.* **38,** 197–208.

Pyati, U. J., Cooper, M. S., Davidson, A. J., Nechiporuk, A., and Kimelman, D. (2006). Sustained Bmp signaling is essential for cloaca development in zebrafish. *Development* **133,** 2275–2284.

Qin, H., Rosenbaum, J. L., and Barr, M. M. (2001). An autosomal recessive polycystic kidney disease gene homolog is involved in intraflagellar transport in *C. elegans* ciliated sensory neurons. *Curr. Biol.* **11,** 457–461.

Raschperger, E., Neve, E. P., Wernerson, A., Hultenby, K., Pettersson, R. F., and Majumdar, A. (2008). The coxsackie and adenovirus receptor (CAR) is required for renal epithelial differentiation within the zebrafish pronephros. *Dev. Biol.* **313,** 455–464.

Reimschuessel, R., and Williams, D. (1995). Development of new nephrons in adult kidneys following gentamicin- induced nephrotoxicity. *Ren. Fail.* **17,** 101–106.

Reimschuessel, R., Bennett, R. O., May, E. B., and Lipsky, M. M. (1990). Development of newly formed nephrons in the goldfish kidney following hexachlorobutadiene-induced nephrotoxicity. *Toxicol. Pathol.* **18,** 32–38.

Reiser, J., Kriz, W., Kretzler, M., and Mundel, P. (2000). The glomerular slit diaphragm is a modified adherens junction. *J. Am. Soc. Nephrol.* **11,** 1–8.

Roselli, S., Gribouval, O., Boute, N., Sich, M., Benessy, F., Attie, T., Gubler, M. C., and Antignac, C. (2002). Podocin localizes in the kidney to the slit diaphragm area. *Am. J. Pathol.* **160,** 131–139.

Rottbauer, W., Baker, K., Wo, Z. G., Mohideen, M. A., Cantiello, H. F., and Fishman, M. C. (2001). Growth and function of the embryonic heart depend upon the cardiac-specific L-type calcium channel alpha1 subunit. *Dev. Cell* **1,** 265–275.

Ruotsalainen, V., Ljungberg, P., Wartiovaara, J., Lenkkeri, U., Kestila, M., Jalanko, H., Holmberg, C., and Tryggvason, K. (1999). Nephrin is specifically located at the slit diaphragm of glomerular podocytes. *Proc. Natl. Acad. Sci. USA* **96,** 7962–7967.

Salice, C. J., Rokous, J. S., Kane, A. S., and Reimschuessel, R. (2001). New nephron development in goldfish (Carassius auratus) kidneys following repeated gentamicin-induced nephrotoxicosis. *Comp. Med.* **51,** 56–59.

Saxén, L. (1987). "Organogenesis of the Kidney." Cambridge University Press, Cambridge.

Schottenfeld, J., Sullivan-Brown, J., and Burdine, R. D. (2007). Zebrafish curly up encodes a Pkd2 ortholog that restricts left-side-specific expression of southpaw. *Development* **134,** 1605–1615.

Schwarz, M., Cecconi, F., Bernier, G., Andrejewski, N., Kammandel, B., Wagner, M., and Gruss, P. (2000). Spatial specification of mammalian eye territories by reciprocal transcriptional repression of Pax2 and Pax6. *Development* **127,** 4325–4334.

Sehnert, A. J., Huq, A., Weinstein, B. M., Walker, C., Fishman, M., and Stainier, D. Y. (2002). Cardiac troponin T is essential in sarcomere assembly and cardiac contractility. *Nat. Genet.* **31,** 106–110.

Seldin, D. W., and Giebisch, G. H. (1992). "The Kidney: Physiology and Pathophysiology." Raven Press, New York.

Serluca, F. C., and Fishman, M. C. (2001). Pre-pattern in the pronephric kidney field of zebrafish. *Development* **128,** 2233–2241.

Serluca, F. C., Drummond, I. A., and Fishman, M. C. (2002). Endothelial signaling in kidney morphogenesis: A role for hemodynamic forces. *Curr. Biol.* **12,** 492–497.

Seufert, D. W., Brennan, H. C., DeGuire, J., Jones, E. A., and Vize, P. D. (1999). Developmental basis of pronephric defects in Xenopus body plan phenotypes. *Dev. Biol.* **215,** 233–242.

Shalaby, F., Rossant, J., Yamaguchi, T. P., Gertsenstein, M., Wu, X. F., Breitman, M. L., and Schuh, A. C. (1995). Failure of blood-island formation and vasculogenesis in Flk-1-deficient mice. *Nature* **376,** 62–66.

Shawlot, W., and Behringer, R. R. (1995). Requirement for Lim1 in head-organizer function. *Nature* **374,** 425–430.

Shmukler, B. E., Kurschat, C. E., Ackermann, G. E., Jiang, L., Zhou, Y., Barut, B., Stuart-Tilley, A. K., Zhao, J., Zon, L. I., Drummond, I. A., Vandorpe, D. H., Paw, B. H., *et al.* (2005). Zebrafish slc4a2/ ae2 anion exchanger: cDNA cloning, mapping, functional characterization, and localization. *Am. J. Physiol. Renal Physiol.* **289,** F835–F849.

Smithers, L., Haddon, C., Jiang, Y., and Lewis, J. (2000). Sequence and embryonic expression of deltaC in the zebrafish. *Mech. Dev.* **90,** 119–123.

Sollner, C., Burghammer, M., Busch-Nentwich, E., Berger, J., Schwarz, H., Riekel, C., and Nicolson, T. (2003). Control of crystal size and lattice formation by starmaker in otolith biomineralization. *Science* **302,** 282–286.

Sprague, J., Bayraktaroglu, L., Bradford, Y., Conlin, T., Dunn, N., Fashena, D., Frazer, K., Haendel, M., Howe, D. G., Knight, J., Mani, P., Moxon, S. A., *et al.* (2008). The Zebrafish Information Network: The zebrafish model organism database provides expanded support for genotypes and phenotypes. *Nucleic. Acids Res.* **36,** D768–D772.

Stainier, D. Y., Weinstein, B. M., Detrich, H. W., III, Zon, L. I., and Fishman, M. C. (1995). Cloche, an early acting zebrafish gene, is required by both the endothelial and hematopoietic lineages. *Development* **121,** 3141–3150.

Stickney, H. L., Imai, Y., Draper, B., Moens, C., and Talbot, W. S. (2007). Zebrafish bmp4 functions during late gastrulation to specify ventroposterior cell fates. *Dev. Biol.* **310,** 71–84.

Sullivan-Brown, J., Schottenfeld, J., Okabe, N., Hostetter, C. L., Serluca, F. C., Thiberge, S. Y., and Burdine, R. D. (2008). Zebrafish mutations affecting cilia motility share similar cystic phenotypes and suggest a mechanism of cyst formation that differs from pkd2 morphants. *Dev. Biol.* **314,** 261–275.

Sun, Z., and Hopkins, N. (2001). vhnf1, the MODY5 and familial GCKD-associated gene, regulates regional specification of the zebrafish gut, pronephros, and hindbrain. *Genes Dev.* **15,** 3217–3229.

Sun, Z., Amsterdam, A., Pazour, G. J., Cole, D. G., Miller, M. S., and Hopkins, N. (2004). A genetic screen in zebrafish identifies cilia genes as a principal cause of cystic kidney. *Development* **131,** 4085–4093.

Tabin, C. J., and Vogan, K. J. (2003). A two-cilia model for vertebrate left-right axis specification. *Genes Dev.* **17,** 1–6.

Taulman, P. D., Haycraft, C. J., Balkovetz, D. F., and Yoder, B. K. (2001). Polaris, a protein involved in left-right axis patterning, localizes to basal bodies and cilia. *Mol. Biol. Cell* **12,** 589–599.

Tavernarakis, N., and Driscoll, M. (1997). Molecular modeling of mechanotransduction in the nematode *Caenorhabditis elegans*. *Annu. Rev. Physiol.* **59,** 659–689.

Thisse, B., Heyer, V., Lux, A., Alunni, V., Degrave, A., Seiliez, I., Kirchner, J., Parkhill, J. P., and Thisse, C. (2004). Spatial and temporal expression of the zebrafish genome by large-scale *in situ* hybridization screening. *Methods Cell Biol.* **77,** 505–519.

Topczewska, J. M., Topczewski, J., Shostak, A., Kume, T., Solnica-Krezel, L., and Hogan, B. L. (2001a). The winged helix transcription factor Foxc1a is essential for somitogenesis in zebrafish. *Genes Dev.* **15,** 2483–2493.

Topczewska, J. M., Topczewski, J., Solnica-Krezel, L., and Hogan, B. L. (2001b). Sequence and expression of zebrafish foxc1a and foxc1b, encoding conserved forkhead/winged helix transcription factors. *Mech. Dev.* **100,** 343–347.

Toyama, R., and Dawid, I. B. (1997). lim6, a novel LIM homeobox gene in the zebrafish: Comparison of its expression pattern with lim1. *Dev. Dyn.* **209,** 406–417.

Tytler, P. (1988). Morphology of the pronephros of the juvenile brown trout, Salmo trutta. *J. Morphol.* **195,** 189–204.

Tytler, P., Ireland, J., and Fitches, E. (1996). A study of the structure and function of the pronephros in the lavvae of the turbot (*Scophthalmus maximus*) and the herring (*Clupea harengus*). *Mar. Fresh. Behav. Physiol.* **28,** 3–18.

Ungar, A., Kelly, G. M., and Moon, R. T. (1995). Wnt4 affects morphogenesis when misexpressed in the zebrafish embryo. *Mech. Dev.* **52,** 1–12.

van Rooijen, E., Giles, R. H., Voest, E. E., van Rooijen, C., Schulte-Merker, S., and van Eeden, F. J. (2008). LRRC50, a conserved ciliary protein implicated in polycystic kidney disease. *J. Am. Soc. Nephrol.* **19,** 1128–1138.

Vize, P. D., Seufert, D. W., Carroll, T. J., and Wallingford, J. B. (1997). Model systems for the study of kidney development: Use of the pronephros in the analysis of organ induction and patterning. *Dev. Biol.* **188,** 189–204.

Vize, P. D., Woolf, A. S., and Bard, J. B. L. (2002). "The Kidney: From Normal Development to Congenital Diseases." Academic Press, Amsterdam, Boston.

Weinstein, B. M., Schier, A. F., Abdelilah, S., Malicki, J., Solnica-Krezel, L., Stemple, D. L., Stainier, D. Y., Zwartkruis, F., Driever, W., and Fishman, M. C. (1996). Hematopoietic mutations in the zebrafish. *Development* **123,** 303–309.

Wingert, R. A., and Davidson, A. J. (2008). The zebrafish pronephros: A model to study nephron segmentation. *Kidney Int.* **73,** 1120–1127.

Wingert, R. A., Selleck, R., Yu, J., Song, H. D., Chen, Z., Song, A., Zhou, Y., Thisse, B., Thisse, C., McMahon, A. P., and Davidson, A. J. (2007). The cdx genes and retinoic acid control the positioning and segmentation of the zebrafish pronephros. *PLoS Genet.* **3,** 1922–1938.

Yang, Y., Jeanpierre, C., Dressler, G. R., Lacoste, M., Niaudet, P., and Gubler, M. C. (1999). WT1 and PAX-2 podocyte expression in Denys-Drash syndrome and isolated diffuse mesangial sclerosis. *Am. J. Pathol.* **154,** 181–192.

Yoder, B. K., Hou, X., and Guay-Woodford, L. M. (2002a). The polycystic kidney disease proteins, polycystin-1, polycystin-2, polaris, and cystin, are co-localized in renal cilia. *J. Am. Soc. Nephrol.* **13,** 2508–2516.

Yoder, B. K., Tousson, A., Millican, L., Wu, J. H., Bugg, C. E., Jr, Schafer, J. A., and Balkovetz, D. F. (2002b). Polaris, a protein disrupted in orpk mutant mice, is required for assembly of renal cilium. *Am. J. Physiol. Renal Physiol.* **282,** F541–F552.

Zhong, T. P., Childs, S., Leu, J. P., and Fishman, M. C. (2001). Gridlock signalling pathway fashions the first embryonic artery. *Nature* **414,** 216–220.

INDEX

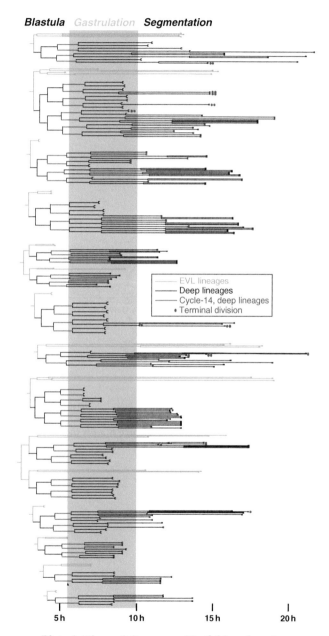

Blastula Gastrulation **Segmentation**

EVL lineages
Deep lineages
Cycle-14, deep lineages
• Terminal division

5 h   10 h   15 h   20 h

Plate 1 (Figure 2.8 on page 22 of this volume)

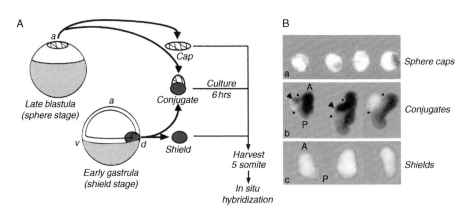

Plate 2 (Figure 6.5 on page 99 of this volume)

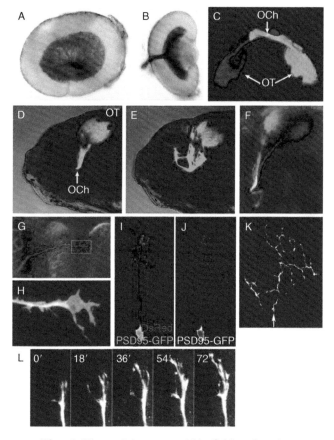

Plate 3 (Figure 9.1 on page 164 of this volume)

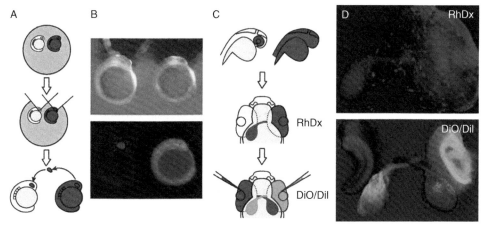

Plate 4 (Figure 9.4 on page 175 of this volume)

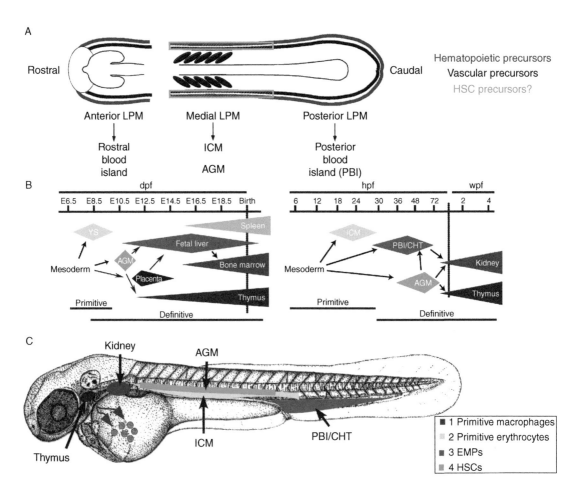

Plate 5 (Figure 11.1 on page 207 of this volume)

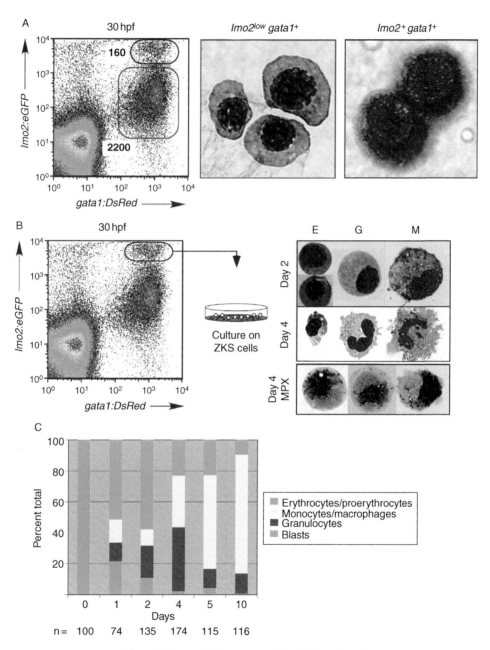

Plate 6 (Figure 11.2 on page 209 of this volume)

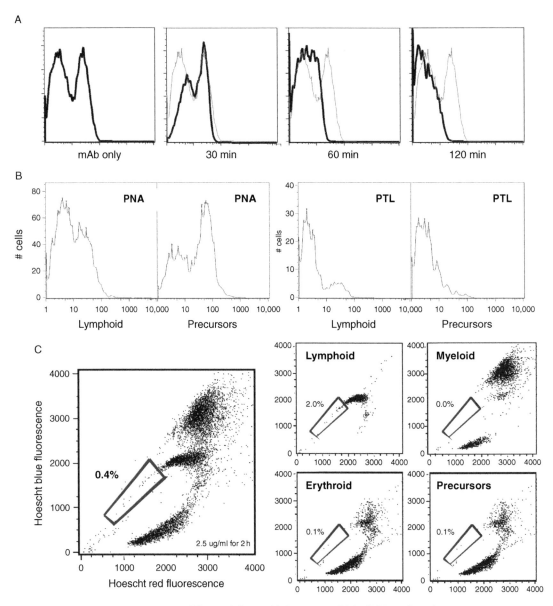

Plate 7 (Figure 11.7 on page 224 of this volume)

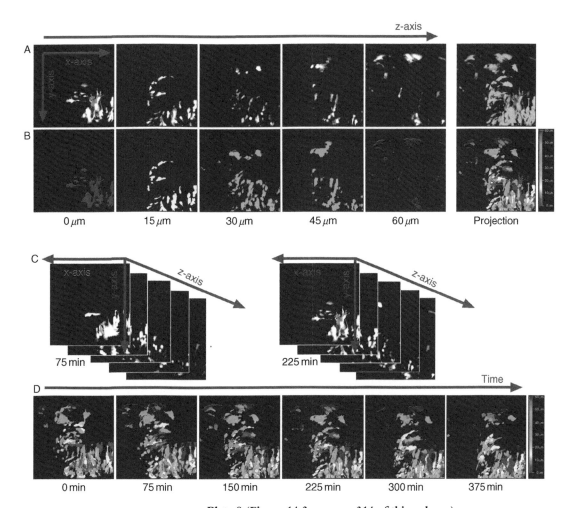

Plate 8 (Figure 14.3 on page 314 of this volume)

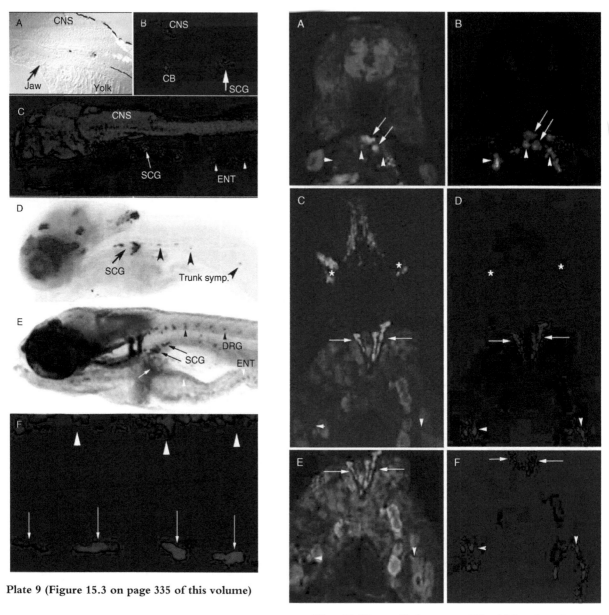

Plate 9 (Figure 15.3 on page 335 of this volume)

Plate 10 (Figure 15.4 on page 338 of this volume)

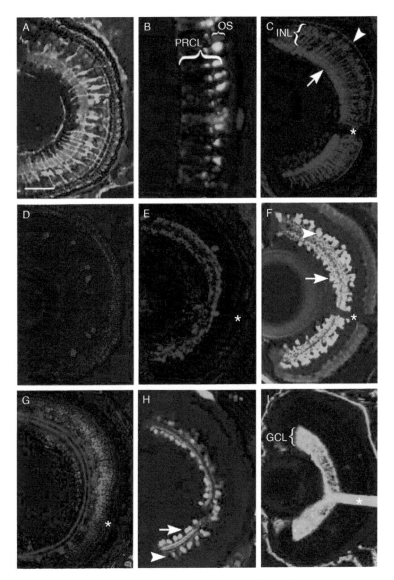

Plate 11 (Figure 16.3 on page 367 of this volume)

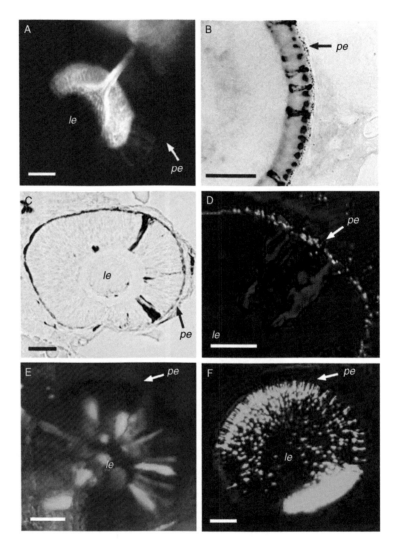

Plate 12 (Figure 16.4 on page 369 of this volume)

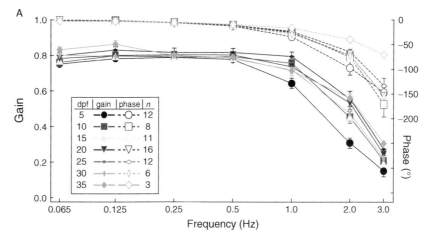

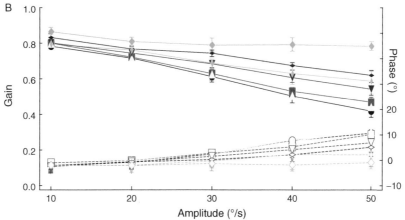

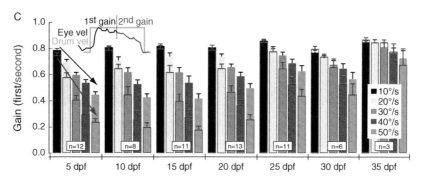

Plate 13 (Figure 17.9 on page 423 of this volume)

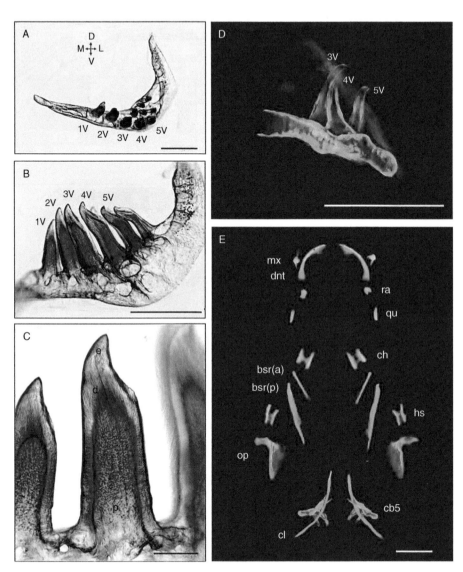

Plate 14 (Figure 19.1 on page 460 of this volume)

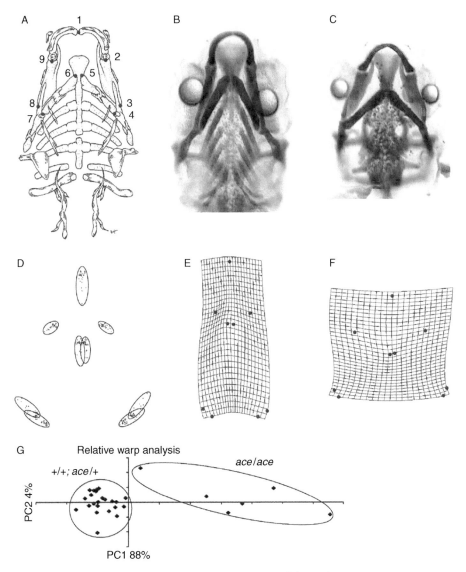

Plate 15 (Figure 19.3 on page 468 of this volume)

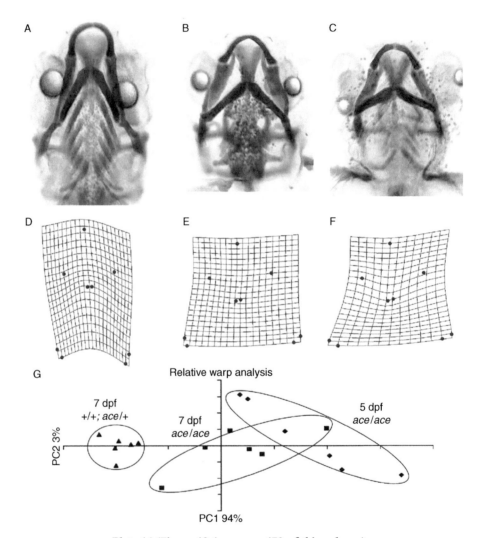

Plate 16 (Figure 19.4 on page 470 of this volume)

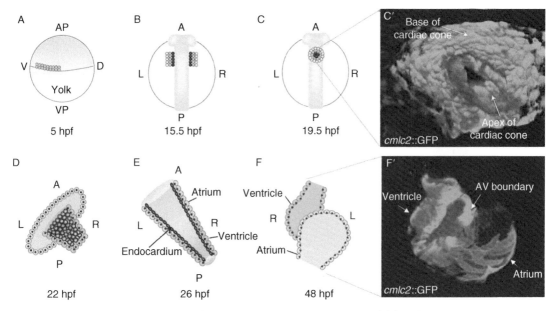

Plate 17 (Figure 20.1 on page 482 of this volume)

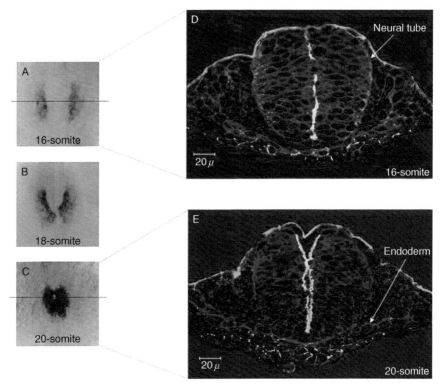

Plate 18 (Figure 20.2 on page 484 of this volume)

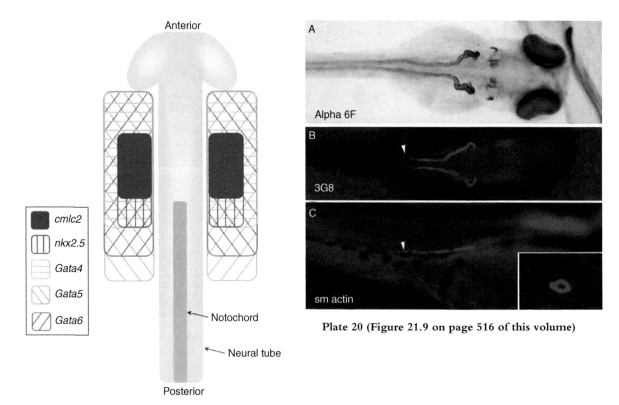

cmlc2

nkx2.5

Gata4

Gata5

Gata6

Anterior

Notochord

Neural tube

Posterior

Plate 19 (Figure 20.4 on page 493 of this volume)

A

Alpha 6F

B

3G8

C

sm actin

Plate 20 (Figure 21.9 on page 516 of this volume)

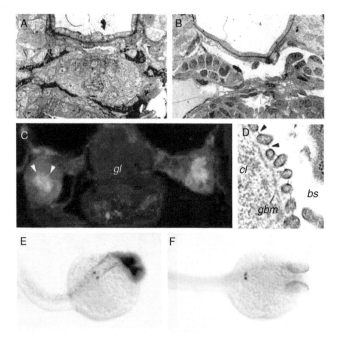

A   B

C   D

gl   cl   gbm   bs

E   F

Plate 21 (Figure 21.12 on page 521 of this volume)

Printed and bound by CPI Group (UK) Ltd, Croydon, CR0 4YY

03/10/2024

01040318-0007